A publication of the
AUSTRALIAN BIOLOGICAL RESOURCES STUDY, CANBERRA

AF785263

Fungi of Australia

Inocybaceae

P. Brandon Matheny & Neale L. Bougher

Australian Government
Department of the Environment and Energy

CSIRO PUBLISHING

© Commonwealth of Australia 2017

This work is copyright. Apart from any use as permitted under the *Copyright Act 1968*, no part may be reproduced or distributed by any process or stored in any retrieval system or data base without prior written permission from the copyright holder. Requests and inquiries concerning reproduction and rights should be addressed to:

Australian Biological Resources Study, GPO Box 787, Canberra, ACT 2601, Australia.

VOLUME EDITOR
Tony Orchard

ASSISTANT EDITORS
Russell Barrett
Patrick M. McCarthy
Zoë Knapp
Brigitte Kuchlmayr
Annabel Wheeler

This work may be cited as:

P.B.Matheny & N.L.Bougher, *Fungi of Australia: Inocybaceae.* ABRS, Canberra; CSIRO Publishing, Melbourne (2017).

This book is available from:

CSIRO PUBLISHING
Locked Bag 10,
Clayton South VIC 3169
Australia

Tel: 1300 788 000 Int: +(61 3) 9545 8400
Fax: (03) 9545 8550 Int: +(61 3) 9545 8550

Email: publishing.sales@csiro.au

Cataloguing-in-Publication entry

Fungi of Australia: Inocybaceae.
P. Brandon Matheny & Neale L. Bougher

Bibliography.

Includes index.

ISBN 978 1 486 30666 4 (hbk.)

ISBN 0 643 05702 1 (hard cover set).

1. Inocybaceae. 2. Fungi - Australia. I. Title. (Series : Fungi of Australia).
579.5930994

Cover Photographs:
Front: *Inocybe mallocyboides* (Photo. R.M.Robinson)
Back: *Inocybe serrata* complex (Photos. P.B.Matheny)
Background: Basidiospores of *Inocybe exigens* (Photo. N.L.Bougher)
Cover designer: Brigitte Kuchlmayr

Published by ABRS, Canberra & CSIRO Publishing, Melbourne
Printed in China by 1010 Printing International Ltd

MIX
Paper from responsible sources
FSC® C016973
www.fsc.org

The paper this book is printed on is in accordance with the rules of the Forest Stewardship Council®. The FSC® promotes environmentally responsible, socially beneficial and economically viable management of the world's forests.

CONTENTS

AUTHORS

Dr P. Brandon Matheny, Department of Ecology and Evolutionary Biology, University of Tennessee, Hesler Biology Building 332, Knoxville, TN 37996-1610, USA.

Dr Neale L. Bougher, Western Australian Herbarium, Science and Conservation Division, Department of Parks and Wildlife, Locked Bag 104, Bentley Delivery Centre 6983, Western Australia.

ILLUSTRATOR

Dr Neale L. Bougher, Western Australian Herbarium, Science and Conservation Division, Department of Parks and Wildlife, Locked Bag 104, Bentley Delivery Centre 6983, Western Australia.

PHOTOGRAPHERS

Dr Timothy J. Baroni, Department of Biological Sciences, 340 Bowers Hall, PO Box 2000, State University of New York – College at Cortland, Cortland, NY 13045, USA.

Dr Matthew D. Barrett, Botanic Gardens and Parks Authority, Fraser Ave, West Perth 6005, Western Australia.

Dr Neale L. Bougher, Western Australian Herbarium, Science and Conservation Division, Department of Parks and Wildlife, Locked Bag 104, Bentley Delivery Centre 6983, Western Australia.

Dr Genevieve M. Gates, School of Plant Science, University of Tasmania, Hobart, Tasmania.

Dr Roy E. Halling, Institute of Systematic Botany, The New York Botanical Garden, 2900 Southern Blvd, Bronx, NY 10458-5126, USA.

Ms Emma Harrower, Department of Ecology and Evolutionary Biology, University of Tennessee, 569 Dabney Hall, Knoxville, TN 37996-1610, USA.

Mr Heino Lepp, PO Box 38, Belconnen 2616, Australian Capital Territory.

Dr P. Brandon Matheny, Department of Ecology and Evolutionary Biology, University of Tennessee, Hesler Biology Building 332, Knoxville, TN 37996-1610, USA.

Dr Richard M. Robinson, Science and Conservation Division, Department of Parks and Wildlife, Brain Street, Manjimup 6258, Western Australia.

Dr Martin Ryberg, Systematic Biology, Evolutionary Biology Centre, Uppsala University, Norbyvägen 18D, 752 36 Uppsala, Sweden.

Ms Katrina Syme, RMB 1020, South Coast Highway, Denmark 6333, Western Australia.

ACKNOWLEDGEMENTS

Funds to support the production of this book were provided by a research grant from the Australian Government's Australian Biological Resources Study National Taxonomy Research Grant Programme (RFL211-31) co-funded by the Western Australian Naturalists' Club Inc. Funds were also contributed by a U.S. National Science Foundation award (DEB-0949517), and from the College of Arts and Sciences at the University of Tennessee. The L.R. Hesler Endowment Fund provided stipend support for undergraduate students at the University of Tennessee. Facilities and administrative support for the project were provided at the Western Australian Herbarium by the Department of Parks and Wildlife (formerly Department of Environment Conservation).

Staff at the following herbaria - AD, BRI, CANB, CNS, CORT, FLAS, HO, M, MEL, NY, OSC, PDD, PERTH, TENN, WTU, ZT - are thanked for arrangement of loans (see Thiers, undated, for details of herbaria). In addition to the herbaria above, the following individuals provided materials for this study via collaborative field work or other significant assistance: Sandra Abell-Davis (James Cook University; Australian Tropical Herbarium), Tim Baroni (SUNY Cortland), Matt Barrett (Botanic Gardens and Parks Authority, W.A.), Kaylene Bransgrove (James Cook University), Judith Curnow (Australian National Herbarium), Pam Catcheside (State Herbarium of South Australia), Nigel Fechner (Queensland Herbarium), Joe Froudist (former president of the Western Australian Naturalists' Club Inc.), Genevieve Gates (University of Tasmania), Roy Halling (New York Botanical Garden), Emma Harrower (University of Tennessee), Roz Hart (past president of the Western Australian Naturalists' Club Inc.), Egon Horak (Eidgenössische Technische Hochschule Zürich), Peter Johnston (Landcare Research, New Zealand Fungal Herbarium), Fernanda Karstedt (Instituto de Botânica, Seçãso de Micologia e Lequenologia, São Paulo, Brazil), Karina Knight (Western Australian Herbarium), Teresa Lebel (National Herbarium of Victoria), Heino Lepp (Australian National Herbarium), Tom May (National Herbarium of Victoria), Götz Palfner (University of Concepción, Chile), Michael Pilkington (U.K.), David Ratkowsky (University of Tasmania), Richard Robinson (formerly Department of Parks and Wildlife, W.A.), Martin Ryberg (Uppsala University), Matt Smith (University of Florida), Katrina Syme (Denmark, W.A.), Kevin Thiele (formerly Department of Parks and Wildlife, W.A.) and staff at the Western Australian Herbarium (PERTH), Jim Trappe (Oregon State University), and Mark True (Two Peoples Bay Nature Reserve, Department of Parks and Wildlife, W.A.).

Particular thanks go to these individuals for their logistical and/or housing support: David Ratkowsky and Genevieve Gates in Hobart, Teresa Lebel and Tom May in Melbourne, Nigel Fechner in Brisbane, Jim Trappe in Canberra, and Peter Johnston in Auckland, New Zealand.

Rosemary Gillen (volunteer, Western Australian Herbarium) is thanked for her vital contributions such as databasing, herbarium procedures and proofing.

The following students and technicians from the University of Tennessee are acknowledged for the generation of sequence data and/or collection maintenance: Katheryn Allen, Joshua Birkebak, Christine Braaten, Sean Dornbush, Emily Giles, Whitaker Hoskins, Hailee Korotkin, Brian Looney, Marisol Sánchez-García, Sarah Slocum, Sarah Sprague, Stacia Warwick, and Aaron Wolfenbarger. Emma Harrower is acknowledged for her SEM work.

We extend special thanks to the Bougher and Matheny families in Western Australia and Tennessee respectively for all their support during the many years leading up to and during the production of this book.

The following collecting license permitting agencies are acknowledged: Government of Western Australia, Department of Parks and Wildlife (permit no. SW011372, CE002808), Wellington Park Management Trust (file no. 07-84-46), Department of Primary Industries, Parks, Water and Environment (DPIPWE), Tasmania (file no. NH-NH-RE-ST-215896), Department of Sustainability and Environment, Victoria (permit no. 10004954, file no. FF383119), and the Department of Environment and Resource Management, Queensland Government for state parks (permit no. TWB/01/2012) and for national parks (WITK10522811).

Kevin Thiele (formerly curator at the Western Australian Herbarium) is especially acknowledged for his coordination of reviews and editorial assistance. We thank Zoë Knapp, Patrick McCarthy, Brigitte Kuchlmayr and Tony Orchard at ABRS for all their work seeing this book through the production process. Edits and reviews were also kindly provided by Brad Kropp (Utah State University), Tom May (Royal Botanic Gardens Melbourne), Patrick McCarthy (formerly ABRS), and one anonymous reviewer. The authors are duly grateful for the careful edits and suggestions to improve this book.

LIST OF KEYS

LIST OF PHYLOGRAMS

INTRODUCTION

The family Inocybaceae Jülich (Agaricales) is a highly diverse, cosmopolitan group of mycorrhizal fungi, including a largely unexplored diversity of endemic native species in Australia. Most Inocybaceae have agaricoid (mushroom-like) fruit bodies. However, they are not renowned for their showiness or allure compared to other families such as the Entolomataceae or Hygrophoraceae. Nevertheless, the Inocybaceae have been subject to thorough descriptive treatment, including elegant watercolour paintings, in classic European texts such as those by Heim (1931), Alessio (1980), and Stangl (1989). In contrast to some other mushroom families, the Inocybaceae also appear to lack any redeeming culinary values. Indeed, many species of the family are poisonous and a few hallucinogenic or deadly (Kosentka *et al.*, 2013). So why study them? Inocybaceae are ubiquitous, highly diverse and difficult to identify at species level. Species of Inocybaceae vary macroscopically and are often exquisite under the microscope, featuring a wide range of anatomical variation. Even in relatively well-studied regions such as Europe, new species continue to be described. Another compelling reason to study the Inocybaceae is because the understanding of species-level diversity and diverse mycorrhizal plant-host associations can contribute to research in the fields of macroevolution, macroecology, and historical biogeography (Matheny *et al.*, 2009; Ryberg & Matheny, 2012). In this sense, the Inocybaceae are a model system to study patterns and processes of evolution on a deep time scale.

Most Inocybaceae can be recognized in the field by a combination of features of their fruit bodies, which are typically brown in colouration, small to medium size with a stipe, a fibrillose to scaly pileus, brownish attached lamellae with whitish fimbriate edges, and dull brown spore deposit. Many feature unusual odours such as spermatic, green corn, of bruised *Pelargonium* leaves, fish, rotten meat, mussels, or spicy cinnamon. Microscopically, the presence of thick-walled and apically crystalliferous hymenial cystidia can confirm their identity, at least for the largest group in the family, the *Inocybe* clade (*Inocybe s. str.*). Many species are medically important because they contain the toxins muscarine or psilocybin (Kuyper, 1986; Kosentka *et al.*, 2013). The Inocybaceae occur in a variety of habitats such as sand dunes, heath-scrub, thickets, woodland, and rainforest, and are an important constituent of alpine areas. Historically the family has been found predominantly in temperate forests, but new species continue to be described from relatively poorly explored tropical regions (Bougher *et al.*, 2012; Matheny *et al.*, 2012a; Matheny *et al.*, 2012b; this volume).

Phylogenetically, the Inocybaceae *sensu* Matheny (2005) and Matheny (2009) form a strongly supported monophyletic group. Currently, seven major clades are recognized in the family, an overview of which is presented in Matheny (2009). In summary, two of these clades have been formally recognised at generic rank, as *Auritella* Matheny & Bougher and *Tubariomyces* Esteve-Rav. & Matheny (Matheny & Bougher 2006a, 2006b; Alvarado *et al.* 2010), while the remaining five major clades (the genus *Inocybe s. str.*, *Nothocybe*, *Pseudosperma*, *Inosperma* and *Mallocybe*) are all currently included in *Inocybe*. Matheny (2009) suggested that all seven clades should at some point be recognised at the genus level. A phylogenetic tree depicting relationships of these clades is presented in Fig. 1, based on findings from a multi-gene analysis of Matheny *et al.* (2012). The classification presented here, however, leaves the genus *Inocybe* as a paraphyletic group. The authors' intentions are to see that the remaining major clades are elevated to generic rank (as discussed in Matheny, 2009). Nevertheless, on-going collaborations with other workers in the field on a world-wide scale will be necessary to forge a consensus on a final generic level classification of the family. Such a move here, in a regional flora, is beyond the scope of this text.

Australia is richer in phylogenetic diversity of Inocybaceae than most other regions of the world. Of the seven major clades in the family, six have now been recorded from Australia (after *Tubariomyces* was discovered here during the current work), more than in South America (two), New Zealand (three), Papua New Guinea (four), North America (four), Europe (five), and Africa (five). India is also an important region for phylogenetic diversity of Inocybaceae, as all major clades, other than *Tubariomyces*, have been recorded there. Several former Gondwanan regions appear to be the richest in Inocybaceae phylogenetic diversity (Africa, Australia, India), yet South America stands as the poorest, with southern South America the least phylogenetically diverse of all.

Few species of Inocybaceae have been documented from Australia. Indeed, only 17 species in the family had been recognized at the onset of the current work (Matheny & Bougher, 2010). In light of this paucity of knowledge, the current study was undertaken to explore and document the diversity of Inocybaceae in the major climatic and vegetation regions of Australia. The study has culminated in recognition of 137 species of Inocybaceae from Australia, 101 of which are described as new (see Appendix 1). Significantly, from a biogeographical and phylogenetic perspective, the study determined that the proportion of species endemic to Australia (121 of 137, 90%) is extremely high.

HISTORY OF THE FAMILY

Inocybe (Fr.) Fr. was first recognized as a tribe of *Agaricus* by Fries (1821), then as a genus later by Fries (1863). Two authors later independently separated the species with angular-nodulose basidiospores into segregate genera, *Astrosporina* Schröt. and *Clypeus* (Britzelm.) P. Karst., both in 1889. *Clypeus* is now considered a later synonym of *Astrosporina*. Earle (1909) recognized four inocyboid genera: *Inocybe* (lectotype: *I. relicina* (Fr. : Fr.) Quél.), *Inocibium* Earle (holotype: *I. lacera* (Fr. : Fr.) P. Kumm.), *Astrosporina* (lectotype: *I. praetervisa* Quél.), and *Agmocybe* Earle (holotype: *I. rimosa* (Bull. : Fr.) P. Kumm. Earle's classification was based on gross morphological differences and was not widely used. Horak (1968, 1977, 1979) recognized two genera, maintaining the smooth-spored species in *Inocybe* and the angular-nodulose-spored species in *Astrosporina*. As with the Earle classification, this was not widely adopted, with most authors accepting a single genus (*Inocybe*) for species of both generalized basidiospore types (e.g., Kühner, 1980). Typification of *Inocybe* has been controversial due to various interpretations of the type species, *I. relicina*, a good summary of which can be found in Kuyper (1986). Moser (1978) finally sorted out the identity of *I. relicina* (a nodulose-spored species endemic to Fennoscandia), which had been previously misinterpreted by earlier authors. Jülich (1982) proposed the family Inocybaceae and recognised both *Inocybe* and *Astrosporina*. However, until Matheny (2005) this move was widely ignored, with most authors (including Singer, 1986) continuing to place *Inocybe* in the Cortinariaceae.

Kuyper (1986) was the first to explicitly test the phylogeny of *Inocybe* through a cladistic analysis using morphological characters. His results suggested the polyphyly of nodulose-spored species. Follow-up studies evaluating *Inocybe* using molecular data have reinforced Kuyper's earlier findings (Matheny *et al.*, 2002; Matheny, 2005; Ryberg *et al.*, 2008; Matheny *et al.*, 2009; Ryberg *et al.*, 2010), confirming *Astrosporina* as a polyphyletic group.

At the family level, Matheny (2005) was the first to suggest resurrecting Inocybaceae because the Cortinariaceae was polyphyletic, a result reinforced by an Agaricales-wide multigene study by Matheny *et al.*, (2006). Some authors have amalgamated the Inocybaceae and Crepidotaceae, which together form a monophyletic group (Matheny *et al.*, 2006). We do not advocate this approach because sufficient differences exist between the two families, as summarized by Matheny (2009). All species of Inocybaceae are considered to be ectomycorrhizal, whereas species of Crepidotaceae are not. Many species of Inocybaceae contain muscarine, which has not been confirmed in any Crepidotaceae. Both families differ in fruit body development as well (Matheny, 2009).

Matheny (2005) began to recognize and name major clades within *Inocybe*, deviating from a traditional taxonomic scheme. Matheny & Bougher (2006a, 2006b) followed up by formally splitting *Inocybe* into multiple genera based on molecular phylogenetic evidence. *Auritella* Matheny & Bougher was established originally for a small suite of species from tropical Africa and Australia united by molecular characters (see also Matheny, 2005) and by the presence of smooth spores, necrobasidia, elongated cheilocystidia, absence of pleurocystidia, and a tricholomatoid or squat habit. Matheny *et al.* (2012b) later added to *Auritella* a species from tropical India with some phenotypic similarities to species from tropical Africa. In this present work we now recognize 12 species of *Auritella*, eight of which are endemic to Australia and three of which are described as new in this work.

INTRODUCTION

Alvarado *et al.* (2010) moved a small group of highly unusual species centered on *Inocybe inexpectata* M. Villarreal, Esteve-Rav., Heykoop & E. Horak (Villarreal *et al.*, 1998) to the genus *Tubariomyces*, originally referred to by Matheny *et al.* (2009) and Matheny (2009) as the "*Mallocybella* clade". Vizzini *et al.* (2013) added a third species to the genus. The current treatment documents two more, including the first species of *Tubariomyces* recorded from Australia. An additional unnamed species occurs in Zambia. Thus, six species (two awaiting formal description) can now be recognized in *Tubariomyces*, a genus of small-sized species with an omphalinoid habit, necrobasidia, smooth basidiospores, presence of cheilocystidia, absence of pleurocystidia, and presence of caulocystidia the entire length of the stipe. To date, *Tubariomyces* is only known from Mediterranean Europe, Zambia, and northern Australia.

The *Mallocybe* clade contains a suite of species common in northern temperate areas, characterized mainly by a tomentose-fibrillose or scaly pileus, reaction of the pileus surface to alkali, necrobasidia, smooth basidiospores, presence of cheilocystidia, and lack of pleurocystidia (Kuyper 1986, as *Inocybe* subgenus *Mallocybe* Kuyp.). The group is absent from the *Nothofagus*-zone of the southern hemisphere (Singer, 1986), observations reinforced by work in New Zealand (Horak, 1977) and here. The group is monophyletic and differs primarily from *Auritella* by the size of the cheilocystidia (<50 μm long in all species except *Inocybe unicolor* Peck) and by molecular characters. Five new species of the *Mallocybe* clade are described here from Australia. None are known from New Zealand and South America. One species, *Inocybe althoffiae* E. Horak, is from Papua New Guinea. However, an ITS sequence (HQ201335) from the type collection suggests that this species may not be a *Mallocybe*. Examination of the type also reveals lack of necropigmented basidia, a feature more consistent with the *Pseudosperma* clade.

Kühner (1980) first conceived of *Inocybe* subgenus *Inosperma* Kuhner (holotype: *Inocybe calamistrata* (Fr. : Fr.) Gillet) when he divided the genus into two groups: species without pleurocystidia in subgenus *Inosperma*; species with pleurocystidia in subgenus *Inocybe*. Kuyper (1986) demonstrated that subgenus *Inocybe* is monophyletic but derived within subg. *Inosperma*. Molecular studies referred to above have confirmed Kuyper's conclusion that subg. *Inosperma* is paraphyletic. However, two elements that make the subgenus paraphyletic have been raised to genus level (viz, *Auritella*, *Tubariomyces*), thus allowing a monophyletic *Inosperma* clade. Based on molecular phylogenetic results (Larsson *et al.*, 2009; Matheny *et al.*, 2009; Kropp *et al.*, 2013), this clade contains the holotype (*I. calamistrata*) and related species of sect. *Cervicolores* Kühner & Romagnesi ex Singer and residual species of sect. *Rimosae* (Fr.) Sacc. (excluding the type, *I. rimosa* (Bull. : Fr.) P. Kumm.). Many species in the *Inosperma* clade have distinctive non-spermatic odours, phaseoliform spores, and/or reddening flesh (Larsson *et al.*, 2009). Five species are reported in this current work from Australia, one of them introduced.

The *Pseudosperma* clade, referred to as *Inocybe* sect. *Rimosae* by Larsson *et al.* (2009), is relatively well represented in Australia, although until this volume no species of the clade had been described from the continent. Most species can be recognized by the following combination of characters: a rimose pileus, white pruinose to furfuraceous stipe apex, elliptic to subphaseoliform spores, mostly cylindric to clavate cheilocystidia, absence of pleurocystidia, absence of necrobasidia, and often olivaceous-tinged lamellae. Five new species are described in this current volume, but several additional species await formal description. Of the new species described here, *I. gracilissima* Matheny & Bougher is common in tropical regions of Australia. *Inocybe pluviorum* Matheny & Bougher occurs in warm-temperate (subtropical) rainforest in Queensland. This species is strongly supported as the sister lineage to the rest of the *Pseudosperma* clade and is thus an important discovery.

The *Nothocybe* clade (or lineage as it is known from a single species) was originally recognized from one collection of an unnamed species reported under *Casuarina* in tropical India (Matheny *et al.*, 2009; Matheny, 2009). The collection is superficially similar to species of the *Pseudosperma* clade. However, *Nothocybe* is the sister group to the *Inocybe* clade. Recent work in India re-discovered the sole species of *Nothocybe*, in association with the genus *Acacia* (Latha *et al.*, 2016). This species has been described as *Inocybe distincta* K.P.D. Latha & Manim. The *Nothocybe* clade appears to be distinguished from species in the *Pseudosperma* clade by the smooth but at times slightly subangular spores and possession of distinct clusters of caulocystidia. *Nothocybe* appears to be a relict lineage known thus far only from tropical India. No species referable to *Nothocybe* have been recorded from Australia.

Of the seven major groups in the Inocybaceae, the *Inocybe* clade is the most diverse in Australia. A general unifying feature from a phylogenetic point-of-view is the presence of pleurocystidia. In addition, the apiculus of the basidiospores is more distinctive compared to the other major clades of Inocybaceae (Kuyper, 1986). In total, 114 species of the *Inocybe* clade are documented in this volume, of which 109 are endemic or extend to other parts of Australasia (e.g., Papua New Guinea, New Zealand). Of these 109 species, 67 feature smooth spores and 42 species have nodulose spores. The phylogenetic backbone of the *Inocybe* clade is poorly resolved, but all taxa with smooth spores or nodulose spores are not monophyletic. At least 16 subgroupings within the *Inocybe* clade were recognized by Ryberg *et al.* (2010) together with several poorly supported subgroups.

SIMILAR GENERA

Species of Inocybaceae are most likely to be confused with species of *Cortinarius* (Pers.) Gray in Australia. Indeed, Grgurinovic (1997) transferred two species originally described in *Cortinarius* (*C. striatulus* Cleland and *C. fibrillosus* Cleland) to *Inocybe* (the latter also described as *I. cystidiocatenata* Grgur.) (Matheny & Bougher, 2010). Gasparini (2006) transferred *I. cystidiocatenata* to *Cortinarius*, a needless action since *C. fibrillosus* holds priority. Molecular analyses confirm that *C. fibrillosus* belongs in *Cortinarius* (Matheny & Ammirati, 2003), in the *Obtusi* clade (Garnica *et al.* 2005). Such confusion is caused by the apparent smooth spores and presence of distinctive cheilocystidia in these unusual species of *Cortinarius*. However, the overall aspect of these species is typical of *Telamonia*-like species of *Cortinarius* from the Northern Hemisphere.

Other more typical species of *Cortinarius* from Australia can be distinguished from specimens of Inocybaceae by their verrucose spores and rust-coloured spore deposit. Most species of *Cortinarius* also lack pleurocystidia and cheilocystidia. Some species of *Hebeloma* (Fr.) P. Kumm. have a similar spore print to Inocybaceae, but are separated most readily by the combination of a viscid pileus surface (which can be easily detected under the microscope by examination of the pileipellis) and typically verrucose spores. Like the Inocybaceae, both *Cortinarius* and *Hebeloma* fruit on soil and are ectomycorrhizal. Species of *Agrocybe* Fayod might be confused with Inocybaceae due to their similar brown spore deposit and occurrence on the ground, but *Agrocybe* typically have distinctive white rhizomorphs at the base of the stipe, often fruit in woody debris such as woodchips, sometimes have an annulus, and all bear a hymeniform type of pileipellis that is lacking in most Inocybaceae (species of *Tubariomyces* have a palisade pileipellis). Species of *Tubaria* (W.G.Sm.) Gillet could be confused with, for example, *Tubariomyces*, but *Tubaria* often fruit on woody debris, have a brighter-coloured spore deposit, lack pleurocystidia and do not have caulocystidia along the entire length of the stipe. Species of *Galerina* Earle are non-mycorrhizal, generally mycenoid to collybioid in habit, often have a hygrophanous pileus, often have verrucose spores with a plage, and have an ochraceous-yellow to rusty brown spore deposit. Species of *Phaeomarasmius* Scherff. and *Flammulaster* Earle are non-mycorrhizal and lack pleurocystidia and necropigmented basidia. *Flammulaster* have a pileipellis that resembles an epithelium, whereas this feature is lacking in the Inocybaceae. Species of *Pholiota* (Fr.) P. Kumm. may share a similar coloured spore deposit to Inocybaceae but can be distinguished because their spores typically have a germ pore and often a lignicolous habit.

ECOLOGY

Most Inocybaceae fruit on soil. All species in the family are considered to be ectomycorrhizal (ECM), that is, they exchange nutrients for photosynthates with plants via specialized root associations (Kuyper, 1986; Singer, 1986). In Australia, Inocybaceae are most commonly found in habitats such as wet and dry sclerophyll forests and cool- or warm-temperate rainforests where ECM eucalypts, *Allocasuarina*, *Acacia*, *Pomaderris*, or *Nothofagus* occur. Among the Myrtaceae, Inocybaceae are most often associated with *Eucalyptus*, *Corymbia*, and *Lophostemon*, but also co-occur with *Leptospermum*, *Melaleuca*, *Kunzea*, *Astartea*, *Chamelaucium*, and possibly other myrtaceous shrubs.

INTRODUCTION

While the overwhelming majority of Inocybaceae in Australia are associated with sclerophyll forests that include eucalypts, the family also associates with other habitat types. As many as seven species have been recorded under *Acacia* where eucalypts are absent. These include *Inocybe acaciae* Matheny & Bougher, *I. froudistii* Matheny & Bougher, *I. alienospora* (Corner & E.Horak) Garridio, *I. alloumbrina* Matheny & Bougher, *I. sejuncta* Matheny, Bougher & M.D.Barrett, *I. subpulverulenta* Matheny, Bougher & Lepp, and *I. luteo-olivacea* Matheny, Bougher & Halling. Several of these species have been recorded also in vegetation where *Acacia* is mixed with eucalypts and other potential mycorrhizal plants. Shrubs of Fabaceae, e.g., *Bossiaea* and *Gastrolobium*, may also be associated with Inocybaceae, but further studies are warranted. *Antidesma* and *Glochidion* (Phyllanthaceae) and *Albizia* (Fabaceae) are other potential plant associates at tropical latitudes. In summary, potential ECM plant families associated with the Inocybaceae in Australia include at least Myrtaceae, Casuarinaceae, Fabaceae, Nothofagaceae, Rhamnaceae, and Phyllanthaceae.

At least nine species of Inocybaceae occur in cool, wet, *Nothofagus* rainforests of Australia. This rainforest habitat is rather specialized due to absence of fire. The nine species we have confirmed from cool-temperate rainforests in Australia are: *I. auritelloides* Matheny, Bougher & G.M.Gates, *I. fragrans* Matheny, Bougher & G.M.Gates, *I. lucifera* Matheny, Bougher & G.M.Gates, *I. meridionalis* Matheny, Bougher & G.M.Gates, *I. tasmanica* Matheny & Bougher, *I. tenax* Matheny & Bougher, *I. pluviosilvestris* Matheny & Bougher, *I. sylvicola* Matheny, Bougher & G.M.Gates, and *I. calamistratoides*. Continued collecting in this habitat type will likely yield additional taxa, given that many species of Inocybaceae have been documented with *Nothofagus* in New Zealand, Papua New Guinea, and southern South America.

Coastal shrub-heath habitat in southwestern Western Australia hosts several novel species. This type of habitat is characterized by presence of ECM shrubs of Myrtaceae and Fabaceae, at times mixed with *Agonis* (Myrtaceae but non-ECM), on sandy soils. Five species have been collected from the shrub-heath habitat: *I. avellaneifolia* Matheny & Bougher, *I. rufescens* Matheny & Bougher, *I. laureata* Matheny & Bougher, *I. exigens* Matheny & Bougher and *I. gomphocephala* Matheny & Bougher. None of these *Inocybe* species are closely related to each other.

As many as twenty species of Inocybaceae have been documented from Australian tropical latitudes (north of 23.5° S). These are typically associated with sclerophyll vegetation (e.g., *Eucalyptus grandis*, *Corymbia*, *Melaleuca*, *Acacia*, *Allocasuarina*). The more common or outstanding of these include *I. humidicola* Matheny & Bougher, *I. gracilissima*, *I. pyrrhopoda* Matheny & Bougher, *I. subangustifolia* Matheny, Bougher & Halling, *I. torresiae* Matheny, Bougher & M.D.Barrett, *I. araneosa* Matheny & Bougher, *Auritella brunnescens* Matheny & Bougher, *A. fulvella* Matheny & Bougher, *Tubariomyces australiensis* Matheny, Bougher & M.D.Barrett, and *T. umbilicatus* Matheny & Bougher, *nom. prov.* Inocybaceae from tropical aseasonal rainforests are not well known and are few. *Inocybe luteo-olivacea* is one such species putatively associated with *Acacia* from tropical rainforest habitats. Two species, *I. boreoaustralis* Matheny & Bougher and *I. rubricosa* Matheny & Bougher, have been documented from warm-temperate rainforest where *Lophostemon* is the only ECM plant present. Several of the tropically occurring species have been collected at higher latitudes in warm-temperate forests. These include *I. torresiae*, *I. flavoalbida* Matheny & Bougher (occurring also in Papua New Guinea), *I. alienospora*, *I. lasseroides* (E.Horak) Garrido, *I. sejuncta*, *I. invadens* Matheny, Bougher & G.M.Gates, *I. bulbinella* Matheny & Bougher, *I. sabulosa* Matheny & Bougher, *I. subtilior* Matheny & Bougher, and *Auritella brunnescens*. Overall, however, climate and latitude (tropical, temperate) are strong factors that govern the known geographical distributions of Inocybaceae in Australia.

Habitat associations can be diverse across groups of Australian Inocybaceae. Among the *I. australiensis* Cleland & Cheel group and its allies, for example, unrelated species occur in cool-temperate *Nothofagus* rainforest, whereas other species in the group are found in temperate eucalypt forests and/or dry sandy woodlands of *Acacia*, *Allocasuarina*, or *Eucalyptus*. *Nothofagus* rainforest associations are highly dispersed throughout the family phylogeny. These associations are likely derived.

As in the case with other macrofungi, seasonal rainfall and temperature conditions influence the fruiting phenology of Australian Inocybaceae. Phenological patterns vary by region as well.

Generally, fruiting coincides with autumn and winter precipitation in southern Australia. However, data about the phenology of most individual species of Australian Inocybaceae are not complete, because most species have not been collected or observed frequently enough. Furthermore, Inocybaceae have been poorly collected, or not collected, in many regions and locations throughout Australia, and therefore current knowledge about the geographical distribution and phenology of individual species is often based on few collections. Records are also biased by the timing of collecting, e.g. most collecting is in autumn/winter in southern Australia, and in the wet months in northern Australia. An individual species may have been recorded so far in few or only one location or region of Australia, perhaps at only one time of the year, but in light of the paucity of collecting effort, it cannot be ruled out that it may fruit elsewhere at the same or a different time.

Several species have been observed fruiting throughout much of the year. *I. cacaocolor* Matheny & Bougher, a widespread species in Queensland, New South Wales, and Western Australia, has been collected between March and July, and again in November. The common and widespread *I. leptospermi* (E.Horak) Garrido has a wide phenology, fruiting between February and November. One of the most common species of *Inocybe* in southern Australia, *I. froudistii*, has been collected between June and August and again between October and November. *Inocybe violaceocaulis* Matheny & Bougher, also common and widespread in southern Australia, has been recorded in May, August and October, and also December and June in Tasmania. *Inocybe viridipes* Matheny, Bougher & G.M.Gates, which occurs in Queensland and Tasmania, has been collected fruiting almost throughout the year, including records in February in Queensland and January, March, and November in Tasmania. *Auritella dolichocystis* Matheny, Trappe & Bougher has been recorded in spring (October-November) and again in early autumn (March–April) in New South Wales. In south-western Western Australia the season is more compressed owing to the Mediterranean climate, with rains coming mainly between May and July, but some Inocybaceae can be collected in April or August and less often in September.

Several species from Tasmania and/or New Zealand, though not as well-known as the species above, fruit in summer (November-January) and winter (May-June). *Inocybe caerulata* Matheny, Bougher & G.M.Gates has been collected in December and January in Tasmania but during May in New Zealand. *Inocybe cyanotrichia* Matheny, Bougher & G.M.Gates generally fruits in summer but has also been collected in May in Tasmania. *Inocybe viscata* (E.Horak) Garrido fruits in November and as late as June. At more northerly latitudes in the tropics, Inocybaceae can be regularly collected between January and April, coinciding with summer precipitation patterns. *Inocybe gracilissima*, common in northern Australia, has been collected between January and April. Another common tropical and warm-temperate species, *I. torresiae*, also fruits between January and April.

HISTORY OF THE STUDY OF AUSTRALIAN INOCYBACEAE

Historically, the Inocybaceae have been poorly studied in Australia and have been under-represented and often erroneously represented in the scientific literature and Australian field guides. Before this current work, approximately 60 unique names for species of Inocybaceae in Australia had been referred to in the literature, which Matheny & Bougher (2010) distilled to 17 accepted species.

In general it is reasonable to conclude that historical application of Northern Hemisphere names of Inocybaceae to indigenous Australian taxa is erroneous, unless confirmed by examination of voucher collections. The application of European names was often based on a broad morphological concept, and given subtle differences between many species of Inocybaceae, it was easy to mistake endemic Australian species for those familiar to European taxonomists, who examined material from Australia.

The first report of the Inocybaceae from Australia was based on a collection identified at the time as *Agaricus lanuginosus* Fr. (non Bull.) made by James Drummond in 1843 from the Swan River, Perth, W.A. This was reported by the Rev. M.J.Berkeley (Drumm. N. 229) in his 'Decades of Fungi' (1844–1856; see Hilton, 1983) and later listed by authors such as Cooke (1881, 1892). *Inocybe lanuginosa* is a Northern Hemisphere species; reports by Berkeley, Cooke and subsequent Australian records from various locations in Australia (see May &

Wood, 1997) probably refer to one or more of the austral species of the *I. lanuginosa* (Bull. : Fr.) P.Kumm. group documented in the current work (see Fig. 115). Of the few other Inocybaceae taxa described from Australian collections during the 19th century, most are now considered to represent other genera (Pegler, 1965). Cooke (1892) applied the European names *I. cincinnata* (Fr. : Fr.) Quél., *I. petiginosa* (Fr. : Fr.) Gillet, *I. plumosa* (Bolton : Fr.) Quél., and *I. flocculosa* Sacc. to Australian species of Inocybaceae, but the collections on which they were based cannot be located at Kew. McAlpine (1895) reported seven species of Inocybaceae from Australia based largely on Cooke's early works. Cleland & Cheel (1915) reported *I. perlata* (Cooke) Sacc. from Australia. Again, this is a European species and the name is likely to be a misapplication (*I. araneosa* Matheny & Bougher, described in the current work, is suggestive of *I. perlata*). Many other early Australian records may also be erroneous, e.g. Baker (1907) listed *I. flocculosa* Berk. (as "*Qnocybe*") as newly recorded in New South Wales. A few years later Cleland & Cheel (1918) reported four additional species from Australia, of which three (*I. albidipes* Cleland & Cheel, *I. australiensis*, and *I. subasterospora* Cleland & Cheel) were described as new. Of these, only *I. australiensis* can be reliably applied today (Matheny & Bougher, 2010).

Cleland (1933) added eight new species (four as *Astrosporina*, four as *Inocybe*) to the Australian Inocybaceae flora. In addition, he described a ninth species, *Naucoria arenacolens* Cleland, which is now accepted as *Auritella arenicolens* (Cleland) Matheny & Bougher (Matheny & Bougher, 2006a). Mathieson (1946) reported eight additional species of Inocybaceae from Australia, three un-named, three with European names (as *I. hirsuta* Lasch., *I. rimosa* Fr. and *I. tomentosa* (Jungh.) Quél.), and two of Cleland's species (*I. serrata* Cleland, *I. cf. murrayana* Cleland). Until 1987, various authors reported species new to the continent but, again, mostly using European names such as *I. asterospora* Quél., *I. casimiri* Velen., *I. fastigiata* (Schaeff. : Fr.) Quél., *I. geophylla* (Bull. : Fr.) P.Kumm., and *I. dulcamara* (Pers.) P.Kumm. The report of *I. posterula* (Britzelm.) Sacc. by Sinnott (1976) refers to an introduced species (*I. sindonia* (Fr.) P.Karst.), as documented in the current work.

Miller & Hilton (1987) described the new species *I. fibrillosibrunnea* O.K.Mill. & R.N.Hilton from Western Australia, and May (1989) reported the New Zealand species *I. scissa* (E.Horak) Garrido from Australia; the latter might represent *I. sylvicola* Matheny, Bougher & G.M.Gates documented here. Syme (1992) referred to two additional species, one as *I. geophylla* var. *lilacina* (Peck) Gillet, which represents the Australian endemic species *I. violaceocaulis* (Matheny & Bougher, 2005). Grgurinovic (1997), studying the collections of J.B. Cleland, added five species to the previously recorded Australian *Inocybe*. Three of these, however, represent species of *Cortinarius* and one a species of *Pholiota* (Matheny & Bougher, 2010). *Inocybe dewrangia* Grgur. was accepted by Matheny & Bougher (2010) and is included in the current work. Grgurinovic & Simpson (2001) re-listed for South Australia the same 14 species names as in Grgurinovic (1997). Studies by Matheny & Bougher (2006a, 2006b) resulted in the designation of the genus *Auritella*, which included *Naucoria arenacolens* and four new species from Australia.

Some erroneous interpretations of species of *Inocybe* have been perpetuated in the Australian literature until recent times. *Inocybe australiensis* has been misapplied in numerous field guides (e.g. Bougher & Syme, 1998; Fuhrer, 2005; Young, 1982; 1994), most likely to a taxon with affinities in the *I. tenax* group or, in the case of McMullan-Fisher *et al.* (2014), to a currently unnamed species. The illustrated reference to *I. australiensis* by Tate (2007) appears erroneous. *Inocybe eutheles* Sacc., as applied by Fuhrer (2005), is now recognized as *I. sindonia* (as documented in the present work). The current name for *I. austrofibrillosa* Grgur. (see in Fuhrer, 2005) is *Cortinarius fibrillosus* Cleland (Matheny & Bougher, 2010). *Inocybe austrofibrillosa* is erroneously presented in Bougher (2009) as a member of the *I. serrata* group. The depiction of *I. fibrillosibrunnea* in Gates & Ratkowsky (2014) is most likely the new species *I. fuscosquarrosa* Matheny, Bougher & G.M.Gates) now described in the *I. fibrillosibrunnea* group. There have been numerous records of *I. patouillardii* Bres. in Australia, and while some of the accounts may be accurate (e.g. Young, 2005), we have not examined any collections acceptable under this name. A listing of *I. patouillardii* from south-eastern Australia in association with pines by Dunstan *et al.* (1998), which was based on an account in Shepherd and Totterdell (1988), is possibly *I. sindonia*. Other Australian reports of *I. patouillardii* (e.g. Cribb, 1987; Tait, 2007) remain unconfirmed.

Based on revisions of type collections of Australian Inocybaceae, Matheny & Bougher (2010) provided an account of excluded and doubtful taxa, and accepted in the family 17 species and two genera, *Auritella* and *Inocybe*. Since that treatment, Bougher & Matheny (2011) documented two exotic species, *I. curvipes* P.Karst. and *I. rufuloides* Bon. The following year Bougher *et al.* (2012) described four new species of *Inocybe* from Australia and provided an expanded description of the species *I. emergens* (Cleland) Grgur. Other additional species of Inocybaceae reported from Australia since 2010 (outside this current work) include the European *I. griseolilacina* J.E.Lange (McMullan-Fisher *et al.*, 2014); based on the photo and description provided, this species appears to be very close to *I. bella* Matheny, Bougher & G.M.Gates described in the current work as new. Also in 2014, Gates & Ratkowsky (2014) reported two new species that are formally described in this current volume as *I. meridionalis* and *I. viridipes*.

In this current work the hitherto poorly studied and often misrepresented phylogenetic diversity of Inocybaceae in Australia is brought to light: 139 taxa, including 137 species, are documented, representing six of the seven major clades in the family, with 101 species described as new (Appendix 1). Numerous additional and undescribed species have been detected during this investigation, but these require further collections and studies to enable formal documentation. We estimate that at least 200 species of Inocybaceae are likely to occur in Australia.

HISTORICAL AND PRESENT-DAY BIOGEOGRAPHY

Evolution of Inocybaceae in Australia, including in the context of other biogeographic regions, is complicated by several factors, but some observations and hypotheses about the accumulation of the family-level diversity in Australia can be discussed at this time. How effectively such hypotheses can be tested is debatable, as the main challenges to infer processes underlying the observed patterns of diversity include dating the origins of such diversity and adequacy of taxon sampling. Despite these limitations, data so far indicate that the evolution of the Inocybaceae in Australia has likely occurred in spurts, with long-distance dispersal a key element, resulting in most of the diversity observed today (Matheny *et al.*, 2009). These observations are supported by the presence of numerous independent Australian lineages in six of the seven major clades of the family and the inference from dated phylogenetic studies that the crown group origins for the majority of these lineages are much too young to be explained by the ancient breakup of Gondwana (Matheny *et al.*, 2009). One exception to this may be the genus *Auritella*.

Auritella - a Gondwanan lineage?

Auritella is a genus now known to comprise at least twelve species, with a geographical distribution in Africa, India, and Australia. A preliminary attempt to date the origin of the genus was presented by Matheny & Bougher (2006a), which suggested an initial split between African and Australian lineages during the late Cretaceous (c. 84 Mya). The later work by Matheny *et al.* (2009) estimated the origin of *Auritella* between 38–105 Mya ago (median age at 68 My). Assuming migration between Africa and Australia remained possible up until 70–95 Mya, an ancient Gondwanan split of *Auritella* cannot be rejected. By contrast, an investigation into the origins of ECM clades of Agaricales by Ryberg & Matheny (2012) suggests a considerably younger origin for *Auritella* between 25–56 Mya. It should be noted that the discovery of *A. foveata* from tropical dipterocarp forests of India (Matheny *et al.*, 2012b) pushes the crown group node of *Auritella* further back in time. Thus, it seems reasonable to suppose that *Auritella* began to diversify during the Cretaceous, followed by splits between Indian, African, and Australian lineages coinciding with the breakup of Gondwana. Additional research is required to test this hypothesis and resolve the dating discrepancies.

Overall, species of *Auritella* are not widely distributed in Australia. In contrast to the *Inocybe* clade, and assuming a Cretaceous origin, the overall net rate of diversification of *Auritella* is quite slow. The same is true for all the other depauperate lineages (i.e., those lacking pleurocystidia) of the family Inocybaceae in Australia. Further phylogenetic work is needed to firmly resolve the evolutionary position of African *Auritella* with respect to those from

Australia. Our present work indicates that the African lineage, containing three species, is nested within the Australian lineage (Fig. 153), but this result has weak support.

Biogeography of depauperate clades of Inocybaceae in Australia

The depauperate clades of Inocybaceae are characterized by the absence of pleurocystidia and presence of smooth spores. Australia is relatively poor in species of the *Inosperma* clade compared to the Northern Hemisphere. Most Australian species (four known to date, all in sect. *Cervicolores*) are associated with sclerophyll vegetation and form a monophyletic group. *Inocybe calamistratoides*, a trans-Tasman species, is the only one found in association with *Nothofagus*. All are found in the eastern states of Australia, and three occur in Tasmania.

Whereas species of the *Inosperma* clade have only been documented from Tasmania and other eastern states of Australia, Western Australia is host to five species of the *Mallocybe* clade. Two of these are also known from Queensland. No species of *Mallocybe* are known from Tasmania or New Zealand, and it is questionable whether any *Mallocybe* clade species occur in New Guinea. As with Australian *Inosperma* species, the five Australian species of *Mallocybe* form a monophyletic group. All are associated with sclerophyll vegetation, and at least two are both temperate and tropical in distribution.

This work presents the first documentation of *Tubariomyces* from Australia. Two species, *T. australiensis* and *T. umbilicatus*, have a tropical distribution in eucalypt woodlands. Both species, however, are known only from single collections, *T. australiensis* from northern Western Australia and *T. umbilicatus* from far north Queensland. In the absence of DNA sequence data, it is not known if the Australian species form a clade and how they are related to the four other species in the genus. *Tubariomyces* now has a known biogeographic distribution in the Mediterranean region, dry tropical Africa (Zambia), and northern Australia. Matheny *et al.* (2009) suggested a late Cretaceous split in *Tubariomyces* (referred to as the "*Mallocybella*" clade), but the potential error around this date is high (95% confidence interval between 33–120 Mya). Ryberg & Matheny (2012) dated the initial split of *Tubariomyces* approximately 30 Mya in the early Oligocene (95% CI: 13–50 Mya).

Unlike the *Inosperma* and *Mallocybe* clades, Australian species of the *Pseudosperma* clade are polyphyletic, indicating independent dispersals into Australia. However, *I. pluviorum*, a warm-temperate rainforest species, is strongly supported as sister to the rest of the *Pseudosperma* clade (Fig. 136). Most Australian species of the *Pseudosperma* clade are tropical in distribution.

Australian species of the Inocybe clade are highly polyphyletic, indicative of long-distance dispersal

At least 24 independent lineages of the *Inocybe* clade have been detected in Australia, encompassing nearly 140 species documented in the current work. Indeed, 51 species are scattered across 23 of these lineages. Various biogeographic patterns among these 23 lineages are summarized below, followed by discussion of a highly diverse clade of austral taxa (STAC) comprising 63 species that have evolved autochthonously across southern South America, temperate Australia, and New Zealand.

Evidence for recent biogeographic relationships between Australia, Papua New Guinea, and south-eastern Asia

Numerous phylogenetic studies indicate a close affinity between Australia and its northern neighbours, most likely of recent origin (Byrne *et al.*, 2011). *Inocybe parvifulva* Matheny & Bougher, a warm-temperate rainforest species from Queensland, forms a sister lineage with *I. papillata* (E.Horak) Garrido from New Guinea (Fig. 134). *Inocybe luteo-olivacea* is most closely related to Thai specimens of *I. sphaerospora* Kobayasi, the latter recently re-documented in Horak, *et al.* (2015). *Inocybe spiniformis* Matheny & Bougher and *I. acanthosperma* Matheny & Bougher, two spinose-spored species together forming a monophyletic group, are the sister lineage to the Thai species *I. thailandica* E.Horak, Matheny & Desjardin, also with spinose spores (Fig. 100). The stellate-spored pair *I. nobilissima* Matheny & Bougher and *I. vagata* Matheny & Bougher (which also occurs in New Zealand) are derived with strong support from south-eastern Asian and northern temperate ancestors (Fig. 93). Australian populations of *I. flavoalbida* are likely just beginning to diverge from populations in New Guinea (Fig. 76). Samples of *I. alienospora* and *I. lasseroides* from Queensland form a robust sister group to Thai and Indian samples of species related to *I. hydrocybiformis* (Corner & E.Horak) Garrido (Fig. 84). The tropical Australian species *I. subangustifolia* is most closely related to Asian samples of *I. angustifolia* (Corner & E.Horak) Garrido and appears to be derived from it. *Inocybe sejuncta* and *I. tumidula* Matheny & Bougher are nested and recently derived from several species in south-eastern Asia and temperate Europe (Fig. 96). The cool-temperate species *I. viscata*, which occurs in Tasmania and New Zealand, is evolutionarily derived from tropical lineages present in Queensland and south-eastern Asia (Fig. 128).

Two species that occur in Papua New Guinea, *I. flavoalbida* and *I. lasseroides*, have now been confirmed in tropical and/or warm-temperate eastern Australia. With additional surveys it is expected that more New Guinean taxa may be found in Australia. Additionally, the finding in this current work of *I. alienospora* in Australia (the type is from Singapore) suggests that such relationships may extend further than New Guinea into other regions of south-eastern Asia.

New Zealand taxa are recently derived from Australian ancestors

In addition to four species from New Zealand demonstrated as derived from Australian ancestors less than 35 Mya ago (Matheny *et al.*, 2009), we identify here additional New Zealand lineages derived from Australian clades. In the *I. assimilata* Britzelm. group, the New Zealand species *I. subclavata* (E.Horak) Garrido is derived from a clade containing the Australian taxa *I. cacaocolor*, *I. alloumbrina*, *I. subferruginea* Matheny & Bougher, *I. solitaria* Matheny, Bougher & G.M.Gates, and *I. cf. neoumbrina* Kobayasi *sensu* E.Horak (Fig. 87). In the *Inosperma* clade, *I. latericia* E.Horak is recently derived (Fig. 142), and the trans-Tasman species *I. vagata*, *I. luteobulbosa* E.Horak, *I. viscata*, *I. serratoides* Matheny & Bougher, and *I. caerulata* also appear derived from Australian ancestors. Relationships of New Zealand taxa to Australian lineages in the *I. leptospermi* group, however, are not resolved.

Pan-temperate biogeographical relationships

Several species exhbit pan-temperate biogeographical relationships. *Inocybe invadens* is strongly supported as derived within the northern temperate lineage containing *I. praetervisa* and allies (Fig. 100). Austral lineages in the *I. assimilata* group are derived from northern temperate ancestors (Fig. 87). *Inocybe violaceocaulis* is the sole Australian lineage nested within the northern temperate *I. flocculosa* Sacc. group (Matheny *et al.*, 2009; Braaten *et al.*, 2014). Given the shallow branch lengths shared among these taxa, it is evident they dispersed recently into Australia. However, the northern temperate *I. lanuginosa* group is unquestionably derived from austral ancestors (Fig. 115). Several other smooth-spored species, *I. austrosquarrosa* Matheny, Bougher & G.M.Gates, *I. bella* and *I. rufescens,* are not well-resolved phylogenetically and have no clear alliance with any other austral taxa. Matheny *et al.* (2009) showed that *I. spadicea* Matheny & Bougher is the sister lineage to the STAC group (discussed below) using nLSU, *rpb1*, and *rpb2* sequences. However, this result is not recovered using ITS, nLSU, and *rpb2* DNA sequences here.

The smooth-spored temperate austral clade (STAC): subantarctic diversification of a group of smooth-spored Inocybe

The STAC group is an endemic Southern Hemisphere clade characterized by high taxonomic diversity (63 species are presented in this current work) and smooth spores. All are exclusively temperate in distribution, occurring mainly in temperate Australia but also found in southern South America and New Zealand. If *I. spadicea*, which has poor support as the sister lineage to STAC, is excluded, then some studies suggest that the group began to diversify in the mid-Eocene about 45 Mya (Matheny *et al.*, 2009), although others (Ryberg & Matheny, 2012) detected a more recent origin about 24 Mya (95% CI: 15–34 Mya). A younger origin may have coincided with the expansion of sclerophyll biomes in Australia between 10–25 Mya as the continent became cooler, dryer, and developed seasonal climates. During this period, eucalypts, some Fabaceae, and Casuarinaceae underwent radiations (Crisp *et al.*, 2004) perhaps promoting co-diversification of the STAC group. The number of *Nothofagus*-associated taxa is considerably less in STAC, which would be consistent with the comparable lack of diversification in *Nothofagus* and contraction and regional extinction of this mesic biome (Crisp *et al.*, 2004; Byrne *et al.*, 2011). Although 63 species in the STAC group are documented here, more are known and awaiting corroboration from additional collections before formal description.

Short-range endemism

May (2002) demonstrated a lack of short-range endemism among macrofungi in Australia, a conclusion that contrasts with the high degree of endemism observed in plants within Australia (Hopper & Gioia, 2004). To assess the degree of short-range endemism among Australian Inocybaceae, we filtered the 139 taxa documented in this current work to eliminate taxa represented by fewer than four collections or known from only one collecting locality. This process resulted in consideration of 63 species of *Auritella* and *Inocybe* in total. Of these 63 species, 26 are restricted to eastern Australia, 14 occur only in Western Australia, one occurs in both Western Australia and South Australia, and 23 species are widespread, occurring both in eastern and western regions of the country. Seven species also have a trans-Tasman distribution (occurring in Australia and New Zealand). Two of the eastern Australian species are exclusively tropical. Based on available data 63% (40 of 63 species) appear to be short-range endemics. If the seven trans-Tasman species are excluded 52% of the 63 species are short-range endemics and known only from Australia. This indicates that nearly half of Australian Inocybaceae may be endemic to sub-regions of the continent.

Eastern Australian and trans-Tasman species of Inocybaceae include: *Auritella brunnescens*, *A. dolichocystis*, *A. serpentinocystis* Matheny, Trappe & Bougher, *Inocybe atrisquamosa* Matheny & Bougher, *I. auritelloides*, *I. caerulata* (trans-Tasman), *I. calamistratoides* (trans-Tasman) *I. cyanotrichia*, *I. fuscosquarrosa*, *I. gracilissima*, *I. humidicola*, *I. lasseroides*, *I. luteobulbosa* (trans-Tasman), *I. meridionalis*, *I. nobilissima*, *I. pallidiflavipes* Matheny, Bougher & G.M.Gates, *I. pluviosilvestris*, *I. parvifulva*, *I. pluvialis* Matheny, Bougher & G.M.Gates, *I. serratoides* (trans-Tasman), *I. sylvicola*, *I. vagata* (trans-Tasman), *I. viridipes*, and *I. viscata* (trans-Tasman). The fourteen species known so far only from Western Australia include: *I. bicornis* Braaten, Bougher & Matheny, *I. campestris* Matheny & Bougher, *I. emergens* (Western Australia and (type) South Australia), *I. farinosipes* Matheny, Bougher & Ryberg, *I. floccosibrunnea* Matheny & Bougher, *I. isabellina* Matheny & Bougher, *I. leucotaenia* Matheny & Bougher, *I. mallocyboides* Matheny, Bougher, R.M.Rob. & K.Syme, *I. nigricans* Matheny, Bougher & Ryberg, *I. sabulosa*, *I. subferruginea*, *I. subtilior*, *I. vittata* Matheny, Bougher & Ryberg, and *I. xanthocystis* Matheny & Bougher. Of these, two (*I. sabulosa*, *I. subtilior*) occur in tropical and temperate regions of Western Australia.

Widespread species confirmed to occur in both eastern and western Australian states include: *I. albofloccosa* Matheny & Bougher, *I. brunneidisca* Matheny, Bougher & G.M.Gates, *I. acaciae* Matheny & Bougher, *I. australiensis*, *I. bulbinella*, *I. calocephala* Matheny & Bougher, *I. calopedes* Matheny & Bougher, *I. cacaocolor*, *I. froudistii*, *I. fulvilubrica* Matheny, Bougher & G.M.Gates, *I. geniculata* Matheny & Bougher, *I. invadens*, *I. leptospermi* (trans-Tasman), *I. nudiuscula* Matheny & Bougher, *I. pyrrhopoda*, *I. redolens* Matheny, Bougher & G.M.Gates, *I. sejuncta*, *I. serrata* (clade I), *I. serrata* (clade II), *I. serrata* (clade III), *I. subflavospora* Matheny & Bougher, *I. torresiae*, and *I. violaceocaulis*.

STUDYING AUSTRALIAN INOCYBACEAE

Collecting fresh fruit bodies

In Australia, as elsewhere, finding Inocybaceae in the field often requires keen eyesight and thorough searching as their fruit bodies are mostly small, brownish, and generally inconspicuous. Many species seem to be found more often in open areas and less often in thick litter. Some occur in sand or bare areas. Other species favour mossy areas (including infrequently on logs) or disturbed areas. All occur near ectomycorrhizal shrubs or trees such as *Acacia*, *Allocasuarina*, *Eucalyptus*, *Gastrolobium*, and *Nothofagus*.

It is not always possible to readily discern and confidently recognise fruit bodies spotted in the field as members of the Inocybaceae. Many species of Australian Inocybaceae do not necessarily have the conical and coarsely radially fibrous pileus typical of the family elsewhere, and therefore may resemble members of other families. Indeed, a wide range of other fungi have similar fruit bodies - often referred to, along with those of the Inocybaceae, as "LBMs - little brown mushrooms". Nevertheless, membership of the Inocybaceae often can be confirmed in the field by careful examination of fruit bodies with a hand lens. Identification can often be achieved to species level based on fruit body characteristics visible on fresh specimens. The best way to recognize and identify Australian Inocybaceae in the field accurately is to become familiar on a first-hand basis with some of the more common species, and to record macroscopic characters accurately. Beyond the field, detailed microscopic examination, and perhaps molecular sequencing, will render the most accurate identifications.

The likelihood of correctly identifying freshly collected specimens of Australian Inocybaceae may be influenced by how the specimens are handled and by the type of data recorded before the fresh specimens are dried and vouchered. Many vouchered specimens of Inocybaceae currently lodged in Australian herbaria may never be identified to species level because they lack adequate descriptive field notes and images. Molecular annotation using DNA sequences is a viable recourse for identification but may be problematic for specimens that were poorly preserved.

The fruit bodies of many species of Inocybaceae have a similar general appearance, and multiple species can sometimes fruit among each other. Thus, a prime consideration when collecting fresh specimens concerns unintentionally collecting representatives of multiple, rather than a single species. Vouchered collections of Inocybaceae in Australian herbaria have often been mixed unintentionally. Some steps can be taken in the field to minimise the possibility of creating mixed collections. When compiling a collection for vouchering it is best to include only specimens that are tightly clustered together and avoid those found at a distance or in other clusters. Discard any specimens that have a different appearance to the majority. Mixed collections may be sorted out later microscopically, and by nucleotide sequence data, but it is desirable to minimize the need.

Before collecting specimens it is desirable to capture close-up images of the fruit bodies *in situ*, preferably showing a number of specimens of various size and maturity. Such images record the positional relationship of the collection with its microhabitat. Also the images can capture delicate features of the specimens that may rapidly dry out, change colour, or be removed by handling (such as any pruinosity on the stipe). In addition, it is desirable to capture images of the surrounding habitat and vegetation. Accompanying notes about the habitat should be recorded, including soil and vegetation type, and nearby plant species.

When uprooting specimens of Inocybaceae from the substrate with a knife or a similar instrument, it is essential to dig well under the specimens to ensure that the base of the stipe is sampled intact. Specimens should not be handled by the stipe as this may remove or disrupt fibrils and pruinosity on the stipe. Any very young fruit bodies and primordia should be examined with a hand lens at this time to determine the presence of a veil (typically as a cortina), as this structure is usually evanescent and easily damaged by handling. The odour of specimens should also be assessed after cutting fruit bodies longitudinally. Rain-soaked specimens should be avoided as they may appear quite different from other specimens, and not yield an otherwise-present odour.

Macroscopic Characters

Size and general appearance

Species of Australian Inocybaceae vary greatly in size, shape, colour and texture. These attributes can be influenced by environmental factors such as moisture, rain, and abrasiveness of the soil/litter against delicate anatomical structures. Many species of the family in Australia have small tricholomatoid or collybioid fruit bodies. Some are omphalinoid or tubarioid as in *Tubariomyces* (Pl. 265, 266). Furthermore, some Australian species closely resemble other fungi. For example, one collection from tropical far north Queensland (PERTH 08362866) related to *Inocybe alienospora* has fruit bodies strongly resembling members of *Cortinarius* subgen. *Telamonia* (Pl. 1E). Other examples include *I. sylvicola*, which may resemble *Bolbitius* Fr. due to its stature and pallid yellowish colours (Pl. 208, 209), *I. viscata* and *I. olivaceohinnulea* Matheny & Bougher that have some attributes typical of *Hebeloma* (a viscid pileus and pinkish brown lamellae respectively; Pl. 212, 213; Pl. 89), and some collections of the *I. serrata* complex have diminutive fruit bodies that resemble *Mycena* (Pers.) Roussel (Pl. 132, 133). A few Australian Inocybaceae have sequestrate (secotioid or truffle-like) fruit bodies, e.g. *I. bicornis* f. *secotioides* Braaten, Bougher & Matheny (Pl. 1D, 43), *Auritella geoaustralis* Matheny & Bougher (Pl. 263), and an undescribed collection from New South Wales related to *I. nudiuscula*.

In some cases, different species with fruit bodies that are similar in stature and overall appearance may not necessarily be closely related. For example, *I. ionocaulis* Matheny & Bougher (Pl. 91, 92) closely resembles members of the *I. serrata* complex in size, stature, the whitish pileus margin, pruinose upper stipe, and vesture of the lower stipe. Microscopically there are also similarities in the size and shape of spores and hymenial cystidia. However, phylogenetically, *I. ionocaulis* is distantly related to the *I. serrata* group and closely allied with the *I. ornatipes* Matheny & Bougher group (see Figs. 7, 45).

Pileus

The pileus of most Australian Inocybaceae is less than 30 mm wide, and some of the smaller species may have pilei that are only several millimetres wide (e.g. *I. serrata* complex, *Tubariomyces*). However, the largest species may have pilei up to *c.* 80 mm wide (e.g. *Auritella robusta* Matheny, Bougher & G.M.Gates). In the taxonomic descriptions we refer to species having "small" fruit bodies as those with a pileus consistently less than 20 (–25) mm diameter and a narrow stipe 1–2 mm in width. "Medium" sized fruit bodies are more than 20 mm diameter but less than 40 mm wide. "Large" (or "robust") fruit bodies are more than 40 mm diameter.

Common pileus shapes include campanulate, conic, convex, or combinations of these at various stages of maturity (Pl. 1). An umbo may or may not be present (Pl. 1A, E, F). The pileus of *Tubariomyces umbilicatus* has a distinct central depression (Pl. 267). However, centrally depressed pilei are not generally common among Australian Inocybaceae, and the fully expanded pileus of many species may not always become markedly upturned at maturity. The strength and persistence of the velipellis may influence pileus shape, because when persistent in dry conditions it may restrict development of an umbo (Kuyper, 1986). In some species the shape of the pileus can be variable and is probably influenced by environmental conditions.

The surface texture of the pileus is particularly informative at the species level. Most species have a dry pileus, but a viscid pileus is characteristic of *I. viscata* (Pl. 212, 213). Some species have a distinctive surface under wet conditions, e.g. the pileus of *I. fulvilubrica* is soapy-lubricous when wet (Pl. 184, 185). Other species such as *I. parvifulva* (Pl. 2J, 221, 222) may appear almost translucent-striate, throughout or near the margin only, when wet (*e.g. I. spiniformis*; Pl. 171, 172).

In some species of Australian Inocybaceae, the surface is predominantly smooth, as in *I. torresiae* (Pl. 2A). However, surface texture in most species is largely influenced by the presence of radially arranged squamules or fibrils derived from the pileipellis. The persistence, size, shape, and density of these surface structures varies between different species (Pl. 2). For example, *I. ornatipes* has conspicuous white, cottony, recurved and erect, sharp-pointed scales (Pl. 2H), whereas *I. clypeata* Matheny & Bougher is densely radially fibrillose (Pl. 1B). Some species have a rimose pileus whereby the surface fibrils split and crack in a radial manner often revealing paler context beneath. The radial fibrils may be coarse, as in *I. calocephala* (Pl. 2B), or fine, as in *I. araneosa* (Pl. 2D, I).

Surface texture of the pileus can also be influenced by the presence of superficial fibrils derived from a universal veil. Such remnants on the pileus are referred to as the velipellis, which closely adheres to the pileus and forms white or greyish, matted-fibrillose or filmy patches. With age it may be conspicuous and persistent on almost the entire pileus as in *I. farinosipes* (Pl. 48–50), persistent only at the centre as in *I. avellaneifolia* (Pl. 66), or evanescent as in *I. fulvilubrica* (Pl. 84, 85). The velipellis may be more conspicuous in dry conditions, as in *I. araneosa* (Pl. 4H) but may be washed off in wet conditions, as in *I. froudistii* (Pl. 4G). Veil remnants at or overhanging the pileus margin are rarely coarse (Pl. 4C) and persistent (Pl. 4F), and when present are usually only visible in very young fruit bodies, e.g. *I. dewrangia* (Pl. 4B), and *I. albofloccosa* (Pl. 4E).

Most Australian Inocybaceae have predominantly brownish fruit bodies (e.g. chestnut, dark brown, ferruginous, fulvous, hazel, olive brown, tan, yellowish brown). All Australian *Auritella* and *Tubariomyces* species are brownish but some of these discolour to blackish brown with age, e.g. *A. robusta* (Pl. 264). However, the following additional pileus colors are also observed: ivory white in *I. eburnea* Matheny, Bougher & G.M.Gates (Pl. 181), bright yellow, greenish yellow or olive-yellow at least when young in species such as *I. luteo-olivacea* (Pl. 139, 140) and *I. subangustifolia* (Pl. 159, 160), straw-yellow to yellow in *I. sylvicola* (Pl. 208, 209), golden-brown in *I. aureorimosa* Matheny & Bougher (Pl. 23, 24), bright rusty brown or bright reddish brown in *I. rufobadia* Matheny & Bougher (Pl. 38, 39), reddish brown or bright brown in *I. spadicea* (Pl. 143), pinkish to reddish brown when young in *I. roseola* Matheny, Bougher & G.M.Gates (Pl. 108), and greyish blue or blue-green tones and a blue margin when young in *I. caerulata* (Pl. 117). In some species, only a portion of the pileus is coloured. For example, *I. gracilissima* co-occurs in Queensland with another member of the *Pseudosperma* clade, *I. araneosa*, but may be distinguished from that species by its deeply conical pileus when young with a pale olive-yellow to yellowish margin (Pl. 228, 229).

For some species, the colour of the pileus changes with maturity. For example, the pileus of *I. araneosa* may be distinctly bi-coloured when young, with a yellowish brown centre and pale brown margin, but uniform at maturity (Pl. 223, 224). In most species with a brown pileus, somewhat subtle changes in shading occur with age, but in species that are not brown the changes may be more conspicuous, e.g. the initially violet pileus of *I. violaceocaulis* becomes predominantly brown at maturity (Pl. 145, 146).

Pileus colour can be influenced by the presence of superficial fibrils derived from the veil. In the *I. serrata* complex, for example, it is not unusual to find adjacent specimens (putatively produced by the same mycelium) with distinctly paler and darker pilei. This difference is largely due to the persistence of a strong velipellis on the pale specimens.

Only a few Australian Inocybaceae are distinctly hygrophanous (e.g. the pileus of *I. roseola*, which is pinkish brown to reddish brown fading to cream; Pl. 108). No Australian species of *Auritella* or *Tubariomyces* are hygrophanous. Albino variants of *Inocybe* species have been reported worldwide (Kuyper, 1986). We have not confirmed such variants in Australia, but *I. roseola* features very pale (hyaline to pale buff) basidiospores. A collection of *I. clypeata* with atypically pale fruit bodies has been found in Western Australia (PERTH 08363706), but it has normal, brown-pigmented spores.

Lamellae

Most Australian Inocybaceae have adnexed, adnate, or receding lamellae. Those of *I. calocephala* may be nearly free, while many species have a notch (sinuate) near the attachement to the stipe. The lamellae may be shortly decurrent in species such as *Tubariomyces australiensis* (Pl. 265, 266). They are unbranched, lack anastomoses, and have lamellulae in several tiers. The lamellar margin of most species is usually finely uneven or fimbriate and whitish when viewed with a hand lens, due to the presence of cheilocystidia and/or paracystidia; this is usually most conspicuous in young specimens. The edges of the lamellae are more coarsely fimbriate in species of *Auritella*, *Tubariomyces*, the *Inosperma* clade, and the *Pseudosperma* clade.

All Australian Inocybaceae have predominantly brown lamellae at full maturity, but subtle colour variations can be significant for characterising some species. *Auritella* species tend to have darker or more sombre brown lamellae at maturity than species of *Inocybe*, largely due to the presence of abundant pigmented necrobasidia. Young lamellae are often whitish, greyish or yellowish, but sometimes appear to be brown from a very early stage as in *I. albofloccosa*. Other colours include light grey in *I. farinosipes* (Pl. 3E), pink or avellaneous becoming brown in *I. avellaneifolia* (Pl. 3F), olive-tinged becoming dark brown with age in *I. alloumbrina* (Pl. 151), greenish to bright yellow becoming pale olive-yellow or dull yellow-olivaceous in *I. subangustifolia* (Pl. 3D), yellow at first in some collections of *I. australiensis* (Pl. 3G), pale first then yellow in *I. sabulosa* (Pl. 3H), pinkish brown becoming dark yellowish brown in *I. olivaceohinnulea* (Pl. 89), and lilac- or violet-tinged when young in *I. bella* (Pl. 4A).

Plate 1. Some pileus and fruit body forms in the Australian Inocybaceae. **A,** Campanulate pileus with well developed umbo - *Inocybe sp.* (PERTH 08318808); **B,** Convex pilei - *Inocybe clypeata* (PERTH 08320268); **C**, Conical pileus - *Inocybe sp.* (PERTH 08373124); **D**, Secotioid (truffle-like) fruit bodies - *Inocybe bicornis* f. *secotioides* (PERTH 08320403); **E**, *Telamonia*-like fruit bodies - *Inocybe sp.* (PERTH 08362866); **F**, Conico-campanulate pilei with umbo - *Inocybe sp.* cf. *tenax* (PERTH 08362971).

Photographs — A to D — N.L. Bougher; E, F — P.B. Matheny.

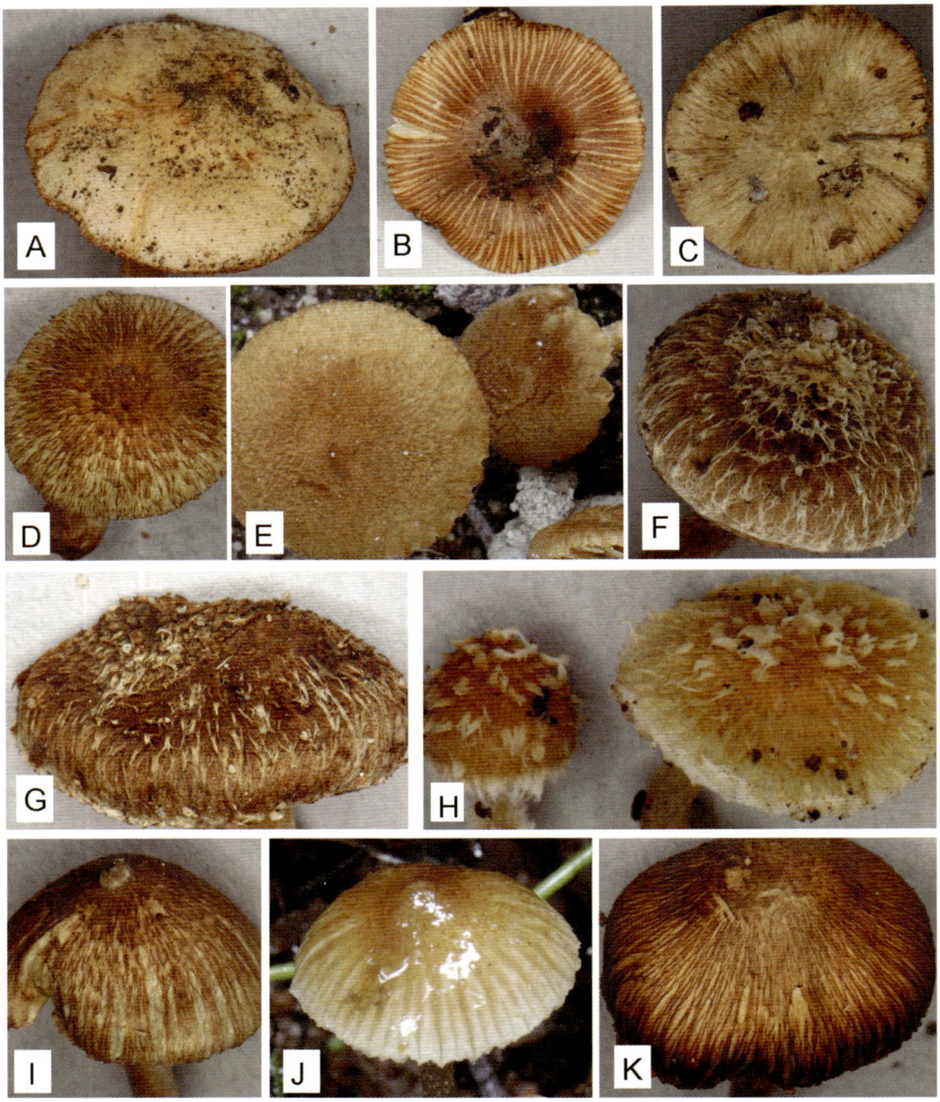

Plate 2. Pileus surfaces in the Australian Inocybaceae. **A,** Smooth pileus - *Inocybe torresiae* (PERTH 08364842); **B**, Strongly rimose pileus - *Inocybe calocephala* (unvouchered); **C**, Finely fibrillose pileus with fibrils adpressed on the pileus - *Inocybe sp.* (unvouchered); **D**, Small squamules at centre merging into coarse radial fibrils which diverge towards the margin - *Inocybe araneosa* (PERTH 08364710); **E**, Finely felty pileus - *Inocybe sp.* (PERTH 08373086); **F**, Densely matted fibrillose-squamulose surface, becoming more fibrillose and appressed towards the margin - *Inocybe fibrillosibrunnea* (PERTH 08318816); **G**, Coarsely radially squarrose pileus with brown or pale-tipped squamules - *Inocybe memorialis* (PERTH 08073317); **H**, Pileus with large recurved and erect sharp-pointed white squamules. - *Inocybe ornatipes* (PERTH 08072906); **I,** Rimose with diverging radial fibrils. - *Inocybe araneosa* (PERTH 08364710); **J**, Translucent-striate pileus in wet conditions - *Inocybe parvifulva* (PERTH 08363064); **K**, Densely radially fibrillose pileus becoming rimose as the fibrils diverge - *Inocybe sp.* (PERTH 08364699).

Photographs — N.L. Bougher, except J — P.B. Matheny.

Plate 3. Fruit bodies context and lamellae in the Australian Inocybaceae. **A,** Brunnescent context, mainly in the stipe - *Auritella brunnescens* (PERTH 08364567); **B**, Rufescent context in the stipe - *Inocybe leucotaenia* (PERTH 08072477); **C**, Rufescent context, mainly in the stipe - *Inocybe rufescens* (PERTH 08383324); **D**, Yellow context - *Inocybe subangustifolia* (PERTH 08364850); **E**, Pale lamellae - *Inocybe farinosipes* (PERTH 08516642); **F**, Pinkish (avellaneous) lamellae - *Inocybe avellaneifolia* (PERTH 08319391); **G**, Yellow lamellae - *Inocybe australiensis* (PERTH 08241724); **H**, Lamellae pale at first becoming yellow - *Inocybe sabulosa* (PERTH 08320292); **I**, Cystidiate lamellae margins - *Auritella brunnescens* (PERTH 08364567).

Photographs — N.L. Bougher, except F, G — P.B. Matheny.

Plate 4. Veil structures in the Australian Inocybaceae. **A,** Thick, intact cortina. - *Inocybe violaceocaulis* (PERTH 08318654); **B**, Cortina rupturing at young stage, with cortinoid fibrils still evident at the pileus margin and stipe base - *Inocybe invadens* (PERTH 08319286); **C**, White velipellis covering entire pileus, and conspicuous appendiculate veil remnants at the pileus margin. - *Inocybe serrata* complex (PERTH 08383294); **D**, Sparse, rapidly evanescent cortina - *Inocybe sp.* (Rimosae) (PERTH 08364761); **E**, Cortina leaving white aggregated fibrils on a young pileus - *Inocybe albofloccosa* (PERTH 08318697); **F**, White pileus margin derived from the cortina - *Inocybe leucotaenia* (PERTH 08363684); **G**, Adpressed whitish velipellis visible on the pileus on the left, washed off the other pileus which was more exposed to rain - *Inocybe froudistii* (PERTH 08320284); **H**, Velipellis visible in dry conditions as an adhering, white, thin central patch. - *Inocybe araneosa* (PERTH 08364729).

Photographs — N.L. Bougher, except B, C — P.B. Matheny.

Plate 5. Stipe base and stipe surface in the Australian Inocybaceae. **A,** Stipe with marginate bulb - *Inocybe emergens* (PERTH 08320462); **B,** Stipe with marginate bulb - *Inocybe nobilissima* (PERTH 08364788); **C,** Stipe entirely pruinose - *Inocybe nobilissima* (PERTH 08364788); **D,** Stipe entirely with thick white, woolly vesture - *Inocybe fibrillosibrunnea* (PERTH 08073937); **E,** Pruinosity conspicuous on the reddish-brown stipe - *Inocybe serrata* complex (PERTH 08320713).

Photographs — N.L. Bougher.

Stipe

Most species of Australian Inocybaceae have a stipe less than 40 mm in length, but many are consistently shorter and some are considerably longer. For example, the stipe in *Inocybe nobilissima* may reach at least 75 mm long. In many species, the size of the stipe varies greatly within a single collection and between different collections. An example of such variation is evident with *I. laureata* where in one collection (PERTH 08318476) the stipe is more elongate in relation to the pileus size than in a second, stockier collection (TENN 066477) (Pl. 73, 74).

The stipe is central in all species, but an occasional fruit body with an eccentric stipe may be encountered. The stipe is usually cylindrical or occasionally compressed but may be contorted or flexuous, especially when emerging through thick litter (e.g. *I. geniculata*; Pl. 81–83). The interior is typically solid or stuffed but with age a central hollow sometimes develops in the upper stipe. The base may be undifferentiated, merely swollen, or distinctly bulbous and marginate as in *I. emergens* (Pl. 5A) and *I. nobilissima* (Pl. 5B).

Assessments of the stipe surface should be carried out using a hand lens and with minimal handling to avoid removing delicate surface features. No species of Australian Inocybaceae have been shown to have a persistent annulus, however, a persistent ring-like zone may be present in *I. hymenocarpa* Matheny & Bougher (Pl. 75). In cortinoid species, the cortina is often rapidly evanescent and usually observed only on very young fruit bodies (Pl. 4A, B, D, E). Similarly, there are rarely any readily visible veil remnants attached to the stipe base, as the attachment is usually only intact in very young fruit bodies (such as in *I. dewrangia*; Pl. 4B).

The texture of the stipe surface is an important taxonomic feature. The stipe of *I. gracilissima* appears smooth other than the presence of furfuraceous fibrils at the extreme apex (Pl. 228, 229). For the majority of species, the stipe is pruinose and/or fibrillose to some extent. Pruinosity is usually due to presence of metuloid caulocystidia; the density and distribution of pruinosity over the stipe surface varies among different species. For example, the entire stipe is pruinose in *I. calocephala* (Pl. 201, 202), *I. nobilissima* (Pl. 5C), *I. tenax* (Pl. 109, 110), and *I. xanthocystis* (Pl. 115, 116). In many other species the pruinosity may not be readily observable with the naked eye, and may occur only on the upper stipe, as in *I. acaciae* (Pl. 64, 65) and *I. acanthosperma* (Pl. 169, 170). In *I. leucotaenia* and *I. alloumbrina*, the stipe is pruinose only at the extreme apex (Pl. 77, 78, 151), while it is not at all pruinose in *I. tomentipes* Matheny & Bougher. For many species pruinosity is best observed under a hand lens; it may be quite conspicuous when overlying a dark-coloured stipe as in some members of the *I. serrata* complex (Pl. 5E). In *I. vittata*, the stipe is conspicuously striatulate due to pruinosity along longitudinal lines (Pl. 112–114).

Many species of Australian Inocybaceae have conspicuous fibrils on the stipe. These are usually longitudinally arranged and whitish, although they may sometimes be coloured (e.g. the rusty or hazel-coloured fibrils occurring just above the stipe base in *I. rufobadia* (Pl. 38, 39). When fibrils are abundant on the stipe they may be referred to as comprising a vesture, as for example in *I. ionocaulis* (Pl. 91, 92). The stipe of *I. floccosibrunnea* has a particularly thick vesture of densely furfuraceous-fibrillose, floccose-fibrillose, or woolly fibrils over its entire length (Pl. 5D). In other species, the stipe is adorned with scales, such as the conspicuous, white, cottony, recurved scales in *I. ornatipes* (Pl. 93, 94).

In many species, the stipe is whitish or brownish but some are otherwise-coloured. For example, *I. ionocaulis* has a vinaceous or deep reddish brown stipe (Pl. 91, 92), and *I. violaceocaulis* has a violet stipe (Pl. 145, 146). Different colours may occur along the length of the stipe. In *I. rufoflavipes* Matheny & Bougher the stipe is reddish brown merging to yellow at the base (Pl. 69, 70). Similarly, colours may be restricted to the apex or base of the stipe (e.g. *I. calocephala* often has pink tinges on the upper stipe but is pallid elsewhere; Pl. 201, 202). *Inocybe viridipes* has greenish or olive tones at the stipe base and is brown elsewhere (Pl. 237, 239). *Inocybe caerulata* has blue tinges at the base (Pl. 117). *Inocybe pallidiflavipes* is usually yellow at the stipe base (Pl. 124, 125). In many species the stipe colour may fade with age, or appear to change with age. This commonly occurs in species with a brownish stipe surface that is initially clothed and obscured by superficial whitish fibrils but becomes exposed during elongation as the fibrils scatter, e.g. some collections of *I. froudistii* (Pl. 55, 56) have a whitish stipe due to persistence of a vesture but others appear more pinkish due to having a disrupted vesture.

In many species of Australian Inocybaceae, the stipe and pileus are concolorous or almost so. However, a contrast in colour between stipe and pileus is evident in several species, e.g. *I. flavoalbida* is so-named in reference to the combination of its honey-yellow to very pale brown pileus and its white stipe (Pl. 138). *Inocybe roseola* has a pink or pale rose pileus that contrasts with its mostly white stipe (Pl. 108). Similarly, the straw-yellow pileus of *I. sylvicola* contrasts with its pallid stipe (Pl. 208, 209).

Basal mycelium

The mycelium attached and immediately below the stipe is usually colourless, whitish or sparse and not conspicuous, including the species with violet fruit bodies such as *I. violaceocaulis*. In *I. pyrrhopoda*, the stipe base is often covered with rusty-reddish fibrils, but the basal mycelium is inconspicuous (Pl. 244).

Context

Most Inocybaceae from Australia have a whitish or pale yellowish to brownish context. However, we have observed some exceptions. In *I. violaceocaulis* the cortex of the stipe and pileus context adjacent to the lamellae is violaceous, particularly in young specimens, but the central stipe and remainder of the pileus context is white (Pl. 4A). In *I. caerulata* the stipe context has light reddish brown to flesh-pink and blue to blue-green colours. In *I. subangustifolia* the context is yellow (Pl. 3D). For many species the context is whiter or paler in the base than in the main length of the stipe (Pl. 3A, B). However, in *I. viridipes* the context is blue to blue-green at the stipe base (Pl. 239). The consistency of the context comprising the stipe may be similar to or more fibrous than that of the pileus.

The context may become dull or slightly change colour with age. For example, in the pileus and stipe of *I. subangustifolia*, the context is pastel yellow to watery greenish yellow, becoming more orange when older, but it does not change from yellow where cut or bruised. In many other species, significant colour changes to the context occur after bruising or being cut and exposed to air. Such colour changes also may be evident on the surface of some fruit bodies after handling or bruising. In many of the species of *Auritella*, the context is brunnescent (e.g. strongly so in *A. brunnescens*; Pl. 3A) or reddish to pinkish brown as in *A. robusta* and *A. serpentinocystis*. Reddening context occurs in some Australian species of *Inocybe* such as *I. rufescens* (Pl. 3C). The stipe context of *I. exigens* is drab creamish, becoming dark reddish after cutting (Pl. 47). Colour changes in some species may occur rapidly after exposure, as in some *Auritella* species, or more slowly (e.g. the stipe context of *I. leucotaenia* slowly becomes pink 3–5 minutes after exposure; Pl. 3B).

Odour and taste

Odour can be a useful diagnostic character for identification. Odour should be assessed in the field on a specimen cut longitudinally, and should be tested again later when describing the specimens. An initial assessment is necessary because the odour may change or dissipate soon after collecting. For example, the odour of *I. spiniformis* is acidulous or faintly musky when first cut, but soon becomes grassy or unremarkable. The peculiar citrine odour of *I. torresiae* may only become apparent some time after collecting.

Australian Inocybaceae exhibit a wide range of odours. The most common distinctive odour in many species of *Inocybe* from temperate regions of the southern hemisphere is an acidulous smell, like that of *Lycoperdon* and *Suillus*, a smell similar to that of a tin can, owing to presence of acetic acid. Other odours include spermatic (like piperidine, Heim, 1931) as in *I. violaceocaulis* or *I. pyrrhopoda*, marzipan (almond-like) as in *I. amygdalina* (E.Horak) Garrido, or bruised geranium (*Pelargonium*) leaves (strongly so in *I. fragrans* and weaker but distinct when cut in *I. pluvialis*). The odour of *I. torresiae* is unusual. Bougher *et al.* (2012) described its odour as sweet or citrine-like. Recent collections of this species have proven that it has a variably strong or weak, sharp and sweet, or fruity-soapy odour similar to orange extract or citrus hand cleaner, that sometimes develops slowly after cutting or after collecting. The honey-like odour of *I. mellita* Matheny, Bougher & Ryberg is distinctive and appears to be diagnostic for this species.

Australian species of *Auritella* also have a wide range of odours such as perfumed rose in *A. robusta*, unpleasant in *A. geoaustralis*, strongly unpleasant (persistently so in dried specimens) in *A. brunnescens*, and musty or very unpleasant like wet socks in *A. serpentinocystis*. Three species have an indistinct odour: *A. arenicolens*, *A. chamaecephala* Matheny, O.K.Mill. & Bougher, and *A. dolichocystis*. No distinctive odours have been recorded for the two known Australian species of *Tubariomyces*. Taste is not considered to be a useful diagnostic character for the family (Kuyper, 1986).

Spore print

Inocybaceae are characterized by dull brown spore deposits, but the print colour is generally not useful for distinguishing species; it is therefore usually unnecessary to make a spore print. Mature discharged spores for microscopic examination can usually be obtained from spore deposits on the stipe, or pileus surface of fruit bodies collected beneath other fruit bodies.

Macrochemical spot tests

Spot tests are not widely used with the Inocybaceae, as they are with some other groups of fungi. The pileus of some Australian Inocybaceae may become brown with potassium hydroxide (KOH), e.g. *A. chamaecephala*, but this is of minimal taxonomic value. In *I. violaceocaulis* the violaceous surface changes to orange with 15% KOH. Dried specimens of *A. serpentinocystis* exude yellow pigment into solutions of KOH (similar to species of the genus *Pleuroflammula* Singer). PDAB (para-dimethylaminobenzaldehyde) has been used overseas with *Inocybe* species as a colour test: e.g. bright turquoise-blue or bright yellow (Matheny *et al.*, 2013).

Vouchered specimens

Dried voucher specimens are primarily used as a source of material for microscopy and DNA extraction, but examination of dried material without using a compound microscope can sometimes be instructive. For example, dried specimens of *A. brunnescens* Matheny & Bougher retain a strong unpleasant odour, unlike the other species of that genus. Dried specimens of *I. violaceocaulis* Matheny & Bougher retain violaceous tinges. The strongly pallid-fimbriate lamellae edges of *I. mallocyboides* are as conspicuous in dried specimens as when fresh. Morphological differences between *I. geniculata* and *I. umbrosa* E.Horak can be readily determined by examination of dried voucher specimens under a hand lens or stereomicroscope. Voucher specimens of *I. geniculata* show isolated clusters of caulocystidia with few fibrils between the clusters. In contrast, voucher specimens of *I. umbrosa* appear clothed with a dense covering of longitudinally arranged fibrils as well as clusters of caulocystidia.

Microscopic Characters

Microscopic structures may be observed in fresh or dried specimens by sectioning tissues or preparing squash mounts. For the latter, it is best to use a hand lens or stereomicroscope when extracting small portions from fruit bodies. This will help to ensure that a particular targeted part of the fruit body is efficiently sampled. For example, the density and prevalence of cheilocystidia in the Australian Inocybaceae can be quite inconsistent among specimens of an individual collection, or in some cases among the lamellae of a single specimen. Under a lens it may be evident that some lamellae margins of a fruit body or certain segments along an individual lamella appear more whitish or finely fimbriate than other parts. It is advantageous to target and pick off the whitest or most conspicuously fimbriate portion of lamellae margin in order to capture a good representation of cheilocystidia.

Under a hand lens or stereomicroscope, a fine forceps can be used to pick off the selected portion and immerse it in a drop of rehydrating fluid on a slide, noting the orientation of the lamella margin on the slide. Before a cover slip is added, the lamella margin will usually expand and unfurl when immersed in a reagent such as 3% KOH. Once the lamella portion has fully unfurled, a coverslip should be gently laid down over the material so as to not squash the preparation. Under the low power objective of a compound microscope it is possible to observe the position of various structures in relation to each other, and the density of structures such as cystidia. After positional relationships have been observed, the cover slip may be tapped to flatten the tissues in order to reveal more of their structure.

Tapping the cover slip not only disrupts the positional relationship of structures comprising various tissues, but also has the potential to distort the shape of structures such as thin-walled cystidia. Also, in some species of Australian Inocybaceae, the crystals dissolve rapidly in reagents like KOH but in others slowly or not at all. If examination of a preparation is delayed after mounting, the crystals may completely dissolve before observation. To ensure accurate observation of cystidial crystals, it is advisable to: (1) make an initial observation by placing a small portion of dried material (less than 1 mm × 1 mm) on a slide without any reagent or coverslip and observing this directly under a compound microscope using a low power objective (although the cystidia and other elements will not be hydrated, the presence of cystidia and apical crystals should be evident); and (2) observe the preparation immediately after adding the reagent and cover slip but before tapping the cover slip. Alternatively, water may be used, which may not dissolve the crystals.

Basidiospores

All species of *Auritella* and *Tubariomyces*, and most species of *Inocybe* in Australia, have smooth spores. However, various forms of angular or nodulose spores also occur among species of *Inocybe*. For example, spores of *I. sinuospora* Matheny & Bougher have an undulating or sinuous outline (Pl. 6A, Fig. 114), and many spores of *I. sejuncta* are minimally angular (Pl. 6B, Fig. 97). *Inocybe imbricata* (Cleland) Garrido has spores with inconspicuous nodules around a rectangular to trapeziform outline (Pl. 7H, Fig. 120). Blunt, moderate-sized nodules occur in many species such as *I. brunneidisca* (Pl. 6C, Fig. 132). The spores of *I. nobilissima* are stellate with 8–10 conical projections each to 2.0–3.0 µm high (Pl. 7I, 8D, Fig. 94). Larger saddled-shaped and crested nodules occur in spores of *I. lasseroides* (Pl. 6D, Fig. 86) and conic-tapering spines 2.0–2.5 µm high occur in spores of several species such as *I. acanthosperma* (Pl. 6E, 8B, Fig. 100).

Spore shape is variable among the smooth-spored species of Inocybaceae in Australia as well. Amygdaliform or subamygdaliform spores occur in a large number of species, e.g. *I. fibrillosibrunnea* (Pl. 6G, Fig. 26) and *I. spadicea* (Pl. 6H, Fig. 82). *Inocybe atrorubripes* Matheny, Bougher & Ryberg has oblong-amygdaliform to cylindric spores with a slight adaxial depression (Pl. 6I, Fig. 55). A few species feature some citriniform spores, e.g. *I. luteo-olivacea* (Pl. 6J, Fig. 78). Elliptic spores occur in others, e.g. *I. fissurata* Matheny & Bougher (elliptic to oblong-elliptic or oblong-reniform; Pl. 7A, Fig. 138) and *I. pluviorum* (broadly elliptic to elliptic; Pl. 7B, Fig. 141). *Auritella geoaustralis* has ovoid-ellipsoid spores (Fig. 159). *Inocybe exigens* Matheny & Bougher has cylindric (less often oblong-amygdaliform) spores (Pl. 7F, Fig. 24). The spores of *A. arenicolens* are distinctively narrowly cylindric to oblong-cylindric (Pl. 7G, 8A, Fig. 154).

Occasionally, spore shape can be variable within individual collections. The spores of the secotioid *I. bicornis* f. *secotioides* are highly variable in size, wall thickness and shape (oblong, subcylindrical, lacrymoid, ovoid, or broadly elliptic; Pl. 7J, Fig. 22). Spores of *I. avellaneifolia* are oblong-amygdaliform to amygdaliform, on occasion with pronounced undulating or subangular outline, less often elliptic with apices conical to bluntly pointed or less often obtuse (Fig. 34).

The apiculus (hilar appendage) may be very small and inconspicuous as in *I. pluviorum* (Fig. 141), tapered as in *I. parvifulva* (Fig. 135), not tapering and wider as in *I. fraudistii* (Fig. 28), or broad and truncate as in *I. spiniformis* (Fig. 101). The hilum is of the open-pore type.

The spores of most species of Australian Inocybaceae appear yellowish brown when viewed under a compound microscope in mountants such as water or KOH. Slight variations in spore colour are sometimes noticeable between and within Australian species, but variability in lighting conditions limits the taxonomic utility of such observations. Consistently pale spores are notable for some species when compared to closely related species, e.g. *I. subflavospora* has paler spores than other Australian species of the *Mallocybe* clade. *Inocybe rosella* and *I. paucicolor* Matheny & Bougher have subhyaline or buff-coloured spores respectively.

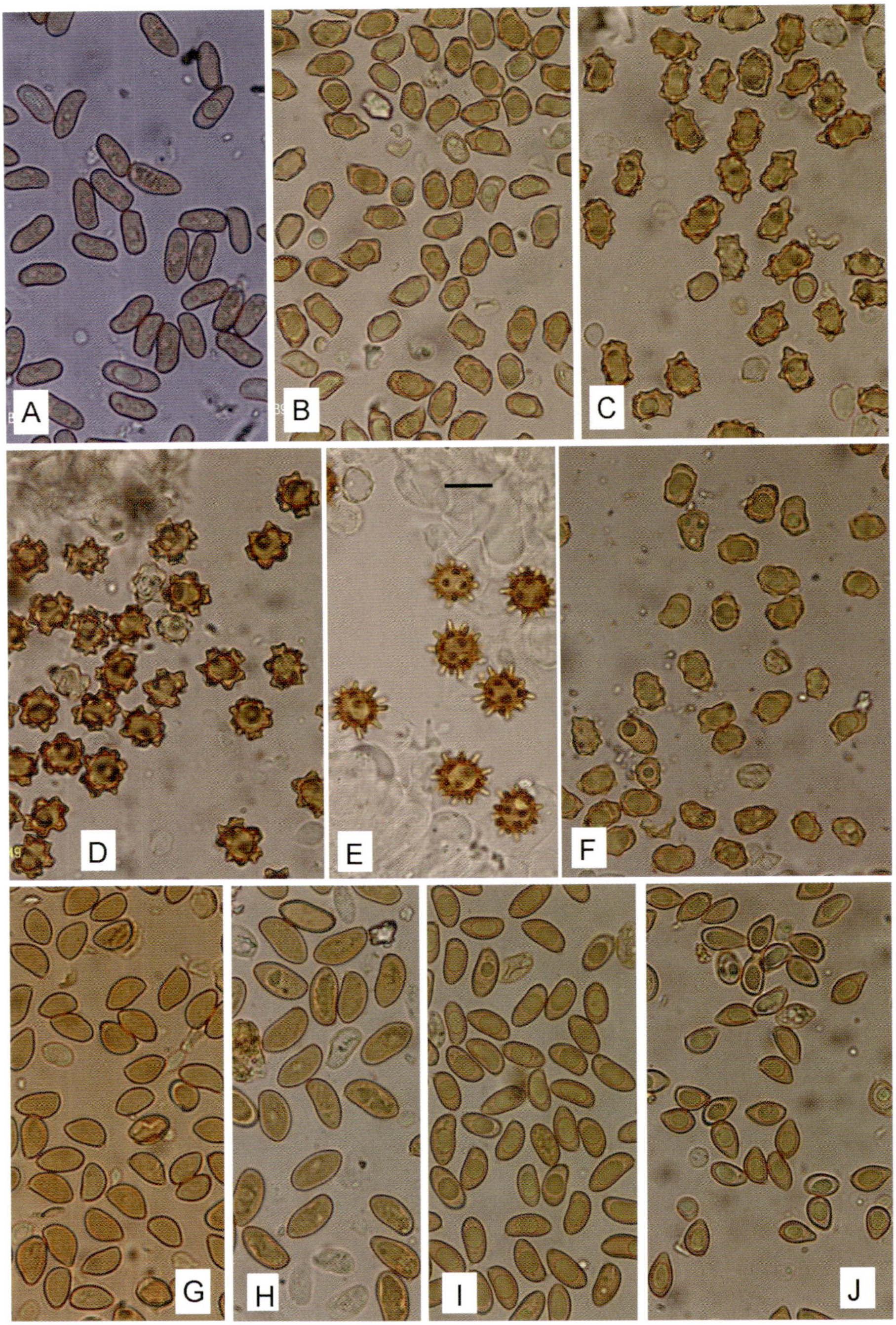

Plate 6. Basidiospores of Australian Inocybaceae. **A,** *Inocybe sinuospora* (PERTH 08320578); **B**, *Inocybe sejuncta* (PERTH 08383103); **C,** *Inocybe brunneidisca* (PERTH 08094217); **D**, *Inocybe lasseroides* (PERTH 08363323); **E,** *Inocybe acanthosperma* (PERTH 08363145); **F**. *Inocybe mellita* (PERTH 08306621); **G**, *Inocybe fibrillosibrunnea* (PERTH 00909904); **H**, *Inocybe spadicea* (PERTH 07678444); **I**, *Inocybe atrorubripes* (PERTH 08306869); **J**, *Inocybe luteo-olivacea* (BRI 794189).

Scale bar (shown in E) = 10 μm. Photographs — N.L. Bougher.

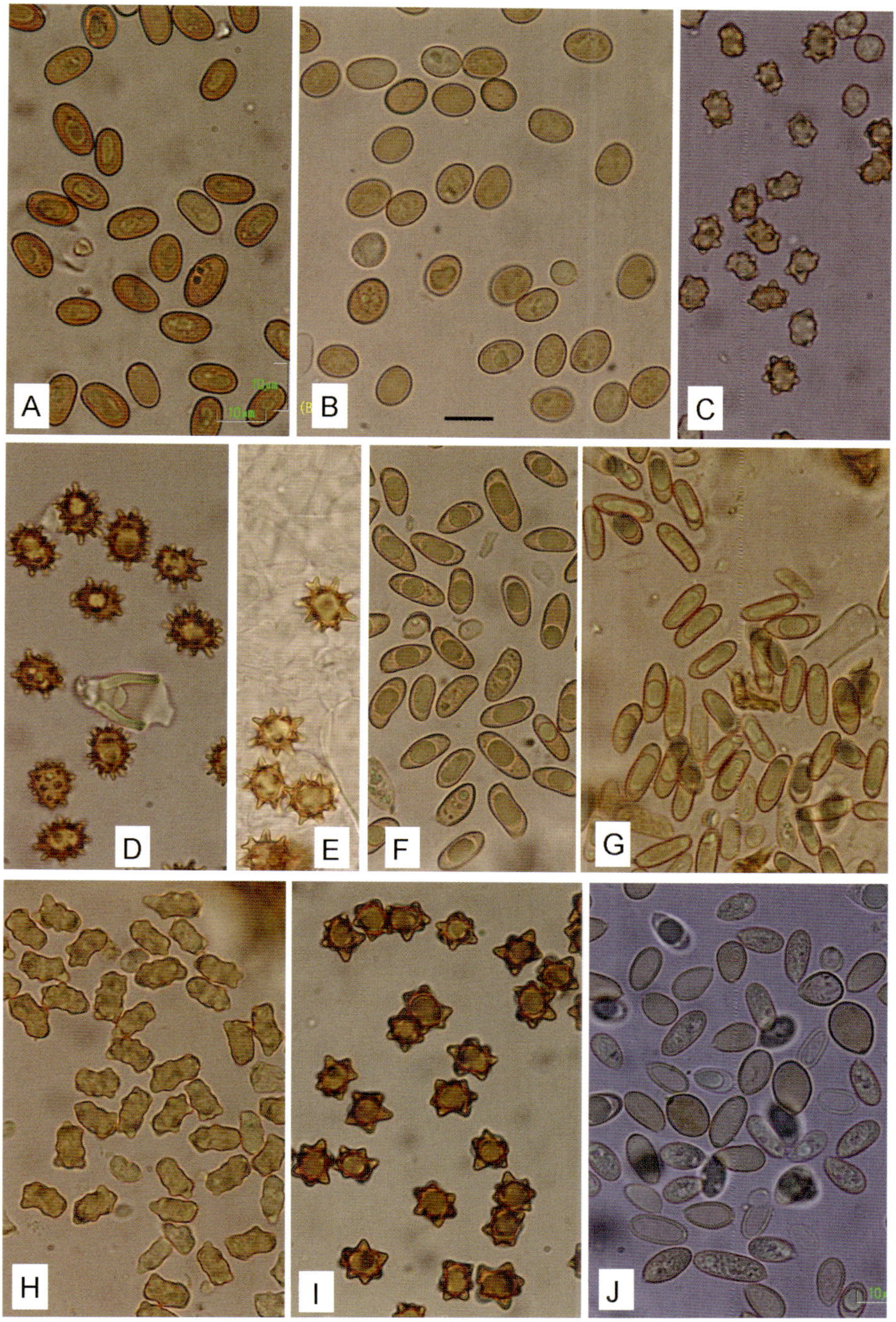

Plate 7. Basidiospores of Australian Inocybaceae. **A,** *Inocybe fissurata* (PERTH 07676824); **B**, *Inocybe pluviorum* (BRI AQ794010); **C,** *Inocybe redolens* (PERTH 07676654); **D**, *Inocybe spiniformis* (PERTH 08363277); **E,** *Inocybe tertia* (PERTH 08362882). **F**, *Inocybe exigens* (PERTH 08383375); **G**. *Auritella arenicolens* (PERTH 08047081); **H**, *Inocybe imbricata* (AD 12227); **I**, *Inocybe nobilissima* (PERTH 08363153); **J**, *Inocybe bicornis* f. *secotioides* (PERTH 08320403).

Scale bar (shown in B) = 10 µm. Photographs — N.L. Bougher.

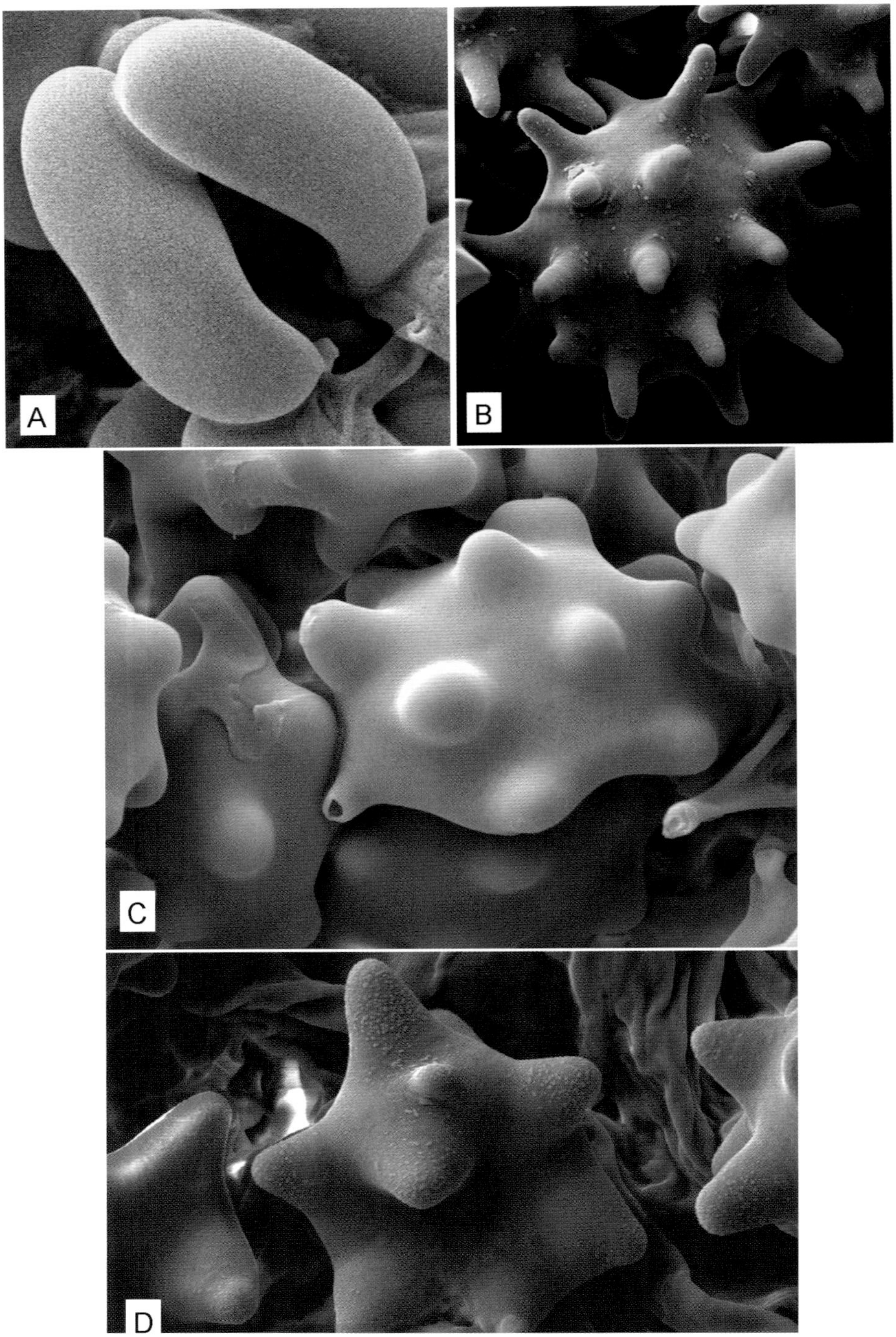

Plate 8. Basidiospores of Australian Inocybaceae under SEM (Scanning Electron Microscope). **A,** *Auritella arenicolens* (TENN 066775); **B,** *Inocybe acanthosperma* (TENN 067005); **C,** *Inocybe redolens* (TENN 067764); **D,** *Inocybe nobilissima* (TENN 067751).

Photographs — E. Harrower.

Cystidia

A diverse range of cystidia occurs on the lamellae and stipe, and these are often of taxonomic and phylogenetic value. Both pleurocystidia and cheilocystidia are typically present in almost all species of the *Inocybe* clade, but pleurocystidia are absent in the *Nothocybe*, *Inosperma*, *Pseudosperma*, and *Mallocybe* clades, and are absent from the genera *Auritella* and *Tubariomyces*. Hymenial cystidia arise from the hymenium or subhymenium except for members of *Mallocybe* and *Auritella* where the cheilocystidia arise from the long, narrow hyphae of the hymenophoral trama.

In species of *Inocybe* in which pleurocystidia are present, the hymenial cystidia are usually metuloid (thick-walled, and with apical crystals of calcium oxalate; Pl. 9, 10A, B), and usually have a basal clamp connection (Pl. 10C). Some, however, may be thin-walled. When they are thick-walled, the cheilocystidia and pleurocystidia may appear yellow, e.g. *I. xanthocystis* has very thick-walled hymenial cystidia [4.0–10.0 (–15.0) µm thick] with the walls often intensely bright yellow (Pl. 10D, Fig. 64). Cheilocystidia often resemble the pleurocystidia, although often the former are somewhat shorter or more lageniform rather than fusiform or ventricose. Among Australian species of the *Inocybe* clade, pleurocystidia are only rarely absent. Australian species in the clade that lack pleurocystidia (or that have very rare pleurocystidia) include *I. nigricans*. Species in the *I. alloumbrina* complex (*I. alloumbrina*, *I. subferruginea*, and *I. cacaocolor*) have thin-walled to very slightly thick-walled (–1.0 µm) hymenial cystidia.

Among all clades lacking pleurocystidia, the cheilocystidia are not metuloid. Within *Inocybe*, for example, members of the *Inosperma* and *Pseudosperma* clades typically possess densely arranged cheilocystidia that are cylindrical, thin-walled, and hyaline, often with at least some multi-septate and branching forms, e.g. *I. araneosa* (Pl. 10G), *I. fissurata* (Fig. 138), and *I. viridipes* (Fig. 146). Cheilocystidia in species of *Inocybe* in the *Mallocybe* clade are also densely arranged, thin-walled and hyaline, are usually more broadly clavate, pyriform, or vesiculose (less often cylindrical) and arise from long, narrow sometimes finely encrusted hyphae, e.g. *I. pyrrhopoda* (Pl. 10E, Fig. 149) and *I. isabellina* (Pl. 10F, Fig. 148). In the genus *Auritella* the cheilocystidia (absent in the secotioid *A. geoaustralis*) are narrow, cylindrical or flexuous or slender-clavate with or without swollen apices, often >50 µm long (to 154 µm in *A. robusta*), and are usually thin-walled but sometimes slightly thick-walled, hyaline or ochraceous (Pl. 11A, B). In Australian species of *Tubariomyces* the cheilocystidia are densely arranged on lamella margins and resemble elements of the pileipellis: they are cylindrical or nearly so, thin-walled to slightly thick-walled and hyaline but sometimes rusty-ochraceous (e.g. *T. australiensis*; Pl. 11C; Fig. 162).

Unlike the pleurocystidia, which are scattered singly on the sides of the lamellae, cheilocystidia are often crowded and intermixed with paracystidia, the latter usually clavate or pyriform, often shorter, thin-walled and hyaline (Pl. 11E). They are generally uniform in shape and crowded, as in *I. mallocyboides* of the *Inocybe* clade (Pl. 11D, Fig 47). Paracystidia have not been observed in *I. alienospora*.

The presence or absence and distribution of caulocystidia are important taxonomic features (Kuyper, 1986). Caulocystidia are absent in several major clades of Australian Inocybaceae, viz, the *Inosperma* and *Mallocybe* clades, but may be present and/or poorly differentiated in some Australian species of the *Pseudosperma* and *Auritella* clades. Occasionally, in some of these species, a small area of caulohymenium may be present at the extreme apex of the stipe, e.g. in *A. robusta*. In other Inocybaceae (*Inocybe* and *Tubariomyces*) caulocystidia may occur along part of the stipe or along the entire length of the stipe (as in *Tubariomyces*), particularly when a cortina is absent. In cortinoid species of the *Inocybe* clade, caulocystidia may occur above the (often widely variable) insertion point of the cortina. In such cases, the caulocystidia tend to resemble cheilocystidia and occur with paracystidia-like cauloparacystidia in an arrangement that matches that on the lamellar margin (Pl. 11G, compare with Pl. 11D). This similarity is due to early-stage homologous development with the hymenium at a time when the lamellar margin is tightly pressed against the stipe in very young fruit bodies (Kuyper, 1986). Unlike the contiguous hymenium, the caulocystidia and cauloparacystidia often occur together in crowded clusters, probably as a result of stipe elongation after their development.

Plate 9. Apical crystals of hymenial cystidia in Australian Inocybaceae, viewed by Scanning Electron Microscope (SEM); **A,** *Inocybe paucicolor* (TENN 069517); **B,** *Inocybe paucicolor* (TENN 069517); **C,** *Inocybe xanthocystis* (TENN 067713). Photographs— E. Harrower.

The stipe of some species may have caulocystidioid hairs below the insertion point of the cortina, sometimes with abundant, superficial, undifferentiated hyphae that comprise a vesture. Like caulocystidia, caulocystidioid hairs are differentiated sterile end-cells but differ from the former by not being developmentally homologous with the cheilocystidia. These often occur more or less evenly scattered rather than in clusters, and are not accompanied by paracystidia, unlike caulocystidia (Kuyper, 1986). Caulocystidioid hairs are often also morphologically differentiated from hymenial cystidia in shape. Some may be thick-walled, quite narrow, long and flexuous, and bear apical crystals (e.g. *I. fuscosquarrosa*; Fig. 29). Although caulocystidioid hairs are not necessarily thick-walled or abundant, metuloid caulocystidioid hairs sometimes occur on the stipe where, unlike cheilocystidia and caulocystidia, they are not accompanied by paracystidia in a hymenial-like arrangement. At least in some species, the metuloid caulocystidioid hairs probably develop on the stipe late in development of the fruit bodies and therefore independently in time compared to the hymenial metuloids that develop earlier (as discussed by Kuyper 1986, with reference to the caulocystidioid hairs of *I. sindonia*). In young representatives of the Australian species *I. clypeata*, there are numerous, differentiated, thin-walled end-cells of varying shape (clavate, sphaeropedunculate, fusiform, cylindrical) just below the pileal insertion and often also on the lower part of the mature stipe. However, stipes of expanded specimens of *I. clypeata* show numerous metuloids (caulocystidioid hairs, presumably late-developing) intermixed with variably-shaped thin-walled cells (Fig. 23).

The presence of caulocystidia without paracystidia along the entire length of the stipe is diagnostic for all known species of *Tubariomyces*, and readily separates *Tubariomyces* from species in the *Mallocybe* clade, with which they may be most readily confused.

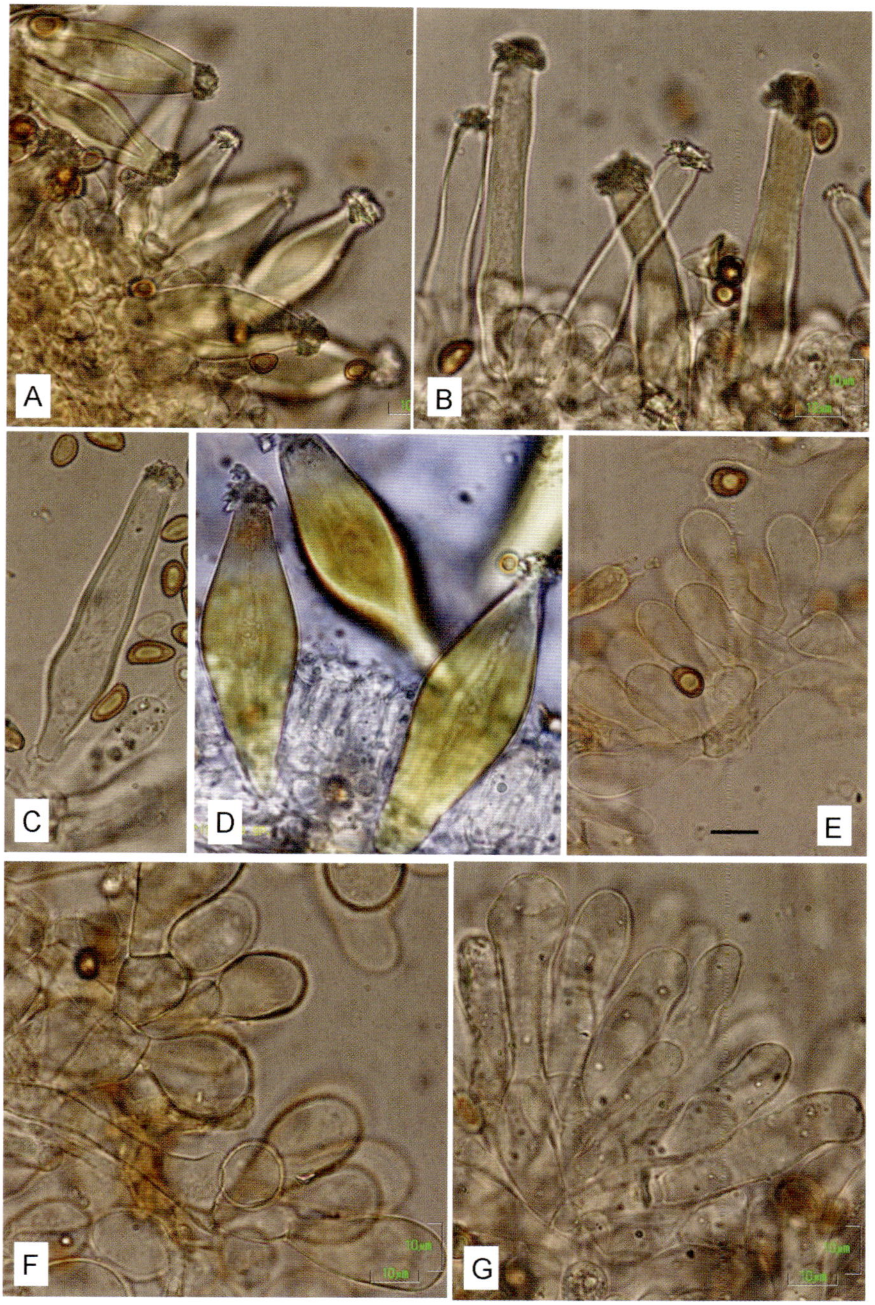

Plate 10. Cystidia of Australian Inocybaceae. **A,** Cheilocystidia - *Inocybe tenax* (MEL 2335645A); **B,** Cheilocystidia - *Inocybe memorialis* (PERTH 08073317); **C,** Pleurocystidium with basal clamp connection - *Inocybe memorialis* (PERTH 08073317); **D,** Cheilocystidia - *Inocybe xanthocystis* (PERTH 08242100); **E,** Cheilocystidia - *Inocybe pyrrhopoda* (PERTH 08363250); **F,** Cheilocystidia - *Inocybe* cf. *isabellina* (PERTH 08309078); **G,** Cheilocystidia - *Inocybe araneosa* (PERTH 08363285).

Scale bar (shown in E) = 10 μm. Photographs — N.L. Bougher.

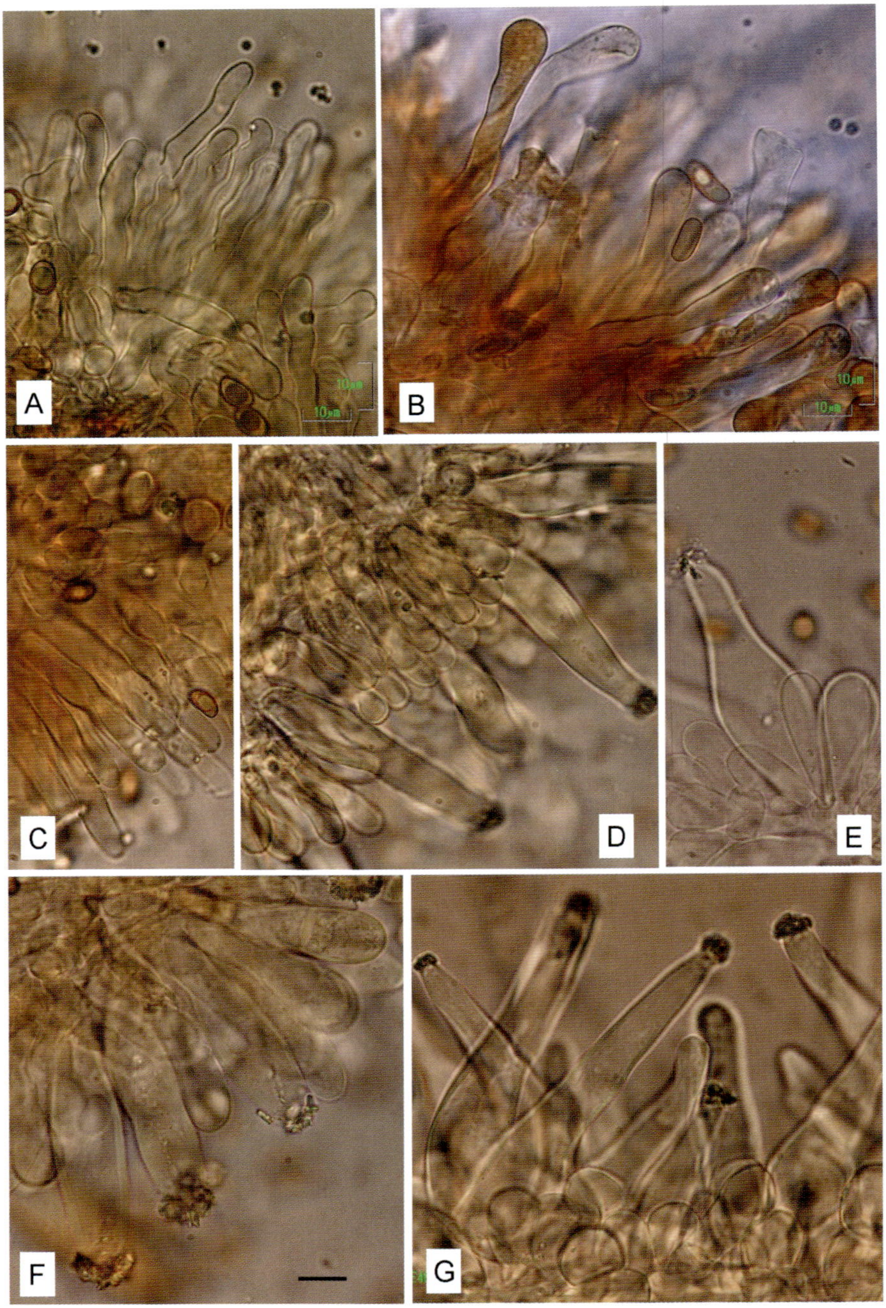

Plate 11. Cystidia of Australian Inocybaceae. **A,** Cheilocystidia - *Auritella serpentinocystis* (PERTH 07536755); **B,** Cheilocystidia - *Auritella arenicolens* (PERTH 05429692); **C,** Cheilocystidia - *Tubariomyces australiensis* (PERTH 08383189); **D,** Cheilocystidia and small, crowded paracystidia - *Inocybe mallocyboides* (PERTH 08242240); **E,** Cheilocystidium and paracystidia - *Inocybe olivaceohinnulea* (PERTH 08319626); **F,** Cheilocystidia and paracystidia - *Inocybe spadicea* (PERTH 07678444); **G,** Caulocystidia and cauloparacystidia - *Inocybe mallocyboides* (PERTH 08242240).

Scale bar (shown in F) = 10 µm. Photographs — N.L. Bougher.

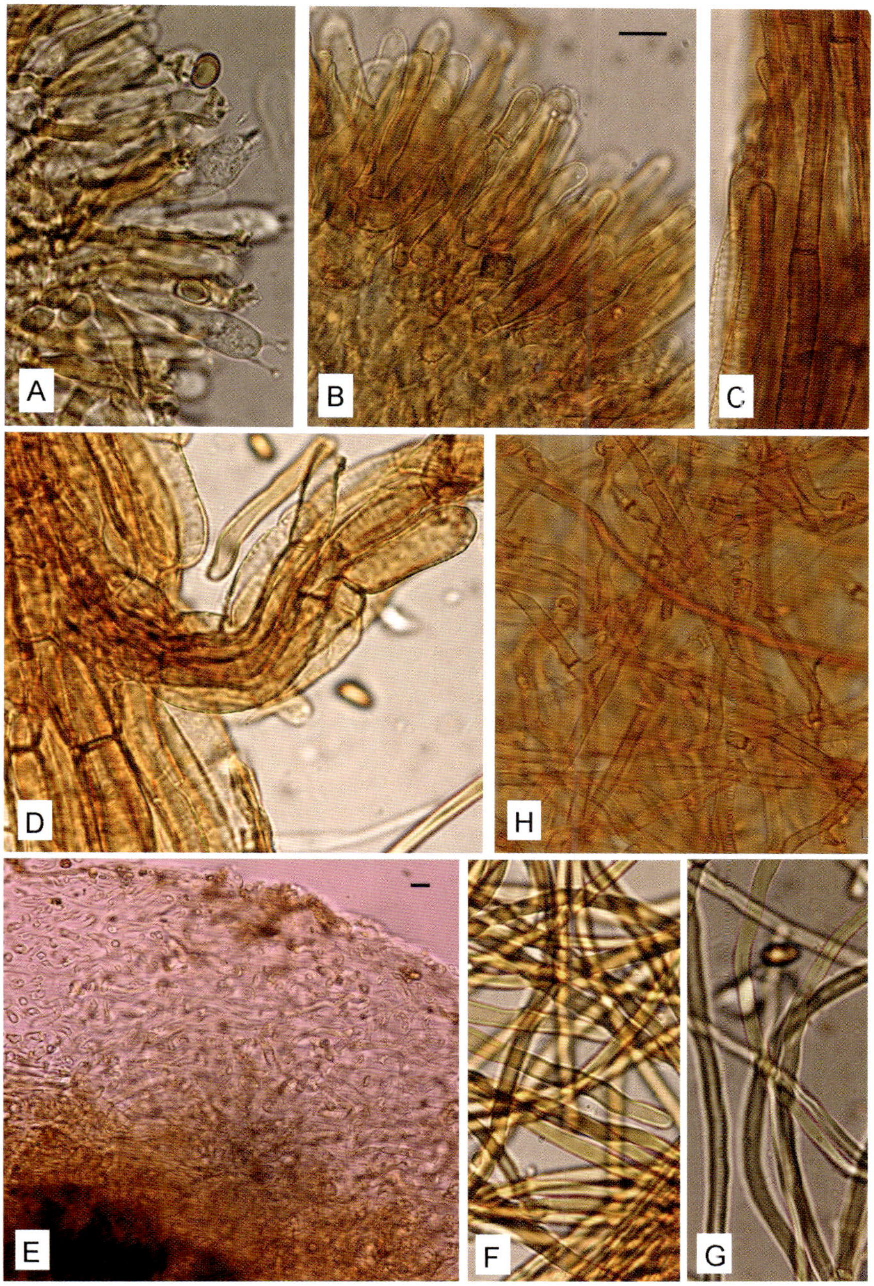

Plate 12. Basidia and pileipellis structures in the Australian Inocybaceae. **A,** Hyaline young basidia and necropigmented mature basidia - *Auritella serpentinocystis* (PERTH 07536755); **B,** Pileipellis - *Tubariomyces australiensis* (PERTH 08383189); **C,** Incrusted pileipellis hyphae - *Inocybe isabellina* (PERTH 08072442); **D,** Pileipellis with emerging bundle of hyphae - *Inocybe pluvialis* (PERTH 08517150); **E,** Broad gelatinised epicutis - *Inocybe viscata* (PERTH 08516871); **F,** Golden velipellis hyphae - *Inocybe pluvialis* (PERTH 08517150); **G,** Pale velipellis hyphae - *Inocybe xanthocystis* (PERTH 08318158); **H,** Rusty hyphae at base of stipe - *Inocybe pyrrhopoda* (PERTH 08363250).

Scale bar shown in B for all (except E) = 10 μm; bar in E = 10 μm. Photographs — N.L. Bougher.

Basidia

Most Australian Inocybaceae have predominantly or exclusively 4–sterigmate basidia; *I. bicornis* is the only known species of the family in Australia that has exclusively 2-sterigmate basidia (Fig. 21). The secotioid form of this species, *Inocybe bicornis* f. *secotioides*, has both 2- and 4-sterigmate basidia (Fig. 22). In other Australian species, 1- or 2-sterigmate basidia are occasionally observed (e.g. *I. flavoalbida*; Fig. 77). As can be the case in other groups of Agaricales, 1- and 2-sterigmate basidia in otherwise predominantly 4-sterigmate species of Inocybaceae may produce exceptionally large spores.

The basidia of most species are clavate or cylindrico-clavate and less than 50 µm long. However, a few species have long and narrow basidia reaching over 60 µm long (e.g. *A. arenicolens*; Fig. 154). Most Australian species have basidia that are thin-walled and always hyaline, or perhaps develop some oily inclusions with age. However, all *Auritella, Tubariomyces*, and *Inocybe* species of the *Inosperma* and *Mallocybe* clades are characterised by the presence of necrobasidia – basidia that are hyaline at first but soon become ochraceous or golden with age and collapse apparently after spore dispersal (Pl. 12A).

Hymenophoral trama

The arrangement of hyphae in the lamellar trama is parallel in all Inocyabaceae, and thus of no taxonomic value within the family.

Pileipellis

The structure of the pileipellis is taxonomically important in some groups of Inocybaceae. The pileipellis of *Tubariomyces*, including the Australian species, is composed of a crowded palisade of mostly cylindrical, rusty-ochraceous terminal cells (Pl. 12B). In contrast, most Australian species of *Auritella* and *Inocybe* have an undifferentiated cutis, often composed of hyphae with encrusted (pigmented ferrugineous, golden or ochraceous) walls (Pl. 12C). Species of the *Mallocybe* clade from Australia tend to have a pileipellis with coarsely encrusted hyphal walls. The hyphae are usually broader in the lower pellis and grade into narrower hyphae towards the surface rather than having a distinct subpellis. At the surface the hyphae may emerge in bundles of trichodermially arranged hyphae comprising squamules in those species that have a scaly pileus, but the consistency of emergence is often variable from the pileus centre to the margin and may be influenced by developmental stages (Pl. 12D). The pileus of some Australian *Inocybe* species may be greasy or soapy-lubricous in wet conditions, e.g. *I. fulvilubrica*, but gelatinisation is usually minimal. *Inocybe viscata* is unusual in having a broad, gelatinised epicutis to 200 µm thick (Pl. 12E).

In some species of *Inocybe*, morphologically distinct velipellis hyphae overlay the pileipellis. This is well demonstrated by species of the *I. tenax* group, which often have an interwoven layer of narrow, thick-walled, rarely branched and infrequently septate velipellis hyphae (e.g. *I. pluvialis*; Pl. 12F). The velipellis hyphae are golden in *I. pluvialis* but much paler in the closely related *I. xanthocystis* (Pl. 12G). In some cases, it is evident that some of the vellipellis hyphae are joined to the pileipellis hyphae.

Hyphal pigments

Encrusting pigments are the most dominant pigment type in Australian Inocybaceae. The pileipellis of many species is composed of hyphae with brown encrusted walls (Pl. 12C). Other tissues may be encrusted in some species, such as hyphae subtending the cheilocystidia in *Auritella fulvella* Matheny & Bougher (Fig. 158). Internal pigments, however, occur in some species. The bright colouration of *I. violaceocaulis* results from a plasmatic pigment. Cheilocystidia in *I. alienospora* have dull yellowish plasmatic pigments or yellow occlusions, while the lamellar edges of *I. lasseroides* are noticeably yellow in mass (as seen under low power microscope) when cystidia with yellow intracellular pigmented occlusions (Fig. 86) are abundant. The cheilocystidia of *A. arenicolens* are at times ochraceous due to an internal pigment (Pl. 11B, Fig. 154). Numerous coarse, rust-coloured, interhyphal pigment deposits occur in the pileipellis of *I. sejuncta*. Rehydrated context tissue of *A. serpentinocystis* exudes a yellow or ochraceous pigment when placed in an alkaline solution, similar to the genus *Pleuroflammula* (Singer, 1986).

INTRODUCTION

Clamp connections

Clamp connections occur at most septa in all fruit body tissues of Australian Inocybaceae.

Biochemical Characters

Muscarine

Muscarine is the most important secondary metabolite found in many species of *Inocybe* worldwide, including in Australia. Muscarine is a quaternary ammonium compound that stimulates the parasympathetic nervous system in humans and other vertebrates. A unique assemblage of symptoms occurs rapidly after muscarine consumption including pupil contraction, salivation, lacrymation, urges to urinate and defecate, gastrointestinal distress, and emesis (Benjamin, 1995). Atropine is a successful antidote. The toxin is produced by several other groups of unrelated agarics (Benjamine, 1995). Introduced species from the Northern Hemisphere such as *I. curvipes* have been implicated in at least one intoxication event in Australia (Matheny & Bougher, 2011). In addition, *I. violaceocaulis* was thought to have been involved in another intoxication event (see discussion under *I. violaceocaulis* for further details).

Phylogenetically, muscarine has evolved on multiple occasions in the Inocybaceae occurring in species of the following major clades: *Mallocybe*, *Inosperma*, *Pseudosperma*, and *Inocybe* (Kosentka *et al.*, 2013). However, muscarine is also absent, in some cases evolutionarily lost, in many species in these same clades. It appears to be an ancestral trait not for the family as a whole, but for an inclusive grouping including, at least, the *Pseudosperma* and *Inocybe* clades, and ostensibly *Nothocybe*, but the latter has not been assayed. Based on limited sampling, muscarine is absent from *Tubariomyces* and Australian species of *Auritella* (viz, *A. brunnescens* and *A. serpentinocystis*). Of native Australian species of *Inocybe*, muscarine has been detected in *I. froudistii* (as *I. aff. fibrillosibrunnea*) and *I. leptospermi* (as *I. cf. graveolens*); muscarine was not detected in *I. serrata* (clade I) (as *I. granulosipes*), *I. pallidiflavipes* (as *I. granulosipes*), *I. cerulata* and *I. viscata* (Kosentka *et al.*, 2013).

Psilocybin and aeruginascin

Psilocybin, a hallucinogen, is a tryptamine alkaloid that acts on serotonin receptor sites (Griffiths *et al.*, 2006). It is one of several co-related psychotropic compounds that tend to co-occur among positive-containing species. Aeruginascin is an indole alkaloid and trimethylammonium analogue of psilocybin, which has been documented in the European species *I. aeruginascens* (Jensen *et al.*, 2006). Psilocybin has been documented from five described species of *Inocybe* in Europe. However, no confirmations of psilocybin- or aeruginascin-containing species have been confirmed in Australian species of Inocybaceae. A species described here as *I. caerulata* from Tasmania and New Zealand is a candidate to assay for presence of hallucinogenic compounds owing to the presence of bluish colours on the pileus and stipe. Muscarine has been confirmed as absent from *I. caerulata*, an apparent evolutionary prerequisite to evolution of hallucinogenic compounds (Kosentka *et al.*, 2013). A hydroxylated indole compound may be present in the northern temperate *I. chrondoderma*, a species that produces a bright blue-green pigment in solutions of PDAB (Matheny *et al.*, 2013).

METHODS

Field sites and collections

First-hand collecting of Australian Inocybaceae by the authors began during multiple expeditions throughout the 1980s and 1990s in temperate and tropical regions Australia-wide (Bougher and colleagues). Expeditions specifically targeting the Australian Inocybaceae were undertaken by Matheny, Bougher and colleagues in 2001 and 2009–2013, primarily in southern temperate regions during autumn and winter months, with the main collecting areas covering diverse habitats in south-western Western Australia and Tasmania. In 2012 the authors and colleagues also undertook an expedition in far north and south-eastern Queensland. Our collecting was opportunistic elsewhere, particularly in New South Wales and Victoria. In addition, a network of collaborators provided numerous collections from northern Western Australia, New South Wales, Northern Territory, Tasmania, and Queensland (see Acknowledgements). To increase sample size and further decrease regional bias, we also directly examined and/or drew upon data from *c.* 1,000 herbarium specimens lodged at AD, BRI, CANB, CNS, CORT, FLAS, HO, M, MEL, NY, OSC, PDD, PERTH, TENN, and ZT [herbarium abbreviations follow *Index Herbariorum*].

Field collections were photographed *in situ* and/or under laboratory conditions. Gross morphological and ecological data were taken from fresh specimens and in the field. "L" refers to the number of lamellae that extend from the pileus margin and reach the stipe. Latitude and longitude were recorded for most collections. Colour guides (Ridgway, 1912; Munsell, 1954; Kornerup & Wanscher, 1967) were used to note colours when fresh. Odours were noted but taste was not; taste is not considered to be of taxonomic significance in the Inocybaceae. Specimens were preserved after air-drying on food dehydrators. Dried specimens were then split between Australian herbaria (mostly PERTH) and TENN. Holotypes are lodged at PERTH, BRI, HO and CANB, and isotypes at TENN, CORT and NY.

We determined that it is best practice not to take for granted that any given vouchered collection of Inocybaceae specimens is necessarily homogenous, i.e. composed of only a single species. We examined as many individual fruit bodies as possible from the same vouchered collection to determine their homogeneity or otherwise.

Faced with large numbers of vouchered collections, we prioritized our effort on vouchers with the following attributes: (1) numerous fruit bodies in good condition; (2) accompanying descriptive notes/diagrams, including morphology and vegetation/nearby plants; (3) accompanying images of fresh specimens.

Microscopy and line drawings

Dried fruit body tissues were sectioned and rehydrated in 3–5% KOH. Occasionally, 1% phloxine was used to enhance contrast, or Congo Red was used to stain hyphal walls. Measurements of basidiospores and other features were made with a Nikon Eclipse 80i compound microscope using Nikon NIS elements D software. Spore measurements were done on spores deposited on the stipe or pileipellis, or by measuring mature spores from lamella squash mounts. Drawings for the figures were constructed in a way to ensure a high degree of accuracy, not only detailing the individual elements in various tissues but also the relative abundance and positional relationships of the elements.

Cheilocystidia are illustrated as they appear along the edge of a lamella, that is, a portion of lamella laid on its side under a coverslip was drawn. Initially, a coverslip was gently rested over the lamella in order to assess the un-squashed distribution and abundance of elements on the lamella, the positions of those elements in relation to each other, and the prevalence and magnitude of apical crystals on cystidia. Microscope images were taken immediately to capture apical crystals, which in at least some *Inocybe* species tend to dissolve and may entirely disappear in KOH by the time a drawing is done. Also, any other change, such as colour change, was monitored. Images were taken at various magnifications including at low power for later reference. The preparation was then gently tapped to separate out some of the elements in order to draw them more clearly and, for example, expose their bases. Images were taken again at various magnifications of the view for later reference. Raw drawings were made in pencil using a camera lucida, carefully noting various attributes such as the positional relationship between the various elements. Final inked drawings were precisely based on the original raw pencil drawings. Pertinent notes and various microscopic images were consulted during this process to help verify and where necessary make slight adjustments to the drawings.

Pleurocystidia were treated in a similar way to the cheilocystidia. Caulocystidia were drawn from squash mounts of stipe tissue. Observations were recorded for most species of the upper and lower stipe, but drawings are usually not presented for all regions of the stipe due to page size limitations.

DNA extraction, PCR, and Sanger sequencing

Details and protocols for DNA extraction, PCR, and Sanger sequencing follow that of Judge *et al.* (2010). For collections older than twenty years, we used a DNA 'high performance' kit, details of which are provided in Baroni & Matheny (2011). Three nuclear loci were targeted for phylogenetic analyses and molecular confirmation: ITS, nLSU, and *rpb2*. Primers used to amplify and sequences these gene regions are provided in Baroni & Matheny (2011).

Phylogenetic analyses

Twenty-five datasets were compiled taking a 'divide and conquer' approach rather than reconstructing a single tree from >2000 DNA sequences. In-house and curated nLSU and *rpb2* family level phylogenies were used to guide dataset assembly for nested monophyletic groups. Outgroups were chosen based on these unpublished large family-level datasets and previously published works. Generally, for closely related taxa we concatenated three-gene supermatrices utilizing ITS, nLSU, and *rpb2* gene sequences. For more inclusive groups (e.g., *Auritella*, the *Pseudosperma* clade) we found the ITS region to be too variable to align, and thus assembled nLSU+*rpb2* DNA sequence datasets only. Sequences analyzed or produced for this study are provided in Appendix 3. Newly generated DNA sequences have been submitted to NCBI GenBank.

Alignments were performed using ClustalX v2.0 (Larkin *et al.*, 2007) and adjusted manually in MacClade (Maddison & Maddison, 2005). Regions ambiguously aligned were omitted before phylogenetic analyses, all of which were done using the Maximum Likelihood (ML) criterion in RAxML 7.2.8 (Stamatakis, 2006; 2014). After concatenation of individual gene alignments, we partitioned the data by rRNA gene region (combining ITS and nLSU in a single partition) and *rpb2* codon positions in the large and more inclusive clades. For smaller data sets of more closely related taxa, we applied a single model of the DNA substitution process. For all partitions we employed a GTRGAMMA model as recommended by Stamatakis (2006). Five hundred bootstrap replicates were performed in RAxML to measure the degree of statistical support for nodes within each phylogenetic tree. Bootstrap proportions >50% are indicated. All alignment and tree files are available on request from the lead author.

Species concepts and recognition

Species were delimited by a combination of morphological, ecological, and phylogenetic observations. Ultimately, gene phylogenies were used to aid circumscription of species and to guide searches for differences in morphology and ecology.

In some cases, habitat type proved useful to distinguish species, as in the case of *I. pluvialis* (wet sclerophyll) and its sister species *I. tenax* (cool-temperate rainforest). In another example, *I. pluviosilvestris* is the only species of the *I. serrata* group in Australia to be found with *Nothofagus* in cool-temperate rainforest. *Inocybe meridionalis* is morphologically indistinguishable from *I. australiensis*, however the former occurs in cool-temperate rainforest whereas the latter and its allies are associated with sclerophyll vegetation. However, habit type is not absolute. Perhaps the most widespread species in Australasia, *I. leptospermi*, primarily occurs with sclerophyll vegetation but has (rarely) been found in association with *Nothofagus*. *Inocybe luteobulbosa* associates with *Nothofagus* in New Zealand, but in Australia it has been recorded in wet sclerophyll forest. Plant association may be phylogenetically conserved in some groups: of the eight Australian species of *Auritella*, none are associated with *Nothofagus*. *Tubariomyces* is likely a non-*Nothofagus*-associated clade as well.

In other instances, clear phylogenetic differences were found between lineages that led to formal recognition of different species. For example, *I. cacaocolor*, *I. subferruginea*, and *I. alloumbrina*, which form a monophyletic group, cannot be reliably distinguished morphologically. However, *I. subferruginea* has been found only in Western Australia, while *I. cacaocolor* and *I. alloumbrina* are primarily eastern Australian species. The delimitation of *I. sylvicola* as new was based on phylogenetic differences between it and New Zealand samples of *I. scissa*, despite the lack of differences in habitat preference and only few and subtle morphological differences. The various clades within the *I. serrata* complex can be distinguished based on molecular differences, but morphologically are almost impossible to separate.

INOCYBACEAE

Inocybaceae Jülich, *Bibliotheca Mycol.* 85: 374 (1982).

Type: *Inocybe* (Fr.) Fr.

Basidioma mostly small to medium, rarely robust; tricholomatoid, collybioid, or omphalinoid; centrally stipitate, fleshy fibrous, rarely tough. *Pileus* conical, convex, pulvinate, campanulate, or plane, sometimes umbonate, margin incurved or mostly decurved; surface dry (rarely viscid or lubricous), fibrillose, tomentose, fibrillose-squamulose, squarrose, with or without a velipellis, rarely hygrophanous, rimose in some speces. *Lamellae* mostly adnexed or sinuate, adnate or subdecurrent, rarely free, generally yellowish brown, brown, or with olivaceous tones at maturity; edges pallid-fimbriate. *Stipe* central, terete or compressed, even or base swollen or bulbous; cortina (partial veil) present or absent, not annulate or rarely with a cortinate ring-zone; surface dry, finely fibrillose, fibrillose, scaly, flocculose, pruinose, or with a vesture of densely aggregated fibrils. *Universal veil* present or absent; if present, then as a velipellis. *Spore print* dull brown or yellowish brown.

Basidiospores smooth, sinuous, angular, nodulose, stellate, or spinose, with slightly thickened wall, often yellowish brown in KOH (rarely subhyaline or pale brown), nondextrinoid, inamyloid, multinucleate, germ pore rarely present, hilum of the open-pore type, apiculus typically indistinct in smooth-spored species lacking pleurocystidia. *Basidia* 4-sterigmate, rarely exclusively 2-sterigmate, hyaline or necropigmented. *Pleurocystidia* absent or present as metuloids, often thick-walled and with crystalliferous apices. *Cheilocystidia* present (rarely absent) as metuloids and mixed with paracystidia or thin-walled and paracystidia absent. *Caulocystidia* absent or when present similar to cheilocystidia. *Pileipellis* a cutis, trichoderm, or palisade of upright elements, hyphae typically encrusted, thin- or thick-walled. *Hymenophoral trama* parallel. *Clamp connections* present throughout the fruit bodies.

Between c. 500–700 species worldwide. Approximately 140 species recognized in Australia. Associating with a wide range of plant families including Pinaceae, Gretaceae, Fagaceae, Juglandaceae, Betulaceae, Casuarinaceae, Nothofagaceae, Orchidaceae, Myrtaceae, Fabaceae, Dipterocarpaceae, Poaceae, Salicaceae, Polygonaceae, Nyctaginaceae, Cistaceae, Ericaceae, Tiliaceae, and Phyllanthaceae.

KEY TO GENERA AND MAJOR CLADES IN AUSTRALIA

1 Pleurocystidia present, if absent, then basidiospores not smooth .... **1. INOCYBE**

1: Pleurocystidia absent 2

2 Pileus rimose (rarely cracked-scaly); lacking necropigmented basidia .... **2. (INOCYBE) PSEUDOSPERMA CLADE**

2: Pileus scaly, tomentose, or fibrillose, not rimose; necropigmented basidia present 3

3 Context rubescent .... **3. (INOCYBE) INOSPERMA CLADE**

3: Context not rubescent (may be brunnescent) 4

4 Cheilocystidia short (<50 µm) and caulocystidia absent; fruit bodies not secotioid .... **4. (INOCYBE) MALLOCYBE CLADE**

4: Cheilocystidia long (>50 µm); if shorter, then caulocystidia present the entire length of the stipe; or fruit bodies secotioid 5

5 Cheilocystidia long (>50 µm); or fruit bodies often tricholomatoid or rarely secotioid; caulocystidia absent or restricted to the extreme apex of the stipe; pileipellis a cutis .... **5. AURITELLA**

5: Cheilocystidia short (reaching up to 50 µm) and fruit bodies omphalinoid, not secotioid; caulocystidia present the entire length of the stipe; pileipellis a palisade of clavate or subcylindrical terminal cells .... **6. TUBARIOMYCES**

SUMMARY PHYLOGENY OF THE INOCYBACEAE

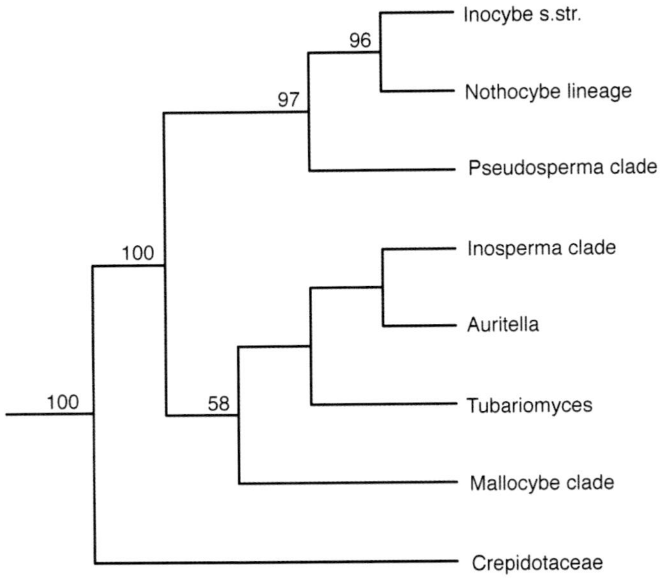

Figure 1. Summary phylogram of the seven major clades and lineages of the Inocybaceae and its sister group, the Crepidotaceae. The tree was reconstructed using Maximum Likelihood (ML) from analysis of nLSU, *rpb1*, and *rpb2* DNA sequence data in Matheny *et al.* (2012b). Numbers above branches represent bootstrap values derived from 1000 bootstrap replicates in that study.

INOCYBE

Inocybe (Fr.) Fr., 1836; Greek *ino-* (fibre), *-cybe* (head), in reference to the fibrous pileus surface.

Type: *I. relicina* (Fr. : Fr.) Quél.

Agaricus tribus *Inocybe* Fr., *Syst. Mycol.* 1: 254 (1821).

Astrosporina Schröt. in Cohn, *Kryptog.-fl. Schlesien, Pilze*: 576 (1889).

Clypeus (Britzelm.) Fayod, *Ann. Sc. Nat.* VII. 9: 362 (1889).

Agaricus subgenus *Clypeus* Britzelm., *Hymen. Südbayern* 3a: 4 (1882) ad int.; 3b (*Nat. Hist. Ver. Augsb., Ber.* 27: 149 (1883).

Agmocybe Earle, Bull. N.Y. Bot. Gard. 5: 439 (1909) (illegit. ICN 52.1).

Inocibium Earle, *Bull. N.Y. Bot. Gard.* 5: 439 (1909).

Inocybella Zerova in Zerova & Peresipkin, *Viznachnik Ukraïnï 5 Basidiomycetes*: 347 (1979) (inval., ICN 41.1).

Pileus fibrillose, rimose, or scaly. *Lamellae* edges indistinctly fimbriate. *Stipe* fibrillose, scaly, or partially pruinose at the apex or pruinose the entire length.

Basidiospores smooth, sinuous, angular, nodulose, stellate, spinose, at times with saddle-shaped or wedge-shaped nodules. *Basidia* hyaline, not necropigmented. *Pleurocystidia* present, typically thick-walled and with crystalliferous apices, rarely absent. *Cheilocystidia* similar to pleurocystidia but often shorter, mixed with paracystidia. *Caulocystidia* present or absent, variably distributed across the stipe, often similar to cheilocystidia and mixed with cauloparacystidia; at times the stipe covered with caulocystidioid hairs, these dissimilar from cheilocystidia, not mixed with cauloparacystidia.

KEY TO SPECIES OF THE INOCYBE CLADE IN AUSTRALIA

1 Under introduced north temperate trees (oak, elm, cottonwood, willow, birch, pine), including pine plantations 2

1: Under native vegetation 6

2 Spores nodulose or angular-nodulose 3

2: Spores smooth 4

3 Pileus dark brown to brown, stipe light brown becoming dark brown below, nowhere pruinose, bulb absent .... **1. I. curvipes**

3: Pileus honey-yellow to brownish yellow, stipe white, entirely pruinose, with a distinct marginate bulb .... **2. I. mixtilis**

4 Fruit bodies with lilac tinges at stipe apex, stipe with distinct brown fibrils.... **3. I. aff. pusio**

4: Fruit bodies lacking any lilac tinges, stipe without distinct brown fibrils 5

5 Pileus and stipe pallid, buff, to pale brown, spores mostly <10 µm long .... **5. I. sindonia**

5: Pileus dark brown, stipe ochraceous to pinkish brown; spores mostly 11–12.5 × 5.5–6.5 µm .... **4. I. rufuloides**

6 Fruit bodies secotioid 7

6: Fruit bodies agaricoid 8

7 Hymenial cystidia broadly clavate to sphaeropedunculate, 30–40 × 18–24 µm .... **18. I. bicornis f. secotioides**

7: Hymenial cystidia fusiform to ventricose, 60–70 × 15–21 µm .... **44. I. nudiuscula**

8 Spores smooth 9

8: Spores sinuous, angular, nodulose, stellate, spinose, or with multiple saddle-shaped nodules 83

9 Stipe entirely pruinose (caulocystidia and cauloparacystidia present below the mid-region of the stipe) 10

9: Stipe pruinose at the apex only or not at all (caulocystidia and cauloparacystidia absent below midregion of stipe) 26

10 Fruit bodies small; pileus <20 (25) mm wide 11

10: Fruit bodies medium; pileus >25 mm wide 23

11 Stipe wiry, tough 12

11: Stipe not wiry and tough 16

12 Pleurocystidia very thick-walled (4–10 µm), bright yellow; SW W.A..... **55. I. xanthocystis**

12: Pleurocystidia moderately thick-walled (3–4 µm), hyaline to pale yellow; Tas. & Qld..... 13

13 In rainforest under *Nothofagus* .... **52. I. tenax**

13: In sclerophyll forest under *Eucalyptus* or in warm temperate rainforest under *Lophostemon* ... 14

14 Odour acidulous; pileus with dark conspicuous scales when fresh .... **53. I. tenuis**

14: Odour of *Pelargonium* or sweet; pileus fibrillose or scaly only when dry 15

15 Pileus fibrillose, lacking dark conspicuous scales, Tas. .... **49. I. pluvialis**

15: Pileus becoming scaly when dry, Qld. .... **47. I. boreoaustralis**

16 Pileus (very) dark brown to dark yellowish brown and scaly 17

16: Pileus not as dark as above, scaly or fibrillose 19

17 Stipe dark reddish brown, in sand under *Allocasuarina* .... **46. I. atrorubripes**

17: Stipe light pinkish cinnamon to pale brown or reddish brown, on soil in leaf litter in wet sclerophyll 18

18 Pileus ≤10 mm wide, often bicolorous, fuscous to reddish brown at the centre, pale brown towards the margin; stipe 0.5–1.0 mm wide .... **50. I. pygmaeformis**

18: Pileus 10–25 mm wide, uniformly dark brown; stipe 2–3 mm wide .... **37. I. geniculata**

19 Pileus with pink tinges, lamellae pink to reddish brown; spores subhyaline............. **51. I. roseola**

19: Pileus lacking pink tinges; lamellae pale brown, strong brown, or yellowish brown; spores yellowish brown 20

20 Stipe noticeably striatulate due to longitudinal raised lines of pruinosity .... **54. I. vittata**

20: Stipe pruinose but not striatulate 21

21 Stipe white .... **66. I. flavoalbida**

21: Stipe coloured 22

22 Stipe avellaneous to cinnamon-brown .... **48 I. fragilis**

22: Stipe ochraceous-salmon .... **30. I. persicinipes**

23 Pileus matted-fibrillose, densely tomentose, or scaly; spores cylindric or oblong, 8–11 × 4–5 µm; caulocystidia bladder-shaped to short-clavate .... **39. I. mallocyboides**

23: Pileus fibrillose; spores amygdaliform or elliptic; caulocystidia not bladder-shaped or short clavate 24

24 Spores >5.5 µm wide .... **21. I. farinosipes**

24: Spores <5.5 µm wide 25

25 Pileus fulvous mixed with shades of brown or tawny; spores 7–9 × 4.5–5.5 µm; Tas. .... **14. I. luteobulbosa**

25: Pileus honey-yellow to very pale brown; spores 6.5–8.5 × 4.0–5.0 µm; Qld. & P.N.G .... **66. I. flavoalbida**

26 Stipe pruinose at the apex or upper half, fibrillose with a vesture or scaly below (lower part of stipe may have caulocystidioid hairs or cells) 27

26: Stipe nowhere pruinose (caulocystidia absent or restricted to extreme apex of stipe, terminal cells of stipe hyphae not differentiated) 67

27 Fruit bodies blueing or with a violet stipe apex 28

27: Fruit bodies not blueing and nowhere violet 29

28 Fruit bodies blueing; pileus blackish brown over the disc, yellowish brown with bluish tones towards the margin; stipe white mixed with blue in lower part... ..... **56. I. caerulata**

28: Fruit bodies not blueing; pileus reddish brown; stipe with lilac hues at apex .... **65. I. bella**

29 Pileus scaly with appressed or recurved scales 30

29: Pileus fibrillose, tomentose, or coarsely fibrillose 49

30 Pileus vivid reddish brown, rusty brown, or dark reddish brown 31

30: Pileus some other colour 32

31 Pileus bright reddish brown to rusty brown; spores 8.0–10.0 × 4.0–5.0 µm.. ..... **16. I. rufobadia**

31: Pileus dark reddish brown; spores 10.0–13.0 × 4.5–6.0 µm .... **20. I. exigens**

32 Fruit bodies squat or peg-like 33

32: Fruit bodies not squat or peg-like 37

33 In sandy soil under shrubs in coastal heath 34

33: Habitat not as above 35

34 Stipe densely pruinose above the cortinate insertion, spores amygdaliform, metuloids from stipe surface 58–135 × 14–17 µm .... **26. I. gomphocephala**

34: Stipe furfuraceous-fibrillose, not obviously pruinose as above, spores oblong-amygdaliform to amygdaliform (some subangular in outline), metuloids from stipe surface 43–77 × 14–19 µm .... **29. I. avellaneifolia**

35 Stipe light pinkish to ochraceous; in lawns or parks .... **27. I. memorialis**

35: Stipe pinkish brown at first, becoming brown, reddish brown, copper brown, or dark brown; often in eucalypt woodland or forest 36

36 Stipe even or with a slight swollen base, lacking any brown fibrils .... **10. I. brevicula**

36: Stipe base distinctly bulbous, often with brown fibrils .... **11. I. campestris**

37 Pileus and stipe with white delicate floccose scales .... **43. I. ornatipes**

37: Pileus and stipe without white floccose scales 38

38 Fruit bodies robust: pileus 14–30 mm wide, stipe 16–30 × 3–8 mm 39

38: Fruit bodies slender: pileus 15–25 mm wide, stipe 15–33 × 2–4 mm 45

39 Spores 6.0–8.0 × 4.5–5.5 µm; in cool-temperate rainforest under *Nothofagus* .. **9. I. auritelloides**

39: Spores 8.0–11.0 × 4.5–7.0 µm; in sclerophyll communities or under planted eucalypts 40

40 Caulocystidia and cauloparacystidia occurring over upper half of stipe, caulocystidioid hairs absent 41

40: Caulocystidia and cauloparacystidia restricted to the extreme apex of stipe, if at all, soon mixed with caulocystidioid hairs descending entire length of stipe 42

41 Stipe base bulbous with brown fibrils .... **11. I. campestris**

41: Stipe even or tapered, the lower part with white longitudinally aggregated floccules ....

.... **27. I. memorialis**

42 Pleurocystidia 60–85 µm long; often under *Acacia* .... **24. I. froudistii**

42: Pleurocystidia 70–115 µm long; often under eucalypts and/or *Allocasuarina* 43

43 Stipe with a pruinose appearance in upper third and reddish brown; Tas. **25. I. fuscosquarrosa**

43: Stipe pruinose at extreme apex only, and brown to yellowish brown above, dark to dark brown below; W.A. ....

44. Tips of scales on pileus whitish .... **22. I. fibrillosibrunnea**

44: Tips of scales on pileus very dark brown to dark brown .... **23. I. floccosibrunnea**

45 Associated with *Nothofagus* .... **15. I. meridionalis**

45: Associated with eucalypts, *Melaleuca, Allocasuarina, Acacia* 46

46 Lower part of stipe with greyish to greyish brown vesture of woolly fibrils **41. I. fibrillosipes**

46: Lower part of stipe with brown to dark brown vesture of woolly fibrils or with dark brown fibrils above the base 47

47 Spore mean length >8.0 µm .... **38. I. eriocaulis**

47: Spore mean length <8.0 µm, ... 48

48 Lower half of stipe woolly-fibrillose, becoming very dark brown .... **7. I. australiensis**

48: Brown fibrils restricted to the very base of the stipe .... **6. I. atrisquamosa**

49 Pileus <25 mm wide; stipe <3(4) mm wide 50

49: Pileus >25 mm wide, or stipe >3 mm wide 62

50 Odour strongly of *Pelargonium* .... **12. I. fragrans**

50: Odour acidulous or not distinctive 51

51 In temperate *Nothofagus* rainforest 52

51: In temperate sclerophyll forests or warm-temperate rainforest with *Lophostemon* 53

52 Pileus smooth to finely fibrillose, spores 6.5–8 × 4–5 µm, cystidia yellow in KOH .... **61. I. pluviosilvestris**

52: Pileus tomentose to coarsely fibrillose, spores 9–10 × 5–6 µm, cystidia hyaline in KOH .... **45. I. tasmanica**

53 Spores very pale (light buff to ochre-buff) in KOH, some cylindric or minimally angular in appearance .... **69. I. paucicolor**

53: Spores yellowish brown in KOH, amygdaliform or elliptic, not angular or cylindrical 54

54 Stipe deep vinaceous brown .... **42. I. ionocaulis**

54: Stipe coloured differently 55

55 Stipe smooth or pruinose above, with few fibrils or with a thin vesture below 56

55: Stipe with a well developed vesture at least on the lower part 58

56 Caulocystidioid hairs absent from lower half of the stipe .... **44. I. nudiuscula**

56: Caulocystidioid hairs present on the lower half of the stipe 57

57 Stipe brown to pale brown; spores mostly 9.0–12.0 × 4.5–6.0 µm .... **68. I. murryana**

57: Stipe salmon or ochraceous; spores mostly 8.0–9.5 × 4.5–5.0 µm .... **30. I. persicinipes**

58 Caulocystidioid hairs absent (terminal cells of vesture hyphae not differentiated) 59

58: Caulocystidioid hairs present and conspicuous below the mid-region of the stipe 60

59 Stipe ground colour light yellowish brown to yellowish brown; spores 7.0–8.5 × 4.0–5.0 µm .... **32. I. vestipedes**

59: Stipe with a reddish brown ground colour; spores 8.5–10.5 (–11.0) × 4.5–5.5 µm .... **31. I. rufoflavipes**

60 Stipe base pale yellow; edge of pileus without white veil remnants .... **60. I. pallidiflavipes**

60: Stipe base not pale yellow; edge of pileus with white veil remnants 61

61 Upper ⅛–⅙ of stipe densely pruinose, remainder with light fibrillose vesture, not mealy .... **62. I. serratoides**

61: Upper ¼–½ of stipe lightly pruinose, below with a dense scurfy or floccose-fibrillose vesture .... **I. serrata complex (63. I. serrata; 58. I. fulvo-olivacea; 59. I. granulosipes;** .... **see also 57. I. alboflocosa)**

62 Basidia exclusively 2-sterigmate .... **17. I. bicornis**

62: Basidia mostly 4-sterigmate (occasionally some 2-sterigmate basidia present) 63

63 Spores on average <10 µm long.. 64

63: Spores on average >10 µm long.. 66

64 Stipe white, associated with *Acacia* .... **28. I. acaciae**

64: Stipe coloured, associated with eucalypts 65

65 Pileus coarsely fibrillose throughout, becoming scaly with age; stipe not densely pruinose, yellow, brownish yellow to yellowish brown .... **8. I. aureorimosa**

65: Pileus smooth at the centre, finely fibrillose elsewhere; stipe densely white pruinose at least above, with a pale brown to pinkish buff ground colour .... **21. I. farinosipes**

66 Pileus convex at first, but soon plano-convex and then plane, reddish brown to bright brown; lamellae pale yellow becoming olive brown; stipe dull yellowish to off-white........ .... **71. I. spadicea**

66: Pileus convex for long duration, pale brown to dark brown; lamellae pale yellowish brown becoming dull brown; stipe pinkish brown .... **19. I. clypeata**

67 Fruit bodies violet, at least on the stipe.. 68

67: Fruit bodies without violet anywhere.. 69

68 Odour spermatic; spores elliptic with rounded apices .... **72. I. violaceocaulis**

68: Odour acidulous; spores amygdaliform with pointed apices .... **65. I. bella**

69 Context of stipe rubescent or becoming pink where cut or bruised 70

69: Context of stipe not changing colour where cut or bruised.. 71

70 Pileus matted-felty to fibrillose, lacking marginal streaks of white fibrils; flesh rubescent; odour spermatic .... **70. I. rufescens**

70: Pileus densely fibrillose, margin streaked with white fibrils; flesh becoming pink; odour not distinctive ....**35. I. leucotaenia**

71 Pileus squamulose to squarrose.. 72

71: Pileus fibrillose (if scaly, only so on older specimens)... 75

72 Pileus brownish yellow to yellowish brown; stipe with well-developed cortinate ring-zone; habit slender; spores amygdaliform to citriniform .... **34. I. hymenocarpa**

72: Pileus tawny or dark brown to blackish brown; stipe without a cortinate ring-zone; habit robust; spores not citriniform 73

73 Pileus dark brown; spores 8.0–10.0 × 4.5–6.5 µm, amygdaliform; in sclerophyllous forest or urban settings under eucalypts .... **64. I. austrosquarrosa**

73: Pileus dark brown to blackish brown or tawny (rusty brown); spores shorter or not as wide as above; in rainforest under *Nothofagus*; 74

74 Spores 6–8 × 4–6 µm, broadly elliptical to sub-amygdaliform .... **9. I. auritelloides**

74: Spores 7–9 × 4–5 µm, amygdaliform .... **13. I. lucifera**

75 Pileus conical or conico-umbonate; fruit bodies slender and small (pileus <25 mm wide, stipe <3 mm thick) 76

75: Pileus hemispherical, convex or plane; if conical, then fruit bodies larger than above; fruit bodies robust, slender or squat 78

76 Spores 7.0–8.5 × 4.0–5.0 µm; stipe with light yellowish brown to yellowish brown ground colour .... **32. I. vestipedes**

76: Spores 8.5–10.5 × 4.5–5.5 µm; stipe cinnamon-buff to clay brown or reddish brown 77

77 Stipe cinnamon-buff to clay-brown; on sand in coastal heath-scrub .... **29. I. avellaneifolia**

77: Stipe with a reddish brown ground colour; on soil in eucalypt forest .... **31. I. rufoflavipes**

78 Fruit bodies robust; stipe to 7 mm wide or squat; spores mostly with conical to bluntly pointed apices (infrequently with mucronate apices) 79

78: Fruit bodies slender; stipe 2–3 mm wide; spores at times with mucronate apices 82

79 Context slowly becoming pink after cutting; young lamellae pinkish .... **35. I. leucotaenia**

79: Context not changing colour where cut; young lamellae not pinkish 80

80 Lamellae avellaneous, becoming brown; pileus brown; short squat habit **29. I. avellaneifolia**

80: Lamellae light gray or pale yellow, becoming olivaceous; pileus orange-brown, fulvous, or yellowish brown; fruit bodies not squat 81

81 Lamellae light grey, becoming olivaceous; spores 9.5–11.5 × 6.0–7.0 µm; associated with temperate vegetation .... **33. I. laureata**

81: Lamellae pale yellow, becoming yellowish green or olivaceous; spores 7.5–9.5 × 4.5–6.0 µm; associated with tropical rainforest .... **67. I. luteo-olivacea**

82 Pileus pale brown to brown-streaked; lamellae brown to chestnut-brown; spores 8.0–10.0 × 5.0–6.5 µm .... **36. I. tomentipes**

82: Pileus olive brown to dark yellowish brown, not streaked, lamellae pinkish brown to yellowish brown; spores 7.0–9.0 × 4.5–5.5 µm .... **40. I. olivaceohinnulea**

83 Spores with multiple saddle-shaped nodules; cheilocystidia with yellow plasmatic pigment or occlusions and dissimilar from pleurocystidia 84

83: Spores spinose, stellate, sinuous, angular, or nodulose; cheilocystidia lacking yellow plasmatic pigment 85

84 Spores 8.0–10.5 × 6.5–8.5 µm .... **73. I. alienospora**

84: Spores 9.5–12.0 × 7.5–10.0 µm .... **74. I. lasseroides**

85 Spores spinose 86

85: Spores stellate, sinuous, angular, or nodulose 89

86 Spores globose (Q 1.00–1.11; Q mean <1.10); stipe with a rounded basal bulb to 8 mm wide .... **84. I. acanthosperma**

86: Spores broadly elliptical to subglobose (Q 1.05–1.32; Q mean >1.10); stipe wide with a slightly swollen base to 3 mm wide 87

87 Spores with 13–16 sharp conical projections .... **86. I. tertia**

87: Spores with 16–24 sharp conical projections 88

88 Spores 10.5–13.5 × 9.0–11.5 (–12.0) µm, pileus scaly .... **85. I. spiniformis**

88: Spores 9.0–11.0 × 8.0–9.5 µm, pileus fibrillose .... **85. I. cf. spiniformis**

89 Spores stellate 90

89: Spores sinuous, angular, or nodulose 91

90 Pileus dark brown to strong brown; stipe long, (to 75 mm); habit slender and stately .... **80. I. nobilissima**

90: Pileus brown to yellowish brown; stipe not as long as above (to 30 mm); habit more robust .... **81. I. vagata**

91 Stipe pruinose below the mid-region and often the entire length (caulocystidia and cauloparacystidia present on lower part of stipe) 92

91: Stipe pruinose at the apex only or not at all (caulocystidia and cauloparacystidia not descending below mid-region of stipe) 114

92 Spores sinuous (some appearing smooth in outline) .... **97. I. sinuospora**

92: Spores angular or nodulose 93

93 Pileus white, yellow, or honey-yellow 94

93: Pileus yellowish brown (brownish yellow), brown, reddish brown, or dark brown 96

94 Pileus 10–25 mm wide, yellow to olive-yellow; lamellae yellow to olivaceous; odour spermatic .... **79. I. subangustifolia**

94: Pileus 20–45 mm wide, whitish or honey-yellow; lamellae not yellow or olivaceous; odour not remarkable or acidulous 95

95 Pileus whitish due to heavy velipellis with yellowish undertones, not rimose; odour mild; in sclerophyllous forests .... **90. I. eburnea**

95: Pileus honey- or straw-yellow, strongly rimose; odour none or slightly acidulous; primarily in *Nothofagus* rainforest .... **108. I. sylvicola**

96 Pileus strongly viscid .... **110. I. viscata**

96: Pileus dry or weakly viscid 97

97 Odour of marzipan or like almond extract .... **88. I. amygdalina**

97: Odour not as above 98

98 Stipe base with a distinct marginate bulb 99

98: Stipe base without a distinct marginate bulb (stipe base tapered, swollen, or rounded bulbous) 108

99 Pileus centre brown, with brown streaks towards the paler margin .... **112. I. brunneidisca**

99: Pileus lacking brown streaks 100

100 Pileus reddish brown or dark brown to brown 101

100: Pileus yellowish brown (fulvous) to brownish yellow 105

101 Pileus dark brown to brown; hymenial cystidia tapered at the bases .... **107. I. leptospermi**

101: Pileus reddish brown; hymenial cystidia sessile or tapered at the base 102

102 Spores oblong-angular to angular with 5–7 low nodules or corners......... ...... **82. I. sejuncta**

102: Spores not oblong and not angular, with 7–10 distinct or low nodules 103

103 Stipe whitish with pale yellow to brownish tinges, spores 9.0–10.5 (–11.0) × 6.5–7.5 µm .... **105. I. calopedes**

103: Stipe whitish with pinkish tinges, spores smaller than above 104

104 Pileus dark reddish brown; spores mostly 7.5–9.5 × 5.5–7.0 µm with 7–10 conspicuous nodules; odour none or weakly acidulous .... **104. I. calocephala**

104: Pileus bronze-brown to light reddish brown; spores mostly 6.0–8.5 × 5.0–6.5 µm with 7–9 low obtusely conical nodules; odour weakly spermatic .... **106. I. chalcoceps**

105 Pleurocystidia walls 2–3 µm thick 106

105: Pleurocystidia walls 3–6 µm thick 107

106 Stipe 3–7 mm wide; pileus brownish yellow to yellowish brown; universal veil remnants present as white fibrils on the margin of the pileus and edge of stipe base .... **89. I. dewrangia**

106: Stipe 1–3 mm wide; pileus strong brown to near yellowish brown; universal veil remnants absent .... **83. I. tumidula**

107 Spores with 12–15 nodules .... **93. I. fulvilubrica**

107: Spores with 7–11 nodules .... **87. I. invadens**

108 Pileus imbricate scaly and dark brown; stipe with a rounded bulb .... **101. I. subpulverulenta**

108: Pileus not as above; stipe tapered, even, or swollen 109

109 Spores oblong-angular .... **91. I. emergens**

109: Spores not oblong 110

110 Stipe tapered towards the base; pleurocystidia <55 µm long, cylindric to subfusiform; spores with 7–11 small nodules .... **96. I. nothopus**

110: Stipe not tapered towards the base (slightly bulbous); pleurocystidia and spores not as above 111

111 Stipe with a pale orange-brown fibrillose zone above the slightly bulbous base ... **94. I. gilviceps**

111: Stipe lacking pale orange-brown fibrils above the base 112

112 Odour not remarkable... **92. I. exigua**

112: Odour like honey or slowly becoming sweet after cutting 113

113 Odour like honey; pleurocystidia with long slender basal pedicel; rare, SW W.A. .. **95. I. mellita**

113: Odour slowly becoming sweet after cutting; hymenial cystidia sessile, deeply embedded in the hymenium; northerly latitudes in tropical sclerophyllous and warm temperate forests .... **109. I. torresiae**

114 Stipe base distinctly bulbous 115

114: Stipe even or with a slightly swollen or slightly enlarged base 117

115 Spores <10 µm; pleurocystidia walls c. 1 µm thick .... **77. I. solitaria**

115: Spores >10 µm long; pleurocystidia walls 2–4 µm thick 116

116 Spores 8–11 × 5.5–7 µm, caulocystidia infrequent .... **111. I. bulbinella**

116: Spores 9–13 × 7–9.5 µm, caulocystidia frequent............................ **101. I. subpulverulenta**

117 Pileus fibrillose, at most becoming excoriate scaly 118

117: Pileus appressed-squamulose or squarrose 120

118 Spores with 8–10 conical nodules .... **113. I. humidicola**

118: Spores with 10–14 nodules.. 119

119 Pileus 6–12 mm wide; stipe 1–2 mm wide; nodules conical............................ **114. I. parvifulva**

119: Pileus 13–22 mm wide; stipe 2–3 mm wide; nodules small and convex or rounded .... **103. I. trachysperma**

120 Stipe with woolly dark brown to brown fibrils or scales, or spores >10 µm long 121

120: Stipe fibrillose, spores <10 µm long.. 123

121 Pleurocystidia absent .... **99. I. nigricans**

121: Pleurocystidia present 122

122 Pleurocystidia thin-walled, spores with 9–14 small nodules .... **98. I. mallopoda**

122: Pleurocystidia thick-walled, spores with 5–8 (–10) small nodules or corners .... **102. I. imbricata**

123 Pleurocystidia thin-walled (<1.5 μm thick); odour none or weakly of *Pelargonium* ....**I. alloumbrina complex (75. I. alloumbrina, 76. I. cacaocolor, 77. I. subferruginea)**

123: Pleurocystidia thick-walled (1.5–2.5 μm thick), odour strongly of *Pelargonium* .... **100. I. redolens**

INOCYBE: INTRODUCED SPECIES

1. Inocybe curvipes P.Karst., *Hedwigia* 29: 176 (1890)

Type: Tammela, Mustiala, Myllyperä, Finland, 22 Aug. 1890, *P.A.Karsten*; holo: *n.v.*

Inocybe radiata Peck, *Bull. Torrey Bot. Club* 22: 488 (1895). Type: Massachusetts, U.S.A., in open grassy ground, Aug. and Sept., *W.D.Jackson*; iso: WTU.

Inocybe lepidocephala Speg., *Anales Mus. Nac. Buenos Aires* 6: 126 (1898). Type: Buenos Aires, Argentina; holo: *n.v.*

Inocybe variabillima Speg., *Anales Mus. Nac. Buenos Aires* 6: 125 (1899). Type: Buenos Aires, Argentina; holo: *n.v.*

Inocybe decipientoides Peck, *Bull. Torrey Bot. Club* 34: 100 (1907); *Astrosporina decipientoides* (Peck) Pearson, *Trans. Brit. Mycol. Soc.* 22: 28 (1938). Type: Massachusetts, U.S.A.; holo: *n.v.*

Inocybe astoriana Murrill, *Mycologia* 3: 104 (1911). Type: New York City, U.S.A.; iso: WTU.

Inocybe jamaicensis Murrill, *Mycologia* 4: 82 (1912). Type: Cinchona, Jamaica, wet mountains area, alt. 4500–5200 ft, in a clay road, 25 Dec. 1908–8 Jan. 1909, *W.A.Murrill & E.L.Murrill WAM & ELM 595*; iso: WTU.

Inocybe ochraceoscabra G.F.Atk., *Amer. J. Bot.* 5: 214 (1918). Type: Cornell Heights, Ithaca [New York], U.S.A.; holo: *n.v.*

Inocybe globocystis Velen., *České Houby* 368 (1920); *Inocybe curvipes* var. *globocystis* (Velen.) Bon, *Docs. Mycol.* 28(109–110): 10 (1998). Type: Bachrändern, Bohemia [Czech Republic]; holo: *n.v.*

Inocybe lanuginella sensu auct. div., non (J.Schröt.) Konrad & Maubl. *Icon. Select. Fung.* 6: 137 (1937). Type: not indicated [Poland].

Inocybe cinerascentipes Huijsman, *Fungus* 25: 26 (1955). Type: Nijmegen, Netherlands; holo: *n.v.*

Inocybe sericeopoda Furrer-Ziogas, *Beih. Sydowia* 10: 62 (1995). Type: Îles d'Hyères, France; holo: *n.v.*

[*Inocybe cicactricata auct. non* Ellis & Everh. *sensu* R.Heim, *Encylop. Mycol.*, 1 Le Genre *Inocybe* (Paris) 1: 351 (1931)]

[*Inocybe carpta* (Scop. : Fr.) P.Kumm. *sensu* Ricken, *Die Blätterpilze*: 101 (1915)]

[*Inocybe lanuginella* (J.Schröt.) Konrad & Maubl.: W.A.Dunstan, B.Dell & N.Malajczuk, *Mycorrhiza* 8: 73 (1998)]

Pileus to 27 mm wide, conical, hemispherical or obtusely conical when immature, becoming subcampanulate, obtusely conical or ±convex upon expansion; margin incurved when young, becoming decurved; velipellis absent, surface dry, appressed-fibrillose, with an unbroken centre when young, later finely appressed-scaly on and around the centre and fibrillose towards the margin; margin torn and at times radially rimulose; dark brown (10YR 3/3; Mummy Brown to bistre) throughout or brown (10YR 4/3; Snuff Brown or Saccardo's Umber) or shading brown towards the margin, occasionally the edge of the margin yellowish brown (10YR 5/4; Tawny-Olive). Context pallid or white, not changing colour where bruised. *Lamellae* adnate to adnexed, moderately close, 34–48 L with several tiers of lamellulae, becoming ventricose with age, to 4 mm broad, light grey when immature to light yellowish brown or brown, eventually becoming dark yellowish brown; edges pallid and fimbriate. *Stipe* 20–40 mm × 3–7 mm at the apex, even or with a slightly swollen base; veil cortinate, fugacious; surface nowhere pruinose, finely fibrillose to fibrillose or twisted-fibrillose, generally pallid above, shading to light brown (Avellaneous) below, but brown to dark brown (bistre) at the base and upwards with age; context solid, mostly pallid, occasionally with a subtle pinkish brown cortex. *Odour* spermatic when first cut, later mixed with an acidulous component. Pl. 13.

Basidiospores (8.5–) 9–12 × 5–6.5 µm [mean 10.3 × 5.6 µm; Q 1.36–1.90; Q mean 1.66 (*n* = 60/2)], angular-nodulose about a trapeziform outline with 6–10 moderately sized to small nodules projecting up to 2 µm, at times these merely corners, usually with a distinct apical nodule, yellowish brown; apiculus distinct. *Basidia* 26–34 × 8–11 µm, mainly 4-sterigmate, less often 2- or 1-sterigmate, then with long sterigmata up to 15 µm long, clavate, cylindrical or ventricose (particularly when 1- or 2-sterigmate). *Pleurocystidia* 47–73 × 19–27 µm, broadly fusiform or saccate and noticeably tapered towards the apex which is usually acute (not obtuse or well-rounded) and crystalliferous, with a long slender basal pedicel, thick-walled, the walls to 2.5 µm thick, hyaline. *Cheilocystidia* 48–66 × 15–24 µm, similar to pleurocystidia but more variable in shape, with abundant clavate pyriform and vesiculose paracystidia among the larger cystidia. *Caulocystidia* absent; sparse vesture hyphae observed on lower part of the stipe. *Pileipellis* a cutis of cylindrical hyphae 4–15 µm wide, many with brown encrusting pigment, some emerging in scattered bundles and with cystidioid terminal elements. *Clamp connections* present. Fig. 2.

Occurs on soil among leaf litter under introduced *Quercus* or *Pinus radiata* and possibly *Salix* in W.A., S.A., A.C.T. and Vic. Introduced into Australia and Papua New Guinea. Fruiting Feb. and Apr.–Aug. in southern Australia.

Specimens examined: W.A.: Carine, Perth, 31°51'S, 115°46'E, gregarious on soil under *Quercus* in suburban backyard garden, 31 Aug. 2001, *B.Dunstan PBM 2210* (PERTH 07676859, formerly CSIRO E7058); Botany Department, campus of University of Western Australia, Crawley, Perth, 31°57'S, 115°52'E, in rose garden, 6 June 1973, *D.Waldie s.n.* (PERTH 00763578). S.A.: Belair Natl Park, unrecorded habitat and plant association, 12 May 1917, *J.B.Cleland s.n.* (AD 12143). A.C.T.: CSIRO, Black Mtn, Canberra, 35°17'S, 149°07'E, in grass at base of *Pinus radiata* near the herbarium, 14 Feb. 1990, *E.Cooper EC4* (CANB 404928; JN035293-nLSU, JN035289-ITS). Vic.: Templestowe, Melbourne, 37°45'S, 145°8'E, growing in lawn in rings under weeping willow (presumably *Salix*), 29 May 1963, *P.Jenkins s.n.* (MEL 1054882); grounds of Institute of Social Welfare Centre, Watsonia, 37°45'S, 145°00'E. on turf under *Quercus*, 25 Apr. 1976, *N.H.Sinnott & D.M.Kershaw NHS1996* (CANB 359621).

Extralimital specimens examined: Papua New Guinea: near Massy, along road under *Pinus*, 27 July 2003, *R.Treu TR37–03* (M 0138268; JN035294-nLSU, JN035290-ITS). U.S.A.: Washington: next to Burke–Gilman Trail, near the Wall of Death, University District, Seattle, on disturbed soil along side of road near *Corylus*, *Prunus* and *Populus*, 18 June 1998, *P.B.Matheny PBM1007* (WTU); *loc. id.*, on grassy lawn under introduced *Arbutus unedo* with *Pinus* in the vicinity, 28 Oct. 1998, *P.B.Matheny PBM1328* (WTU); Green Lake, Seattle, King County, on ground under *Betula*, 15 Sept. 2002, *J.F.Ammirati & M.T.Seidl PBM2491* (WTU; QY239022-nLSU, AY337414-*rpb2*, AY239032-*rpb1*).

Diagnostic features: On soil under introduced northern temperate ectomycorrhizal trees. Fruit bodies medium, brown, dark brown or yellowish brown; stipe generally darker towards the base. Odour spermatic. Hymenial cystidia broadly fusiform or ventricose. Spores angular-nodulose, on average >10 µm long, typically with 6–10 low nodules and/or corners.

Inocybe curvipes is widespread in the Northern Hemisphere temperate zone having been described numerous times in Europe, eastern and western North America and Central America. ITS sequences of Australian collections of *I. curvipes* match (97–99% similarity) representatives of this species from Europe (Matheny & Bougher, 2011). In Australia, *I. curvipes* has a restricted habitat as it occurs in gardens and suburban plantings of *Quercus* and *Pinus* and possibly *Salix* (Matheny & Bougher, 2011); it has not been found in natural vegetation, or in exotic plantations of *Pinus*. The species was likely introduced into Australia together with imported northern temperate ectomycorrhizal plants. It has been similarly introduced into other Southern Hemisphere regions, including with *Pinus* in South Africa [Dunstan *et al.*, 1998; as *I. lanuginella* (J.Schröt.) Konrad & Maubl.], and it is also widespread in South America under *Pinus* (Cortez & Coelho, 2005). *Inocybe curvipes* was recognised for the first time in Australia by Matheny & Bougher (2010), but it has been present in Australia for at least a century; it was collected by J.B.Cleland in South Australia (AD 12143) in 1917. That specimen was originally unidentified by Cleland, and later incorporated, together with *Astrosporina asterospora sensu* Cleland, and merged into the protologue of *I. dewrangia* Grgur. (Grgurinovic, 1997; Matheny & Bougher, 2010). Since publication of Matheny & Bougher (2011) blastn analyses of ITS sequences of *I. curvipes* have revealed the species also occurs with *Pinus* in Cameroon, Zambia, Japan and China.

Inocybe curvipes is toxic to humans and can occur in densely populated urban habitats near non-native, northern temperate ectomycorrhizal trees. One collection from Victoria in the 1960s in Vic. (originally identified as *Inocybe* sp.; MEL 1054882) is annotated "cause of muscarin poisoning in woman who gathered it to eat it in mistake for mushroom" (Bougher & Matheny, 2011).

The presence of 1- and 2-sterigmate basidia is not a consistent feature of this species, but they are present in some collections, including one of the two Western Australian collections examined (PERTH 07676859).

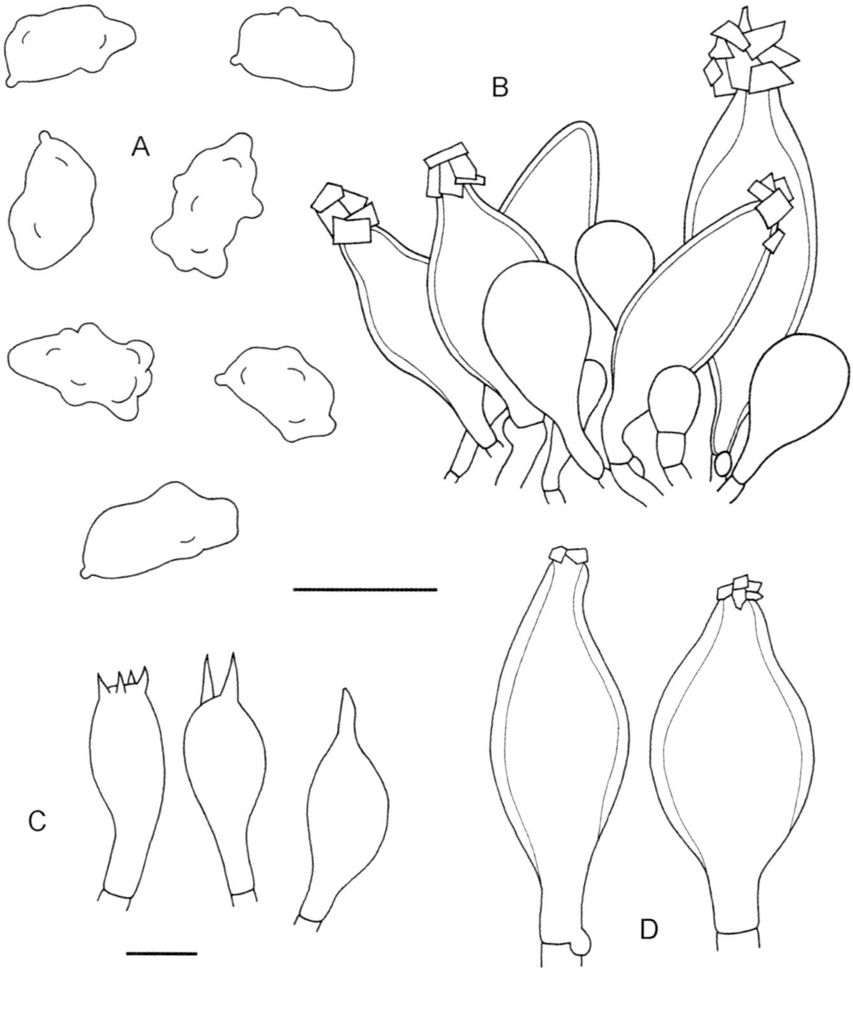

Figure 2. *Inocybe curvipes.* All PERTH 07676859. **A,** Basidiospores; **B,** Cheilocystidia and paracystidia; **C,** Basidia; **D,** Pleurocystidia. Scale bars: 10 μm (longer bar for spores only).

Plate 13. *Inocybe curvipes*. PERTH 07676859. Photograph — N.L.Bougher.

2. Inocybe mixtilis (Britzelm.) Sacc., *Syll. Fung*. 5: 780 (1887)

Agaricus mixtilis Britzelm., *Ber. Naturhist. Vereins Augsburg* 28: 152 (1885). Type: Bavaria, Germany; neo: *n.v.*, *fide* J.Stangl, *Int. J. Mycol. Lichenol*. 1: 332 (1984).

Pileus up to 20 mm wide, conical to campanulate, at times with a broad umbo; margin incurved to decurved, with a satiny or silky sheen; velipellis absent; surface tacky viscid, finely fibrillose; brownish orange (6C7 to 6C8). *Lamellae* adnexed to almost free, moderately close, initially white, becoming greyish brown. *Stipe* 20–35 × 3–6 mm, white; base with a marginate bulb; surface densely pruinose the entire length; cortinate veil remnants observed on a very young specimen, ostensibly connected to the base of the stipe. *Odour* spermatic. Pl. 14.

Basidiospores 7.5–9.5 × 5.0–7.0 µm [mean 8.1 × 5.6 µm; Q 1.21–1.70; Q mean 1.47 (*n* = 18/1)], angular-nodulose with mostly 8–11 obtuse nodules (or corners at times); nodules sometimes quite pronounced, especially at the apex, yellowish brown; apiculus distinct. *Basidia* 24–27 × 8–9 µm, clavate, hyaline, 4-sterigmate. *Pleurocystidia* 43–60 x 17–25 µm, mostly broadly ventricose or broadly utriform; apices obtuse or well-rounded, usually with crystals, thick-walled, the walls 2–4 µm thick, hyaline. *Cheilocystidia* similar to pleurocystidia, often shorter, mixed with clavate to pyriform paracystidia. *Caulocystidia* similar to cheilocystidia and descending in scattered clusters along the entire length of the stipe, mixed with clavate to pyriform cauloparacystidia. *Pileipellis* a cutis, yellowish brown in mass; hyphae cylindrical, mostly 5–12 µm wide, encrusted, their walls slightly thickened; overlain by a thin superficial hyaline layer of very narrow hyphae, these mostly 3–4 µm wide, thin-walled. *Clamp connections* present. Fig. 3.

Occurs on soil among litter under introduced *Pinus* and *Ulmus* in the A.C.T.

Specimens examined: A.C.T.: Yarralumla, 3.5 km WNW of Capital Hill, Canberra, 35°18'20"S, 149°05'00"E, alt. 560 m, on soil under thick layer of pine needles under *Pinus* and *Ulmus*, 23 May 1998, *H.Lepp 2002* (CANB 574539, PERTH 08516669; KP170952-nLSU, KP308781-ITS, KM406195-*rpb2*); *loc. id.*, *H.Lepp 2003* (CANB 574543).

Extralimital specimens examined: France: Pas-de-Calais, Hardelot, dunes d'Ecaut, under *Pinus pinaster* with *Betula* and *Populus tremula*, 1 Nov. 2007, *P.-A.Moreau PAM07110104* (TENN 063575; HQ641115-nLSU, HQ586872-ITS, KM406196-*rpb2*). Norway: Oppland, Lunner, S. Oppdalen, UTM 93 85, gregarious to scattered on soil in eutrophic *Picea* forest on forest edge, middle boreal zone, alt. 400 m, 20 Aug. 2002, *G.Gaarder PBM2379* (WTU).

Diagnostic features: Fruit bodies medium; pileus brownish orange, finely fibrillose, tacky-viscid; stipe white, pruinose along its entire length, with a distinct marginate bulb. Under introduced trees, e.g. *Pinus*. Spores nodulose, <10 µm long.

This is the first report of this introduced, northern temperate species from Australia. Some specimens (e.g. CANB 574543) co-occur and fruit simultaneously with the smooth-spored exotic species *I. sindonia*.

The diagnostic features listed above separate it from similar species native to Australia and associated with native plants, *viz. I. fulvilubrica*, *I. dewrangia* and *I. invadens*.

Collections of *I. mixtilis* from the Australian Capital Territory correspond well with samples studied from Norway and France, but the Norwegian specimen has more coarsely nodulose spores (with mostly 10–12 moderate-sized nodules). There is substantial genetic diversity within *I. mixtilis* apparent from differences observed in DNA sequences (ITS, LSU, *rpb2*) among European and North American populations. Genetically, an Australian sample is identical with the French collection cited above, which was also collected under *Pinus* [see *I. mixtilis* (as outgroup) in Fig. 93].

There is at least one other exotic nodulose-spored species of *Inocybe* associated with *Pinus* in Australia [in 'Agri' area, pines, Donnybrook Sunklands, Western Australia, associated with *Thelephora terrestris* in four-year old pines (*Pinus*), 21 May 1980, *R.N.Hilton UWA 2391* (PERTH 0904589)]. The spores and cystidia are similar to those of *I. mixtilis* (CANB 574539), but the dried specimens are much more delicate. In the absence of descriptive notes and a DNA sequence, further collections are necessary to elucidate its identity.

Plate 14. *Inocybe mixtilis*. CANB 574539. Photograph — H.Lepp.

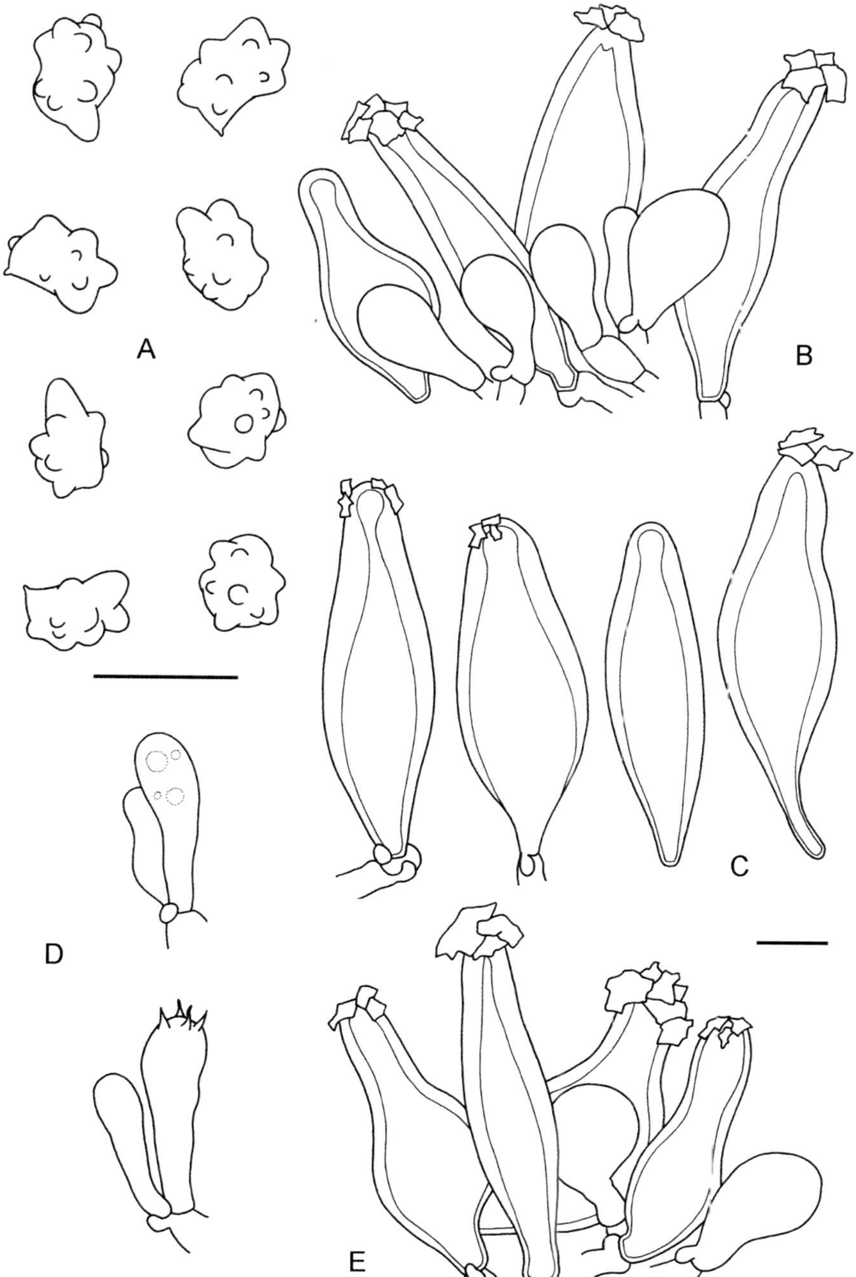

Figure 3. *Inocybe mixtilis.* All CANB 574539. **A,** Basidiospores; **B,** Cheilocystidia and paracystidia; **C,** Pleurocystidia; **D,** Basidia; **E,** Caulocystidia and cauloparacystidia from upper stipe. Scale bars = 10 µm (longer bar for spores only).

3. Inocybe *aff.* pusio P.Karst., *Bidr. Känned. Finl. Nat. Folk* 48: 465 (1889)

[*Inocybe pusio* P.Karst., *Bidr. Känned. Finl. Nat. Folk* 48: 465 (1889)

Type: Tavastia australis, Tammela, Syrjä, Finland, 8 Aug. 1889, *P.A.Karsten 1608*; holo: H, *n.v.*.

Inocybe fechtneri Velen., *České Houby* 372 (1920). Type: [Czech Republic]; holo: *n.v.*

Inocybe obscura var. *obscurissima* R.Heim, *Genre Inocybe*: 257 (1931). Type: not known.

Inocybe pusio f. *velata* Reumaux, *Docs. Mycol.* 12(48): 20 (1983). Type: France; holo: *n.v.*]

Pileus 10–16 mm wide, conical to plane, often with a nipple-like umbo; margin decurved; velipellis absent; surface dry, shiny, fibrillose, occasionally torn; dark yellowish brown (10YR 3/3) to brown (10YR 4/3). Context thin, watery brown to pale brown. *Lamellae* adnexed, moderately close, with several tiers of lamellulae, brown to yellowish brown (10YR 5/3 to 5/4), colour in youth not observed, ventricose; edges white-fimbriate. *Stipe* 24–30 × 1.5–2.0 mm, terete, even, at most with a slightly swollen base; surface pruinose at the extreme apex only, with scattered brown fibrils on the lower part; apex with a distinctive lilac sheen, elsewhere very pale brown to pale brown (10YR 7/3 to 6/3). Context solid. *Odour* spermatic. Pl. 15.

Basidiospores 8.5–10.5 (–11.0) × 4.5–5.5 µm [mean 9.3 × 5.0 µm; Q 1.64–2.10; Q mean 1.85 (*n* = 20/1)], smooth, amygdaliform, with bluntly pointed or slightly attenuated apices, yellowish brown; apiculus small but distinctive. *Basidia* 26–34 × 8–9 µm, clavate, hyaline, 4-sterigmate. *Pleurocystidia* 62–80 × 13–17 µm, fusiform or narrowly fusiform to fusiform-ventricose (at times aberrant in shape); apices obtuse or sometimes subcapitate, often bearing crystals; slightly thick-walled, the walls mostly 1–2 µm thick, hyaline or pale yellow. *Cheilocystidia* similar to pleurocystidia, mixed with hyaline thin-walled paracystidia. *Caulocystidia* scattered in small clusters at the stipe apex, similar to hymenial cystidia, mixed with caulocystidioid cells. Brown encrusted superficial hyphae present on lower part of stipe. *Pileipellis* a cutis of cylindrical to inflated hyphae, these mostly 7–17 µm wide, dark brown in mass, encrusted, ±thin-walled. *Clamp connections* present. Fig. 4.

Occurs on soil under planted *Pinus radiata*. Introduced to Tas.; also in *Pinus radiata* stands in New Zealand. Fruiting May.

Specimen examined: Tas.: near Scottsdale, on soil under *Pinus radiata* at head of track to Cuckoo Falls, 24 May 2010, *F.Karstedt PBM3408* (TENN 068282; KC305373-nLSU).

Diagnostic features: Pileus dark brown to dark yellowish brown, fibrillose, with a nipple-like umbo; stipe apex lilac, with scattered brown fibrils below. Occurring under introduced Monterey pine.

No authenticated collections of *I. pusio* have been examined for comparison with the Tasmanian collection described here. Likewise, no DNA sequences of European *I. pusio* are available for comparison. DNA sequences of the Tasmanian collection clearly place it in the *I. cincinnata* (Fr. : Fr.) Quél. group (data not presented here) together with European and North American sequences of *I. cincinnata*, *I. aff. cincinnata*, *I.* aff. *pseudorhacodes* Tak. Kobay. & Courtec. from Malaysia, and North American material labeled *I. pusio* (which may be questioned; see Matheny, 2005), *I. cinnamomea* A.H.Sm., and two unclarified taxa. Of these, the Tasmanian material is most similar to sequences of the North American *I. pusio*. Morphologically, the Tasmanian specimen differs from descriptions of *I. pusio* (e.g. Kupyer, 1986) by having a nipple-like umbo, the lack of a rimose or rimulose pileus surface, and the presence of scattered brown fibrils on the stipe. Microscopically, the hymenial cystidia of the Tasmanian collection are more variable in shape and more often with a subcapitate apex than those of *I. pusio*. It differs from *I. cincinnata* in the lack of scales on the pileus, the presence of the nipple-like umbo and the absence of brownish encrusted paracystidia. DNA sequences of *I. cincinnata* from Europe and North America are clearly distinct from those of the Tasmanian collection. Because of the lack of a thorough molecular taxonomic treatment of taxa in the *I. cincinnata* group, we refer to the introduced Tasmanian (and New Zealand; see below) species as *I.* aff. *pusio*. While the Tasmanian material was collected under *Pinus radiata*, in Europe the species is almost always collected under hardwoods, but has been found at least once with *Pinus* (Kuyper, 1986).

The ITS sequence of the Tasmanian collection is the same (99% similar) as two sequences (GQ267475, GQ267476) from unidentified fruit bodies collected in *Pinus radiata* plantations in

New Zealand (Walbert *et al.*, 2010), and one sequence (JX989972) produced from an *Alnus* ectomycorrhiza from France (Roy *et al.*, 2013).

Several other European species with lilac or violaceous tinges to the stipe have been described. Both *I. hygrophana* Glowinski & Stangl and *I. ionochlora* Romagn. are distinguished by their hygrophanous and pallescent pileus, while *I. griseolilacina* J.E.Lange is characterised by its *Pelargonium* odour. *Inocybe huijsmanii* Kuyp. has a pale brown pileus, longitudinally white-fibrillose stipe, and a *Pelargonium* smell, and *I. amethystina* Kuyper has a dark reddish brown pileus, lacks brown fibrils or scales on the stipe, and has yellow lageniform pleurocystidia. A key to several infraspecific taxa of *I. pusio* was provided by Bon (1997).

The synonymy above is for *I. pusio s. str.* and follows Kuyper (1986) and the literature references therein.

Plate 15. *Inocybe* aff. *pusio*. TENN 068282. Photograph — P.B.Matheny.

4. **Inocybe rufuloides** Bon, *Docs. Mycol.* 14(53): 28 (1984)

Type: Cayeux-sur-Mer (Somme), France, 19 May 1983, *Bon, Vast & Claus*; holo: *n.v.*

Pileus 5–35 mm wide, convex to hemispherical when young, becoming conico-campanulate, maintaining a broad central umbo, applanate with age; margin with evanescent whitish appendiculate fibrils (suggestive of a veil); surface dry, radially fibrillose with appressed pale silvery fibrils some fusing into narrow appressed squamules; centre of pileus more matted-fibrillose, overlying the mouse brown (near 6E4) to dull bronze-brown (near 6E5) then later darker brown (near 7F8) pileal surface. Context pallid, up to 5 mm thick. *Lamellae* broadly adnexed, crowded, then closely spaced with abundant lamellulae in at least 3 tiers, initially pale cream, becoming pale fawn (near 5B4), finally dull brown (near 5D6 to 6D6) or dull greyish brown (near 6D4), never dark or rusty brown, slightly ventricose, broad, up to 7 mm deep; edges minutely fimbriate (but not conspicuous to the eye), paler or concolorous with the faces.

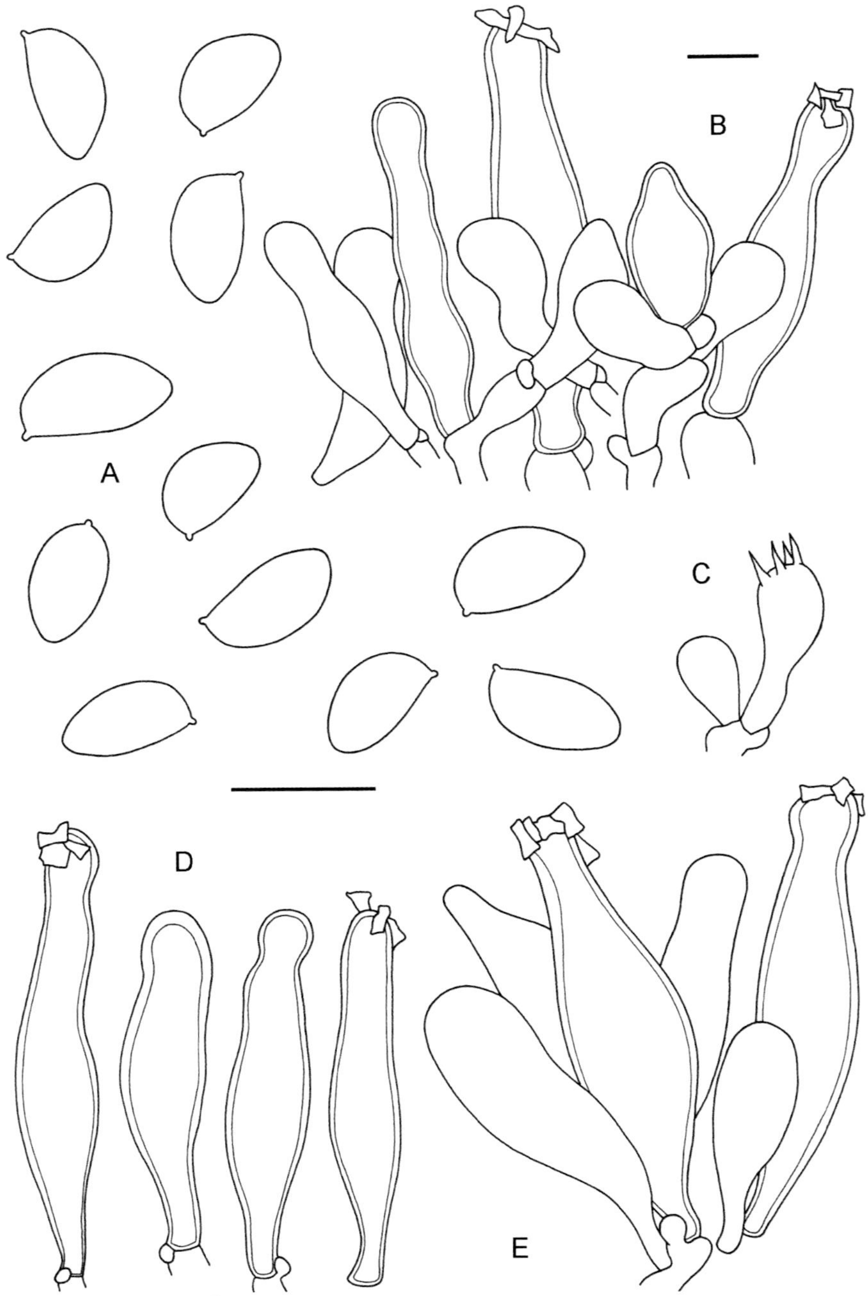

Figure 4. *Inocybe* aff. *pusio*. All TENN 068282. **A,** Basidiospores; **B,** Cheilocystidia and paracystidia; **C,** Basidia; **D,** Pleurocystidia; **E,** Caulocystidia and caulocystidioid cells. Scale bars = 10 µm (longer bar for spores only).

Stipe up to 50 × 8 mm, cylindrical with a slightly swollen base or clavate; veil sparse, rapidly evanescent and composed of whitish fibrils; surface dry, not obviously pruinose, longitudinally silky-fibrillose with whitish appressed fibrils over the entire length (easy to remove on handling); fibrils breaking up into minute floccules near the apex; dull whitish with a slight pinkish tinge especially towards the apex in some specimens, darker and dull when old; basal mycelium white but not copious or conspicuous. Context solid, pale cream. *Odour* grassy, spermatic. Pl. 16, 17.

Basidiospores 11.0–12.5 (–13.0) × 5.5–6.5 µm [mean 11.3 × 6.1 µm; Q 1.36–1.90; Q mean 1.66 (*n* = 60/1)], smooth, ovate to subamygdaliform in profile, occasionally with a slight adaxial depression, elliptic to ovate in face view, yellowish brown; apiculus small but prominent. *Basidia* 25–37 × 8.5–11.5 µm, clavate to narrowly ventricose, hyaline, 4-sterigmate. *Pleurocystidia* 53–75 × 14–17 µm, lageniform, subfusiform or utriform, crystalliferous, thick-walled, the walls to 2 µm thick, hyaline. *Cheilocystidia* 20–56 × 11–16 µm, similar to pleurocystidia, crowded, forming sterile lamella edge, mixed with thin-walled paracystidia sometimes in short chains. *Caulocystidia* few, similar to hymenial cystidia at the extreme stipe apex; end cells otherwise undifferentiated. *Pileipellis* a cutis of cylindrical hyphae up to 10 µm wide, many with brown encrusting pigment; subpellis not well differentiated but with more tightly packed hyphae. *Clamp connections* present. Fig. 5.

Occurs in sandy soil of the coastal plain in south-western W.A.; an introduced species that is often abundant under or near planted *Pinus radiata*. Fruiting June to August.

Specimens examined: W.A.: Modong Nature Reserve, Jandakot Regional Park, Thomas Rd, Oakford, Perth, 32°13'49"S, 115°54'14"E, in gregarious troops in disturbed gravelly roadside edge near *Pinus* plantation (*Melaleuca preissiana* and *Adenanthos* sp. also present), 9 July 2006, *N.L.Bougher, N.Goldsborough & J.Keeble E8353* (PERTH 07700598, as *Inocybe* sp. in Bougher *et al.*, 2006a; JN035295-nLSU, JN035292-ITS); Bold Park, Perth, 31°56'19"S, 115°46'57"E, abundant in litter under *Pinus radiata*, adjacent to Ecology Centre, just S of Oceanic Drive, 10 July 2008, *N.L.Bougher NLB479* (PERTH 08073147); *loc. id.*, 1 June 2010, *N.L.Bougher NLB618* (PERTH 08305978; JN035291-ITS); *loc. id.*, abundant along edge of limestone track in banksia woodland with *Pinus radiata* plantation nearby (30 m to nearest pine tree), 29 July 2009 *N.L.Bougher NLB541* (PERTH 08305951); *loc. id.*, 300 m SE of previous location (20 m to nearest pine tree), 31°57'15"S, 115°46'20.5"E, 1 June 2010, *N.L.Bougher NLB913* (PERTH 08320233); Kensington, Technology Park, Perth, gregarious troops in *Pinus* plantation, including areas with surface limestone, 8 Aug. 2012, *N.L.Bougher NLB1063* [fruiting at same location in late June 2014] (PERTH 08363692).

Diagnostic features: Growing near *Pinus*. Fruit bodies medium, the pileus dull bronze-brown, radially fibrillose; lamellae pale; stipe not pruinose, with a swollen base, slight pinkish tinges on the upper part, lacking lilac. Odour grassy/spermatic. Spores smooth, large, mostly 11.0-12.5 × 5.5–6.5 µm; hymenial cystidia present as metuloids.

Inocybe rufuloides is an introduced species in *Pinus* plantations in Australia (Matheny & Bougher, 2011) where it was first recorded and illustrated as *Inocybe* sp. specimen ID: 2377 (Bougher *et al.*, 2006a). It is an ectomycorrhizal associate of *Pinus* in Europe, e.g. *P. maritima* (Kuyper, 1986) and *P. pinea* (Iotti *et al.*, 2005). Collections of *I. rufuloides* from France listed by Bon (1984) indicated a possible association with *Pinus* and other conifers such as *Picea*. In Australia, *I. rufuloides* is an exotic species apparently exclusively associated with *Pinus*. Matheny & Bougher (2011) identified the Western Australian collections cited here as *I. rufuloides* based on a comparison with literature descriptions, while a molecular sequence matched material identified by J.Vauras as *I. rufuloides* (Ryberg, 2010). The Australian specimens also agree with the circumscription provided by Vesper & Richter (2010) in their comparison of morphological data for *I. rufuloides* published by eight other authors.

The European collections occur in sand (Vesper & Richter, 2010) or dune-sand (Kuyper, 1986), and, similarly, the Australian specimen grew in nutrient-poor sand. However, they are less reddish brown overall than specimens illustrated as *I. aff. rufuloides* by Vesper & Richter (2010).

Plate 16. *Inocybe rufuloides* (PERTH 08073147). Photograph — N.L.Bougher.

Plate 17. *Inocybe rufuloides* (PERTH 08320233). Photograph — N.L.Bougher.

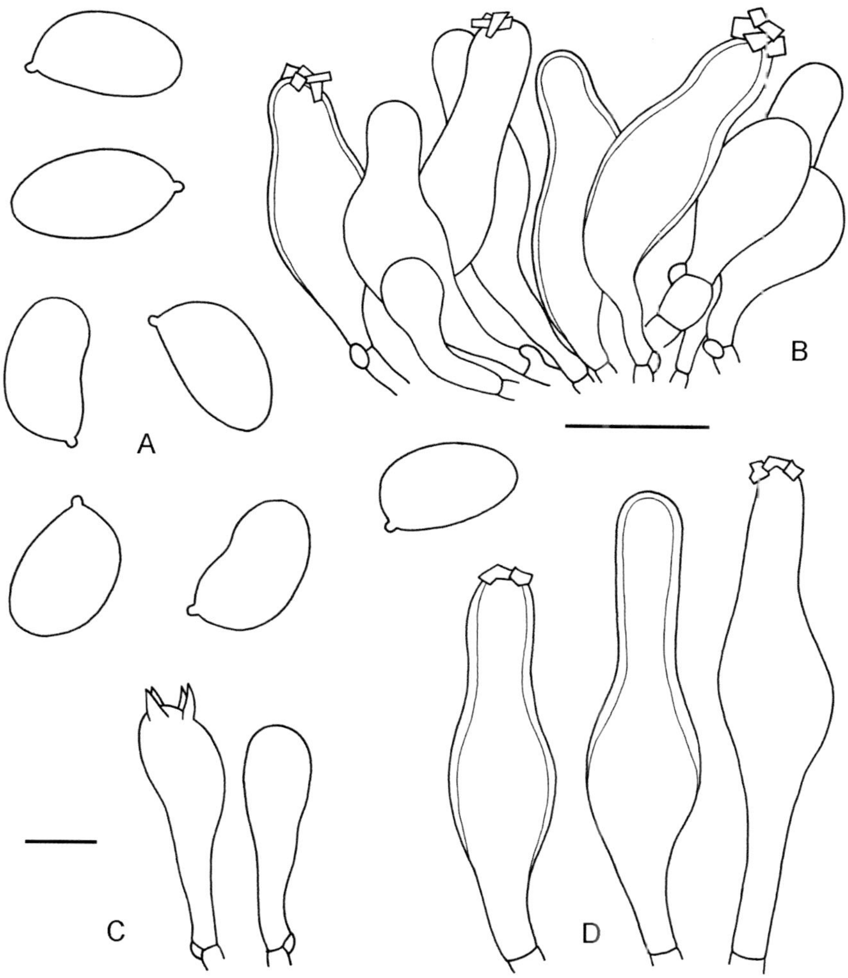

Figure 5. *Inocybe rufuloides*. All PERTH 08073147. **A,** Basidiospores; **B,** Cheilocystidia and paracystidia; **C,** Basidia; **D,** Pleurocystidia. Scale bars: 10 µm (longer bar for spores only).

5. Inocybe sindonia (Fr.) P.Karst., *Bidr. Känned. Finl. Nat. Folk* 32: 465 (1879)

Agaricus muticus Fr., *Monogr. Hymenomyc. Sueciae* 2: 346 (1863); *Inocybe mutica* (Fr.) Sacc., *Syll. Fung.* 5: 769 (1887). Type: Not designated.

Inocybe cortinata Rolland, *Bull. Soc. Mycol. Fr.* 17: 177 (1901). Type: Europe; not located.

Inocybe uliginosa Velen., *České Houby* 374 (1920). Type: Europe; holo: *n.v.*

Inocybe commutabilis Furrer-Ziogas, *Schweiz. Z. Pilzk.* 30: 127 (1952). Type: Europe; holo: *n.v.*

Inocybe lucifuga f. *gralla* Furrer-Ziogas, *Schweiz Z. Pilzk.* 30: 134 (1952). Type: Europe; holo: *n.v.*

Inocybe kuehneri Stangl & Veselský, *České Mykol.* 28: 199 (1974). Type: Czechoslovakia; holo: *n.v.*

Inocybe lepiotoides Reumaux, *Docs. Mycol.* 12(48): 10 ('1983') [1982]. Type: France; holo: *n.v.*

Inocybe eutheles var. *claricolor* Reumaux, *Bull. Trimest. Soc. Mycol. Fr.* 100: 205 (1984). Type: France; holo: *n.v.*

Inocybe eutheles var. *fusoideicystis* Reumaux, *Bull. Trimest. Soc. Mycol. Fr.* 100: 206 (1984). Type: France; holo: *n.v.*

Pileus 10–40 mm wide, conical or deeply so when young, becoming obtusely conical, campanulate or plano-umbonate; margin inrolled when young and raggedly fibrillose due to presence of a partial veil; surface dry, fibrillose and tending to split or become rimose, occasionally becoming appressed-scaly towards the centre, exceptionally with recurved scales at the centre; pale brown or fawn-beige, near buff but paler or with brownish shades (near 6B5 to 6B4, 6C4 to 6C5). *Lamellae* sinuate, close, with several tiers of lamellulae, whitish at first, becoming a little paler or concolorous with the pileus or pale greyish brown. *Stipe* 30–75 × 4–6 mm, even, dry, with a sparse covering of longitudinal fibrils, pruinose in appearance, dull off-white with occasional brown tints of pileal colour. Context solid. *Odour* pronounced, spermatic, or unpleasant, less often not noticeable. Pl. 18.

Basidiospores 8.0–10.0 (–11.0) × 4.5–5.5 µm [mean 9.2 × 5.0 µm; Q 1.55–2.11; Q mean 1.82 (n = 31/3)], smooth, amygdaliform with pointed or bluntly pointed apicies, not adaxially depressed, yellowish brown in deposit. *Basidia* 25–31 × 8–10 µm, clavate, hyaline, 4-sterigmate. *Pleurocystidia* 55–84 × 12–15 µm, fusiform or narrowly fusiform, with obtuse apices often crystalliferous, thick-walled, the walls mostly 2–3 µm thick, yellow to hyaline. *Cheilocystidia* similar to pleurocysidia, mixed with crowded small clavate paracystidia. *Caulocystidia* similar to hymenial cystidia, hyaline, scattered but mixed with cauloparacystidia at least down to stipe centre; some caulocystidia quite large (96 × 19 µm), absent above the stipe base. *Pileipellis* a cutis of cylindrical hyphae, these buff or very pale yellowish brown in mass, mostly 6–17 µm wide, smooth or faintly wrinkled in appearance, ±thin-walled. *Clamp connections* present. Fig. 6.

This species, native to the Northern Hemisphere, occurs in W.A., A.C.T. and Vic.; on soil, typically in lawns under planted *Pinus radiata*, *Quercus* and/or *Ulmus*, but also in *Pinus* plantations, in south-western and south-eastern Australia. Fruiting Apr. to July.

Specimens examined: W.A.: Mundaring Weir pine forest, E of Perth, 31°58'00"S, 116°10'00"E, 16 June 1974, *R.Rainbird UWA 1942* (PERTH 00771422); Bartons Mill Pine Plantation, 29 June 1983, *R.N.Hilton UWA 2877* (PERTH 0916226, TENN 068201). A.C.T.: Canberra, City Hill, 3 km N of Capital Hill, 35°17'S, 149°07'E, alt. 580 m, numerous and solitary in grass under planted pine and cypress, 2 June 1987, *H.Lepp 53* (CANB 574548, NY, PERTH 08516707); Reid, 3 km NE of Capital Hill, Canberra, 35°17'S, 149°08'E, alt. 600 m, numerous in cultivated lawn under *Quercus*, 1 May 1988, *H.Lepp 216* (CANB 574552, PERTH 08516782); 5 km NW of Capital Hill, Canberra, 35°18'S, 149°04'E, alt. 630 m, in *Quercus suber* (Cork oak) plantation, 23 Apr. 1989, *H.Lepp 312* (CANB 574564, PERTH 08516715); Macquarie, 8 km NW of Capital Hill, Canberra, 35°15'S, 149°03'E, alt. 600 m, on soil in open grassy area near tall *Pinus*, 3 June 1992, *H.Lepp 833* (CANB 574572, PERTH 08516790); Acton, 4 km NNW of Capital Hill, Canberra, 35°16.5'S, 149°06.5'E, alt. 580 m, a single cluster on bare soil near mature *Ulmus* (elm), 15 July 1993, *H.Lepp 1003* (CANB 574531, PERTH 08516723); Yarralumla, 3.5 km WNW of Capital Hill, Canberra, 35°18'20"S, 149°05'00"E, alt. 560 m, on soil among *Pinus* and *Ulmus*, 23 May 1998, *H.Lepp 2003* (CANB 574543, PERTH 08516731). Vic.: Western Highlands region, Mount Macedon Forest Park, 37°23'S, 144°37'E, alt. 900 m, on ground under *Pinus*, 11 Apr. 1976, *D.M.Kershaw, T.Mochanuk, K.H.Sinnott & N.H.Sinnott NHS1959* (CANB 359624, K, PERTH 08516812); *loc. id.*, 11 Apr. 1976, *D.M.Kershaw, T.Mochanuk, K.H.Sinnott & N.H.Sinnott NHS1960* (CANB 359627, K, PERTH 08516804, VPRI); Western Highlands region, slopes of Mount Macedon, 37°23'S, 144°37'E, alt. 300 m, on ground under *Pinus*, 16 May 1976, *K.H.Sinnott, J.H.Willis et al. NHS2009* (CANB 359644, PERTH 08516758).

Diagnostic features: Pileus conical to campanulate, buff, pale brown to fawn; flesh typically with a spermatic odour; stipe whitish, at times with tints of the pileus colour. Under introduced trees (pine, oak or elm).

A composite description is presented here based on collections made in south-eastern Australia, while the list of taxonomic synonyms is reproduced from Kuyper (1986). This is the first report of *I. sindonia* from Western Australia, where it occurs in pine plantations. Walbert *et al.* (2010) reported it from plantations of *Pinus radiata* in New Zealand, and we have produced DNA sequence data from a New Zealand collection [Bottle Lake, Christchurch, 9 Apr. 2004, J.A.Cooper (PERTH 08516766, PDD 80239; KC305363-nLSU, KP636833-ITS)]. A record from south-eastern Australia, associated with pines and listed by Dunstan *et al.* (1998) as *I. patouillardii* Bres. is possibly *I. sindonia*, while in some Australian localities *I. sindonia* co-occurs and fruits simultaneously with the nodulose-spored exotic species *I. mixtilis*.

Inocybe sindonia has had a confusing taxonomic history with numerous names, e.g. *I. eutheles sensu* Kühner, *I. lucifuga* f. *gralla* Furrer-Ziogas and *I. kuehneri* Stangl & Veselsky (Kuyper,

1986). We have determined that several Australian Capital Territory collections in CANB, labelled *I. posterula* (Britzelm.) Sacc. and determined by D.N.Pegler, actually represent *I. sindonia*.

Kuyper (1986) reported a smaller spore size for European collections of *I. sindonia* (7.0–9.5 × 4.0–5.5 µm), those of south-eastern Australia specimens tending to be slightly longer, including some up to 11 µm long.

Gates & Ratkowsky (2014) included a photograph of *I. sindonia* from Tasmania, with a more whitish pileus than is usual for this species.

Plate 18. *Inocybe sindonia*. CANB 574572. Photograph — H.Lepp.

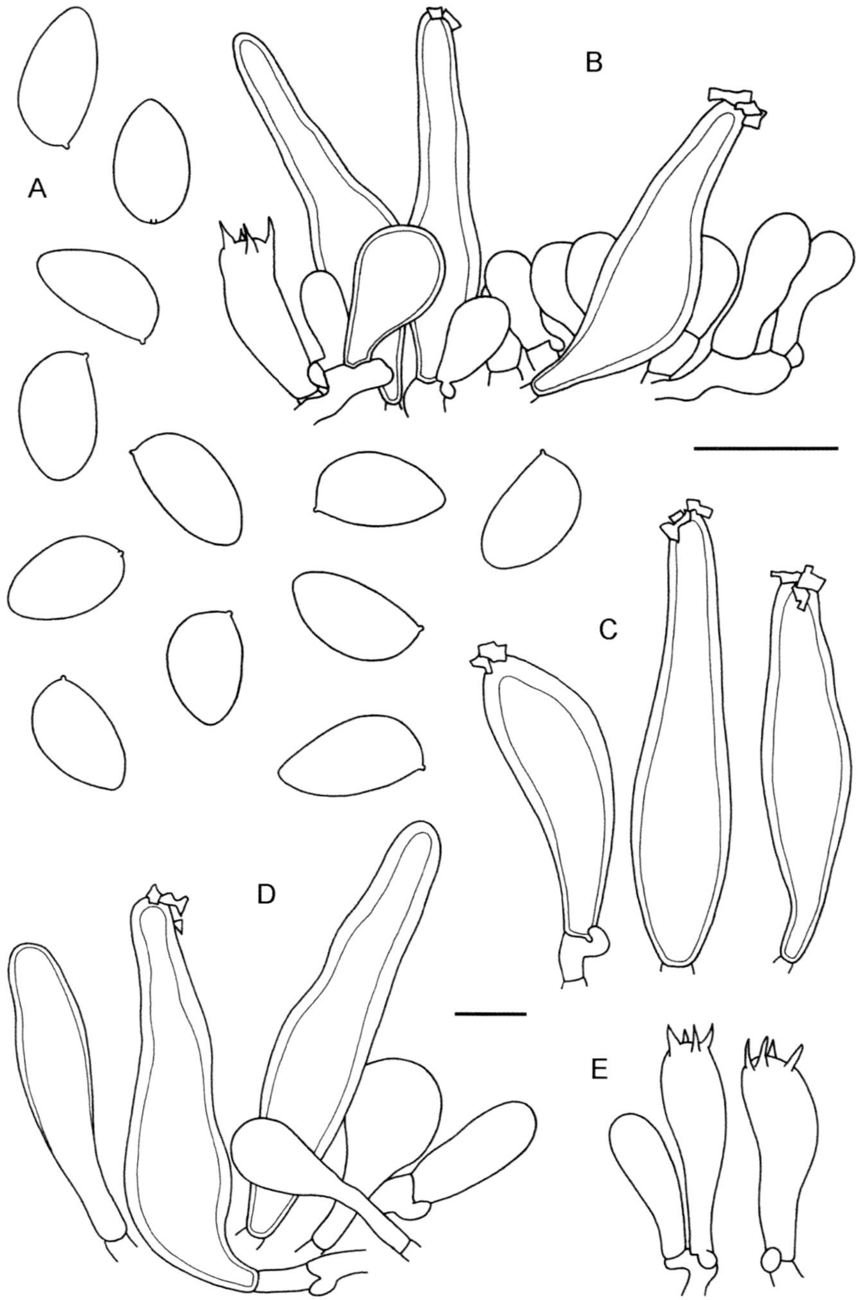

Figure 6. *Inocybe sindonia.* All CANB 574572. **A,** Basidiospores; **B,** Cheilocystidia and paracystidia; **C,** Pleurocystidia; **D,** Caulocystidia and cauloparacystidia; **E,** Basidia. Scale bars = 10 µm (longer bar for spores only).

SMOOTH-SPORED SPECIES OF THE *INOCYBE* CLADE

Seventy two species with smooth spores and pleurocystidia (viz, the *Inocybe* clade *p.p.*) are documented below for Australia. Sixty three of these species occur in a hyper-diverse clade with a subantarctic geographic distribution (southern South America, temperate regions of Australia, and New Zealand). This group is referred to for convenience as the "smooth-spored, temperate austral clade" (STAC). The remaining nine species occupy independent lineages elsewhere in the *Inocybe* clade. Smooth-spored species in the *Inocybe* clade do not form a monophyletic group (Matheny *et al.*, 2002; Matheny, 2005; Matheny *et al.*, 2009; Ryberg *et al.*, 2010).

STAC GROUP

The STAC group (smooth-spored, temperate austral clade) represents the most diverse indigenous clade of austral Inocybaceae. Sixty three species from Australia are documented here, and numerous others await formal description or are known to date only from environmental sequences. All species feature smooth basidiospores, typically with an indistinct apiculus, possess pleurocystidia, and are constrained geographically to temperate regions of the southern hemisphere. The most commonly encountered species include members of the *I. serrata* group (and complex), the *I. fibrillosibrunnea* group, and the *I. australiensis* group. In addition to these three groups, five additional groups are presented below.

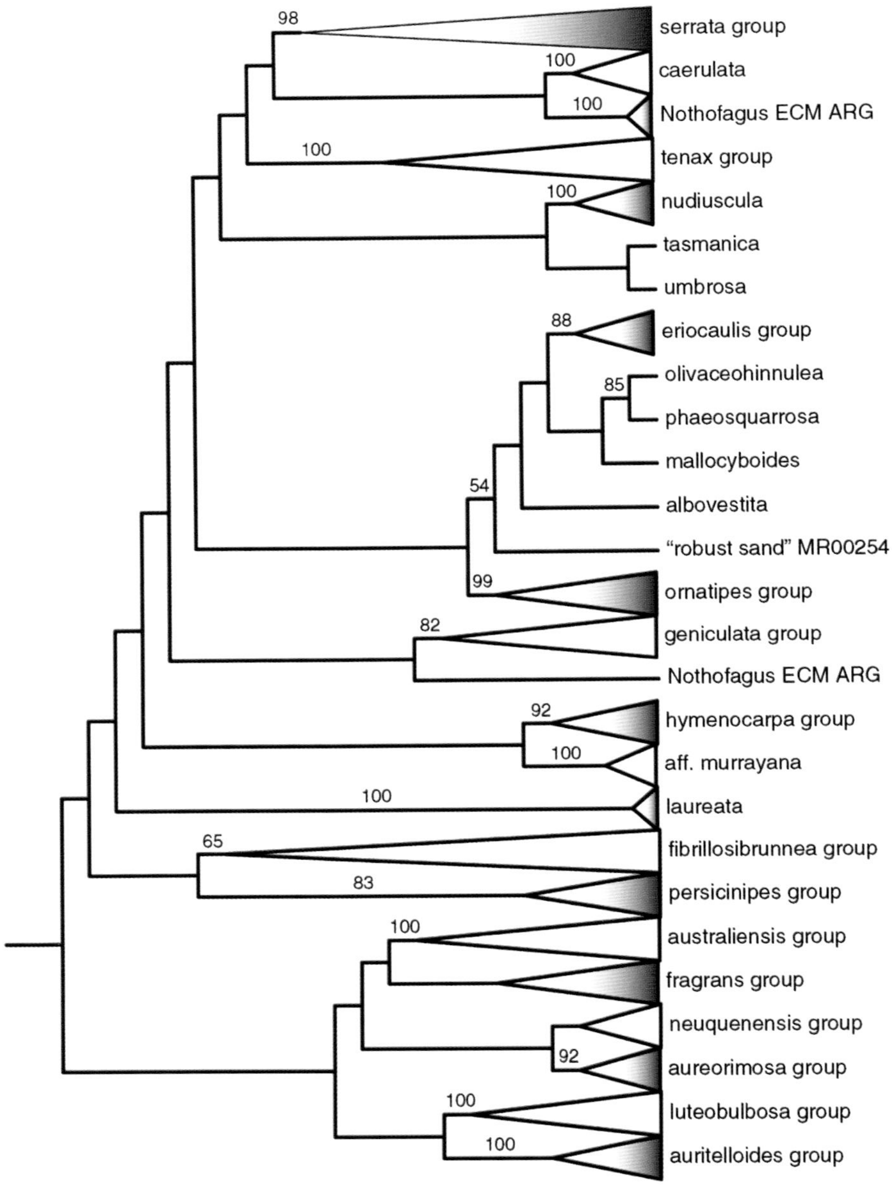

Figure 7. Maximum Likelihood phylogram of the smooth-spored temperate austral clade (STAC) of *Inocybe*. Bootstrap values are shown above or below branches. The tree was inferred from combined ITS, nLSU, and *rpb2* gene regions and includes 2892 total sites and 296 taxa. Clades are collapsed to indicate general groupings. Outgroups include *I. stellatospora*, *I. subochracea*, and *I. tubarioides* following Matheny *et al.* (2009). Two Australian taxa are placed outside the STAC group: *I. austrosquarrosa* and *I. spadicea*. In Matheny *et al.* (2009) using nLSU, *rpb1*, and *rpb2* sequences, *I. spadicea* is the sister lineage to the remainder of the STAC group.

INOCYBE AUSTRALIENSIS GROUP AND ALLIES

This is an inclusive grouping of 28 species including the *I. australiensis* group and numerous other lineages, many of which are known only from environmental samples [e.g. *Nothofagus* roots; see Nouhra *et al.* (2013)]. The grouping as a whole is not well supported by phylogenetic analysis, but the majority of species (16) are found in association with *Nothofagus*. Detailed descriptions are provided below for ten species from Australia.

6. **Inocybe atrisquamosa** Matheny & Bougher, *sp. nov.*

MycoBank: MB 809544

Type: near Central Station, Fraser Island, Qld, gregarious on sandy soil under *Eucalyptus pilularis*, *E. microcorys* and *Syncarpia*, 19 May 2011, *T.J.Baroni & N.Fechner TJB10483*; holo: BRI AQ798333; iso: CORT, PERTH 08516898, TENN 069505; KJ729907-nLSU, KJ729881-ITS, KJ729938-*rpb2*).

Pileus 9–24 mm wide, conical, convex, becoming broadly convex and nearly plane with age, at times with a broad obtuse umbo; margin decurved; surface dry; velipellis absent; centre with densely erect dark brown scales, which are thick and conical, erect on the disc, becoming recurved towards the margin to appressed at the very margin; margin also with dark brown appressed radiate fibrils; surface fibrous, splitting lacinate, and subtly striate or rimulose near the margin; centre dark brown (6F5–F8, 6F4–7F4, 7F3–8F3; Mummy Brown) shading to brown (5D4 to 6D3; Snuff Brown) towards the margin; context pale brown, whitish or sordid white, thin and c. 1–2 mm thick. *Lamellae* adnate, moderately close (24–28 L) with several tiers of lamellulae, initially with yellow hues (near 4A3–B3, 4C4), becoming light yellowish brown to greyish brown (5D4) or dark grey brown (6E4); edges pallid-fimbriate, broad, up to 3 mm deep. *Stipe* 10–27 × 1.5–3.0 mm, terete, subeven or with a swollen base 2.5–6.0 mm wide; surface densely pruinose over the upper third but less so downwards or pruinose-pubescent near the base; paler brown than the pileus: pale tan-brown to cinnamon (5C4 to 5D5–D6), becoming brown (6D4) with a darker brown (6F5) or reddish brown base, dark brown fibrils present just above the base; context solid, brown. *Odour* absent. Pl. 19, 20.

Basidiospores 6.0–8.0 (–9.5) × 4.0–5.0 µm [mean 7.4 × 4.5 µm; Q 1.40–1.80 (–1.90); Q mean 1.64 (*n* = 61/3)], smooth, amygdaliform, broadly amygdaliform or ovate elliptic, rarely distally elongated; apices conical to bluntly pointed or obtuse, yellowish brown; apiculus small and not distinctive. *Basidia* 20–26 × 6–8 µm, clavate, hyaline, 4-sterigmate. *Pleurocystidia* 40–53 × 12–15 µm, fusiform or narrowly so; apices obtuse, crystalliferous; base pedicellate to broadly rounded, thick-walled, the walls mostly 2.5–4.0 µm thick, hyaline to pale yellow. *Cheilocystidia* similar to pleurocystidia, often shorter and more ventricose, mixed with paracystidia and some basidia. *Caulocystidia* 40–98 × 14–26 µm, some larger than hymenial cystidia, many lageniform or fusiform, mixed with paracystidia in discrete crowded clusters at least down to the centre of the stipe as well as large thin- to thick-walled broadly clavate cells, hyaline; caulocystidioid cells or hairs present on the lower part of the stipe, in BRI AQ798334 arising as terminal thick-walled hairs of vesture hyphae, mixed with the shorter thin-walled clavate to subcylindrical cells. *Pileipellis* composed of facsicles of upturned cylindrical hyphae, dark yellowish brown in mass, these mostly 7–20 µm wide, encrusted, ±thin-walled. *Clamp connections* present. Fig. 9.

Occurs in Qld, N.S.W. and Tas.; grows on soil under *Eucalyptus* and/or *Acacia* in (warm) temperate forests, including wet sclerophyll forest. Fruiting May–June and Sept.–Nov.

Additional specimens examined: Qld: Apple Tree Park, Springbrook Natl Park, gregarious or scattered on sandy soil under *Eucalyptus*, 2 June 2011, *T.J.Baroni & N.Fechner TJB10553* (BRI AQ798334, CORT, PERTH 08516901, TENN 069506; KJ729909-nLSU, KJ729883-ITS, KJ729940-*rpb2*). N.S.W.: 4 km E of Visitors' Centre, Warrumbungle Natl Park, under *Eucalyptus*, 3 Sept. 2010, *T.Elliott T35059* (OSC; KJ729908-nLSU, KJ729882-ITS, KJ729939-*rpb2*). Tas.: Chauncy Vale, on soil in dry sclerophyll forest under *Eucalyptus* and *Acacia*, 29 Nov. 2001, *G.Gates & D.Ratkowsky 129* (Herb. G.M.Gates, TENN 065712; KJ729910-nLSU, KJ729884-ITS).

Etymology: From the Latin *ater* (black) and *squamosus* (coarsely scaly), in reference to the dark erect-scaly pileus.

Diagnostic features: Fruit bodies small. Pileus dark brown to brown, disc with erect scales; stipe pruinose or pubescent for much of its length; odour absent. Spores small, mostly 6.0–8.0 × 4.0–5.0 µm, amygdaliform to broadly amygdaliform or ovate elliptic; caulocystidia much larger than hymenial cystidia. On soil under eucalypts in eastern Australia.

Inocybe atrisquamosa can be confused with *I. australiensis*, but the latter is characterised by the presence of dark woolly fibrils over the lower half of the stipe. Molecular phylogenetic results support the separation of the two species, both of which are reciprocally monophyletic with good support (Fig. 8). *Inocybe campestris* forms a monophyletic group with both *I. atrisquamosa* and *I. australiensis* (Fig. 8), but has more robust fruit bodies and the pileus surface is not squarrose.

Inocybe meridionalis is similar to *I. atrisquamosa* morphologically, but occurs in *Nothofagus*-dominated rainforest and is phylogenetically distantly related (Fig. 8), while *I. eriocaulis* differs in the distinct woolly-fibrillose lower half of the stipe and its larger and more decidedly amygdaliform spores. Lastly, *I. fibrillosipes* can be distinguished from *I. atrisquamosa* by its pinkish to reddish brown stipe covered with greyish brown superficial fibrils on the lower part, the end cells of which are not differentiated.

Inocybe atrisquamosa has also been detected (based on ITS blastn results of GenBank) from hyphal ingrowth bags in field sites dominated by *Eucalyptus pilularis* in wet sclerophyll in south-eastern Queensland (Peachester S.F., Qld, 26°50'S, 152°53'E; Bastias *et al.*, 2006; GenBank accession numbers DQ388826, DQ388837, DQ388843).

7. **Inocybe australiensis** Cleland & Cheel, *Trans. Roy. Soc. S. Austalia* 42: 109 (1918)

Type: Neutral Bay, Sydney, N.S.W., 9 May 1915, *coll. unknown*; lecto: AD 5455, *fide* P.B.Matheny & N.L.Bougher, *Muelleria* 28: 90 (2010).

Pileus 15–25 mm wide, conical to obtusely conical, becoming plano-convex or plane, at times with a low obtuse umbo, velipellis not observed; margin initially slightly incurved, soon decurved; surface dry, the disc squarrose-scaly to squamulose with weakly recurved tips, coarsely fibrillose to rimulose towards the margin; very dark to blackish brown or dark bistre (10YR 2/2) to dark yellowish brown (Mummy Brown) at the centre, shading to dark brown to brown (Prout's Brown) towards the margin; context whitish, not changing colour upon exposure. *Lamellae* adnexed, moderately close, 30–38 L with several tiers of lamellulae, light grey to (very) pale brown (10YR 6/3) or yellowish when young (10YR 7/2 to 7/3), becoming light yellowish brown (2.5Y 6/4) to (dark) yellowish brown with age (10YR 5/4 to 4/4); edges indistinctly pallid-fimbriate. *Stipe* 15–33 × 2–4 mm, often slightly swollen at the base, upper half pruinose, woolly fibrillose or densely fibrillose below; apex yellowish red (Cinnamon) shading to brown (Sayal Brown) or dark brown below, eventually yellowish brown (Snuff Brown) throughout, with age dark reddish brown at the apex and becoming bistre or very dark brown on lower part or throughout; context with pinkish cinnamon exterior and whitish interior. *Odour* not remarkable or acidulous. Pl. 21, 22.

Basidiospores 6.5–9.0 × 4.5–5.0 µm [mean 7.7 × 4.6 µm; Q 1.44–1.89; Q mean 1.70 (n = 60/2)], smooth, amygdaliform to subamygdaliform, with pointed to bluntly pointed apices, yellowish brown; apiculus small but distinct. *Basidia* 23–25 × 7–9 µm, clavate, hyaline, 4-sterigmate. *Pleurocystidia* 48–75 × 10–15 µm, fusiform to narrowly fusiform, thick-walled, the walls mostly 2–4 µm thick, hyaline or pale yellow; apices obtuse, occasionally with crystals. *Cheilocystidia* similar to pleurocystidia, at times shorter or thin-walled, scattered and mixed with paracystidia and basidia. *Caulocystidia* 30–108 × 15–22 µm, larger than hymenial cystidia, typically fusiform, cylindrical or lageniform but occasionally irregular in outline; slightly thick-walled, the walls hyaline; apices obtuse; mixed with cauloparacystidia, including some multiseptate forms, above the stipe centre (cauloparacystidia not observed above the base). *Pileipellis* a trichoderm, composed of broad tufts of ascending hyphae, these dark yellowish brown in mass, encrusted, with slightly thickened walls, mostly cylindrical to inflated, 10–17 µm wide. *Clamp connections* present. Fig. 10.

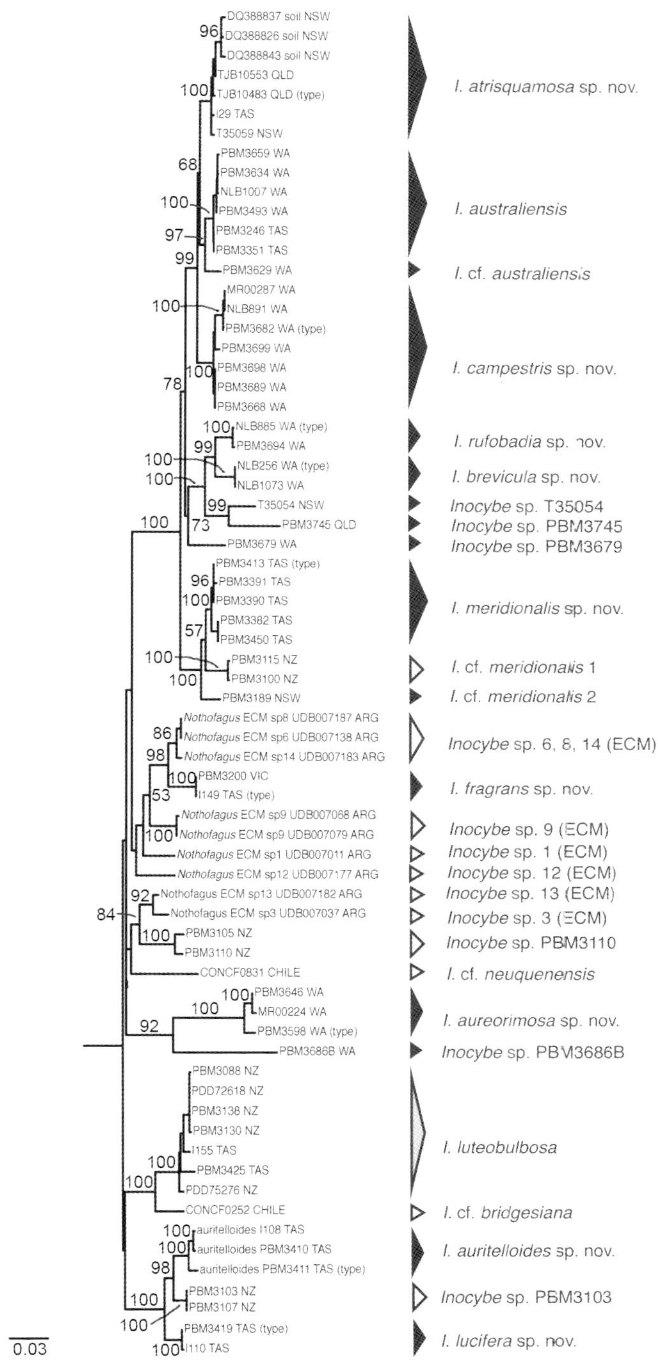

Figure 8. Maximum Likelihood phylogram of the *Inocybe australiensis* group and allied taxa produced from the analysis of the STAC group shown in Fig. 7. Australian taxa are indicated by black-filled triangles. Grey-filled triangles illustrate species occurring in Australia and New Zealand. Clear triangles indicate New Zealand or southern South American taxa. Bootstrap values are shown above the branches.

Plate 19. *Inocybe atrisquamosa,* holotype (BRI AQ798333). Photograph — T.J.Baroni.

Plate 20. *Inocybe atrisquamosa* (BRI AQ798334). Photograph — T.J.Baroni.

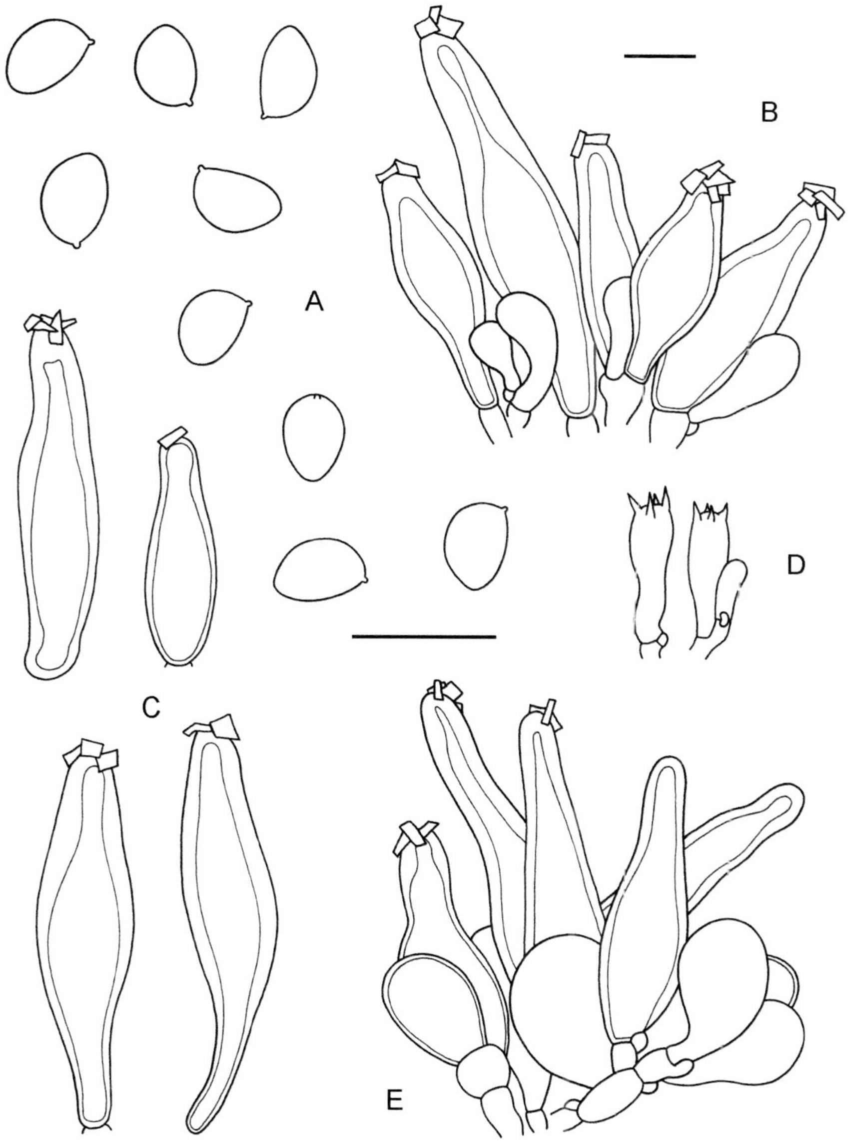

Figure 9. *Inocybe atrisquamosa*, isotype (PERTH 08516898). **A,** Basidiospores; **B,** Cheilocystidia and paracystidia; **C,** Pleurocystidia; **D,** Basidia; **E,** Caulocystidia and cauloparacystidia. Scale bars = 10 µm (longer bar for spores only).

Occurs in W.A., N.S.W. and Tas. on soil or sand in a range of vegetation types including mixed tall forests of *Eucalyptus*, *Corymbia* and/or *Allocasuarina*, closed canopy *Melaleuca* thicket mixed with *Agonis flexuosa*, under planted *Eucalyptus gomphocephala*, and in wet sclerophyll forest including a mixture of *Eucalyptus*, *Acacia* and *Pomaderris*. Fruiting May–July.

Additional specimens examined: W.A.: Yeagerup Dunes area, Ritter Rd, D'Entrecasteaux Natl Park, 34°33'0.4"S, 115°51'47.7"E, scattered singly on sandy soil in Jarrah-Marri-*Allocasuarina* forest, 9 June 2010, *P.B.Matheny PBM3493* (PERTH 08241724, TENN 065713; KJ756435-nLSU, KJ756470-ITS, KJ729955-*rpb2*); Bibbulmun Track and Shelley Beach Rd, West Cape Howe Natl Park, 35°05'56.8"S, 117°37'44.1"E, scattered singly on soil in a closed canopy *Melaleuca* thicket mixed with *Agonis flexuosa*, 18 July 2011, *P.B.Matheny PBM3634* (PERTH 08319529, TENN 066588; KJ756436-nLSU, KJ756471-ITS, KJ729956-*rpb2*); Gull Rock Natl Park, 34°59'48.3"S, 117°58'19.7"E, on mossy soil in open woodland dominated by *Corymbia calophylla* and *Agonis flexuosa*, 19 June 2012, *N.L.Bougher & M.True NLB1007* (PERTH 08383359; KJ801160-nLSU, KJ756472-ITS); Cheynes Beach Town outside Waychinicup Natl Park, 34°53'5.2"S, 118°24'25.3"E, gregarious or scattered singly or in pairs in sand under planted *Eucalyptus gomphocephala* and *Acacia*, 20 July 2011, *P.B.Matheny PBM3659* (PERTH 08318905, TENN 066527; KJ756437-nLSU, KJ756473-ITS); *loc. id.*, in sandy soil under *Eucalyptus gomphocephala*, 20 July 2011, *M.Ryberg MR00264* (PERTH 08256268, TENN 066491); c. 150 m S along track off Rosa Brook Rd, near Margaret River, 33°58'21.0"S, 115°06'34.4"E, on side of track in forest of *Corymbia calophylla* and *Eucalyptus marginata*, 1 July 2008, *N.L.Bougher NLB468* (PERTH 08073023). N.S.W.: Chatswood, Sydney, 21 May 1916, *J.B.Cleland* (AD 5456; watercolour 96 by Miss Clarke, *n.v.*); Sydney, "pre-1920", *J.B.Cleland* (AD 5457). Tas.: Chauncy Vale Wildlife Sanctuary, scattered singly on soil in small pocket of wet sclerophyll forest under *Eucalyptus globulus*, *Acacia* and *Pomaderris*, 11 June 2009, *G.Gates, D.Ratkowsky & P.B.Matheny PBM3246* (TENN 065714; KJ756433-nLSU, KJ756468-ITS, KJ729953-*rpb2*); Lower Kermandie, scattered singly on soil along track in wet sclerophyll forest under *Eucalyptus*, *Acacia dealbata* and *Pomaderris*, 19 May 2010, *F.Karstedt & G.Gates PBM3351* (PERTH 08574731, TENN 065715; KJ756434-nLSU, KJ756469-ITS, KJ729954-*rpb2*).

Diagnostic features: Fruit bodies small. Pileus scaly, blackish brown to dark brown; stipe pruinose above the woolly to densely fibrillose lower part, base swollen. Spores typically less than 8 µm long. Widespread in southern-temperate Australia.

The foregoing description is based on specimens collected in the south-west of Western Australia and in Tasmania, and these collections match the type (Matheny & Bougher, 2010) remarkably well in outward appearance as well as microscopically. *Inocybe australiensis* is characterised by the relatively small size of the fruit bodies, the very dark and often squarrose-scaly pileus, the woolly or densely fibrillose lower part of the stipe, and the small spores. Some collections can have yellowish lamellae at younger stages.

The *I. australiensis* group includes at least 11 species-level lineages or clades (Fig. 8), six of which are documented here as species. *Inocybe campestris* is characterised by its larger fruit bodies and an association with *Allocasuarina* and/or *Acacia*. *Inocybe atrisquamosa*, widely distributed in the Queensland, New South Wales and Tasmania, is very similar to *I. australiensis* but differs in its amygdaliform to broadly amygdaliform spores and the absence of brownish woolly fibrils on the lower part of the stipe. The Tasmanian *I. meridionalis* occurs in rainforest and wet sclerophyll habitats containing *Nothofagus*, and while phylogenetic analysis supports it as a separate species, it is morphologically similar to *I. australiensis*. *Inocybe rufobadia* and *I. brevicula* are known only from the south-west of Western Australia. The former is easily distinguished from *I. australiensis* by its bright rusty brown or vivid reddish brown pileus, while *I. brevicula* is characterised by its short, squat fruit bodies and the absence of dark woolly fibrils on the lower part of the stipe. Several unnamed and mostly single-stem lineages are apparent in the phylogenetic tree figure (Fig. 8); this group clearly merits further scrutiny, and it is likely that additional new species will come to light.

We also note at least two instances of strongly supported gene conflict between *rpb*2 and nuclear rRNA gene regions in the *I. australiensis* group. Given what are likely several recent speciation events in the group, and with multiple species exhibiting broad geographic distributions, incomplete lineage sorting seems a likely explanation for the gene conflict.

Inocybe phaeosquarrosa E.Horak, first described from New Zealand, can be distinguished from *I. australiensis* by the scaly stipe and presence of thin-walled cystidia, while the broadly similar *I. umbrosa* E.Horak is characterised by its entirely pruinose stipe. Both of these species are phylogenetically distantly related from *I. australiensis* (data not shown). Horak (1980) suggested that *I. serrata* Cleland and *I. granulosipes* Cleland might be conspecific with

I. australiensis, leading to the misapplication of the name *I. australiensis* to members of what probably is the *I. serrata* group by Grgurinovic (1997; indeed, a species with ellipsoidal spores is illustrated), and by Bougher & Syme (1998). The photograph labelled *I. australiensis* in Fuhrer (2009) depicts specimens without the woolly fibrillose lower part of the stipe, and these appear to be pruinose the entire stipe length, possible a species in the *I. tenax* group. Furthermore, a photograph of "*Inocybe australiensis*" in McMullan-Fisher *et al.* (2014) is doubtful since the background colour of the pileus is pale brown and the stipe is almost white.

8. Inocybe aureorimosa Matheny & Bougher, *sp. nov.*

MycoBank: MB 809262

Type: 1 km N of Catterick Rd, near Greenbushes, W.A., 33°50'50.6"S, 116°04'50.3"E, gregaricus to scattered on soil in Jarrah forest under *Eucalyptus marginata* and *Corymbia calophylla*, 12 July 2011, *P.B.Matheny PBM3598*; holo: PERTH 08319871, iso: TENN 066614 (GenBank DNA accession no. KF915274-nLSU, KF977198-ITS, KF991371-*rpb2*).

Pileus 15–60 mm wide, plano-convex even when immature, with a large obtuse umbo, becoming plane; margin deeply decurved, at times undulate; velipellis absent or present over the disc; surface dry, coarsely fibrillose overall; disc unbroken or with a few appressed scales when young, but becoming more obviously appressed-squamulose with recurved tips or excoriate, rimose or rimulose towards the margin that becomes torn with age; yellowish brown to brownish yellow (10YR 5/6, 6/6 to 5/8; mainly Buckthorn Brown mixed with Yellow Ochre) or golden brown (6E8 to 6D8, 5E8 to 5D7; Sudan Brown to Burnt Sienna); context whitish with buff tinges, not changing colour upon exposure, not confluent with stipe. *Lamellae* sinuate or ascending-adnate to, occasionally, adnate, very close to crowded, 36–60 L, with several tiers of lamellulae; initially light grey (2.5Y 7/2) to pale yellow (2.5Y 7/4 to 6/4), becoming light yellowish brown to light olive-brown (2.5Y 6/4 to 5/4) or, later, dark yellowish brown (10YR 4/4), rather broad, 2–8 mm deep, ventricose; edges indistinctly pallid-fimbriate. *Stipe* 20–50 × 3–11 mm, terete even, slightly swollen, or tapered slightly at the base, not bulbous; partial veil whitish and fugacious, surface pruinose at the apex or not so obviously pruinose at least on older specimens, elsewhere fibrillose, yellow when immature (2.5Y 7/6 to 6/4), becoming coloured red like the pileus; brownish yellow to yellowish brown (10YR 6/6 to 5/6); context solid, pallid or with isabelline tones, becoming yellowish brown to olivaceous brown. *Odour* not remarkable. Pl. 23, 24.

Basidiospores 8.5–11.0 (–11.5) × 5.0–7.0 µm [mean 9.8 × 6.0 µm; Q 1.38–1.88 (–2.00); Q mean 1.64 (*n* = 68/4)], smooth, mostly subamygdaliform to amygdaliform with bluntly pointed to pointed apices, some more broadly amygdaliform, yellowish brown; apiculus small, typically not very distinctive. *Basidia* 26–45 × 8–11 µm, clavate, hyaline, 4-stergimate. *Pleurocystidia* 70–94 (–113) × 12–24 (–28) µm, mostly fusiform, typically widest above the short basal pedicel, occasionally more cylindrical (*E5186* and *PBM3598* are at the wider end of the range, whereas *PBM3646* and *MR00224* are at the narrow end of the range); apices obtuse but occasionally swollen, often bearing crystals, usually thick-walled, the walls 1.5–3.0 µm thick, hyaline, sometimes only thick-walled apically or thin-walled. *Cheilocystidia* heteromorphic, composed of cells similar to pleurocystidia but also variably-shaped cells, these broadly lageniform, clavate or saccate; hyaline or ochraceous, mixed with paracystidia.

Caulocystidia 36–130 × 20–45 µm, often larger than cystidia in the hymenium, descending to the stipe centre in scattered clusters, shaped more like cheilocystidia including large lageniform cells mixed with saccate ventricose cylindrical or clavate cells, and cauloparacystidia that are variable in abundance, thin- and thick-walled forms present, hyaline or ochraceous; apical crystals present on some. *Pileipellis* a cutis giving rise to scattered fascicles of recurved hyphae, these bright rusty brown in mass, mostly cylindrical to inflated, 7–25 µm wide, encrusted, thin-walled to slightly thickened. *Clamp connections* present. Fig. 11.

Known only from south-western W.A.; on soil in in Jarrah forest under *Corymbia calophylla* and *Eucalyptus marginata*, in Karri forest under *E. marginata* and *E. diversicolor*; also in coastal shrub under *Melaleuca*, *Agonis*, *Spyridium*, *Acacia* and *Westringia*. Fruiting May–July.

Plate 21. *Inocybe australiensis* (PERTH 08241724). Photograph — P.B.Matheny.

Plate 22. *Inocybe australiensis* (PERTH 08319529). Photograph — P.B.Matheny.

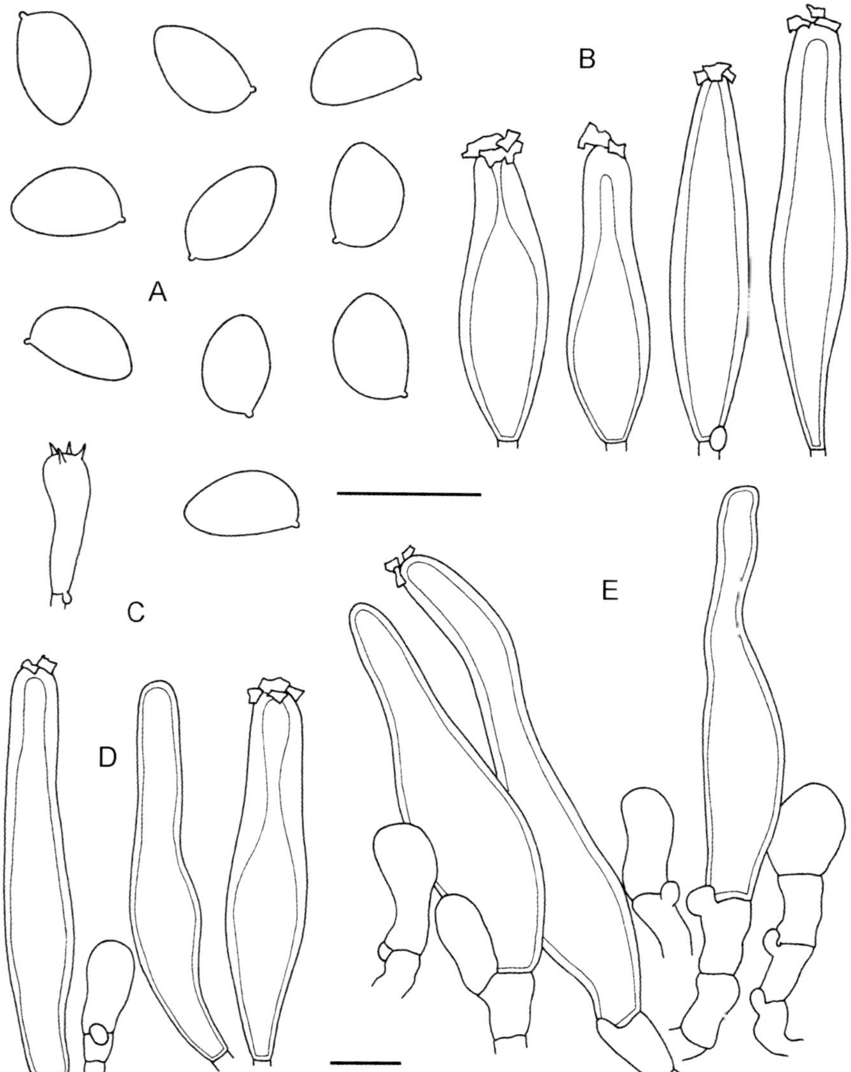

Figure 10. *Inocybe australiensis* (A, C–E, PERTH 08319529; B, PERTH 08241724). **A,** Basidiospores; **B,** Pleurocystidia; **C,** Basidium; **D,** Cheilocystidia and paracystidium; **E,** Caulocystidia and cauloparacystidia (from stipe apex). Scale bars = 10 µm (longer bar for spores only).

Additional specimens examined: W.A.: Two Peoples Bay Nature Reserve, Rocky Point Track, under *Melaleuca thymoides*, *Agonis flexuosa*, *Spyridium globulosum* (Rhamnaceae) and *Westringia* (Lamiaceae), 28 May 1992, *K.Syme KS539/92* (PERTH 05429684, TENN 068228); the Knoll, near Walpole, on lateritic soil under *Eucalyptus marginata* and *E. diversicolor*, 14 July 1994, *N.L.Bougher E5186* (PERTH 07655304); type locality, on lateritic soil with gravel under *Corymbia calophylla* and *Eucalyptus*, 12 July 2011, *M.Ryberg MR00223* (PERTH 08320144, TENN 066583); type locality, on lateritic soil in Jarrah forest, 12 July 2011, *M.Ryberg MR00224* (PERTH 08320136, TENN 066654; KF915273-nLSU, KF991370-*rpb2*); Two Peoples Bay Nature Reserve, 30 km E of Albany, Sinkers Reef Track, 34°58'54.2"S, 118°09'58.8"E, on sandy soil in coastal shrub under *Melaleuca*, *Agonis*, *Spyridium* and *Acacia*, 19 July 2011, *P.B.Matheny PBM3646* (PERTH 08319405, TENN 066458; KF915272-nLSU, KF977197-ITS, KF991369-*rpb2*).

Etymology: From the Latin *aureus* (golden) and *rimosus* (cracked), in reference to the pileus.

Diagnostic features: Fruit bodies medium sized to robust. Pileus golden-brown, brownish yellow to yellowish brown, coarsely fibrillose to squamulose, rimulose. Spores smooth, mostly subamygdaliform to amygdaliform, 8.5–11.0 × 5.0–7.0 µm; caulocystidia and hymenial cystidia large and conspicuous.

Inocybe aureorimosa appears to be most closely related to an unidentified species based on *PBM3686B* (PERTH 08318999) (Fig. 8). This collection has comparatively large spores (mostly c. 12–17 × 7 µm), and it was collected in sand under *Allocasuarina* in Kunjin in the Western Australian wheatbelt region.

Inocybe spadicea is similar to *I. aureorimosa* in outward appearance but, at a microscopic level, it has large clavate cells mixed among the caulocystidia that descend to just above the stipe base, while the spores are longer (mostly 9.0–14 × 5.5–7.0 µm) and more distinctly amygdaliform than in *I. aureorimosa*. Phylogenetically, *I. spadicea* and *I. aureorimosa* are rather distantly related in contrast to the unresolved *PBM3686B* mentioned above. *Inocybe rufobadia*, also Western Australian and with a bright reddish or golden-brown pileus, differs from *I. aureorimosa* by its smaller cylindrical spores measuring 8–10 × 4–5 µm. Moreover, the pileus of *I. rufobadia* is more vivid reddish brown than in *I. aureorimosa*.

Future genetic studies may eventually see several other collections being referred to *I. aureorimosa*. Thus, PERTH 07677731 [*PBM2131*, *E6766*, N of Madfish Bay Rd, William Bay Natl Park, W.A. under *Eucalyptus*, 23 June 2001] has longer and more oblong-amygdaliform spores (9–12 × 5.5–6.0 µm) and noticeably long, narrowly fusiform pleurocystidia (90–122 × 13–18 µm). Another collection [Vinegar Well off Barber Rd, near Kunjin, W.A., under *Eucalyptus loxophleba*, *Allocasuarina huegeliana* and *Acacia acuminata* (PERTH 07659652, *E6179*)] has more robustly shaped pleurocystidia (82–95 × 15–21 µm; like some of the specimens cited here), but it also has longer spores (12.0–13.0 × 6.0–6.5 um) and has caulocystidia on the upper half of the stipe that are similar in shape to the pleurocystidia. All of these collections share with *I. aureorimosa* bright fulvous or golden-brown colours, lamellae with an olivaceous cast, and a pale yellowish stipe.

Inocybe aureorimosa is now the correct name for PERTH specimens referred to as "*Inocybe longissima*, nom. prov." in records up to 2001. It also resembles a species illustrated by Fuhrer (2005) as "*Inocybe* sp." (no. 143) but from forests of *Nothofagus cunninghamii*.

Plate 23. *Inocybe aureorimosa*, holotype (PERTH 08319871). Photograph — P.B.Matheny.

Plate 24. *Inocybe aureorimosa* (PERTH 08319405). Photograph — P.B.Matheny.

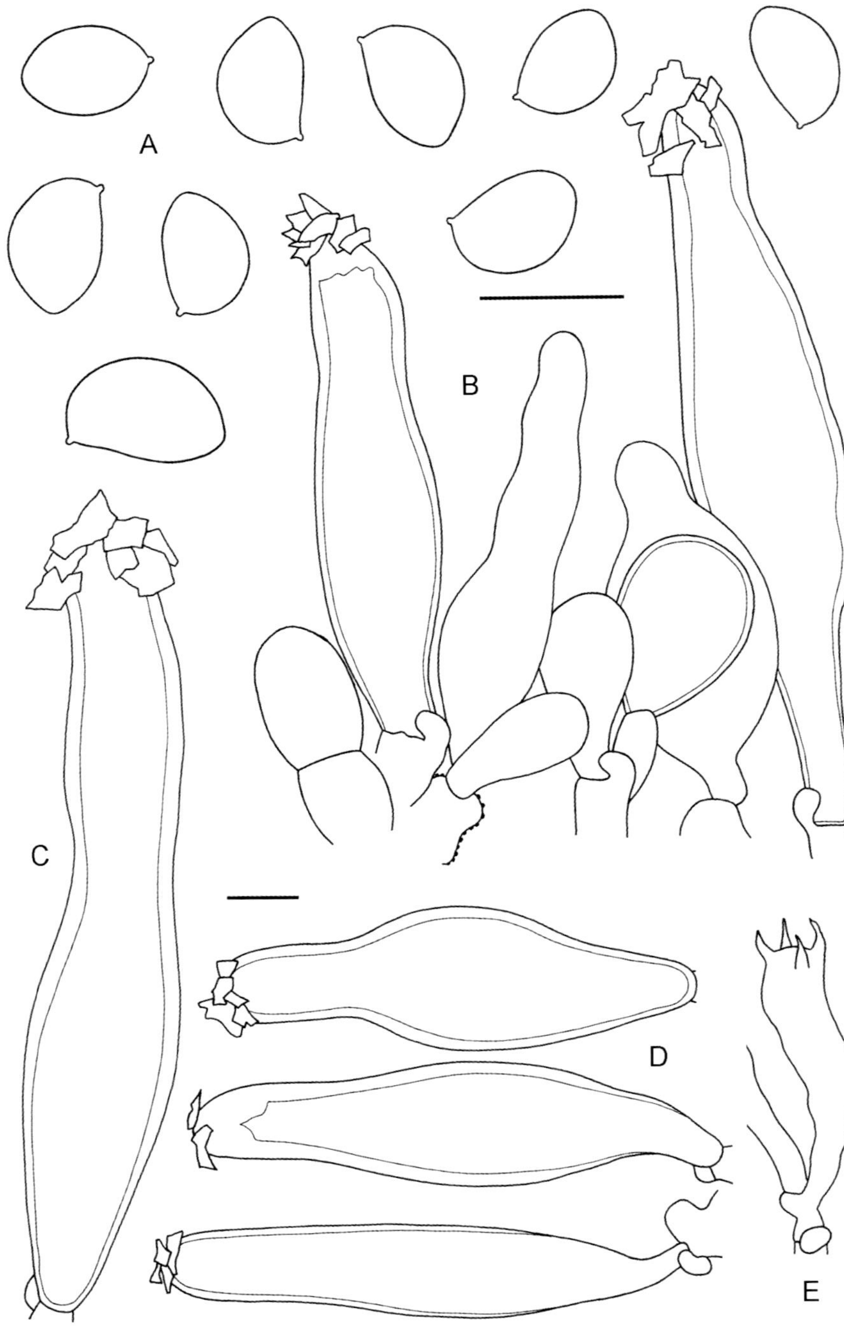

Figure 11. *Inocybe aureorimosa* (holotype). **A,** Basidiospores; **B,** Cheilocystidia and paracystidia; **C,** A large caulocystidium from upper stipe; **D,** Pleurocystidia; **E,** Basidium. Scale bars = 10 μm (longer bar for spores only).

9. **Inocybe auritelloides** Matheny, Bougher & G.M.Gates, *sp. nov.*

MycoBank: MB 809263

Type: Vale of Belvoir, 30 km W of Gowrie Park, Tas., scattered singly on soil in cool-temperate rainforest under *Nothofagus cunninghamii*, 25 May 2010, *G.Gates PBM3411*; holo: PERTH 08516928; iso: TENN 068294 (KJ729911-nLSU, KJ729885-ITS, KJ729941-*rpb2*).

Pileus 25–42 mm wide, broadly convex, becoming plano-convex or plane with a slightly obtuse umbo; margin decurved but slightly uplifted; velipellis absent; surface dry, the disc squarrose-scaly to coarsely-fibrillose, coarsely-fibrillose towards the margin with fibrils diverging near the margin, with appressed imbricate scales around the centre; blackish brown at the centre (possibly darker due to high moisture content), shading to dark brown (7.5YR 3/2) or dark yellowish brown (10YR 3/4) towards the margin, lighter or pale brown between the spreading fibrils, or with reddish brown or tawny fibrils and scales (*1108*); context tough, dull brown or sombre brown, not changing colour where exposed, up to 4 mm thick under the disc. *Lamellae* adnexed to sinuate, very close, 40–50 L, with several tiers of lamellulae, yellowish brown to brown (10YR 5/4 to 4/3), becoming dark yellowish brown (10YR 4/4) or dark reddish brown (5YR 3/3), 4–6 mm deep; edges pallid-fimbriate or indistinctly so, or concolorous with the faces. *Stipe* 30–75 × 5–8 mm at the apex, even to tapered or slightly enlarged at the base; pruinose at the (extreme) apex, if at all, elsewhere the surface fibrillose; light reddish brown (5YR 6/3) at the extreme apex, below reddish brown (5YR 4/4 to 7.5YR 4/4; Chestnut), becoming very dark brown (10YR 3/4; Bay Brown); context solid, light yellowish brown, with very dark brown areas developing towards the base. *Odour* like *Pelargonium* (*PBM3410*) or not remarkable (*PBM3411*, *1108*). Pl. 25, 26.

Basidiospores 6.0–8.0 (–9.0) × 4.0–6.0 (–6.5) µm [mean 7.1 × 5.0 µm; Q 1.23–1.70 (–2.00); Q mean 1.42 (*n* = 54/3)], smooth, mostly broadly elliptic to broadly subamygdaliform, not adaxially depressed; apices rounded to bluntly pointed; apiculus small. *Basidia* 26–29 × 7–10 µm, clavate, hyaline, 4-sterigmate (rarely 1-sterigmate). *Pleurocystidia* 55–80 × 14–17 µm, fusiform-ventricose to fusiform, sometimes with a distinct basal pedicel; apices subacute or obtuse, with crystals or more often lacking crystals; thin- or thick-walled, the walls often thickened apically to 1–3 µm, hyaline. *Cheilocystidia* similar to pleurocystidia but of variable size and shape (utriform or lageniform), occasionally with a subcapitate apex and a broad truncate base, mixed with clavate ellipsoidal and variously-shaped mainly thin-walled paracystidia, and some basidia. *Caulocystidia* present but infrequent at the extreme apex, similar to cheilocystidia, variable in shape (lageniform, fusiform or slender-clavate), mixed with thin-walled and thick-walled cauloparacystidia of similar shape to the paracystidia; stipe hyphae encrusted, russet to rusty brown in KOH. *Pileipellis* a cutis where sampled between the margin and the disc, with some upturned elements, dark yellowish brown in mass, hyphae cylindrical, encrusted, mostly 6–14 µm wide, thin-walled or slightly thickened. *Clamp connections* present. Fig. 12.

Occurs on soil under *Nothofagus cunninghamii* in cool-temperate rainforest in Tas. Fruiting Feb.–May.

Additional specimens examined: Tas.: Growling Swallet, Mount Field Natl Park, 17 Feb. 2005, *G.Gates & D.Ratkowsky 1108* (PERTH 08516936, TENN 068292; KJ729912-nLSU, KJ729886-ITS); Vale of Belvoir (type locality), 30 km W of Gowrie Park, solitary along track in rainforest under *Nothofagus cunninghamii*, 25 May 2010, *F.Karstedt PBM3410* (TENN 068293; KJ729913-nLSU, KJ729887-ITS, KJ729942-*rpb2*).

Etymology: The epithet *auritelloides* alludes to its similarity to species of *Auritella* having sombre brown colouration and tough flesh.

Diagnostic features: Fruit bodies robust with firm flesh. Pileus dark brown, blackish brown or tawny, the centre squarrose-scaly, the stipe fibrillose. Spores broadly elliptical to subamygdaliform, mostly 6–8 × 4–6 µm; caulocystidia infrequent and restricted to the stipe apex.

Collections of *Inocybe auritelloides* vary in several ways. Two specimens (*PBM3410*, *PBM3411*) have a very dark blackish brown pileus centre (possibly due to excessive moisture), whereas a third (*1108*) is distinctly red-brown or tawny. Furthermore, *PBM3410* featured a *Pelargonium* smell, whereas *PBM3411* and *1108* lacked any odour. Some variation in microscopic characters is also evident, with the holotype having a greater proportion of subamygdaliform spores with thicker walls and a larger apiculus than in *1108*. The latter often has larger hymenial cystidia with longer necks and thicker walls than the holotype. However, the spores and cystidia of these collections overlap in size and form, and in overall terms the three collections are quite similar microscopically. Moreover, their DNA sequences are almost identical (Fig. 8), and they share the same association with *Nothofagus cunninghamii*.

Inocybe lucifera, also endemic to *Nothofagus*-dominated rainforest in Tasmania, is a closely related but distinct species. The two overlap in overall colour (the former is similar to *1108* having a reddish brown or tawny pileus), and both tend to have distinctive or pungent odours. However, the spores of *I. lucifera* are longer and narrower (mean 8.0 × 4.5 µm) compared to *I. auritelloides* (mean 7.1 × 5.0 µm).

Inocybe auritelloides is also similar to *I. austrosquarrosa*, but the latter occurs in wet sclerophyll forest or lawns and has an association with eucalypts. Moreover, the spores of *I. austrosquarrosa* have a more amygdaliform outline and are longer and narrower than in *I. auritelloides*, and caulocystidia appear to be entirely absent.

Inocybe cerea E.Horak, from under *Nothofagus* in New Zealand (Horak, 1977), shares the small, mostly broadly elliptic spore outline of *I. auritelloides*, but differs in other respects, for example its non-scaly pileus and yellow, wax-yellow or brownish yellow stipe with pink tints, and its thin-walled cheilocystidia.

10. **Inocybe brevicula** Matheny & Bougher, *sp. nov.*

MycoBank: MB 809268

Type: West Bay, Leeuwin-Naturaliste Natl Park, Augusta, W.A., 34°17'16.8"S, 115°09'18.6"E, on sandy track in low lying area near sedges and myrtaceous shrubs in Jarrah-Marri woodland under *Eucalyptus marginata* and *Corymbia calophylla*, 17 June 2007, *N.L.Bougher NLB256*; holo: PERTH 08092338; iso: TENN 068645 (JN974925-nLSU, KJ756478-ITS, KJ729950-*rpb2*).

Pileus 8–20 mm wide, hemispherical when immature, becoming convex to plano-convex with age; umbo absent; margin initially incurved, becoming decurved, non-appendiculate; surface dry; velipellis uncommon, a greyish covering of the disc, coarsely radially fibrillose when young, not striate, becoming appressed-squamulose at the centre and towards the margin; centre dark brown to dark yellowish brown or "Bay Brown" (near 7.5YR 2/3), except where covered by greyish velipellis, yellowish brown to dull brown between the spreading fibrils or fibrillose scales; context pallid or cream, dulling slowly after being cut. *Lamellae* adnexed, close with 1 or 2 tiers of lamellulae between each pair of lamellae, broad, 2–4 mm deep (quite deep for the size of the pileus), edges entire or inconspicuously fimbriate; at first pinkish cream, soon becoming near isabelline (2.5Y 5/8) then dull dark brown, not bruising. *Stipe* 8–15 × 2–3 mm, quite short for its size to pileus width, terete, even but with a slightly swollen non-marginate base in some specimens, without any conspicuous basal mycelium; veil not evident; surface with pinkish tinges when young, later becoming dark brown like the pileus, pruinose at least at the apex; context solid, coloured like the pileus. *Odour* not distinctive. Pl. 27, 28.

Basidiospores (7.0–) 7.5–9.5 (–10.0) × 4.0–5.0 µm [mean 8.5 × 4.6 µm; Q 1.56–2.18 (–2.25); Q mean 1.86 (n = 39/2)], smooth, amygdaliform to oblong-amygdaliform; apices bluntly pointed, yellowish brown; apiculus small. *Basidia* 25–31 × 7–8 µm, clavate to cylindro-clavate, hyaline, yellow, or ochraceous, 4-sterigmate. *Pleurocystidia* 60–72 (–80) × 10–15 (–21) µm, narrowly lageniform, narrowly fusiform to cylindrical; apices obtuse and crystalliferous, thick-walled, the walls 3–4 µm thick, hyaline or ochraceous. *Cheilocystidia* similar to pleurocystidia and variable in shape (lageniform, narrowly fusiform or cylindrical), mixed with short clavate paracystidia and basidia, these hyaline but at times ochraceous or yellow. *Caulocystidia* 48–60 × 14–17 µm, in scattered clusters, descending the entire length of the stipe, mostly lageniform, also fusiform or cylindrical, thick-walled, mostly 2–3 µm thick, hyaline or ochraceous; mixed with cauloparacystidia and caulocystidioid cells of varying shape and size. *Pileipellis* composed of cylindrical to somewhat inflated hyphae mostly 7–19 µm wide and with thickened walls, dark yellowish brown in mass, smooth in appearance or at least not distinctly encrusted. *Clamp connections* present. Fig. 13.

Plate 25. *Inocybe auritelloides*, holotype (PERTH 08516928). Photograph — P.B.Matheny.

Plate 26. *Inocybe auritelloides* (PERTH 08516936). Photograph — G.M.Gates.

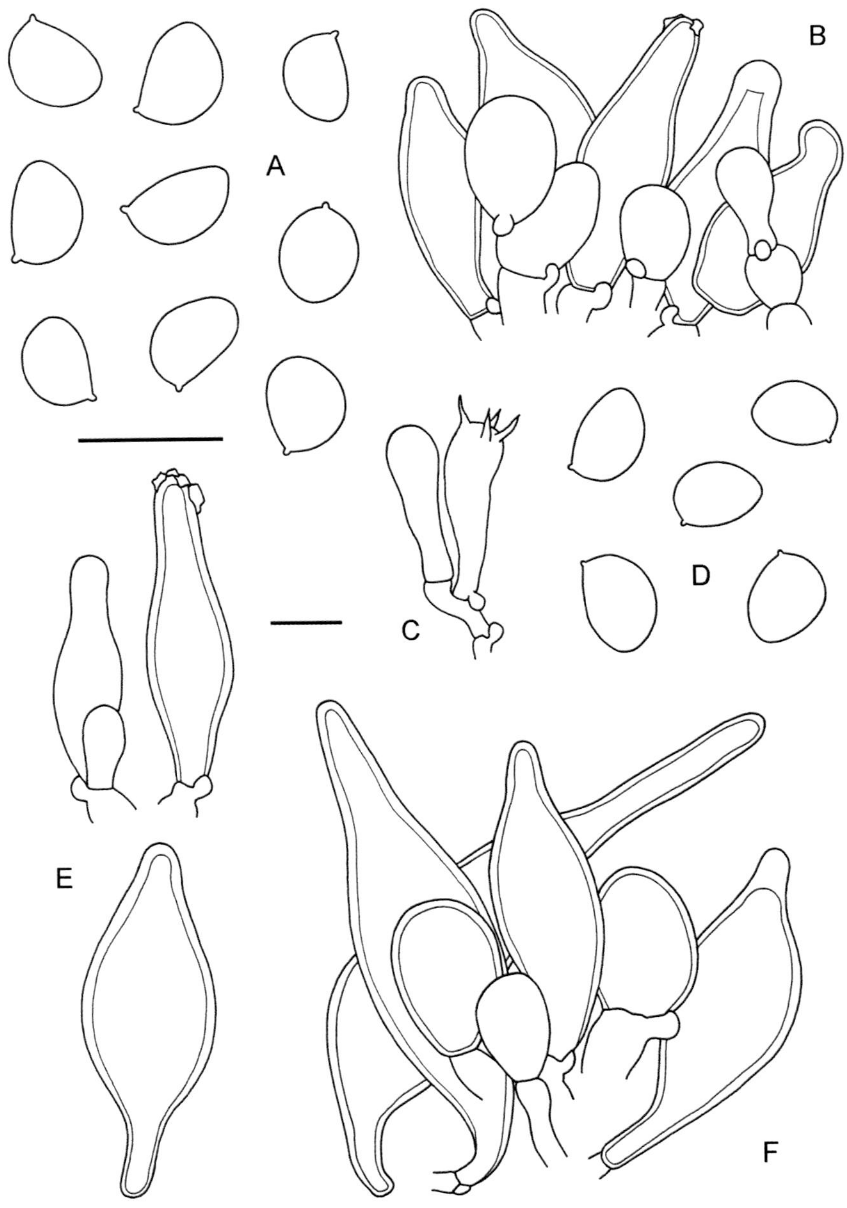

Figure 12. *Inocybe auritelloides* (A–C, E, holotype; D, F, TENN 068292). **A,** Basidiospores; **B,** Cheilocystidia and paracystidia; **C,** Basidia; **D,** Basidiospores; **E,** Pleurocystidia; **F,** Caulocystidia and cauloparacystidia (upper stipe). Scale bars = 10 μm (longer bar for spores only).

Plate 27. *Inocybe brevicula*, holotype (PERTH 08092338). Photograph — N.L.Bougher.

Plate 28. *Inocybe brevicula* (PERTH 08516634). Photograph — N.L.Bougher.

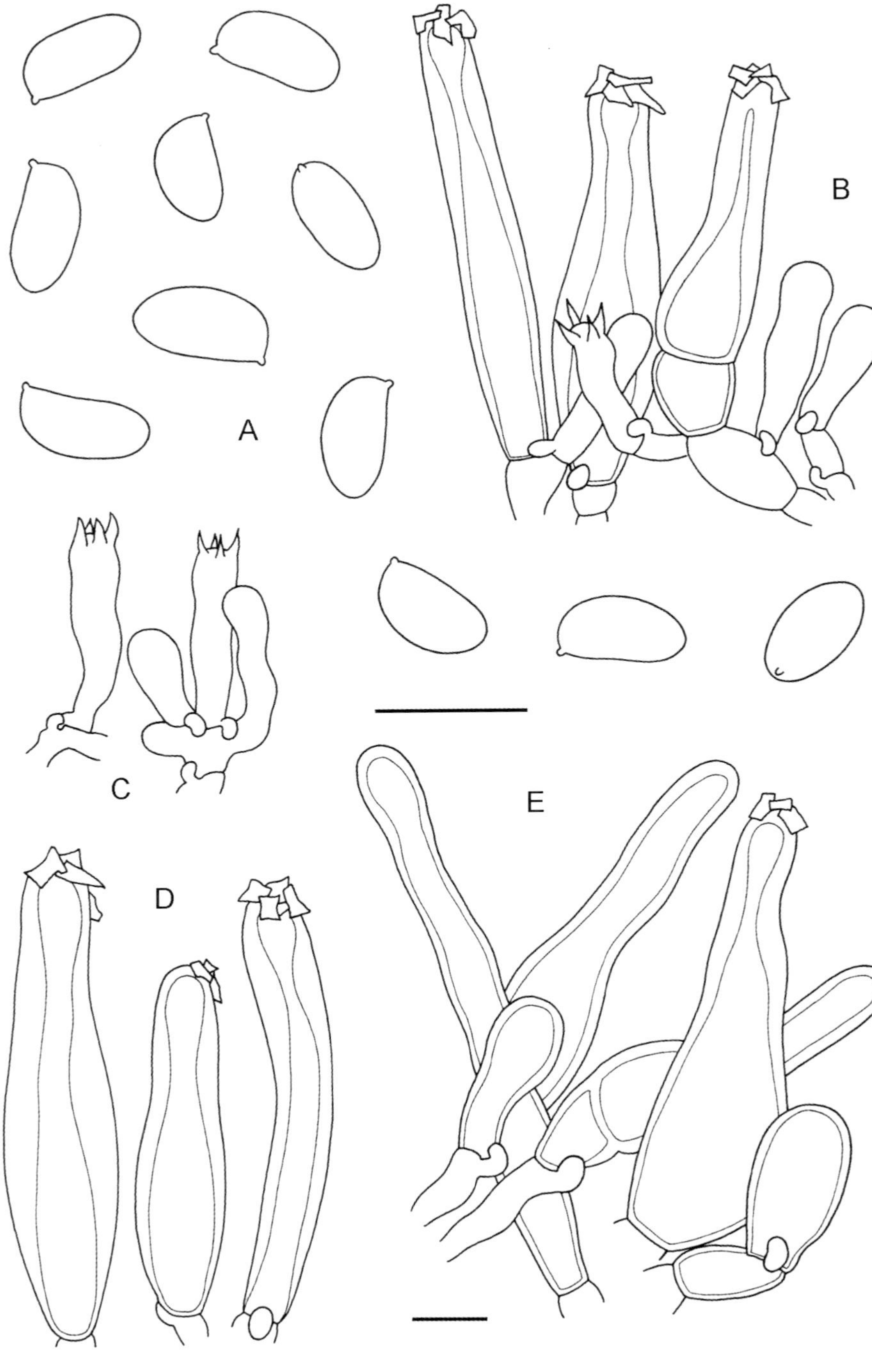

Figure 13. *Inocybe brevicula*, holotype (PERTH 08092338). **A,** Basidiospores; **B,** Cheilocystidia and paracystidia/basidia; **C,** Basidia; **D,** Pleurocystidia; **E,** Caulocystidia and cauloparacystidia/caulocystidioid cells.

Scale bars = 10 μm (longer bar for spores only).

Known only from soil in south-western W.A.; grows under or near eucalypts, including *Eucalyptus marginata*, *Corymbia calophylla* and/or *E. diversicolor*. Fruiting June.

Additional specimen examined: W.A.: Perup Nature Reserve, near Perup Ecology Centre, 34°10'36.9"S, 116°35'44.7"E, on mossy track in between Karri (*Eucalyptus diversicolor*) plantation and Jarrah-Marri forest of *E. marginata* and *Corymbia calophylla*, 2 June 2013, *N.L.Bougher NLB1073* (PERTH 08516634, TENN 068646; KJ756443-nLSU, KJ756479-ITS, KJ756404-*rpb2*).

Etymology: From the Latin *brevis* (short) and the suffix *-cula* (indicating somewhat), in reference to the rather squat fruit bodies.

Diagnostic features: Fruit bodies short and squat. Pileus dark brown, appressed-squamulose to coarsely fibrillose; stipe pruinose at least at the apex, becoming coloured like the pileus; odour not distinctive. Spores narrow (4–5 µm wide).

A member of the *I. australiensis* group (Fig. 8), the available collections suggest *I. brevicula* produces short, squat fruit bodies similar in stature to those of *I. gomphocephala*; see also an image of the holotype collection in Bougher *et al.* (2007a) as specimen ID: 2865 (incorrectly as "ID: 2865"). However, *I. gomphocephala* is distantly related, in the *I. fibrillosibrunnea* group; morphologically, it has broader spores (5–7 µm) and an association with shrubs of Myrtaceae, Rhamnaceae and Fabaceae in coastal shrub-heath. *Inocybe campestris* usually produces larger (and not squat) fruit bodies, the stipe is longer and wider (16–30 × 3–7) mm, and DNA sequences readily separate the two species (Fig. 8). Finally, *I. avellaneifolia*, which occurs in similar coastal shrub-heath habitat, has lighter brown colours and the presence of caulocystidioid cells only on the lower part of the stipe.

11. Inocybe campestris Matheny & Bougher, *sp. nov.*

MycoBank: MB 809274

Type: 14 km W of Corrigin, Old Kunjin Railway, wheatbelt region, W.A., 32°21'25.5"S, 117°44'10.6"E, gregarious to scattered singly on sandy soil under planted *Allocasuarina* with nearby *Eucalyptus* spp., 26 July 2011, *P.B.Matheny PBM3682*; holo: PERTH 08319030; iso: TENN 066462 (GenBank DNA accession no. KJ756458-nLSU, KJ756416-*rpb2*).

Pileus 14–30 mm wide, obtusely conical to conico-campanulate or convex, becoming plane with age, occasionally with a low broad umbo; margin decurved and at times undulate; velipellis typically absent but present in some specimens as somewhat lighter brown portion over the disc; surface generally smooth and unbroken at the centre or with appressed fibrillose squamules, fibrillose towards the margin, not rimose; predominantly Bay Brown or dark yellowish brown to Snuff Brown or brown (10YR 3/2 to 4/3) to dark brown (7.5YR 3/2), at times the disc a lighter shade of brown; context thick under the disc, to 6 mm deep, bright white to pallid or dingy pale brown, colour not changing upon exposure, not confluent with the stipe. *Lamellae* ascending-adnate, adnate to uncinate, close to moderately close, 26–48 L, with several tiers of lamellulae, brown (10YR 5/3) to dark yellowish brown (10YR 4/4) or umber or strong brown with age (7.5 YR 4/4), up to 4 mm deep, edges not distinctly pallid-fimbriate. *Stipe* 16–30 × 3–7 mm, terete or slightly compressed, most often with a slight rounded, submarginate to oblique bulb up to 8 mm wide, occasionally the bulb is conspicuous; veil not directly observed; only the apex pruinose, or pruinose along the upper half to three-quarters or with a pruinose appearance almost the entire length, fibrillose above the base, overall with a satiny appearance for most of its length; when young pinkish to reddish yellow (5YR 7/3, 7/6, 6/8), becoming strong brown to brown or umber (7.5YR 5/6, 5/4, 4/4; 6E8 to 7E8) or dark reddish brown or near copper-brown (7.5YR 4/4; 6F6) throughout, often darker above the wider base, occasionally with clear distinct dark brown fibrils along the rim of the bulbous base; context solid with a pale interior, white in the bulb. *Odour* absent. Pl. 29–31.

Basidiospores 7.5–10.0 × 4.5–5.5 µm [mean 8.4 × 4.9 µm; Q 1.45–2.00; Q mean 1.72 (n = 47/3)], smooth, amygdaliform to subamygdaliform, occasionally more elliptic and with a slight ventral depression; apices bluntly pointed or less often obtuse, yellowish brown; apiculus small and not distinctive. *Basidia* 24–34 × 7–9 µm, 4-sterigmate, clavate to cylindrical, hyaline. *Pleurocystidia* 50–65 × 10–13 µm, narrowly fusiform to cylindrical; apices obtuse; bases truncate or pedicellate, crystalliferous, thick-walled, the walls generally 2.5–5.0 µm thick, hyaline to pale yellow. *Cheilocystidia* similar to pleurocystidia, at times shorter, occasionally thin-walled, scattered along the lamellar edge, mixed with paracystidia and basidia. *Caulocystidia* similar to hymenial cystidia but with fewer or no apical crystals, mixed with cauloparacystidia, occurring in scattered but frequent clusters at least down to the centre of the stipe, as clusters or rare individual cells above the stipe base. *Pileipellis* a cutis with scattered slightly ascending tufts of hyphae, these smooth to faintly wrinkled (encrusting pigment not evident), not thick-walled, cylindrical to inflated, mostly 8–15 µm wide, brown or dark yellowish brown in mass; end cells not differentiated. *Clamp connections* present. Fig. 14.

Occurs in sandy soil under a mixture of *Allocasuarina*, *Acacia* and/or *Eucalyptus* in Perth and the wheatbelt region of W.A. Fruiting July.

Additional specimens examined: W.A.: former site of W.A. Herbarium, corner of Hayman Rd and George St, Kensington, Perth, gregarious to scattered on grey sand under native planted *Eucalyptus* spp., 22 July 2011, *N.L.Bougher PBM3668* (PERTH 08319197, TENN 066544; KJ778840-ITS); Vinegar Well, off Barber Rd, Kunjin, W of Corrigin, 32°02'31.7"S, 117°40'06.5"E, in litter under *Acacia acuminata* and *Allocasuarina huegeliana* (no eucalypts nearby), 27 July 2011, *N.L.Bougher NLB891* (PERTH 08320381, TENN 066719; KJ756459-nLSU, KJ778841-ITS, KJ756417-*rpb2*); *loc. id.*, scattered singly on compact sand track under *Allocasuarina* and *Acacia acuminata*, 27 July 2011, *P.B.Matheny PBM3689* (PERTH 08318964, TENN066552; KJ756460-nLSU, KJ778842-ITS, KJ756418-*rpb2*); *loc. id.*, gregarious on sandy soil under *Allocasuarina* spp. and *Acacia acuminata* (no eucalypts nearby), 27 July 2011, *P.B.Matheny PBM3698* (PERTH 08318867, TENN 066539; KJ756461-nLSU, KJ756419-*rpb2*); *loc. id.*, scattered singly or in pairs on sand under *Allocasuarina* spp., 27 July 2011, *P.B.Matheny PBM3699* (PERTH 08318859, TENN 066533; KJ756462-nLSU, KJ778843-ITS, KJ756420-*rpb2*); *loc. id.*, gregarious on sandy soil under *Eucalyptus melanoxylon* and *Acacia acuminata*, 27 July 2011, *M.Ryberg MR00287* (TENN 066547; KJ756463-nLSU, KJ778844-ITS, KJ756421-*rpb2*); Old Kunjin Railway, c. 14 km W of Corrigin, along main road, 32°21'25.9"S, 117°44'17.1"E, in sandy soil under planted eucalypts and *Allocasuarina* bushes, 26 July 2011, *N.L.Bougher NLB876* (PERTH 08320543, TENN 066801); *loc. id.*, but c. 400 m distant, in sandy soil under planted eucalypts and *Allocasuarina* bushes, 26 July 2011, *N.L.Bougher NLB877* (PERTH 08320535).

Etymology: From the Latin *campestris* (pertaining to plains or flat areas), in reference to its occurrence in flat sandy regions of W.A.

Diagnostic features: Fruit bodies medium-sized and robust. Pileus fibrillose-squamulose; stipe pinkish brown, reddish brown or coppery brown by maturity, the base broadly bulbous and with brown fibrils.

Inocybe campestris is locally abundant in the Corrigin area of the wheatbelt region of Western Australia under a variety of putative mycorrhizal hosts including *Allocasuarina*, *Acacia* and *Eucalyptus*. It has also been recorded in sand under native planted eucalypts in the Perth metropolitan area. The brown fibrils at the stipe base are not always evident due to the aggregation of sandy soil around the base of the stipe when collected. In contrast, the bulb at the stipe base can be quite conspicuous at other times.

Inocybe australiensis and *I. atrisquamosa* differ from *I. campestris* in having smaller fruit bodies, a woolly brown fibrillose covering on the lower part of the stipe (of *I. australiensis*), and smaller spores. *Inocybe umbrosa* E.Horak, which occurs in New Zealand, differs in its smaller size, more slender habit, even and entirely pruinose stipe, and the presence of ephemeral, white fibrillose veil remnants at the base of the stipe.

Squat or less mature fruit bodies of *I. campestris* can be confused with *I. brevicula*, which has been recorded in Jarrah-Marri woodland and at the edges of Karri forest. However, the fruit bodies of *I. campestris* are generally more robust with a longer and thicker stipe. DNA sequences readily separate the two species. *Inocybe rufobadia* differs in having a bright reddish brown to rusty brown pileus.

Plate 29. *Inocybe campestris*, holotype (PERTH 08319030). Photograph — P.B.Matheny.

Plate 30. *Inocybe campestris* (PERTH 08310381). Note brown fibrils at the base of the stipe. Photograph — N.L.Bougher.

Plate 31. *Inocybe campestris* (PERTH 08318867). Photograph — P.B.Matheny.

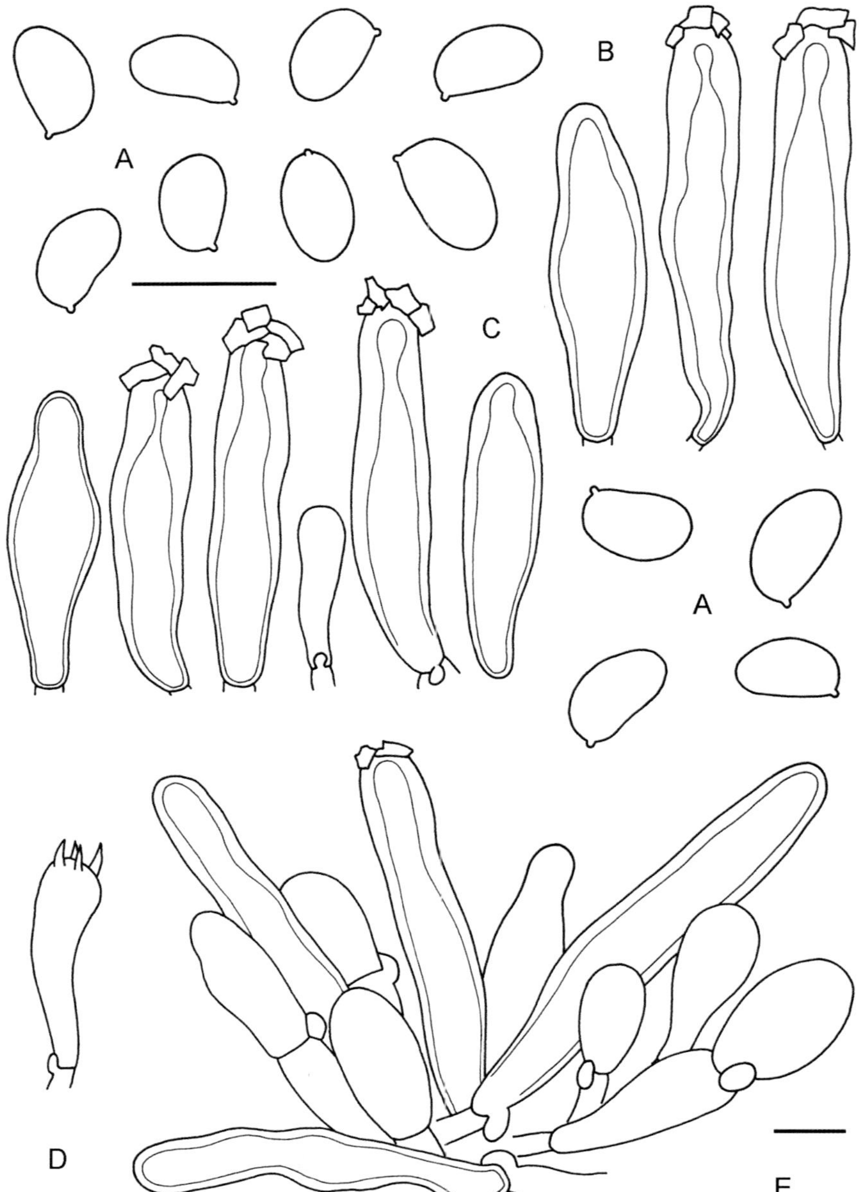

Figure 14. *Inocybe campestris,* holotype (PERTH 08319030). **A,** Basidiospores; **B,** Pleurocystidia; **C,** Cheilocystidia and a paracystidium; **D,** Basidium; **E,** Caulocystidia and cauloparacystidia. Scale bars = 10 μm (longer bar for spores only).

12. Inocybe fragrans Matheny, Bougher & G.M.Gates, *sp. nov.*

MycoBank: MB 809329

Type: Arve Road, Leap Loop, Tas., 43°09'43.93"S, 146°46'43.41"E, on soil under *Nothofagus cunninghamii*, *Eucalyptus*, *Eucryphia* and other rainforest understory plants, 5 Feb. 2011, *G.M.Gates 1149*; holo: HO 562696; iso: TENN 067060 (GenBank accession DNA No. KF915278-nLSU, KF977202-ITS, KF991375-*rpb2*).

Pileus 10–18 mm wide, conical with an acute apex, expanding to broadly conical, becoming plane with age, developing a small obtuse umbo; margin decurved to slightly so, at times uplifted after expansion; velipellis absent; surface smooth and unbroken on the disc, occasionally with small appressed scales at the centre, radially fibrillose to rimulose towards the margin; centre fulvous or brownish yellow (5C7), shading cream or chamois (5B5–4A3) or light yellowish brown (2.5Y 6/4) towards the margin, or the centre strong brown (7.5YR 4/4); margin Buckthorn Brown or yellowish brown (10YR 5/8), becoming light olive-brown (2.5Y 5/4) around the centre with age, not hygrophanous; context thin-fleshed, somewhat fragile, light yellowish brown. *Lamellae* adnexed, uncinate, moderately close, with at least 1 tier of lamellulae, pale brown (10YR 6/3) to yellowish brown (10YR 5/6), ventricose, medium, c. 2.5 mm deep; edges lightly pallid-fimbriate. *Stipe* 19–30 × 1.5–3 mm, even, terete, not bulbous, but occasionally swollen at the base, pruinose over the upper ⅙–¼, lightly silky-fibrillose below, with thin silky whitish fibrils over a pale yellow (2.5Y 7/4) ground colour; cortina not confirmed but presumably fugacious; context with a central pallid pith between the pale yellow cortex. *Odour* strongly of bruised *Pelargonium* leaves. Pl. 32.

Basidiospores 7.0–8.5 × 4.0–5.0 [mean 7.7 × 4.7 µm; Q 1.40–1.88; Q mean 1.64 (n = 31/2)], smooth, mostly elliptic to less often subamygdaliform, with obtuse to bluntly pointed apices, some ovate in face view, yellowish brown; apiculus small and not conspicuous. *Basidia* 19–24 × 6–7 µm, clavate to cylindroclavate, hyaline, 4-sterigmate. *Pleurocystidia* 44–58 (–65) × 11–18 µm, fusiform to lageniform (the ventricose median portion with a short cylindrical neck), tapered towards the base, thick-walled, the walls 1.0–2.5 µm thick, hyaline or yellow; apices obtuse, crystalliferous. *Cheilocystidia* similar to pleurocystidia but often shorter, fusiform, lageniform, or ventricose, mixed with paracystidia. *Caulocystidia* absent, caulocystidioid hairs much larger than hymenial cystidia (48–112 × 13–25 µm), frequent at the apex and numerous at the mid-region of the stipe where they resemble differentiated end cells, cauloparacystidia not observed; mostly fusiform, thin- to thick-walled, the walls to 2 µm thick, hyaline or pale yellow; occasionally with crystalliferous apices. *Pileipellis* a cutis of cylindrical hyphae mostly 5–19 µm wide, weakly encrusted, brownish yellow in mass. *Clamp connections* present. Fig. 15.

Occurs in Vic. and Tas.; grows on soil under *Nothofagus cunninghamii* in rainforest or in mixed rainforest vegetation including *Eucalyptus*. Fruiting Feb.–May.

Additional specimen examined: Vic.: Acheron Way, Somers Park, Yarra Ranges Natl Park, on soil in rainforest under *Nothofagus cunninghamii*, 29 May 2009, *P.B.Matheny PBM3200* (TENN 063632; KF915279-nLSU, KF977203-ITS).

Etymology: From the Latin *fragrans* (fragrant), in reference to the pleasant smell of bruised *Pelargonium* leaves given off by this species.

Diagnostic features: The combination of the brownish yellow to chamois pileus giving a strong odour of *Pelargonium*, the small, smooth, mostly elliptic spores, and the occurrence in temperate rainforest with *Nothofagus cunninghamii*.

Inocybe cerea E.Horak, from New Zealand, is similarly coloured and also occurs in *Nothofagus* forests, but it differs by having subglobose to ovate spores and an acidulous odour (Horak, 1977). *Inocybe luteobulbosa* E.Horak can be similar in colour to *I. fragrans*, but it retains the sulphur-yellow veil remnants on the pileus, it has an entirely pruinose stipe, lacks a distinctive odour (Horak, 1977), and is distantly phylogenetically related (Fig. 8). *Inocybe fragrans* occupies a unique branch in the STAC '*australiensis* group' (Fig. 8).

Plate 32. *Inocybe fragrans*, holotype (HO 562696). Photograph — G.M.Gates.

13. Inocybe lucifera Matheny, Bougher & G.M.Gates, *sp. nov.*

MycoBank: MB 809387

Type: Vale of Belvoir, c. 30 km W of Gowrie Park, Tas., on soil in rainforest under *Nothofagus cunninghamii*, 25 May 2010, *F.Karstedt PBM3419*; holo: PERTH 08517029; iso: TENN 065791 (KP170939-LSU, KP308763-ITS, KJ811595-*rpb2*).

Pileus 14–32 mm wide, obtusely conical, expanding with age and developing a pronounced bluntly conical umbo; margin decurved; surface dry; velipellis absent; centre smooth and unbroken but becoming appressed-squamulose or developing recurved scales; margin fibrillose to coarsely so and becoming rimose, revealing whitish flesh underneath, orange-brown to tawny, paler between the spreading fibrils towards the margin. *Lamellae* adnate, close, brown to yellowish brown, with several tiers of lamellulae, broad, up to 4 mm deep; edges not conspicuously fimbriate. *Stipe* 25–45 × 3–5 mm, even, terete, flexuous; surface finely fibrillose (nowhere pruinose) but with small tawny-coloured tufts of fibrils above the base, elsewhere pale brown and contrasting in colour to the pileus; context whitish, solid, becoming hollow. *Odour* pungent. Pl. 33.

Basidiospores (6.5–) 7.0–9.0 × 4.0–5.0 µm [mean 8.0 × 4.5 µm; Q 1.50–2.00; Q mean 1.79 (n = 42/2)], smooth, mostly amygdaliform with bluntly pointed apices, occasionally cylindrical, elliptic, or subamygdaliform with an adaxial (ventral) depression imparting a subreniform appearance, yellowish brown to light yellowish brown (many pale brownish yellow and more variable in shape from lamellar mounts of the type); apiculus small and not distinctive. *Basidia* 22–26 × 7–8 µm, clavate, hyaline, 4-sterigmate. *Pleurocystidia* 36–68 × 10–15 µm, fusiform to fusiform-ventricose, only slightly thick-walled with the walls up to 1.5 µm thick apically, hyaline; apices obtuse or less often subcapitate, crystalliferous. *Cheilocystidia* similar to pleurocystidia but often shorter, mixed with short clavate hyaline paracystidia, these occasionally with slightly thickened walls. *Caulocystidia* similar to hymenial cystidia but often more flexuous and mixed with cauloparacystidia, observed only at the apex of the stipe; stipe hyphae with scattered golden inclusions or pigment deposits.

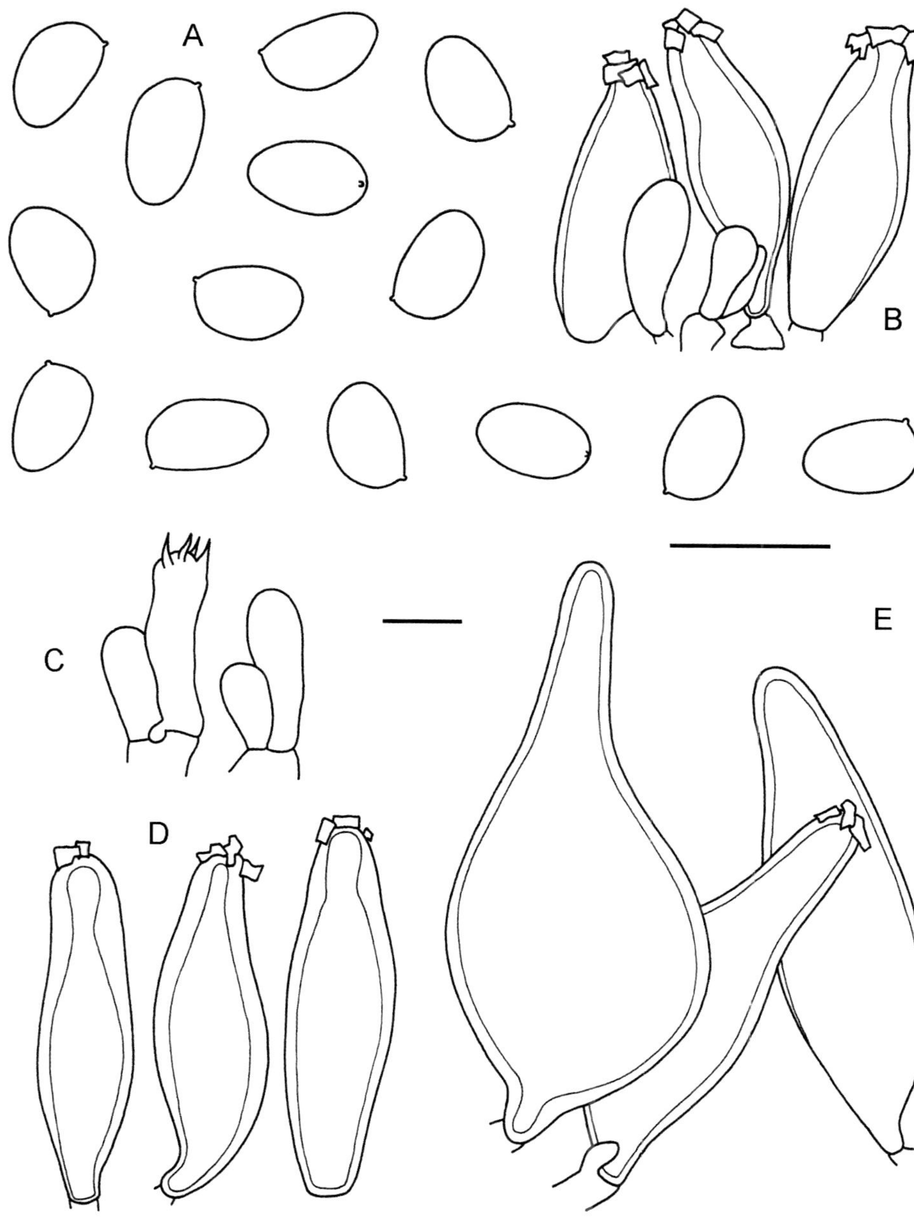

Figure 15. *Inocybe fragrans* (A, B, D, E, holotype (HO 562696); C, (TENN 063632)). **A,** Basidiospores; **B,** Cheilocystidia and paracystidia; **C,** Basidia; **D,** Pleurocystidia; **E,** Caulocystidioid hairs from stipe apex. Scale bars = 10 μm (longer bar for spores only).

Pileipellis composed of fascicles of erect hyphae (when sampled from disc with recurved scales), very dark brown in mass; hyphae cylindrical, mostly 7–15 µm wide, encrusted but not conspicuously so. *Clamp connections* present. Fig. 16.

Known only from Tas.; grows on soil in cool-temperate rainforest or mixed wet sclerophyll under *Nothofagus cunninghamii.* Fruiting Feb.–May.

Additional specimen examined: Tas.: Upper Florentine Valley, Timbs Track, in old wet sclerophyll forest with rainforest understorey under *Nothofagus cunninghamii* and *Eucalyptus* spp., 24 Feb. 2005, *G.Gates* & *D.Ratkowsky 1110* (HO 571944, TENN 068496 (KP170940-nLSU, KP308764-ITS).

Etymology: From the Latin *lucifera* (bearer of light), in reference to the bright tawny to reddish brown colour of the pileus.

Diagnostic features: Pileus bright tawny to reddish brown, with a bluntly conical umbo; margin rimose; centre becoming appressed-scaly with age. Stipe pale brown with tawny tufts of fibrils above the stipe base, nowhere pruinose.

Inocybe lucifera is one of three closely related species associated with *Nothofagus* in Tasmania and New Zealand. *Inocybe auritelloides* has a darker brown pileus and stipe, at least under wet conditions. The two can be difficult to distinguish, but the spores of *I. auritelloides* are shorter and broader (mean 7.1 × 5.0 µm). An undescribed species from New Zealand (TENN 063680, TENN 063683) is also morphologically similar to *I. lucifera*, and all three are readily distinguished by ITS, LSU, and *rpb2* gene sequences and form a strongly supported clade (Fig. 8). For morphological comparison with other species, see the discussion under *I. auritelloides*.

Plate 33. *Inocybe lucifera* (HO 571944). Photograph — G.M.Gates.

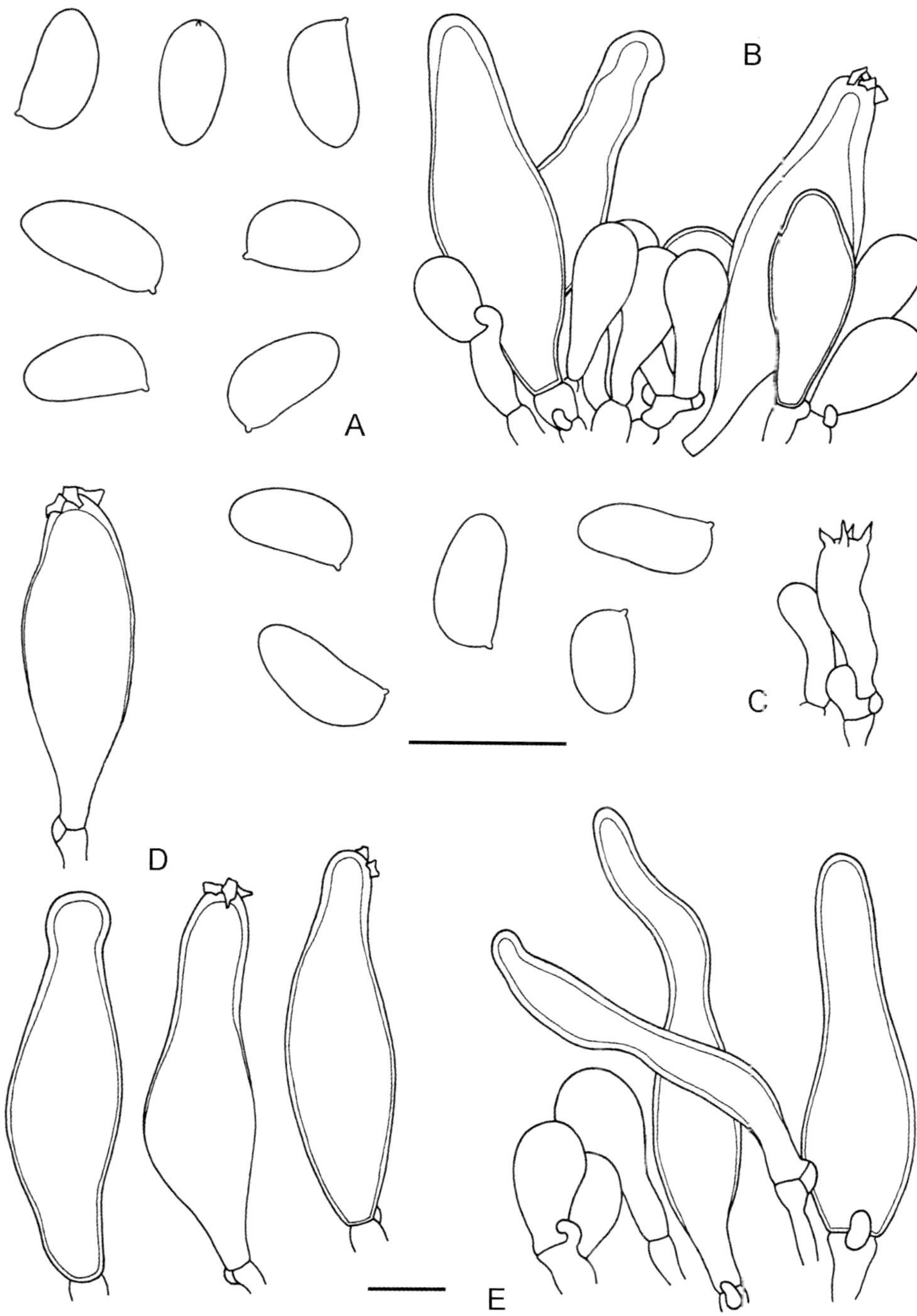

Figure 16. *Inocybe lucifera*, all holotype (PERTH 08517029). **A,** Basidiospores; **B,** Cheilocystidia and paracystidia; **C,** Basidia; **D,** Pleurocystidia; **E,** Caulocystidia and cauloparacystidia. Scale bars = 10 µm (longer bar for spores only).

14. Inocybe luteobulbosa E.Horak, *New Zealand J. Bot.* 15: 725 ('1978') [1977]

Type: Tawa Track, Urewera Natl Park, Gisborne area, North Island, New Zealand, on the ground under *Nothofagus fusca*, 24 June 1968, *E.Horak*; holo: PDD 27129.

Pileus 10–25 mm wide, conical to obtusely conical; umbo absent or small and obtuse; velipellis absent; margin decurved; surface at times viscid but not glutinous, fibrillose, nowhere scaly, not rimose; margin appendiculate, with a narrow marginal flap or with denticulate fibrils; disc fulvous (10YR 5/8 to 5/6; Buckthorn Brown) to reddish brown (5YR 4/4), strong brown (7.5YR 4/4) or tawny; margin light yellowish brown (Honey Yellow) to pale yellow or stramineous (2.5Y 6/4 to 7/4), uniformly coloured when immature; context thin, pallid to isabelline. *Lamellae* adnexed, moderately close, c. 26–30 L, with several tiers of lamellulae; pallid to light grey or pale brown, becoming light yellowish brown to olive-yellow (2.5Y 6/4 to 6/6) or brown (10YR 6/3 to 5/4) with age, ventricose; edges indistinctly fimbriate. *Stipe* 17–37 × 2–4 mm at the apex, base rounded-bulbous or swollen, up to 6 mm wide; surface pruinose over the upper half, fibrillose to finely fibrillose or smooth below, whitish or pallid to light yellowish buff or with a very pale yellowish cast: context solid, becoming hollow, buff, but white in the base, not changing colour when exposed. *Odour* acidulous or not distinctive. Pl. 34, 35.

Basidiospores 7.0–9.0 × 4.5–5.5 µm [mean 8 × 5 µm; Q 1.36–1.84 (–2.00); Q mean 1.60 (n = 35/2)], amygdaliform with bluntly pointed apices, yellowish brown; apiculus small. *Basidia* 26-28 × 7–8 µm, clavate, hyaline, 4-sterigmate. *Pleurocystidia* 47–60 × 13–19 µm, fusiform, thick-walled, the walls 2–3 µm thick, hyaline; apices crystalliferous. *Cheilocystidia* similar to pleurocystidia but often shorter, mixed with paracystidia that resemble basidiole-like cells, these narrowly clavate and hyaline, basidia also detected on gill edge. *Caulocystidia* numerous at the apex, mostly fusiform, thick-walled and with crystalliferous apices but longer than pleurocystidia (to 90 µm long), mixed with cauloparacystidia, hyaline, and at times mixed with cells of intermediate size, shape and wall thickness; descending below the centre of the stipe. *Pileipellis* a cutis of cylindrical hyphae, these mostly 5–12 µm wide, tawny in mass and lightly encrusted.

Occurs on soil in wet sclerophyll forest in Tas. and in *Nothofagus* forest in New Zealand. Fruiting Mar.–May (Australia).

Additional specimens examined: Tas.: Duckhole Lake Trail, near Dover, on soil among bryophytes in wet sclerophyll forest under *Eucalyptus*, *Pomaderris* and *Acacia*, 29 May 2010, *P.B.Matheny PBM3425* (PERTH 08574758, TENN 067034; KP170941-nLSU, KP308765-ITS, KJ811596-*rpb2*); Pipeline Track, Mt Wellington, on soil in gravel along roadside in wet sclerophyll forest under *Pomaderris apetala* and *Eucalyptus obliqua*, 20 Mar. 2011, *G.Gates 1155* (HO 562695; KP170942-nLSU, KP308766-ITS, KJ811597-*rpb2*).

Extralimital specimens examined: New Zealand: North Island, Tongariro Natl Park, Mt Ruapehu, Whakapapa, on soil in *Nothofagus* forest, 27 Apr. 2001, *E.Horak & A.Horak* (PDD 72618; HM060318-ITS); North Island, Kaimanawa Forest Park, Kiko Rd, Lake Taupo district, on soil along track in *Nothofagus* forest, 7 May 2009, *P.B.Matheny PBM3088* (PDD 97897, TENN 063665; KP170943-nLSU, KP308768-ITS, KJ811599-*rpb2*); North Island, Kaimanawa Forest Park towards Tree Trunk Gorge, Kaimanawa, 15 km S of Turangi, on soil under *Nothofagus*, 10 May 2009, *P.B.Matheny PBM3130* (TENN 063872; KP170944-nLSU, KP308769-ITS, KJ811600-*rpb2*); *loc. id.*, under *Nothofagus*, 10 May 2009, *P.B.Matheny PBM3138* (TENN 63707, PDD 97902; KP170945-nLSU, KP308770-ITS, KJ811601-*rpb2*). South Island, Takaka Hill Track, Nelson, on soil under *Nothofagus menziesii*, 17 Apr. 2004, *P.Leonard* (PDD 75276, as *I. cerea*; JQ313573-nLSU, KP308767-ITS, KJ811598-*rpb2*).

Diagnostic features: Fruit bodies medium sized. Pileus generally fulvous to tawny at the centre and yellowish towards the margin; lamellae with olivaceous tones; stipe with a pruinose apex and a swollen to rounded-bulbous base. Spores smooth, amygdaliform; caulocystidia present below the mid-region of the stipe.

This description of *Inocybe luteobulbosa* is based on specimens collected in Tasmania and New Zealand, and because all collections sequenced so far form a monophyletic group (Fig. 8), we recognise them as belonging to a single species. However, some ecological differences between the Australian and New Zealand specimens have been noticed. For example, *I. luteobulbosa* occurs in *Nothofagus* forests in New Zealand, whereas in Tasmania it is found thus far in wet sclerophyll forest under *Eucalyptus*, *Pomaderris* and *Acacia*. Sequences obtained show minor divergence between New Zealand and Australian populations (Fig. 8), but neither forms a monophyletic group.

The New Zealand specimens cited here closely match material (PDD72618) determined by E.Horak as *I. luteobulbosa* in terms of DNA sequence similarity. However, we note several morphological traits from the protologue that are inconsistent with our field observations of New Zealand and Tasmanian material. Our specimens feature a stipe that is not obviously pruinose for its entire length, a white to yellowish buff stipe (not beige, and in no case turning pink or reddish brown), and we did not observe a dense covering of yellow fibrils at the stipe base, a feature that can be missed on older specimens. Pending the availability of further specimens we consider those cited above from both sides of the Tasman Sea to represent a broad morphological species.

Horak (1977) described *I. luteobulbosa* f. *volvata* E.Horak from New Zealand with a volvate stipe base; we have not observed this form in Australia. Furthermore, we have determined that material at PDD mislabelled *Inocybe cerea* E.Horak (PDD75276) represents *I. luteobulbosa* based on morphology and DNA sequence similarity. *Inocybe cerea* differs from *I. luteobulbosa* by the even and non-bulbous stipe and smaller ovate spores (6.0–7.0 × 4.5–5.0 µm).

Phylogenetically, *I. luteobulbosa* is nested in a poorly supported clade containing the *I. australiensis* group and its allies (Fig. 8). Material labelled *Inocybe* cf. *bridgesiana* Singer from Chile is most closely related to *I. luteobulbosa*.

15. **Inocybe meridionalis** Matheny, Bougher & G.M.Gates, *sp. nov.*

MycoBank: MB 809392

Type: Vale of Belvoir, 30 km W of Gowrie Park, Tas., scattered singly on soil in rainforest on track under seedlings of *Nothofagus cunninghamii*, 25 May 2010, *F.Karstedt & P.B.Matheny PBM3413*; holo: PERTH 08517037; iso: TENN 065708 (KF915281-nLSU, KF977204-ITS, KF991377-*rpb2*).

Pileus 10–21 mm wide, obtusely conical to convex, becoming plano-convex, lacking an umbo, not at all distinctive or, occasionally, the margin decurved; surface dry; velipellis absent, squamulose but with a squarrose-scaly disc, coarsely fibrillose towards the margin, fibrils sometimes diverging somewhat to reveal a lighter colour; dark brown or very dark brown or Mummy Brown on the disc and a shade lighter (dark yellowish brown or Snuff Brown) towards the margin; context thin, whitish to pale brown. *Lamellae* subsinuate to adnexed, moderately close, 20–30 L, with several tiers of lamellulae, pale brown to yellowish brown (10YR 6/3 to 5/3), in some cases (e.g. *PBM3390*) olive-yellow (2.5Y 6/6) to light olive-brown (2.5Y 5/6) or isabelline, becoming dark yellowish brown (10YR 4/4) with age, medium, ventricose; edges pallid, indistinctly fimbriate. *Stipe* 18–30 × 1–3 mm at the apex, terete, occasionally flexuous, the base swollen but not bulbous, at most up to 3 mm wide; surface with a pruinose appearance at least over upper third to half, with a weakly developed vesture of brownish fibrils on the lower half, becoming glabrous with age; partial veil fugacious, reddish brown to dusky red (5YR 4/3 to 2.5YR 3/2) or vinaceous brown, at times darker towards the base, the extreme base whitish; context solid, becoming hollow. *Odour* acidulous or not remarkable. Pl. 36, 37.

Basidiospores 6.5–8.5 × 4.0–5.5 µm [mean 7.4 × 4.7 µm; Q (1.30–) 1.36–1.84 (–1.88); Q mean 1.60 (n = 63/3)], smooth, usually subamygdaliform to amygdaliform and with bluntly pointed apices, at times elliptic, more variably shaped in lamella mounts of the type (*PBM3413*), smaller broadly elliptic spores occasionally observed from lamella mounts, yellowish brown; apiculus small. *Basidia* 17–29 × 7–9 µm, clavate, hyaline or becoming yellowish, 4-sterigmate. *Pleurocystidia* 43–55 (–65) × 10–16 µm, fusiform, the apices obtuse and bearing crystals, thick-walled; walls 2–3 µm thick, hyaline. *Cheilocystidia* similar to pleurocystidia but often shorter and more lageniform, at times thin-walled, mixed with clavate to pyriform paracystidia. *Caulocystidia* mixed with cauloparacystidia (these thick-walled at times) along the upper half of the stipe, occurring in numerous scattered clusters, similar to pleurocystidia or cheilocystidia but often larger (70–90 × 15–26 µm) and/or lageniform; younger specimens lacking well-developed caulocystidioid cells, but these can be numerous on other specimens arising as terminal elements of the dark yellowish brown superficial hyphae and variously shaped (cylindrical and flexuous, fusiform, lageniform or clavate), at times bearing apical crystals. *Pileipellis* composed of broad pyramidal fascicles of upturned hyphae, these very dark yellowish brown in mass, mostly 6–15 µm wide, with slightly thickened walls, encrusted; terminal cells not differentiated. *Clamp connections* present. Fig. 18.

Plate 34. *Inocybe luteobulbosa* (TENN 067034). Photograph — P.B.Matheny.

Plate 35. *Inocybe luteobulbosa* (HO 562695). Photograph — G.M.Gates.

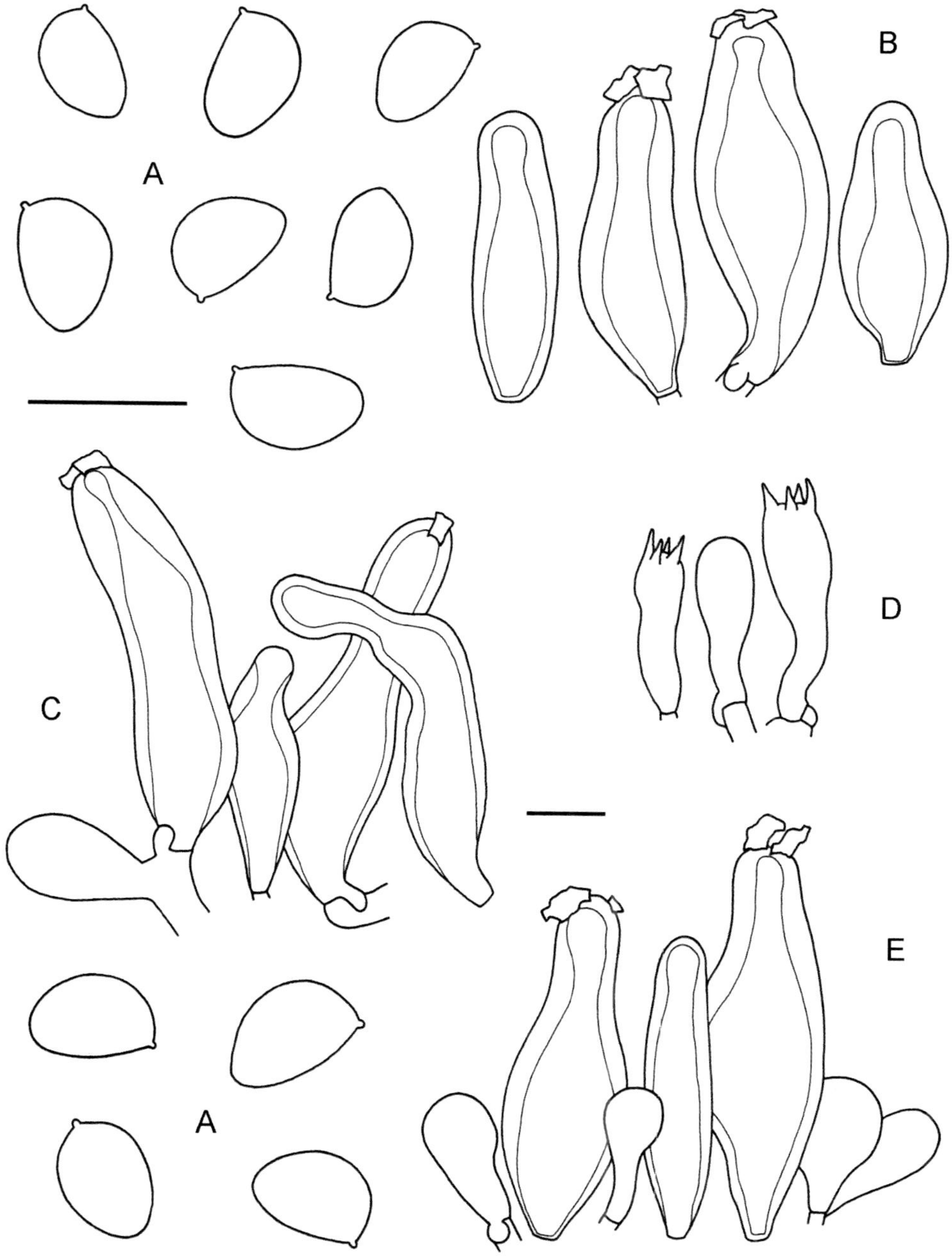

Figure 17. *Inocybe luteobulbosa*, all PERTH 08574758. **A,** Basidiospores; **B,** Pleurocystidia; **C,** Caulocystidia; **D,** Basidia; **E,** Cheilocystidia and paracystidia. Scale bars = 10 µm (longer bar for spores only).

Occurs in soil under *Nothofagus cunninghamii* in cool-temperate rainforest in Tasmania and in wet sclerophyll forest containing *N. cunninghamii*. Fruiting May and June.

Additional specimens examined: Tas.: Blue Tier Forest Reserve, Goblin Track, scattered singly on soil in rainforest under *Nothofagus cunninghamii* and *Leptospermum*, alt. 800 m, 22 May 2010, *F.Karstedt & P.B.Matheny PBM3382* (PERTH 08517045, TENN 065709; KF915284-nLSU, KF977207-ITS, KF991380-*rpb2*); *loc. id.*, same habitat, 22 May 2010, *F.Karstedt & P.B.Matheny PBM3383* (TENN 068202); Blue Tier Forest Reserve, Mount Michael Track, on soil in rainforest under *Nothofagus cunninghamii*, 22 May 2010, *G.Gates, D.Ratkowsky & P.B.Matheny PBM3390* (PERTH 08517053, TENN 065705; KF915282-nLSU, KF977205-ITS, KF991378-*rpb2*); *loc. id.*, same habitat, 22 May 2010, *G.Gates, D.Ratkowsky & P.B.Matheny PBM3391* (TENN 065705; KF915283-nLSU, KF977206-ITS, KF991379-*rpb2*); West Pyengana, Ralphs Falls Track, alt. 800 m, on soil in rainforest under *Nothofagus cunninghamii*, 23 May 2010, *P.B.Matheny PBM3396* (TENN 068203); Mount Field Natl Park, track to Lady Barron Falls, gregarious on ground among bryophytes in wet sclerophyll forest under *Eucalyptus* and *Nothofagus cunninghamii*, 1 June 2010, *G.Gates, D.Ratkowsky & P.B.Matheny PBM3450* (PERTH 08517061, TENN 065707; KF915285-nLSU, KF977208-ITS, KF991381-*rpb2*); *loc. id.*, on soil in wet sclerophyll forest under mixture of *Nothofagus cunninghamii*, *Eucalyptus*, *Pomaderris* and *Acacia*, 1 June 2010, *G.Gates & P.B.Matheny PBM3454* (TENN 068204).

Etymology: From the Latin *meridionalis* (southern), in reference to its occurrence in Tas. in association with Southern Beech (*Nothofagus*).

Diagnostic features: Differs from *Inocybe australiensis* by its association with *Nothofagus cunninghamii* or in mixed vegetation with *Nothofagus* and other ectomycorrhizal plants.

Inocybe meridionalis is morphologically similar to *I. australiensis* Cleland & Cheel, but phylogenetic analyses (Fig. 8) support the separation of multiple lineages within the *I. australiensis* group (see discussion under *I. australiensis*). We apply the name *I. australiensis s. str.* to a widespread lineage that occurs in habitats with eucalypts or other Myrtaceae, but not with *Nothofagus*. We also note that Cleland & Cheel (1918) described *I. australiensis* from a suburb of Sydney, New South Wales, a locality likely lacking *Nothofagus*. Phylogenetic analyses support the recognition of *I. meridionalis* for fungi morphologically similar to *I. australiensis* but occurring under *Nothofagus*. Morphologically, hymenial cystidia of the type of *I. meridionalis* tend to be somewhat shorter (mostly 43–55 × 10–16 μm) than in *I. australiensis* (48–75 × 10–15 μm). An anomalous collection from eucalypt forest in New South Wales (e.g. *PBM3189*, TENN 063642) could be assigned to *I. australiensis*. However, its sequences place it as the sister lineage to *I. meridionalis* and an unclarified lineage from New Zealand with strong statistical support. Another sclerophyll-associated collection from New South Wales (*PBM3190*, TENN 063643) exhibits longer spores and some 2-sterigmate basidia. It is hoped that the availability of further collections will help to resolve the relationship of *I. australiensis* and *I. meridionalis*.

Some variants of *I. meridionalis* feature yellowish or yellow-olive-tinged immature lamellae, whereas in others such lamellae are whitish (TENN 065705; Pl. 37). However, molecular analyses do not support their taxonomic separation.

A photograph of *I. meridionalis* is included under this name in Gates & Ratkowsky (2014).

16. Inocybe rufobadia Matheny & Bougher, *sp. nov.*

MycoBank: MB 809453

Type: Corrigin Wildflower Drive, 5 km W of Corrigin, W.A., 32°20'15.1"S, 117°49'22.7"E, in yellow sand under *Allocasuarina campestris* and heath plants (no eucalypts present), 27 July 2011, *N.L.Bougher NLB885*; holo: PERTH 08320454; iso: TENN 066725 (KF915290-nLSU, KF977213-ITS, KF991385-*rpb2*).

Pileus 7–25 mm wide, broadly convex or hemispherical when immature, becoming broadly convex to plano-convex or applanate, at times with a low broad umbo or becoming depressed in the centre; margin decurved; velipellis absent or occasionally present; surface with fibrillose squamules at the centre with recurved tips or matted-scaly, coarsely fibrillose towards the margin, predominantly bright rusty brown or vivid reddish brown (6E8 to 6D8 or hazel; Antique Brown to Sudan Brown); context firm, tinged yellow or white, up to 3 mm thick under the disc. *Lamellae* ascending, adnate to adnexed, or adnate with a slight decurrent tooth, close, 30–40 L, with several tiers of lamellulae; pale yellow (4A5) to yellowish when young, light yellowish

Plate 36. *Inocybe meridionalis*, holotype (PERTH 08517037). Photograph — P.B.Matheny.

Plate 37. *Inocybe meridionalis* (TENN 065705). Photograph — P.B.Matheny.

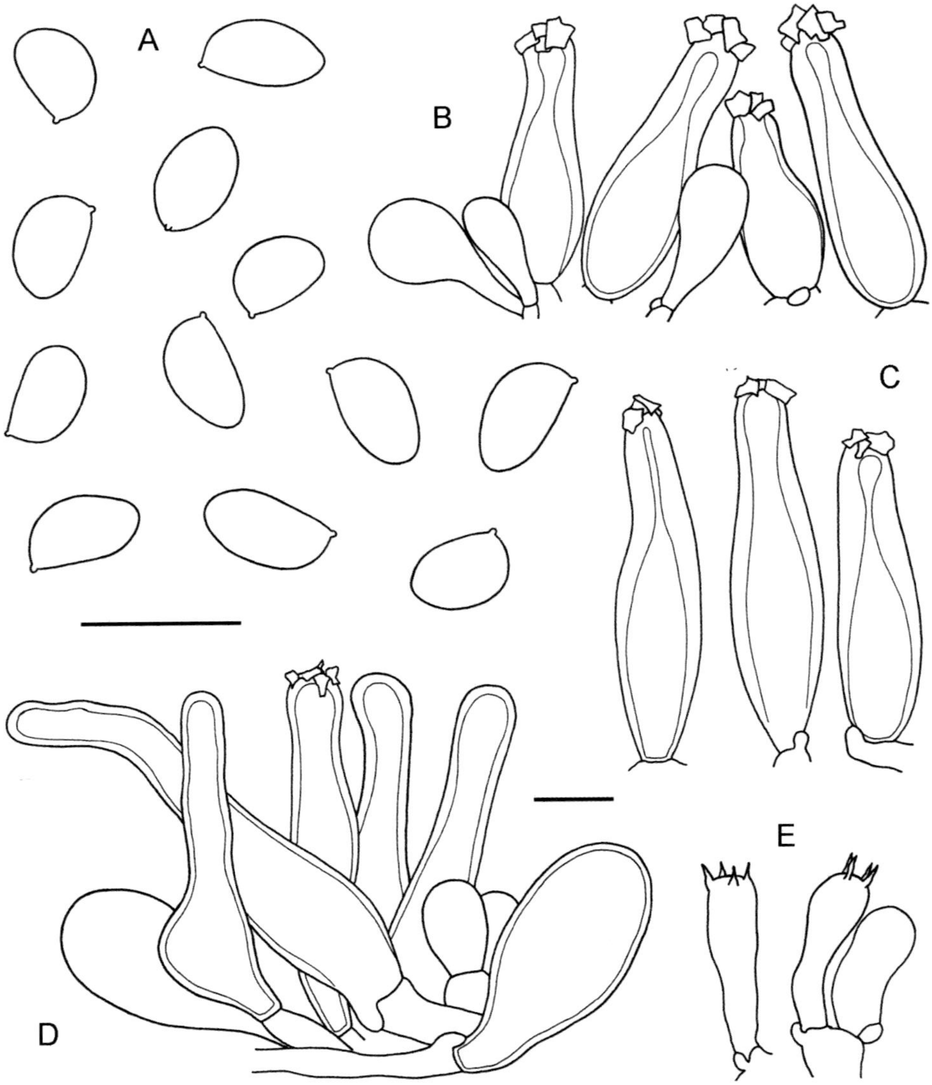

Figure 18. *Inocybe meridionalis*, holotype (PERTH 08517037). **A,** Basidiospores; **B,** Cheilocystidia and paracystidia; **C,** Pleurocystidia; **D,** Caulocystidia and cauloparacystidia; **E,** Basidia. Scale bars = 10 μm (longer bar for spores only).

brown to olive-yellow (2.5Y 7/4, 6/4, 6/6), becoming dark yellowish brown or Dresden Brown, broad, up to 3 mm deep; edges not distinctly pallid-fimbriate. *Stipe* 12–35 × 1.5–4.0 mm, even, terete, with or without a slightly swollen base, at least the upper half weakly pruinose under a hand lens (with a slight satiny appearance), pruinose the entire length, or with minute rusty-coloured scales in contrast to the yellow ground colour, with rusty or hazel-coloured fibrils just above the stipe base; pale yellow, yellow to bright yellow (3A6; 2.5Y 7/6 to 7/4; Chamois), or light yellowish brown (2.5Y 7/4 to 6/4), in age yellowish brown or near Buckthorn Brown (10YR 5/8); context becoming hollow, yellow or pallid. *Odour* none. Pl. 38, 39.

Basidiospores 8.0–10.0 × 4.0–5.0 µm [mean 8.9 × 4.6 µm; Q 1.70–2.18 (-2.25); Q mean 1.94 (*n* = 60/3)], smooth, mostly cylindrical, narrow, oblong-amygdaliform, less often elliptic, some with a slight adaxial depression, with obtuse or bluntly pointed apices, yellowish brown; apiculus small and often not very distinctive. *Basidia* 29–40 × 8–10 µm, narrowly clavate to clavate, hyaline or ochraceous, 4-sterigmate. *Pleurocystidia* 51–72 × 10–17 µm, fusiform to (sub)cylindrical, the apices obtuse, some bearing crystals, thick-walled, walls 1–3 µm thick, occasionally only thick-walled apically, hyaline to pale yellow. *Cheilocystidia* similar to pleurocystidia but often shorter, hyaline to ochraceous, mixed with paracystidia. *Caulocystidia* descending below the stipe centre, mainly clavate (29–43 × 15–19 µm), mixed also with fusiform cells, typically thick-walled, the walls 1.5–3.0 µm thick, occasionally thin-walled, hyaline to ochraceous, numerous, mixed with cauloparacystidia. *Pileipellis* a cutis of cylindrical hyphae, these bright rust-coloured in mass, mostly 6–19 µm wide, encrusted, thin-walled or with slightly thickened walls. *Clamp connections* present. Fig. 19.

Occurs in sand under *Allocasuarina* and *Acacia* in the wheatbelt region of W.A. Fruiting July.

Additional specimens examined: W.A.: Kunjin, Vinegar Well, off Barber Rd, 32°23'3 .7"S, 117°40'06.5"E, scattered singly on sandy soil under *Allocasuarina* and *Acacia acuminata* (no eucalypts present), 27 July 2011, *P.B.Matheny PBM3685* (PERTH 08319006, TENN 066496); *loc. id.*, same habitat, 27 July 2011, *P.B.Matheny PBM3694* (PERTH 08318905, TENN 066467; KF915291-nLSU, KF977214-ITS, KF991386-*rpb2*).

Etymology: From the Latin *rufobadia* (red bay-brown), in reference to the bright reddish brown or rusty brown pileus colour.

Diagnostic features: Pileus bright rusty brown or reddish brown. Lamellae yellowish when young. Stipe bright to pale yellow with minute but conspicuous rust-brown scales with hazel or rusty fibrils above the stipe base. Spores narrow, mostly cylindrical. Caulocystidia mostly clavate. In sand under *Allocasuarina* and *Acacia*.

Inocybe rufobadia is a striking new species from the Western Australian wheatbelt region where it occurs in sandy soils in association with *Allocasuarina* and *Acacia*, and in the absence of eucalypts. In addition to the vivid reddish brown or rusty brown pileus and the yellow stipe with rusty brown scales, it also has distinctively narrow and mostly cylindrical spores and the presence of clavate caulocystidia. Hazel or rusty fibrils are also present above the stipe base at least on younger specimens.

Two other species of *Inocybe* from southern Western Australia are somewhat similar to *I. rufobadia* in having strikingly bright reddish or golden brown pilei. One of these, *I. spadicea* differs from *I. rufobadia* by its larger amygdaliform spores (11.5–12.5 × 5.0–6.0 µm), and its pileus is not as vivid reddish brown. The second species, *I. aureorimosa*, has a golden brown to yellowish brown pileus, larger spores (mean c. 10 × 6 µm) and larger cheilocystidia and caulocystidia. Moreover, *I. spadicea* and *I. aureorimosa* have been found near eucalypts in woodland and forest, unlike *I. rufobadia* which occurs in shrubland and heath vegetation in the absence of eucalypts. Phylogenetically, all three species are rather distantly related (Fig. 8).

Inocybe brevicula, collected under eucalypts in the south-west of Western Australia, is similar to *I. rufobadia*, but it has dark brown pileus, squat stature, and a short, minutely pruinose, dark brown stipe. Similarly, *I. campestris* differs primarily in its dark brown colouration.

Plate 38. *Inocybe rufobadia*, holotype (PERTH 08320454). Photograph — N.L.Bougher.

Plate 39. *Inocybe rufobadia* (PERTH 08318905). Photograph — P.B.Matheny.

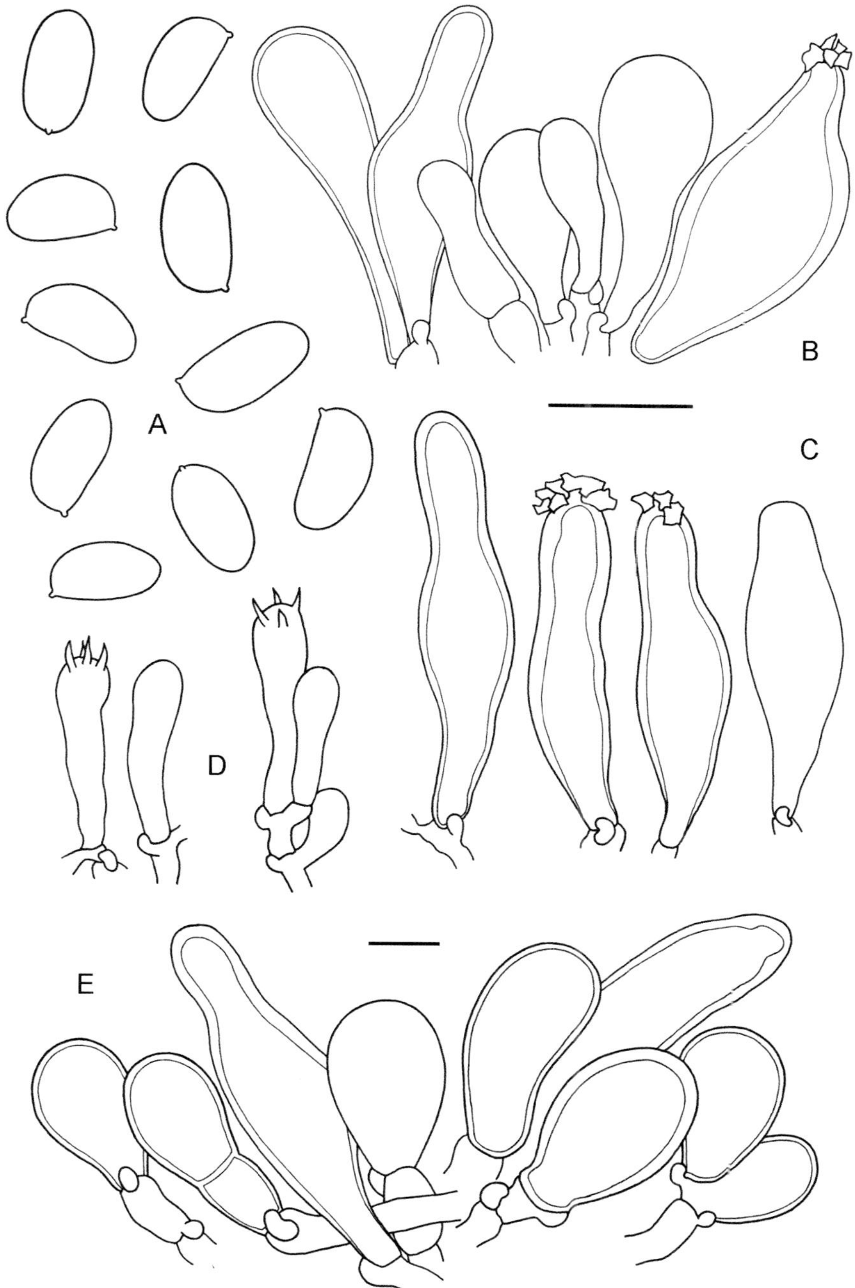

Figure 19. *Inocybe rufobadia*, holotype (PERTH 08320454). **A,** Basidiospores; **B,** Cheilocystidia and paracystidia; **C,** Pleurocystidia; **D,** Basidia; **E,** Caulocystidia and cauloparacystidia. Scale bars = 10 μm (longer bar for spores only).

INOCYBE FIBRILLOSIBRUNNEA GROUP

This is a moderately well supported clade containing 16 known species from Australia. *Inocybe froudistii* is the most widespread and common species, but *I. fuscosquarrosa* is common in Tas. Many species in the group are currently known only from southwestern W.A. These include *I. gomphocephala*, *I. fibrillosibrunnea*, *I. floccosibrunnea*, *I. farinosipes*, *I. memorialis*, *I. bicornis*, *I. bicornis* f. *secotioides*, *I. exigens*, and *I. clypeata*. Several species are known only from single collections or await further documentation. Most have medium-sized fruit bodies and are characterized by a squamulose pileus surface, presence of a partial veil, and have a furfuraceous-fibrillose to floccose-fibrillose stipe surface. Exceptions, however, are known (e.g. the densely pruinose stipe of *I. farinosipes*; the densely fibrillose pileus surface of *I. clypeata*).

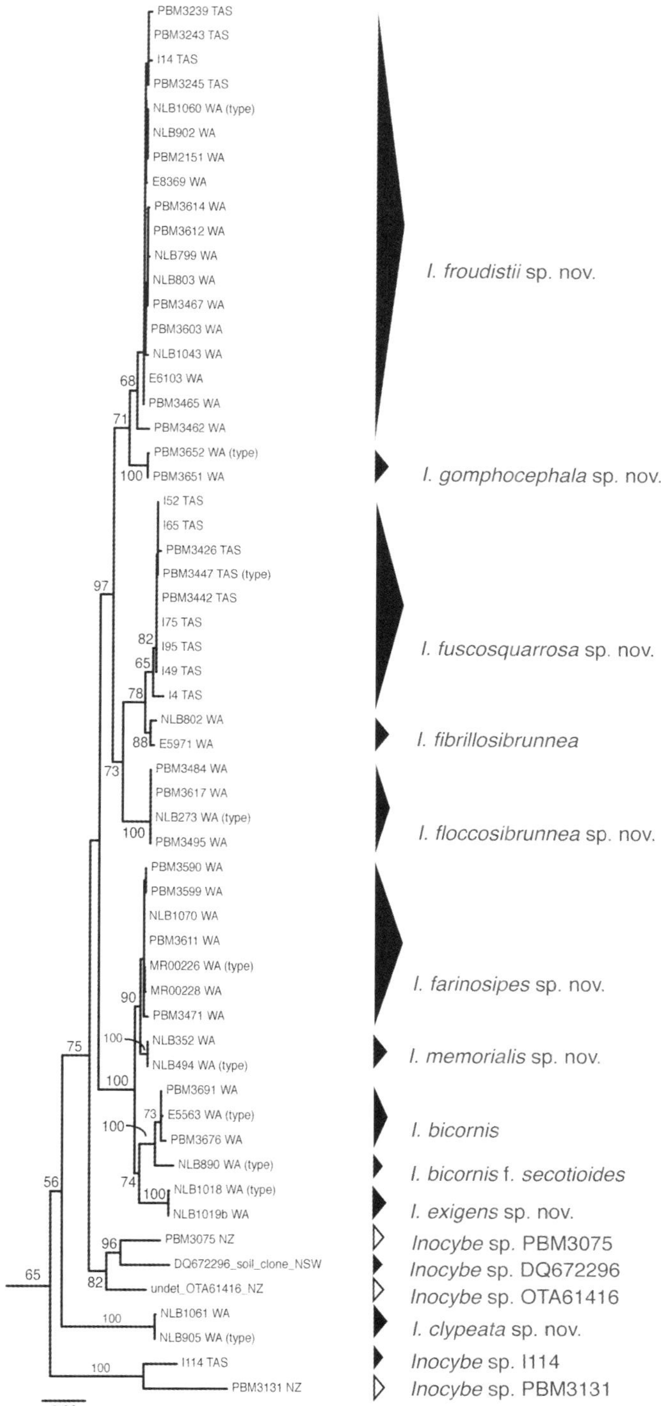

Figure 20. Maximum Likelihood phylogram of the *Inocybe fibrillosibrunnea* group produced from the analysis of the STAC group shown in Fig. 7. Australian taxa are indicated by black-filled triangles. White triangles indicate New Zealand taxa. Bootstrap values are shown above the branches.

17. Inocybe bicornis f. **bicornis** Braaten, Bougher & Matheny, *in* C.C.Braaten, P.B.Matheny, D.L.Viess, M.G.Wood, J.H.Williams & N.L.Bougher, *Botany* 92: 19 (2014)

Type: Maurice Barnes' paddock, near Letchford, N of Kellerberrin, W.A., on the ground under *Eucalyptus wandoo* in revegetation site, 20 July 1995, *W.Dunstan E5563*; holo: PERTH 07605684; iso: TENN 068364 (EU555447-nLSU, KC305486-ITS, KC305448-*rpb2*).

Pileus 14–25 mm wide, convex to broadly convex when immature, becoming plane, lacking an umbo; velipellis occasionally present over the disc; margin decurved, often undulate on older specimens; surface shaggy to coarsely fibrillose, at times with appressed fibrillose squamules, fibrillose towards the margin where the fibrils diverge before maturity, brownish yellow or fulvous to yellowish brown or brown (near 6D6; Buckthorn Brown to Russet); context white or whitish, up to 4 mm thick under the disc, not changing colour where exposed. *Lamellae* adnexed to uncinate, moderately close to close with 32–40 L and several tiers of lamellulae, yellowish brown or brown (10YR 5/3 to 6/3; Avellaneous or 6C4 to Wood Brown) when immature, becoming dark yellowish brown with age; edges obscurely pallid-fimbriate or appearing concolorous with the faces; ventricose, medium, up to 3 mm deep. *Stipe* 15–35 × 3–6 mm, terete, occasionally squat when under soil but elongating by maturity; base swollen or weakly bulbous but not distinctly marginate; partial veil not observed but presumably present due to floccose fibrils above the stipe base; surface pruinose at the apex or at least visibly so towards the centre, at times appearing furfuraceous, lower part with scattered dense clusters of whitish floccules or fibrils, these more densely arranged above the base; light brown (Cinnamon Buff to Tawny-Olive); context dull-coloured or brownish to reddish brown, white in the slightly bulbous base. *Odour* none. Pl. 40–42.

Basidiospores 10.0–15.5 (–16.0) × 5.5–7.5 µm [mean 12.6 × 6.3 µm; Q 1.67–2.41 (–2.58); Q mean 2.01 (n = 42/3)], smooth, very variable in shape (oblong-amygdaliform to cylindrical with bluntly pointed to obtuse apices, at times reniform due to a kinked or depressed ventral side, less often subangular or sinuous in outline), apiculus at times quite prominent, yellowish brown. *Basidia* 31–41 × 9–11 µm, clavate, hyaline, 2-sterigmate. *Pleurocystidia* 62–92 × 11–15 (–19) µm, narrowly fusiform to narrowly lageniform, at times nearly cylindrical and occasionally shorter and ventricose to subclavate; apices obtuse, swollen or sometimes indistinctly subcapitate; bases narrow-pedicellate or truncate, crystalliferous or bare; walls 1–3 µm thick, hyaline. *Cheilocystidia* similar to pleurocystidia or shorter and more variable in shape, mixed with abundant hyaline paracystidia and cystidioid cells of variable size and form (clavate, ventricose or cylindrical). Many cylindrical cells multi-septate and arising from a long narrow subtending hypha, extending at least to the same level as the cheilocystidia (sometimes beyond). *Caulocystidia* 45–141 × 12–18 µm, very variable in shape, at the apex similar to pleurocystidia, but also flexuous or sublageniform to cylindrical, thin- or thick-walled; at apex and centre of the stipe mixed with shorter thin-walled hyaline cystidioid cells of various shapes (e.g. clavate, subcylindrical), these dissimilar to sterile elements on the lamellar edge, end cells undifferentiated or occasionally fusiform above the base. *Pileipellis* a cutis of lightly encrusted hyphae, these mostly cylindrical, up to 11 µm wide, yellowish brown in mass. *Clamp connections* present. Fig. 21.

Occurs on sandy soil or completely hidden in sandy soil under planted or revegetated sites with *Eucalyptus wandoo*, *Eucalyptus* sp. and/or *Allocasuarina* in south-western W.A. Fruiting July.

Additional specimens examined: W.A.: Old Kunjin Railway, 14 km W of Corrigin, 32°21'25.9"S, 117°44'17.1"E, emerging from compact coarse sandy soil or pushing up plaques of wet sand under *Allocasuarina* (Black Tamma), 26 July 2011, *P.B.Matheny PBM3676* (PERTH 08319103, TENN 066454; KC305409-nLSU, KC305487-ITS, KC305448-rpb2); *loc. id.*, under *Allocasuarina*, 26 July 2011, *P.B.Matheny PBM3680* (PERTH 08319057, TENN 066488); *loc. id.*, scattered completely under sandy soil under planted *Eucalyptus* and *Allocasuarina*, 26 July 2011, *N.L.Bougher NLB873* (PERTH 08320586, TENN 066711); Corrigin Wildflower Drive, 5 km W of Corrigin, 32°20'19.1"S, 117°40'20.0"E, scattered singly on yellow sand under *Allocasuarina*, 27 July 2011, *P.B.Matheny PBM3691* (PERTH 08318948, TENN 066640; KJ756442-nLSU, KJ756477-ITS, KJ729952-rpb2).

Diagnostic features: Forms a monophyletic group with *Inocybe bicornis* f. *secotioides* but differs in the agaricoid habit and exclusively 2-sterigmate basidia.

Inocybe bicornis is rather similar to *I. sinuospora* and *I. emergens* which co-occur in the wheat-belt region of south-western Western Australia, and also *I. farinosipes* from Jarrah forests in the south of the state. However, all three of those species have 4-sterigmate basidia. Moreover, *I. emergens* has nodulose spores and is closely related to *I. sinuospora*, both of which are phylogenetically quite distant from *I. bicornis* (Braaten *et al.*, 2014). *Inocybe farinosipes* is the sister taxon to *I. bicornis* (Fig. 20) but, in addition to its 4-sterigmate basidia, it also has smaller spores.

Exclusively 2-sterigmate basidia are extremely rare in the Inocybaceae (Kuyper, 1986), *I. bicornis* being the only example in the Australian mycota. *Inocybe aberrans* (E.Horak) Garrido, described from *Nothofagus* forest in Papua New Guinea (Horak, 1979a), has 2-sterigmate basidia but is readily distinguished by its large stellate spores and slender scaly habit.

The secotioid *I. bicornis* f. *secotioides* shares very similar LSU and ITS sequences with agaricoid forms of *I. bicornis* (Fig. 20). It has 2- and 4- sterigmate basidia, clavate-sphaeropedunculate hymenial cystidia, and differently shaped spores (see below). Both the secotioid and agaricoid forms of *I. bicornis* have been collected under *Allocasuarina* in the wheat-belt region of Western Australia.

18. Inocybe bicornis f. **secotioides** Braaten, Bougher & Matheny, *in* C.C.Braaten, P.B.Matheny, D.L.Viess, M.G.Wood, J.H.Williams & N.L.Bougher, *Botany* 92: 20 (2014)

Type: Vinegar Well, off Barber Rd, Kunjin, W of Corrigin, W.A., 32°23'31.7"S, 117°40'06.5"E, under cracks in sand or remaining hidden beneath or barely emerging above the surface of the sand, near *Allocasuarina huegeliana* (no eucalypts nearby), 27 July 2011, *P.B.Matheny & N.L.Bougher NLB890*; holo: PERTH 08320403; iso: TENN 066804 (KC305410-nLSU, KC305488-ITS, KC305447-*rpb2*).

Pileus 7–12 mm wide, the margin appressed and clasping the stipe, convoluted button-shaped with strongly incurved margins, showing no evidence of a veil; surface with some appressed radial fibrils; colour dull yellowish (near 5C5) to yellowish brown and mixed with paler areas; context isabelline. *Lamellae* a convoluted lamella-like gleba, pale avellaneous to yellowish brown. *Stipe* 12–15 × 5 mm, even but slightly swollen at the base, with some pruinosity below the insertion of the pileal margin, otherwise finely fibrillose, pallid or almost white; context solid. *Odour* none. Pl. 43 (see also Pl. 1D).

Basidiospores 7.5–13.5 × 5.0–7.5 (–8.4) µm [mean 10.5 × 5.9 µm; Q 1.46–2.43; Q mean 1.80 (n = 33/1)], oblong, subcylindrical, lacrymoid, ovoid or broadly elliptic, smooth, yellowish brown, with an oblique apiculus, highly variable in size, shape, wall thickness and appearance of the cytoplasm (granular to hyaline, with or without a large guttule); apiculus variable, a truncate peg up to 1.2 µm long or shorter, broader and obtuse. *Basidia* 17–26 × 6–8 µm, clavate to cylindrical, hyaline, 2- or 4- sterigmate, ballistosporic. *Hymenial cystidia* 31–60 (–65) × 18–24 (–29) µm, broadly clavate to sphaeropedunculate with a long narrow pedicel, thick-walled, the walls mostly 2–3 µm thick, yellowish brown, infrequent and unevenly distributed. *Caulocystidia* occasionally present on the upper stipe among small scattered patches of fertile hymenium. *Pileipellis* a cutis of loosely interwoven septate hyphae mostly 8–11 µm wide, slightly pigmented pale yellow. *Clamp connections* present. Fig. 22 (see also Pl. 7J).

Plate 40. *Inocybe bicornis*, holotype (PERTH 07605684). Photograph — N.L.Bougher.

Plate 41. *Inocybe bicornis* f. *bicornis* (PERTH 08319103). Photograph — P.B.Matheny.

Plate 42. *Inocybe bicornis* f. *bicornis* (PERTH 08318948). Photograph — P.B.Matheny.

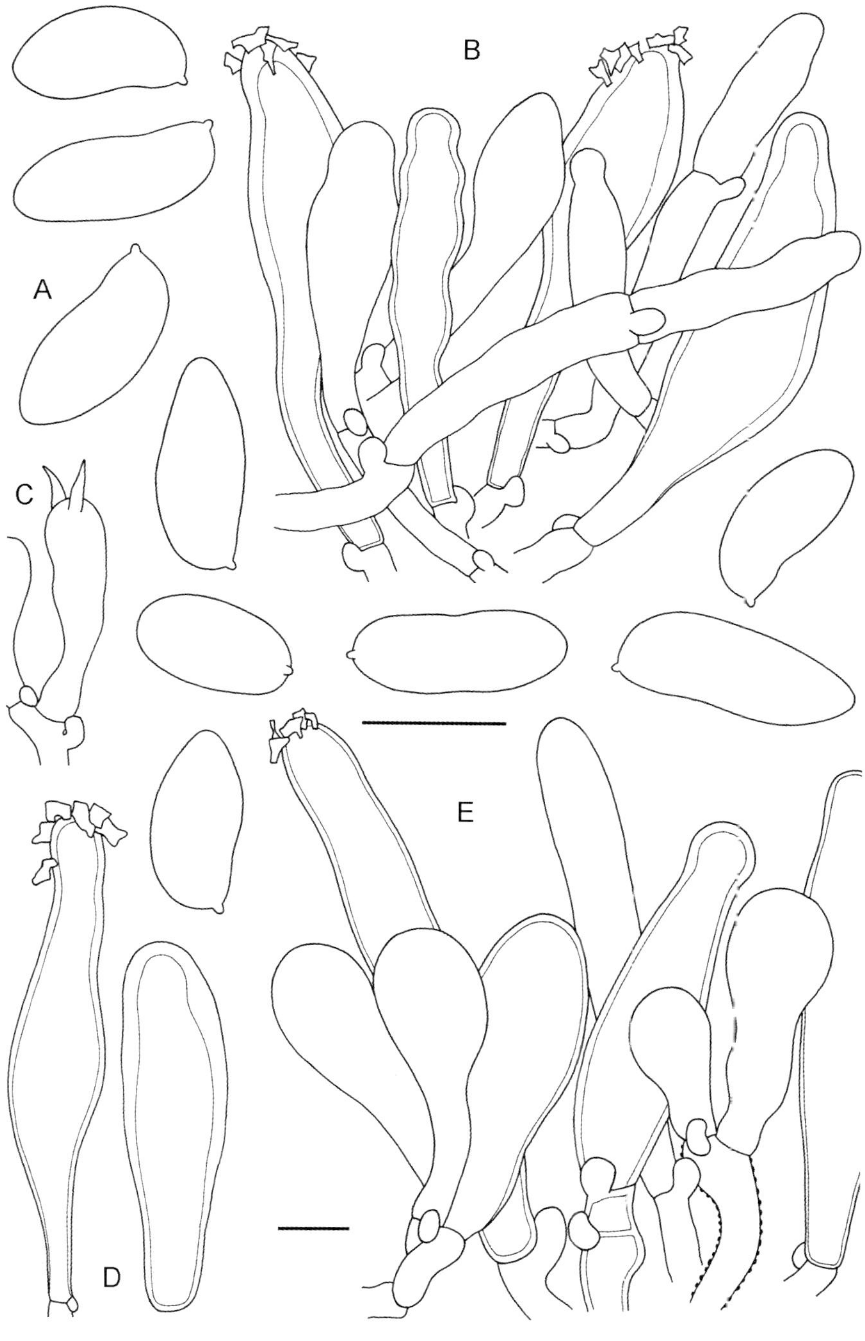

Figure 21. *Inocybe bicornis*, holotype (PERTH 07605684). **A,** Basidiospores; **B,** Cheilocystidia and paracystidia/cystidioid cells; **C,** Basidia; **D,** Pleurocystidia; **E,** Caulocystidia and cauloparacystidia/cystidioid cells. Scale bars = 10 μm (longer bar for spores only).

Known only from the type locality in south-western W.A. Occurs below cracks in sand on a track (hypogeous), remaining hidden or barely emerging above the surface, near *Allocasuarina huegeliana* with no eucalypts nearby. Fruiting July.

Diagnostic features: Forms a monophyletic clade with *I. bicornis* but differs in having a secotioid habit and some 4-sterigmate basidia.

Molecular data from two loci confirm the taxonomic affinity of *I. bicornis* f. *secotioides* and agaricoid forms of *I. bicornis* (Fig. 20; Braaten *et al.*, 2014). We suspect secotioid individuals are likely to have struggled to emerge from compacted sand along a track, resulting in the compressed, secotioid habit. Nevertheless, formal taxonomic distinction is appropriate, the secotioid form of *I. bicornis* differing from the agaricoid form in having both 2-sterigmate and 4-sterigmate basidia, clavate-sphaeropedunculate hymenial cystidia, and spores of a different shape (Braaten *et al.*, 2014). Furthermore, *I. bicornis* f. *secotioides* exhibits much greater variation in spore size and shape, perhaps a result of the deformed lamellae.

A second secotioid form of *Inocybe* attributable to *I. nudiuscula* has been recorded from Victoria (J.Trappe, unpublished). More data are required to determine their morphological differences, but phylogenetic analysis of DNA sequences clearly sets them apart.

Plate 43. *Inocybe bicornis* f. *secotioides*, holotype (PERTH 08320403). Photograph — N.L.Bougher.

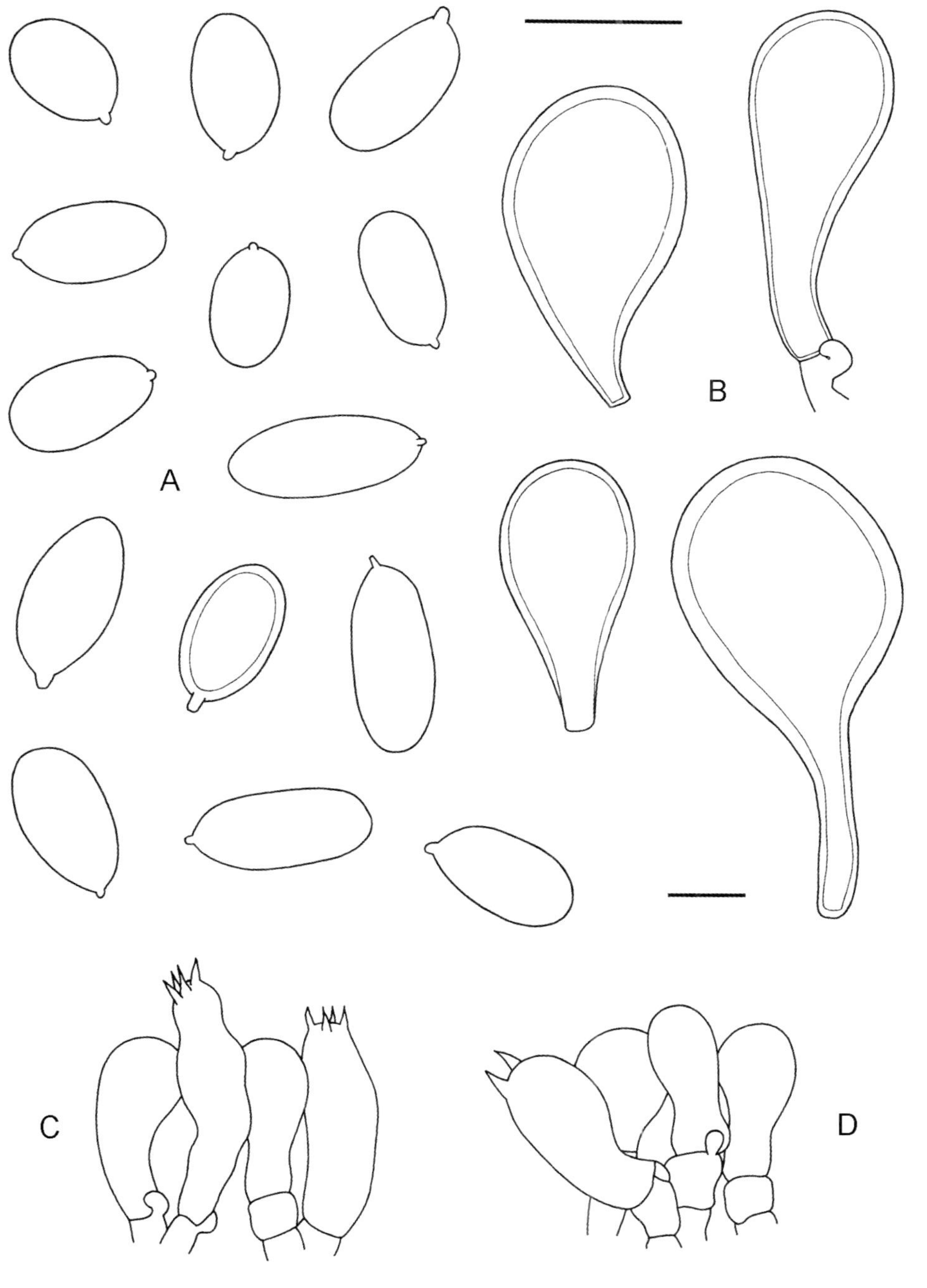

Figure 22. *Inocybe bicornis* f. *secotioides*, holotype (PERTH 08320403). **A,** Basidiospores; **B,** Hymenial cystidia; **C,** Hymenium with 4-sterigmate basidia; **D,** Hymenium with 2-sterigmate basidia.
Scale bars = 10 μm (longer bar for spores only).

19. Inocybe clypeata Matheny & Bougher, *sp. nov.*

MycoBank: MB 809317

Type: N of path, Bold Park, eastern slope of Reabold Hill, Perth, W.A., 31°56'18.0"S, 115°46'48.7"E, on ground among litter over limestone rock in shrubland about 3 m tall, under *Acacia* spp., *Templetonia retusa*, *Melaleuca huegelii*, *Spyridium globulosum* and scattered *Eucalyptus gomphocephala*, 4 Aug. 2011, *N.L.Bougher NLB905*; holo: PERTH 08320268; iso: TENN 066812 (KF853405-nLSU, KF871781-ITS, KF891958-*rpb2*).

Pileus 20–30 mm wide, convex and often remaining so to maturity, eventually applanate or with an upturned margin that is barely decurved; surface smooth to densely matted-fibrillose at the centre, elsewhere densely and coarsely radially fibrillose, with imbricate bundles of fibrils becoming separate (rimose) only at the extreme margin of the pileus, becoming ragged/split with age; overall colour pale brown (near 6D5) at first due to prominence of pale and silky fibrils when young, becoming darker brown (near 6F8) when mature, the margin mostly concolorous with pileus or paler at the extreme margin; context white. *Lamellae* adnexed, closely spaced, c. 28 L, with several tiers of lamellulae, very pale yellowish brown (near 5B3) when young, maturing to dull brown (near 6D5), up to 8 mm deep (very deep considering pileus width), not ventricose; edges minutely fimbriate but not distinctly paler than the face. *Stipe* 20-35 × 4–5 mm, cylindrical or slightly widening towards the base but without a swollen or marginate bulb, lowest part of stipe covered with sand and soil; surface densely pruinose or branny-furfuraceous almost its entire length, densely fibrillose above the base; overall colour pinkish brown (near 9D4) but can be faint in some specimens; context dull, white at the base, solid or hollow near the apex. *Odour* not distinctive. Pl. 44, 45.

Basidiospores 10.0–11.5 × 5.5–6.5 µm [mean 10.5 × 6.0 µm; Q 1.54–1.91; Q mean 1.75 (n = 31/2)], smooth, amygdaliform to broadly so, with pointed to bluntly pointed apices, at times with a suprahilar depression, yellowish brown; apiculus not conspicuous. *Basidia* 31–34 × 9–10 µm, clavate, hyaline, 4-sterigmate. *Pleurocystidia* 70–91 × 15–19 µm, fusiform-ventricose, numerous, often with a tapered or well-developed neck and a narrow basal pedicel up to 25–30 µm long, thick-walled, the walls 1.5–3.0 µm thick, hyaline to pale yellow; apices obtuse, crystalliferous. *Cheilocystidia* shorter than pleurocystidia, mostly lacking well-developed necks, thin- to thick-walled, hyaline or yellow, mixed with paracystidia. *Caulocystidia* consisting of mixture of metuloids and variously shaped thin-walled cystidioid cells, in dense clusters, found above and below the stipe centre; metuloids similar to pleurocystidia; caulocystidioid cells short-clavate, clavate, sphaeropedunculate, saccate, ventricose, fusiform, or cylindrical, these mostly 35–72 × 14–31 µm; metuloids rare at the apex on immature specimens, common and readily observed in older material, and thus may represent caulocystidioid cells. *Pileipellis* a cutis of cylindrical hyphae mostly 5–12 µm wide, light yellowish brown in mass, thin-walled, at most weakly encrusted. *Clamp connections* present. Fig. 23.

Known only from the type locality in western Perth, W.A.; in shallow soil over limestone among various shrubs including *Acacia*, *Templetonia*, *Melaleuca* and scattered trees of *Eucalyptus gomphocephala*. Fruiting July and Aug.

Additional specimens examined: W.A.: type locality, 4 Aug. 2011, *N.L.Bougher NLB904* (PERTH 08320276, TENN 066811); *loc. id.*, 12 July 2012, *N.L.Bougher NLB1061* (PERTH 08363706; KF853406-nLSU, KF871782-ITS, KF891959-*rpb2*).

Etymology: From the Latin *clypeatus* (shield-like), in reference to the often convex mature pileus.

Diagnostic features: Pileus often convex at maturity, the surface densely fibrillose. Stipe pruinose to branny-furfuraceous almost its entire length, pinkish in some specimens. Spores smooth, amygdaliform, some adaxially depressed. Caulocystidioid cells present for the entire length of the stipe.

Inocybe clypeata is described as new based on collections made on limestone soil in shrubland in Perth, Western Australia. To date the species has only been found where fabaceous shrubs (*Acacia* and *Templetonia*) are mixed with scattered trees of *Eucalyptus*. It can be distinguished from others in the greater *I. fibrillosibrunnea* group by the convex, densely fibrillose pileus, pruinose or branny-furfuraceous stipe (due to numerous, densely arranged caulocystidioid cells), and its unique phylogenetic placement (Fig. 20).

Macroscopically, *I. leucotaenia*, which occurs with *I. clypeata* at the same locality (Bold Park), can be confused with *I. clypeata* in wet conditions when the distinctive white pileal margin of *I. leucotaenia* might not be conspicuous. However, pleurocystidia are less abundant in *I. leucotaenia* and long pedicels are less common. Also, its spores are shorter (mean 9.1 × 5.8 µm) than those of *I. clypeata* and not adaxially depressed. *Inocybe clypeata* and *I. leucotaenia* are well separated phylogenetically.

20. Inocybe exigens Matheny & Bougher, *sp. nov.*

MycoBank: MB 809322

Type: Two Peoples Bay Nature Reserve, sand track parallel to Little Beach Rd, W.A., 34°58'31.3"S, 118°11'04.6"E, in sand on the track in low woodland of *Agonis flexuosa*, *Allocasuarina* and myrtaceous shrubs, 20 June 2012, *N.L.Bougher NLB1018*; holo: PERTH 08383367; iso: TENN 068647 (KJ801164-nLSU, KJ778854-ITS, KJ756426-*rpb2*).

Pileus 7–15 mm wide, convex, lacking an umbo, flatter with age; margin strongly incurved when young, with sparse hyaline whitish evanescent veil remnants; surface dry, recurved-squamulose at and near the central disc, fibrillose only near the margin, scales becoming appressed over the centre with age; entirely dark reddish brown (near 5YR 3/2–3/3) overall, tips of squamules light grey; context paler than in the stipe. *Lamellae* adnexed to sinuate, close, with several tiers of lamellulae, to 3 mm deep, ventricose, edges paler than face and conspicuously fimbriate (umber 7.5YR 4/4). *Stipe* 12–25 × 2–4 mm, even, terete, without a swollen base; surface pruinose or almost furfuraceous to at least half way down the stipe, sparsely and finely fibrillose below; dull reddish brown (near 2.5 YR 2/4), paler towards the apex; context solid, drab creamish, or dulling dark reddish after being cut. *Odour* none. Pl. 46, 47.

Basidiospores (9.5–) 10.0–13.0 × 4.5–6.0 µm [mean 11.3 × 5.2 µm; Q (1.82–) 1.87–2.44; Q mean 2.17 (n = 43/2)], smooth, cylindrical (less often oblong-amygdaliform) with obtuse or bluntly pointed apices, occasionally with a pronounced adaxial depression, yellowish brown; apiculus small. *Basidia* 28–34 × 8–10 µm, 4-sterigmate, clavate, hyaline. *Pleurocystidia* 65–95 × 10–15 µm, cylindrical to narrowly fusiform, at times with an undulate outline, with obtuse or indistinctly subcapitate apices often crystalliferous, thick-walled (1.0–2.5 µm thick), hyaline. *Cheilocystidia* often shorter than pleurocystidia, fusiform to ventricose, often with an undulate outline, crowded with variously shaped cystidioid elements including forms with subcapitate or capitate apices and occasionally lecythiform, also associated with clavate, thin-walled, hyaline paracystidia. *Caulocystidia* in dense clusters in upper stipe, fusiform to cylindrical, some longer than hymenial cystidia (some >110 µm long), at times flexuous with an undulate outline; apices obtuse or subcapitate, occasionally capitate, infrequently crystalliferous, thick-walled (to 3 µm), hyaline or pale yellowish, or some developing yellowish brown contents, mixed with clavate cauloparacystidia and larger variously shaped caulocystidioid cells. *Pileipellis* composed of cylindrical to inflated hyphae mostly 10–30 µm wide, brown in mass, with slightly thickened walls, weakly encrusted. *Clamp connections* present. Fig. 24 (see also Pl. 7F).

Occurs in sand in coastal low woodland or shrubland in southern W.A., associated with *Agonis flexuosa*, *Allocasuarina*, *Spyridium*, *Leucopogon* and myrtaceous shrubs (no eucalypts nearby). Fruiting June.

Additional specimens examined: W.A.: Two Peoples Bay Nature Reserve, Sinker Reef Track (c. 20 m from end of gravel road), 34°58'51.9"S, 118°09'58.0" E, in sand and mossy ground in shrubland with emergent *Agonis flexuosa*, as well as *Spyridium globulosum*, *Leucopogon* and other shrubs (no eucalypts nearby), 20 June 2012, *N.L.Bougher NLB1019b* (PERTH 08383375, TENN 068648; KJ801165-nLSU, KJ778855-ITS, KJ756427-*rpb2*).

Etymology: From the Latin *exigens* (scanty), in reference to the small size.

Diagnostic features: Fruit bodies small. Pileus dark reddish brown and recurved-squamulose; stipe pruinose or furfuraceous at least in the upper half. Spores large and cylindrical (10.0–13.0 × 4.5–6.0 µm); pleurocystidia often cylindrical to narrowly fusiform. On sand in woodland or coastal scrub heath.

Inocybe exigens is most closely related to several species in the *I. farinosipes* group (Fig. 20). *Inocybe rufobadia*, described from sand in the wheatbelt region in south-western Western Australia, might be confused with *I. exigens*, but the latter has larger fruit bodies, a vivid reddish brown to rusty brown pileus, and shorter spores.

Plate 44. *Inocybe clypeata*, holotype (PERTH 08383375). Photograph — N.L.Bougher.

Plate 45. *Inocybe clypeata* (PERTH 08320276). Photograph — N.L.Bougher.

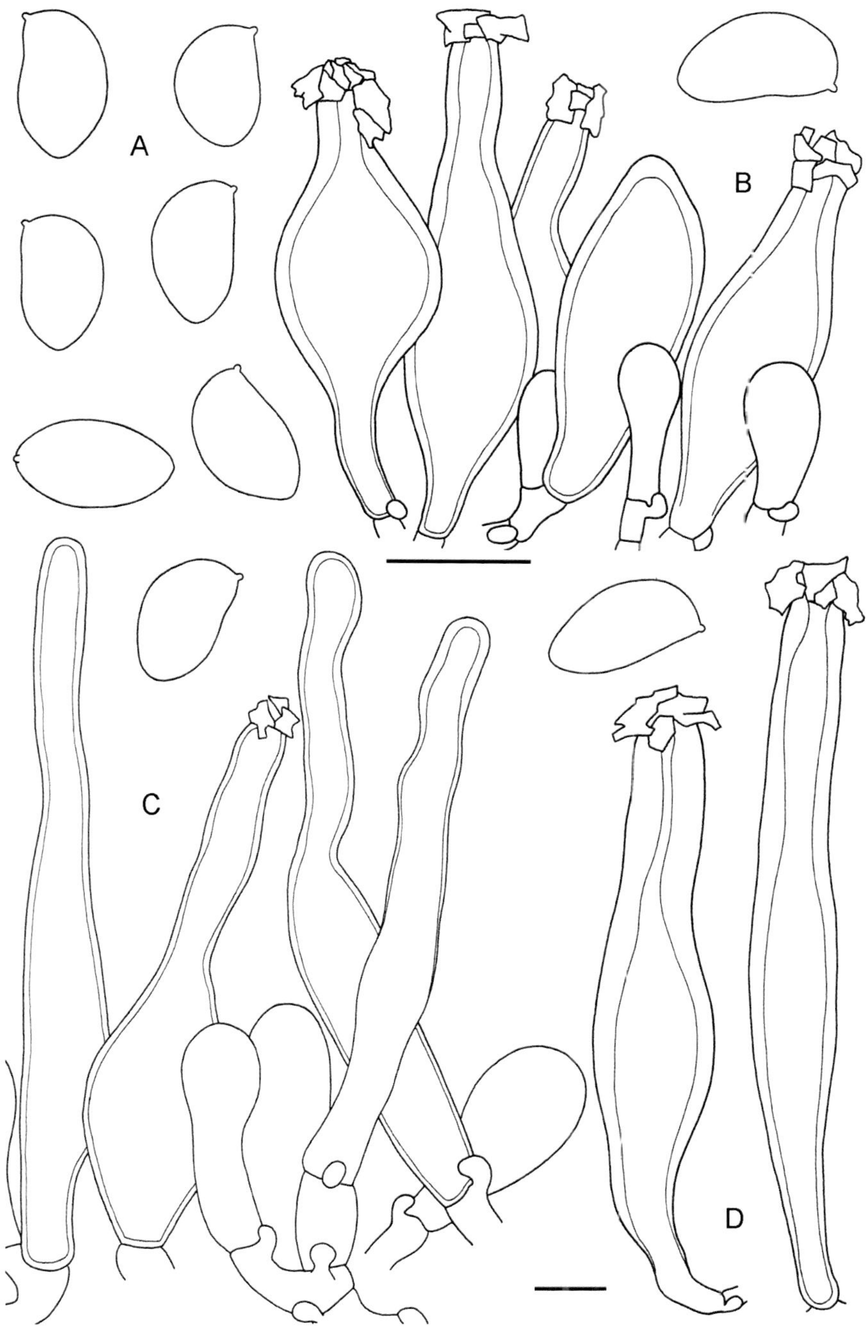

Figure 23. *Inocybe clypeata,* holotype (PERTH 08383375). **A,** Basidiospores; **B,** Cheilocystidia and paracystidia; **C,** Caulocystidia and caulocystidioid cells (from mature stipe apex); **D**, Pleurocystidia. Scale bars = 10 µm (longer bar for spores only).

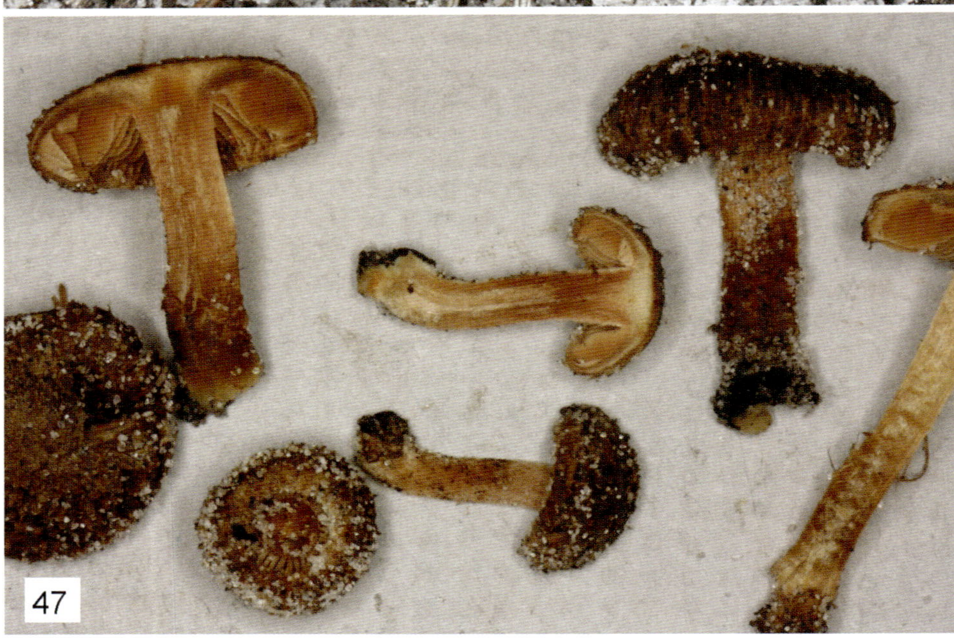

Plate 46. *Inocybe exigens*, holotype (PERTH 08383367). Photograph — N.L.Bougher.

Plate 47. *Inocybe exigens*, holotype (PERTH 08383367). Note the reddening flesh. Photograph — N.L.Bougher.

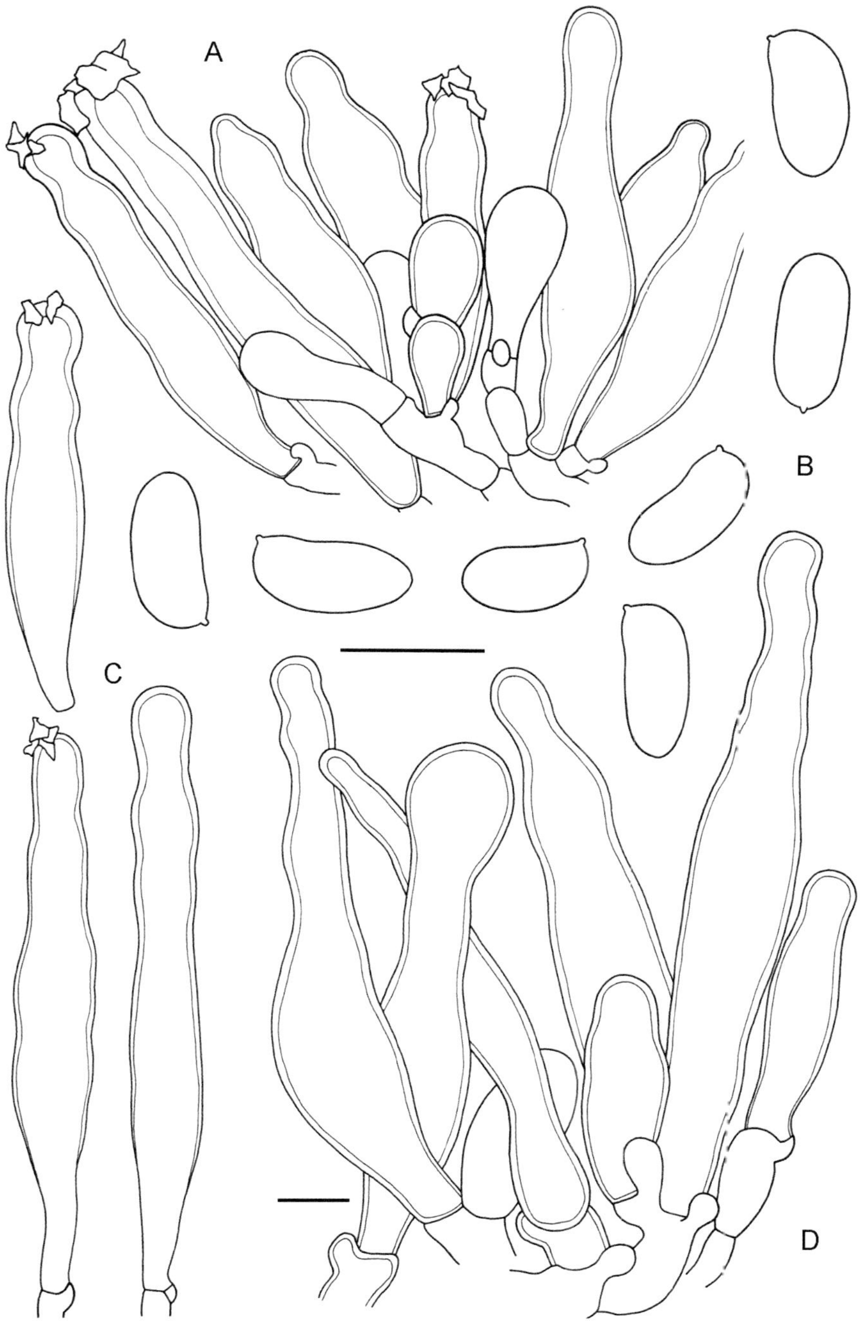

Figure 24. *Inocybe exigens,* holotype (PERTH 08383367). **A,** Cheilocystidia and paracystidia; **B,** Basidiospores; **C,** Pleurocystidia; **D,** Caulocystidia and cauloparacystidia. Scale bars = 10 µm (longer bar for spores only).

21. Inocybe farinosipes Matheny, Bougher & Ryberg, *sp. nov.*

MycoBank: MB 809323

Type: 1 km N of Catterick Rd, near Greenbushes, W.A., 33°50'50.6"S, 116°04'50.3"E, on lateritic soil in recently burnt Jarrah (*Eucalyptus marginata*) forest, 12 July 2011, *M.Ryberg MR00226*; holo: PERTH 08320101; iso: TENN 066461 (KC305405-nLSU, KC305483-ITS, KC305445-*rpb2*).

Pileus 10–37 mm wide, parabolic, conical, broadly convex, plano-convex to plane; umbo absent or inconspicuous; margin decurved to slightly uplifted with age, occasionally undulate, the margin edge occasionally with sparse fibrillose remnants; surface dry, not at all scaly, smooth and unbroken in the centre, radially fibrillose to coarsely so towards the margin or throughout, becoming weakly rimulose but not revealing the underlying context; colour variable: without velipellis dark brown (10YR 3/3) in the centre, shading to brown (10YR 4/3; Snuff Brown or Saccardo's Umber) towards the margin; with velipellis the centre starkly whitish, elsewhere umbrinous or brown to yellowish brown (10YR 5/4 to 4/3) or chamois at the centre and shading yellowish brown with olivaceous tones (Isabella Colour) towards the margin, in some cases pallid superficial fibrils impart a light greyish brown colour throughout; context whitish, not confluent with the stipe, not changing colour where bruised. *Lamellae* adnexed to ascending-adnate or sinuate, close to crowded with 32–50 L and several tiers of lamellulae, ventricose, occasionally narrow or medium, 2–4 mm deep; light grey or pale brown, becoming light yellowish brown or yellowish brown (10YR 6/4 to 5/4); edges pallid, indistinctly fimbriate. *Stipe* 18–32 × 3–7 mm, even, terete or compressed, somewhat swollen or widened at the base; base never marginate, but up to 9 mm wide; veil remnants rarely directly observed, when present, seen attached close to the base of the stipe (as in MR00226); surface densely pruinose, with a powdered appearance on the upper half and, at times, appearing pruinose down to the base, on other occasions the lower part of the stipe is densely covered by white fibrils; ground colour varying from very pale brown (10YR 8/4 to 7/3; Warm Buff) to ±salmon-coloured, pinkish buff or reddish yellow (7.5YR 6/6) or strong brown (7.5YR 5/6), white at the base; context solid, fleshy-fibrous, pale ochre-buff, or coloured like the surface. *Odour* none. Pl. 48–50 (see also Pl. 3E).

Basidiospores 8.0–10.5 (–11.0) × (5.0–) 5.5–6.5 (–7.0) µm [mean 9.1 × 5.9 µm; Q 1.33–1.80 (-1.91); Q mean 1.56 (*n* = 90/5)], smooth, mostly broadly amygdaliform (with bluntly pointed apices) to elliptic or broadly elliptic with rounded apices (especially in the type), occasionally with an adaxial (vental) depression, light yellowish brown; apiculus not distinct. *Basidia* 24–34 × 8–11 µm, clavate to cylindroclavate, hyaline, 4-sterigmate. *Pleurocystidia* 65–105 × 11–18 µm, narrowly fusiform, cylindrical to fusiform-ventricose, typically with a short basal pedicel; some apices bearing crystals, obtuse to subacute; wall hyaline, variable in thickness (to 1.5 µm thick in the type, other collections with thin-walled forms mixed with walls to 2.0–3.5 µm thick). *Cheilocystidia* similar to pleurocystidia but occasionally shorter, mixed with slender paracystidia. *Caulocystidia* abundant and in isolated dense clusters on the upper and lower parts of the stipe, similar to hymenial cystidia, but at times with subcapitate apices, mixed with cauloparacystidia and cells of intermediate size, shape and wall thickness; walls hyaline or sometimes with brown pigment. *Pileipellis* a cutis of cylindrical hyphae that are golden brown or brownish yellow in mass, cylindrical to inflated, mostly 5–20 µm wide; in some collections the cutis contains a compact interwoven layer of thin-walled and mostly 7–14 µm wide hyphae forming a hyaline zone above the pigmented layer. *Clamp connections* present. Fig. 25.

Occurs in south-western W.A.; on soil among mossy or burnt ground in or near eucalypt forest dominated by *Eucalyptus marginata*, *Corymbia calophylla* or *E. diversicolor*. Fruiting June to July.

Additional specimens examined: W.A.: Asquith Rd, Surface Forest Block, 28.3 km NNE from Collie, ForestCheck monitoring site 17, 37°07'22"S, 116°15'43"E, on soil in forest with associated *Eucalyptus marginata*, *Corymbia calophylla*, *Banksia grandis* and *Allocasuarina fraseriana*, 9 July 2009, *R.M.Robinson & K.Syme FC1498* (PERTH 06672914); Boranup Drive, Leeuwin-Naturaliste Natl Park, 34°10'11.5"S, 115°04'04.7"E, on soil along track in wet sclerophyll in Karri forest under *E. diversicolor*, *C. calophylla*, *Acacia* and *Agonis flexuosa*, 7 June 2010, *P.B.Matheny PBM3471* (PERTH 08242054, TENN 067776; KC305407-nLSU); type locality, *M.Ryberg MR00225* (PERTH 08320128, TENN 066517); *loc. id.*, 12 July 2011, *M.Ryberg MR00227* (PERTH 08320098, TENN 066604); *loc. id.*, 12 July 2011, *M.Ryberg MR00228* (TENN 06659; KC305408-nLSU, KC305485-ITS, KC305446-*rpb2*); Westralia Conservation Park, near Collie, scattered singly on soil on cut road bank in Jarrah forest under *E. marginata* and *C. calophylla*, 11 July 2011, *P.B.Matheny PBM3590* (PERTH 08319944, TENN 066492; KC305403-nLSU, KC305481-ITS); type locality, scattered singly or in pairs on unburnt soil in Jarrah forest under *E. marginata* and *C. calophylla*, 12 July 2011, *P.B.Matheny PBM3599* (PERTH 08319863, TENN066494); near Greenbushes, 33°51'08.6"S, 116°06'34.4"S, scattered singly on soil in c. 3 year-old burn beside blue gum plantation, 13 July 2011, *P.B.Matheny PBM3611* (PERTH 08319758, TENN 066628; KC305406-nLSU, KC305484-ITS, KC305444-*rpb2*); Woylie Walk, c. 500 m from Perup Ecology Centre, Perup Nature Reserve, among moss on burnt mound in tea-tree shrubland in dampland (some sedges present) c. 10 m from Jarrah (*E. marginata*) and Marri (*C. calophylla*) forest, 2 June 2013, *N.L.Bougher NLB1070* (PERTH 08516642, TENN 068380; KJ801166-nLSU, KJ778857, KJ756428-*rpb2*).

Etymology: From the Latin *farinosus* (powdery) and *pes* (a foot), in reference to the conspicuously pruinose upper half of the stipe and its overall densely powdered appearance.

Diagnostic features: Fruit bodies often robust. Pileus at times with a conspicuous velipellis; stipe densely powdered with a widened base but lacking a distinctive bulb, with pinkish tinges above but white below. Spores smooth, broadly subamygdaliform to elliptic; caulocystidia present the entire length of the stipe.

The following suite of traits is characteristic of *I. farinosipes*: the robust stipe, the absence of a marginate bulb, a powdered surface at least on the upper half of the stipe, and pinkish or salmon tinges on the upper part of the stipe, the base being white. Some collections have a distinct velipellis, or the ground colour of the pileus can be obscured by a whitish fibrillose superficial layer. *Inocybe farinosipes* is closely allied to *I. bicornis* that differs mainly in its 2-sterigmate basidia, although it has a similar stipe with a pruinose-furfuraceous appearance or scattered dense pruina.

Inocybe farinosipes is most closely related to *I. memorialis* (Fig. 20), although the latter is distinguished by a scaly pileus and its occurrence in lawns and parks. *Inocybe exigens*, which also has a scaly pileus surface, differs from *I. farinosipes* genetically (ITS locus 6% dissimilar) and ecologically; it grows in sand in shrubland with *Spyridium*, *Leucopogon* and *Agonis flexuosa*.

22. Inocybe fibrillosibrunnea O.K.Mill. & R.N.Hilton, *Sydowia* 39: 132 (1987)

Type: 10 km E of Kalamunda, W.A., on ground under *Eucalyptus marginata* and *Corymbia calophylla*, 31 July 1977, *B.Dell UWA2204*; holo: PERTH 00777234.

Pileus 5–40 mm wide, convex, convex-umbonate, or, in smaller collections, conical and then campanulate with a pronounced umbo; margin slightly incurved, eventually splitting and eroded, at first adorned with whitish fibrils; surface dry; with or without a distinct finely matted velipellis at the centre, the disc densely matted fibrillose-squamulose with recurved and often whitish tips, becoming more appressed and fibrillose towards the margin and often rimulose to rimose at the margin where the radial fibrils diverge; dark brown to dark yellowish brown (10YR3/3 to 3/4 or 10YR4/4, also 7.5YR3/2; Dresden Brown to Snuff Brown), concolorous or a shade lighter between spreading fibrils; context pallid. *Lamellae* ascending-adnate to sinuate or adnexed, subdistant to moderately close, with several tiers of lamellulae; initially light brown (7.5YR 6/4), becoming brown (7.5YR 5/4) and eventually dark yellowish brown (10YR 4/4) with age, up to 3–4 mm deep, the edges pallid-fimbriate. *Stipe* 10–60 mm × 1–8 mm at the apex, terete, even, sometimes with a slight bulb; veil cortinate, evanescent; surface dry, never pruinose, densely furfurfaceous-fibrillose, initially contiguous, later forming numerous small scattered tufts; stipe base densely floccose-fibrillose; upper part pinkish brown to yellowish brown (10YR 5/3 to 5/4), shading to dark brown or brown (10YR 4/3) below; context solid, with a brown cortex and a pale interior. *Odour* faintly acidulous. Pl. 51, 52 (see also Pl. 2F, 5D).

Basidiospores 9.0–11.0 × 5.0–6.0 µm [mean 9.9 × 5.4 µm; Q 1.63–2.20, Q mean 1.83 (n = 20/1)], smooth, amygdaliform with bluntly pointed apices, not reniform but occasionally with a suprahilar depression, yellowish brown, the wall to 0.7 µm thick; apiculus small but distinct. *Basidia* 26–30 × 8–10 µm, clavate, hyaline, 4-sterigmate. *Pleurocystidia* 70–115 × 11–20 µm, lageniform with long cylindrical to tapered necks and a short basal pedicel, thick-walled (2.5–4.0 µm), (pale) yellow in KOH; apices obtuse, sparsely crystalliferous.

Plate 48. *Inocybe farinosipes*, holotype (PERTH 08320101). Photograph — M.Ryberg.

Plate 49. *Inocybe farinosipes* (PERTH 08319944). Photograph — P.B.Matheny.

Plate 50. *Inocybe farinosipes* (PERTH 08516642). Photograph — N.L.Bougher.

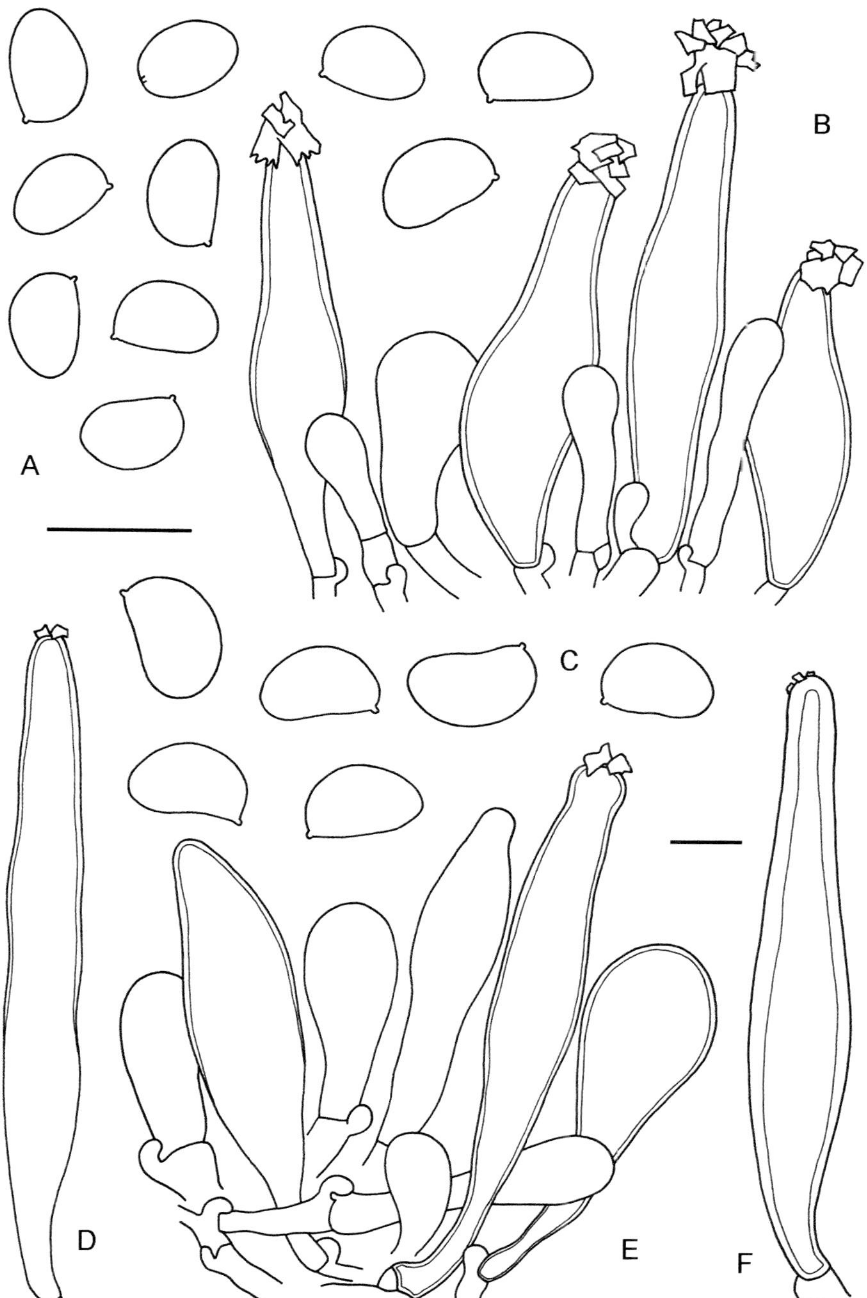

Figure 25. *Inocybe farinosipes*. **A,** Basidiospores, holotype; **B,** Cheilocystidia and paracystidia, holotype (PERTH 08320101); **C,** Basidiospores (PERTH 08319944); **D,** Pleurocystidium (PERTH 08319863); **E,** Caulocystidia and cauloparacystidia from upper stipe, holotype (PERTH 08320101); **F,** Pleurocystidium (PERTH 08319944). Scale bars = 10 µm (longer bar for spores only).

Cheilocystidia similar to pleurocystidia but often shorter and mixed with abundant clavate thin-walled hyaline paracystidia and some basidia. *Caulocystidia* and cauloparacystidia absent, but caulocystidioid hairs forming dense clusters similar to pleurocystidia at the stipe apex, at times with lanceolate necks, mixed with smaller cells of intermediate size and shape (cylindrical, fusiform or clavate), thin- or thick-walled, hyaline (not homologous with paracystidia), intergrading with the longer cells stemming from vestiture hyphae just below the extreme apex, these usually filiform, thick-walled, continuing down to the base of the stipe. *Pileipellis* a cutis of regular hyphae giving rise to broad ascending fascicles of cylindrical to inflated hyphae 5-28 µm wide, thin-walled or with slightly thickened walls, brown-pigmented in mass, encrusting pigments very faint; terminal cells filiform to cylindrical, thin- to thick-walled, often flexuous and tapered distally. *Clamp connections* present. Fig. 26 (see also Pl. 6G).

Occurs on soil in eucalypt and sheoak forests in south-western W.A., in association with *Eucalyptus*, *Corymbia* and *Allocasuarina*. Fruiting June and July.

Additional specimens examined: W.A.: Julimar Forest, in open woodland of *Eucalyptus wandoo* and *E. accedens*, 12 July 1992, *N.L.Bougher*, *R.N.Hilton*, *K.Syme*, *K.Clarke* & *M.Brundrett E4793* (PERTH 07660375); "Site 18, ungrazed", 31 km along Bencubbin–Kellerberrin road, N of Kellerberrin, 31°22'S, 117°45'E, under *Eucalyptus loxophleba* and *E. wandoo*, 29 June 1998, *I.C.Tommerup & S.Q.Bolsenbroek E5970* (PERTH 07608101); Durokoppin Nature Reserve, 26 km along Bencubbin–Kellerberrin road, N of Kellerberrin, 31°24'S, 117°46'E, under *Allocasuarina*, 29 June 1998, *I.C.Tommerup & S.Q.Bolsenbroek E5971* (PERTH 07607032; EU307848-nLSU, KC305473-ITS, EU307849-*rpb1*, EU307850-*rpb2*); Cascades Rd, near Greenbushes, 33°53'24.4"S, 116°08'09.7"E, in leaf litter on soil in Jarrah-Marri forest under *Eucalyptus marginata* and *Corymbia calophylla*, 13 July 2011, *N.L.Bougher NLB802* (PERTH 08318816, KC305397-nLSU, KC305474-ITS, KC305439-*rpb2*); Old Kunjin railway, c. 14 km W of Corrigin, along main road, 32°21'25.5"S, 117°44'10.6"E, in sandy soil under *Allocasuarina*, 26 July 2011, *N.L.Bougher NLB875* (PERTH 08320551, TENN 066721).

Diagnostic features: Fruit bodies medium-sized, occasionally robust. Pileus coarsely matted fibrillose-squamulose with recurved scales; scale tips often whitish in contrast to the dark brown to dark yellowish brown surface; stipe pruinose only at the extreme apex, furfuraceous-woolly elsewhere. Spores smooth, amygdaliform; pleurocystidia very long (some >100 µm); stipe surface with clusters of caulocystidioid hairs (cauloparacystidia absent), these typically filiform or similar to pleurocystidia, mixed with shorter cells of variable shape and wall thickness; hairs present throughout the stipe.

The foregoing description is based on several recent collections and expands on those provided by Miller & Hilton (1987) and Matheny & Bougher (2010), which were based solely on the holotype. *Inocybe fibrillosibrunnea* is now known from multiple localities in south-western Western Australia, but it is not as commonly encountered as two other closely related species, *I. froudistii*, which is widely distributed in southern Australia, and *I. fuscosquarrosa* which appears to be common in Tasmania *Inocybe gomphocephala*, another member of the *I. fibrillosibrunnea* alliance (Fig. 20), is a short-squat species known only from coastal scrub heath habitat in south-western Western Australia.

We consider the circumscription of *I. fibrillosibrunnea* to include the type and collections clustered in a clade with *E5971* for historical reasons (Kropp *et al.*, 2010, Fig. 20). Unfortunately we failed to sequence DNA from the holotype. We note considerable sequence variation in the greater *I. fibrillosibrunnea* group, hence the recognition of five phylogenetic species (Fig. 20). One of these five species, *I. floccosibrunnea* from Western Australia., is phylogenetically distinct from *I. fibrillosibrunnea* with strong support (Fig. 20) and can be distinguished by the very dark brown tips of the scales on the pileus.

A photograph, supposedly of *I. fibrillosibrunnea*, included in Gates & Ratkowsky (2014), probably represents *I. fuscosquarrosa*, a common Tasmanian species.

Plate 51. *Inocybe fibrillosibrunnea*, air-dried holotype (PERTH 00777234). Photograph — N.L.Bougher.

Plate 52. *Inocybe fibrillosibrunnea* (PERTH 07607032). Photograph — N.L.Bougher.

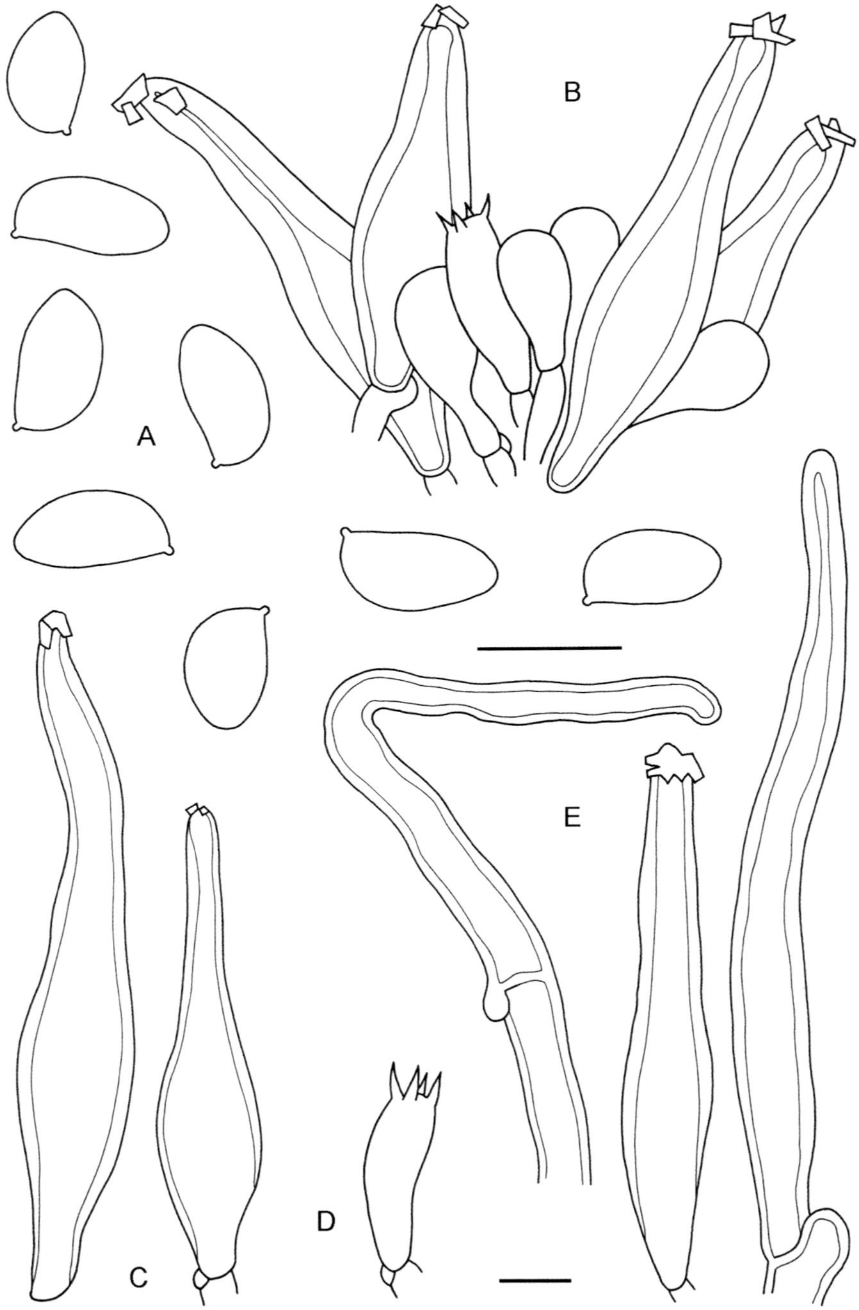

Figure 26. *Inocybe fibrillosibrunnea.* **A,** Basidiospores (holotype); **B,** Cheilocystidia with paracystidia and basidium (PERTH 07607032); **C,** Pleurocystidia, holotype (PERTH 00777234); **D,** Basidium (PERTH 07607032); **E.** Caulocystidioid hairs from stipe apex (PERTH 07607032). Scale bars = 10 µm (longer bar for spores only).

23. **Inocybe floccosibrunnea** Matheny & Bougher, *sp. nov.*

MycoBank: MB 809327

Type: Mt Clare, Walpole-Nornalup Natl Park, W.A., 34°59'39"S, 116°39'18"E, in tall forest of *Eucalyptus diversicolor* and *E. guilfoylei*, 19 June 2007, *N.L.Bougher NLB273*; holo: PERTH 08073937; iso: TENN 069538 (JN974918-nLSU, KC305477-ITS, KC305440-*rpb2*).

Pileus (8–) 16–32 mm wide, at first deeply obtusely conical but becoming conical or campanulate and later expanding to plane with an obtuse umbo; margin remaining incurved for some time, later decurved; surface dry, velipellis absent, squarrose-scaly with recurved tips on the scales over the disc and outwards towards the margin, the tips as dark as the scales, coarsely fibrillose near the margin, rimulose to rimose at the edge where fibrils diverge; colour very dark brown shading to dark brown (10YR 3/3) or dark brown (10YR 3/3) shading to brown (10YR 4/3) towards the margin; context pallid with an olive-yellow tone, firm when fresh, not confluent with the stipe. *Lamellae* adnexed to sinuate, moderately close, c. 26–30 L, with several tiers of lamellulae; light yellowish brown (2.5Y 6/4) when young (pinkish tinges absent), becoming light olive-brown (2.5Y 5/4; near Dresden Brown) with age, ventricose, generally up to 4 mm deep, the edges pallid but often indistinctly fimbriate. *Stipe* 25–60 × 3–5 mm at the apex, terete, gradually enlarged and up to 7 mm wide towards the base; surface dry, veil (cortina) fugacious, nowhere obviously pruinose (possibly at the extreme apex), densely furfuraceous-fibrillose, floccose-fibrillose or woolly, with fibrils scattered over the entire length of the stipe, fibrils more densely accumulated towards the base; brown to yellowish brown above (10YR 5/3 to 5/4) or dark brown to brown (10YR 4/3) below; context solid, flesh pallid with reddish brown to brown tones, the cortex brown. *Odour* absent or weakly acidulous. Pl. 53, 54.

Basidiospores 8.5–11.0 × 5.0–6.0 μm [mean 9.7 × 5.5 μm; Q 1.42–2.15 (–2.20); Q mean 1.79 (*n* = 42/2)], smooth, mostly amygdaliform to broadly amygdaliform, apices often conical (occasionally appearing subangular), at times with a slight adaxial depression, yellowish brown; apiculus small but conspicuous. *Basidia* 28–32 × 8–10 μm, clavate, hyaline, 4-sterigmate. *Pleurocystidia* 79–102 × 10–17 μm, narrowly sublageniform to subfusiform, necks long and cylindrical or slightly tapered towards the obtuse and often crystalliferous apices, thick-walled (walls up to 3 μm), pale yellow. *Cheilocystidia* similar to pleurocystidia but shorter, mixed with abundant hyaline paracystidia and some basidia. *Caulocystidia* and cauloparacystidia not observed, terminal cells of vesture hyphae often thick-walled (1.0–1.5 μm), mostly long (to 180 μm), filiform or narrowly cylindrical or irregularly so, occasionally tapered towards the apices, less often fusiform or rarely clavate, apices infrequently crystalliferous. *Pileipellis* a cutis giving rise to large broad fascicles of trichodermal elements, dark yellowish brown in mass; hyphae mostly cylindrical to inflated, 8–25 μm wide, thick-walled (1.0–1.5 μm), smooth or weakly encrusted; terminal cells largely undifferentiated, typically with tapered apices, occasionally fusiform. *Clamp connections* present. Fig. 27.

Occurs in south-western W.A.; on soil in litter in Karri (*Eucalyptus diversicolor*) forest, or mixed Karri/Tingle (*E. guilfoylei*) forest, or Karri/Jarrah/Marri (*E. marginata* / *Corymbia calophylla*) forest, sometimes with *Allocasuarina* also present. Fruiting June and July.

Additional specimens examined: W.A.: Warren River Loop Trail, Warren Natl Park, 34°30'36.6"S, 115°57'33.6"E, gregarious, solitary or in pairs on soil in Karri forest under *Eucalyptus diversicolor* and *Allocasuarina*, 8 June 2010, *P.B.Matheny PBM3484* (PERTH 08241813, TENN 069523; KC305400-nLSU, KC305478-ITS, KC305442-*rpb2*); Shedley Drive intersection with Tingle Drive, Walpole-Nornalup Natl Park, 34°59'24.5"S, 116°39'20.6"E, on soil in Tingle-Karri (*Eucalyptus guilfoylei*, *E. diversicolor*) forest mixed with *Allocasuarina*, 9 June 2010, *P.B.Matheny & N.L.Bougher PBM3495* (PERTH 08242305, TENN 065924; KC305399-nLSU, KC305476-ITS); Southern Forest Sculpture Walk, Northcliffe, 34°37'47.8"E, 116°07'41.0"S, scattered singly on sandy soil under Jarrah in mixed forest of *Eucalyptus marginata*, *E. diversicolor* and *Corymbia calophylla*, 14 July 2011, *P.B.Matheny PBM3617* (PERTH 08319685, TENN 066576; KC305401-nLSU, KC305479-ITS, KC305441-*rpb2*).

Etymology: From the Latin *floccosus* (having tufts of soft hairs) and *brunnea* (brown), in reference to the floccose-fibrillose stipe and the brown pileus.

Diagnostic features: Fruit bodies medium-sized, not squat. Pileus brown to very dark brown, squarrose-scaly over the disc, scale tips concolorous with surface; lamellae lacking pink tinges;

stipe furfuraceous-fibrillose, floccose-fibrillose or woolly-fibrillose, pruinose at the extreme apex. Spores smooth, 8.5–11.0 × 5.0–6.0 µm, with conical apices. Caulocystidia and cauloparacystidia absent; terminal cells of vesture hyphae cystidioid due to their thick walls, lacking crystals, often narrowly cylindrical and flexuous, hyaline or pale ochraceous.

Inocybe floccosibrunnea comprises a distinct monophyletic group of collections from south-western Western Australia that are genetically distinct from *I. fibrillosibrunnea* (Fig. 20). Collections of both species match the holotype of *I. fibrillosibrunnea* in almost all macroscopic and microscopic attributes. Prior to the current work we had been applying the name *I. fibrillosibrunnea* to all collections morphologically matching the type of that species (Matheny & Bougher, 2010). One distinction between the two species that may be significant is a difference in the colour of scale tips on the pileus surface: white in *I. fibrillosibrunnea* but concolorous with the rest of the scales in *I. floccosibrunnea*.

In our type studies (Matheny & Bougher, 2010), we reported the presence of caulocystidia similar to pleurocystidia at the stipe apex in the holotype of *I. fibrillosibrunnea* but not the presence of cauloparacystidia. Other collections designated as *I. fibrillosibrunnea* in the current work lack caulocystidia. However, we have not observed caulocystidia in any of the collections assigned to *I. floccosibrunnea*. Stipes of the latter are dominated by caulocystidioid hairs including some very long terminal cells occasionally with apical crystals.

Inocybe floccosibrunnea is also morphologically similar to several other genetically closely related species, *viz. I. froudistii*, *I. memorialis* and *I. fuscosquarrosa*. The pileus scales of *I. floccosibrunnea* are uniformly brown, unlike *I. froudistii* and *I. memorialis* which, like *I. fibrillosibrunnea*, have distinctively whitish or pallid scale tips. *Inocybe fuscosquarrosa* can be distinguished most reliably by the pruinose upper part of the stipe and the lightly pallid fibrillose vesture elsewhere on the reddish brown stipe. Long pleurocystidia and the stipe characters separate the last species from *I. australiensis* and *I. umbrosa* E.Horak, a New Zealand endemic.

24. **Inocybe froudistii** Matheny & Bougher, *sp. nov.*

MycoBank: MB 809330

Type: Bold Park, eastern slope of Reabold Hill, Perth, W.A., 31°56'18.0"S, 115°46'48.7"E, emerging from soil in sparse litter over limestone rock in shrubland with *Melaleuca huegelii*, *Acacia* sp. and *Templetonia retusa*, 12 July 2012, *N.L.Bougher, M.J.Bougher & R.E.Bougher NLB1060*; holo: PERTH 08363714; iso: TENN 067784 (KJ801168-nLSU, KJ756429-*rpb2*).

Pileus 10–27 mm wide, initially pulvinate, becoming conical (rarely deeply conical), conico-campanulate to convex, with or without a distinct but low central umbo, almost plane with age; margin initially incurved, but soon becoming decurved, occasionally with white appendiculate fibrils; surface dry, radially fibrillose-squarrose with small recurved squamules that are present when young or develop later (some with pale tips), scales in some collections overlain by an irregular network of pale greyish brown fibrils over the entire pileus, coarsely fibrillose throughout or fibrillose and rimulose towards the margin; dark yellowish brown, dark brown (near 5YR 3/3) or bay-brown (near 10YR 3/4), a few collections dark reddish brown (5YR 2/2 to 5/4); context white to watery brown, up to 2 mm thick under the disc, firm, not fragile. *Lamellae* adnexed to sinuate, moderately close to close, 22–30 L, with several tiers of lamellulae, whitish when very young, becoming light yellowish brown (near 10YR 6/4) with a slight pinkish tinge, eventually yellowish brown (10YR 5/6) or brown (7.5YR 5/4 to 4/4); ventricose or even, 2–4 mm deep; edges indistinctly pallid-fimbriate. *Stipe* 15–50 × 2–7 mm at the apex, terete or compressed, with an even or slightly enlarged or swollen base; partial veil observed only on very young specimens, composed of fine whitish silky fibrils, fugacious; surface densely pruinose to scurfy-pruinose at the apex, with silky superficial fibrils below, or becoming densely pruinose to at least halfway along the stipe, fibrillose-floccose near the base; white, dull whitish to pinkish brown (near 8D5), at times light brown, brown to umber (7.5YR 6/4 or 5/6 to 4/4 or 5/4); context white at the stipe base, more dull in the interior, solid, becoming hollow. *Odour* absent, acidulous, weakly of *Pelargonium* or farinaceous. Pl. 55, 56.

Plate 53. *Inocybe floccosibrunnea*, holotype (PERTH 08073937). Photograph — N.L.Bougher.

Plate 54. *Inocybe floccosibrunnea* (PERTH 08241813). Photograph — P.B.Matheny.

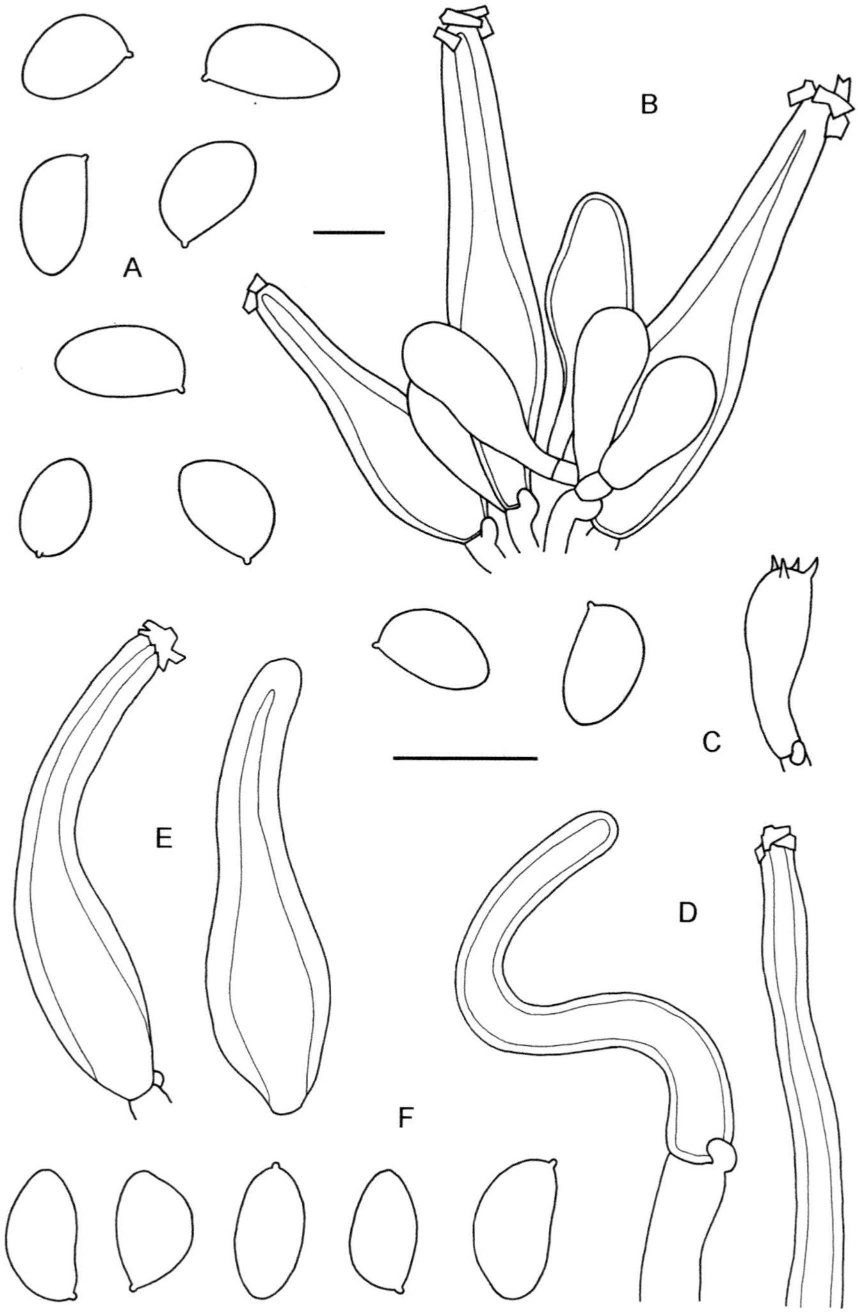

Figure 27. *Inocybe floccosibrunnea* (A–D, holotype (PERTH 08073937); E, F, PERTH 008241813). **A,** Basidiospores; **B,** Cheilocystidia and paracystidia; **C,** Basidium; **D,** Caulocystidioid hairs from upper stipe, including the upper part of a long, narrow, filiform hair bearing apical crystal; **E,** Pleurocystidia; **F,** Basidiospores. Scale bars = 10 µm (longer bar for spores only).

Basidiospores 9.0–11.5 (–12.0) × 5.5–7.0 µm [mean 10.3 × 6.1 µm; Q 1.50–1.93 (–2.09); Q mean 1.69 (n = 73/6)], smooth, broadly amygdaliform to amygdaliform with bluntly pointed apices, yellowish brown; apiculus often distinct. *Basidia* 23–36 × 8–12 µm, clavate, hyaline, 4-sterigmate (rarely with 2). *Pleurocystidia* 58–84 (–109) × 12–19 µm, fusiform, narrowly fusiform to fusiform-ventricose, often with distinct necks; apices obtuse and bearing crystals, thick-walled (2–4 µm), hyaline. *Cheilocystidia* similar to pleurocystidia but occasionally lageniform and/or shorter or thin-walled, hyaline to ochraceous, mixed with paracystidia, these rarely hyaline or brownish. *Caulocystidia* in dense clusters mixed with cauloparacystidia at the apex of the stipe, fusiform to irregularly cylindrical or flexuous, thick-walled and bearing apical crystals, also mixed with cystidioid terminal cells of variable shape (short- or long- cylindrical, short-clavate, clavate or fusiform), all descending below the stipe centre, some occurring above the stipe base in some collections; lower part of stipe with thick-walled apical crystal-bearing long-flexuous caulocystidioid hairs to 150 µm long; some collections with only sparse superficial hyphae present on the lower part of the stipe and above the base. *Pileipellis* a cutis giving rise to broad fascicles of upturned hyphae, the outermost layer composed of smooth hyaline hyphae mostly 5–15 µm wide, with terminal end cells tapered at the apices; lower layer composed of dark yellowish brown encrusted hyphae mostly 7–16 µm wide; hyaline superficial layer inconspicuous in some collections. *Clamp connections* present. Fig. 28.

Known from south-western W.A. and Tas.; grows on soil in thickets, shrubland, woodland and in lawns under *Acacia dealbata*, *A. rostellifera*, or near other *Acacia* species mixed with *Templetonia* (Fabaceae) and *Melaleuca huegelii*, eucalypts and/or *Allocasuarina*. Also in dry or wet sclerophyll forest under a mixture of eucalypts and *Acacia*, or in recently burnt eucalypt forest. Locally common and widespread in temperate regions. Fruiting June–Aug., Oct. and Nov.

Additional specimens examined: W.A.: Bold Park, near Christchurch playing fields, near peg 2369, City Beach, Perth, on soil in *Acacia* thicket under *Acacia rostellifera*, 22 June 1999, *N.L.Bougher E6103* (PERTH 07589069, TENN 067786; KC305461-ITS); Bold Park, near Old Skyline drive-in, Oceanic Drive, Perth, on sandy soil and attached to small buried sticks in open woodland of *Eucalyptus decipiens* near transition to area with shrubland on limestone, 5 July 2001, *P.B.Matheny PBM2151, E6944* (PERTH 07678118, TENN 066819; KJ801169-nLSU, KJ811567-*rpb2*); Bold Park, Rochdale Rd sand dunes, Floreat Park, Perth, on sandy soil (limestone) under *Melaleuca* and *Acacia rostellifera* in damp pocket between sand dunes, 5 July 2001, *N.L.Bougher PBM2154, E6945* (PERTH 07678096); *loc. id.*, 5 July 2001, *N.L.Bougher PBM2152, E6946* (PERTH 07678088); Bold Park, just S of the old Skyline drive-in, City Beach, on soil in open woodland of *Eucalyptus decipiens*, 27 June 2002, *N.L.Bougher E7218* (PERTH 07673744); *loc. id.*, 27 June 2002, *N.L.Bougher E7219* (PERTH 07673752); Brownman Swamp, Mount Brown Lake and adjacent bushland, Henderson Naval Base, Perth, among litter in *Acacia* woodland, 19 July 2006, *J.Froudist, N.Goldsborough, V.Wood & P.Spencer E8369* (PERTH 07700261; GB KC305374-nLSU, KC305432-*rpb2*; an *in situ* image, as *Inocybe* sp. specimen ID 2551, was included in Bougher *et al.*, 2006b); Leeuwin-Naturaliste Natl Park, Boranup Forest, Boranup Drive, 34°10'40.4"S, 115°03'44.4"E, scattered singly on soil in wet sclerophyll forest in Karri forest under *Eucalyptus diversicolor*, *Corymbia calophylla*, *Acacia* and *Agonis flexuosa*, 7 June 2010, *P.B.Matheny* PBM3462 (PERTH 08242143, TENN 067794; KC305382-nLSU, KC305460-ITS, KC305433-*rpb2*); *loc. id.*, same habitat, 7 June 2010, *P.B.Matheny & A.Dyson PBM3465* (PERTH 08242119, TENN 067791; KJ729944-nLSU); Leeuwin-Naturaliste Natl Park, Boranup Forest, Boranup Drive, 34°10'11.5"S, 115°04'04.7"E, 7 June 2010, *P.B.Matheny PBM3467* (PERTH 08242097, TENN 067792; KC305379-nLSU, KC305457-ITS, KC305424-*rpb2*); *loc. id.*, 7 June 2010, *P.B.Matheny PBM3468* (PERTH 08242089, TENN 067793; KJ729918-nLSU); near Greenbushes, 33°50'50.6"S, 116°04'50.3"E, scattered to gregarious on unburnt soil in Jarrah-Marri forest under *Eucalyptus marginata* and *Corymbia calophylla*, 12 July 2011, *P.B.Matheny PBM3603* (TENN 066607; KC305376-nLSU, KC305428-*rpb2*); near Greenbushes, 33°51'05.5"S, 116°05'32.9"E, scattered singly on soil in c. three year-old burnt Jarrah-Marri forest, next to Blue Gum plantation, track BB252, 13 July 2011, *P.B.Matheny PBM3612* (PERTH 08319731, TENN 066486; KC305378-nLSU, KC305456-ITS, KC305426-*rpb2*); near Greenbushes, 33°51'08.6"S, 116°06'34.4"E, scattered singly on soil in c. three year-old burnt Jarrah-Marri forest, next to Blue Gum plantation, track BB252, 14 July 2011, *P.B.Matheny PBM3614* (PERTH 08319715, TENN 066557; KC305377-nLSU, KC305455-ITS, KC305425-*rpb2*); Balingup–Bridgetown road, 1 km N of Catterick Rd turnoff, 33°50'50.1"S, 116°04'51.2"E, in charcoal and burnt ground in recently burnt Jarrah-Marri forest under *Eucalyptus marginata* and *Corymbia calophylla*, 12 July 2011, *N.L.Bougher NLB799* (PERTH 08318123, TENN 066750; KC305375-nLSU, KC305454-ITS, KC305423-*rpb2*); Bold Park, Reabold Hill, eastern slope, on soil with some leaf litter in shrubland under *Acacia* spp., *Templetonia* and *Eucalyptus gomphocephala*, 1 Aug. 2011, *N.L.Bougher NLB902* (PERTH 08320284, TENN 066810; KC305380-nLSU, KC305458-ITS, KC305427-*rpb2*); Bold Park, Reabold Hill, eastern slope (just N of path), in shrubland c. 3 m tall with *Acacia* spp., *Melaleuca huegelii* and *Eucalyptus gomphocephala*, 1 Aug. 2011, *N.L.Bougher NLB899* (PERTH 08320314); *loc. id.*, 1 Aug.

2011, *N.L.Bougher NLB906* (PERTH 08320241); Kings Park, Fungi Survey Site 4, Forrest Drive, 31°58'00.3"S, 115°49'23.8"E, in sheoak litter at base of *Allocasuarina fraseriana* in open sheoak woodland including *Banksia* and a few *Corymbia calophylla*, 5 July 2012, *N.L.Bougher & L.Rowland NLB1042* (PERTH 08383391); Kings Park, Fungi Survey Site 2, Aberdare lowlands, 31°57'59.6"S, 115°49'13.5"E, in sheoak litter at base of *Allocasuarina fraseriana* in open sheoak woodland including *Banksia* and few *Eucalyptus marginata* and *Corymbia calophylla*, 5 July 2012, *N.L.Bougher & L.Rowland NLB1043* (PERTH 08383405, TENN 067785; KJ801170-nLSU, KJ778859-ITS, KJ756432-*rpb2*). Tas.: Chauncy Vale Wildlife Sanctuary, on soil in dry sclerophyll forest under mixture of *Acacia dealabata* and *Eucalyptus*, 13 Oct. 2001, *G.Gates & D.Ratkowsky I4* (TENN 067787; KC305398-nLSU, KC305475-ITS); *loc. id.*, same habitat, 14 Nov. 2001, *C.Vincent I14* (TENN 067788; KC305386-nLSU); *loc. id.*, scattered singly on soil in dry sclerophyll forest under *Acacia dealbata* and *Eucalyptus*, 11 June 2009, *G.Gates, D.Ratkowsky & P.B.Matheny PBM3239* (TENN 067789; KC305385-nLSU, KJ756430-*rpb2*); *loc. id.*, gregarious on lawn under *Acacia dealbata*, 11 June 2009, *G.Gates, D.Ratkowsky & P.B.Matheny PBM3243* (TENN 067790; KC305384-nLSU, KJ756431-*rpb2*); *loc. id.*, gregarious on lawn under *Acacia dealbata*, 11 June 2009, *P.B.Matheny, G.Gates & D.Ratkowsky PBM3245* (TENN 067771, formerly TENN 065741; KC305383-nLSU, KC305462-ITS, KC305431-*rpb2*).

Etymology: The name honours Joseph Henryk Stanislaw Froudist (1946–2013), naturalist and supporter of Australian field mycology.

Diagnostic features: Fruit bodies relatively robust. Pileus often dark yellowish or bay-brown, fibrillose-squarrose to coarsely squamulose, often with pale grey scales, at times coarsely fibrillose; stipe pruinose at the apex or at least to centre, sometimes below, elsewhere with pallid-superficial fibrils; lower part of stipe dominated by long conspicuous caulocystidioid hairs. Spores broadly amygdaliform, mean length >10 µm; pleurocystidia often shorter (<80 µm) than in *I. fibrillosibrunnea* and not lageniform.

Inocybe froudistii is one of five species with a close phylogenetic affinity to *I. fibrillosibrunnea* (Fig. 20). Another similar species, *I. gomphocephala*, also occurs in coastal scrub heath, but it has a squat and peg-like habit due to the very short stipe. Microscopically *I. gomphocephala* is very similar to *I. froudistii*, although the spores of *I. gomphocephala* are shorter (mean 9.1 × 6.0 µm vs 10.3 × 6.1 µm). Both rRNA and *rpb2* gene trees support the separation of *I. froudistii* from *I. gomphocephala*, and each is indicated as reciprocally monophyletic (Braaten *et al.*, 2013; Fig. 20). *Inocybe fibrillosibrunnea* and *I. floccosibrunnea*, which we interpret to be less common than *I. froudistii* in Western Australia, are distinguished by longer lageniform pleurocystidia and a pubescent stipe with scattered tufts of fibrils. These differences are difficult to observe, but molecular data readily separate the three species (Fig. 20). *Inocybe memorialis* differs in lacking caulocystidioid hairs on the stipe surface, and it occurs in parks and grass under eucalypts.

Some phenotypic variation has been observed between populations of *I. froudistii* in Western Australia and Tasmania. The coarse pileus scales and white vesture on the stipe surface may not be as conspicuous in samples that are rain-soaked, and the greyish scales of some Western Australian specimens have not been observed in Tasmanian collections. Many of the Western Australian collections, including the holotype, have a more robust/stocky appearance than Tasmanian material, but some tall and slender specimens have been observed in Western Australia (e.g. PERTH 08320284).

Ecologically, *I. froudistii* appears to be most commonly found in habitats where *Acacia* also occurs. Indeed, several specimens were collected in sites with no eucalypts, and only *Acacia* as a putative ectomycorrhizal partner.

Muscarine has been detected at a relatively high concentration in *I. froudistii* (*PMB3245* from Tasmania) by Kosentka *et al.* (2013).

Plate 55. *Inocybe froudistii*, holotype (PERTH 08363714). Note conspicuous white vesture on the stipe. Photograph — N.L.Bougher.

Plate 56. *Inocybe froudistii* (TENN 067789). Note rain-soaked appearance in this Tasmanian collection, e.g. less conspicuous white vesture on the stipe. Photograph — P.B.Matheny.

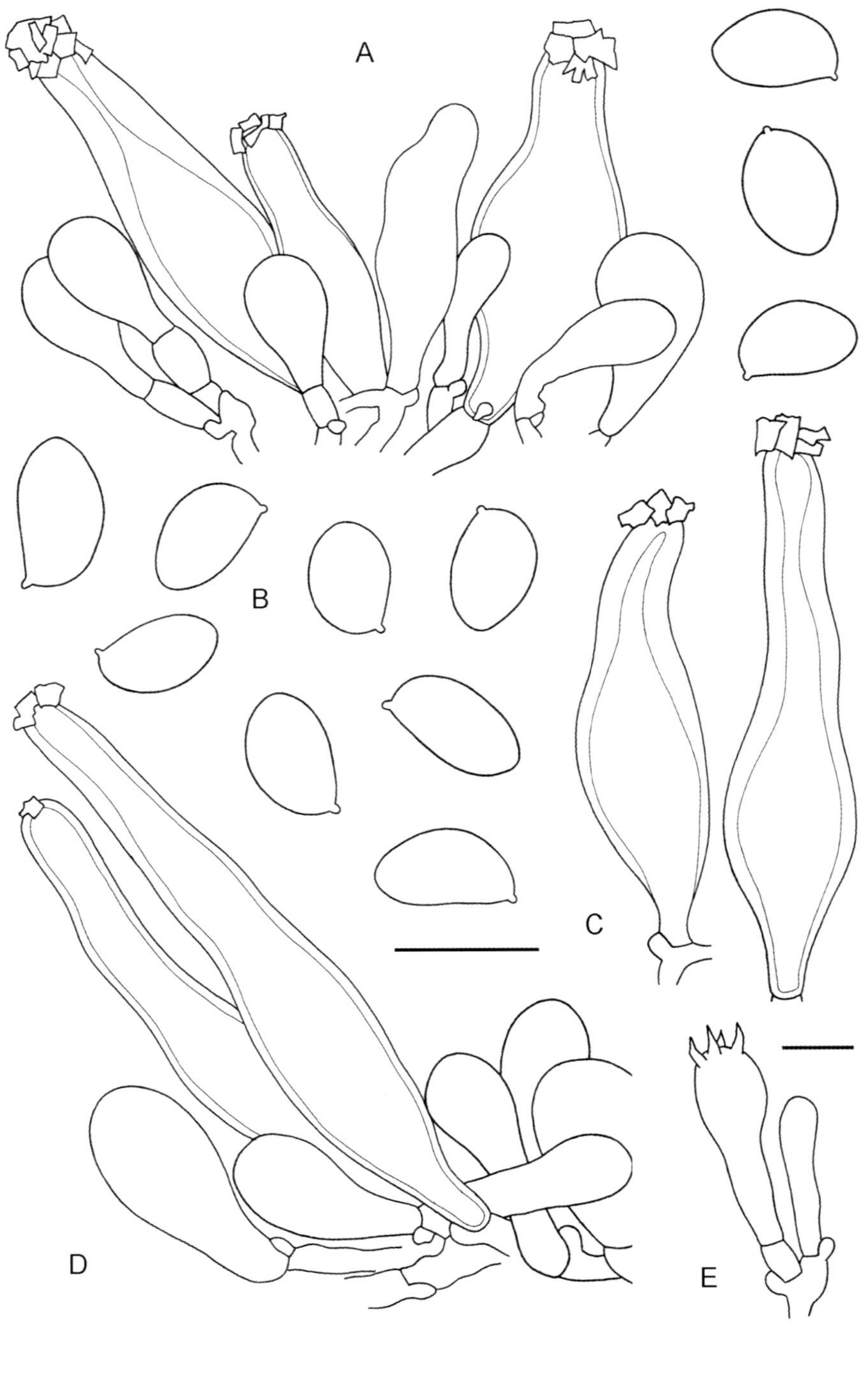

Figure 28. *Inocybe froudistii,* all holotype (PERTH 08363714). **A,** Cheilocystidia and paracystidia; **B,** Basidiospores; **C,** Pleurocystidia; **D,** Caulocystidia with paracystidia from stipe apex; **E,** Basidia. Scale bars = 10 μm (longer bar for spores only).

25. **Inocybe fuscosquarrosa** Matheny, Bougher & G.M.Gates, *sp. nov.*

MycoBank: MB 809331

Type: track to Lady Barron Falls, Mount Field Natl Park, Tas., scattered singly on soil in wet sclerophyll forest under *Eucalyptus regnans*, *E. obliqua*, *Pomaderris* and *Acacia*, 1 June 2010, *G.M.Gates & D.Ratkowsky PBM3447*; holo: PERTH 08556458; iso: TENN 069527 (KC305389-nLSU, KC305465-ITS, KC305437-*rpb2*).

Pileus 10–40 mm wide, conical, obtusely conical or conico-campanulate, or convex to conico-convex, becoming plane and distinctly umbonate with age, margin decurved; surface dry; velipellis absent; the centre squarrose to shaggy-squamulose with slightly recurved scales, imbricate; margin coarsely fibrillose and rimose, edges of the pileus fringed with veil remnants; mostly dark yellowish brown (10YR 3/4; Bay Brown; 6F4) at the centre (dark reddish brown if overly moist) or very dark brown (10YR 2/2), shading to dark brown (10YR 3/3; Sepia or Mummy Brown), dark yellowish brown (10YR 4/4; Dresden Brown) or brown (10YR 5/3 to 4/3) towards the margin, a lighter shade of brown between the spreading fibrils, the scale tips sometimes lighter (pale brown to brown) than the scales themselves; context thin, up to 3 mm thick under the disc, not firm, pallid or brownish. *Lamellae* adnexed, sinuate or subfree, moderately close to subdistant with c. 22–34 L, with several tiers of lamellulae; at first pinkish grey or light reddish brown (5YR 6/4; Fawn), becoming light brown (7.5YR 6/2 to 6/4), reddish brown (5YR 5/3; Avellaneous), brown (7.5YR 5/4 or 10YR 5/3) or strong brown (7.5YR 4/4) with age, ventricose, up to 4 mm deep, the edges pallid and indistinctly fimbriate under a lens. *Stipe* 15–65 mm long, 2–4 mm wide at the apex, up to 6 mm wide at the base, central, terete, even or only slightly swollen at the base, dry; veil cortinate and fugacious; most collections pruinose at the apex (upper third) only, less often with a pruinose appearance just below the stipe centre, with a thin layer of superficial fibrils (vesture) for much of its length when young, but this wearing away with age, becoming weakly fibrillose or with fine scattered pallid silky fibrils below, almost glabrous, base with white (less often yellowish) tomentum; brown with an apricot-pink hue or reddish brown (5YR 5/4 to 4/4), becoming dark reddish brown (5YR 3/4) with age; context solid, concolorous with the surface. *Odour* not remarkable or faintly acidulous. Pl. 57–59.

Basidiospores 8.5–11.0 (–11.5) × 5.0–7.0 µm [mean 9.9 × 6.0 µm; Q 1.43–1.86 (–1.91); Q mean 1.64 (*n* = 44/3)], smooth, amygdaliform, less often elliptic, at times with a slight adaxial depression, yellowish brown; apices conical; apiculus conspicuous. *Basidia* 30–34 × 9–11 µm, clavate, hyaline, 4-sterigmate. *Pleurocystidia* (56–) 67–102 × 9–17 µm, mostly fusiform to fusiform-ventricose or sublageniform, apices obtuse, bases rounded, crystalliferous or not; thick-walled, the walls mostly 2.0–3.0 µm thick, yellow or pale yellow. *Cheilocystidia* similar to pleurocystidia but shorter, some ventricose, clavate or lageniform, mixed with crowded short thin-walled clavate hyaline paracystidia. *Caulocystidia* as caulocystidioid hairs (cauloparacystidia not observed), at times similar to pleurocystidia (mostly 60–125 × 7–13 µm), or long and narrowly cylindrical, with obtuse to subacute apices, these crystalliferous or not; mingled with mainly thick-walled narrowly clavate (35–45 × 9–10 µm) or cylindrical cells of various sizes, dissimilar to the paracystidia on the lamellar edges, these sometimes in dense clusters or flexuous; descending the stipe often in clusters, at times scattered on the lower part of the stipe. *Pileipellis* composed of broad fascicles of upturned cylindrical to inflated hyphae mostly 7–15 µm wide, dark yellowish brown in mass, thick-walled (1–2 µm), smooth, end cells not differentiated. *Clamp connections* present. Fig. 29.

Common in Tas.; in wet and dry sclerophyll forest typically under mixtures of *Eucalyptus*, *Acacia* and/or *Pomaderris*. Fruiting Oct., Jan. and Apr.–June.

Additional specimens examined: Tas.: Chauncy Vale Wildlife Sanctuary, in dry sclerophyll forest under *Eucalyptus* and *Acacia*, 13 Oct. 2001, *G.M.Gates & D.Ratkowsky 13* (TENN 065716); *loc. id.*, 28 Mar. 2002, *G.M.Gates*, *D.Ratkowsky 165* (TENN 069530; KC305391-nLSU, KC305467-ITS); Lower Kermandie, 5 Jan. 2002, *G.M.Gates & D.Ratkowsky 149* (TENN 069528; KC305396-nLSU, KC305472-ITS); Upper Kermandie, 10 Jan. 2002, *G.M.Gates*, *D.Ratkowsky 152* (TENN 069529; KC35890-nLSU, KC305466); Mount Field Natl Park, 9 May 2002, *G.M.Gates 175* (TENN 069531; KC305395-nLSU, KC305471-ITS, KC305435-*rpb2*); Warra, plot 8H, 27 Apr. 2004, *G.M.Gates & D.Ratkowsky 195* (TENN 069532; KC305394-nLSU, KC30547-ITS); Duckhole L., near Dover, scattered and solitary on edge of road under *Eucalyptus*, *Acacia* and *Pomaderris* in wet sclerophyll forest, 29 May 2010, *G.M.Gates PBM3426* (TENN 069526; KC3053920-nLSU, KC305468-ITS, KC305436-*rpb2*); between tracks to Lady Barron Falls and Russell Falls, Mount Field Natl Park, gregarious to scattered singly on bank on soil along road, 1 June 2010, *F.Karstedt PBM3442* (TENN 065787; KC305393-nLSU, KC305469-ITS, KC305438-*rpb2*).

Etymology: From the Latin *fuscus* (dark) and *squarrosa* (upturned scales), in reference to the dark brown squarrose-scaly pileus.

Diagnostic features: Fruit bodies medium-sized to robust. Pileus squarrose to shaggy-squamulose with a rimose margin; stipe often with a pruinose appearance at least at the apex and at times below the stipe centre, elsewhere with a light covering of silky fibrils, reddish brown. Spores smooth, amygdaliform; pleurocystidia often very long [(56–) 67–102 × 9–17 µm]; dense clusters of caulocystidioid hairs similar to hymenial cystidia occurring throughout the stipe; cauloparacystidia absent.

Inocybe fuscosquarrosa, an eastern Australian species, is most closely related to the western Australian *I. fibrillosibrunnea* and *I. floccosibrunnea* (Fig. 20). *Inocybe fuscosquarrosa* can be distinguished from the latter pair by its more strongly developed pruinose stipe (at the apex) and reddish brown colouration to the stipe. Microscopically, these species are virtually indistinguishable. *Inocybe froudistii* is subtly distinguished by having shorter pleurocystidia (mostly 58–84 × 12–19 µm).

In the field, smaller forms of *I. fuscosquarrosa* can be confused with *I. australiensis* or *I. geniculata*, when the whitish fibrils and vesture on the lower part of the stipe are worn away. Otherwise, *I. australiensis* differs by its smaller fruit bodies, a brownish vesture of woolly fibrils on the lower half of the stipe, short elliptic to ovate spores, and much shorter hymenial cystidia. *Inocybe geniculata* is best distinguished by a uniformly densely pruinose stipe that is often bent or flexuous.

Gates & Ratkowsky (2014) provided an illustration of a Tasmanian fungus that they called *I. fibrillosibrunnea*. It is more likely that this represents *I. fuscosquarrosa*.

26. **Inocybe gomphocephala** Matheny & Bougher, *sp. nov.*

MycoBank: MB 809334

Type: E of Gardiner Lake, Two Peoples Bay Nature Reserve, W.A, 34°58'04.2"S, 118°09'39.8"E, on sandy soil in coastal heath-scrub under *Agonis* and unidentified shrubs (Myrtaceae and Fabaceae), 19 July 2011, *P.B.Matheny PBM3652*; holo: PERTH 08517010; iso: TENN 066646; KC305387-nLSU, KC305463-ITS, KC305434-*rpb2*).

Pileus 7–17 mm, pulvinate to broadly convex, at times obtusely conical, becoming plano-convex with age, not umbonate; margin initially inrolled, later decurved; velipellis present over the fibrillose-squamulose to woolly-fibrillose disc, coarsely fibrillose to squamulose towards the margin, edge of margin coarsely fibrillose; colour predominantly dark yellowish brown or Mummy Brown mixed with shades of brown or Prout's Brown; context pale brown to pallid, not changing color where cut. *Lamellae* adnexed, close, broad, very pale brown (10YR 7/3) to avellaneous, edges pallid-fimbriate. *Stipe* to 10 mm long and 3–4 mm wide at the apex, flaring outwards towards the base (to 6 mm wide), squat, densely pruinose above the cortinate insertion, fibrillose below; colour light reddish brown to pale brown or brown; context not observed. *Odour* none. Pl. 60, 61.

Basidiospores 8.0–11.0 × 5.0–7.0 µm [mean 9.1 × 6.0 µm; Q 1.31–1.80 (–1.91); Q mean 1.52 (*n* = 36/1)], amygdaliform with a bluntly pointed apex, often adaxially convex, with a conspicuous apiculus. *Basidia* 30–38 × 8–11 µm, clavate, hyaline, 4-sterigmate. *Pleurocystidia* 58–84 × 12–17 µm, fusiform with distinct necks and obtuse apices, often bearing crystals, only slightly thick-walled (1–2 µm thick), hyaline. *Cheilocystidia* similar to pleurocystidia, but occasionally shorter and more lageniform, mixed with mainly clavate paracystidia. *Caulocystidia* densely arranged in mixtures of metuloids, cauloparacystidia and thin-walled cells of various sizes and shape; metuloids similar to hymenial cystidia but some longer, 58–135 × 14–17 µm, these narrowly lageniform, fusiform or cylindrical, only slightly thick-walled; apices frequently crystalliferous; other terminal cells generally shorter, fusiform, clavate, hyaline and thin-walled to slightly thick-walled, present as a dense palisade on the lower part of the stipe.

Plate 57. *Inocybe fuscosquarrosa*, holotype (PERTH 08556458). Photograph — P.B.Matheny.

Plate 58. *Inocybe fuscosquarrosa* (TENN 065787). Photograph — P.B.Matheny.

Plate 59. *Inocybe fuscosquarrosa* (TENN 069526). Photograph — P.B.Matheny.

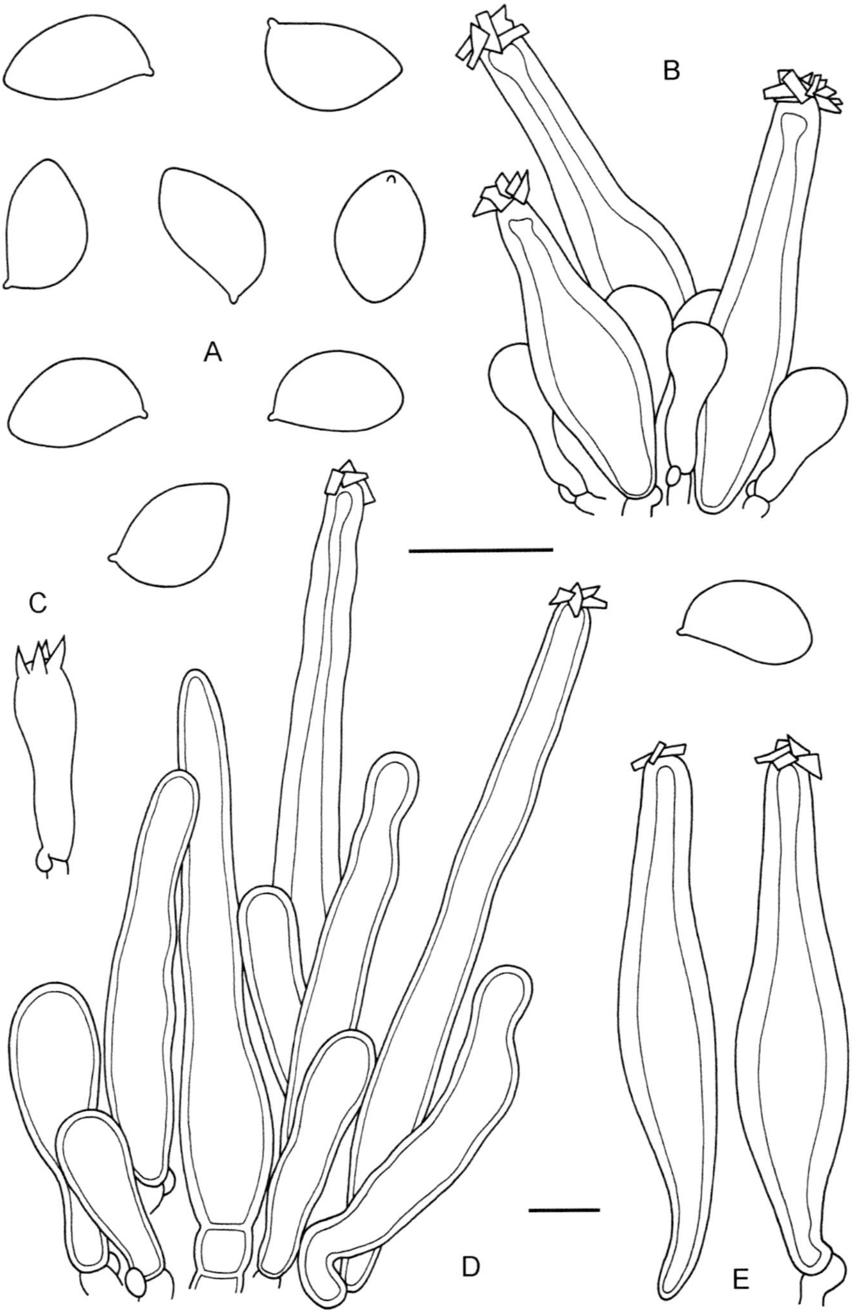

Figure 29. *Inocybe fuscosquarrosa*, all holotype (PERTH 08556458). **A,** Basidiospores; **B,** Cheilocystidia and paracystidia; **C,** Basidium; **D,** Upper stipe - caulocystidioid hairs and associated cells dissimilar to the paracystidia on the lamellar edges; **E,** Pleurocystidia. Scale bars = 10 μm (longer bar for spores only).

Pileipellis composed of broad tufts of ascending hyphae, these regularly arranged, weakly encrusted to smooth, dark yellowish brown to russet-brown in mass, cylindrical, swollen or fusiform, 7–25 µm wide. *Clamp connections* present. Fig. 30.

Known only from the type locality, Two Peoples Bay Nature Reserve in south-western W.A.; grows on sandy soil in coastal scrub heath under *Agonis*, *Spyridium* and other shrubs of Myrtaceae and Fabaceae. Fruiting July.

Additional specimen examined: W.A.: type locality, *P.B.Matheny PBM3651* (PERTH 08319375, TENN 066525; KC305388-nLSU, KC305464-ITS).

Etymology: From the Greek *gomphus* (a nail or peg) and *cephalus* (-headed), in reference to the short squat habit of the fruit bodies.

Diagnostic features: Fruit bodies small, squat, peg-shaped. Pileus <20 mm wide, fibrillose-squamulose. Lamellae very pale brown to avellaneous. Stipe densely pruinose above the cortinate insertion, fibrillose below. Spores 8.0–11.0 × 5.0–7.0 µm. Cystidia only slightly thick-walled.

This is a member of the *I. fibrillosibrunnea* complex and the sister species of *I. froudistii* (labelled "*I. aff. fibrillosibrunnea*" in Braaten *et al.*, 2014). *Inocybe gomphocephala* is distinguished not only by its unique evolutionary position (Fig. 20), but also by the pale brown to avellaneous lamellae, a densely pruinose stipe apex, and its small, squat stature. Its ecology is also unique; it occurs in sandy coastal scrub heath vegetation and probably in a mycorrhizal association with species of Rhamnaceae, Myrtaceae and/or Fabaceae.

Inocybe avellaneifolia (*I. acaciae* group) can be confused with *I. gomphocephala*, but it is characterised by its furfuraceous-fibrillose stipe surface, oblong-amygdaliform to amygdaliform spores (some with a subangular outline), and shorter metuloids on the stipe surface. This species has also been recorded from similar heath-scrub vegetation in south-western Western Australia. *Inocybe brevicula* (*I. australiensis* group) can also be mistaken for *I. gomphocephala*, but it has narrower spores and is associated with eucalypts in woodland and forest.

At least one other unresolved species collected in a similar habitat at Two Peoples Bay in Western Australia could be confused with *I. gomphocephala*. The specimen (TENN 066447; *PBM3649*) has a squat habit and a fibrillose-squamulose pileus, but the lamellae are initially pallid, not pale brown or avellaneous. Another specimen, PERTH 8318476 (*NLB833*), is genetically very similar but it has more robust fruit bodies. Phylogenetic data support their separation from *I. gomphocephala* (data not shown).

A further collection, from Kensington, Perth, Western Australia (PERTH 7679866; *E8130*), also has a squat habit and grows in sand. However, fruit bodies are darker overall and have pale cream lamellae when young. We lack DNA sequence data from this collection, and therefore merely note the possibility of it being referable to *I. gomphocephala*.

27. **Inocybe memorialis** Matheny & Bougher, *sp. nov.*

MycoBank: MB 809391

Type: Tuart Grove section, Pinnaroo Valley Memorial Park, Padbury, W.A., 31°48'28.79"S, 115°46'8.4"E, near isolated trees of *Eucalyptus marginata* and *E. gomphocephala* over grass grazed by kangaroos, 21 July 2008, *N.L.Bougher NLB494*; holo: PERTH 08073317 (KP308778-ITS).

Pileus 10–25 mm wide, convex to bluntly conical, becoming almost applanate and slightly upturned; margin incurved when young, not appendiculate; surface dry, radially scaly with imbricate and recurved brown scales especially near the centre, and coarse fibrillose-appressed scales near the margin; dark brown (near 7F5 to 7F6) when young and unchanging with age or slightly paler; scales often white-tipped and contrastingly paler than the pileus surface, margin occasionally pale or white, becoming radially split with age; context pallid, not changing colour where bruised. *Lamellae* adnexed, close, with c. 30–35 L and several tiers of lamellulae; pale pinkish grey (near 6B2) when young, becoming warm tan (near 5B4), never dark or rust-brown; edges pallid and indistinctly fimbriate; slightly ventricose, up to 5 mm deep.

Plate 60. *Inocybe gomphocephala*, holotype (PERTH 08517010). Photograph — P.B.Matheny.

Plate 61. *Inocybe gomphocephala* (PERTH 08319375). Photograph — P.B.Matheny.

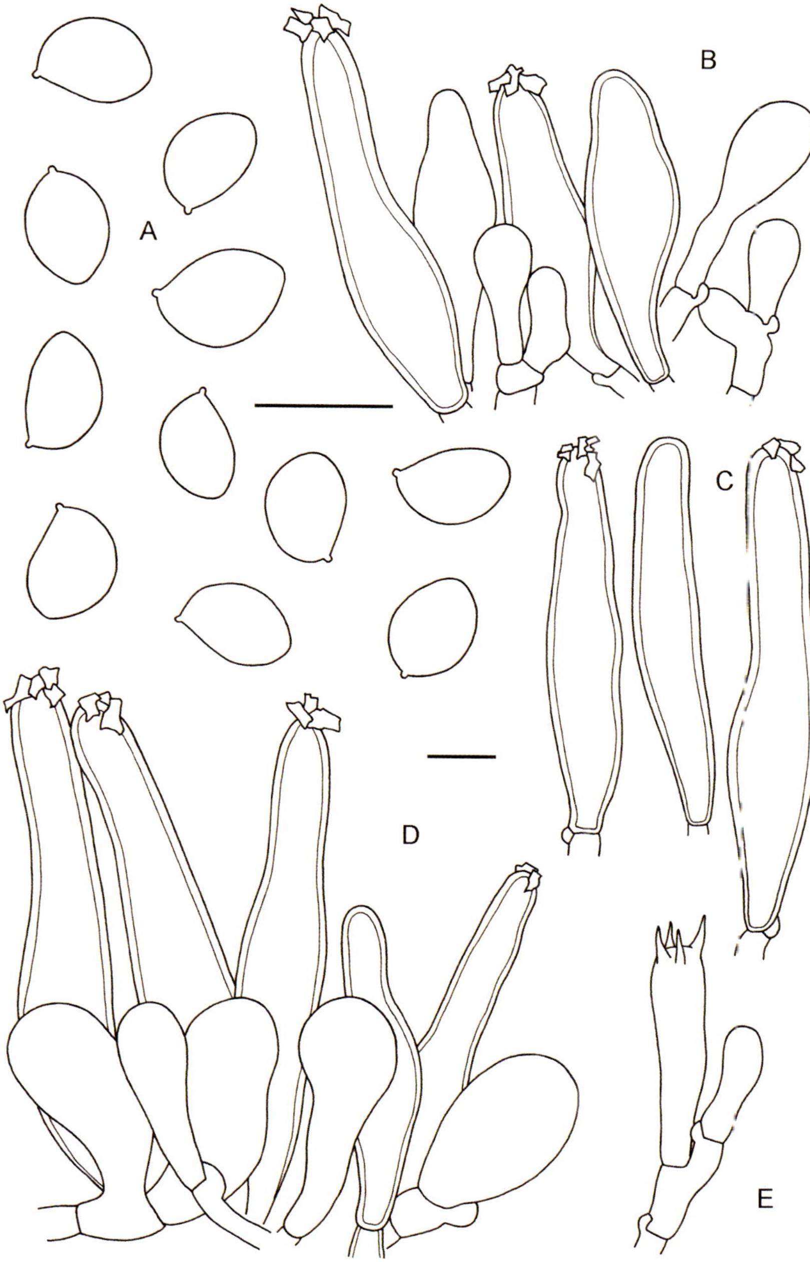

Figure 30. *Inocybe gomphocephala,* all holotype (PERTH 08517010). **A,** Basidiospores; **B,** Cheilocystidia and paracystidia; **C,** Pleurocystidia; D, Caulocystidia and cauloparacystidia (from upper stipe); **E,** Basidia. Scale bars = 10 µm (longer bar for spores only).

Stipe 10–20 × 3–6 mm, terete, even or tapering slightly towards the base; veil sparse, whitish, rapidly evanescent (disappearing before the buttons expand); surface dry, the apex densely hairy-pruinose, otherwise covered by fine whitish fibrils orientated longitudinally but aggregated into floccules (especially when young); ground colour dull ochraceous to pale pinkish brown (near 6B3); context pallid, duller at the stipe base and dull in over-mature specimens, solid but developing vertical fissures near the apex with age. *Odour* not distinctive. Pl. 62, 63 (see also Pl. 2G).

Basidiospores 8.0–10.5 (–11.0) × 5.0–6.0 µm [mean 9.4 × 5.4 µm; Q 1.58–2.00; Q mean 1.74 (*n* = 36/2)], smooth, subelliptic to subamygdaliform; apices bluntly pointed or obtuse, at times with a slight adaxial depression, yellowish brown; apiculus indistinct. *Basidia* 28–36 × 8–9 µm, clavate, hyaline, 4-sterigmate. *Pleurocystidia* 53–84 × 10–14 µm, narrowly fusiform or cylindrical, with a lower ventricose swelling, tapering towards the base; apices obtuse and crystalliferous, thick-walled (walls mostly 1.5–2.5 µm), hyaline. *Cheilocystidia* similar to pleurocystidia, often shorter, at times lageniform, mixed with hyaline clavate paracystidia. *Caulocystidia* abundant, similar to hymenial cystidia but occasionally thin-walled, mixed with cauloparacystidia at the stipe apex and descending to the middle portion of the stipe; caulocystidioid hairs not observed; end-cells of hyphae in the lower stipe short-cylindrical or fusoid. *Pileipellis* a cutis of ascending elements, dark yellowish brown to dark rusty brown in mass, hyphae cylindrical to inflated, mostly 6–20 µm wide, encrusted, thin-walled or with slightly thickened walls, terminal cells similar but hyaline or nearly so. *Clamp connections* present. Fig. 31 (see also Pl. 10B, 10C).

Known from parkland in south-western W.A.; grows on soil under *Eucalyptus marginata* and *E. gomphocephala*. Fruiting July.

Additional specimen examined: W.A.: Perry Lakes, W side near Perry Lakes Drive, among grass in parkland under natural stand of *Eucalyptus gomphocephala*, 9 July 2007, *N.L.Bougher & M.J.Bougher NLB352B* (PERTH 08073775, TENN 069534; JN974919-nLSU, KP308779).

Etymology: From the Latin *memorialis* (of memory), in reference to the type locality being in a cemetery.

Diagnostic features: Fruit bodies medium, with a squat appearance. Pileus dark brown, often with white-tipped erect scales that contrast with the dark brown ground colour of the pileus; lamellae pinkish grey, but becoming pale brown at maturity; stipe apex densely hairy-pruinose, elsewhere with fine floccules or fibrils. Spores smooth; hymenial cystidia <85 µm.

Inocybe memorialis most closely resembles *I. fibrillosibrunnea* and *I. froudistii* in gross morphology, although the three are phylogenetically distinct (Fig. 20). *Inocybe memorialis* differs from the other two species by the absence of caulocystidioid hairs on the stipe surface. Indeed, *I. memorialis* has a mixture of caulocystidia and cauloparacystidia that descends over much of the upper part of the stipe, but the lower part has large undifferentiated end cells. *Inocybe froudistii* also differs from *I. memorialis* by the non-squat habit and predominant association with *Acacia*.

Phylogenetically, *I. memorialis* forms a sister lineage to *I. farinosipes* (Fig. 20), both sharing the densely pruinose stipes or stipe apices. The latter, known from eucalypt forest in south-western Western Australia, is readily distinguished by the absence of scales on the pileus.

Inocybe gomphocephala and *I. avellaneifolia* also have a squat habit, but both occur under shrubby Myrtaceae and Fabaceae in sandy coastal heath habitats in south-western Western Australia.. Moreover, *I. gomphocephala* has very long metuloids on the stipe surface, whereas *I. avellaneifolia* lacks the contrastingly coloured, fibrillose, pileal squamules of *I. memorialis*.

Plate 62. *Inocybe memorialis,* holotype (PERTH 08073317). Photograph — N.L.Bougher.

Plate 63. *Inocybe memorialis,* holotype (PERTH 08073317). Photograph — N.L.Bougher.

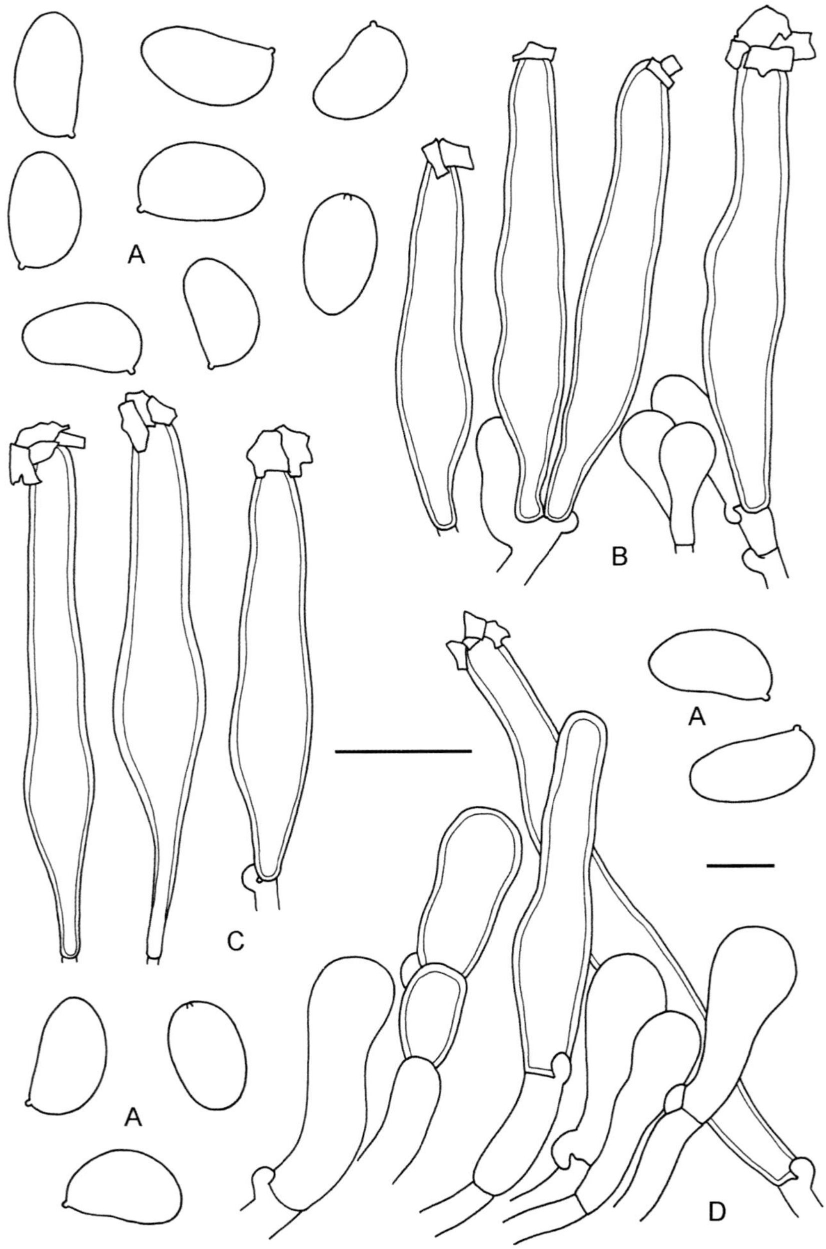

Figure 31. *Inocybe memorialis,* all holotype (PERTH 08073317). **A,** Basidiospores; **B,** Cheilocystidia and paracystidia; **C,** Pleurocystidia; **D,** Caulocystidia and cauloparacystidia. Scale bars = 10 µm (longer bar for spores only).

INOCYBE PERSICINIPES GROUP

This is a well-supported group of eight species from Australia, almost all of which are currently known only from Western Australia. All species possess a partial veil, but no morphological attributes appear unique to the group. All are associated with sclerophyll vegetation. *Inocybe acaciae*, as the epithet suggests, may be a mycorrhizal partner with *Acacia* (Fabaceae). None of the species are commonly encountered and are known presently from three or fewer collections. *Inocybe laureata* appears to be sister to the remainder of the STAC group, but this position is poorly supported.

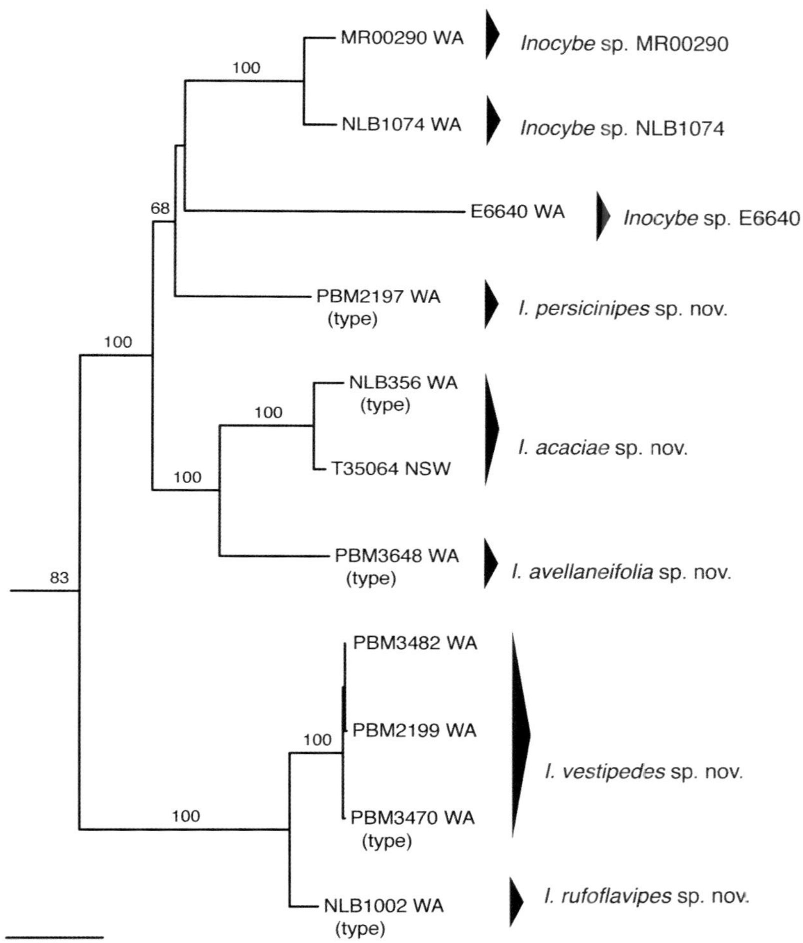

Figure 32. Maximum Likelihood phylogram of the *Inocybe persicinipes* group produced from the analysis of the STAC group shown in Fig. 7. All species shown occur in Australia. Bootstrap values are indicated above the branches.

28. Inocybe acaciae Matheny & Bougher, *sp. nov.*

MycoBank: MB 808390

Type: Port Kennedy Scientific Park, Rockingham Lakes Regional Park, Port Kennedy, W.A., in litter under thicket of *Acacia rostellifera* in *Acacia*-dominated shrubland (eucalypts absent), 17 July 2007, *N.L.Bougher & R.Hart NLB356*; holo: PERTH 08073813 (JN974921-nLSU, KJ729858-ITS, KJ729921-*rpb2*).

Pileus 8–20 mm wide, obtusely conical or convex, becoming broadly convex; margin incurved to decurved, non-appendiculate, entire, not rimose; the surface densely covered with long appressed whitish silky radial fibrils; velipellis absent; dull brown; context pallid or white, not changing colour when exposed. *Lamellae* adnexed, close, 35–45 L, with several tiers of lamellulae, initially pallid, becoming light grey, eventually dull brown (never rusty), to 3 mm deep, ventricose; edges indistinctly fimbriate, concolorous with the faces. *Stipe* 10–30 × 3–6 mm, even or slightly enlarged below, cylindrical or widened towards the apex, partial veil cortinate and fugacious, leaving no trace; surface dry, minutely but densely pruinose near the apex only, elsewhere longitudinally silky-fibrillose; white; context white or pallid, solid, becoming hollow especially near the apex. *Odour* not distinctive. Pl. 64, 65.

Basidiospores 8.0–11.5 × 5.0–6.0 µm [mean 9.9 × 5.5 µm; Q 1.60–2.10; Q mean 1.81 (n = 22/2)], smooth, amygdaliform with conical or bluntly pointed apices, yellowish brown; apiculus small, at times with a slight adaxial depression. *Basidia* 31–36 × 9–10 µm, clavate, hyaline, 4-sterigmate. *Pleurocystidia* 54–80 × 10–16 µm, fusiform, with obtuse apices, crystalliferous, thick-walled (1.0–3.5 µm), hyaline. *Cheilocystidia* similar to pleurocystidia, occasionally shorter and utriform, some subcylindrical, mixed with clavate and various-shaped paracystidia and basidia. *Caulocystidia* similar to hymenial cystidia at the apex, mixed with cauloparacystidia, hyaline or pale yellow; elsewhere with numerous densely arranged caulocystidioid elements up to 105 µm long, thick-walled, hyaline, these also scattered on the lower part of the stipe (not in clusters), absent above the stipe base. *Pileipellis* a cutis, ochraceous-tawny in mass; hyphae cylindrical to inflated, 5–20 µm wide, thin-walled, weakly encrusted if at all, overlain by narrow layer of hyaline cylindrical hyphae. *Clamp connections* present. Fig. 33.

Known from south-western W.A. (in *Acacia* thickets in sand dunes) and northern N.S.W. (dry *Eucalyptus* woodland with *Acacia* spp. in the understorey). Fruiting June, July and Sept.

Additional specimens examined: W.A.: sand dunes along Ritter Rd, near L. Yeagerup, W of Vasse Hwy and Pemberton, 23 June 1985, *N.L.Bougher E3729* (PERTH 07569386, ZT 2738). N.S.W.: 6 km E of Visitors' Centre, Warrumbungle Natl Park, 31°17'32"S, 149°00'29"E, on soil under *Eucalyptus*, 3 Sept. 2010, *T.Elliott T35064* (OSC; KJ729888-nLSU, KJ729860-ITS, KJ729922-*rpb2*).

Etymology: The epithet *acaciae* is in reference to its association with *Acacia* spp.

Diagnostic features: Fruit bodies medium sized . Pileus dull brown, smooth to finely fibrillose; stipe white with a pruinose apex. Spores smooth, 8.0–11.5 × 5.0–6.0 µm; cystidia present as metuloids; caulocystidia present at the stipe apex, dominated by clustered caulocystidioid cells below the stipe apex and scattered on the lower part of the stipe. Occurs in *Acacia* thickets or dry eucalypt woodland.

The collection from northern New South Wales is genetically distinct (ITS 95% similar) from the Western Australian type specimen, its spores being somewhat shorter than the type but overlapping in their range. Notes with the eastern Australian specimen state that it was collected under *Eucalyptus*, but several species of *Acacia* are common understorey plants in dry eucalypt forest in the Warrumbungle National Park, New South Wales. While we include the New South Wales material here, additional collections should clarify the circumscription of *I. acaciae*. Spores of the Lake Yeagerup specimen are also slightly smaller than the type, but overall morphology and a very similar ecology suggest *I. acaciae*.

Inocybe acaciae has similar macroscopic morphology and habitat to *I. farinosipes*, but the latter has an entirely densely pruinose stipe.

An image of the holotype of *I. acaciae in situ* (as *Inocybe* sp., specimen ID: 3343) was included in Bougher *et al.* (2007b).

Plate 64. *Inocybe acaciae*, holotype (PERTH 08073813). Photograph — N.L.Bougher.

Plate 65. *Inocybe acaciae*, holotype (PERTH 08073813). Photograph — N.L.Bougher.

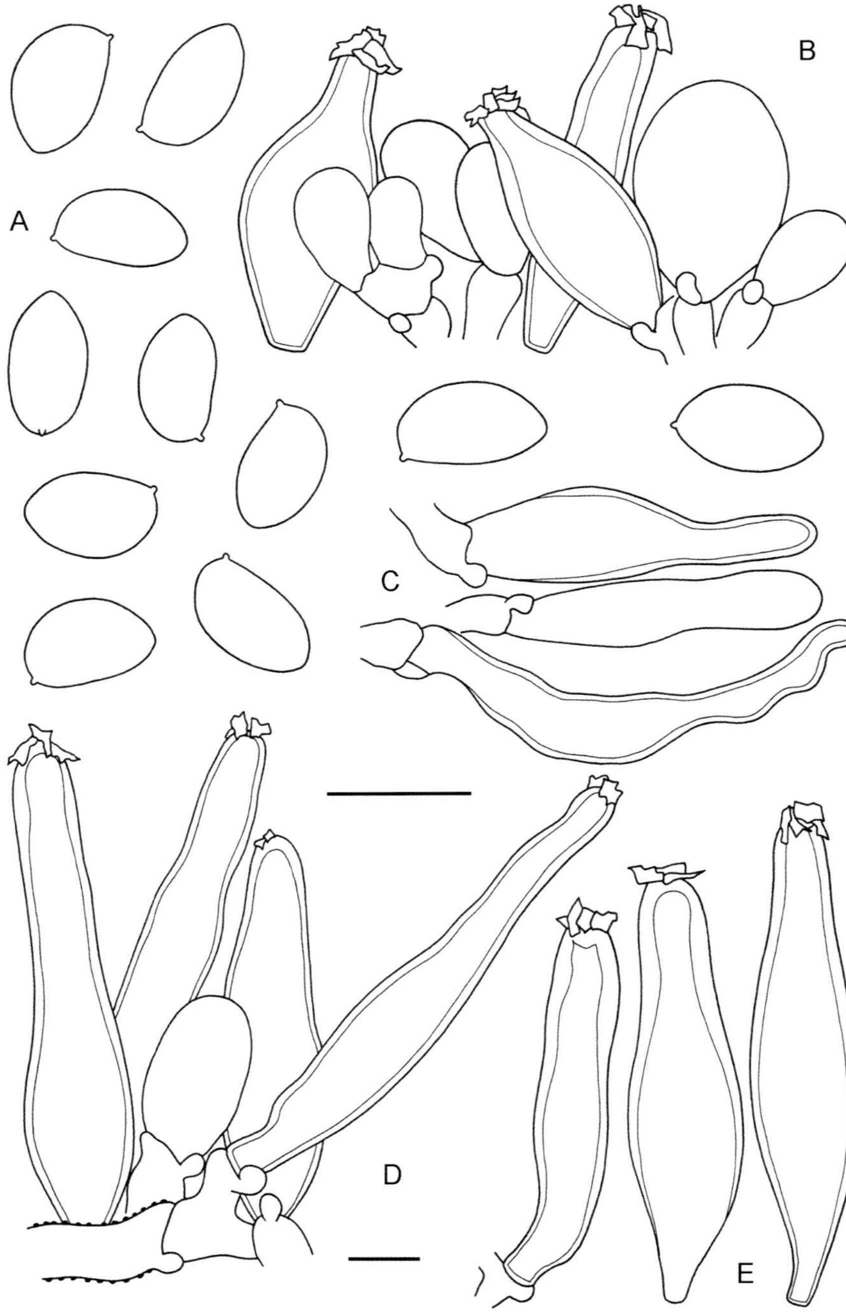

Figure 33. *Inocybe acaciae,* all holotype (PERTH 08073813). **A,** Basidiospores; **B,** Cheilocystidia and paracystidia; **C,** Caulocystidioid elements from lower stipe; **D,** Caulocystidia and cauloparacystidia from upper stipe; **E,** Pleurocystidia. Scale bars = 10 μm (longer bar for spores only).

29. Inocybe avellaneifolia Matheny & Bougher, *sp. nov.*

MycoBank: MB 809265

Type: Sinker Reef Track, Two Peoples Bay Nature Reserve, 30 km E of Albany, W.A., 34°58'54.2"S, 118°09'58.8"E, scattered singly on sand in coastal heath-scrub under *Melaleuca*, *Agonis*, *Spyridium*, *Acacia* and other fabaceous and myrtaceous shrubs, 19 July 2011, *P.B.Matheny PBM3648*; holo: PERTH 08319391; iso: TENN 066593 (KJ756440-nLSU, KJ756475-ITS, KJ729951-*rpb2*).

Pileus 8–18 mm wide, conical to plane, margin initially incurved, later decurved, lacking an umbo; at first the disc and most of surface covered with thin pale greyish brown fibrils of the velipellis, surface fibrillose over the disc and towards the margin, not rimose; centre pale brown when young but brown elsewhere (10YR 4/3), later the centre remaining pale brown due to the obvious velipellis, elsewhere brown (10YR 4/3); context pallid, not changing colour with exposure. *Lamellae* adnexed, moderately close, 34–42 L, with several tiers of lamellulae, rarely forked, initially avellaneous, becoming brown (no olive or yellow tones), edges pallid but not distinctly fimbriate, of moderate depth. *Stipe* 10–15 × 2–4 mm at the apex, terete, flattened, surface with a furfuraceous-fibrillose covering of moderately well-developed vesture, not obviously pruinose, veil (cortina) fugacious, light brown to light yellowish brown (7.5YR 7/4 to 6/4; Cinnamon-Buff to Clay Colour); context solid, coloured like the surface. *Odour* absent. Pl. 66.

Basidiospores 8.5–10.5 × 4.5–5.5 µm [mean 9.9 × 5.2 µm; Q 1.55–2.22; Q mean 1.91 (*n* = 27/1)], smooth, oblong-amygdaliform to amygdaliform, occasionally with a pronounced undulate or subangular outline, less often elliptic, the apices conical to bluntly pointed, less often obtuse, yellowish brown; apiculus small but relatively distinct. *Basidia* 29–36 × 8–9 µm, clavate to cylindroclavate, hyaline, 4-sterigmate. *Pleurocystidia* 53–67 × 12–15 µm, fusiform, apices obtuse and crystalliferous, thick-walled (walls mostly 2–3 µm thick), hyaline. *Cheilocystidia* similar to pleurocystidia, at times shorter or thin-walled, mixed with clavate hyaline paracystidia. *Caulocystidia* forming dense widely spaced clusters on the upper part of the stipe, similar to the hymenial cystidia but sometimes larger and with longer necks, accompanied by caulocystidioid cells of intermediate shape and clavate cauloparacystidia similar to those on the lamellae; lower stipe with scattered solitary caulocystidioid cells arising as terminal cells of the stipitipellis and the superficial hyphae. *Pileipellis* a cutis, dark yellowish brown in mass, hyphae mostly 9–20 µm wide, encrusted, with slightly thickened walls. *Clamp connections* present. Fig. 34.

Known only from the type locality in south-western W.A.; in sand in coastal shrub-heath under *Melaleuca*, *Agonis*, *Spyridium* and *Acacia*. Fruiting July.

Etymology: From the Latin *avellaneus* (hazel-nut brown) and *folia* (a leaf), in reference to the colour of the lamellae at all stages of development.

Diagnostic features: Fruit bodies small. Pileus fibrillose with a greyish velipellis; lamellae avellaneous when young, becoming brown; stipe surface furfuraceous-fibrillose. Spores mostly oblong-amygdaliform to amygdaliform, occasionally with an undulate or subangular outline. On sand under shrubs of Myrtaceae, Rhamnaceae and Fabaceae in coastal shrub-heath

Inocybe avellaneifolia is characterised by its small size, fibrillose pileus, furfuraceous-fibrillose stipe, avellaneous lamellae even when immature, and specialised habitat. *Inocybe acaciae* is most closely related to *I. avellaneifolia* (Fig. 32), but it differs in having larger fruit bodies, pallid lamellae when immature, and a white stipe with a pruinose apex.

Inocybe gomphocephala and *I. avellaneifolia* occur in coastal scrub-heath habitats and share a similar small, squat habit and overall colouration. However, the stipe of *I. gomphocephala* is densely pruinose above the cortinate insertion. Phylogenetically, *I. gomphocephala* is a member of the *I. fibrillosibrunnea* group (Fig. 20). *Inocybe brevicula*, which also produces short, squat fruit bodies, differs subtly by its narrower spores, darker overall colouration, and its association with eucalypts in Jarrah forest or at the margins of Karri forest.

Plate 66. *Inocybe avellaneifolia*, holotype (PERTH 08319391), showing the avellaneous lamellae of this species. Photograph — P.B.Matheny.

30. **Inocybe persicinipes** Matheny & Bougher, *sp. nov.*

MycoBank: MB 809430

Type: Walk-through Tree Trail, Karri Valley Resort, L. Beedelup, near Manjimup, W.A., on soil among moss along track in Karri forest under *Eucalyptus diversicolor*, 19 Aug. 2001, *P.B.Matheny PBM2197/E7044*; holo: PERTH 07676727; iso: TENN 066744 (EU600837-nLSU, KF977215-ITS, EU600836-*rpb2*, EU600835-*rpb1*).

Pileus 10–15 mm wide, conical to obtusely conical, occasionally with a slight broad umbo; margin decurved, the edge initially with a ring of pale brown veil remnants; a pale brownish velipellis visible over the centre and/or towards the margin of some specimens; surface dry, smooth and unbroken in the centre, finely fibrillose towards the margin, not rimose; very pale brown (10YR 7/3; Avellaneous to Wood Brown; 5C4 to 5D4) where covered by the velipellis, elsewhere brown to yellowish brown (10YR 5/3, 5/4, 4/3; Saccardo's Umber); context pallid, up to 2 mm thick under the disc, not changing colour where bruised. *Lamellae* adnexed, moderately close, 24–38 L, with several tiers of lamellulae, pale brown, becoming yellowish brown, ventricose, up to 2 mm deep; edges pallid-fimbriate. *Stipe* 15–30 × 2.5–4.0 mm, even; partial veil cortinate and fugacious; surface densely pallid-pruinose at least over the upper half of the stipe and sometimes below, otherwise lower part of the stipe silky-fibrillose; conspicuously ochraceous or near salmon throughout (7.5YR 6/8; Light Ochraceous-Salmon); context solid, somewhat fragile, concolorous with the surface. *Odour* not remarkable. Pl. 67, 68.

Basidiospores 8.0–9.5 × 4.5–5.0 µm [mean 8.9 × 4.9 µm; Q 1.60–2.04 (–2.11); Q mean 1.80 (n = 20/1)], smooth, amygdaliform with bluntly pointed apices, at times with an adaxial depression, brown to brownish yellow; apiculus small and indistinct. *Basidia* 27–34 × 7–10 µm, clavate, hyaline, 4-sterigmate. *Pleurocystidia* 70–93 × 14–21 µm, lageniform to fusiform, with a long tapered neck and a short pedicel, apices obtuse and bearing crystals, thick-walled, the walls 1.5–3.0 µm thick, hyaline to pale yellow.

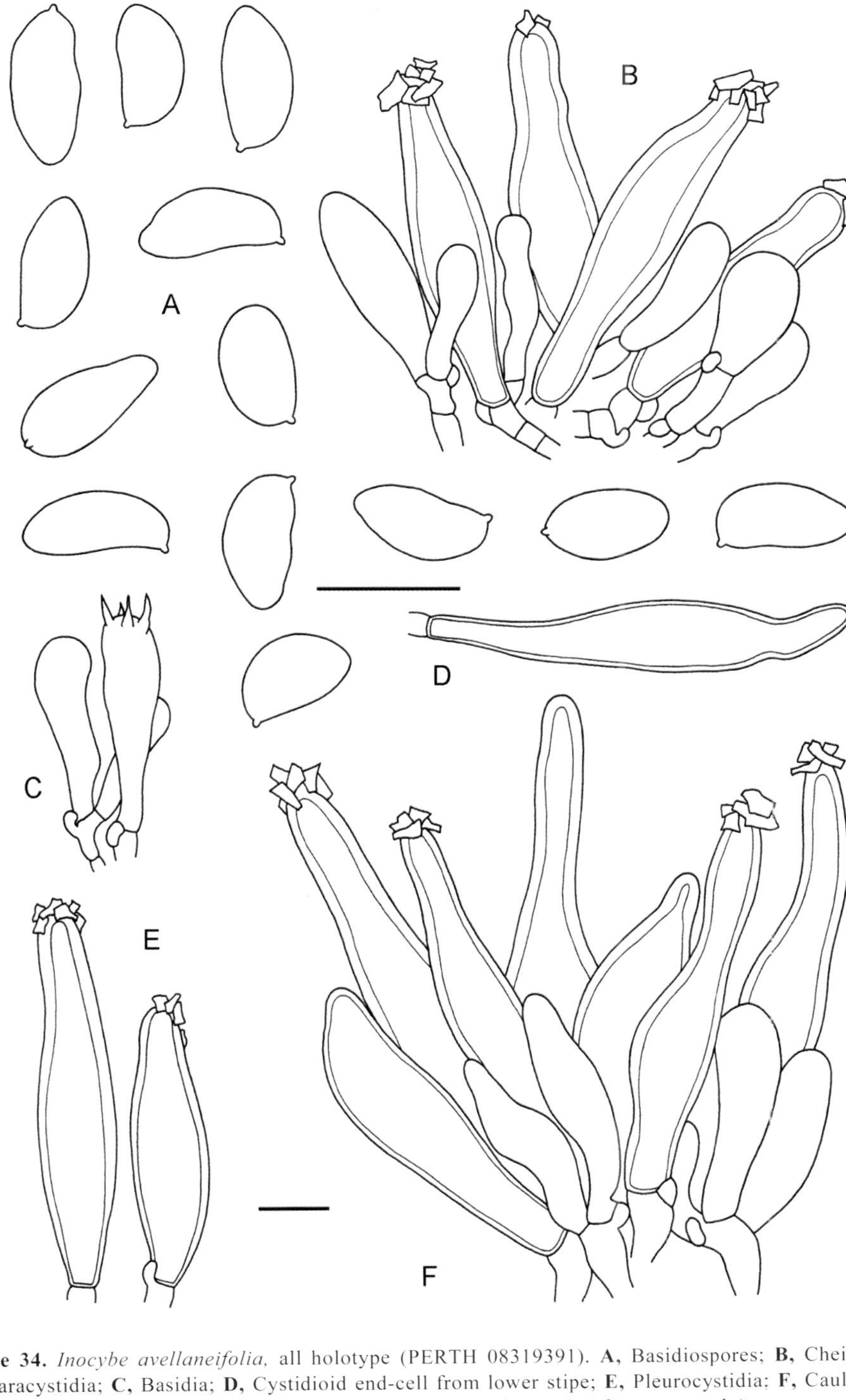

Figure 34. *Inocybe avellaneifolia,* all holotype (PERTH 08319391). **A,** Basidiospores; **B,** Cheilocystidia and paracystidia; **C,** Basidia; **D,** Cystidioid end-cell from lower stipe; **E,** Pleurocystidia; **F,** Caulocystidia and cauloparacystidia from upper stipe. Scale bars = 10 µm (longer bar for spores only).

Cheilocystidia similar to pleurocystidia, often shorter, at times utriform, mixed with clavate, obovate or elliptic paracystidia. *Caulocystidia* descending the entire length of the stipe, mixed with cauloparacystidia and cells of intermediate size and shape, these occasionally quite long (to 110 µm), mostly fusiform, cylindrical, lageniform or utriform, the apices often bearing crystals, thick-walled, hyaline. *Pileipellis* a cutis of cylindrical to inflated hyphae, brownish yellow in mass, mostly 8–20 µm wide, weakly encrusted. *Clamp connections* present. Fig. 35.

Occurs in south-western W.A., on soil in Karri forest under *Eucalyptus diversicolor*. Fruiting Aug.

Additional specimens examined (*I. aff. persicinipes*): W.A.: Kunjin-Jubuk revegetation corridor, Brookton Hwy, c. 14 km W of Corrigin, on soil under various planted eucalypts, *Acacia acuminata* and *Templetonia sulcata*, 20 July 2000, *N.L.Bougher & I.C.Tommerup E6640* (PERTH 07674422, TENN 068466; KP170977-nLSU, KP308808-ITS, KM406215-*rpb2*); Vinegar Well, off Barber Rd, Kunjin, 32°23'31.7"E, 117°40'06.5"S, in sandy soil under *Allocasuarina*, 27 July 2011, *M.Ryberg MR00290* (TENN 066565; KP170978-nLSU, KP308809-ITS, KM406216-*rpb2*).

Etymology: From the Latin *persicinipes* (a peach-coloured foot), in reference to the ochraceous salmon or peach-coloured stipe.

Diagnostic features: Pileus smooth to finely-fibrillose, with veil remnants over the disc and overhanging the margin. Stipe pruinose (at least the upper half), at times silky-fibrillose below, ochraceous to salmon-coloured. Pleurocystidia often lageniform, with long tapered necks. A dense covering of caulocystidia descending the entire length of the stipe.

Inocybe persicinipes is characterised by an ochraceous salmon-coloured stipe that is densely pruinose at least over the upper half, the presence of veil remnants over the pileus and on the margin when young, and its occurrence in Karri forest in south-western Western Australia. It occupies a strongly supported clade with four other lineages (Fig. 32), all of which are rather incompletely known and represented by one or few collections. *E6640* (PERTH 07674422) is morphologically most similar to *I. persicinipes*, but it differs in its phylogenetic position. *MR00290* (TENN 066565) has a reddish or reddish brown stipe and more obtuse and contorted cheilocystidia; it is also associated with *Allocasuarina*. The lineage named as *I. acaciae* has a distinctive white stipe and a frequent association with *Acacia*. *Inocybe avellaneifolia* occurs in coastal shrub-heath under mixed myrtaceous and fabaceous shrubs and differs morphologically by its avellaneous lamellae and squat habit. *Inocybe gomphocephala*, also from coastal shrub-heath and *I. campestris* from Perth and the Wheatbelt region of Western Australia, share the same squat habit and sandy substratum. They are very similar to *I. avellaneifolia* but occur under *Eucalyptus*.

Inocybe murrayana, known only from the type locality in South Australia, is microscopically similar to *I. persicinipes* but differs in having larger oblong-amygdaliform spores (these mostly 10.0–11.0 × 5.0–5.5 µm) and a cinnamon-brown, Buckthorn Brown to Dresden Brown pileus, and tobacco-brown to pale brown stipe lacking salmon or ochraceous tints.

31. **Inocybe rufoflavipes** Matheny & Bougher, *sp. nov.*

MycoBank: MB 809454

Type: Nanga Bridge, Nanga Rd, c. 8 km S of Dwellingup, W.A., on soil among moss and fallen woody debris in Jarrah forest, under *Eucalyptus marginata* (also with nearby *E. camaldulensis*), 17 June 2012, *N.L.Bougher NLB1002*; holo: PERTH 08383316; iso: TENN 067782 (KP170993-nLSU, KP308820-ITS, KM406233-*rpb2*).

Pileus 10–25 mm wide, sharply conical with a decurved margin, with some velar fibrils (pale but not white) persistently adhering, eventually conico-campanulate with a ragged-eroded margin in old age, finely radially silky-fibrillose, but smooth at the central disc; dark greyish brown (near 10YR 4/3) at the margin, reddish brown to dark reddish brown or darker (almost black) towards the centre. *Lamellae* adnexed, to 1 mm deep, closely spaced, edges very minutely fimbriate; umber (7.5YR 4/4), paler when young; lamellulae abundant. *Stipe* 20–35 × 1–3 mm, cylindrical, often flexuous; surface densely pruinose near the apex, sparsely fibrillose below; dull reddish brown (near 5YR 4/4) merging to yellow (near 2.5Y 8/4) at the base; white fibrillose cortina present in button. *Odour* weakly spermatic-grassy. Pl. 69, 70.

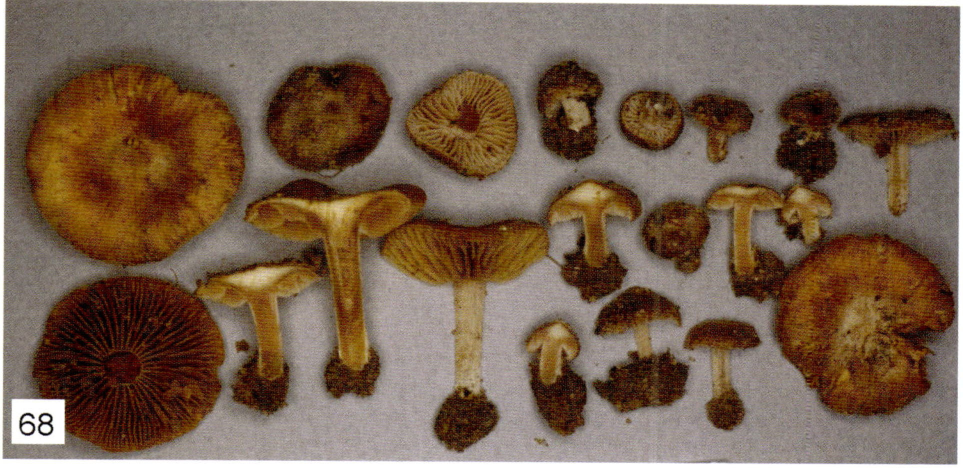

Plate 67. *Inocybe persicinipes*, holotype (PERTH 07676727). Photograph — P.B.Matheny.

Plate 68. *Inocybe* aff. *persicinipes* (PERTH 07674422). Photograph — N.L.Bougher.

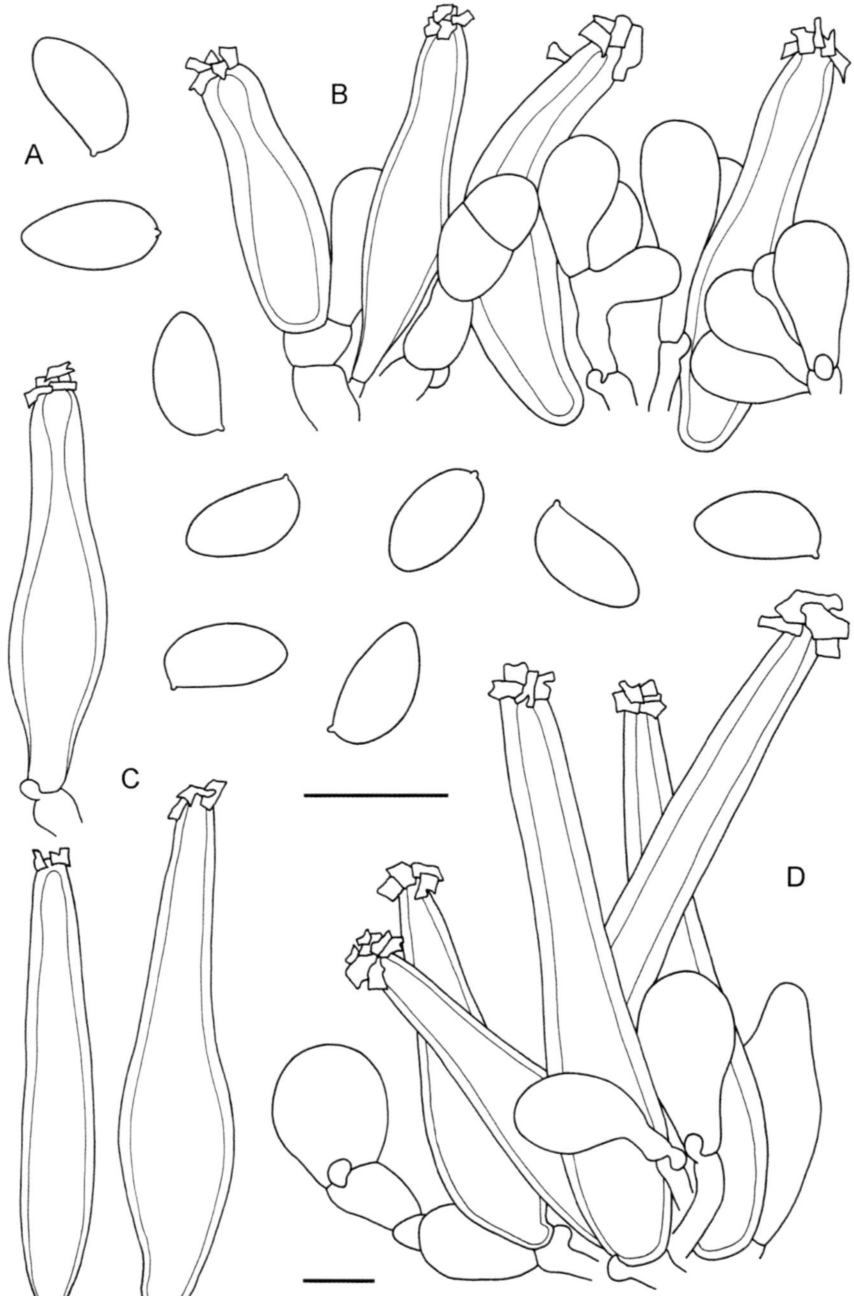

Figure 35. *Inocybe persicinipes,* all holotype (PERTH 07676727). **A,** Basidiospores; **B,** Cheilocystidia and paracystidia; **C,** Pleurocystidia; **D,** Caulocystidia and cauloparacystidia. Scale bars = 10 μm (longer bar for spores only).

Basidiospores 8.5–10.5 (–11.0) × 4.5–5.5 µm [mean 9.2 × 5.1 µm; Q 1.55–2.15 (–2.20); Q mean 1.81 (*n* =30/1)], smooth, amygdaliform, oblong-amygdaliform or fusiform, yellowish brown; apices often attenuated, subacute or more bluntly pointed; apiculus small, pale. *Basidia* 25–32 × 7–9 µm, 4-sterigmate, clavate, hyaline. *Pleurocystidia* 51–70 × 8.5–13.0 µm, narrowly fusiform or ventricose, with obtuse apices often with crystals, thick-walled (to 2 µm), hyaline. *Cheilocystidia* similar to pleurocystidia, abundant but not crowded, mixed with intermediate-shaped elements and small clavate elements (probably basidia). *Caulocystidia* present in the upper stipe in scattered dense clusters, fusiform, cylindrical, some longer than hymenial cystidia, thick-walled (to 2 µm), sometimes mixed with small clavate cauloparacystidia and larger variably shaped caulocystidioid cells. *Pileipellis* a cutis of cylindrical hyphae, these mostly 5–15 µm wide, dark yellowish brown in mass, encrusted. *Clamp connections* present. Fig. 36.

Known only from the type locality: on soil in Jarrah (*Eucalyptus marginata*) forest, in south-western W.A. Fruiting June.

Etymology: From the Latin *rufus* (reddish), *flavus* (yellow) and *pes* (a foot), in reference to the stipe that is reddish brown merging to yellow at the base.

Diagnostic features: Fruit bodies small. Pileus dark and conico-campanulate with persistent pale velar remnants at the margin; stipe reddish brown merging to yellow at the base. Spores smooth, amygdaliform to fusiform. Darker than its sister species *I. vestipedes*.

The fruit bodies of *I. rufoflavipes* resemble those of *I. vestipedes* from Karri forest in south-western Western Australia. Both have a finely fibrillose conico-campanulate pileus with pale or whitish velar remnants adhering to the pileus margin. However, the former is distinguishable by its darker pileus and a reddish brown stipe merging to yellow at the base. The two species also differ microscopically, *I. rufoflavipes* having larger and more amygdaliform/fusiform spores than *I. vestipedes*. The two differ genetically and are phylogenetically distantly related to other species. (Fig. 32).

32. Inocybe vestipedes Matheny & Bougher, *sp. nov.*

MycoBank: MB 809478

Type: Boranup Forest off Boranup Drive, Leeuwin-Naturaliste Natl Park, W.A., 34°10'11.5"S, 115°04'04.7"E, gregarious on soil in wet sclerophyll in Karri forest under *Eucalyptus diversicolor*, *Corymbia calophylla* and *Agonis flexuosa*, 7 June 2010, *N.L.Bougher PBM3470*; holo: PERTH 08242062; iso: TENN 0677809 (KP171092-nLSU, KP641640-ITS, KM656136-*rpb2*).

Pileus 7–23 mm wide, initially deeply conical, less so with age, expanding to broadly conical or plane with a small conical or pointed umbo; margin incurved when immature, becoming decurved; surface dry, finely fibrillose, later somewhat diffracted scaly to fibrillose around the disc but never squarrose or squamulose, radially fibrillose; velipellis absent, shiny; dark brown to umber or near strong brown when moist (7.5YR 3/2 to 4/4 or 5/6), or dark brown (Sepia) at the centre, to brown or yellowish brown (10YR 4/3 to 5/4), towards the margin brownish yellow (10YR 6/6) to yellowish brown (10YR 5/6); context thin, not confluent with the stipe, brownish. *Lamellae* adnexed, moderately close, 24–28 L, with intervening shorter lamellulae; light grey to very pale brown (10YR 7/2 to 7/3), pale brown to light yellowish brown (10YR 6/3 to 6/4), becoming dark yellowish brown (10YR 4/4); edges pallid, indistinctly fimbriate; ventricose. *Stipe* 20–50 × 1.5–2.0 mm, terete, often flexuous; surface not pruinose, or pruinose at the extreme apex, covered with a heavy white partial veil and below with a dense whitish fibrillose vesture, these wearing away with age revealing a light yellowish brown ground colour (10YR 6/4), later near yellowish brown (10YR 5/4), glabrescent but with scattered fibrillose patches especially at the apex and above the base; context solid, coloured like the surface, white in the base. *Odour* acidulous or not remarkable. Pl. 71, 72.

Basidiospores 7.0–8.5 × 4.0–5.0 µm [mean 7.8 × 4.6 µm; Q 1.50–1.78; Q mean 1.69 (*n* = 40/2)], smooth, subamygdaliform to elliptic or occasionally cylindrical-oblong in face view, the apices rounded to bluntly pointed, yellowish brown; apiculus small but generally distinct. *Basidia* 19–27 × 8–9 µm, clavate, hyaline, 4-sterigmate.

Plate 69. *Inocybe rufoflavipes*, holotype (PERTH 08383316). Photograph — N.L.Bougher.

Plate 70. *Inocybe rufoflavipes*, holotype (PERTH 08383316). Photograph — N.L.Bougher.

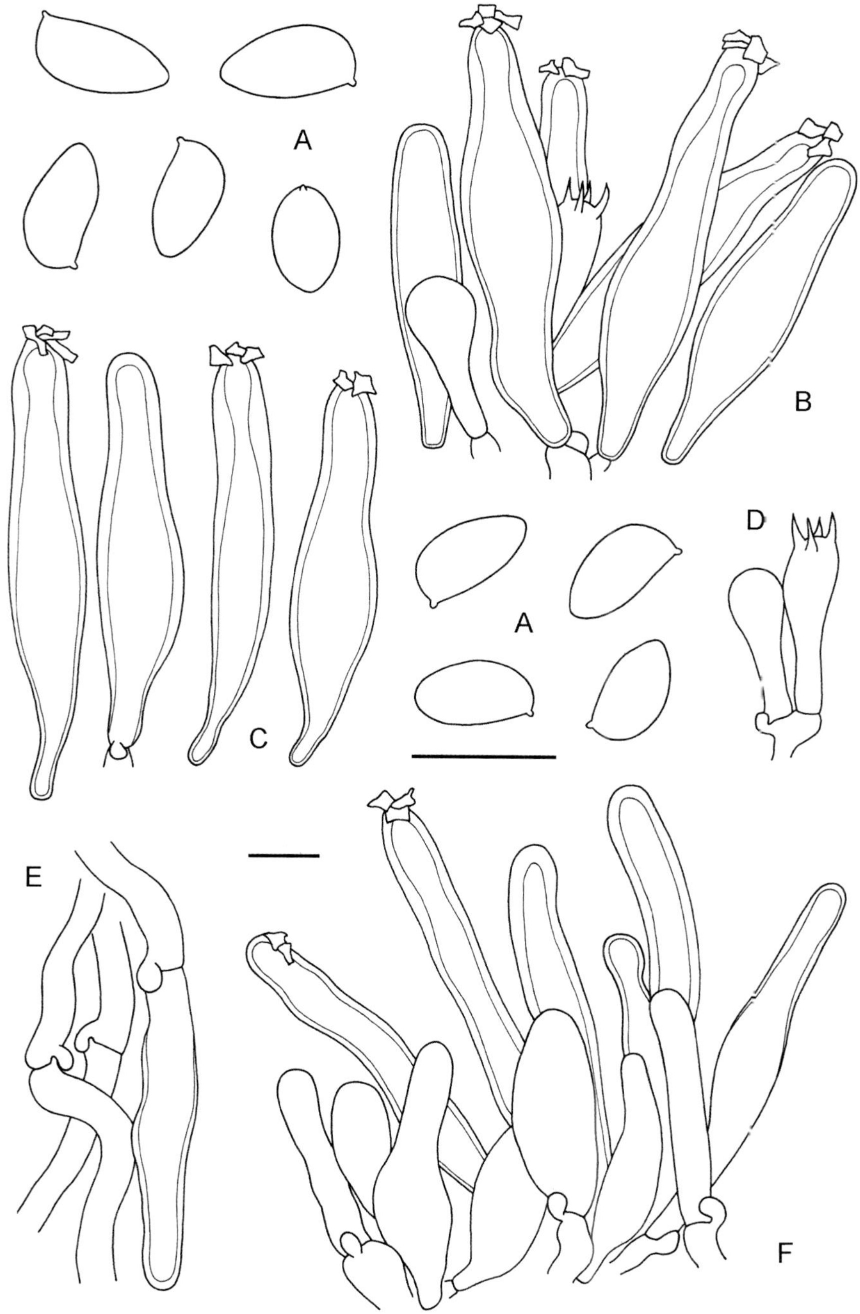

Figure 36. *Inocybe rufoflavipes,* all holotype (PERTH 08383316). **A,** Basidiospores; **B,** Cheilocystidia and paracystidia; **C,** Pleurocystidia; **D,** Basidia; **E,** Superficial hyphae on stipe with caulocystidioid terminal; **F,** Caulocystidia and cauloparacystidia. Scale bars = 10 µm (longer bar for spores only).

Pleurocystidia 58–72 × 10–14 µm, narrowly fusiform, the apices obtuse and often bearing crystals, thick-walled (2–3 µm), hyaline; numerous and frequent. *Cheilocystidia* similar to pleurocystidia, mixed with paracystidia, basidia and cells similar to pleurocystidia but shorter and thin-walled. *Caulocystidia* usually present but occurring only at the extreme apex, similar to cheilocystidia and mixed with infrequent cauloparacystidia (in some specimens caulocystidia were not observed at all), with scattered clusters of caulocystidioid cells, these typically fusiform to cylindrical, thick-walled, mixed with cells of intermediate size and shape with no obvious cauloparacystidia; central and lower portions of the stipe covered with a superficial interwoven layer of hyaline hyphae with undifferentiated terminal cells, these mostly 5–10 µm wide. *Pileipellis* a cutis of cylindrical hyphae 5–20 µm wide, encrusted with pigment, thin- to slightly thick-walled; some sections with a thin superficial layer of hyaline interwoven cylindrical hyphae. *Clamp connections* present. Fig. 37.

Occurs in south-western W.A.; on soil in Karri forest under *Eucalyptus diversicolor* or in mixed Karri forest including *Allocasuarina*. Fruiting June–Aug.

Additional specimens examined: W.A.: Walk-through Tree Trail, Karri Valley Resort, L. Beedelup, 22 km NE of Pemberton, on soil in Karri forest under *Eucalyptus diversicolor*, 19 Aug. 2001, *P.B.Matheny PBM2199 E7046* (PERTH 07676743; EU6008590-nLSU, KP641641-ITS, EU600858-*rpb2*, EU600857-*rpb1*); Heartbreak Trail, Warren Natl Park, 34°29'46.0"S, 115°57'08.8"E, scattered singly on soil in mixed Karri forest including *Allocasuarina* and *Eucalyptus diversicolor*, 8 June 2010, *N.L.Bougher PBM3482* (PERTH 08241848, TENN 067781; KP171093-nLSU, KP641642-ITS, KM656137-*rpb2*).

Etymology: From the Latin *vestitus* (clothed) and *pes* (a foot), in reference to the heavy vesture of superficial hyphae on the stipe.

Diagnostic features: Fruit bodies slender and small. Pileus conical or plane, with a conical umbo at maturity, not scaly, dark brown in the centre, yellowish brown to brownish yellow towards the margin. Stipe with a heavy superficial layer of white fibrils with a light yellowish brown to yellowish brown ground colour. Spores small, 7.0–8.5 × 4.0–5.0 µm.

Inocybe vestipedes is a small-spored species with slender fruit bodies, a fibrillose conical pileus and a stipe with a heavy white vesture. It differs from members of the *I. serrata* group by the limited presence of caulocystidioid cells and/or caulocystidia to the stipe apex.

Inocybe rufoflavipes, is most closely related to *I. vestipedes*, but the former has larger and amygdaliform to fusiform spores, and it also differs genetically (Fig. 32). *Inocybe albovestita* E.Horak (Horak, 1977) resembles *I. vestipedes* in outward appearance, but it has larger spores (8.0–11.0 × 4.5–5.5 µm), more pronounced caulocystidia, a trichodermal pileipellis, and it occurs under *Nothofagus* in New Zealand. Phylogenetically, *I. albovestita* (based on PDD 71309; det. Horak) is somewhat distant from *I. vestipedes* and *I. rufoflavipes*.

33. Inocybe laureata Matheny & Bougher, *sp. nov.*

MycoBank: MB 809385

Type: Sinker Reef Track, Two Peoples Bay Nature Reserve, W.A., 34°56'54.2"S, 118°09'58.8"E, in litter in heath-scrub under *Agonis flexuosa* near *Acacia*, *Melaleuca*, *Olax*, *Spyridium* and other shrubs of Myrtaceae and Fabaceae, 19 July 2011, *N.L.Bougher NLB833*; holo: PERTH 08318476; iso: TENN 066737 (KP170929-nLSU, KP308751-ITS, KM245997-*rpb2*).

Pileus 7–20 mm wide, at first pulvinate to broadly convex, becoming convex to bluntly campanulate, margin initially incurved, becoming strongly decurved; surface dry; velipellis sparse or absent, smooth at the centre, surrounded by matted or woolly fibrils, these occasionally forming small fibrillose squamules; fibrils more radially oriented near the extreme margin, becoming rimulose; uniformly fulvous or orange-brown (10YR 5/6) to yellowish brown (10YR 5/4), sometimes light grey over the disc when the velipellis is present; context whitish to pale yellow, not changing colour where cut or bruised. *Lamellae* adnate, ascending-adnate to broadly adnexed with a brief decurrent tooth or shallow sinuation, close, with several tiers of lamellulae, light grey, very pale brown to pale brown when young (not avellaneous), becoming light olive-brown to olive-brown (near 2.5Y 5/6), broad, up to 3 mm deep, edges pallid and densely fimbriate.

Plate 71. *Inocybe vestipedes*, holotype (PERTH 08242062). Photograph — P.B.Matheny.

Plate 72. *Inocybe vestipedes* (PERTH 08241848). Photograph — P.B.Matheny.

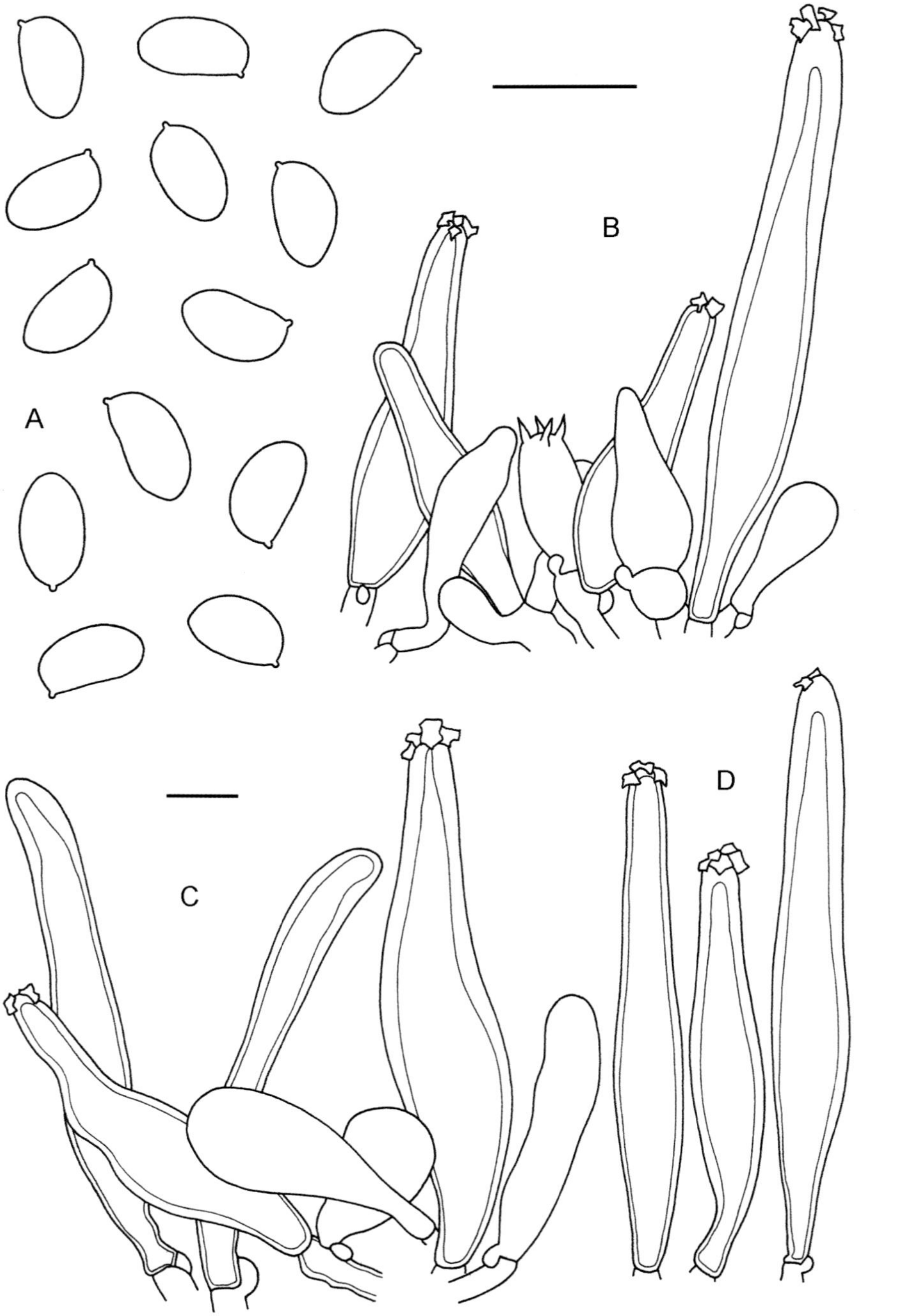

Figure 37. *Inocybe vestipedes,* all holotype (PERTH 08242062). **A,** Basidiospores; **B,** Cheilocystidia and paracystidia; **C,** Caulocystidia and cauloparacystidia; **D,** Pleurocystidia. Scale bars = 10 μm (longer bar for spores only).

Stipe 10–30 × 3–8 mm, even, cylindrical, or more often tapered towards the base which is not swollen or bulbous; surface dry, never pruinose, sparsely furfuraceous to finely fibrillose; fibrils more densely accumulated above the base, at times becoming tattered, dull whitish with some brownish or isabelline undertones, contrasting in colour with the pileus; context solid, dull straw. *Odour* none. Pl. 73, 74.

Basidiospores 9.5–11.5 × 6–7 µm [mean 10.6 × 6.5 µm; Q 1.50–1.77; Q mean 1.63 (n = 26/2)], smooth, amygdaliform to broadly amygdaliform, light yellowish brown; apices bluntly pointed. *Basidia* 33–40 × 8–11.5 µm, 4-sterigmate, narrowly clavate, at times widest below the apex, hyaline. *Pleurocystidia* 55–75 (–110) × 12–17 µm, narrowly fusiform or subcylindrical; bases rounded or pedicellate; apices obtuse, crystalliferous, mostly thick-walled (1–3 µm thick), occasionally thin-walled, hyaline. *Cheilocystidia* often shorter and more variable in shape than the pleurocystidia, including some broader fusiform ventricose and subcapitate forms, mostly hyaline, occasionally ochraceous, mixed with abundant clavate thin-walled paracystidia. *Caulocystidia* similar to cheilocystidia but often longer (to 115 µm long), narrowly fusiform, cylindrical or sometimes lageniform, hyaline or ochraceous, numerous at the extreme apex of the stipe and mixed with abundant clavate thin-walled or slightly thick-walled cauloparacystidia; none observed below the mid-region of the stipe. *Pileipellis* a cutis of cylindrical to somewhat inflated hyphae mostly 6–14 µm wide, smooth or with weak encrustations, thin-walled or occasionally with thick walls, pale brown in mass. *Clamp connections* present. Fig. 38.

Known only from litter in shrub-heath at Two Peoples Bay Nature Reserve in south-western W.A. Fruiting July.

Additional specimen examined: W.A.: type locality, scattered to gregarious on sand in coastal shrub under *Melaleuca* (shrub), *Agonis*, *Spyridium*, *Acacia* and other fabaceous and myrtaceous shrubs, 19 July 2011, *P.B.Matheny PBM3649* (TENN 066477; KP170930-nLSU, KP308752-ITS, KM245998-*rpb2*).

Etymology: From the Latin *laureatus* (crowned with a laurel as a sign of victory, triumphant) - the last named species of this work.

Diagnostic features: Fruit bodies medium sized, robust or squat. Pileus orange-brown, fulvous or yellowish brown, mainly smooth or matted-fibrillose, but it can develop fibrillose squamules; lamellae pale grey, becoming olivaceous or olivaceous-brown; stipe solid, tapering towards the base. Spores smooth, amygdaliform, comparatively large, 9.5–11.5 × 6–7 µm; pleurocystidia narrowly fusiform to cylindrical, thin- or thick-walled.

Inocybe laureata inhabits the shrub-heath in south-western Western Australia along with at least three similar species. One, *I. avellaneifolia*, differs in having avellaneous lamellae when immature, a darker pileus and narrower spores. *Inocybe rufescens* has a reddish brown pileus and rubescent flesh, while *I. leucotaenia*, from the Perth area, is characterised by its pinkish lamellae when young, stipe context that becomes pink after exposure, and bands of white fibrils radiating towards the pileal margin, in contrast to the umbrinous or reddish brown surface colour. Phylogenetic analysis supports the autonomous position of *I. laureata* (Fig. 7).

Plate 73. *Inocybe laureata*, holotype (PERTH 08318476). Photograph — N.L.Bougher.

Plate 74. *Inocybe laureata* (TENN 066477). Photograph — P.B.Matheny.

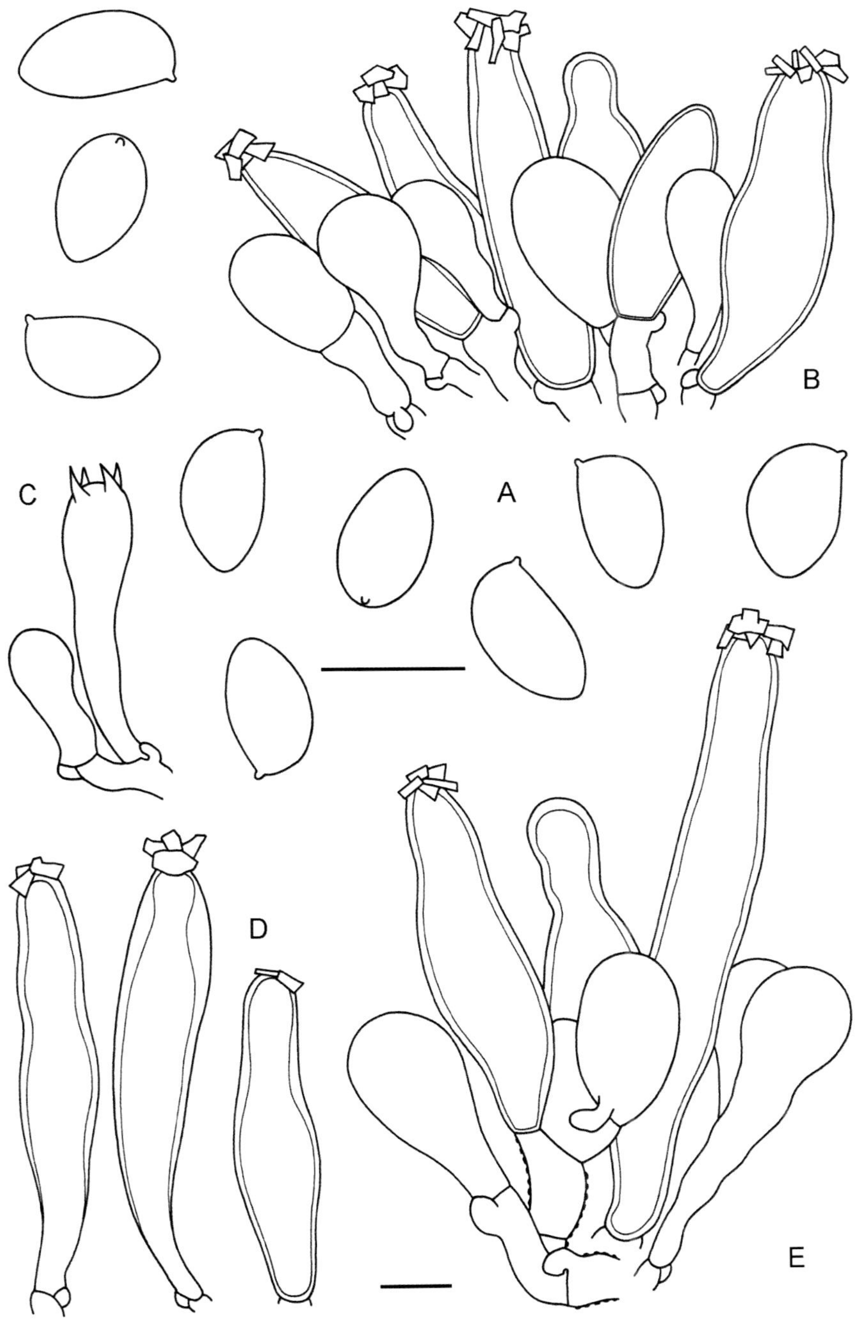

Figure 38. *Inocybe laureata,* all holotype (PERTH 08318476). **A,** Basidiospores; **B,** Cheilocystidia and paracystidia; **C,** Basidia; **D,** Pleurocystidia; **E,** Caulocystidia and cauloparacystidia. Scale bars = 10 µm (longer bar for spores only).

INOCYBE HYMENOCARPA GROUP

This group is a small robust clade of three species from Australia. In all three species, the pileus is hemispherical when young, the stipe coarsely fibrillose, and caulocystidia are absent or found only at the extreme apex. The *I. hymenocarpa* group is sister to a fourth but unclarified lineage. All are presently known only from southwestern Western Australia.

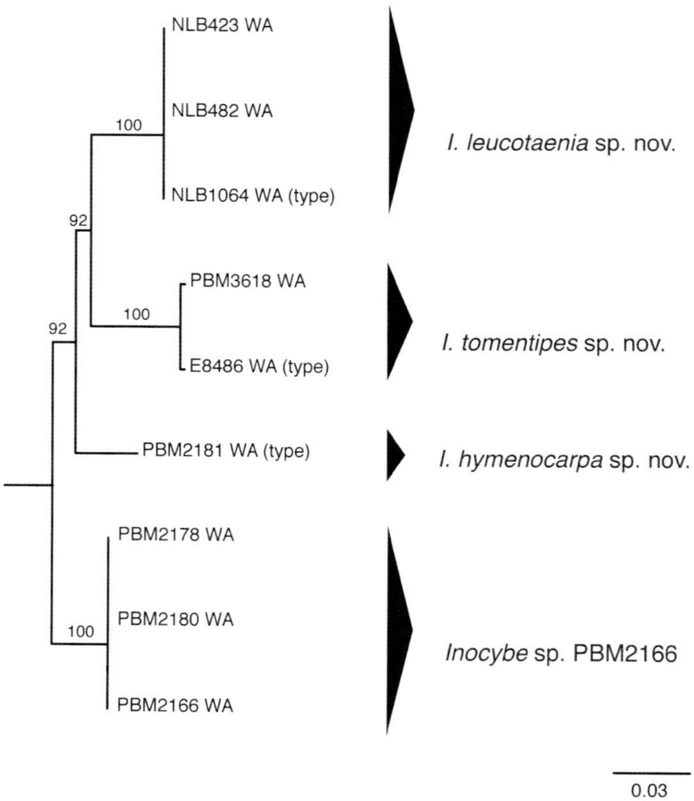

Figure 39. Maximum Likelihood phylogram of the *Inocybe hymenocarpa* group produced from the analysis of the STAC group shown in Fig. 7. All species shown occur in Australia. Bootstrap values are indicated above branches.

34. **Inocybe hymenocarpa** Matheny & Bougher, *sp. nov.*

MycoBank: MB 809380

Type: Walk-through Tree Trail, Karri Valley Resort, L. Beedelup, 22 km NE of Pemberton, W.A., gregarious to scattered on soil along trail in Karri forest under *Eucalyptus diversicolor*, *Trymalium spathulatum* and *Chorilaena quercifolia*, 18 Aug. 2001, *P.B.Matheny PBM2181, E7030*; holo: PERTH 07676603; iso: TENN 066684 (EU307851-nLSU, KP171132-ITS, EU307852-*rpb1*, EU307853-*rpb2*).

Pileus 13–40 mm wide, parabolic to conical when young, becoming broadly conical, the margin incurved to decurved, edge of margin fringed with denticulate fibrils on expanded specimens, lacking an umbo; surface dry; velipellis not observed, centre at first unbroken but later becoming cracked-areolate, areolate-scaly or coarsely squarrose with age, fibrillose and rimulose towards the margin, the centre fulvous (10YR 5/8; Sudan Brown to Ochraceous-Tawny), but shading yellowish brown (10YR 7/6 or 5/6; Tawny-Olive, Clay Colour, or Cinnamon-Buff) towards the margin; context pale brown, up to 2 mm thick under the disc; all tissues darkening with KOH (PERTH 7606974). *Lamellae* narrowly adnate, subdistant, 20–22 L, with several tiers of lamellulae; light brown (7.5YR 6/4; Avellaneous) when young, becoming yellowish brown (10RY 5/4) with age; ventricose, up to 3 mm deep; edges pallid, indistinctly fimbriate. *Stipe* 30–55 × 2–3 mm, even, terete, often flexuous; surface with a prominent cortina that can leave a defined ring zone in a superior position, whitish cottony fibrillose to fibrillose-pruinose above the cortinate insertion, fibrillose below; very pale brown (10YR 7/4 to 7/6; Cinnamon-Buff to Tawny-Olive); context solid, coloured ±like the surface. *Odour* unremarkable. Pl. 75, 76.

Basidiospores 7.5–9.5 (–10.0) × 5.0–6.0 µm [mean 8.4 × 5.3 µm; Q 1.42–1.82 (–1.90); Q mean 1.58 (*n* = 31/2)], smooth, amygdaliform, occasionally almost citriniform; apices conical, bluntly pointed, less often mucronate, yellowish brown; apiculus small. *Basidia* 25–30 × 8–9 µm, clavate, hyaline, 4-sterigmate. *Pleurocystidia* 68–93 × 11–15 µm, narrowly fusiform, with elongated necks and a slender basal pedicel, apices obtuse and typically bearing crystals, walls mostly 1–3 µm thick. *Cheilocystidia* similar to pleurocystidia, somewhat shorter or lageniform, scattered among paracystidia and some basidia. *Caulocystidia* absent; stipitipellis composed of interwoven superficial hyphae, these mostly 5–20 µm wide, terminal cells generally tapered towards the apices but largely undifferentiated, thin- to slightly thick-walled; subtending cells encrusted at times. *Pileipellis* a cutis giving rise to broad fascicles of interwoven to trichodermally arranged hyphae that are yellowish brown or Ochraceous-Tawny in mass; hyphae cylindrical to inflated, to 23 µm wide; walls thin to slightly thick-walled, at most faintly encrusted. *Clamp connections* present. Fig. 40.

Occurs in south-western W.A.; on soil in Jarrah and Karri forest under *Eucalyptus marginata* and *Corymbia calophylla* or *E. diversicolor* with a *Trymalium* (Rhamnaceae) and *Chorilaena* (Rutaceae) understorey. Fruiting June–Aug.

Additional specimens examined: W.A.: Plavins forest block, Murray River Rd, Dwellingup, on soil in Jarrah forest under *Eucalyptus marginata* and *Corymbia calophylla*, 26 June 1997, *S.Bolsenbroek E5885* (PERTH 07608489, TENN 068359); *loc. id.*, same habitat, 26 June 1997, *S.Bolsenbroek E5886* (PERTH 07608497, TENN 068360); *loc. id.*, Plavins (burnt) forest block, Murray River Rd, Dwellingup, on soil in Jarrah forest under *Eucalyptus marginata* and *Corymbia calophylla*, 23 July 1998, *S.Bolsenbroek & J.Catchpole E6023* (PERTH 07606974, TENN 068362).

Etymology: From the Greek *hymen* (a membrane) and *karpos* (a fruit), in reference to the well-developed stipe vesture and partial veil that covers the immature hymenium (fertile layer).

Diagnostic features: Fruit bodies medium-sized but with a slender habit. Pileus hemispherical and with a denticulate margin when young; coarse scales developing over the pileus centre with age; partial veil well-developed on the stipe, at times forming a ring-zone. Pleurocystidia long and narrow.

Inocybe hymenocarpa forms a monophyletic group with *I. leucotaenia* and *I. tomentipes* (Fig. 39). It differs from *I. leucotaenia* in having a context that does not stain pink, longer pleurocystidia, and the absence of distinct whitish fibrils overlain along the pileus margin. *Inocybe tomentipes* has a more fibrillose-streaked (never scaly) pileus, slightly broader spores, and it lacks a cortinate ring-zone. It is also somewhat similar to *I. cingulata* E.Horak, an associate of Fagaceae in Papua New Guinea (Horak, 1980), which has a distinct cortinate ring-zone and amygdaliform-mucronate spores. However, the latter has a spermatic to fruity odour. *Inocybe scabriuscula* E.Horak, from New Zealand (Horak, 1977), appears outwardly similar but differs in the woolly to scaly stipe covering, smaller spores (6.5–8.5 × 4.0–5.0 µm), and an association with *Nothofagus*. *Inocybe albovestita* E.Horak, also from New Zealand (Horak, 1977) and superficially similar, differs in having a reddish brown stipe and an association with *Nothofagus*. Sequences of the last two species represented by PDD 88624 (*I. scabriuscula*) and PDD 71309 (*I. albovestita*) are also different to *I. hymenocarpa*. *Inocybe scabriuscula* is part of the *I. serrata* group (Fig. 66), whereas *I. albovestita* is allied elsewhere in a poorly supported phylogenetic position in the STAC group (Fig. 7).

An additional collection from a different locality in W.A. that could be referred to as *I. hymenocarpa* is PERTH 00910449 (*UWA 2635*). However, the specimen is more robust, has an umbonate pileus, and gill edges with large (to 50 µm diam.), vesiculose or saccate cystidia mixed with more typical cheilocystidia. DNA sequences of this collection and from additional collections such as *E5898* (PERTH 07608616, TENN 068361) will be required to measure the significance of morphological variation in *I. hymenocarpa*.

75

76

Plate 75. *Inocybe hymenocarpa*, holotype (PERTH 07676603). Photograph — P.B.Matheny.

Plate 76. *Inocybe hymenocarpa* (PERTH 07608497). Photograph — N.L.Bougher.

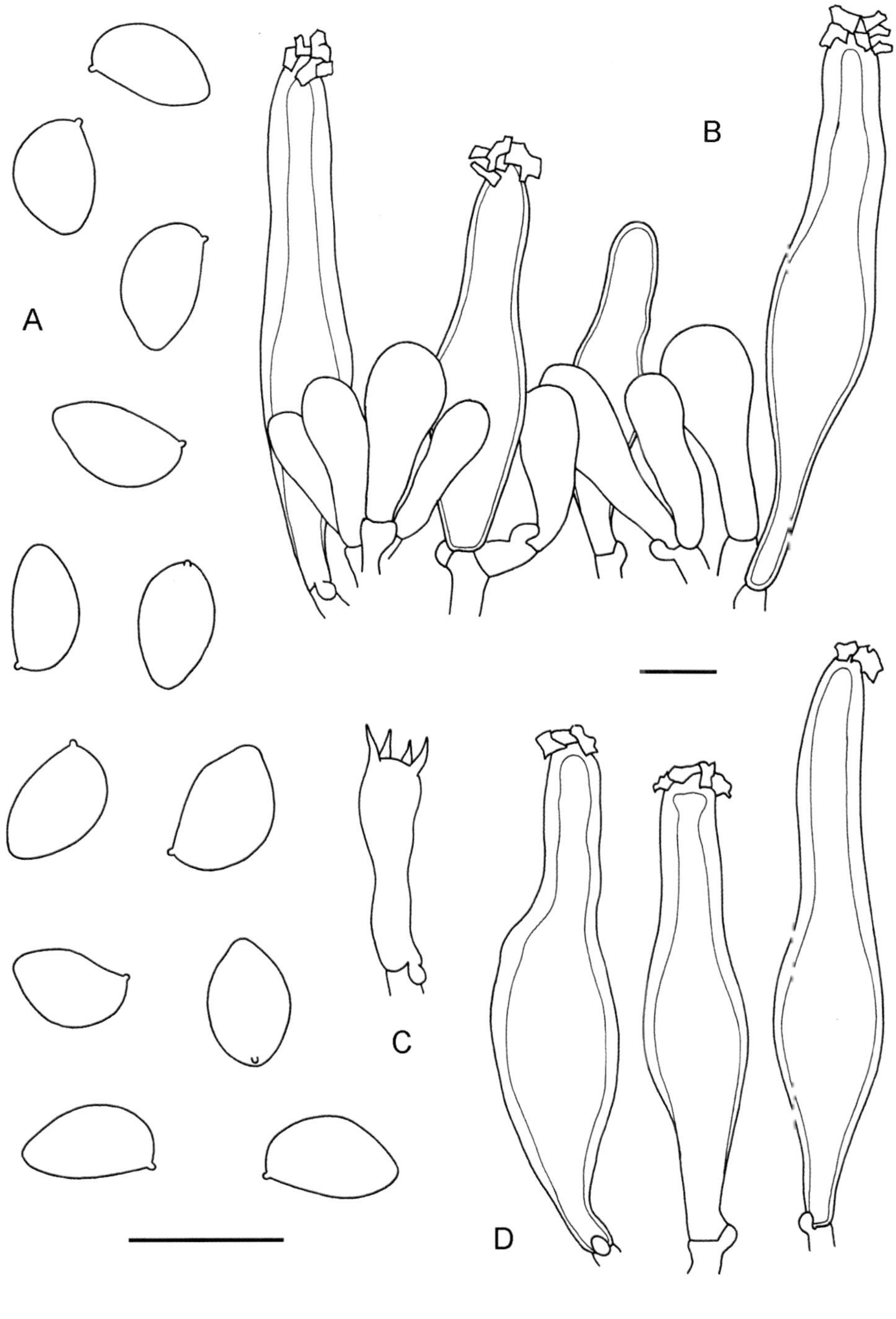

Figure 40. *Inocybe hymenocarpa,* all holotype (PERTH 07676603). **A,** Basidiospores; **B,** Cheilocystidia and paracystidia; **C,** Basidium; **D,** Pleurocystidia. Scale bars = 10 μm (longer bar for spores only).

35. **Inocybe leucotaenia** Matheny & Bougher, *sp. nov.*

MycoBank: MB 809386

Type: N side of Tuart carpark, Bold Park, Bold Park Drive, City Beach, Perth, W.A., solitary or more often in tight or caespitose clusters, on soil among litter in open woodland near *Eucalyptus gomphocephala* and *Corymbia calophylla*, 8 Aug. 2012, *N.L.Bougher NLB1064*; holo: PERTH 08363684; iso: TENN 067770 (KF853408-nLSU, KF871784-ITS, KF891961-*rpb2*).

Pileus 5–35 mm wide, at first hemispherical to oblong, becoming convex with a persistently incurved margin, eventually broadly convex to applanate, occasionally with an undulate margin; umbo low or absent; surface densely fibrillose; fibrils whitish near the margin when young; centre smooth; margin entire or occasionally splitting, with whitish matted-silky fibrils from the evanescent cortina, later developing small appressed and imbricate scales, with the centre becoming scurfy-areolate or recurved-scaly; young buttons greyish silvery brown (near 5YR 4/4) where whitish fibrils are abundant, or darker brown (umbrinous, near 7.5YR 4/4) where less fibrillose or more water-soaked, similarly coloured with age or reddish brown (near 5YR 3/4), margin often paler except in extreme age, due to the accumulation of fibrils; centre not paler than elsewhere. *Lamellae* adnexed to more broadly attached, close, with numerous lamellulae, dull pinkish, light reddish brown or pale tan with a pinkish tinge (near 5YR 5/4 or 6/4) when young, becoming umber-brown (near 7.5YR 4/4) or rich chocolate-brown with age, 4–5 mm deep, ventricose, the edges pallid and minutely fimbriate. *Stipe* 15–45 × 2–7 mm, even or mostly cylindrical, at times tapering towards the base, bent or contorted; surface densely longitudinally fibrillose over the whole length, partial veil present as a whitish silky cortina but leaving no trace by maturity, pruinose at the extreme apex only, whitish and becoming duller with age as fibrils become more sparse; base white, often tightly binding a soil clod; context solid, dull in stipe but turning pinkish within 3–5 minutes of being cut. *Odour* not distinctive. Pl. 77, 78 (see also Pl. 3B, 4F).

Basidiospores (7.5–) 8.0–10.5 × 5.5–6.5 µm [mean 9.1 × 5.8 µm; Q (1.25–) 1.36–1.76 (–1.82); Q mean 1.56 (*n* = 50/2)], smooth, yellowish brown, broadly amygdaliform or amygdaliform, less often elliptic, the apices pointed or bluntly so; apiculus not distinct. *Basidia* 27–32 × 7–10 µm, clavate, hyaline, 4-sterigmate. *Pleurocystidia* 53–75 × 11–15 µm, mostly fusiform, occasionally skittle-shaped (especially in the type collection), less often utriform, typically with a tapered basal pedicel, only slightly thick-walled (1–2 µm), hyaline; apices occasionally bearing crystals, sometimes swollen or indistinctly subcapitate (especially in the type). *Cheilocystidia* similar to pleurocystidia but at times shorter, mixed with paracystidia. *Caulocystidia* absent, the end cells mostly undifferentiated except at the extreme apex where some end cells of interwoven superficial hyphae are fusiform-ventricose, thin-walled, with or without apical crystals; cauloparacystidia absent. *Pileipellis* a cutis of cylindrical hyphae, these golden brown to yellowish brown in mass, mostly 7–12 µm wide, thin-walled, sometimes with faint encrustations. *Clamp connections* present. Fig. 41.

Known only from City Beach, Perth, W.A., on soil or among moss in open woodland near *Eucalyptus gomphocephala* and *Corymbia calophylla*. Fruiting June–Aug.

Additional specimens examined: W.A.: Bold Park, area to S of Reabold Hill, City Beach, Perth, on the ground in woodland near *Eucalyptus gomphocephala* and *Corymbia calophylla*, 11 June 2008, *N.L.Bougher & J.Keeble NLB423* (PERTH 08072477; KF853409-nLSU, KF871785-ITS, KF891962-*rpb2*); Bold Park, N side of Tuart carpark, Bold Park Drive, City Beach, Perth, on soil in woodland of *Eucalyptus gomphocephala*, 15 July 2008, *N.L.Bougher NLB482* (PERTH 08073171, TENN 067771; KF853407-nLSU, KF871783-ITS, KF891960-*rpb2*); Bold Park, Reabold Hill, City Beach, Perth, among moss on almost bare limestone soil, near *Acacia* spp. and *Melaleuca huegelii* with *Eucalyptus gomphocephala* nearby, 1 Aug. 2011, *N.L.Bougher NLB900* (PERTH 08320306).

Etymology: From the Greek *leukos* (white) and *tainia* (a ribbon), in reference to the conspicuous white superficial fibrils towards the margin of the pileus.

Diagnostic features: Pileus hemispherical when young, with appressed, matted white fibrils towards the margin; lamellae initially pinkish-tinged; stipe entirely fibrillose (non-pruinose), its context slowly becoming pink after exposure.

Plate 77. *Inocybe leucotaenia*, holotype (PERTH 08363684). Photograph — N.L.Bougher.

Plate 78. *Inocybe leucotaenia* (PERTH 08072477). Photograph — N.L.Bougher.

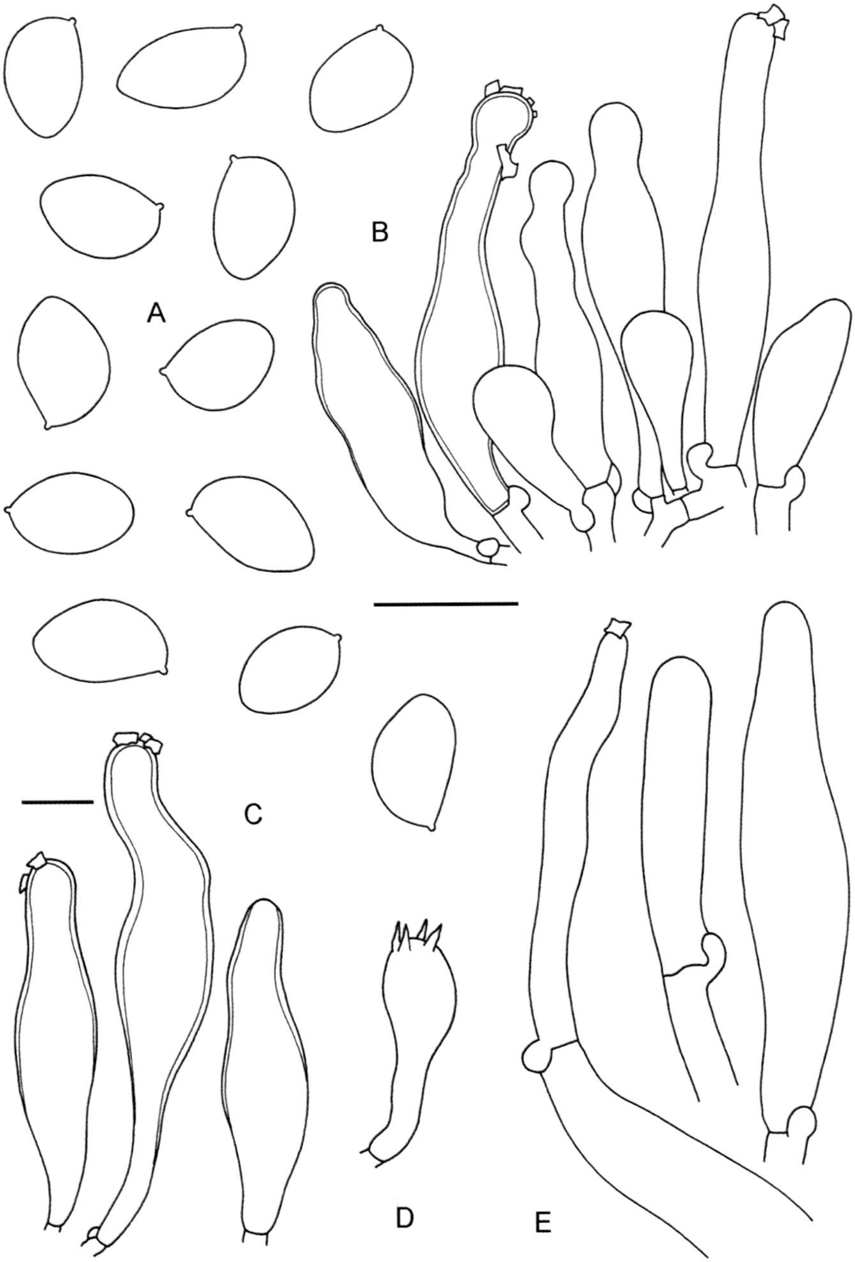

Figure 41. *Inocybe leucotaenia,* all holotype (PERTH 08363684). **A,** Basidiospores; **B,** Cheilocystidia and paracystidia; **C,** Pleurocystidia; **D,** Basidia; **E,** Caulocystidioid and terminal cells from stipe apex. Scale bars = 10 µm (longer bar for spores only).

Three other species, *I. tomentipes*, *I. hymenocarpa* and *I. clypeata*, also have hemispherical immature pilei, a fibrillose to coarsely fibrillose stipe, and either lack caulocystidia or have them only at the extreme apex of the stipe. However, *I. tomentipes* differs in cystidial morphology (cheilocystidia lack capitate apices, pleurocystidia are more slender and cystidial walls are thicker), and by spores that tend to be mucronate. Furthermore, its variegated-streaked pileus lacks conspicuous white marginal fibrils. *Inocybe hymenocarpa* has more slender fruit bodies, larger pleurocystidia, a denticulate pileal margin, and a cortinate ring zone, while *I. clypeata* has longer spores (mean 10.5 × 6.0 µm), pleurocystidia with a long basal pedicel, and its context does not stain pink. Neither *I. tomentipes* nor *I. hymenocarpa* have been observed with context that slowly becomes pink upon exposure, but this trait should be examined more carefully. Phylogenetically *Inocybe leucotaenia*, *I. tomentipes* and *I. hymenocarpa* form a monophyletic clade sister to *I. clypeata* (Fig. 39). DNA sequences (ITS, nLSU and *rpb2* regions) distinguish all four species (Fig. 39).

36. Inocybe tomentipes Matheny & Bougher, *sp. nov.*

MycoBank: MB 809474

Type: Brixton Street Wetland, Kenwick, Perth, W.A., in litter in open woodland of *Corymbia calophylla*, 27 June 2007, *N.L.Bougher E8486*; holo: PERTH 08093334 (KF853411-nLSU, KF871787-ITS).

Pileus 15–20 mm wide, obtusely conical to convex, lacking an umbo; margin decurved; surface dry, smooth to finely fibrillose, without recurved or erect scales, not hygrophanous, lacking a velipellis; brown mixed with pale brown or pale greyish brown areas or with radial streaks (10YR 5/3 or 6/3; 6E7 to 6F7); context whitish, not changing colour upon exposure, c. 2–3 mm thick under the disc, not thin-fleshed. *Lamellae* adnexed or uncinate, subdistant to moderately close, 22–30 L, with several tiers of lamellulae; brown (7.5YR 5/4) to chestnut-brown (near 6E6; Tawny-Olive to Sayal Brown), broad, to 5 mm deep, the edges distinctly or indistinctly pallid-fimbriate. *Stipe* 30–45 × 2–3 mm, terete, occasionally filiform, a partial veil not directly observed but presumably present when immature; surface nowhere pruinose, rather densely longitudinally fibrillose with soft pale brown fibrils; ground colour brown to yellowish brown (10YR 5/3 to 5/4) or Clay Color, white at the extreme base; context solid, becoming hollow; cortex concolorous with the surface, interior pallid and pithy, bright yellow-cream in *E8486*. *Odour* not remarkable. Pl. 79, 80.

Basidiospores 8.0–10.0 × 5.0–6.5 µm [mean 9.2 × 5.9 µm; Q (1.31–) 1.35–1.73; Q mean 1.55 (n = 25/1)], smooth, mostly broadly amygdaliform, with pointed apices, some citriniform due to mucronate apices, others elliptic with more rounded apices, yellowish brown; apiculus distinct. *Basidia* 28–32 × 8–9 µm, clavate, hyaline, 4-sterigmate. *Pleurocystidia* 68–77 × 10–12 µm, narrowly fusiform to narrowly lageniform, numerous; apices obtuse, not swollen or subcapitate, often bearing crystals, thick-walled (2–3 µm), pale yellow to hyaline. *Cheilocystidia* similar to pleurocystidia, often shorter, mixed with paracystidia. *Caulocystidia* absent; stipitipellis composed of superficial hyphae that are cylindrical, hyaline, smooth, often flexuous, with slightly thickened walls (to 1.5 µm thick); terminal cells not differentiated, mostly 6–12 µm wide and cylindrical, fusiform or irregularly cylindrical, often with subacute apices. *Pileipellis* with a superficial layer of interwoven hyphae that are smooth, mostly 5–10 µm wide and thin-walled, subtended by a dark yellowish brown pigmented zone of encrusted cylindrical hyphae mostly 6–15 µm wide and ±thin-walled. *Clamp connections* present. Fig. 42.

Occurs among litter on unburnt soil in woodland of *Corymbia calophylla* and in Jarrah forest (*Eucalyptus marginata*, *C. calophylla*, *Agonis* spp.) mixed with *Allocasuarina* and *Astartea* (myrtaceous shrub) in south-western W.A. Fruiting June and July.

Additional specimen examined: W.A.: 1 km N of Weld R., Southwest Hwy, 34°41'16.6"S, 116°31'11.5"E, scattered singly on unburnt soil in Jarrah forest under *Allocasuarina*, *Agonis* and *Astartea*, 14 July 2011, *P.B.Matheny PBM3618* (PERTH 08319677, TENN 066570; KF853410-nLSU, KF871786-ITS, KF891963-*rpb2*).

Etymology: From the Latin *tomentum* (a densely interwoven hair covering) and *pes* (a foot), in reference to the coarsely fibrillose stipe.

Plate 79. *Inocybe tomentipes*, holotype (PERTH 08093334). Photograph — N.L.Bougher.

Plate 80. *Inocybe tomentipes* (PERTH 08319677). Photograph — P.B.Matheny.

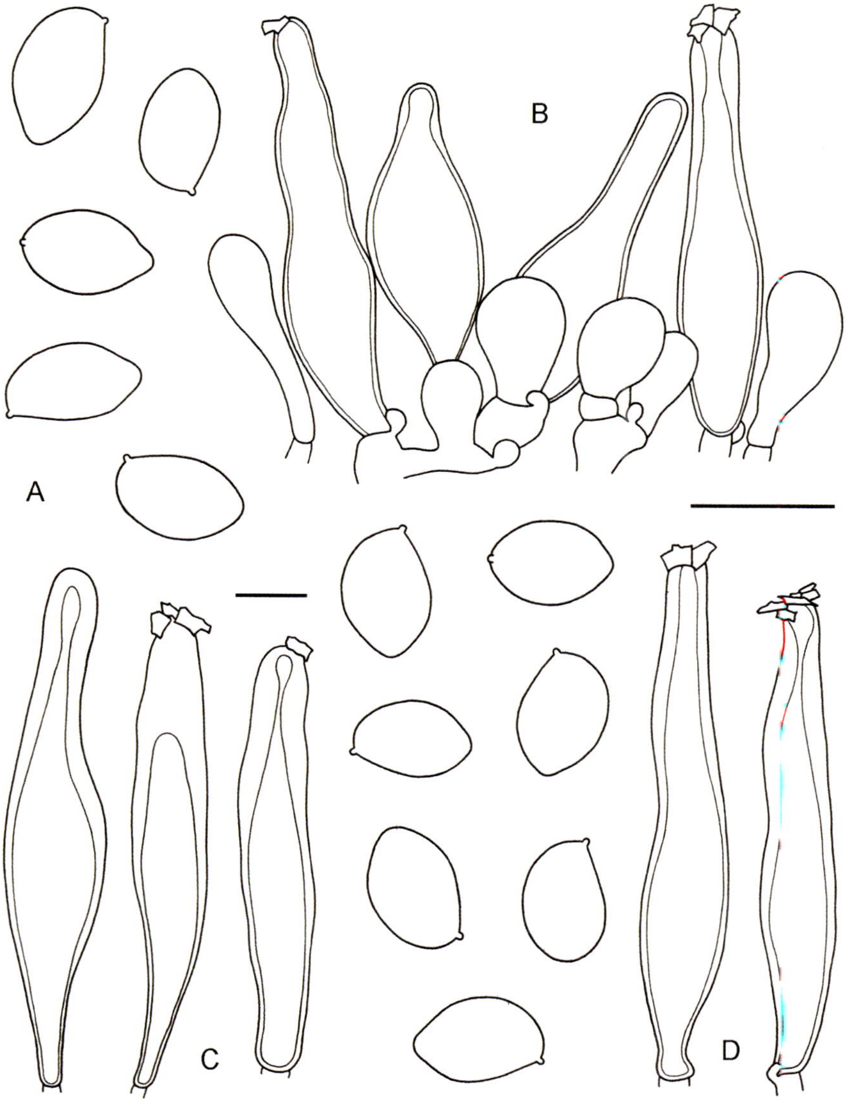

Figure 42. *Inocybe tomentipes*, all holotype (PERTH 08093334), except D (PERTH 08319677). **A,** Basidiospores; **B,** Cheilocystidia and paracystidia; **C,** Pleurocystidia; **D,** Pleurocystidia.

Scale bars = 10 µm (longer bar for spores only).

Diagnostic features: Fruit bodies with slender stipe; pileus pale brown to brown-streaked, smooth to finely fibrillose (not scaly); stipe coarsely fibrillose, never pruinose. Basidiospores smooth, broad, sometimes citriniform; pleurocystidia very slender; caulocystidia absent.

Inocybe tomentipes is one of three closely related but distinct taxa (Fig. 39) from south-western Western Australia that have a heavy partial veil or a coarsely fibrillose stipe covering and comparatively broad spores. It could be confused with *I. olivaceohinnulea*, but the latter has an olivaceous pileus, pinkish brown lamellae, a stipe that becomes dark towards the base, and somewhat smaller spores. *Inocybe leucotaenia* differs in having a pileus covering with distinct whitish fibrils adorning the margin, a tendency to become scaly, young pinkish to avellaneous lamellae, a stipe context that becomes slowly pink upon exposure, and its occurrence in Tuart (*Eucalyptus gomphocephala*) urban bushland. *Inocybe hymenocarpa* also has a pileus surface that becomes scaly, but sometimes with a distinctive cortinate ring-zone. Moreover, the spores are somewhat smaller than those of *I. tomentipes* and *I. leucotaenia*, although the size ranges do overlap.

INOCYBE GENICULATA GROUP

This small group with robust phylogenetic support comprises five species. Only one, *I. geniculata*, is described here, the only one which occurs in Australia, where it is present in Tasmania, Victoria and south-western Western Australia. The other four species occur either in southern South America or New Zealand. The group is unified by the entirely pruinose stipe and presence of caulocystidia and cauloparacystidia down to the base of the stipe.

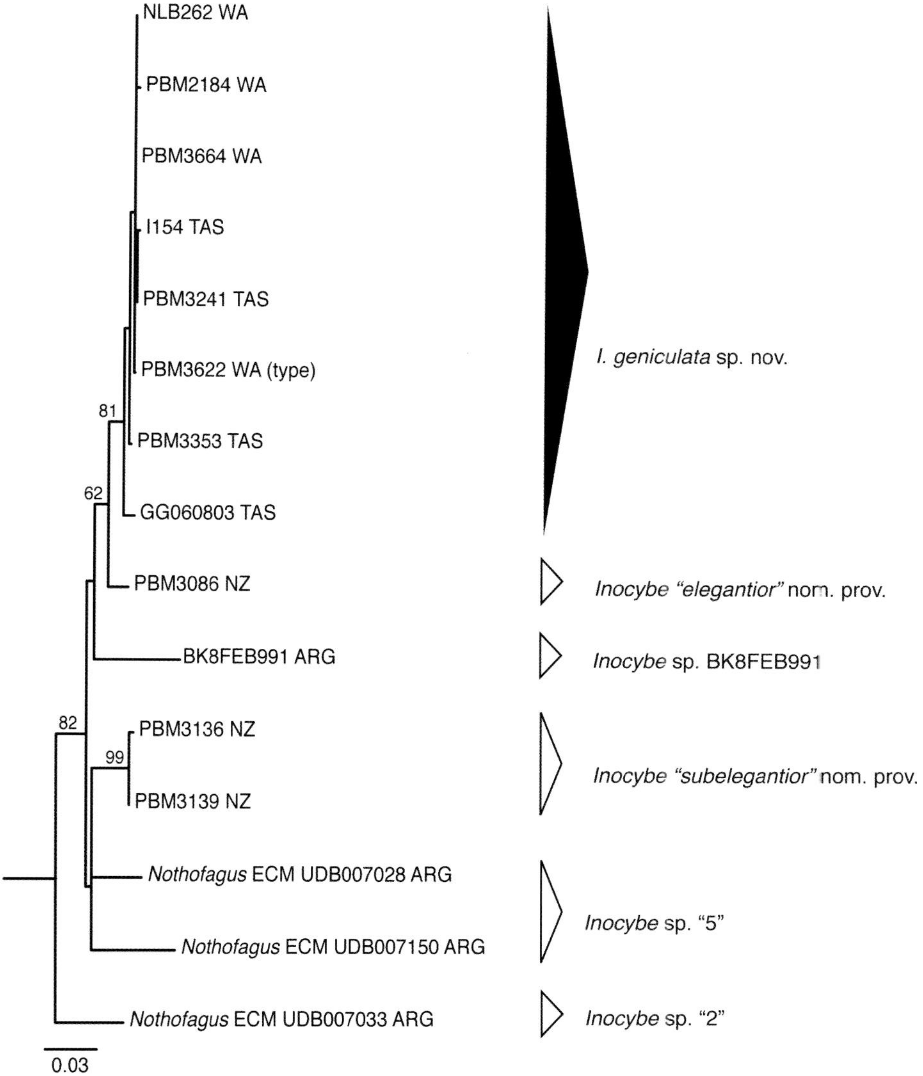

Figure 43. Maximum Likelihood phylogram of the *Inocybe geniculata* group produced from the analysis of the STAC group shown in Fig. 7. Australian taxa are indicated by black-filled triangles. White triangles indicate lineages from New Zealand or Argentina. Bootstrap values are shown above branches.

37. Inocybe geniculata Matheny & Bougher, *sp. nov.*

MycoBank: MB 809332

Type: Campbell Rd and Inlet Drive, Denmark, W.A., 34°59'11.2"S, 117°20'57.2" E, on soil in leaf litter in Karri forest under *Eucalyptus diversicolor* and *Agonis flexuosa*, 16 July 2011, *P.B.Matheny PBM3622*; holo: PERTH 08319642; iso: TENN 066464 (KJ801171-nLSU; KJ778860-ITS).

Pileus 10–22 mm wide, initially conical, expanding to plano-convex or plane with a low obtuse umbo; margin decurved, becoming undulate or raised, surface dry, disc with upturned scales concolorous with the pileus surface, radially fibrillose towards the margin, occasionally rimulose but often torn; dark yellowish brown (10YR 3/4; Bay Brown) to very dark brown, becoming excoriate and brown beneath the torn surface; context thin-fleshed, dull brown where moist. *Lamellae* adnexed, subdistant, 24–28 L, with several tiers of lamellulae, when young pink to light brown (7.5YR 7/4 to 6/4; Light Pinkish Cinnamon to Cinnamon), becoming yellowish brown (10YR 5/3) with age, ventricose, medium to broad, up to 4 mm deep, edges not distinctly pallid-fimbriate. *Stipe* 25–35 × 2–3 mm, often bent or flexuous, even to slightly swollen at the base, veil absent, pruinose its entire length, reddish yellow (7.5YR 7/6 to 6/6; Ochraceous-Tawny to Cinnamon) to strong brown (7.5YR 5/6), becoming dark brown at the base with age; context solid, coloured like the surface, with a pallid interior. *Odour* acidulous. Pl. 81–83.

Basidiospores 7.5–10.5 × 4.5–6.0 µm [mean 9.0 × 5.3 µm; Q (1.36–) 1.46–1.91; Q mean 1.70 (n = 71/4)], smooth, amygdaliform to subamygdaliform with pointed to bluntly pointed apices, yellowish brown; apiculus small and indistinct. *Basidia* 23–28 × 7–10 µm, clavate, hyaline, 4-sterigmate. *Pleurocystidia* 56–72 × 11–18 µm, fusiform to lageniform, thick-walled (walls 2.5–5.0 µm), hyaline or pale yellow; base tapered or sessile and without a distinct basal pedicel; apices obtuse and crystalliferous. *Cheilocystidia* 42–66 × 11–21 µm, similar to pleurocystidia but often shorter and lageniform, mixed with intermediate-sized cells with thick walls and paracystidia, in clusters. *Caulocystidia* descending the entire length of the stipe, similar to cheilocystidia but occasionally longer (to 91 µm), in dense clusters mixed with cauloparacystidia. *Pileipellis* composed of trichodermal bundles of hyphae mostly 7–20 µm wide, repent towards the margin, fuscous or dark yellowish brown in mass, with encrusted somewhat thickened walls. *Clamp connections* present. Fig. 44.

Known from W.A., Vic. and Tas.; on soil, typically in leaf litter in wet sclerophyll forest or tall wet forest under *Eucalyptus*, *Corymbia*, *Pomaderris* and/or *Allocasuarina*, in temperate regions. Fruiting Mar.–Aug.

Additional specimens examined: W.A.: K. & J.Ling property, Corrigin, 32°38.3'33.3"S, 117°07.1'06.6"E, under *Allocasuarina campestris*, 2 Aug. 2000, *N.L.Bougher & I.C.Tommerup E6648* (PERTH 07674511); Walk-through Tree Trail, Karri Valley Resort, L. Beedelup, 22 km NE of Pemberton, on soil in Karri forest under *Eucalyptus diversicolor*, 17 Aug. 2001, *P.B.Matheny PBM2200 (E7047)* (PERTH 07676751); *loc. id.*, 18 Aug. 2001, *P.B.Matheny PBM2184, E7033* (PERTH 07676646; EU555468-nLSU, KJ778864-ITS, EU555467-*rpb2*); Old Vasse Rd, Warren Natl Park, 34°49'33.3"S, 115°96'58.3"E, on soil among litter in tall wet forest of *C. calophylla* and *E. diversicolor*, 18 June 2007, *N.L.Bougher NLB262* (PERTH 08092257; KJ801175-nLSU, KJ778865-ITS, KJ811571-*rpb2*); Woodlands Rd, Porongurup Natl Park, 34°40'06.4"S, 117°52'53.6"E, on soil in leaf litter on 4 year-old burn in tall forest of *C. calophylla*, *E. diversicolor* and *E. marginata*, 21 July 2011, *P.B.Matheny PBM3664* (TENN 066592, PERTH 08319235; KJ801176-nLSU, KJ778866-ITS, KJ811572-*rpb2*). Vic.: Bonang Hwy at Cooney Ridge Rd, N of Orbost, in recently burnt habitat under *E. sieberi*, 29 Apr. 1995, *N.L.Bougher & R.Vigalys E5262* (PERTH 07710453, TENN 067074). Tas.: Chauncy Vale, on soil under *Eucalyptus* and *Pomaderris*, in pocket of wet sclerophyll forest, 11 June 2009, *P.B.Matheny, G.Gates & D.Ratkowsky PBM3241* (TENN 067071; KJ801172-nLSU, KJ778861-ITS, KJ811568-*rpb2*); Lower Kermandie, on soil in wet sclerophyll forest under *Eucalyptus* and *Pomaderris*, 19 May 2010, *F.Karstedt PBM3353* (TENN 067072; KJ801173-nLSU, KJ778862-ITS, KJ811569-*rpb2*); Pipeline Track, Mt Wellington, on soil under *Pomaderris apetala*, 20 Mar. 2011, *G.Gates I154* (HO 562694, TENN 067073; KJ801174-nLSU, KJ778863-ITS, KJ811570-*rpb2*).

Etymology: From the Latin *geniculatus* (bent abruptly like a knee), in reference to the distinctively bent stipe.

Diagnostic features: Fruit bodies mostly small. Pileus dark yellowish brown to very dark brown with recurved scales; lamellae pink to light brown when young; stipe often bent or flexuous, conspicuously pruinose over its entire length, ochraceous-tawny to cinnamon. Spores smooth.

Inocybe geniculata can be readily identified in the field by its recurved-scaly, dark brown pileus and an ochraceous-tawny to cinnamon stipe that is often bent or flexuous and conspicuously pruinose (lacking longitudinal fibrils). It has been recorded consistently in leaf litter typically under eucalypts in wet sclerophyll forest or wet tall forest in south-eastern and south-western Australia.

Those entities most closely related to *I. geniculata* include two undescribed taxa from New Zealand with entirely pruinose stipes, *I.* "*elegantior*" and *I.* "*subelegantior*", two provisionally named taxa. The former is similar microscopically to *I. geniculata*, but it has a pallid-incarnate stipe and occurs under *Nothofagus*. *Inocybe* "*subelegantior*" is indistinguishable microscopically from *I. geniculata* and *I.* "*elegantior*", but it has a pale brown to brown pileus and a pallid stipe with a pale brown undertone; it occurs with *Nothofagus* and *Kunzea*.

Phylogenetic analysis (Fig. 43) supports the separation of *I. geniculata*, *I.* "*elegantior*" and *I.* "*subelegantior*". We lack molecular data from authentic specimens of *I. umbrosa* E.Horak, which is most similar to this trio among previously described taxa. However, *I. umbrosa* has a fuscous (or dark brown) pileus and stipe and somewhat smaller spores [6.5–8.5 (–9.0) × 4.0–5.0 µm; Horak, 1977]. Morphological differences between *I. geniculata* and *I. umbrosa* can also be observed by examination of dried specimens under a hand-lens or stereomicroscope. Thus, *I. geniculata* has isolated clusters of caulocystidia with little or no covering of fibrils between them. In contrast, *I. umbrosa* (including the holotype PDD 27131) is clothed with a dense covering of longitudinally-arranged fibrils as well as clusters of caulocystidia. Material identified by E.Horak as *I. umbrosa* (PDD 72521) indicates a species that may be more closely related to *I. strigiceps* E.Horak *nom. inval.* [*non I. strigiceps* (Fr.: Fr.) Sacc.], renamed in the current work as *Inocybe tenax* Matheny & Bougher, *nom. nov. Inocybe tenax* is phylogenetically distantly related to the clade containing *I. geniculata*, *I.* "*elegantior*" and *I.* "*subelegantior*". *Inocybe geniculata* also possesses a unique 16-base-pair insertion (conserved in both eastern and western populations) towards the 5′ end of the nuclear large subunit ribosomal RNA gene, an insertion also lacking in *I.* "*elegantior*" and *I.* "*subelegantior*".

Most collections of *I. geniculata* have a distinctive, recurved-scaly and dark brown pileus. However, one specimen from Porongurup National Park in south-western Western Australia (PERTH 08319235) has a non-scaly (fibrillose), yellowish brown pileus. We initially referred to this collection as a possible Australian form of *I.* "*subelegantior*", but sequence data confirm it to be *I. geniculata*.

Inocybe australiensis Cleland & Cheel superficially resembles *I. geniculata*, but it can be distinguished in the field by the presence of woolly-tomentose fibrils on the lower part of the stipe.

Plate 81. *Inocybe geniculata*, holotype (PERTH 08319642). Photograph — P.B.Matheny.

Plate 82. *Inocybe geniculata* (PERTH 08092257). Photograph — N.L.Bougher.

Plate 83. *Inocybe geniculata* (PERTH 08319235). Photograph — P.B.Matheny.

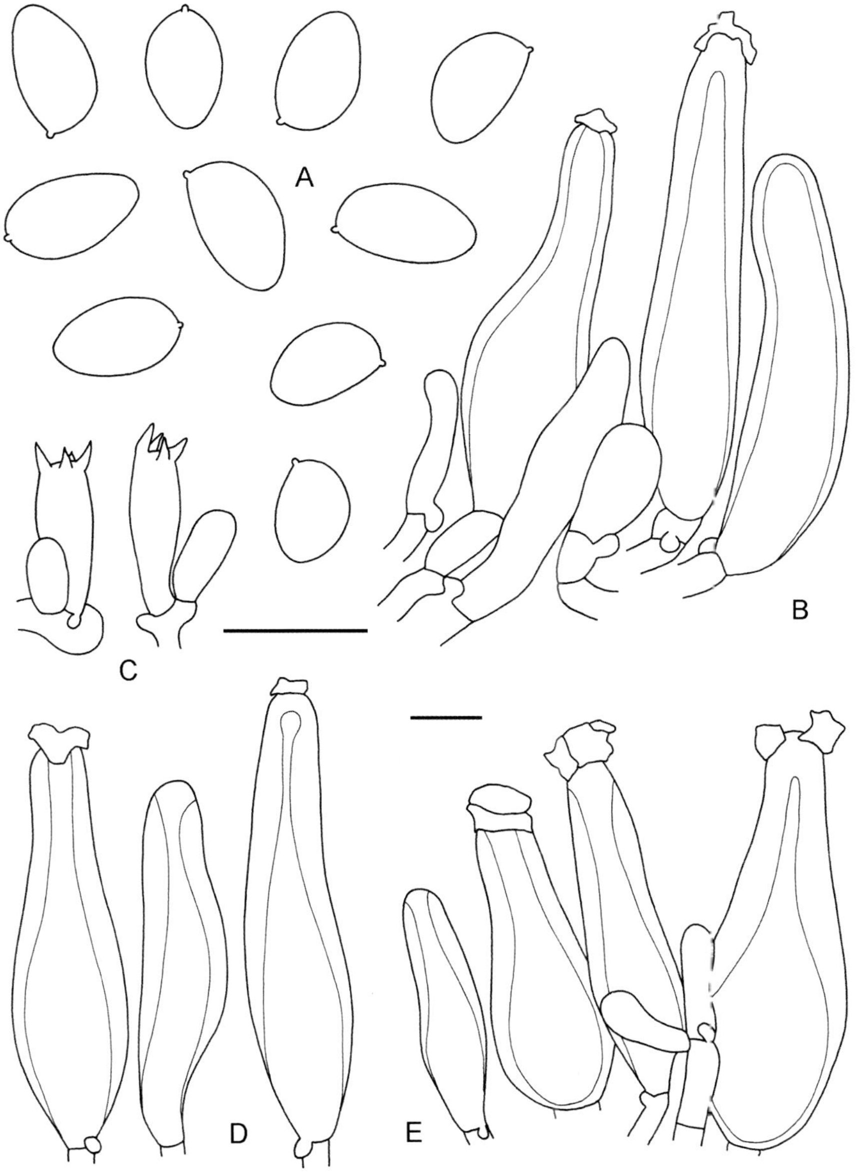

Figure 44. *Inocybe geniculata,* all holotype (PERTH 08319642), except B (PERTH 08092257). **A,** Basidiospores; **B,** Caulocystidia and cauloparacystidia; **C,** Basidia; **D,** Pleurocystidia; **E,** Cheilocystidia and paracystidia. Scale bars = 10 µm (longer bar for spores only).

INOCYBE ERIOCAULIS, *I. ORNATIPES* GROUP AND ALLIES

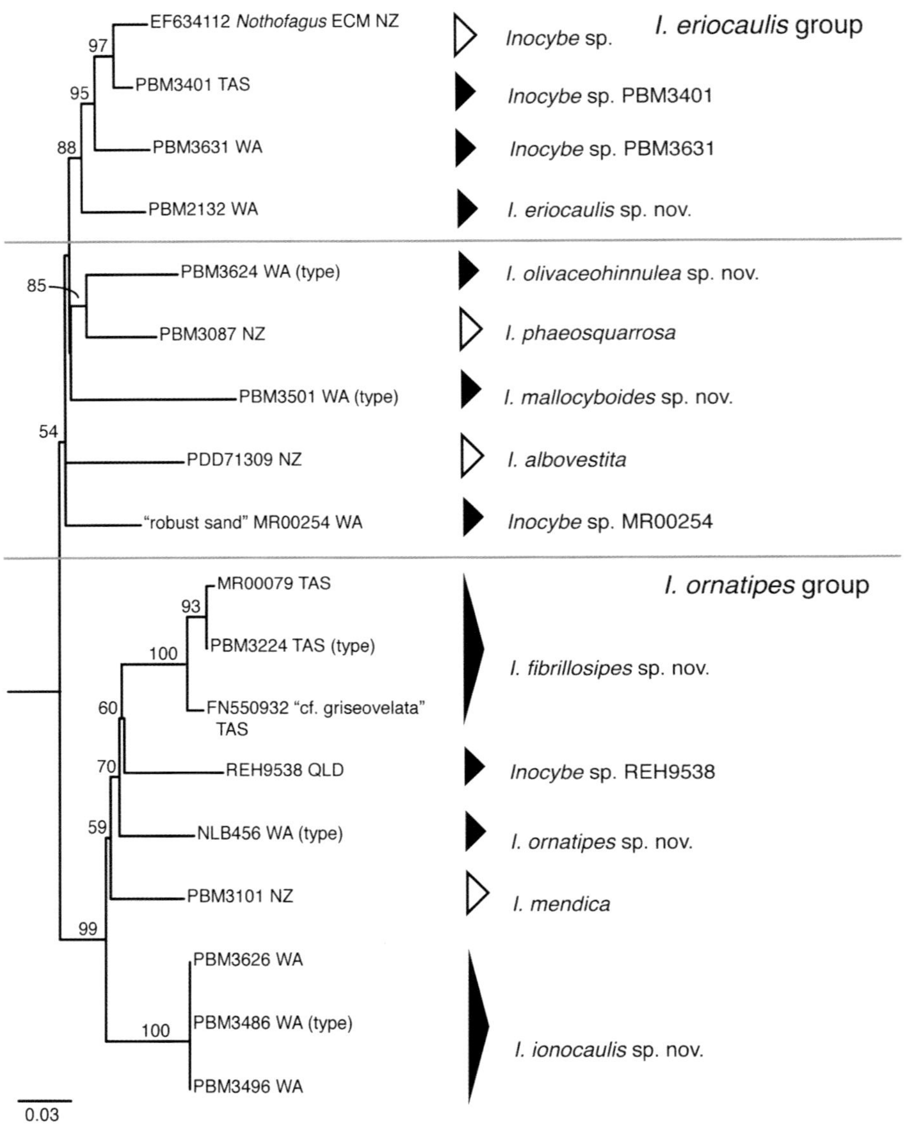

Figure 45. Maximum Likelihood phylogram of the *Inocybe eriocaulis* group, the *I. ornatipes* group, and allied taxa. The topology was produced from the analysis of the STAC group shown in Fig. 7. Australian taxa are indicated by black-filled triangles. White triangles indicate taxa from New Zealand. Bootstrap values are shown above branches.

Fourteen Australian species can be found in this alliance, but the overall grouping is weakly supported, and no obvious morphological features unite them all. The *I. eriocaulis* group comprises four lineages, three of which are not very well known and require further study. One is known only from a *Nothofagus* root sample in New Zealand. The other three lineages share in common the upper pruinose portion of the stipe and an often distinct woolly-fibrillose lower portion of the stipe. Five lineages, four of which have been named, form a weakly supported paraphyletic group giving rise to the *I. eriocaulis* group. The four named species include two new species described from Western Australia. and two known species from New Zealand. The grouping with strong support between the Western Australian species *I. olivaceohinnulea* and the New Zealand species *I. phaeosquarrosa* is rather surprising as these two species differ in macroscopic appearance. The *I. ornatipes* group is strongly supported and contains five species, four of which are from Australia. *Inocybe fibrillosipes* is described as new from Tasmania (and actually has more gross morphological affinities with the *I. eriocaulis* group). *Inocybe ornatipes* is unique in the group by virtue of its floccose-scaly appearance. *Inocybe ionocaulis* resembles species in the *I. serrata* group but is clearly allied phylogenetically with *I. ornatipes* and relatives.

38. **Inocybe eriocaulis** Matheny & Bougher, *sp. nov.*

MycoBank: MB 809321

Type: along old railway line road, Plantagenet Loc. 3298, 12 km W of Denmark, W.A., 34°58'59"S, 117°15'00"E, on the ground in Karri forest along overgrown track under *Eucalyptus patens*, *E. diversicolor*, *Corymbia calophylla* and *Agonis*, 24 June 2001, *P.B.Matheny PBM2129 (E6762)*; holo: PERTH 07677685; iso: TENN 068949.

Pileus 6–15 mm wide, convex to plano-convex, becoming plane, at times with a low obtuse umbo; margin decurved; surface dry; velipellis absent, centre squarrose to squamulose with recurved scales, at times lacerate-scaly; fibrillose-squamulose, matted-fibrillose or fibrillose towards the margin; dark brown (10YR 3/4; Mummy Brown) at the centre, shading to brown (10YR 4/3; Snuff Brown) towards the margin; context white, thin, unchanging where bruised. *Lamellae* narrowly adnate to adnexed, distant, 14–22 L, with several tiers of lamellulae, light brown to brown (7.5YR 6/4, 5/4, 4/4), becoming dark brown (10YR 3/4), edges pallid-fimbriate, ventricose, up to 2 mm deep. *Stipe* 15–25 × 1–2.5 mm, terete, even or slightly swollen at the base, veil (cortina) fugacious, upper third to half densely pruinose, below this densely tomentose-fibrillose or woolly-fibrillose, more densely so when young; brown to dark brown like the pileus, apex at times pinkish grey to light reddish brown or reddish grey (5YR 6/2, 6/3, 5/2; near Avellaneous); context pale brown above, dark brown below, solid. *Odour* not remarkable. Pl. 84–86.

Basidiospores 7.0–10.5 (–11.0) × 4.5–5.5 µm [mean 8.6 × 5.0 µm; Q 1.43–2.10; Q mean 1.72 (n = 30/3)], smooth, amygdaliform with conical apices, rarely mucronate or ovate-elliptic, yellowish brown to brownish yellow; apiculus small but distinct. *Basidia* 20–28 × 7–9 µm, 4-sterigmate, subclavate, hyaline. *Pleurocystidia* 55–78 × 14–17 µm, fusiform, apices obtuse and crystalliferous, thick-walled (2–3 µm), pale yellow to hyaline. *Cheilocystidia* similar to pleurocystidia but occasionally lageniform and shorter, mixed with clavate, thin-walled, hyaline paracystidia. *Caulocystidia* often lageniform and scattered or in clusters mixed with cauloparacystidia and/or cells intermediate in size and shape, these descending to the centre of the stipe (absent on the lower half) but intergrading with caulocystidioid cells; lower half of the stipe composed of interwoven vesture hyphae that are cylindrical, thick-walled, smooth, pale brownish to hyaline, and up to 10 µm wide. *Pileipellis* a trichoderm of cylindrical hyphae that are mostly 7–22 µm wide, thick-walled, encrusted, dark brown in mass, with undifferentiated end cells. *Clamp connections* present. Fig. 46.

Occurs in south-western W.A. in tall wet Karri forest, Jarrah woodland and *Agonis* woodland mixed with *Melaleuca*, generally under *Eucalyptus patens*, *E. diversicolor*, *E. marginata*, *Corymbia calophylla*, *Melaleuca*, *Allocasuarina*, *Oxylobium*, *Agonis* or a mixture of these. Fruiting June, July and Sept.

Additional specimens examined: W.A.: Firebreak Gully Track, Two Peoples Bay Nature Reserve, 34°59'04"S, 118°10'59"E, in sand under *Melaleuca* and *Eucalyptus marginata*, 9 July 1991, *K.Syme KS250/91* (PERTH

03978834, TENN 066694); Firebreak Gully Rd, Two Peoples Bay Nature Reserve, 35°00'S, 118°12' E, under *Melaleuca* and *Agonis*, 11 Sept. 1991, *K.Syme, M.Hart et al. KS 360/91* (PERTH 05302625, TENN 068952); *loc. id., K.Syme, M.Hart et al. KS 361/91* (PERTH 05302633, TENN 068951); just outside William Bay Natl Park, near L. Byleveld, 15 km W of Denmark, 35°00'S, 117°15'E, on soil under *Eucalyptus*, 23 June 2001, *P.B.Matheny, N.L.Bougher & R.Agerer PBM2121 (E6744)* (PERTH 07651937, TENN 068950); corner of SW Highway and William Bay Rd, 34°59'S, 117°13'E, on soil in grey sand along track under *Corymbia calophylla*, *Eucalyptus marginata* and *Allocasuarina* sp., 23 June 2001, *T.Lebel PBM2132 (E6768)* (PERTH 07651619, TENN 066895; EU569843-nLSU, KJ778853-ITS, EU569842-*rpb2*, EU569841-*rpb1*).

Etymology: From the Greek *erion* (wool) and *kaulos* (a stalk), in reference to the distinct woolly lower half of the stipe.

Diagnostic features: Fruit bodies small and slender. Pileus dark brown, squarrose to squamulose; stipe coloured like the pileus, densely pruinose over the upper third or half, woolly-fibrillose below. Spores smooth and amygdaliform.

Inocybe eriocaulis is morphologically similar to *I. australiensis*, which differs most noticeably by its smaller spores. In addition, *I. eriocaulis* features reddish grey (or vinaceous) tones to the stipe surface. Surprisingly, the two species are not very closely related. Indeed, several other cryptic lineages from south-western Western Australia, Tasmania and New Zealand are more closely related to *I. eriocaulis* (Fig. 45). These are all rather uniform in terms of their morphology, and the group is in need of further analysis.

39. Inocybe mallocyboides Matheny, Bougher, R.M.Rob. & K.Syme, *sp. nov.*

MycoBank: MB 809388

Type: London Forest Block, 1.8 km S of Mountain Rd, along Renzo Rd extension, W.A., 34°46'40.0"S, 116°55'34.0"E, in Jarrah forest under *Eucalyptus marginata*, *Kunzea* and other myrtaceous shrubs, 10 June 2010, *R.M.Robinson & K.Syme PBM3501*; holo: PERTH 08242240; iso: TENN 069669 (KP170947-nLSU, KP308773-ITS).

Pileus 10–30 mm wide, initially pulvinate, becoming broadly convex or obtusely conical to plano-convex, often flattened or occasionally slightly depressed in the centre, less often with a low obtuse umbo; margin at first incurved, soon becoming decurved; surface dry, dull, often covered with sand particles or grains, the centre matted-fibrillose when young, tomentose-fibrillose towards the margin, soon areolate-scaly on the disc to almost squamulose, scales not prominently recurved, remaining tomentose-fibrillose towards the margin; colour overall fulvous or yellowish brown (10YR 5/6; near Hazel or Ochraceous-Tawny), becoming dark yellowish brown (10YR 4/4) mixed with shades of strong brown or brown (7.5YR 4/4), the centre occasionally brown and darker than the margin; context whitish to pale yellow (5Y 8/3), with isabelline areas, firm, rather tough for the small size, colour not changing where cut or bruised, up to 3 mm thick under the centre, not confluent with the stipe. *Lamellae* adnate to ascending-adnate or subsinuate, close to crowded, c. 38–48 L with lamellulae of varying lengths, pallid or very pale brown when young, becoming light yellowish brown to olive-brown (2.5Y 6/4 to 5/4); edges pallid-fimbriate, conspicuously so even in dried specimens, ventricose, to 4 mm deep. *Stipe* 15–35 × 3–4 mm, even, not bulbous, base with adhering sandy soil; partial veil not observed, surface initially striatulate, appearing pruinose over most of the length under a hand lens, bicolorous with whitish upper half and yellowish brown lower part, becoming velutinous in appearance, with yellowish brown fibrils, these eventually covering the entire stipe in age and becoming dark yellowish brown or reddish brown (5YR 5/4 to 4/4; near Russet or Cinnamon-Brown); context yellowish brown to brown, solid, firm. *Odour* mild to somewhat acidulous. Pl. 87, 88.

Basidiospores 8.0–11.0 × 4.0–5.0 µm [mean 9.5 × 4.6 µm; Q (1.60–) 1.65–2.38; Q mean 2.07 (n = 45/4)], smooth, cylindrical or oblong-elliptic to oblong-amygdaliform, with bluntly pointed or obtuse apices, yellowish brown; apiculus small but distinct. *Pleurocystidia* 56–90 × 11–14 µm, narrowly fusiform to narrowly sublageniform or ±cylindrical, with distinct tapered necks, apices obtuse, occasionally bearing crystals, thick-walled (2–3 µm), hyaline to pale yellow. *Cheilocystidia* similar to pleurocystidia, numerous, mixed with crowded hyaline clavate paracystidia.

Plate 84. *Inocybe eriocaulis*, holotype (PERTH 07677685). Photograph — P.B.Matheny.

Plate 85. *Inocybe eriocaulis*, holotype (PERTH 07677685). Photograph — P.B.Matheny.

Plate 86. *Inocybe eriocaulis* (PERTH 07651937). Photograph — N.L.Bougher.

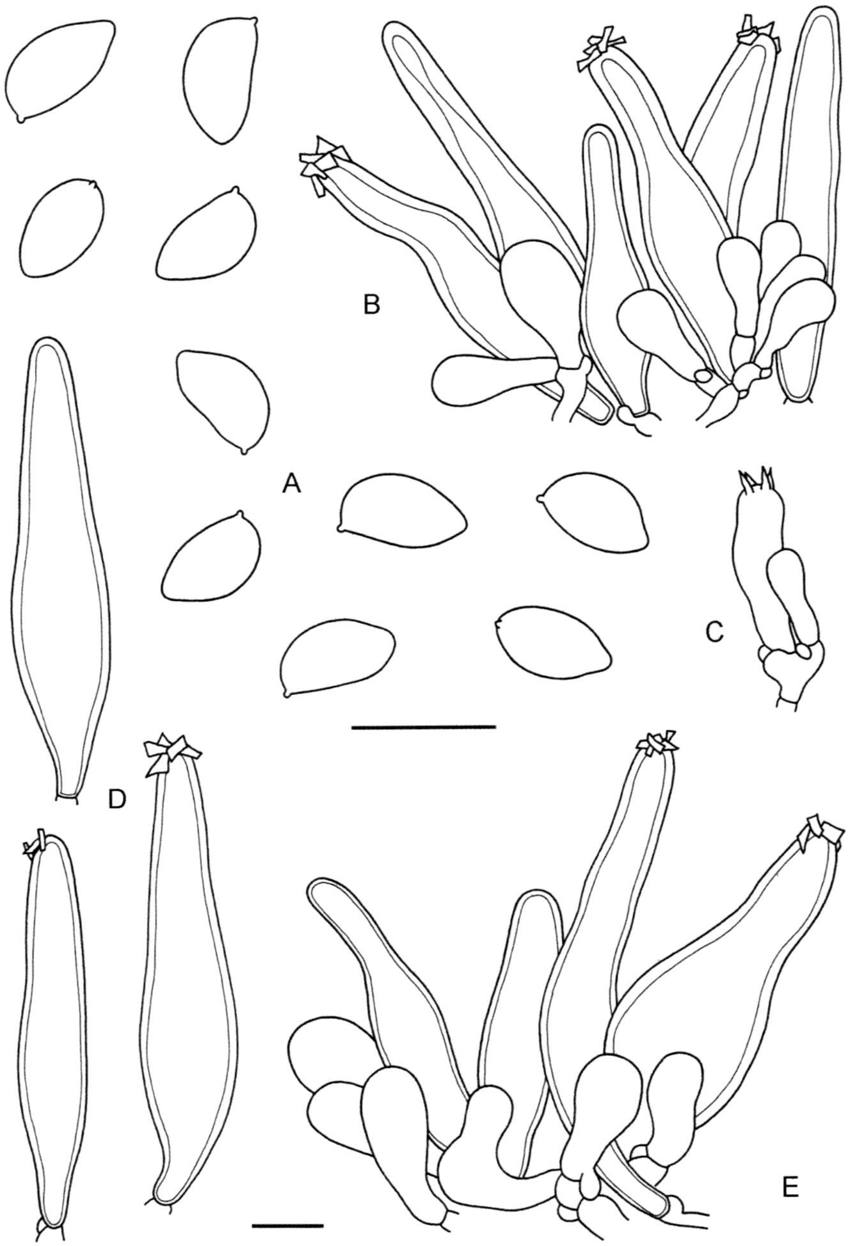

Figure 46. *Inocybe eriocaulis,* all holotype (PERTH 07677685) except E and C (PERTH 07651619). **A,** Basidiospores; **B,** Cheilocystidia and paracystidia; **C,** Basidia; **D,** Pleurocystdia; **E,** Caulocystidia and cauloparacystidia. Scale bars = 10 μm (longer bar for spores only).

Caulocystidia scattered and mixed in dense clusters of bladder-shaped to short clavate cells similar to paracystidia, but these often (in the holotype) or sometimes (other collections) thick-walled and ochraceous to brown-pigmented, subtending hyphae and stipitipellis hyphae often encrusted; caulocystidia and bladder-shaped thick-walled cells descending the entire length of the stipe with the latter densely arranged. *Pileipellis* a cutis giving rise to broad fascicles of erect hyphae that are narrow, smooth, cylindrical, hyaline, up to 5 µm wide and thin-walled; subtending layer composed of rich rusty-ochraceous hyphae encrusted and with slightly thickened walls, often inflated, up to 32 µm wide, frequently septate. *Clamp connections* present. Fig. 47 (see also Pl. 11D, 11G). Occurs on sandy soil in Jarrah (*Eucalyptus marginata*) forest in south-western W.A.; found near other eucalypts or *Melaleuca*, and in black, peaty, sandy soil in shrubland or heath with *Taxandria linearifolia* and *Melaleuca preissiana*. Fruiting Apr.–June.

Additional specimens examined: W.A.: Syme property, Denmark, 34°57'S, 117°21'E, 18 May 1984, *R.N.Hilton UWA2963* (PERTH 00917788); Two Peoples Bay Nature Reserve, 34°58'S, 118°09'E, gregarious on soil under *Melaleuca baxteri*, 23 Apr. 1992, *K.Syme 418/92* (PERTH 05371988); Cemetery Rd, Walpole-Nornalup Natl Park, 34°58'S, 116°45'E, on soil under *Eucalyptus marginata*, *Corymbia calophylla*, *Agonis* spp., *Allocasuarina fraseriana* and *Melaleuca*, 3 June 1992, *K.Syme, M.Hart & N.L.Bougher E4636* (PERTH 07581769); near Denmark, on sandy soil under *Eucalyptus patens*, 24 June 2001, *R.M.Robinson PBM2730* (E6765, PERTH 07677723, TENN 069635); Surprise Forest Block, Western Rd, 1.5 km N of Mountain Rd, 34°46'02"S, 116°55'34"E, in black peaty sandy soil in shrubland to heath with associated *Taxandria linearifolia*, *Beaufortia sparsa* and *Melaleuca preissiana*, 31 May 2005, *R.M.Robinson & K.Syme WFM133* (PERTH 06664784); London Forest Block, 1 km W of Nornalup Rd, along gravel pit track, 34°48'35"S, 116°58'32"E, on sandy soil in forest to woodland with *Allocasuarina fraseriana*, *Eucalyptus marginata*, *Xanthorrhoea preissii*, *Banksia grandis* and *Taxandria parviceps*, 12 June 2009, *R.M.Robinson & K.Syme WFM379* (PERTH 06674216); London Forest Block, 1.8 km S of Mountain Rd, along Renzo Rd extension, 34°46'40"S, 116°55'34"E, on soil in unburnt sedge, 18 June 2009, *R.M.Robinson & K.Syme WFM431* (PERTH 06674690); London Forest Block, intersection of Mountain Rd and Renzo Rd extension, on sandy soil in Jarrah forest under *Eucalyptus marginata*, *Allocasuarina*, *Kunzea* other myrtaceous shrubs, *Jacksonia*, and with *Allocasuarina* in the vicinity, 10 June 2010, *P.B.Matheny, K.Syme, R.M.Robinson & N.L.Bougher PBM3502* (PERTH 08242232, TENN 065774).

Etymology: The epithet *mallocyboides* alludes to the fruit bodies being similar in appearance to those of species in the *Mallocybe* clade.

Diagnostic features: Fruit bodies medium-sized. Pileus tomentose-fibrillose, fulvous to hazel, with a pallid to pale yellow context; stipe velutinous. Spores narrow; caulocystidia descending the entire length of the stipe, abundant, thin- or thick-walled, often mixed with pigmented bladder-shaped cells.

Inocybe mallocyboides is a distinctive new species that is best recognised in the field by its tomentose pileus, velutinous stipe, and its occurrence in sandy soil. Microscopically, it is further characterised by its narrow spores, some of which are cylindrical, and the unusual stipe covering composed of caulocystidia mixed with thin- to thick-walled, bladder-shaped cells along the entire length of the stipe. Its closest relative currently is unresolved, but the species is part of the STAC group (Fig. 7).

An additional collection from Two Peoples Bay (PERTH 03978583, *KS228/91*) is similar to *I. mallocyboides*, but it has broader spores (mostly 10.0–11.5 × 5.5–7.0 µm) and more robust fruit bodies. It may represent a distinct species, but further collections are required before that can be determined.

Plate 87. *Inocybe mallocyboides,* holotype (PERTH 08242240). Photograph — P.B.Matheny.

Plate 88. *Inocybe mallocyboides* (PERTH 06674216). Photograph — R.M.Robinson.

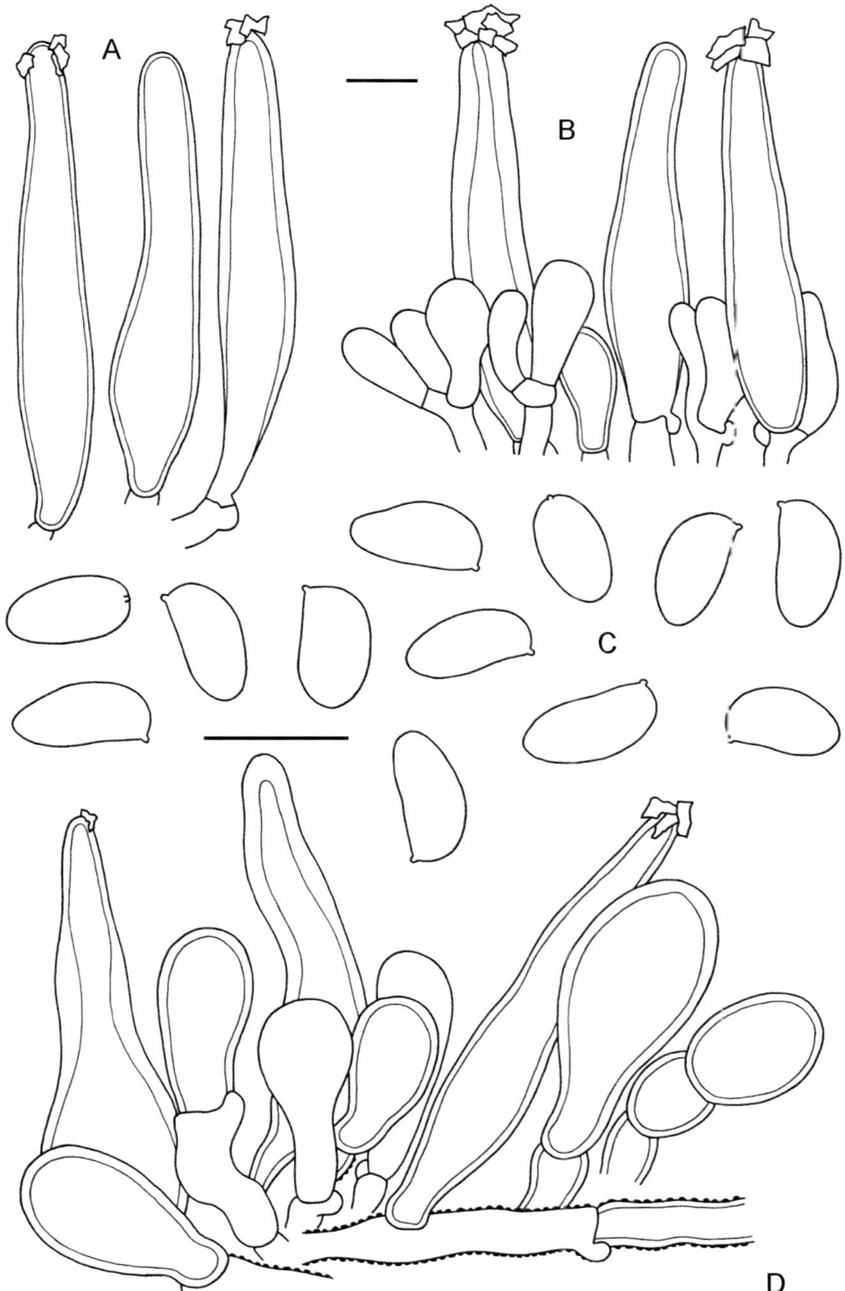

Figure 47. *Inocybe mallocyboides,* all holotype (PERTH 08242240). **A,** Pleurocystidia; **B,** Cheilocystidia and paracystidia; **C,** Basidiospores; **D,** Caulocystidia and cauloparacystidia from mid-stipe. Scale bars = 10 µm (longer bar for spores only).

40. **Inocybe olivaceohinnulea** Matheny & Bougher, *sp. nov.*

MycoBank: MB 809397

Type: Campbell Rd and Inlet Drive, Denmark, W.A., 34°59'11.2"S, 117°20'57.2"E, scattered singly on soil in disturbed Karri forest under *Eucalyptus diversicolor* and *Agonis flexuosa*, 16 July 2011, *P.B.Matheny PBM3624*; holo: PERTH 08319626; iso: TENN 066490 (KP170965-nLSU, KP308797-ITS, KM406205-*rpb2*).

Pileus 13–23 mm wide, plano-convex to almost plane, lacking an umbo; margin decurved; surface dry; velipellis absent, dull, initially fibrillose throughout, with age the disc finely appressed squamulose but fibrillose towards the margin, not rimose; olive-brown to dark yellowish brown (2.5Y 4/4 to 10YR 4/4), disc a shade darker than elsewhere; context coloured like the surface when moist but with pallid areas after drying out, not confluent with the stipe, fleshy-fibrous, not fragile. *Lamellae* adnexed, moderately close to subdistant, 24–28 L, with several tiers of lamellulae, edges conspicuously pallid-fimbriate, broad, to 4 mm deep, pinkish brown to brown or yellowish brown (10YR 5/3 to 5/4; Fawn to Wood Brown), becoming dark yellowish brown. *Stipe* 30–46 × 2–3 mm, even; veil not observed; the surface completely non-pruinose, fibrillose to densely so; apex pale brown (10YR 6/3); lower part dark yellowish brown (10YR 4/4); context solid, becoming hollow; lower part can become very dark brown to almost blackish brown. *Odour* absent. Pl. 89.

Basidiospores 7.0–9.0 × 4.5–5.5 µm [mean 8.2 × 5.1; Q 1.40–1.80; mean Q 1.59 (n = 27/1)], smooth, amygdaliform to broadly amygdaliform; apices conical or occasionally submucronate, yellowish brown; apiculus small. *Basidia* 24–26 × 7–9 µm, clavate to cylindroclavate, hyaline, 4-sterigmate. *Pleurocystidia* 53–61 × 12–20 µm, mostly fusiform, occasionally lageniform, apices obtuse and at times bearing crystals, thick-walled (1.5–2.5 µm), hyaline. *Cheilocystidia* similar to pleurocystidia, some shorter, mixed with hyaline clavate thin-walled paracystidia and occasional larger thick-walled clavate cells. *Caulocystidia* not observed even at the apex of the stipe; surface of stipe composed of a vesture of smooth cylindrical slightly thick-walled hyphae, these interwoven, light yellowish brown in mass but subhyaline or very pale brown when viewed individually, mostly 5–8 µm wide, with an almost gelatinous appearance; vesture hyphae rusty-tawny in mass towards the base of the stipe. *Pileipellis* composed of cylindrical to inflated repent hyphae mostly 5–15 µm wide, with thickened walls, brownish to tawny in mass but encrusting pigments not observed. *Clamp connections* present. Fig. 48 (see also Pl. 11E).

Known only from the type locality, on soil in disturbed Karri forest in south-western W.A. under *Eucalyptus diversicolor* and *Agonis flexuosa*. Fruiting July.

Etymology: From the Latin *olivaceus* (olive-green) and *hinnuleus* (fawn-coloured), in reference to the olive-brown pileus and fawn or pinkish brown lamellae.

Diagnostic features: Fruit bodies slender. Pileus mainly fibrillose, olive-brown; lamellae pinkish brown at first; stipe densely fibrillose, not pruinose, pale brown above, dark yellowish brown below; odour absent.

Based on phylogenetic analysis (Figs 7, 45), *I. olivaceohinnulea* is most closely related to *I. phaeosquarrosa* E.Horak from New Zealand (Horak, 1977), but the two species differ in several morphological and ecological characteristics. Thus, *I. phaeosquarrosa* has a dark brown, squarrose pileus, a warty and scabrous stipe, thin-walled hymenial cystidia, ovate spores and it fruits on rotten wood and bark of *Nothofagus*. Both species share the same cystidial morphology and a darker stipe base, although the cystidia of *I. olivaceohinnulea* are more thick-walled than in *I. phaeosquarrosa* (Horak, 1977).

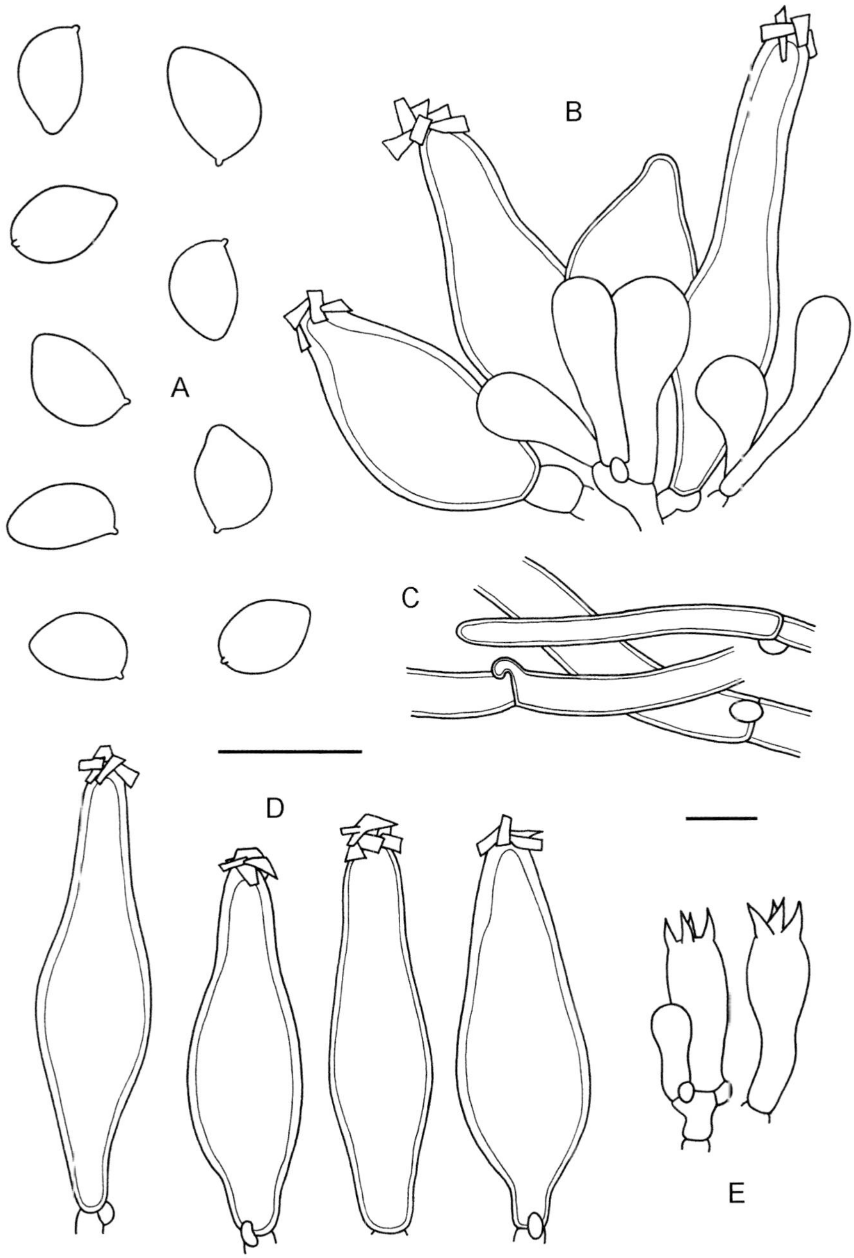

Figure 48. *Inocybe olivaceohinnulea,* all holotype (PERTH 08319626). **A,** Basidiospores; **B,** Cheilocystidia and paracystidia; **C,** Vesture hyphae from upper stipe; **D,** Pleurocystidia; **E,** Basidia. Scale bars = 10 µm (longer bar for spores only).

Plate 89. *Inocybe olivaceohinnulea*, holotype (PERTH 08319626). Photograph — P.B.Matheny.

41. Inocybe fibrillosipes Matheny, Bougher & G.M.Gates, *sp. nov.*

MycoBank: MB 809324

Type: Lady Barron Track, Mount Field Natl Park, Tas., 42°40'S, 146°43'E, gregarious to scattered in wet sclerophyll forest along bank of track under *Eucalyptus regnans*, *Acacia* and *Pomaderris*, 8 June 2009, *D.Ratkowsky, G.Gates & P.B.Matheny PBM3224*; holo: PERTH 08517002; iso: TENN 068200 (KJ801167-nLSU, KJ778858-ITS).

Pileus 7–15 mm wide, conical, expanding with age, lacking velipellis and umbo; margin decurved to nearly straight; surface dry, squarrose over the disc; scales often with recurved tips, fibrillose towards the margin; dark yellowish brown or very dark brown (10YR 3/4 to 2/2) or Mummy Brown, at times brown (10YR 4/3), with tips of scales lighter in colour or greyish brown; context thin-fleshed, not changing colour upon exposure. *Lamella* adnexed, moderately close, with several tiers of lamellulae, initially pale brown (10YR 6/3), becoming yellowish brown (10YR 5/4) with age; edges pallid but indistinctly fimbriate; ventricose, of medium depth. *Stipe* 15–45 × 1.0–2.5 mm, terete, often flexuous, even, partial veil not observed directly but presumably evanescent; upper half distinctly pruinose, lower half covered with a conspicuous vesture of greyish to greyish brown fibrils, elsewhere pinkish brown to reddish brown (7.5YR 5/4 to 5YR 4/4); context solid. *Odour* not remarkable. Pl. 90.

Basidiospores 6.5–8.5 × 4.5–5.0 µm [mean 7.6 × 4.7 µm; Q 1.40–1.78; Q mean 1.60 (n = 35/1)], smooth, mostly subamygdaliform to amygdaliform with bluntly pointed apices, not usually adaxially depressed, yellowish brown; apiculus not distinct. *Basidia* 22–25 × 6–8 µm, clavate, hyaline, 4-sterigmate. *Pleurocystidia* 50–60 × 10–14 µm, fusiform to narrowly fusiform, with or without a pedicel, base often rounded, apex obtuse and bearing crystals, thick-walled (1.5–3.0 µm), pale yellow. *Cheilocystidia* similar to pleurocystidia but often shorter and sometimes lageniform, yellow to ochraceous, mixed with paracystidia that become slightly thick-walled and/or ochraceous. *Caulocystidia* 75–91 × 20–26 µm, fusiform to lageniform, at times with flexuous necks, larger than hymenial cystidia, frequent on the upper part of the stipe, forming scattered clusters mixed with cauloparacystidia; lower part of stipe composed of a dense interwoven superficial layer of hyaline cylindrical hyphae, these mostly 5–12 µm wide, thin-walled, smooth; terminal cells not differentiated other than being tapered towards the apices. *Pileipellis* composed of tufts of upturned fascicles of hyphae, these cylindrical to inflated and mostly 7–25 µm wide, walls noticeably thickened (1–2 µm) on subtending hyphae; terminal cells more thin-walled, smooth, lighter in colour and typically tapered towards the apices; subtending hyphae dark yellowish brown in mass, with smooth to wrinkled or indistinctly encrusted walls; septa frequent. *Clamp connections* present. Fig. 49.

Occurs on soil in wet sclerophyll forest in Tas., under a mixture of *Eucalyptus regnans*, *Acacia* and *Pomaderris* and/or *Nothofagus*. Fruiting June–Aug.

Additional specimens examined: Tas.: Warra Long-Term Ecological Research site, plot 1898 E3 and A1, 43°5.5'S, 146°38.8'E, on soil in mixed forest including *Eucalyptus*, *Nothofagus cuninghamii* and *Pomaderris*, 1 Aug. 2006, *M.Ryberg MR00074* (FN550917-nLSU + ITS); *loc. id.*, 3 Aug. 2006, *G.Gates 060803* (FN550932-nLSU + ITS).

Etymology: From the Latin *fibrillosus* (fibril-bearing) and *pes* (a foot), in reference to the dense superficial layer of greyish fibrils on the lower half of the stipe.

Diagnostic features: Fruit bodies small and slender. Pileus dark yellowish brown to dark brown, with paler recurved scales over the disc; stipe pinkish brown to reddish brown and distinctly pruinose on the upper half, covered with a greyish fibrillose vesture on the lower half. Spores amygdaliform, <9 µm long; caulocystidioid hairs absent from the lower half of the stipe.

Inocybe fibrillosipes is characterised by the dark brown, squarrose pileus, the scales of which are greyish at the tips, and a stipe that is pinkish to reddish brown, pruinose above but starkly greyish fibrillose below. The spores are small and smooth, metuloid hymenial cystidia are present, and the species belongs phylogenetically to the STAC group where it is most closely related to two Western Australian taxa, *I. ornatipes* and *I. ionocaulis*, as well as *I. mendica* E.Horak from New Zealand (Fig. 45). *Inocybe ornatipes* is readily distinguished by its bicolorous warm brown to pale brownish yellow pileus and conspicuous, floccose, white cottony scales on the pileus and stipe. *Inocybe ionocaulis* has a fibrillose pileus and vinaceous stipe and, like *I. fibrillosipes*, it is pruinose along the upper half of the stipe and has a whitish, fibrillose vesture below. *Inocybe mendica* has smaller fruit bodies without any trace of scales (Horak, 1977).

Plate 90. *Inocybe fibrillosipes*, holotype (PERTH 08517002). Photograph — P.B.Matheny.

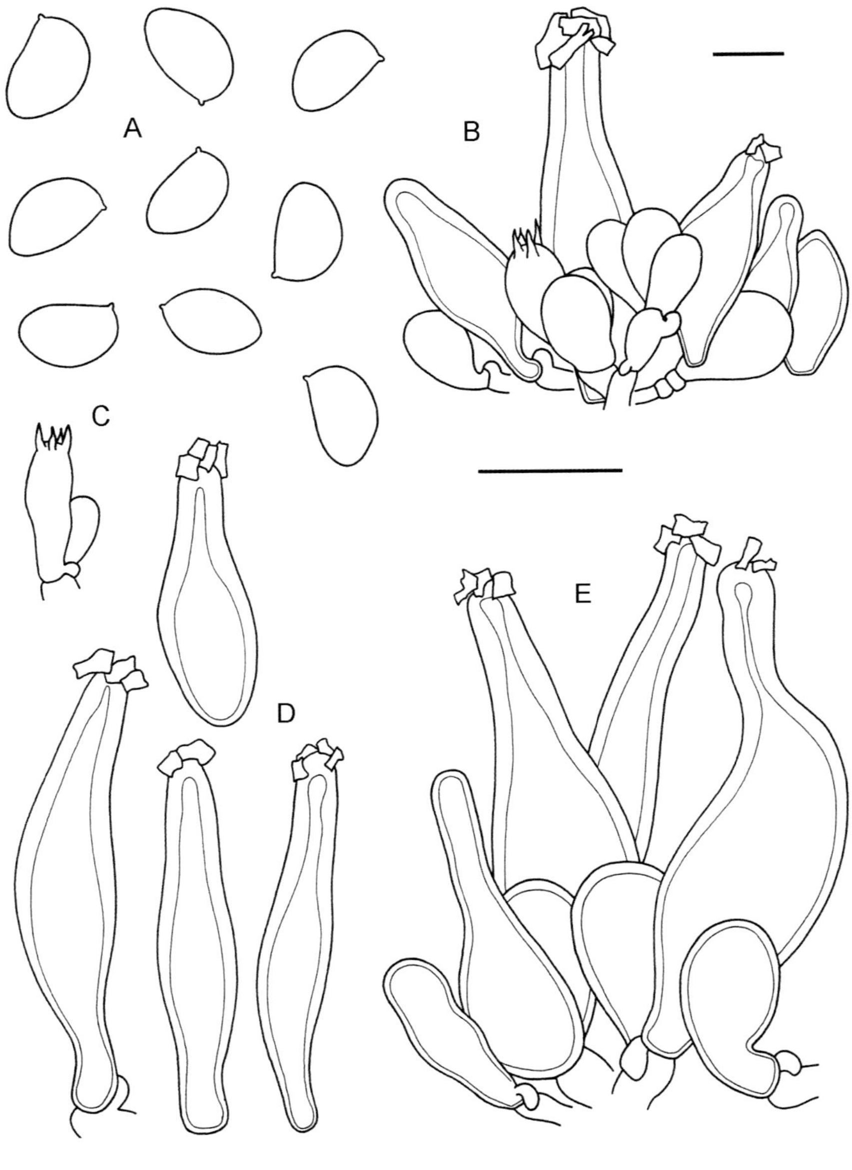

Figure 49. *Inocybe fibrillosipes,* all holotype (PERTH 08517002). **A,** Basidiospores; **B,** Cheilocystidia and paracystidia; **C,** Basidium; **D,** Pleurocystidia; **E,** Caulocystidia and cauloparacystidia. Scale bars = 10 µm (longer bar for spores only).

42. **Inocybe ionocaulis** Matheny & Bougher, *sp. nov.*

MycoBank: MB 809383

Type: Warren River Loop Walk, Warren Natl Park, W.A., 34°30'36.6"S, 115°37'33.6"E, gregarious or scattered singly or in pairs in soil in Karri forest under *Eucalyptus diversicolor* and *Allocasuarina*, 8 June 2010, *P.B.Matheny PBM3486*; holo: PERTH 08241791; iso: TENN 068287 (KJ801188-nLSU, KP171134-ITS, KJ811579-*rpb2*).

Pileus 7–21 mm wide, conical, bell-shaped or convex, becoming plane, at times with a small obtuse umbo at maturity, margin decurved or straight; velipellis absent; surface dry; disc unbroken, diffract-scaly or areolate, finely fibrillose towards the entire or rimulose margin, somewhat glossy; edge fringed with pallid finely silky-fibrillose veil remnants; centre dark brown to very dark brown, shading to paler brown towards the margin (10YR 2/2, 3/2, 4/3) when young, becoming dark brown (10YR 3/3) at the centre with a yellowish brown (10YR 4/4) margin; context coloured like the surface when moist, thin, fragile overall. *Lamellae* ascending, adnate to sinuate or adnexed, subdistant to moderately close, 18–30 L, with at least 3 tiers of lamellulae, initially light brown to brown (7.5YR 6/4, 5/4), becoming brown, yellowish brown or dark yellowish brown (10YR 5/6, 4/3, 4/4); edges pallid and indistinctly fimbriate; ventricose, 2–3 mm deep. *Stipe* 20–50 × 1.5–3.0 mm at the apex, even or with a slightly swollen base, pruinose above the pileal insertion (upper half to upper third), below with dense white fibrils, often broken up in areas or worn away but compact at the the base and upwards; reddish brown or vinaceous brown (5YR 5/3, 4/3; Army Brown) at the apex and as a ground colour beneath the vesture below; context solid, becoming hollow, with a light reddish brown interior; cortex coloured like ground colour. *Odour* mild or acidulous. Pl. 91, 92.

Basidiospores 7.5–9.5 × 4.5–5.5 µm [mean 8.3 × 5.1 µm; Q 1.45–1.85 (–1.90); Q mean 1.64 (*n* = 31/3)], smooth, amygdaliform, with pointed to bluntly pointed apices, yellowish brown; apiculus small. *Basidia* 24–29 × 7–8 µm, clavate, hyaline, 4-sterigmate. *Pleurocystidia* 53–82 × 10–20 µm, fusiform, fusiform-ventricose or lageniform, some (especially in PBM3626) with a slender basal pedicel; apices obtuse and typically bearing crystals, bases truncate, rounded or slightly flaring, thick-walled (2.5–4.0 µm), yellow. *Cheilocystidia* similar to pleurocystidia, sometimes shorter and lageniform, crowded, mixed with scattered paracystidia (not abundant in the type collection) and basidia. *Caulocystidia* in scattered dense clusters at the stipe apex, these lageniform to fusiform (to 120 µm long), some with long narrow necks, hyaline, mixed with cauloparacystidia; caulocystidioid cells at the stipe centre fusiform, lageniform or clavate; cystidia not observed in area just above the stipe base. *Pileipellis* a cutis of cylindrical to inflated hyphae mostly 6–21 µm wide, dark yellowish brown to dark brown in mass, some thin-walled, others with slightly thickened walls, encrusted. *Clamp connections* present. Fig. 50.

Occurs in Karri forest or mixed Tingle-Karri forest in south-western W.A. under *Eucalyptus diversicolor*, *E. jacksonii* and *Allocasuarina*. Fruiting June and July.

Additional specimens examined: W.A.: Shedley Drive intersection with Tingle Drive, Walpole-Nornalup Natl Park, 34°59'24.5"S, 116°39'20.6"E, on sandy soil in Tingle-Karri forest under *Eucalyptus jacksonii*, *E. diversicolor* and *Allocasuarina*, 9 June 2010, *P.B.Matheny PBM3496* (PERTH 08242291, TENN 065773; KP171135-ITS); Campbell Rd and Inlet Drive, Denmark, 34°59'11.2"S, 117°20'57.2"E, scattered singly on soil in leaf litter in disturbed Karri forest under *Eucalyptus diversicolor* and *Agonis flexuosa*, 16 July 2011, *P.B.Matheny PBM3626* (PERTH 08319561, TENN 066636; KJ801189-nLSU, KP171136-ITS, KJ811580-*rpb2*).

Etymology: From the Greek *ion* (violet) and *kaulos* (a stalk), in reference to the vinaceous or deep reddish brown stipe.

Diagnostic features: Fruit bodies small; habit slender and elongate. Pileus dark brown fibrillose, the edge with whitish veil remnants; stipe vinaceous brown or reddish brown, pruinose at the apex, but covered with whitish vesture below. Spores smooth; hymenial cystidia yellow.

Discussion: *Inocybe ionocaulis* resembles species of the *I. serrata* group, such as *I. granulosipes* and *I. albofloccosa*. All have fruit bodies of similar size and stature, the upper stipe is pruinose and the lower stipe is white-fibrillose. However, *I. ionocaulis* has a deep reddish brown or vinaceous brown stipe and a dark brown pileus. In contrast, *I. albofloccosa* has a pale brown to brown pileus and a stipe that is generally pinkish grey to yellowish brown (exceptionally vinaceous-brown). That species and other members of the *I. serrata* complex are more common than *I. ionocaulis*, which is known only from Karri forests in the south-west of Western Australia.

Phylogenetically, *I. ionocaulis* is only quite distantly related to the *I. serrata* group; indeed, it is closer to the *I. ornatipes* group (Fig. 45).

Plate 91. *Inocybe ionocaulis*, holotype (PERTH 08241791). Photograph — P.B.Matheny.

Plate 92. *Inocybe ionocaulis* (PERTH 08319561). Photograph — P.B.Matheny.

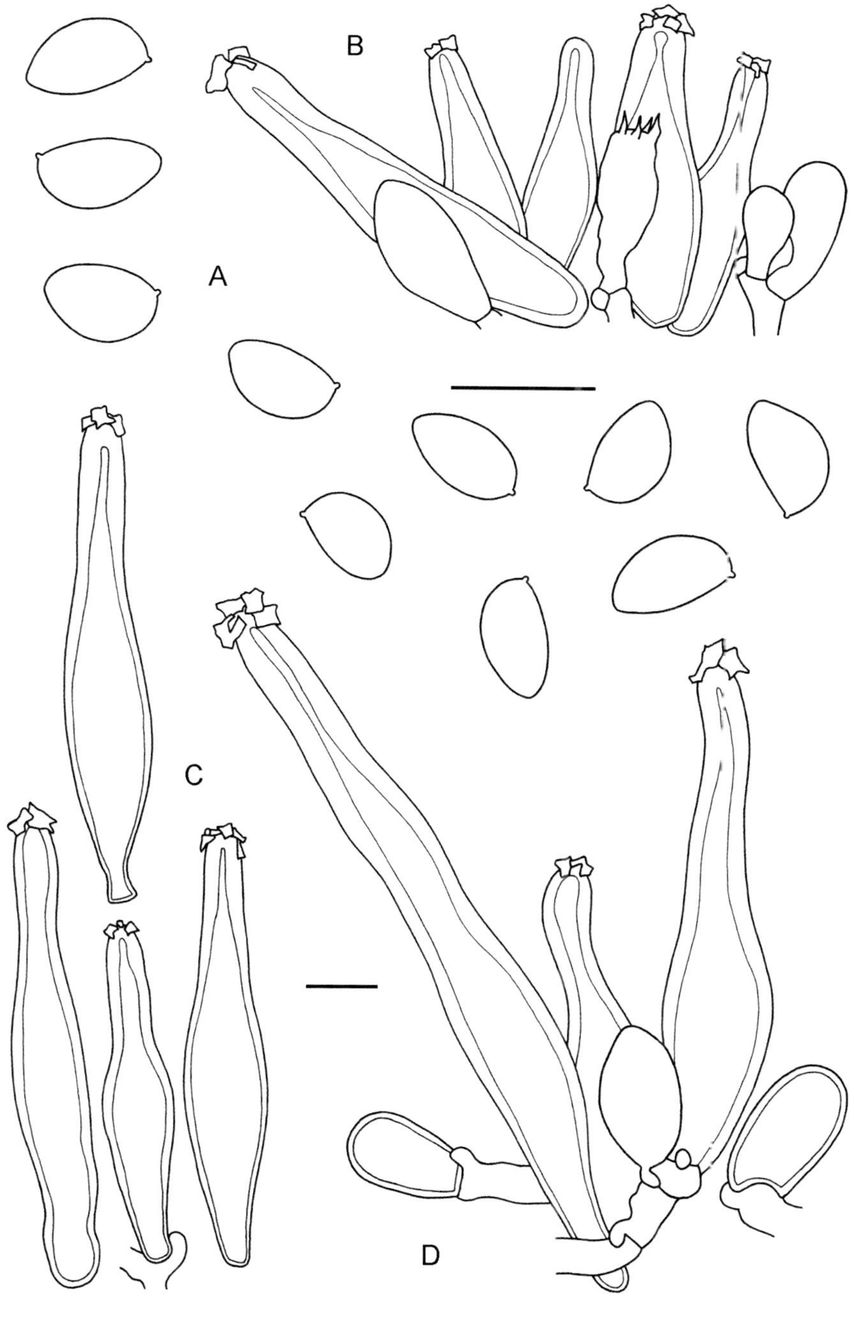

Figure 50. *Inocybe ionocaulis,* all holotype (PERTH 08241791). **A,** Basidiospores; **B,** Cheilocystidia, paracystidia, and basidium; **C,** Pleurocystidia; **D,** Caulocystidia and cauloparacystidia from upper stipe. Scale bars = 10 µm (longer bar for spores only).

43. **Inocybe ornatipes** Matheny & Bougher, *sp. nov.*

MycoBank: MB 809545

Type: Boranup Drive, Leeuwin-Naturaliste Natl Park, W.A., among litter but attached to soil in tall eucalypt forest dominated by Karri (*Eucalyptus diversicolor*), 30 June 2008, *N.L.Bougher NLB456*; holo: PERTH 08072906 (KP170966-nLSU, KP308798-ITS, KM406206-*rpb2*).

Pileus 5–15 mm wide, at first bluntly conical with a strongly incurved margin attached to a white cortinoid partial veil, later becoming campanulate, expanding fully to nearly applanate, with a slightly raised centre and a thin but not translucent-striate margin (some pale veil remnants adhering to the margin in some specimens); surface dry, densely covered with recurved and erect sharp-pointed scales, becoming radially arranged, with underlying surface minutely radially fibrillose; scales slightly paler than the pileal surface; bright warm tan to brownish yellow (near 6E6) when young, becoming paler and more yellowish towards the margin with age (between 4B5 and 5B5); context concolorous, dull, watery, unchanging when cut, thin over the stipe apex (less than 1 mm). *Lamellae* adnexed, closely spaced, with at least 1 tier of lamellulae, to 2 mm deep, slightly ventricose; edges faintly fimbriate, pallid or concolorous with the faces, pale cream in button stage, remaining pale but becoming pale yellowish brown (near 4B4), stained brown with patchy spores. *Stipe* 15–45 × 1–2 mm, narrowly cylindrical, often flexuous or bent; solid; surface dry, upper part covered with white squamules when young, later pruinose; lower part with white silky longitudinal fibrils that are easily removed by handling, underlying colour dull yellowish; annular zone absent, base not differentiated; basal mycelium not conspicuous, white; context solid. *Odour* not distinctive. Pl. 93, 94.

Basidiospores 7.6–9.2 × 4.6–6.1 µm [mean 8.5 × 5.3 µm; Q 1.44–1.74; Q mean 1.59 (n = 30)], smooth, subamygdaliform to amygdaliform, with bluntly pointed apices, yellowish brown; apiculus small. *Basidia* 25–28 × 8–10 µm, clavate, hyaline, 4-sterigmate. *Pleurocystidia* 52–67 × 10–16 µm, fusiform to lageniform, at times with long tapering necks; apices obtuse, bearing crystals, thick-walled (to 2 µm), hyaline to pale yellowish. *Cheilocystidia* similar to pleurocystidia, mixed with small basidia-like paracystidia and some basidia, abundant, at times in crowded clusters. *Caulocystidia* similar to hymenial cystidia but occasionally longer, clustered and mixed with cauloparacystidia at the stipe apex, these clavate, vesiculose, pyriform, and thin- or thick-walled; terminal cells on lower part of stipe with thickened walls but not differentiated. *Pileipellis* a cutis of hyphae inflated up to 20 µm wide and with finely encrusted brown walls, giving rise to reflexed and erect bundles of paler mainly cylindrical and smooth hyphae. *Clamp connections* present. Fig. 51.

Occurs among litter but attached to the soil in tall eucalypt forest dominated by Karri (*Eucalyptus diversicolor*) in south-western W.A. Fruiting June.

Etymology: From the Latin *ornatus* (adorned) and *pes* (a foot), in reference to the conspicuous cottony squamules on the pileus and stipe.

Diagnostic features: Fruit bodies small; habit slender and elongate. Pileus and stipe with whitish or pale cottony recurved or pointed scales; ground colour of the pileus warm tan to brownish yellow, yellowish on the stipe. Spores smooth; pleurocystidia thick-walled.

Discussion: Although known only from the type locality, *I. ornatipes* is well characterised by morphology and multiple gene sequences. Thus, species known from New Zealand with a scaly pileus and stipe include *I. strobilomyces* E.Horak, which has more robust, brown fruit bodies, while the stipe, unlike that of *I. ornatipes*, is not floccose-squamulose (Horak, 1977). Moreover, *I. phaeosquarrosa* E.Horak differs in its dark brown colours, thin-walled cystidia, and its occurrence on woody debris and bark in *Nothofagus* forests (Horak, 1977).

Species most closely related to *I. ornatipes* include *I. fibrillosipes* Matheny, Bougher, & G.M. Gates, *I. ionocaulis* Matheny & Bougher, and *I. mendica* E.Horak, the latter from New Zealand (Fig. 45), but they differ in their surface textures and coloration. Thus, the Tasmanian *I. fibrillosipes* has a dark brown squarrose pileus with a pruinose upper stipe and a greyish, densely fibrillose lower stipe. *Inocybe ionocaulis* differs in its fibrillose texture and the vinaceous colours of the stipe, while *I. mendica* has diminutive fruit bodies that lack squamules.

Plate 93. *Inocybe ornatipes*, holotype (PERTH 08072906). Photograph — N.L.Bougher.

Plate 94. *Inocybe ornatipes*, holotype (PERTH 08072906). Note conspicuous squamules on the pileus at all stages, and on the stipe when young. Photograph — N.L.Bougher.

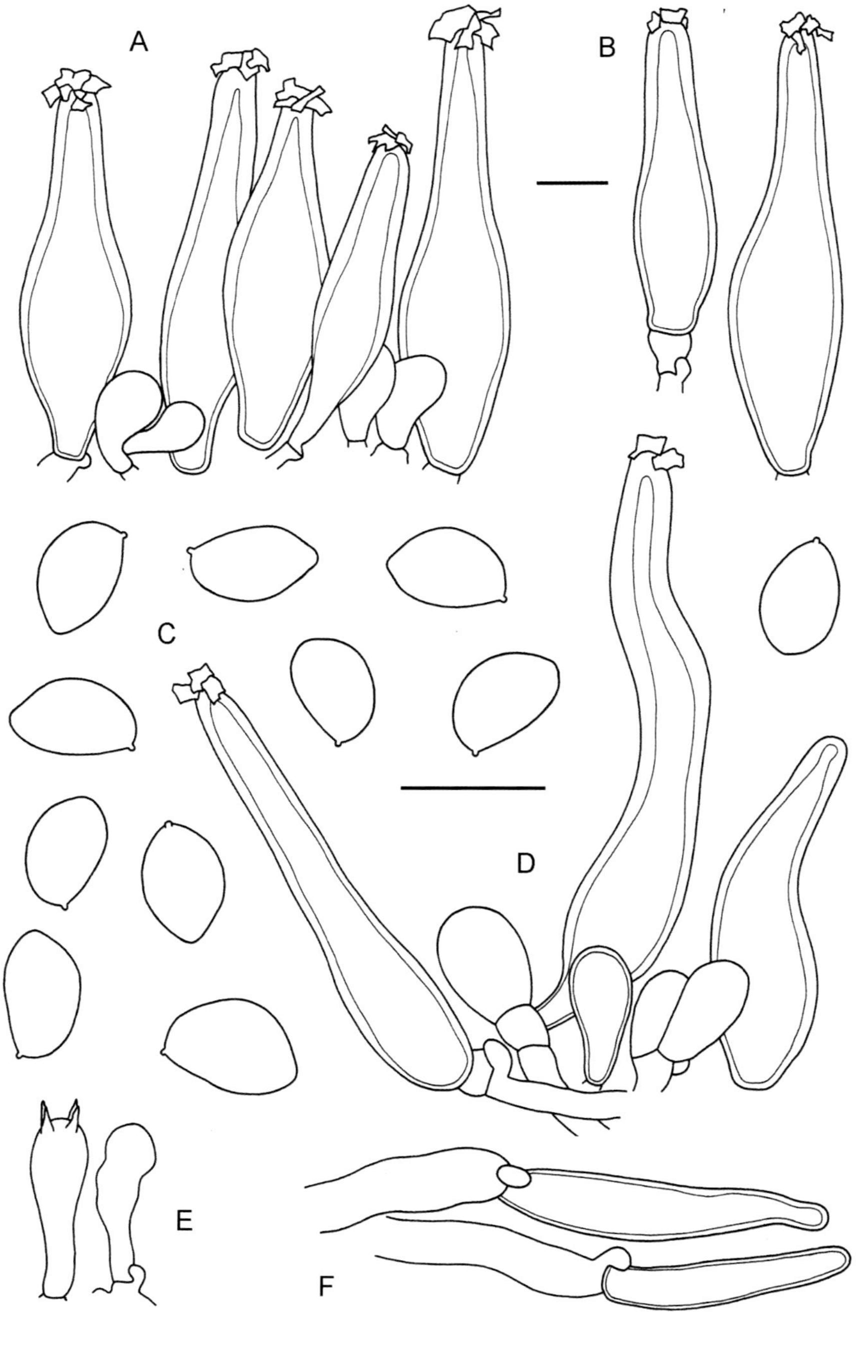

Figure 51. *Inocybe ornatipes*, holotype (PERTH 08072906). **A,** Cheilocystidia and paracystidia; **B,** Pleurocystidia; **C,** Basidiospores; **D,** Caulocystidia from stipe apex; **E,** Basidia; **F,** Caulocystidioid end-cells from the lower stipe. Scale bars = 10 μm (longer bar for spores only).

44. Inocybe nudiuscula Matheny & Bougher, *sp. nov.*

MycoBank: MB 809396

Type: 1 km N of Catterick Rd, near Greenbushes, W.A., 33°50'50.6"S, 116°04'50.3"E, scattered singly on unburnt soil in Jarrah forest under *Eucalyptus marginata* and *Corymbia calophylla*, 12 July 2011, *P.B.Matheny PBM3602*; holo: PERTH 08319839; iso: TENN 066573 (KP170962-nLSU, KP308793-ITS, KC305451-*rpb2*).

Pileus 12–25 mm wide, conical, obtusely conical or plano-convex, becoming plane with a small nipple-like umbo; margin initially incurved, soon decurved, becoming uplifted; velipellis absent or indistinct; surface dry, glossy; disc generally smooth and unbroken, occasionally with a few appressed scales, radially fibrillose and minutely rimulose towards the margin; very pale brown over the umbo when a velipellis is present, yellowish brown, umbrinous or dark brown (10YR 4/3 to 3/2; Snuff Brown to Prout's Brown) at the centre, shading to brown or yellowish brown (10YR 4/3 to 5/4; Tawny-Olive) towards the margin, much paler between the spreading fibrils; context whitish or yellowish brown when moist, not changing colour where bruised or exposed, thin except under the disc. *Lamellae* adnexed to sinuate, moderately close to close, 26–40 L, with several tiers of lamellulae, pale brown, light yellowish brown or yellowish brown (2.5Y 6/4 to 10YR 6/3), medium, ventricose; edges pallid and indistinctly fimbriate. *Stipe* 25–37 × 1.5–3.0 mm, even or with a slightly swollen base, slightly flexuous; veil cortinate, fugacious; surface lacking a well-developed vesture, upper third to half lightly pruinose, silky-fibrillose below, more densely fibrillose above the base; buff, pale ochraceous, to ochraceous-buff (Warm Buff, Light Ochraceous-Buff to Light Ochraceous Salmon) or very pale brown (10YR 7/3 to 7/4) with a buff or ochraceous-buff tone; context coloured like the surface, solid. *Odour* absent, acidulous or weakly of *Pelargonium*. Pl. 95, 96.

Basidiospores 7.5–9.5 (–10.0) × 4.5–5.5 µm [mean 8.4 × 5.0 µm; Q 1.45–1.90; Q mean 1.68 (*n* = 52/3)], smooth, amygdaliform, with pointed to bluntly pointed apices, yellowish brown; apiculus small and often indistinct. *Basidia* 24–26 × 7–8 µm, clavate, hyaline, 4-sterigmate. *Pleurocystidia* 53–65 × 10–19, narrowly fusiform, cylindrical, fusiform or sublageniform; apices obtuse, crystalliferous; walls thick (2–4 µm), especially apically, hyaline to pale greenish. *Cheilocystidia* similar to pleurocystidia but more variable in shape (utriform, ventricose or broadly fusiform) and often shorter, mixed with clavate paracystidia, thick-walled, the walls hyaline to yellow. *Caulocystidia* similar to pleurocystidia but often longer and occasionally subcylindrical, mixed with cauloparacystidia, forming scattered clusters over the upper part of the stipe, these most dense towards the apex, absent below the stipe centre. *Pileipellis* a cutis of cylindrical hyphae that are dark yellowish brown in mass, mostly 4–12 µm wide, encrusted, thin-walled, overlain on the disc by a superficial layer of interwoven hyaline thin-walled cylindrical hyphae. *Clamp connections* present. Fig. 52.

Occurs in south-western W.A., south-eastern Qld and Vic.; grows in soil under eucalypts or a mixture of eucalypts and *Allocasuarina* or *Acacia*, or under *Lophostemon*. Fruiting Nov., Feb.–July.

Additional collections examined: Qld: Cunninghams Gap, Main Range Natl Park, 28°02'59"S, 152°23'37.3"E, alt. 776 m, on soil in sclerophyll forest under *Eucalyptus* and *Allocasuarina*, 29 Feb. 2012, *R.E.Halling PBM3774* (PERTH 08363137, TENN 066573; KP170963-nLSU, KP308794-ITS, KC305449-*rpb2*); *loc. id.*, alt. 770 m, scattered singly on soil on bank of track in warm-temperate rainforest under *Lophostemon confertum*, 3 Mar. 2012, *P.B.Matheny PBM3782* (PERTH 08363072, TENN 067014; KP170964-nLSU, KP308795-ITS, KC305450-*rpb2*). Vic.: Benambra–Corryong road, 2 km S of Sipthorp Track, Claridge site No. 98, Alpine Natl Park, on a sheltered slope under *Acacia dealbata*, *Eucalyptus dalrympleana* and *E. radiata*, 25 May 1996, *A.Jumpponen* (*Trappe 18260*) (OSC, but material extinguished; EU555457-nLSU, KP308796-ITS); Black Mountain Rd, 3.2 km E of Junction Rams Horn Track, Claridge site No. 81, Alpine Natl Park, in a gully under *Acacia dealbata*, *Eucalyptus dalrympleana* and *E. pauciflora*, 7 Nov. 1996, *E.C.Ázares* (*Trappe 20013*; OSC).

Etymology: From the Latin *nudiusculus* (somewhat naked), in reference to the bare stipe.

Diagnostic features: Fruit bodies small, agaricoid or secotioid. Pileus umbrinous or brown, becoming rimose; stipe (where present) pruinose over the upper third, with a thin, pallid vesture below, ochre-buff, with a veil. Spores smooth, to 10 µm long, amygdaliform. On soil under eucalypts or mixture of eucalypts and *Allocasuarina* or *Acacia*, or under *Lophostemon* (Myrtaceae).

Plate 95. *Inocybe nudiuscula*, holotype (PERTH 08319839). Photograph — P.B.Matheny.

Plate 96. *Inocybe nudiuscula* (PERTH 08363137). Photograph — P.B.Matheny.

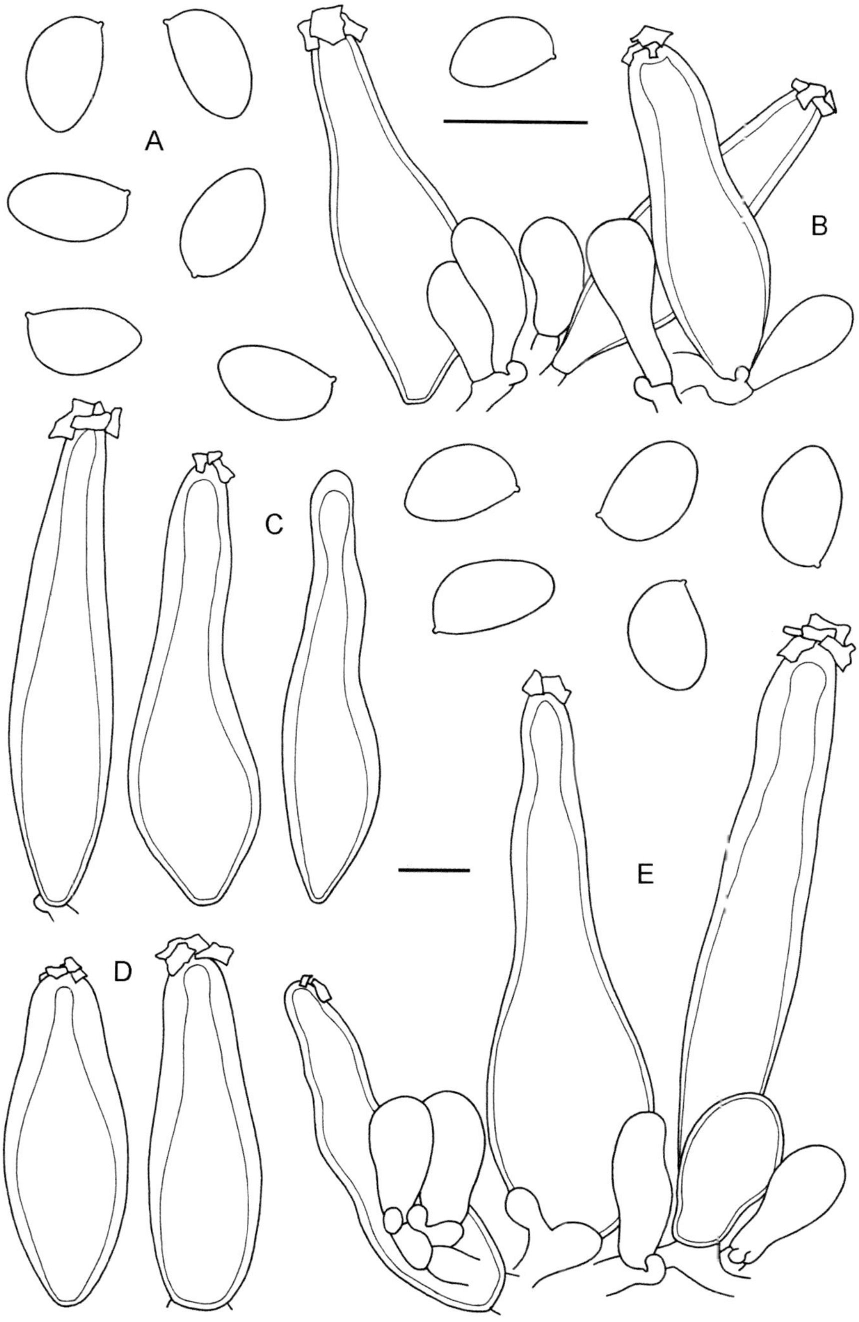

Figure 52. *Inocybe nudiuscula,* all holotype (PERTH 08319839), except D (PERTH 08363137). **A,** Basidiospores; **B,** Cheilocystidia and paracystidia; **C,** Pleurocystidia; **D,** Pleurocystidia; **E,** Caulocystidia and cauloparacystidia. Scale bars = 10 µm (longer bar for spores only).

Inocybe nudiuscula occupies a unique lineage in the STAC group in an unresolved phylogenetic position (Fig. 7). Collections have been identified from warm-temperate south-eastern Queensland and the south-west of Western Australia, and secotioid forms have been collected in Victoria. The Queensland collections have hymenial cystidia that are shorter, stockier, have a less slender, attenuated upper part, and are slightly thicker-walled than the Western Australia holotype. All collections cited here form a robust monophyletic group, but within it some genetic differences are evident between those sampled from the different localities (data not shown). However, all other characteristics are comparatively uniform, including overall coloration, size, texture, spores, and its association with various Myrtaceae.

One collection from Victoria (*Trappe 18260*; and see Matheny *et al.*, 2009; Braaten *et al.*, 2014) represents a secotioid form that groups phylogenetically with the other collections of *I. nudiuscula* cited above. However, due to inadequate material and documentation, we are unable to shed further light on the form. *Trappe 18260* is extinguished, and a second secotioid collection from Victoria (*Trappe 20013*) has only one small fruit body and lacks any molecular data at this time. Its basidia are statismosporic, 4-sterigmate, and spore size is variable: 9.0–10.5 (–12.0) × 5.0–6.0 (–7.0) µm.

45. **Inocybe tasmanica** Matheny & Bougher, *sp. nov.*

MycoBank: MB 809471

Type: Ralphs Falls, West Pyengana, Tas., alt. 800 m, gregarious on soil in cool-temperate rainforest under *Nothofagus cunninghamii*, 23 May 2010, *P.B.Matheny PBM3395*; holo: PERTH 08517231; iso: TENN 068285 (KF808340-nLSU, KF830028-ITS, KF830046-*rpb2*).

Pileus 3–11 mm wide, convex to plano-convex, usually lacking an umbo; margin incurved to decurved, entire; velipellis absent; surface dry, tomentose-fibrillose to coarsely fibrillose, not at all scaly and not rimose; dark brown at the centre shading to yellowish brown (Snuff Brown) towards the margin; context very thin. *Lamellae* adnexed to subfree, moderately close, with several tiers of lamellulae, ventricose, pinkish grey-brown (7.5YR 6/4; Avellaneous) to light brown; edges pallid, indistinctly fimbriate. *Stipe* 11–14 × 1.5–2.0 mm, even, often flexuous; partial veil cortinate, fugacious; upper half pruinose, with a light whitish vesture below; ground colour light brown to vinaceous grey-brown (Avellaneous), becoming brown (7.5YR 6/4 to 5/4); context solid, concolorous with the surface. *Odour* acidulous. Pl. 97.

Basidiospores 8.5–10.0 × 5.0–6.0 µm [mean 9.4 × 5.5 µm; Q 1.55–1.93 (–2.00); Q mean 1.72 (n = 21/1)], smooth, ranging from amygdaliform with barely attenuated apices to limoniform with attenuated-mucronate apices, yellowish brown; apiculus small. *Basidia* 24–26 × 7–9 µm, clavate, hyaline, 4-sterigmate. *Pleurocystidia* 53–62 × 11–14 µm, fusiform or ventricose with a slender basal pedicel, apices obtuse, often bearing crystals, thick-walled, the walls 1.5–3.0 µm, hyaline. *Cheilocystidia* similar to pleurocystidia, at times shorter or more variable in shape (broadly clavate to subcylindrical), mixed with small slender clavate paracystidia (often in dense clusters) and basidia. *Caulocystidia* similar to pleurocystidia but larger (67–115 × 15–19 µm), descending at least to the stipe centre, mixed with cells of various shapes and sizes, including cauloparacystidia and caulocystidioid terminal cells of vesture hyphae (the latter often cylindrical and thick-walled), lower half of stipe composed of interwoven vesture hyphae. *Pileipellis* a cutis giving rise to some upturned elements, dark brown to yellowish brown in mass, hyphae weakly encrusted, frequently septate, cylindrical to inflated, mostly 6–20 µm wide, thick-walled (walls 1.0–1.5 µm), terminal elements more narrow and smooth with fewer septa. *Clamp connections* present. Fig. 53.

On soil in cool-temperate rainforest under *Nothofagus cunninghamii*. Known only from the type collection. Fruiting May.

Etymology: The epithet *tasmanica* is in reference to the type locality of the new species.

Diagnostic features: Fruit bodies small; pileus less than 15 mm wide, surface tomentose to coarsely fibrillose. Lamellae pinkish grey-brown to light brown. Stipe pruinose over upper half, covered with thin whitish vesture below. Spores mostly 9–10 × 5–6 µm, many limoniform. In cool-temperate rainforest under *Nothofagus*.

Inocybe tasmanica is known from only a single collection, but it warrants recognition as a new species due to its unique diagnostic features. Moreover, it also occupies a unique phylogenetic position, being isolated from all the major groups in the STAC group (Fig. 7).

Several other small species of *Inocybe* could be confused with *I. tasmanica*. *Inocybe pluviosilvestris* (*I. serrata* group) also occurs in *Nothofagus* forests in Tasmania, but it has a smooth pileus, yellow hymenial cystidia, and smaller spores (6.5–8.0 × 4.0–5.0 µm). The New Zealand species *I. mendica* E.Horak has slightly smaller spores (8.0–9.5 × 5.0–5.5 µm), a light brown pileus, and a distinct phylogenetic position nested in the *I. ionocaulis* group (Fig. 45).

Members of the *I. serrata* group, with the exception of *I. pluviosilvestris* (*I. serrata*, *I. granulosipes*, *I. fulvo-olivacea*, *I. destruens* and *I. scabriuscula*), could be confused with *I. tasmanica*, but all occur in non-rainforest habitats, or differ by their scaly appearance.

97

Plate 97. *Inocybe tasmanica*, holotype (PERTH 08517231). Photograph — P.B.Matheny.

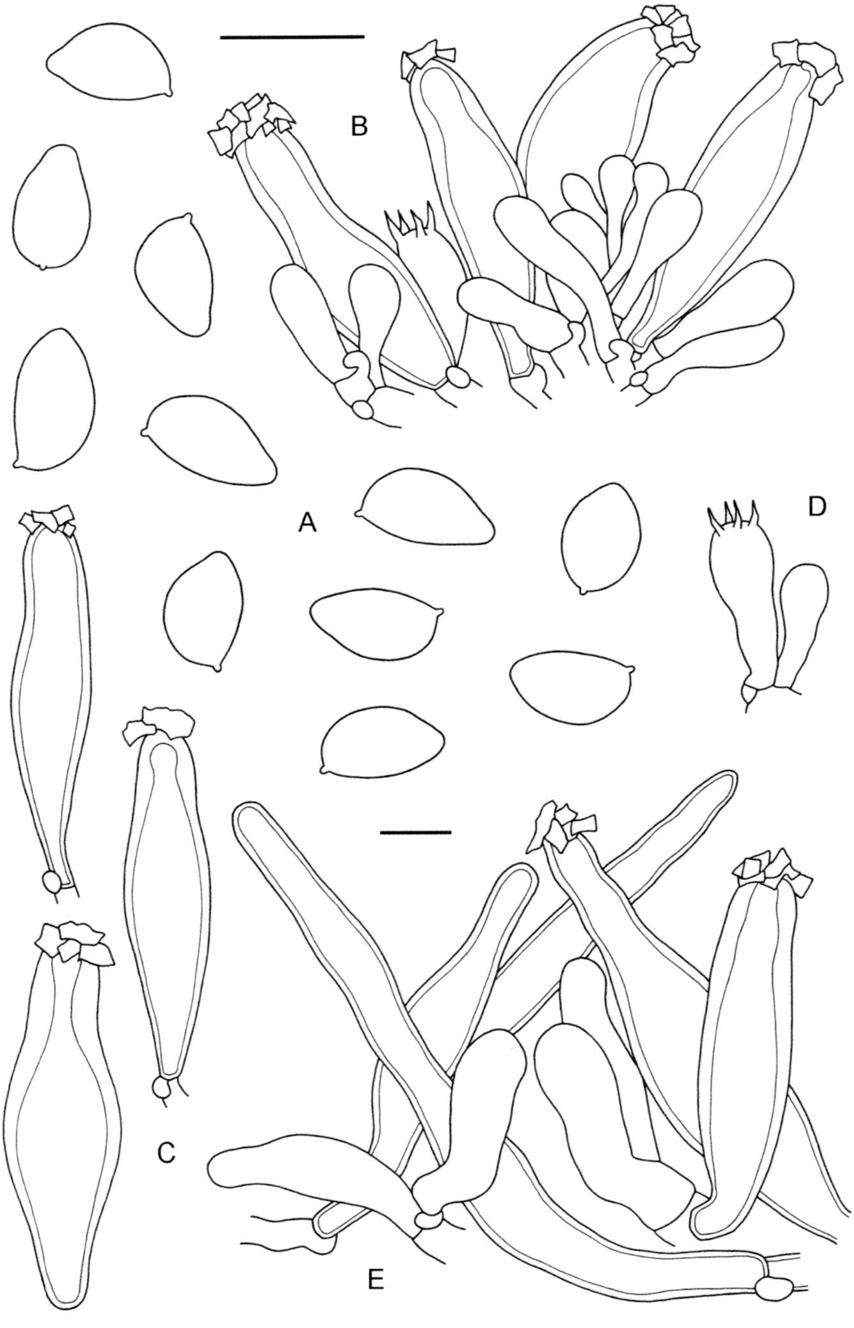

Figure 53. *Inocybe tasmanica,* all holotype (PERTH 08517231). **A,** Basidiospores; **B,** Cheilocystidia and paracystidia; **C,** Pleurocystidia; **D,** Basidia; **E,** Caulocystidia and caulocystidioid hairs, and cauloparacystidia, Scale bars = 10 µm (longer bar for spores only).

INOCYBE TENAX GROUP

The *I. tenax* group is a strongly supported monophyletic group containing 14 species, ten of which are documented here from Australia. Species in the group are generally geographically restricted, other than *I. tenax* itself which contains distinct eastern Australian and New Zealand populations. Thirteen of the 14 species occur only in Australia. One of these has been introduced in eucalypt plantations in Argentina (Fig. 54). Many of the species in the *I. tenax* group are small, and all feature a distinct pruinose and even stipe.

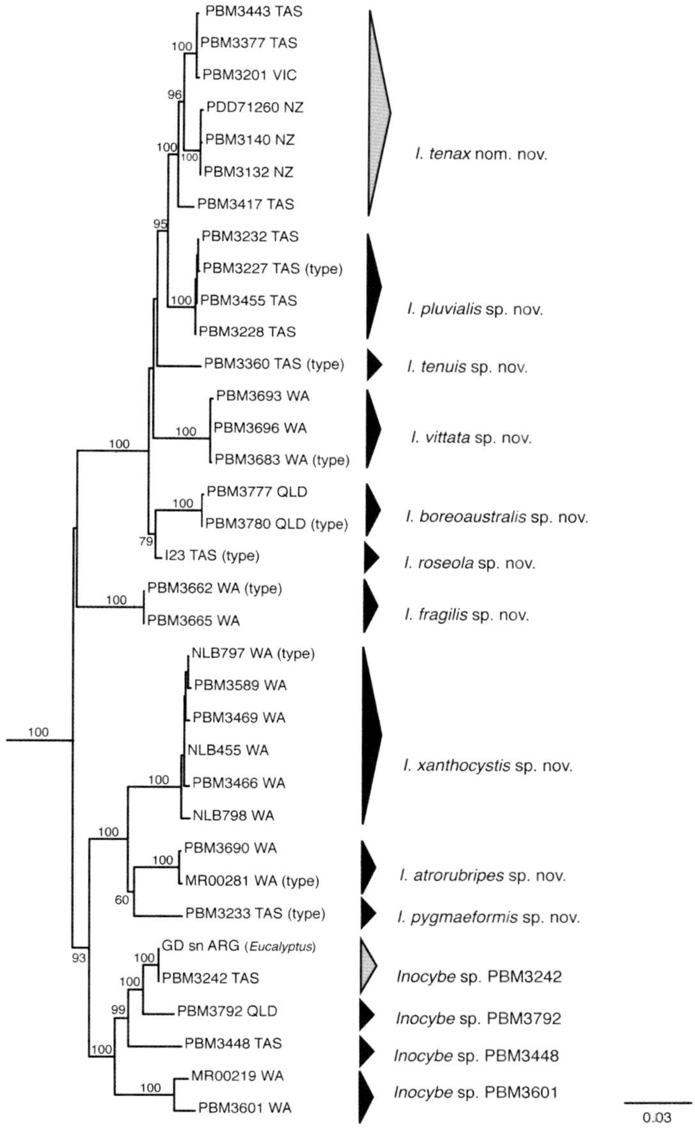

Figure 54. Maximum Likelihood phylogram of the *Inocybe tenax* group, the topology of which was produced from the analysis of the STAC group shown in Fig. 7. Australian taxa are indicated by black-filled triangles. Grey-filled triangles indicate taxa that occur in Australia and New Zealand or Australia and Argentina. Bootstrap values are shown above branches.

46. Inocybe atrorubripes Matheny, Bougher & Ryberg, *sp. nov.*

MycoBank: MB 808683

Type: Corrigin Wildflower Drive, near airfield, 5 km W of Corrigin, W.A., 32°20'19.1"S, 117°49'20.0"E, solitary or in small clusters on sandy to lateritic soil in open space under *Allocasuarina campestris*, 27 July 2011, *M.Ryberg MR00281*; holo: PERTH 08306869; iso: TENN 066456 (KF808350-nLSU, KF830038-ITS, KF830054-*rpb2*).

Pileus 6–22 mm wide, parabolic to conical, becoming plano-convex to plane with age, not umbonate; margin initially incurved, soon decurved; surface dry; velipellis often present over the disc; surface coarsely fibrillose, at times breaking up into small appressed squamules or floccules on and around the disc; mostly dark yellowish brown (10YR 3/3; Bay Brown); disc often dark or greyish black (bistre); margin dark brown (7.5YR 3/2) to brown (10YR 5/3); edges pallid due to veil remnants; context thin, somewhat fragile, dark reddish brown, not changing colour where bruised but becoming white after drying. *Lamellae* adnate to adnexed, subdistant, with c. 20 L and several tiers of lamellulae, brown, dark reddish brown, to dark brown or very dark brown with age (5.YR 3/2 to 3/4 or 7.5YR 5/4, 4/4, 3/2), ventricose, 1.0–3.5 mm deep, the edges indistinctly pallid-fimbriate. *Stipe* 8–29 × 1–3 mm, even, terete, at times slightly compressed or grooved, with base sometimes slightly swollen, central or occasionally excentric; surface dry, entirely pruinose, occasionally the pruina sparse above the base, dark reddish brown (2.5YR 2/4 to 3/4; 5YR 3/4; Rood's Brown to Walnut Brown) but sometimes ferruginous (2.5YR 5/8; "red") when very young, usually with a pallid base; context hollow in the stipe, dark reddish brown. *Odour* absent. Pl. 98–100.

Basidiospores 9.0–11.5 × 4.5–5.5 µm [mean 10.1 × 5.0 µm; Q 1.80–2.30; Q mean 2.03 (*n* = 27/2)], smooth, oblong-amygdaliform or cylindrical, at times with a slight adaxial depression, rarely slightly angular in face view; apices often pointed; apiculus small, indistinct. *Basidia* 29–34 × 7–10 µm, ±clavate or narrowly so, generally widest near the middle, hyaline, 4-sterigmate. *Pleurocystidia* 60–84 × 12–14 µm, mostly narrowly fusiform to lageniform, with a long tapered neck; apices obtuse to subacute and crystalliferous; walls mostly 2–3 µm thick, pale yellow. *Cheilocystidia* similar to pleurocystidia, mixed with dense clusters of short clavate hyaline paracystidia. *Caulocystidia* descending the entire length of the stipe, densely arranged, dimorphic: (1) long lageniform thick-walled cells with subacute to obtuse apices and lanceolate necks, mostly 80–105 × 13–20 µm; and (2) short pyriform, saccate or broadly clavate cells, mostly 25–33 × 17–20 µm, with walls 1.0–1.5 µm thick, hyaline, mixed with occasional cauloparacystidia. *Pileipellis* a cutis giving rise to clusters of ascending hyphae, dark yellowish brown in mass; hyphae cylindrical, mostly 6–12 µm wide, encrusted to coarsely so, with walls slightly thickened; velipellis hyphae present above the cutis, these compactly interwoven, smooth, mostly 5–10 µm wide, thin-walled to slightly thick-walled, frequently septate. *Clamp connections* present. Fig. 55 (see also Pl. 6I).

Found in the wheat-belt region of south-western W.A.; in yellow sand or sandy soil under *Allocasuarina* and/or *Acacia*. Fruiting July.

Additional specimens examined: W.A.: Old Kunjin railway, c. 14 km W of Corrigin, along main road, 32°21'25.5"S, 117°44'10.6"E, in sandy soil under planted eucalypts and *Allocasuarina* bushes, 26 July 2011, *N.L.Bougher NLB879* (PERTH 08320519, TENN 066763); type locality, on yellow sand under *Allocasuarina campestris* (Black Tamma), 27 July 2011, *P.B.Matheny PBM3690* (PERTH 08318956, TENN 066551; KF808349-nLSU, KF830037-ITS, KF830053-*rpb2*); off Barber Rd, Vinegar Well, Kunjin, W of Corrigin, 32°23'31.7"S, 117°40'6.5"E, on soil in open woodland of *Acacia acuminata* and *Allocasuarina huegeliana*, 27 July 2011, *N.L.Bougher NLB894* (PERTH 08320357); *loc. id.*, *N.L.Bougher NLB897* (PERTH 08320330).

Etymology: From the Latin *atroruber* (dark red) and *pes* (a foot), in reference to the dark reddish brown stipe.

Diagnostic features: Fruit bodies small. Pileus very dark brown (greyish black) to dark yellowish brown or brown, appressed-squamulose; velipellis present; lamellae quite dark; stipe dark reddish brown, pruinose for its entire length, at times compressed or grooved, hollow. Spores smooth; hymenial cystidia as metuloids. In woodland under *Allocasuarina* and/or *Acacia*.

Inocybe atrorubripes is most similar to *I. vittata*, which shares the similar habitat and an apparent preference for *Allocasuarina* (and/or *Acacia)*, but the pileus, lamellae and stipe are significantly darker in *I. atrorubripes*. Phylogenetic data also support their separation (Fig. 54). *Inocybe nigricans* can be mistaken for *I. atrorubripes* as it is a very dark-capped species also occurring in *Allocasuarina-Acacia* shrubland or woodland in the Western Australian wheatbelt region. However, the former has a fibrillose (not pruinose), brown (not dark reddish brown) stipe and nodulose spores. *Inocybe geniculata* is also similar, but it has a recurved-scaly pileus, a light cinnamon to pale brown stipe, and it grows in leaf litter in wet sclerophyll forest in Tasmania.

The caulocystidia of *I. atrorubripes* are noteworthy due their density and distribution (not in scattered clusters) on the stipe surface and their dimorphism, with long, lageniform or lanceolate cells and slightly thick-walled pyriform, saccate or broadly clavate cells.

Plate 98. *Inocybe atrorubripes*, holotype (PERTH 08306869). Photograph — M.Ryberg.

Plate 99. *Inocybe atrorubripes* (PERTH 08320330). Photograph — N.L.Bougher.

Plate 100. *Inocybe atrorubripes* (PERTH 08318956). Photograph — P.B.Matheny.

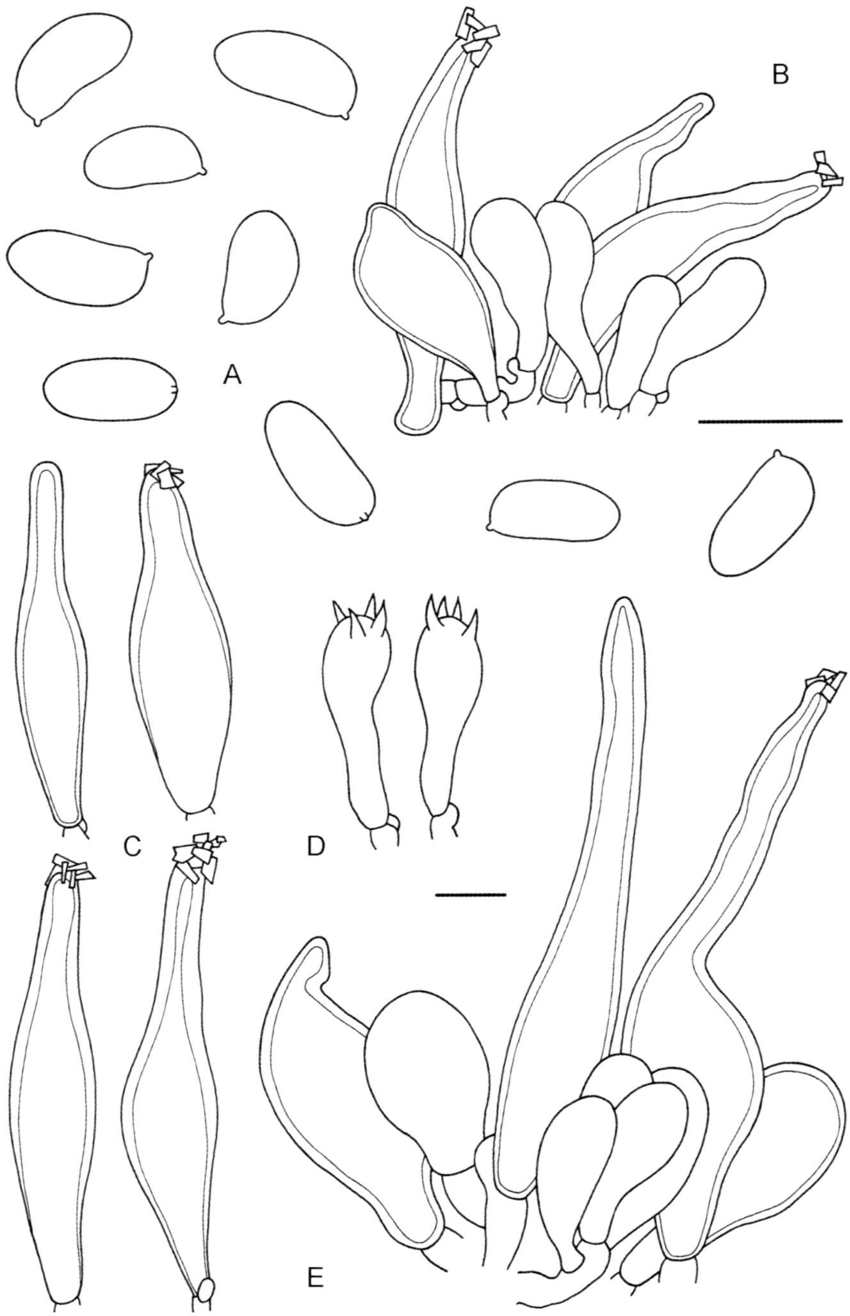

Figure 55. *Inocybe atrorubripes,* all holotype (PERTH 08306869). **A,** Basidiospores; **B,** Cheilocystidia and paracystidia; **C,** Pleurocystidia; **D,** Basidia; **E,** Caulocystidia and cauloparacystidia. Scale bars = 10 µm (longer bar for spores only).

47. **Inocybe boreoaustralis** Matheny & Bougher, *sp. nov.*

MycoBank: MB 809267

Type: Rainforest Circuit Track, Cunninghams Gap, Main Range Natl Park, Qld, 28°02'59"S, 152°23'37.3"E, alt. 770 m, scattered singly on soil under *Lophostemon* in warm-temperate rainforest, 3 Mar. 2012, *P.B.Matheny PBM3780*; holo: PERTH 08363080; iso: TENN 067012 (KF808359-nLSU, KF830044-ITS, KF830060-*rpb2*).

Pileus 4–12 mm wide, conical or obtusely so, expanding with age to broadly conical or ±plane, often with a small nipple-like or obtuse umbo; margin decurved; surface dry; velipellis present over the centre, occasionally inconspicuous; disc smooth and unbroken, matted-fibrillose to fibrillose or with small appressed scales towards the margin, at times excoriate around the centre; margin entire; colour variable depending on field conditions: when moist the centre dark brown or chestnut-brown (7.5YR 3/2), with margin brown (10YR 4/3 to 3/2; Snuff Brown or Sepia), and centre at times light grey over the umbo due to velipellis; when dry, the disc light yellowish brown due to velipellis (2.5Y 6/4), elsewhere light olive-brown overall (2.5Y 5/4) to olive-brown (2.5Y 4/4) and paler between the small squamules; context very thin, pallid or concolorous with the pileus surface when moist, not changing colour where cut or bruised. *Lamellae* adnexed, subdistant, 16–18 L, with few tiers of lamellulae, light yellowish brown to yellowish brown, the edges not noticeably pallid-fimbriate except finely so under a hand lens, ventricose, medium. *Stipe* 9–25 × 0.5–1.0 mm, terete, even, base slightly swollen, no veil observed, surface densely pruinose throughout; reddish brown (5YR 4/4) to dark reddish brown or dark brown (5YR 3/4, 5/3 or 7.5YR 4/2), pallid at the extreme base; context solid, becoming hollow, pallid at the base, reddish brown elsewhere. *Odour* like *Pelargonium*, sweet, or acidulous and the *Pelargonium* odour developing after being cut. Pl. 101, 102.

Basidiospores (7.0–) 8.0–9.5 (–10.0) × 4.5–5.5 (–6.0) µm [mean 8.7 × 5.2 µm; Q 1.50–1.91 (-2.00); Q mean 1.67 (n = 31/2)], smooth, amygdaliform with conical to bluntly pointed apices (at times submucronate), yellowish brown; apiculus small and indistinct. *Basidia* 21–30 × 7–8 µm, clavate or cylindroclavate, hyaline, 4-sterigmate. *Pleurocystidia* 40–65 × 13–20 µm, fusiform, apices obtuse and crystalliferous, bases often rounded; walls thick (2–4 µm), yellow. *Cheilocystidia* similar to pleurocystidia, mixed with crowded hyaline clavate paracystidia. *Caulocystidia* descending the entire length of the stipe, generally larger and more often lageniform than hymenial cystidia, hyaline, apices obtuse to subacute, mixed with hyaline clavate cauloparacystidia. *Pileipellis* a cutis of cylindrical hyphae mostly 5–14 µm wide, russet-brown to dark yellowish brown in mass, with thickened walls, weakly encrusted, superimposed by an interwoven layer of velipellis hyphae, these narrowly cylindrical, 3–5 µm wide, thick-walled, smooth, pale yellowish; individual cells extending for considerable lengths without septa. *Clamp connections* present. Fig. 56.

Endemic to south-eastern Qld; occurs in warm-temperate rainforest under *Lophostemon* or among sclerophyll vegetation (dominated by *Eucalyptus* and *Allocasuarina*) adjacent to rainforest. Fruiting Feb. and Mar.

Additional specimen examined: Qld: Cunninghams Gap, Main Range Natl Park, 28°02'59"S, 152°23'37"E, alt. 776 m, scattered singly on soil in sclerophyll forest under *Eucalyptus* and *Allocasuarina*, 29 Feb. 2012, *R.E.Halling & N.Fechner PBM3777* (PERTH 08363110, TENN 067009; KF808358-nLSU, KF830045-ITS, KF830061-*rpb2*).

Etymology: From the Latin *borealis* (northern) and *australis* (southern), in reference to the occurrence of this species in a northern part of the Southern Hemisphere.

Diagnostic features: Fruit bodies small, the pileus usually with a greyish velipellis, fibrillose or becoming appressed-squamulose; stipe densely pruinose for its entire length, the base slightly swollen; odour of *Pelargonium* or becoming so after being cut.

Inocybe boreoaustralis is one of a complex of species from temperate Australia and New Zealand related to *I. tenax* (Fig. 54, and see comments under *I. tenax* and *I. tenuis*). Based on current data, *I. boreoaustralis* is distinguished by its occurrence in south-eastern Queensland and a distinct phylogenetic position (Fig. 54). It resembles several other species in the complex with narrowly cylindrical, thick-walled, velipellis hyphae, and hymenial cystidia typically with rounded-truncate bases.

101

102

Plate 101. *Inocybe boreoaustralis*, holotype (PERTH 08363080). Photograph — P.B.Matheny.

Plate 102. *Inocybe boreoaustralis* (PERTH 08363110). Photograph — P.B.Matheny.

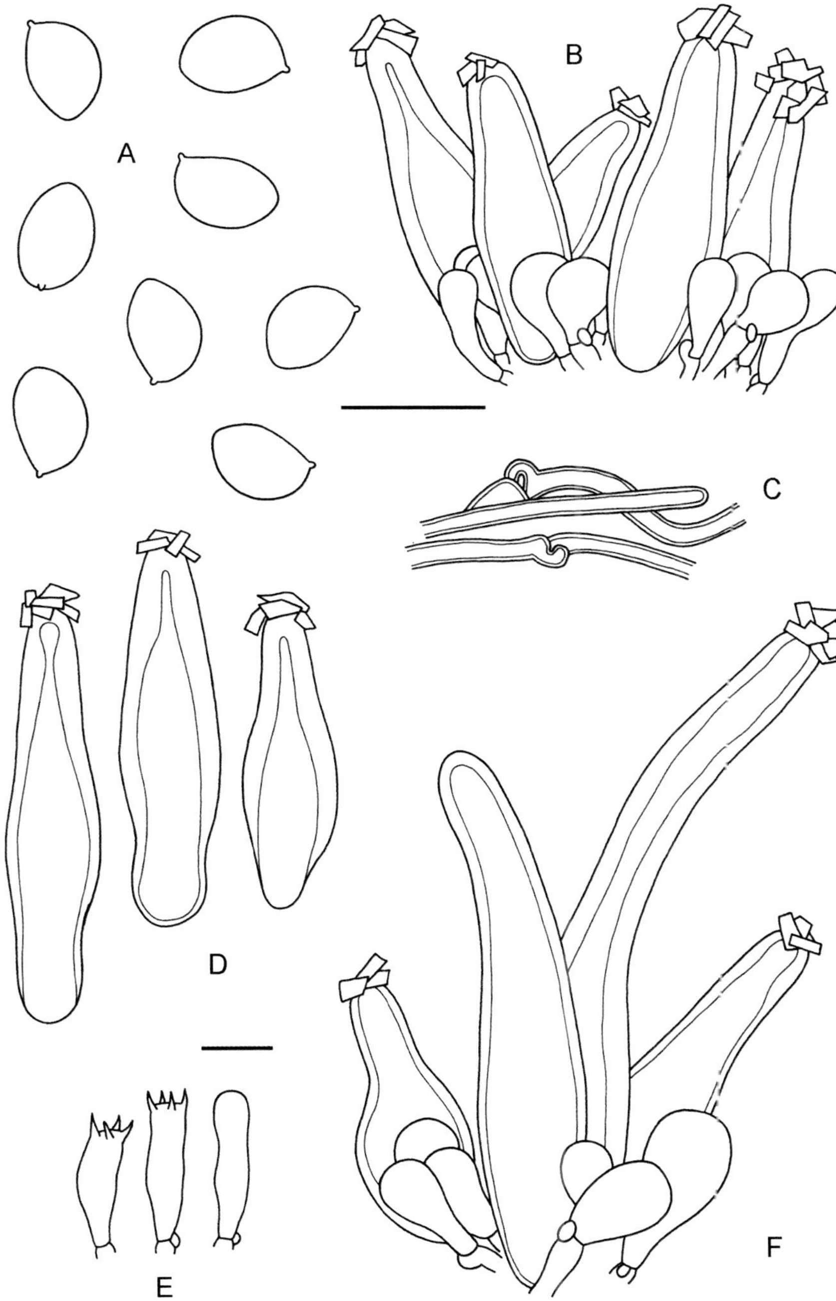

Figure 56. *Inocybe boreoaustralis,* all holotype (PERTH 08363080). **A,** Basidiospores; **B,** Cheilocystidia and paracystidia; **C,** Velipellis hyphae; **D,** Pleurocystidia; **E,** Basidia; **F,** Caulocystidia and cauloparacystidia. Scale bars = 10 µm (longer bar for spores only).

48. Inocybe fragilis Matheny & Bougher, *sp. nov.*

MycoBank: MB 809328

Type: Woodlands Rd, Porongurup Natl Park, W.A., 34°40'06.4"S, 117°52'53.6"E, gregarious, solitary or in pairs, on compact soil in litter in tall wet Karri forest under *Corymbia calophylla*, *Eucalyptus diversicolor* and *E. marginata* last burnt 4 years previously, 21 July 2011, *P.B.Matheny PBM3662*; holo PERTH 8319251; iso: TENN 066507 (KF808341-nLSU, KF830029-ITS, KF830047-*rpb2*).

Pileus 4–12 mm wide, initially conical or conico-convex, soon plane with a small obtuse umbo, margin decurved or straight to uplifted with age, surface fibrillose throughout, slightly excoriate or becoming coarsely fibrillose towards the margin, dry; disc often lighter in colour due to the velipellis; overall colour mostly fulvous (yellowish brown) or Buckthorn Brown mixed with shades of Dresden Brown (dark yellowish brown), typically fulvous at the centre and throughout or shading a little darker towards the margin; context pallid, thin. *Lamellae* adnexed, subdistant, 20–22 L, with several tiers of lamellulae, at first pale brown to light yellowish brown, yellowish brown with age, edges indistinctly pallid-fimbriate, ventricose, not narrow. *Stipe* 6–18 × 0.75–1.00 mm, even, flexuous, pruinose throughout, Avellaneous or reddish brown when young, with a white base, becoming Cinnamon Buff to Clay Colour; context solid. *Odour* acidulous. Pl. 103, 104.

Basidiospores 7.5–9.5 (–10.5) × 4.0–5.0 µm [mean 8.4 × 4.6 µm; Q 1.60–2.11 (–2.33); Q mean 1.83 (n = 41/2)], smooth, amygdaliform, sometimes with an attenuated apex; apiculus small. *Basidia* 22–24 × 6–8 µm, clavate or cylindroclavate, 4-sterigmate, hyaline. *Pleurocystidia* 50-60 × 10–18 µm, fusiform, with a tapered or cylindrical neck, the apices subacute, crystalliferous, the bases sometimes rounded, thick-walled (2–3 (–5) µm), hyaline to pale yellowish. *Cheilocystidia* similar to pleurocystidia but occasionally shorter; paracystidia basidiole-like, crowded. *Caulocystidia* similar to cheilocystidia and pleurocystidia but more often lanceolate, also lageniform or fusiform, 26–74 × 10–17 µm; apices subacute, mostly lacking crystals; descending the entire length of the stipe in numerous scattered clusters mixed with cells of intermediate size and shape, including cauloparacystidia, but these difficult to distinguish on the lower part of the stipe. *Pileipellis* a cutis of cylindrical hyphae mostly 7-20 µm wide and thin-walled, weakly encrusted, golden yellow in mass, subtending an interwoven layer of velipellis hyphae, these long, rarely branched, cylindrical, narrow, the walls 1.0–1.5 µm thick, with infrequent septa, hyaline or pale yellowish. *Clamp connections* present. Fig. 57.

Known only from the type locality in Porongurup National Park, south-western W.A.; grows on soil in leaf litter under *Corymbia calophylla*, *Eucalyptus diversicolor* and *E. marginata* in Karri forest. Fruiting July.

Additional specimens examined: W.A.: type locality, same habitat and collection date, *P.B.Matheny PBM3665* (PERTH 8319227, TENN 066504; KF808342-nLSU, KF830030-ITS, KF830048-*rpb2*); *loc. id.* (c. 300 m distant), same habitat and collection date, *N.L.Bougher NLB856* (PERTH 8320764).

Etymology: From the Latin *fragilis*, in reference to the small, soft fruit bodies.

Diagnostic features: Similar to *I. tenax* in colour, spores and pileipellis, but the fruit bodies very small (pileus <15 mm wide) and fragile, not tough. Pileus with a whitish disc in contrast to the yellowish brown margin. Spores 4–5 µm wide.

Inocybe fragilis is a species that is strongly phylogenetically supported in the *I. tenax* group (Fig. 54). It shares with this assemblage of temperate taxa the comparatively small stature, a flexuous pruinose stipe with a white base, the presence of a velipellis, smooth spores, and a preference for leaf litter habitats. The velipellis hyphae are similar to those of *I. tenax* and *I. pluvialis* and others in being interwoven, thick-walled and infrequently septate. Unlike other species in the group, the fruit bodies of *I. fragilis* are soft and fragile.

Plate 103. *Inocybe fragilis*, holotype (PERTH 8319251). Photograph — P.B.Matheny.

Plate 104. *Inocybe fragilis* (PERTH 08319227). Photograph — P.B.Matheny.

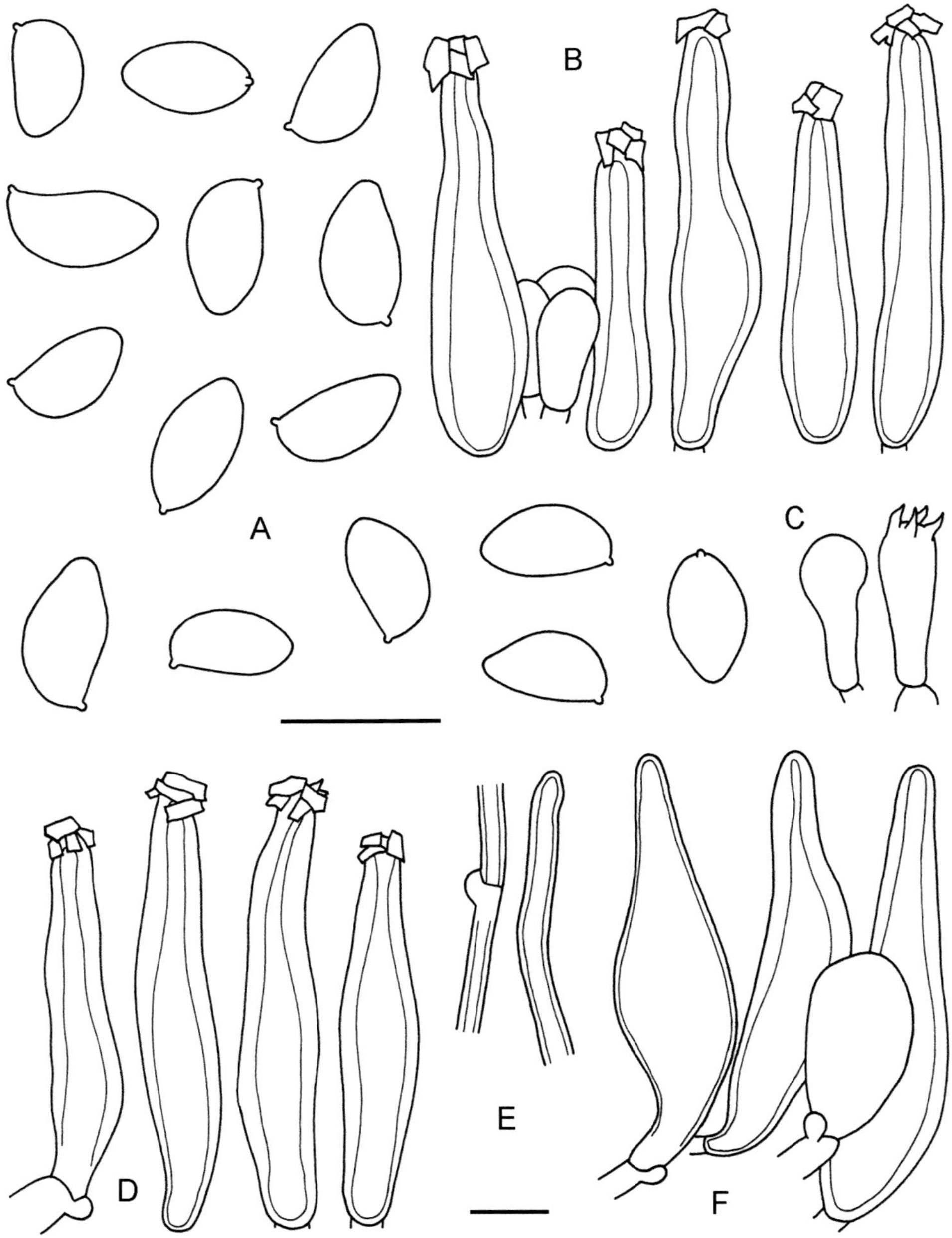

Figure 57. *Inocybe fragilis,* all holotype (PERTH 8319251). **A,** Basidiospores; **B,** Cheilocystidia and paracystidia; **C,** Basidia; **D,** Pleurocystidia; **E,** Velipellis elements; **F,** Caulocystidia and a cauloparacystidium. Scale bars = 10 µm (longer bar for spores only).

49. **Inocybe pluvialis** Matheny, Bougher & G.M.Gates, *sp. nov.*

MycoBank: MB 809431

Type: Lady Barron Track, Mount Field Natl Park, Tas., 42°40'S, 146°43'E, on soil on bank of track in wet sclerophyll forest under *Eucalyptus regnans*, *Acacia* and *Pomaderris*, 8 June 2009, *G.Gates*, *D.Ratkowsky & P.B.Matheny PBM3227*; holo: PERTH 08517150; iso: TENN 067041 (KF853398-nLSU, KF871774-ITS, KF891951-*rpb2*).

Pileus 8–16 mm wide, obtusely conical to conical, becoming convex to plane, at times with a small obtuse umbo; velipellis absent or present as a greyish area over the disc; disc at first smooth and unbroken, soon breaking up and becoming finely appressed-scaly, eventually excoriate around the disc, not rimose, fibrillose or finely fibrillose towards the margin; occasionally the margin translucent-striate or weakly so; disc dark brown (7.5YR 3/2), shading to strong brown or umber (7.5YR 4/4) towards the margin, colour more fulvous or Sayal Brown (7.5YR 5/6) after moisture loss; context thin to very thin, coloured like the surface. *Lamellae* adnexed, moderately close to subdistant, with several tiers of lamellulae, yellowish brown (10YR 5/4) to strong brown (7.5YR 4/4) or brown, ventricose, broad; edges indistinctly pallid-fimbriate. *Stipe* 35–40 × 1.0–1.5 mm at the apex, even but often flexuous, not at all bulbous, wiry, stiff; partial veil not observed; lightly pruinose throughout, but the base with a thin covering of pallid tomentum; reddish brown (5YR 5/4 to 4/4) to dark reddish brown (5YR 3/4 to 3/3); context solid, concolorous with the surface. *Odour* not strong but distinctly of *Pelargonium* when cut, less often acidulous. Pl. 105, 106.

Basidiospores 8.0–9.5 × 4.5–6.0 µm [mean 8.5 × 5.1 µm; Q 1.45–1.90; Q mean 1.66 (*n* = 40/2)], smooth, amygdaliform with pointed apices, yellowish brown; apiculus indistinct. *Basidia* 25–26 × 7–9 µm, cylindroclavate, 4-sterigmate, hyaline. *Pleurocystidia* 49–75 × 13–15 µm, fusiform to lageniform with a rounded-truncate base (as in *I. tenax*); base wider than the apex, the latter often subacute and/or crystalliferous, thick-walled (walls to 3–4 µm), hyaline or pale yellow. *Cheilocystidia* similar to pleurocystidia but often shorter, mixed with hyaline paracystidia. *Caulocystidia* descending the entire length of the stipe, similar to hymenial cystidia but longer (e.g. 96 × 15 µm) and lanceolate or narrowly fusiform. *Pileipellis* bilayered, the uppermost of hyaline narrow smooth- and thick-walled (to 2 µm) velipellis hyphae up to 3-6 µm wide, golden, aseptate for long lengths; subtending layer (pileipellis proper) of wider weakly encrusted dark brown hyphae. *Clamp connections* present. Fig. 58 (see also Pl. 12D, 12F).

Known only from Mount Field Natl Park, Tas.; on soil in cut banks along tracks in wet sclerophyll forest under *Eucalyptus regnans*, *Pomaderris* and *Acacia*. Fruiting June.

Additional specimens examined: Tas.: Mount Field Natl Park, on bank under *Eucalyptus regnans*, 8 June 2009, *G.Gates & D.Ratkowsky PBM3228* (TENN 067042; KF853401-nLSU, KF871777-ITS, KF891954-*rpb2*); *loc. id.*, *G.Gates, P.B.Matheny, & D.Ratkowsky PBM3232* (TENN 067043; KF853399-nLSU, KF871775-ITS, KF891952-*rpb2*); track to Lady Barron Falls, Mount Field Natl Park, on soil in wet sclerophyll forest under *Eucalyptus regnans* and *Pomaderris*, 1 June 2010, *P.B.Matheny PBM3438* (PERTH 08517169, TENN 067044); *loc. id.*, *P.B.Matheny PBM3455* (TENN 067045; KF853400-nLSU, KF871776-ITS, KF891953-*rpb2*).

Etymology: From the Latin *pluvialis* (of rain), in reference to the wet sclerophyll forest habitat.

Diagnostic features: Fruit bodies small. Stipe even, pruinose for its entire length except for a pallid tomentum at the base, stiff, wiry, easily plucked from the substratum with the base intact, reddish brown. Odour typically of *Pelargonium*. Spores smooth, amygdaliform. Pleurocystidia thick-walled, often with a rounded-truncate base. Velipellis hyphae golden, thick-walled and aseptate for long lengths.

Inocybe pluvialis is a member of the *I. tenax* group (Fig. 54); species characterised by their small size, entirely pruinose and often tough and wiry stipe, and smooth spores. It can also be characterised by the presence of narrowly cylindrical and thick-walled velipellis hyphae on the pileus surface. Within the group, *I. pluvialis* is strongly supported as a lineage associated with wet sclerophyll forest habitat, in contrast to *I. tenax* that is associated with *Nothofagus* and rainforest habitats in south-eastern Australia and New Zealand. Phylogenetically, *I. pluvialis* forms a monophyletic group sister to *I. tenax* (Fig. 54), and morphologically, it shares with *I. tenax* thick-walled velipellis hyphae, hymenial cystidia with rounded-truncate bases, and an overall dark reddish brown colouration. However, the lamellae of *I. pluvialis* can lack the reddish brown tones of *I. tenax*. Collections of *I. pluvialis* typically exhibit a subtle odour of *Pelargonium* when cut, whereas those of *I. tenax* from Australia and New Zealand have an initial strong acidulous odour that can be followed by a *Pelargonium*-like odour in Australian collections. We also observed some specimens of *I. pluvialis* with a translucent-striate pileal margin, a character never seen in *I. tenax*. These last three morphological traits are rather subtle and should be rigorously assessed for their consistency and overlap. Otherwise, habitat association is the best indicator for *I. pluvialis* which appears to fruit in the absence of *Nothofagus*.

Inocybe pluvialis is most likely to be confused with *I. tenuis*, also described from wet sclerophyll forest in Tasmania. The latter differs subtly in having pleurocystidia that are tapered towards the base (not rounded-truncate) and an acidulous odour, but it is also distinguished by its phylogenetic position (Fig. 54). *Inocybe xanthocystis*, known only from south-western Western Australia, differs from *I. pluvialis* in its bright yellow pleurocystidia with very thick walls [4–10 (–15) µm].

50. Inocybe pygmaeformis Matheny, Bougher & G.M.Gates, *sp. nov.*

MycoBank: MB 809446

Type: Lady Barron Track, Mount Field Natl Park, Tas., 42°40'S, 146°43'E, scattered on soil in wet sclerophyll forest under *Eucalyptus regnans*, *Acacia* and *Pomaderris*, 8 June 2009, *P.B.Matheny, G.Gates & D.Ratkowsky PBM3233*; holo: PERTH 08556431; iso: TENN 067736 (KF808348-nLSU, KF830036-ITS).

Pileus 4–10 mm wide, conical to obtusely so, expanding to broadly plano-convex with a slight obtuse umbo, margin decurved; surface dry, velipellis present over the disc (coarsely fibrillose) and towards the margin (fibrillose) at least when young, velipellis remnants more noticeable towards the margin with age; centre becoming recurved-squamulose, margin developing few appressed scales by maturity, not rimose; centre very dark brown (Fuscous) or dark reddish brown, often sharply grading to brown or pale brown at the margin in most specimens; context thin-fleshed. *Lamellae* adnexed, subdistant with c. 20–24 L and several tiers of lamellulae; pale brown, becoming light yellowish brown to brown, edges pallid but indistinctly fimbriate, ventricose, medium to broad. *Stipe* 12–25 × 0.5–1.0 mm, thin, slender and flexuous; even or slightly swollen at the base (not bulbous), pruinose for its entire length, cortina absent, light reddish brown to reddish brown (5YR 6/4 to 5/4), the base at times whitish; context not observed. *Odour* weakly of *Pelargonium* or absent. Pl. 107.

Basidiospores 7.5–10.0 × 4.5–5.0 µm [mean 8.5 × 4.8 µm; Q 1.60–2.00; Q mean 1.76 (n = 23/1)], smooth, amygdaliform, with bluntly pointed or conical apices, rather thick-walled (c. 0.5 µm), yellowish brown; apiculus small. *Basidia* 20–25 × 6–8 µm, subcylindrical to clavate, hyaline, 4-sterigmate. *Pleurocystidia* 53–75 × 14–18 µm, fusiform, with obtuse and sparsely crystalliferous apices, tapering towards the base which at times is rounded-truncate; walls yellow, rather thick (3–6 µm). *Cheilocystidia* similar to pleurocystidia but often shorter and more ventricose, mixed with hyaline and short-clavate paracystidia. *Caulocystidia* descending the entire length of the stipe, similar to cheilocystidia but often longer (to 90 µm), in scattered clusters mixed with abundant thin-walled broadly clavate, pyriform and sphaeropedunculate cauloparacystidia that are mostly larger than those associated with the cheilocystidia. *Pileipellis* a cutis giving rise to ascending elements, tawny to reddish orange-brown in mass, the hyphae cylindrical, mostly 5–13 µm wide, with walls 0.5–1.0 µm thick, weakly encrusted; velipellis hyphae superimposed over the cutis hyphae, hyaline in mass, narrowly cylindrical and mostly 3–5 µm wide, smooth, interwoven or erect, the walls 1–2 µm thick, sparingly septate, hyaline to pale yellow when viewed individually. *Clamp connections* present. Fig. 59.

Occurs on soil in wet sclerophyll forest in Tas., and is known only from the type collection; grows under *Eucalyptus*, *Acacia* and *Pomaderris*. Fruiting in June.

Etymology: From the Latin *pygmaeus* (a dwarf) and *-formis* (-formed), in reference to the very small and slender fruit bodies.

Plate 105. *Inocybe pluvialis*, holotype (PERTH 08517150). Photograph — P.B.Matheny.

Plate 106. *Inocybe pluvialis*. TENN 067042. Photograph — P.B.Matheny.

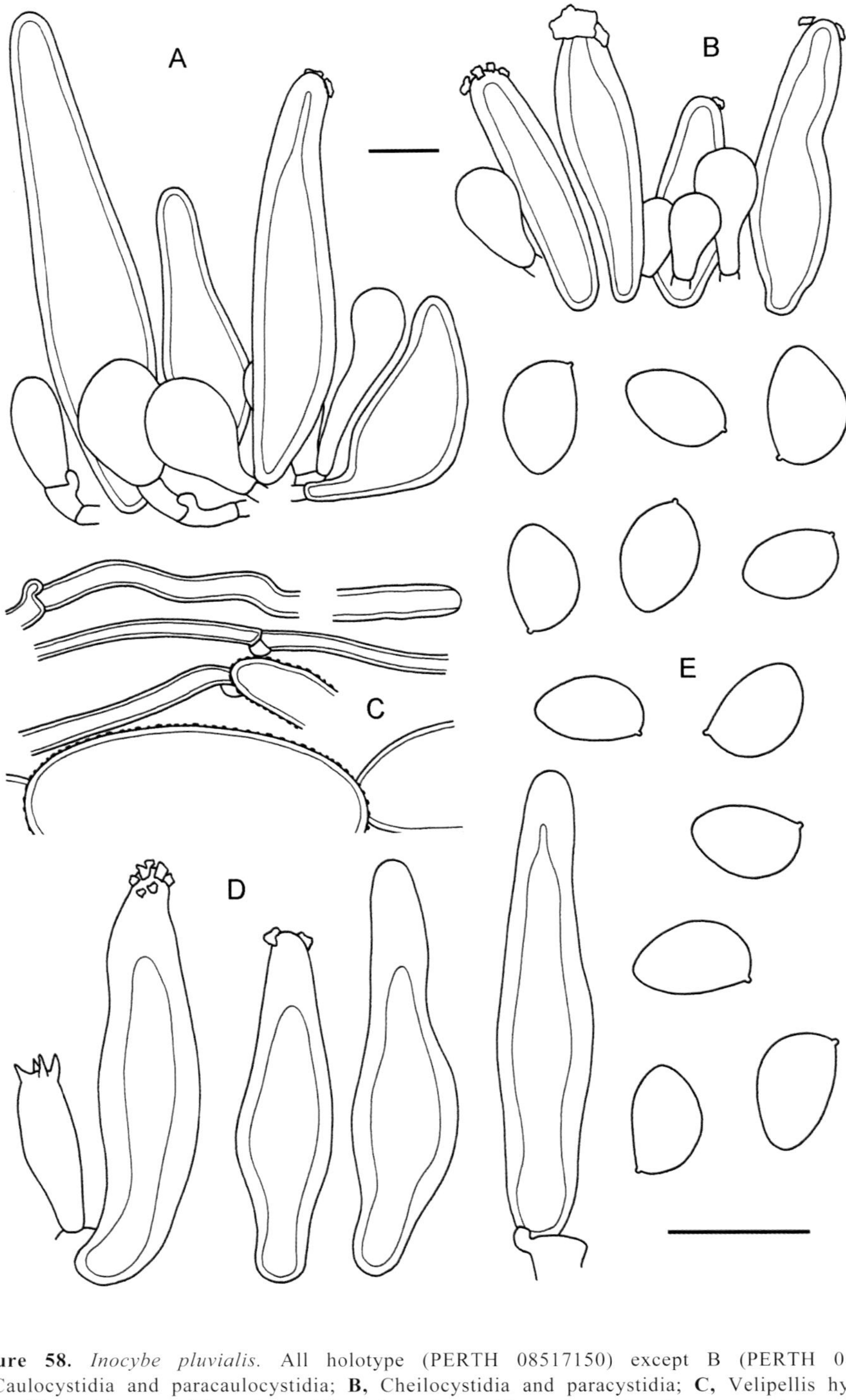

Figure 58. *Inocybe pluvialis.* All holotype (PERTH 08517150) except B (PERTH 08517169). **A,** Caulocystidia and paracaulocystidia; **B,** Cheilocystidia and paracystidia; **C,** Velipellis hyphae and pileipellis; **D,** Pleurocystidia and a basidium. Scale bars = 10 μm (longer bar for spores only).

Diagnostic features: Fruit bodies very small and slender. Pileus with a very dark brown centre that becomes erect-scaly; margin often noticeably paler; velipellis present; stipe pruinose for its entire length, light reddish brown to reddish brown, not bulbous. Spores smooth, to 10 µm long, amygdaliform. Hymenial cystidia fusiform to ventricose, thick-walled, yellow. Velipellis hyphae narrowly cylindrical, thick-walled, sparingly septate.

Inocybe pygmaeformis is phenotypically most similar to *I. strigiceps* f. *pygmaea* from Papua New Guinea (Horak, 1980), which is recombined below as *I. tenax* f. *pygmaea*. Although the taxa appear very similar, they differ by the longer pleurocystidia and larger spores in *I. pygmaeformis*, as well as ecologically and in their grographical distribution.

Plate 107. *Inocybe pygmaeformis*, holotype (PERTH 08556431). Photograph — P.B.Matheny.

Plate 108. *Inocybe roseola*, holotype (HO 569186). Photograph — G.M.Gates.

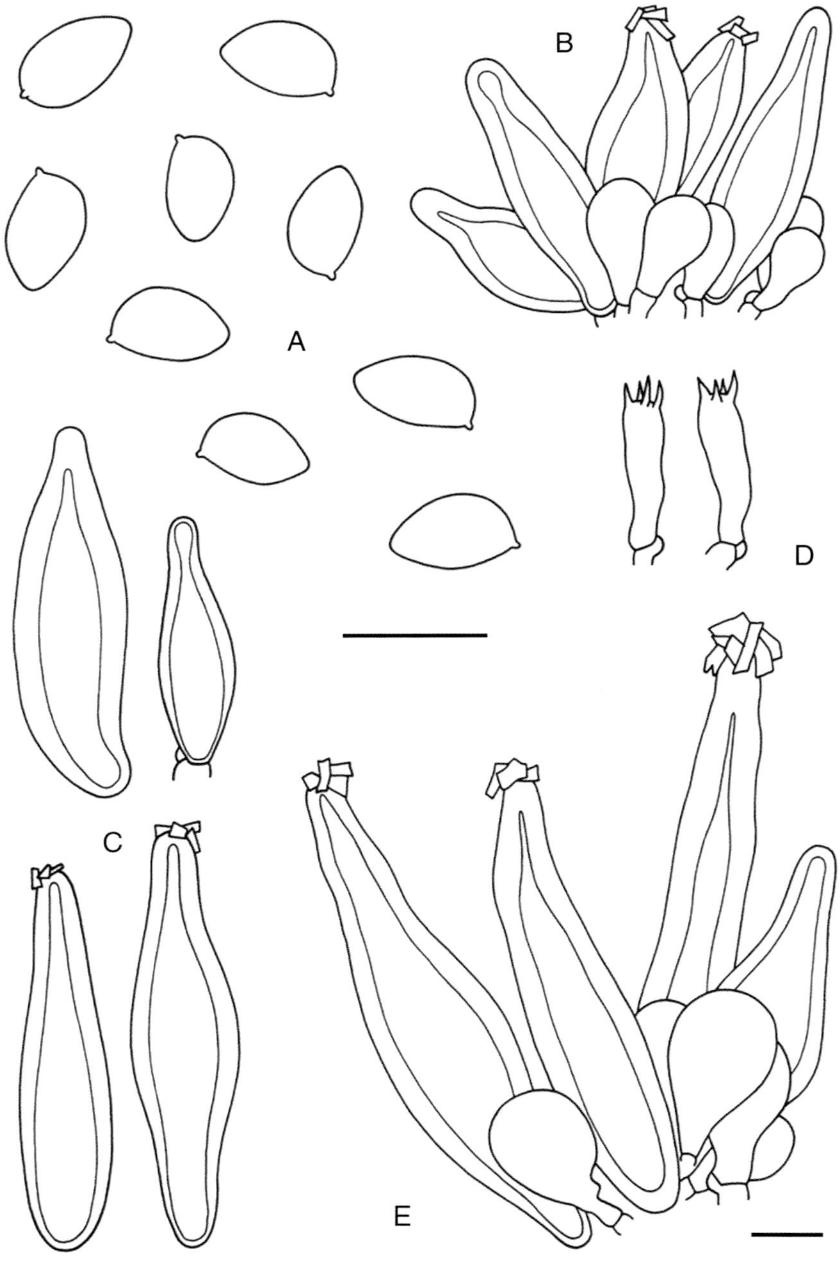

Figure 59. *Inocybe pygmaeformis.* All holotype (PERTH 08556431). **A,** Basidiospores; **B,** Cheilocystidia and paracystidia; **C,** Pleurocystidia; **D,** Basidia; **E,** Caulocystidia and cauloparacystidia. Scale bars = 10 μm (longer bar for spores only).

51. Inocybe roseola Matheny, Bougher & G.M.Gates, *sp. nov.*

MycoBank: MB 809449

Type: Chauncy Vale Wildlife Sanctuary, 40 km N of Hobart, Tas., on the ground in dry sclerophyll forest under *Eucalyptus* and *Acacia*, 20 Nov. 2001, *G.M.Gates & D.Ratkowsky 123*; holo: HO 569186; iso: TENN 068271 (KF808357-nLSU).

Pileus 3–8 mm wide, conical, with a blunt apex, expanding with age; margin incurved, becoming decurved; surface dry, some specimens with a whitish smooth velipellis over the disc, finely fibrillose, dry; margin entire, the edge with scattered white veil remnants, pinkish brown to reddish brown (2.5YR 4/4; near Avellaneous or Fawn Colour), creamy white over the disc when a velipellis present, hygrophanous, fading to cream; context white, thick under the disc. *Lamellae* ascending-adnate to adnexed, moderately close, with several tiers of lamellulae, pale rose or reddish brown (2.5YR 5/4 to 5YR 5/4), ventricose, 2 mm deep; edges pallid, indistinctly fimbriate. *Stipe* 7–12 mm × 1–2 mm, even, terete, at times flexuous, lacking a bulb; surface pruinose for the entire length, white, at times with pinkish tinges at the apex, becoming patchily dull yellow; context solid, becoming hollow, white or coloured like the surface. *Odour* absent. Pl. 108.

Basidiospores 7.0–9.0 (–9.5) × 5.0–6.0 µm [mean 7.8 × 5.2 µm; Q 1.33–1.73; Q mean 1.52 (*n* = 25/1)], smooth, mostly amygdaliform to broadly amygdaliform, with bluntly pointed or slightly attenuated apices, less often elliptic, (sub)hyaline to pale buff or completely hyaline and without guttules; apiculus small, indistinct. *Basidia* 22–31 × 6–9 µm, clavate, hyaline, 4-sterigmate. *Pleurocystidia* 50–65 × 12–17 µm, mostly fusiform, occasionally fusiform-ventricose; apices obtuse and bearing crystals, tapered towards the rounded base, the walls 1.5-2.5 µm thick, hyaline. *Cheilocystidia* similar to pleurocystidia, at times shorter or more ventricose, mixed with clavate to subcylindrical paracystidia. *Caulocystidia* in dense clusters descending the entire length of the stipe, similar to pleurocystidia or somewhat longer, mixed with cells of various shapes and sizes, including cauloparacystidia. *Pileipellis* with a superficial layer of interwoven narrowly cylindrical hyphae that are smooth and mostly 4–7 µm wide, thin- to slightly thick-walled, hyaline, occasionally flexuous; subtending layer dull yellowish in mass, hyphae mostly 5–17 µm wide, thin-walled, smooth in appearance. *Clamp connections* present. Fig. 60.

Known only from the type locality in southern Tas.; in dry sclerophyll forest under *Eucalyptus* and *Acacia*. Fruiting Nov.

Etymology: From the Latin *roseola* (pink or pale rose), in reference to the colour of the lamellae.

Diagnostic features: Fruit bodies small and delicate. Pileus pinkish to reddish brown; lamellae pale rose to reddish brown, in contrast to the mostly white, entirely pruinose stipe. Spores smooth, hyaline or very pale buff.

Phylogenetic analyses support the inclusion of *I. roseola* in the *I. tenax* group (Fig. 54). Morphological attributes also confirm this relationship (small size, entirely pruinose stipe, the presence of a velipellis, smooth spores). *Inocybe roseola* forms a sister lineage to the Queensland species *I. boreoaustralis*, the latter differing by its appressed-squamulose pileus and lack of pink colours, as well as ecology and distribution.

The spores of *I. roseola* are distinctive due to their very pale or buff colour. A low percentage of non-guttulate spores are completely hyaline.

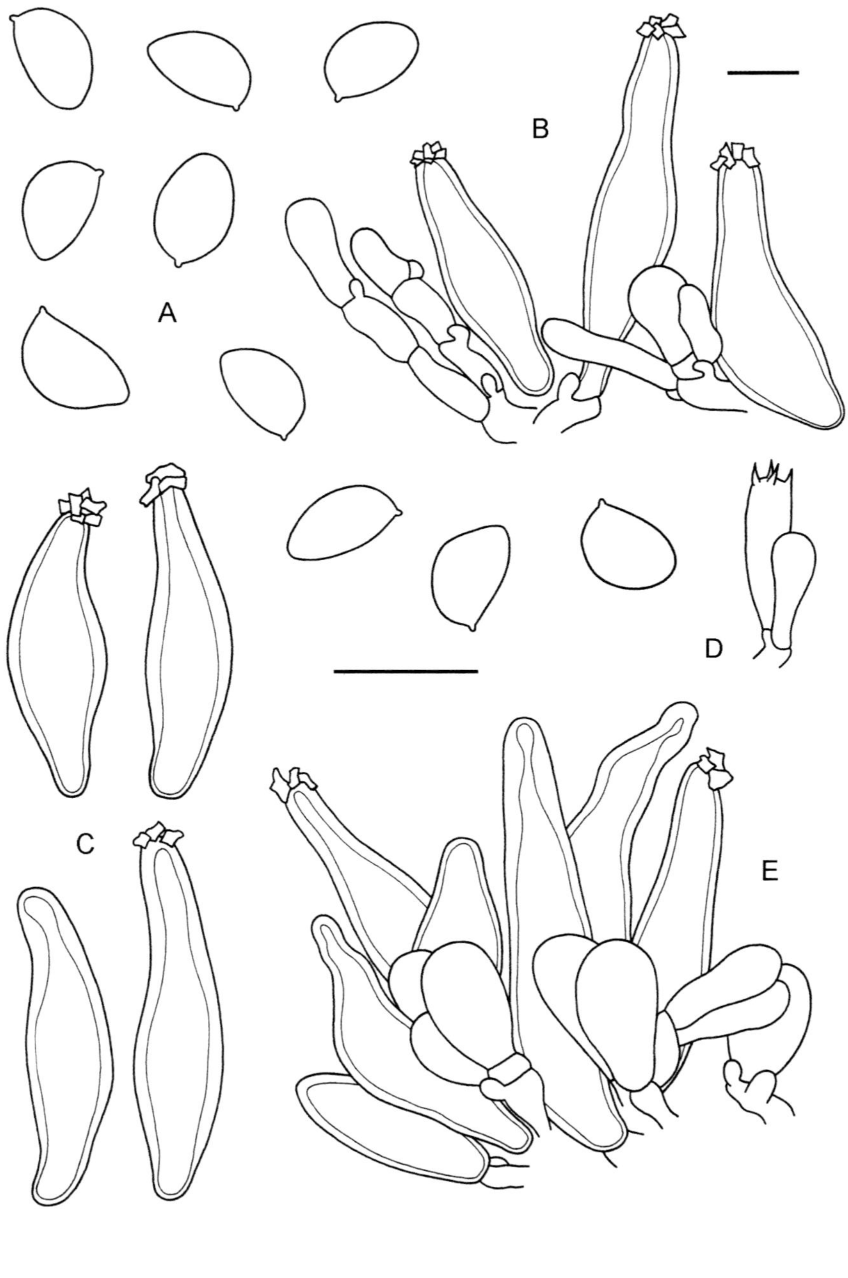

Figure 60. *Inocybe roseola.* All holotype (HO 569186). **A,** Basidiospores; **B,** Cheilocystidia and paracystidia; **C,** Pleurocystidia; **D,** Basidia; **E,** Caulocystidia and cauloparacystidia. Scale bars = 10 µm (longer bar for spores only).

52. Inocybe tenax Matheny & Bougher, *nom. nov.*

MycoBank: MB 809546

Basionym: *Inocybe strigiceps* E.Horak, *Persoonia* 11: 22 (1980), *nom. inval.*, *non Inocybe strigiceps* (Fr.: Fr.) Sacc., *Syll. Fung. (Abellini)* 5: 791 (1887). Type: Hut Track, Mt Michael, Frigano, Eastern Highlands, Papua New Guinea, alt. 2400 m, on soil in *Nothofagus* forest, 8 Dec. 1971, *E.Horak*; holo: ZT 71/400 *n.v.*

Pileus 9–22 mm wide, conical to obtusely so, expanding with age to plano-convex, at times with a small obtuse umbo; margin decurved, dry; velipellis absent or present as a greyish area over the disc, not rimose; disc matted-fibrillose to matted-fibrillose-scaly, at times even squarrose or recurved-scaly, fibrillose to coarsely fibrillose towards the entire (not striate) margin; edge of margin occasionally denticulate; dark yellowish brown (10YR 3/4; Bay Brown) shading to dark yellowish brown (10YR 4/3) or brown (10YR 5/3) towards the margin; context thin-fleshed, not changing colour where cut or bruised. *Lamellae* adnexed, moderately close, with several tiers of lamellulae, strong brown to reddish brown (7.5YR 5/4 to 4/4 and 5YR 4/4), eventually yellowish brown or rusty, ventricose, edges pallid-fimbriate. *Stipe* 15–45 × 1.0–2.5 mm at the apex, terete, even but with a slightly swollen base less than 4 mm wide, not bulbous, tough, wiry; surface pruinose throughout; reddish brown to vinaceous brown (5YR 5/3 to 4/3) or weak red to dusky red (2.5YR 4/2 to 3/2); base with a dense white mycelium; context solid, becoming hollow, with a pale reddish brown pithy interior. *Odour* acidulous or strongly acidulous when first cut, later with a *Pelargonium* component, occasionally nondescript. Pl. 109, 110.

Basidiospores 7.5–9.5 × 5.0–6.0 µm [mean 8.5 × 5.3 µm; Q 1.45–1.80; Q mean 1.62 (*n* = 27/3)], amygdaliform to broadly so, with pointed apices, ovate in face view, yellowish brown, apiculus indistinct. *Basidia* 25–30 × 7–8 µm, clavate, hyaline, 4-sterigmate. *Pleurocystidia* 57-67 × 13–18 µm, fusiform, with a rounded-truncate base, pedicel absent, apices subacute, very thick-walled (4–5 µm) and yellow, arising from subhymenium. *Cheilocystidia* similar to pleurocystidia, sometimes shorter and obclavate, mixed with hyaline paracystidia. *Caulocystidia* descending the entire length of the stipe, similar to hymenial cystidia but at times noticeably longer (to 90 µm), in dense clusters mixed with cauloparacystidia and cells of intermediate size and shape. *Pileipellis* a cutis of brown-pigmented encrusted hyphae. Pellis overlain with a layer of velipellis hyphae, these narrow, cylindrical, hyaline to golden, 4–5 µm wide, smooth and with walls to 2 µm thick, rarely branching and with distant septa. *Clamp connections* present. Fig. 61 (see also Pl. 10A).

Occurs in Vic. and Tas.; on soil in rainforest under *Nothofagus cunninghamii*, either in mixed forest including *Leptospermum*, or in forest that includes *Acacia*, *Pomaderris* and/or *Eucalyptus*. Also in Papua New Guinea and in New Zealand. Fruiting May and June in Australia.

Specimens examined: Vic.: Somers Park, Acheron Way, Yarra Ranges Natl Park, on wet muddy soil in rainforest gully under *Nothofagus cunninghamii* with nearby *Acacia*, 29 May 2009, *T.May PBM3201* (MEL 2335645A, as *I. umbrosa*, TENN 063633; KF853393-nLSU, KF871769-ITS, KF891946-*rpb2*). Tas.: Goblin Track, Blue Tier, alt. 800 m, on soil in mixed forest of *Nothofagus cunninghamii* and *Leptospermum*, 22 May 2010, *F.Karstedt & P.B.Matheny PBM3377* (TENN 067035; KF853392-nLSU, KF871768-ITS, KF891945-*rpb2*); Ralphs Falls Track, West Pyengana, alt. 800 m, on soil in rainforest under *Nothofagus cunninghamii*, 23 May 2010, *G.Gates, D.Ratkowsky & P.B.Matheny PBM3397* (TENN 067036); Vale of Belvoir, on soil in rainforest under *Nothofagus cunninghamii*, 25 May 2010, *F.Karstedt PBM3417* (TENN 067037; KF853395-nLSU, KF871771-ITS, KF891948-*rpb2*); track to Duckhole L., near Dover, on soil in wet sclerophyll forest under *Nothofagus cunninghamii* with nearby *Eucalyptus* and *Acacia*, 29 May 2010, *P.B.Matheny PBM3428* (TENN 067038); *loc. id.*, under *Nothofagus cunninghamii*, 29 May 2010, *P.B.Matheny PBM3429* (TENN 067039); track to Lady Barron Falls, Mount Field Natl Park, on soil in wet sclerophyll forest under *Nothofagus cunninghamii* and *Pomaderris*, 1 June 2010, *P.B.Matheny PBM3443* (TENN 067040; KF853394-nLSU, KF871770-ITS, KF891947-*rpb2*).

Extralimital specimens examined: New Zealand: North Island, Taupo District, Whakapapa Village, Ridge Track, in broad-leaved forest with *Nothofagus solandri* var. *cliffortioides*, 8 Apr. 2005, *J.A.Cooper JAC9423* (PDD 80851); North Island, 15 km S of Turangi, Kaimanawa Forest Park, along Kaimanawa Rd towards Tree Trunk Gorge, on soil under *Nothofagus* and *Leptospermum*, 10 May 2009, *P.B.Matheny PBM3132* (PDD 97903, TENN 063877; KF853397-nLSU, KF871773-ITS, KF891950-*rpb2*); *loc. id.*, on soil under *Nothofagus*, 10 May 2009, *P.B.Matheny PBM3140* (PDD 97907, TENN 063709; KF853396-nLSU, KF871772-ITS, KF891949-*rpb2*). South Island, Otago Lakes, Haast Pass, Cameron Flat, on soil under *Nothofagus menziesii* in beech forest, 3 Apr. 2000, *E.Horak & A.Horak* (PDD71260; KF977216-ITS).

Etymology: From the Latin *tenax* (tough), in reference to the tough, wiry stipe.

Diagnostic features: Fruit bodies small. Stipe tough, wiry, pruinose throughout and with a swollen but non-bulbous base. Hymenial cystidia very thick-walled, yellow, typically with a rounded-truncate base. Hyphae of velipellis thick-walled, rarely branched, with distant septa.

Australian collections of *I. tenax* are microscopically very similar to specimens from New Zealand. All exhibit pleurocystidia with a characteristic rounded-truncate base, and they share the small size, tough and wiry stipe, overall coloration, acidulous odour and occurrence with *Nothofagus*. However, Australian samples have slightly longer spores with more attenuated apices and, frequently, a scaly pileus with recurved scales on the disc. We also noticed that the velipellis hyphae (suprapellis) are golden in KOH, unlike New Zealand material in which these hyphae are hyaline to pale greenish yellow. Phylogenetic data support the separation of Australian and New Zealand populations, with high bootstrap support (Fig. 54). However, one Tasmanian collection (*PBM3417*, under *Nothofagus cunninghamii*) renders the Australian group paraphyletic, from within which the New Zealand group is derived. Morphologically and ecologically, we cannot separate the latter collection from the others, although geography appears to be an important feature. We conservatively recognise all three lineages as a single species, *I. tenax*, and suggest the New Zealand element recently arrived there via long-distance dispersal from Australia. Missing from this explanation, however, are phylogenetic data from New Guinea collections.

Several taxa form a robust monophyletic group with *I. tenax* (Fig. 54). The closest of these, *I. pluvialis*, differs primarily by its occurrence in a wet sclerophyll forest habitat without *Nothofagus* in Tasmania. It also exhibits subtle morphological differences, *viz.* a translucent-striate pileus margin and a bruised-*Pelargonium* odour. Another species, *I. tenuis*, occurs in a similar habitat in Tasmania, but it has pleurocystidia that are tapered towards the base.

More distantly related lineages in the *I. tenax* group (Fig. 54) include *I. boreoaustralis*, known from montane sclerophyll forest and montane warm-temperate rainforest under *Lophostemon confertus* in south-eastern Queensland. *Inocybe boreoaustralis*, which resembles typical forms of *I. tenax* with a matted-fibrillose to squamulose pileus and a distinct odour, is probably a cryptic species that has been isolated geographically. A second species, *I. roseola*, occurs in dry sclerophyll forest (*Eucalyptus* and *Acacia*) in Tasmania, and possesses several noteworthy morphological features: the pileus is often bicolorous with a pinkish buff to avellaneous margin and a cream-coloured disc; the lamellae are pinkish buff; and the stipe is entirely pruinose but white and becoming yellowish with age. A third species, *I. vittata*, appears to be geographically isolated in the wheatbelt region of south-western Western Australia where it associates with *Allocasuarina* and/or *Acacia*; it is morphologically unique due to the striatulate-pruinose stipe.

Horak (1980) also described an infraspecific taxon, *I. strigiceps* f. *pygmaea*, from Papua New Guinea for a smaller form with shorter cystidia. In accordance with the newly proposed species name, *I. tenax*, we rename this form as:

Inocybe tenax f. **pygmaea** Matheny & Bougher, *nom. nov.*

MycoBank: MB 809548

Basionym: *Inocybe strigiceps* f. *pygmaea* E.Horak, *Persoonia* 11: 24 (1980), *nom. inval.*. Type: Okapa Track, Mt Michael, Frigano, Eastern Highlands, Papua New Guinea, alt. 2400 m, 9 Dec. 1971, *E.Horak*; holo: ZT 71/410 *n.v.*

Based on the description in the literature (Horak, 1980), this taxon appears most similar to *I. pygmaeformis* but it differs in its smaller spores, shorter hymenial cystidia, and its association with *Nothofagus* and/or *Lithocarpus*.

Plate 109. *Inocybe tenax*. TENN 067035. Photograph — P.B.Matheny.

Plate 110. *Inocybe tenax*. TENN 067040. Photograph — P.B.Matheny.

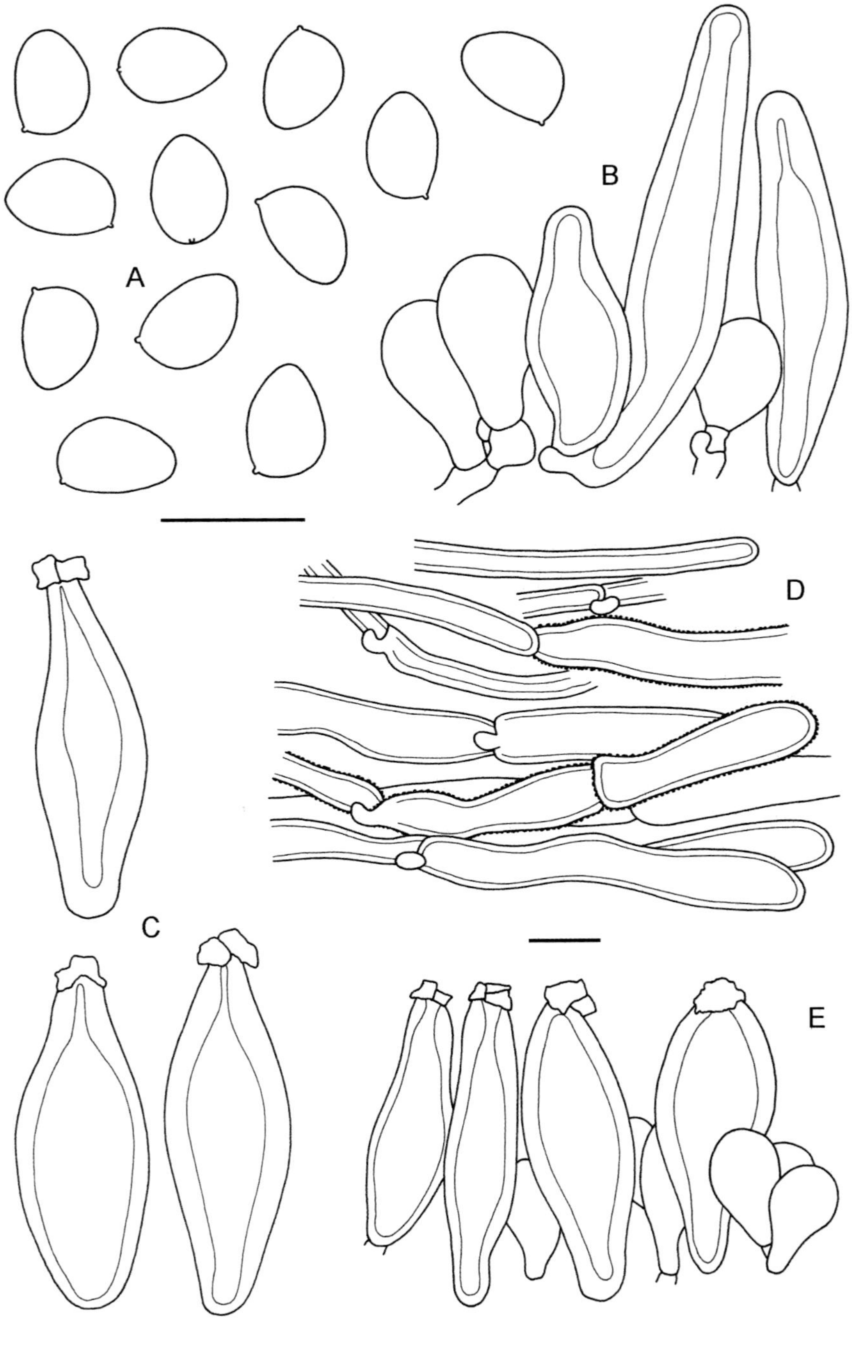

Figure 61. *Inocybe tenax.* All TENN 063633. **A,** Basidiospores; **B,** Caulocystidia and cauloparacystidia; **C,** Pleurocystidia; **D,** Pileipellis with overlying velipellis hyphae; **E,** Cheilocystidia and paracystidia. Scale bars = 10 μm (longer bar for spores only).

53. **Inocybe tenuis** Matheny & Bougher, *sp. nov.*

MycoBank: MB 809472

Type: Lower Kermandie, Tas., 43°12'S, 146°52'E, scattered singly on soil in wet sclercphyll forest under *Eucalyptus* and *Pomaderris*, 19 May 2010, *P.B.Matheny PBM3360*; holo: PERTH 08517258; iso: TENN 065790 (KF853404-nLSU, KF871780-ITS, KF891957-*rpb2*).

Pileus 10–23 mm wide, conical but soon expanding with a broad obtuse umbo, becoming ±plane; margin uplifted in age; surface dry; velipellis present but sometimes indistinct, recurved-squamulose around the disc and outwards towards the margin, becoming entire or translucent-striate and torn or eroded with age; centre dark yellowish brown (10YR 3/4; Bay Brown) shading to yellowish brown (10YR 4/4 to 4/3; Saccardo's Umber) or brown; context thin-fleshed, not changing colour where bruised. *Lamellae* adnexed, moderately close to almost subdistant, 20–26 L, with several tiers of lamellulae, pale brown, brown or yellowish brown (10YR 6/3, 5/3, 5/4), edges pallid and indistinctly fimbriate, ventricose. *Stipe* 23–35 × 1–2 mm, thin, wiry, flexuous, pruinose the entire length, base swollen slightly and whitish, elsewhere reddish brown (5YR 4/3); context solid, coloured like the surface. *Odour* acidulous. Pl. 111.

Basidiospores 8.0–10.0 (–10.5) × 4.5–5.5 µm [mean 9.0 × 4.9 µm; Q 1.60–2.11; Q mean 1.86 (*n* = 31/1)], smooth, amygdaliform, with conical or bluntly pointed apices, yellowish brown, apiculus indistinct. *Basidia* 24–28 × 7–9 µm, clavate to cylindroclavate, hyaline, 4-sterigmate. *Pleurocystidia* 53–70 × 14–19 µm, fusiform-ventricose to fusiform, often tapered towards the apex and base (some with rounded bases); apices obtuse to subacute, crystalliferous, thick-walled, the walls mostly 3–5 µm thick, pale yellow. *Cheilocystidia* similar to pleurocystidia but shorter, and often ventricose or utriform, mixed with dense clusters of hyaline clavate paracystidia. *Caulocystidia* descending the entire length of the stipe in scattered clusters, similar to hymenial cystidia, frequently fusiform or subcylindrical with tapered apices that tend to be subacute, thick-walled (2–3 µm), hyaline and mixed with clavate hyaline cauloparacystidia. *Pileipellis* composed of cylindrical to inflated hyphae, these frequently septate and cylindrical to inflated, 5–12 µm wide, tawny to rust-brown in mass, slightly thick-walled, weakly encrusted, giving rise to narrowly cylindrical velipellis hyphae (mostly 4–6 µm wide) with fewer septa, with walls 1.0–1.5 µm thick and with ochre-tawny pigment. *Clamp connections* present. Fig. 62.

Known only from wet sclerophyll forest at the type locality in southern Tas.; grows under *Eucalyptus* and *Pomaderris*. Fruiting May.

Etymology: From the Latin *tenuis* (thin, fine), in reference to the small, slender fruit bodies.

Diagnostic features: Fruit bodies small and slender, with a thin wiry stipe. Pileus recurved-scaly, the margin becoming translucent-striate. Stipe reddish brown with a white swollen base, pruinose throughout.

Inocybe tenuis belongs to the *I. tenax* group of species (Fig. 54), several of which (but not all) exhibit a thin wiry or tough stipe, smooth amygdaliform spores, and distinct velipellis hyphae on the pileus surface. Within this clade resides a complex of at least six taxa that are phylogenetically distinct and can differ ecologically, but often only have minor differences in morphology. One, *I. tenax*, is best characterised by its occurrence in *Nothofagus* rainforest vegetation. *Inocybe pluvialis* is most similar to *I. tenuis* as both species occur in wet sclerophyll forests of Tasmania. It differs from *I. tenuis* in its *Pelargonium* odour (in contrast to the acidulous smell of *I. tenuis*), brighter brown pileipellis hyphae and in the absence of conspicuous dark scales against the pale pileus background. However, more collections of *I. tenuis* are needed to determine the constancy of such differences.

Three other species in the *I. tenax* complex (Fig. 54) are readily separable from *I. tenuis*. Thus, *I. vittata*, from south-western Western Australia, is characterised by its non-wiry, striatulate stipe. The Tasmanian *I. roseola* has a pileus with pinkish tinges, pink to reddish brown lamellae, a white stipe and subhyaline spores, while *I. boreoaustralis*, known from montane forest in south-eastern Queensland, has a distinct *Pelargonium* or sweet smell. Finally, *I. xanthocystis*, from Western Australia, is macroscopically similar to *I. tenuis* and other members of the *I. tenax* complex, but it has pleurocystidia with very thick (4–10 (–15) µm wide), bright yellow walls.

In conclusion, members of the *I. tenax* complex have probably undergone recent isolation with many sorting out as independent phylogenetic lineages, which we recognise at the species level based on habitat type and geography (Fig. 54).

Plate 111. *Inocybe tenuis*, holotype (PERTH 08517258). Photograph — P.B.Matheny.

54. **Inocybe vittata** Matheny, Bougher & Ryberg, *sp. nov.*

MycoBank: MB 809480

Type: Old Kunjin Railway, 14 km W of Corrigin, W.A., 32°21'25.5"S, 117°44'10.6"E, scattered singly on sandy soil under *Allocasuarina obesa* (Black Tamma) with eucalypts in the vicinity, 26 July 2011, *P.B.Matheny PBM3683*; holo: PERTH 08319022; iso: TENN 066487 (KF853403-nLSU, KF871779-ITS, KF891956-*rpb2*).

Pileus 5–25 mm wide, initially conical, bluntly conical or hemispherical, remaining bluntly conical or becoming plano-convex, usually lacking an umbo (if present, then low and obtuse), with a greyish and usually conspicuous velipellis initially covering the entire surface, with remnants over the centre and towards the margin or the centre only with age; margin at first incurved, becoming decurved or almost straight, with margin initially rimmed with whitish veil remnants; surface dry, typically coarsely fibrillose, floccose-fibrillose or fibrillose-squamulose, usually scales present around the disc or areolate with age; (dark) yellowish brown to brown or fulvous (10YR 5/4, 4/4, 5/3, or 4/3) or varying mixture of these shades, or strong brown to umber (7.5YR 5/6 to 4/4; Buckthorn Brown, Ochraceous-Tawny or Antique Brown (5YR 4/4)) mixed with shades of Brussels Brown, most stages with scattered pale brown to pallid veil remnants especially towards the margin giving a lighter colour (pale brown), at least in younger stages; context thin, whitish to pale brown, not changing colour when exposed. *Lamellae* adnexed to uncinate, subdistant, 22–26 L, with several tiers of lamellulae, greyish in buttons, soon becoming strong brown or umber (7.5YR 5/4, 5/6, 4/4) even when young, then yellowish brown to reddish brown (10YR 5/4 to 5YR 4/4) with age, medium to broad, up to 2 mm deep, ventricose, the edges distinctly pallid-fimbriate under a hand lens.

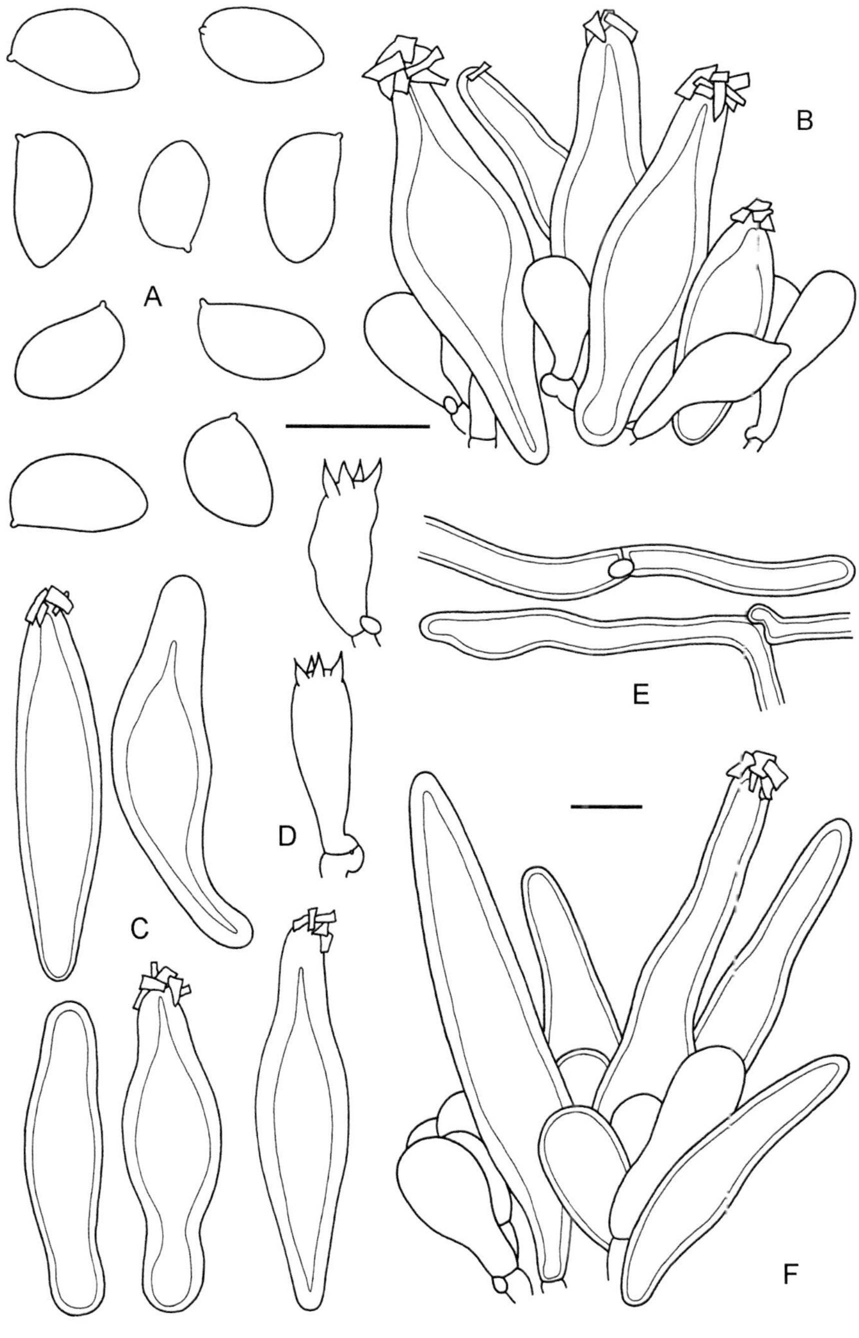

Figure 62. *Inocybe tenuis.* All holotype (PERTH 08517258). **A,** Basidiospores; **B,** Cheilocystidia and paracystidia; **C,** Pleurocystidia; **D,** Basidia; **E,** Velipellis hyphae; **F,** Caulocystidia and cauloparacystidia. Scale bars = 10 μm (longer bar for spores only).

Stipe 12–32 × 1–3 mm, even or slightly swollen at the base, with a fibrillose veil connecting the margin of the pileus to the edge of the stipe base; surface striatulate (due to pruinosity running in longitudinal lines) usually for the entire stipe length, densely pruinose, with vesture apparent just above the base, light yellowish brown, becoming brown or strong brown (7.5YR 5/4) or light reddish brown or reddish brown (10YR 6/4, 7.5YR 5/6 to 4/4, 5YR 6/3 to 5/3; Fawn Colour or Avellaneous); base pallid; context solid, becoming hollow, coloured like the surface, with a pale brown interior. *Odour* acidulous or not remarkable. Pl. 112–114.

Basidiospores 8.5–12.0 × 5.0–6.0 µm [mean 10.0 × 5.4 µm; Q 1.55–2.20 (–2.40); Q mean 1.84 (*n* = 48/3)], smooth, amygdaliform to oblong-amygdaliform with conical or bluntly pointed apices, yellowish brown; apiculus small, indistinct. *Basidia* 24–33 × 8–10 µm, clavate to subcylindrical, hyaline, 4-sterigmate. *Pleurocystidia* 60–86 × 12–14 µm, narrowly fusiform, apices typically subacute and bearing crystals, with or without a tapered pedicel, thick-walled, the walls 3–4 µm thick apically, thinner elsewhere, pale yellow or sometimes hyaline. *Cheilocystidia* similar to pleurocystidia but more variable in shape (mostly narrowly fusiform, also cylindrical or lageniform), densely arranged, mixed with thin-walled clavate hyaline paracystidia and short thick-walled clavate cells. *Caulocystidia* descending the entire length of the stipe in very dense clusters, similar to cheilocystidia, mixed with cauloparacystidia and cells of varied shape and size, including short thick-walled clavate cells. *Pileipellis* a cutis of cylindrical to inflated hyphae mostly 8–22 µm, tawny in mass, encrusted and with thickened walls; overlain by an interwoven layer of velipellis hyphae, these pale or hyaline, narrow, mostly 3–5 µm wide, with walls 1–2 µm thick with few septa. *Clamp connections* present. Fig. 63.

Grows in sandy soil under *Allocasuarina* or *Allocasuarina* mixed with *Acacia* in the wheatbelt region of W.A. Fruiting July.

Additional specimens examined: W.A.: type locality, 26 July 2011, *N.L.Bougher BOU872* (PERTH 08320608, TENN 066759); Corrigin Wildflower Drive, 5 km W of Corrigin, 32°20'19.1"S, 117°49'20.0"E, scattered singly on sandy soil under *Allocasuarina obesa* (Black Tamma), 27 July 2011, *P.B.Matheny PBM3684* (PERTH 08319014, TENN 066575); Vinegar Well, off Barber Rd, Kunjin, 32°23'31.7"S, 117°40'06.5"E, scattered singly in sand under *Allocasuarina* spp. and *Acacia acuminata*, 27 July 2011, *P.B.Matheny PBM3693* (PERTH 08318913, TENN 066620; KP171098-nLSU, KP641650-ITS, KM656140-*rpb2*); *loc. id.*, same habitat and date of collection, *P.B.Matheny PBM3696* (PERTH 08318883, TENN 066615) (KF853402-nLSU, KF871778-ITS, KF891955-*rpb2*); *loc. id.*, in litter under *Acacia acuminata*, *Allocasuarina huegeliana*, 27 July 2011, *N.L.Bougher NLB889* (PERTH 08320411, TENN 066690).

Etymology: From the Latin *vittatus* (bound with a ribbon, plaited or striped), in reference to the longitudinally striatulate stipe surface.

Diagnostic features: Fruit bodies small; pileus often with a conspicuous velipellis and veil remnants at the margin. Stipe striatulate and pruinose for its entire length, not wiry or tough. Associated with *Allocasuarina*.

Inocybe vittata is part of the *I. tenax* group (Fig. 54), having veil remnants on the pileus surface consisting of narrow, hyaline, thick-walled and sparingly septate hyphae. However, its characteristically striatulate stipe is diagnostic, being caused by a dense longitudinal arrangement of caulocystidia. Furthermore, the stipe of *I. vittata* is not tough or wiry, unlike, for example, *I. tenax* and *I. pluvialis*.

55. **Inocybe xanthocystis** Matheny & Bougher, *sp. nov.*

MycoBank: MB 809481

Type: Bridgetown Rd, 1 km N of Catterick Rd turn-off, near Balingup, W.A., 33°50'50.6"S, 116°04'50.3"E, among charcoal and burnt ground in Jarrah forest under *Eucalyptus marginata* and *Corymbia calophylla*, 12 July 2011, *N.L.Bougher NLB797*; holo: PERTH 08318158; iso: TENN 066753 (KF808352–nLSU, KF830040-ITS, KF830056-*rpb2*).

Pileus 7–14 mm wide, hemispherical, becoming conical to obtusely conical, expanding with age to plano-convex; margin initially incurved but soon decurved, usually lacking an umbo; centre often with a pale brownish to silvery grey velipellis layer, this more apparent as specimens lose moisture; disc unbroken or at most appressed-scaly, fibrillose towards the margin, not at all rimose; edge of margin with strigose hairs (veil remnants) when very young; dark reddish brown (5YR 3/3–3/4), shading at times to reddish brown towards the margin (5YR 4/3–4/4) but centre typically pallid, pale brown or greyish due to the velipellis; other collections more yellowish brown with scattered floccules of whitish velipellis remnants, at times mottled; context very thin, reddish brown, not changing colour upon exposure. *Lamellae* adnexed, moderately close to subdistant, about 24 L, with several tiers of lamellulae; light brown (7.5YR 6/4) to light yellowish brown or reddish brown when young, becoming umber (7.5YR 4/4) with age; edges pallid and indistinctly fimbriate, ventricose, at most up to 2 mm deep. *Stipe* 15–40 × 1–2 mm, terete, cylindrical, even, not at all bulbous, often flexuous, tough (easy to remove intact from the substratum), pruinose over its entire length; pinkish brown to reddish brown (10R 4/4) or strong brown, brown to umber (7.5YR 5/6–5/4–4/4) with age, white-mycelioid at the base; context solid, concolorous with the surface. *Odour* weakly acidulous. Pl. 115, 116.

Basidiospores 7.0–9.0 × 4.5–5.5 µm [mean 8.0 × 5.1 µm; Q 1.36–1.80; Q mean 1.58 (*n* = 48/3)], smooth, amygdaliform, with bluntly pointed apices, yellowish brown, with a small apiculus. *Basidia* 24–29 × 7–9 µm, 4-sterigmate, clavate, hyaline. *Pleurocystidia* 46–74 × 14–21 µm, fusiform, often with a tapered pedicel or rounded-truncate; apices obtuse, crystalliferous; thick- to very thick-walled [4–10 (–15) µm], the walls often intensely bright yellow in KOH. *Cheilocystidia* similar to pleurocystidia but at times lageniform or shorter or wider (to 27 µm wide), mixed with hyaline paracystidia. *Caulocystidia* scattered in dense clusters descending the entire length of the stipe, mostly fusiform with blunt or subacute apices, mixed with cells of intermediate size and shape; thick-walled (2–4 µm), the walls hyaline, not yellow in KOH. *Pileipellis* composed of 2 layers: an outer layer of velipellis hyphae that are interwoven to trichodermally arranged, long (>200 µm) and unbranched, narrow (3–4 µm wide), cylindrical, flexuous, thin-walled or slightly thick-walled (to 1.5 µm), with infrequent septa; inner subtending layer composed of broader (5–15 µm wide) frequently septate hyphae that are brownish yellow or golden brown in mass, smooth or weakly encrusted. *Clamp connections* present. Fig. 64 (see also Pl. 9C, 10D, 12G).

Occurs on soil in litter under *Eucalyptus* and/or *Corymbia* in Jarrah and Karri forest in south-western W.A. Fruiting June–Aug.

Additional specimens examined: W.A.: Boranup Forest off Boranup Drive, Leeuwin-Naturaliste Natl Park, 34°10'40.4"S, 115°03'44.4"E, on roadside soil in Karri forest under *Eucalyptus diversicolor*, *Corymbia calophylla*, *Acacia* and *Agonis flexuosa*, 7 June 2010, *P.B.Matheny PBM3466* (PERTH 08242100, TENN 0667712; KF808355-nLSU, KF830042-ITS, KF830058-*rpb2*); *loc. id.*, 34°10'11.5"S, 115°04'04.7"E, 7 June 2010, *P.B.Matheny PBM3469* (PERTH 08242070, TENN 067713; KF808353-nLSU, KF830041-ITS, KF830057-*rpb2*); near "Pipeline site 3", Wellington Natl Park, on lateritic soil in Jarrah Forest under *Eucalyptus marginata* and *Corymbia calophylla*, 11 July 2011, *P.B.Matheny PBM3589* (PERTH 08319952, TENN 066546; KF808351-nLSU, KF830039-ITS, KF830055-*rpb2*); Boranup Drive, Leeuwin-Naturaliste Natl Park, in litter attached to soil and decaying leaf litter in Karri forest under *Eucalyptus diversicolor*, 30 June 2008, *N.L.Bougher NLB455* (PERTH 08072892; KF808354-nLSU); type locality, same habitat and date of collection, *N.L.Bougher NLB798* (PERTH 08318131, TENN 066733; KF808356-nLSU, KF830043-ITS, KF830059-*rpb2*); Love Spring Track, Boranup Forest, Leeuwin-Naturaliste Natl Park, 34°06'21.4"S, 115°02'49.2"E, scattered singly in litter near edge of track in *Eucalyptus diversicolor* forest, 2 July 2008, *N.L.Bougher NLB475* (PERTH 08073104); Shelley Beach Rd, intersection with Bibbulmun Track, West Cape Howe Natl Park, 35°05'56.8"S, 117°37'44.1"E, among litter under *Melaleuca* sp. in *Agonis flexuosa* woodland, 18 July 2011, *N.L.Bougher NLB826* (PERTH 0318549).

Etymology: From the Greek *xanthos-* (yellow) and *cystis* (a bladder), in reference to the striking yellow and very thick-walled hymenial cystidia.

Diagnostic features: Fruit bodies small. Pileus centre pale brownish to silvery grey due to velipellis, margin reddish brown to yellowish brown and with strigose hairs; stipe even, entirely pruinose, tough, pinkish brown to reddish brown. Odour weakly acidulous. Hymenial cystidia very thick-walled (4–15 µm), intensely bright yellow; velipellis hyphae ±thin-walled and very long without septa (more than 200 µm long). In eucalypt forests in south-western W.A.

Plate 112. *Inocybe vittata*, holotype (PERTH 08319022). Note the conspicuous pale velar remnants on the pileus margin. Photograph — P.B.Matheny.

Plate 113. *Inocybe vittata*. PERTH 08320608. Note the striatulate stipe. Photograph — N.L.Bougher.

Plate 114. *Inocybe vittata*. PERTH 08320411. Note how young specimens are covered with a conspicuous greyish velipellis. Photograph — N.L.Bougher.

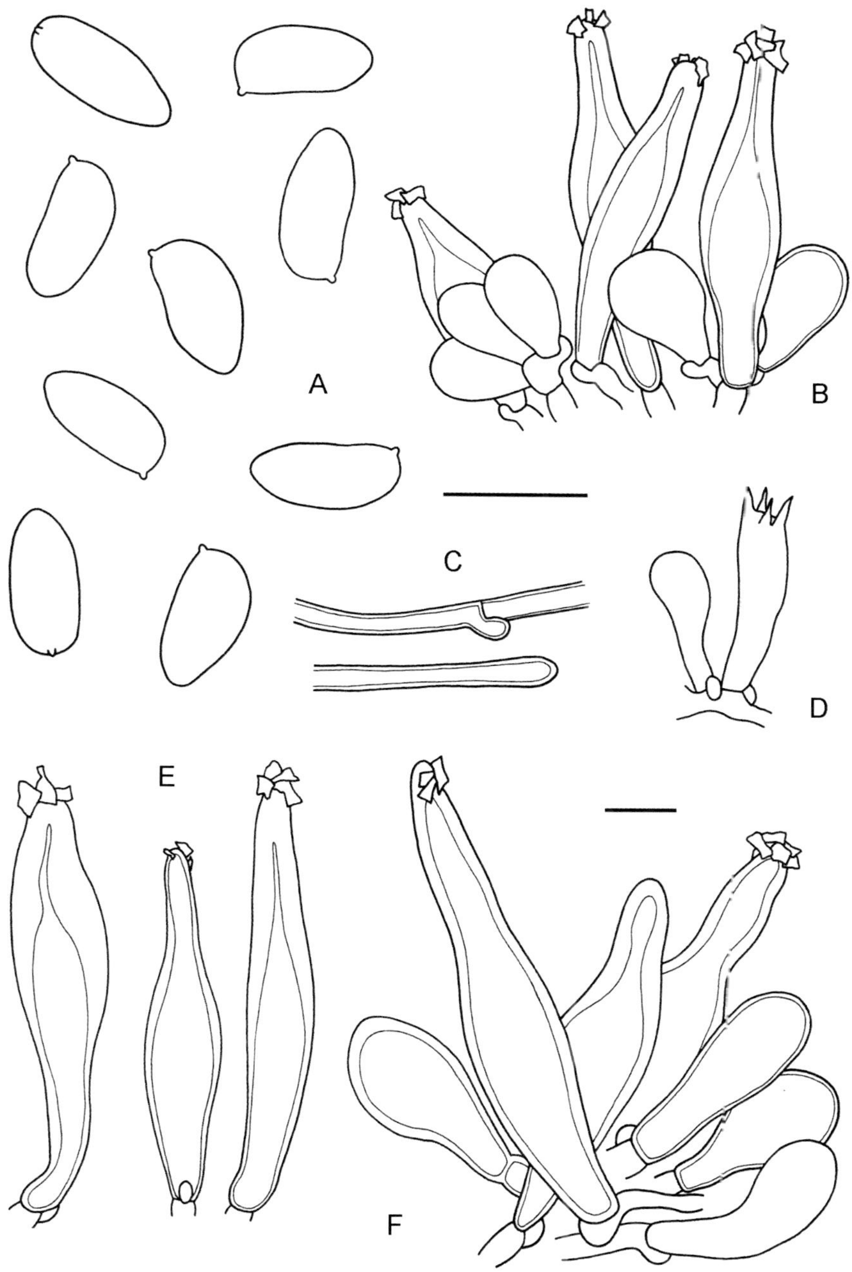

Figure 63. *Inocybe vittata.* All holotype (PERTH 08319022). **A,** Basidiospores; **B,** Cheilocystidia and paracystidia; **C,** Velipellis hyphae; **D,** Basidia; **E,** Pleurocystidia; **F,** Caulocystidia and cauloparacystidia. Scale bars = 10 µm (longer bar for spores only).

Inocybe xanthocystis is one of numerous species (perhaps as many as 14) in the *I. tenax* group, most of which are endemic to temperate Australia and New Zealand (Fig. 54). It is most closely related to *I. pygmaeformis* and *I. atrorubripes*, the three forming a robustly supported monophyletic group (Fig. 54). *Inocybe pygmaeformis* occurs under *Eucalyptus obliqua*, *Acacia* and *Pomaderris* in Tasmanian wet sclerophyll forest, while *I. atrorubripes* is associated with *Allocasuarina* in the wheatbelt of Western Australia. The latter has a dark brown pileus disc with a dark yellowish brown margin, a vinaceous stipe, and a more robust habit.

Like many other species in the *I. tenax* group, *I. xanthocystis* is rather small in stature, has a tough, even, entirely pruinose stipe, universal veil remnants as a distinct velipellis over the pileus surface and strigose hairs on the edge of the pileus margin when young. This taxon is unusual in having intensely bright yellow hymenial cystidia with comparatively thick walls [4-10 (–15) µm], predominantly thin-walled velipellar hyphae, a weak acidulous odour, and a habitat in eucalypt forests in south-western Western Australia. By contrast, *I. tenax* appears to be restricted to *Nothofagus* rainforest habitats in New Zealand and south-eastern Australia, while the closely related *I. pluvialis* differs from *I. tenax* by its occurrence in wet sclerophyll forest in Tasmania as well as genetic divergences (Fig. 54). Both of these species have thick-walled velipellis hyphae and hymenial cystidia that are not as thick-walled and intensely yellow as in *I. xanthocystis*. Moreover, the bases of their hymenial cystidia are more consistently rounded-truncate than those of *I. xanthocystis* which are often tapered. The wall thickness of pleurocystidia and cheilocystidia in *I. xanthocystis* is the thickest we have observed in *Inocybe*. We have also noted that the shape of the hymenial cystidia can vary between extremely thick-walled and stocky (PBM3466) and very thick-walled but often more slender (PBM3469, NLB455).

56. **Inocybe caerulata** Matheny, Bougher & G.M.Gates, *sp. nov.*

MycoBank: MB 809272

Type: Tongariro Crossing, Tongariro Natl Park, North Island, New Zealand, on soil in tea-tree forest under *Kunzea ericoides*, 9 May 2009, *P.B.Matheny PBM3127*; holo: PDD 97860; iso: TENN 063699 (JQ313560-nLSU, KJ756490-ITS, KJ756411-*rpb2*).

Pileus 10–25 mm wide, conical, obtusely conical or plano-convex, umbonate, margin decurved; surface dry, disc fibrillose or fibrillose-squamulose, unbroken, margin fibrillose, not rimose, velipellis not observed, centre blackish brown shading yellowish brown towards the margin, at times with greyish blue or blue-green tones, margins blue when immature; context pale brown, not obviously discolouring when cut or bruised. *Lamellae* narrowly adnate, moderately close, with several tiers of lamellulae, brown to yellowish brown (10YR 5/3 to 5/4), the edges pallid-fimbriate, ventricose, 4–5 mm deep. *Stipe* 25–75 × 3–5 mm, even, terete, base slightly swollen, cortina fugacious, densely white-pruinose above its insertion, fibrillose to lightly floccose-fibrillose below, brown (7.5YR 5/4) to reddish brown or yellowish red (5YR 5/4 to 5.6) above, lower half densely fibrillose, white mixed with blue; context not obviously changing colour, light reddish brown (2.5Y 6/4) to flesh pink, darker below, blue to blue-green at the base, solid, becoming hollow. *Odour* mild. Pl. 117.

Basidiospores 7.0–9.5 × 5.0–6.0 µm [mean 8.3 × 5.4 µm; Q 1.27–1.87; Q mean 1.46–1.63 (n = 53/3)], smooth, broadly amygdaliform with bluntly pointed apices, yellowish brown, with a small or indistinct apiculus. *Basidia* 29–32 × 7–9 µm, subclavate, hyaline, 4-sterigmate. *Pleurocystidia* 48–65 × 13–18 µm, fusiform, with or without a short basal pedicel, apices crystalliferous; walls 1.5–3.0 µm thick apically, hyaline. *Cheilocystidia* similar to pleurocystidia but shorter and more broadly fusiform to utriform, often thin-walled or slightly thick-walled, mixed with hyaline paracystidia. *Caulocystidia* at the stipe apex narrowly fusiform to cylindrical; some with crystalliferous apices, thin-walled to slightly thick-walled; lower part of the stipe lacking or with occasional caulocystidioid cells that are thin-walled to slightly thick-walled, fusiform, irregularly cylindrical or sinuous. *Pileipellis* a cutis of cylindrical hyphae, mostly 5–10 µm wide, with weak encrustations, yellowish brown in mass. *Clamp connections* present. Fig. 65.

Grows on soil in wet sclerophyll forest in southern Tas. and in tea-tree forest in northern New Zealand. Fruiting Dec. and Jan. in Tas.; May in New Zealand.

Plate 115. *Inocybe xanthocystis*, holotype (PERTH 08318158). Photograph — N.L.Bougher.

Plate 116. *Inocybe xanthocystis*. PERTH 08072892. Photograph — N.L.Bougher.

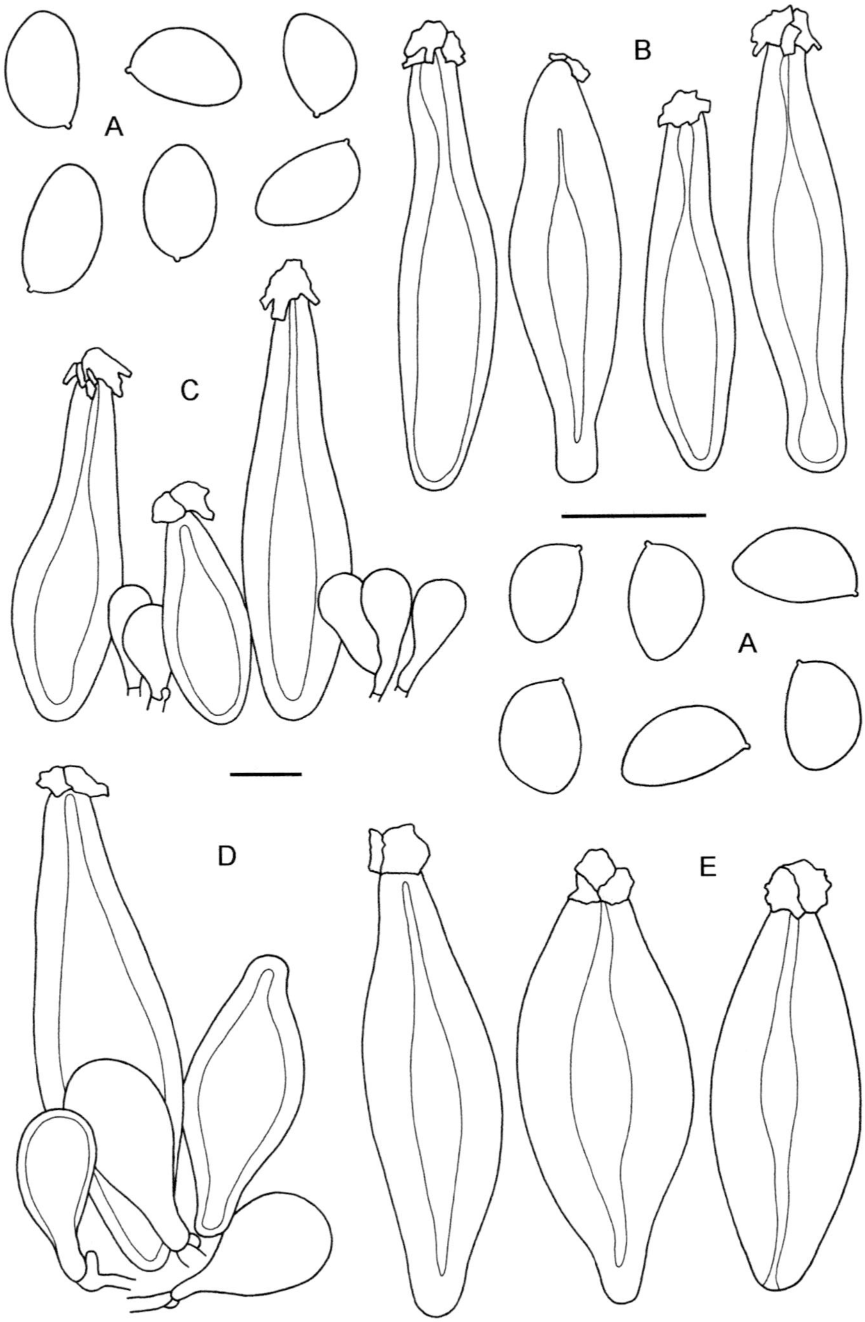

Figure 64. *Inocybe xanthocystis*. All holotype (PERTH 08318158), except E (PERTH 08242100). **A,** Basidiospores; **B,** Pleurocystidia; **C,** Cheilocystidia and paracystidia; **D,** Caulocystidia and cauloparacystidia; **E,** Pleurocystidia. Scale bars = 10 µm (longer bar for spores only).

Additional specimens examined: Tas.: Bermuda Rd, wet sclerophyll forest, 13 July 2000, *G.M.Gates 157a* [erroneously labelled I16] (HO 567548; KJ756452-nLSU, KJ756491-ITS, KJ756412-*rpb2*); Mt Arthur, wet sclerophyll forest, 19 Jan. 2002, *G.M.Gates & D.Ratkowsky I57* (HO 567549; KJ756453-nLSU, KJ756492-ITS); Myrtle Forest, wet sclerophyll forest, 11 Dec. 2008, *G.M.Gates & D.Ratkowsky I141* (HO 567550; KJ756454-nLSU, KJ756493-ITS).

Etymology: From the Latin *caeruleus* (blue) and *-ata* (possession of), referring to the blue-tinged fruit bodies.

Diagnostic features: Fruit bodies medium-sized. Pileus centre and/or margin and stipe base tinged blue; veil fugacious. Spores smooth; metuloid cystidia present as cheilocystidia, pleurocystidia and caulocystidia at the stipe apex.

Inocybe caerulata, a distinctive new species from Tasmania and New Zealand, can be readily recognised in the field due to the blue colours on the pileus and/or stipe. An assay for muscarine in collection *PMB3217* from New Zealand (Kosentka *et al.*, 2013) proved negative. All psilocybin-containing species from the Northern Hemisphere lack muscarine, but psilocybin assays have not been performed on *I. caerulata*. ITS and LSU sequences of *PBM3127* from New Zealand and three collections from Tasmania are sufficiently similar to confirm their conspecificity. Some variation was observed in spore size, particularly those from I141 (Tas.), which are somewhat larger than in the holotype, but their ranges overlap. This fungus grows under *Kunzea ericoides* in New Zealand, and although its associated vegetation in Tasmania is not known, a relationship with Myrtaceae is likely.

Plate 117. *Inocybe caerulata*, holotype (PDD 97860). Photograph — P.B.Matheny.

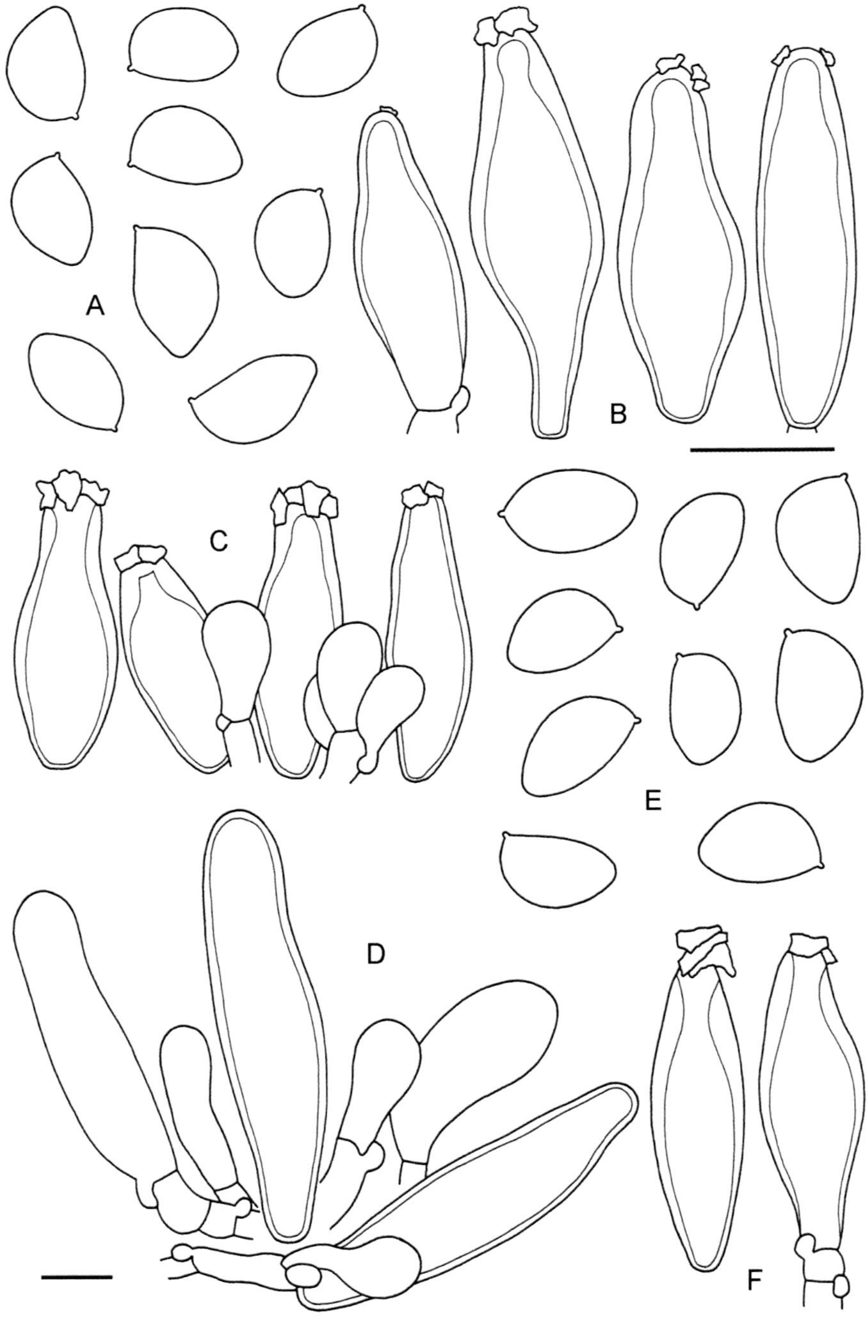

Figure 65. *Inocybe caerulata.* All holotype (PDD 97860), except E, F (HO 567550). **A,** Basidiospores; **B,** Pleurocystidia; **C,** Cheilocystidia and paracystidia; **D,** Caulocystidia and cauloparacystidia; **E,** Basidiospores; **F,** Pleurocystidia. Scale bars = 10 μm (longer bar for spores only).

The blue tinges at the stipe base and along the pileal margin when young may indicate the presence of a universal veil in *I. caerulata*. With age, dark blackish blue fibrils can be detected on the disc of the pileus. Additional specimens, including young buttons, will be necessary to confirm such a developmental sequence. The context does not obviously undergo any colour changes to red or blue, but these traits warrant further attention. Only a small number of austral taxa exhibit blue-green, olive-green or blue colouration. For example, *I. calamistratoides* E.Horak (1977), described from New Zealand, has a blackish green or olive-tinged stipe base. It is readily distinguished from *I. caerulata* by its squarrose to squamulose pileus and stipe, fruity or *Pelargonium*-like odour, and the absence of pleurocystidia. *Inocybe viridipes*, from eastern Australia, also has green or olive tinges at the stipe base but rubescent context, very dark brown lamellae, and it lacks pleurocystidia and an obvious odour. The Tasmanian *I. cyanotrichia* is reddish brown with dark brown lamellae, a fishy odour, and no pleurocystidia; however, it has bluish tinges on the pileal disc and at the stipe base. *Inocybe calamistratoides*, *I. viridipes* and *I. cyanotrichia* are members of *Inocybe* section *Cervicolores*, a monophyletic grouping within the *Inosperma* clade.

The sister group to *I. caerulata* is not known with confidence, and the species appears to occupy a rather isolated position within the STAC group (Fig. 7).

INOCYBE SERRATA GROUP

Several species in the *Inocybe serrata* group are common and widespread in temperate regions of Australia. Nine phylogenetic species can be recognized with seven occurring in Australia. The New Zealand species, *I. scabriuscula* and *I. destruens*, are also placed in the group. While it is not clear what morphological features, if any, unite this cluster of smooth-spored species, the Australian taxa share the presence of a typically well-developed vesture of white fibrils on the stipe that give rise to distinctive caulocystidioid hairs. The *I. serrata* complex is composed of three distinct clades, but future studies will be required to resolve the status of *I. fulvo-olivacea* and *I. granulosipes* with respect to the *I. serrata* complex.

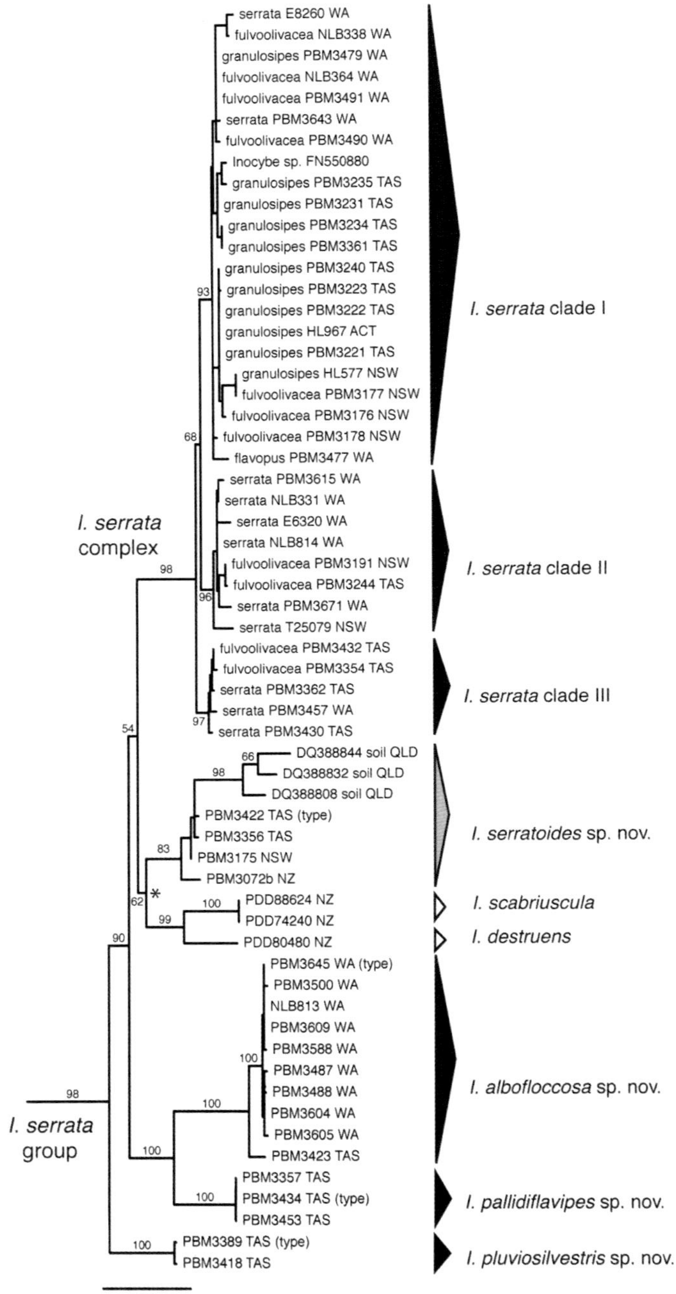

Figure 66. Maximum Likelihood phylogram of the *Inocybe serrata* group, the topology of which was produced from the analysis of the STAC group shown in Fig. 7. Australian taxa are indicated by black-filled triangles. Grey-filled triangles indicate taxa that occur in Australia and New Zealand. White triangles indicate New Zealand species. Bootstrap values are shown above the branches. The asterisk (*) indicates an internode that strongly conflicts between individual gene region phylogenies.

57. **Inocybe albofloccosa** Matheny & Bougher, *sp. nov.*

MycoBank: MB 808681

Type: near boat ramp area, Heritage Trail, Two Peoples Bay Nature Reserve, W.A., 34°58'31"S, 118°11'04"E, on sandy soil under *Melaleuca* shrubs in Peppermint (*Agonis flexuosa*) woodland, 19 July 2011, *P.B.Matheny PBM3645*; holo: PERTH 08319413; iso: TENN 066502 (KJ729891-nLSU, KJ729923-ITS, KJ729924-*rpb2*).

Pileus 8–25 mm wide, conical to obtusely conical, expanding with age to plane and developing a small obtuse umbo; veil remnants a superficial silky fibrillose layer over the centre and towards the margin when young with fringed appendiculate remains on the edge; veil remnants or velipellis generally less conspicuous with age; margin incurved, becoming decurved; surface dry, at most scurfy or excoriate on the disc or around the centre, fibrillose towards the margin; pale brown to brown (10YR 6/3 to 5/3) or paler whitish when young, becoming yellowish brown (10YR 5/4) and shading pale brown (10YR5/3) towards the margin, or the centre brownish yellow (10YR 6/6) shading to light yellowish brown (10YR 6/4) towards the margin, with age generally fulvous or yellowish brown (10YR 5/6 to 5/8), at times dark brown (sepia or bistre) on the disc; context pale brown to isabelline, up to 2.5 mm thick under the disc, thin elsewhere; colour not changing upon exposure. *Lamellae* ascending-adnexed to adnexed, subdistant to moderately close, 20–30 L, with several tiers of lamellulae, light yellowish brown (10YR 6/4) to brown (7.5YR5/4) when young, becoming clay-brown to yellowish brown (10YR 5/4 to 5/3) or dark yellowish brown with age; edges pallid-fimbriate or indistinctly so, up to 3 mm deep. *Stipe* 17–45 × 1.5–3.0 mm, terete, even or with a slightly swollen base, often flexuous; cortina fugacious; upper ¼ slightly pruinose, below with a silky to scurfy fibrillose or floccose vesture; ground colour at apex with a pinkish grey to brown cast (7 5 R 6/2–5/2) or light pinkish cinnamon (7.5YR 7/4), shading brown (7.5YR 5/4) or yellowish brown below, at times reddish brown (Wood Brown) above and brown (10YR 4/3 to 4/2) towards the base or throughout, exceptionally vinaceous-brown (2.5YR 4/2) at the apex and dark vinaceous or dusky red (2.5YR 3/2) below; context solid, becoming hollow, interior pallid or light reddish brown, cortex reddish brown at the apex, brown below. *Odour* not remarkable, or faintly acidulous. Pl. 118–121.

Basidiospores (6.5–) 7.0–8.5 (–9.0) × 4.5–5.5 µm [mean 7.8 × 4.8 µm; Q 1.40–1.80; Q mean 1.64 (*n* = 49/4)], amygdaliform, yellowish brown, with a small and indistinct apiculus. *Basidia* 24–26 × 7–8 µm, cylindrical to cylindroclavate, hyaline, 4-sterigmate. *Pleurocystidia* 65–75 × 12–15 µm, fusiform to narrowly fusiform or sublageniform, the walls 3–5 µm thick, bright yellow to pale yellow, rarely hyaline; apices subacute to obtuse, many lacking crystals in KOH mounts, frequent. *Cheilocystidia* similar to pleurocystidia but often shorter or lageniform, mixed with clavate to saccate hyaline paracystidia. *Caulocystidia* similar to pleurocystidia or cheilocystidia, narrowly fusiform or lageniform, thick-walled, present at the apex in scattered clusters mixed with cauloparacystidia and cells of intermediate size, shape and wall thickness, the latter short-fusiform, clavate, or cylindrical and thin- to thick-walled; the remainder of stipe covered with an interwoven layer of the terminal cells that form caulocystidioid hairs 45–110 × 4–7 µm, these narrowly fusiform to cylindrical, often flexuous, narrow, thick-walled, at times with crystalliferous apices. *Pileipellis* composed of 2 distinct layers; upper layer a superficial zone of interwoven cylindrical hyaline hyphae similar to the vesture on the stipe surface, the terminal cells of which are long, unbranched, flexuous, smooth and thick-walled (to 1.5 µm thick), 40–100 × 3–8 µm, arising from a subtending zone of broader and more septate hyphae; lower layer a pigmented zone, russet to tawny in mass, composed of frequently septate encrusted hyphae, 5–20 µm wide. *Clamp connections* present. Fig. 67.

Occurs on soil in Jarrah and Karri forests in south-western W.A., under *Eucalyptus marginata*, *Corymbia calophylla*, *E. diversicolor* and *Allocasuarina*, also in Peppermint (*Agonis flexuosa*) woodland under *Melaleuca* shrubs. Known in mixed rainforest in Tas., including *Nothofagus cunninghamii* and *Eucalyptus*. Fruiting May–July.

Additional specimens examined: Tas.: O'Neils Ck, Gowrie Park, on soil under *Eucalyptus* spp. and *Nothofagus cunninghamii*, 25 May 2010, *F.Karstedt PBM3423* (TENN 067737; KJ729892-nLSU, KJ729864-ITS, KJ729925-*rpb2*). W.A.: Warren River Loop Trail, Warren Natl Park, 34°30'36.6"S, 115°57'33.6"E, on wood and soil in Karri forest under *Eucalyptus* and *Allocasuarina*, 8 June 2010, *P.B.Matheny PBM3487* (PERTH 08241775, TENN 067711; KJ729893-nLSU, KJ729865-ITS); *loc. id.*, *P.B.Matheny PBM3488* (PERTH 08241783, TENN065783; KJ729894-nLSU, KJ729866-ITS, KJ729926-*rpb2*); Ritter Rd, Yeagarup Dunes area, d'Entrecasteaux Natl Park,

34°33'00.4"S, 115°51'47.7"E, on sandy soil in Jarrah forest mixed with *Eucalyptus marginata*, *Corymbia calophylla*, *Allocasuarina*, 9 June 2010, *P.B.Matheny PBM3500* (PERTH 08242259, TENN065780; KJ729895-nLSU, KJ729867-ITS, KJ729927-*rpb2*); 1 km N of Catterick Rd, near Greenbushes, 33°50'50.6"S, 116°04'50.3"E, on unburnt soil in Jarrah forest under *Eucalyptus marginata* and *Corymbia calophylla*, 12 July 2011, *P.B.Matheny PBM3604* (PERTH 08319820, TENN 066500; KJ729896-nLSU, KJ729868-ITS, KJ729928-*rpb2*); *loc. id.*, *P.B.Matheny PBM3605* (PERTH 08319812, TENN 066612; KJ729897-nLSU, KJ729869-ITS, KJ729929-*rpb2*); Cascades Rd, near Greenbushes, N of Bridgetown, 33°53'24.4"S, 116°08'09.7"E, on 2 year-old burn in Jarrah forest under *Eucalyptus marginata* and *Corymbia calophylla*, 13 July 2011, *P.B.Matheny PBM3609* (PERTH 08319774, TENN 066519; (KJ729898-nLSU, KJ729870-ITS, KJ729930-*rpb2*); South Western Hwy, c. 1 km N of Weld River Bridge, 34°41'16.6"S, 116°31'11.5"E, on soil in forest of *Eucalyptus marginata*, *Corymbia calophylla* and blackbutt, 14 July 2011, *N.L.Bougher NLB813* (PERTH 08318697, TENN 066662; KJ729899-nLSU, KJ729871-ITS, KJ729931-*rpb2*).

Etymology: From the Latin *albo* (white) and *floccosa* (with cottony scales), in reference to the white floccose stipe vesture.

Diagnostic features: Fruit bodies small or slender. Pileus margin initially white-fringed; lamellae brown even in youth; stipe covered with a dense white scurfy to floccose vesture of fibrils. Spores small, smooth, amygdaliform, 7.0–8.5 × 4.5–5.5 µm; pleurocystidia fusiform with bright yellow to pale yellow thick walls. Common in Jarrah and Marri forests of south-western Western Australia. Also in mixed rainforest including *Eucalyptus* in Tasmania.

Inocybe albofloccosa belongs to the *I. serrata* group (*I. serrata* complex, with *I. pallidiflavipes*, *I. serratoides*, *I. scabriuscula*, *I. destruens* and *I. pluviosilvestris*) based on phylogenetic evidence (Fig. 66). To date the majority of collections of *I. albofloccosa* have been from temperate south-western Western Australia, where it appears to be common in Jarrah forests. We have also collected this species in Karri forest and under shrubs of *Melaleuca* and *Agonis* trees, as well as one collection from mixed rainforest (*Eucalyptus*, *Nothofagus*) in Tasmania (*PBM3423*) that is morphologically similar to the Western Australian collections but genetically distinct (Fig. 66).

Inocybe albofloccosa appears to be a cryptic species in the *I. serrata* group. In the field we have misidentified it as *I. serrata*, *I. fulvo-olivacea*, *I. granulosipes* and *I. ionocaulis* most likely due to the changes in pileus and stipe colour as the universal veil remnants disappear or weather with age; mature fruit bodies are predominantly fulvous or yellowish brown. However, the young fruit bodies are usually pale or whitish due to an abundance of velar fibrils on the pileus and a dense vesture covering the stipe. We have observed some collections to be extremely pale when young, and persistently so (e.g. PERTH 08241783), whereas some others become brown more rapidly (e.g. the holotype). Colouration of the fruit bodies in *I. albofloccosa* and other members of the *I. serrata* group is likely to be affected by the initial density and abundance of the velar fibrils and conditions during their development, particularly moisture, and the hardness and/or abrasiveness of the substratum from which they emerge.

Within the *I. serrata* group, *I. albofloccosa* occupies an isolated phylogenetic lineage (Fig. 66), hence its description as a new species. The dense whitish stipe vesture, lamellae that are brown when immature, and the white-fringed pileus margin in youth can provide clues to its identity, but fresh, unweathered specimens are essential for an accurate determination. The exclusively Tasmanian *I. pallidiflavipes* is similar, but can be distinguished in the field by its yellow to pale yellow stipe base. *Inocybe ionocaulis* is superficially similar to *I. albofloccosa* having a fringed white pileus margin when young; it is known only from south-western Western Australia in similar habitats as *I. albofloccosa*. Phylogenetically, *I. ionocaulis* is quite distantly related and may best be distinguished in the field by its combination of bistre or dark brown pileus, the lack of velar development on the pileus, and vinaceous to dusky red stipe.

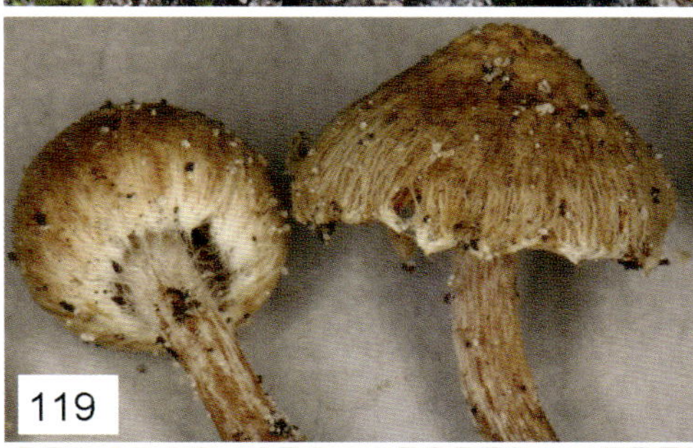

Plate 118. *Inocybe albofloccosa*, holotype (PERTH 08319413). Photograph — P.B.Matheny.

Plate 119. *Inocybe albofloccosa*. PERTH 08318697. Note abundant velar development and cortina. Photograph — N.L.Bougher.

Plate 120. *Inocybe albofloccosa*. PERTH 08241783. Note pale colour of the fruit bodies, white appendiculate pileus margin, and dense vestiture on the stipe. Photograph — P.B.Matheny.

Plate 121. *Inocybe albofloccosa*. PERTH 08319820. Photograph — P.B.Matheny.

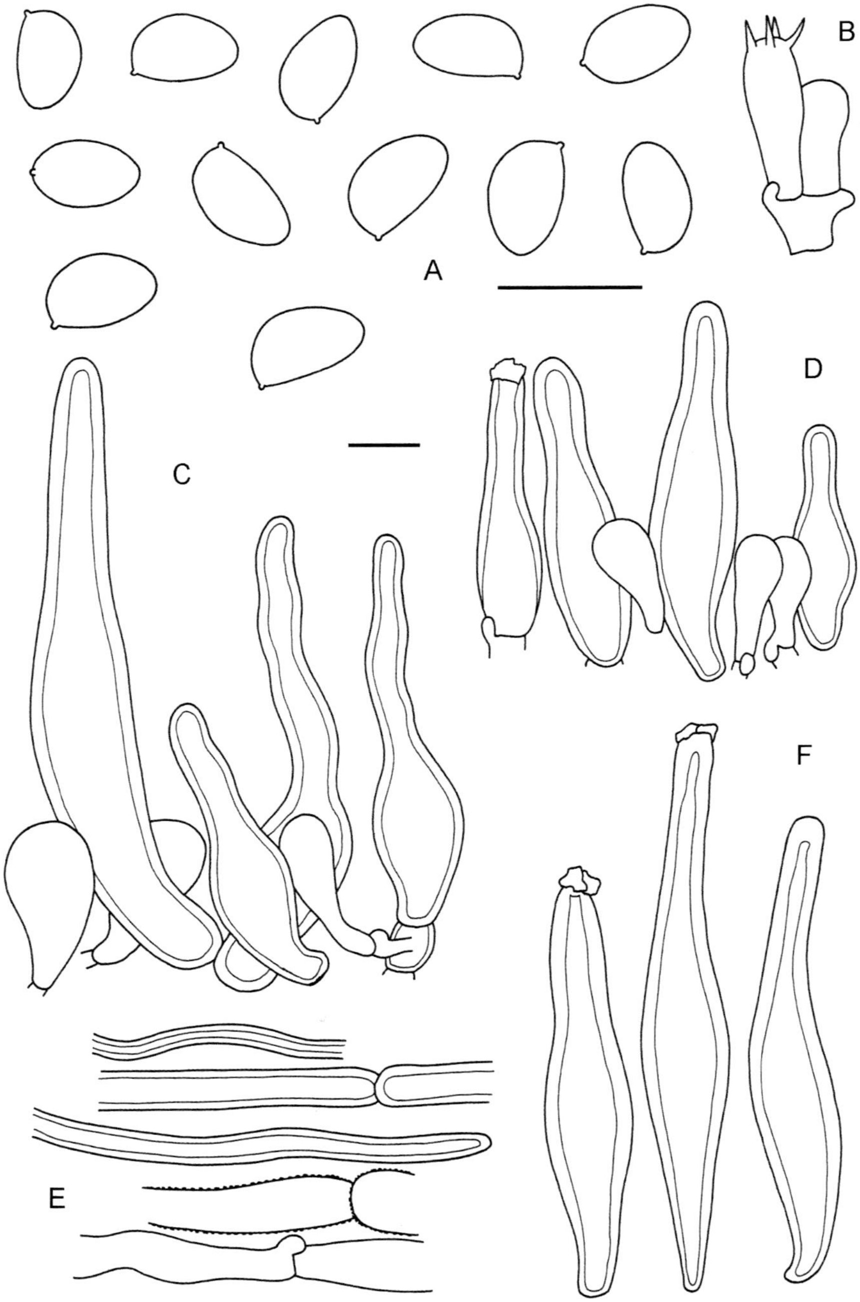

Figure 67. *Inocybe albofloccosa.* All holotype (PERTH 08319413). **A,** Basidiospores; **B,** Basidium; **C,** Caulocystidia and cauloparacystidia; **D,** Cheilocystidia and paracystidia; **E,** Pileipellis and velipellis; **F,** Pleurocystidia. Scale bars = 10 μm (longer bar for spores only).

58. Inocybe fulvo-olivacea Cleland, *Trans. Roy. Soc. S. Australia* 57: 192 (1933)

Type: Belair, S.A., on the ground, 4 Aug. 1928, *J.B.Cleland*; lecto: AD 40427 (ADW 12681), *fide* P.B.Matheny & N.L.Bougher, *Muelleria* 28: 95 (2010).

Pileus 12 mm diam., ±deeply convex, finely fibrillose, yellowish brown (Tawny Olive). *Lamellae* adnate, ascending, moderately close, lighter than dark yellowish brown (Saccardo's Umber). *Stipe* 25 mm × 1–2 mm, slender, granular, slightly hollow, pallid brownish, often a little rosy above. Pl. 122.

Basidiospores 6.5–8.0 × 4.5–5.5 µm [mean 6.9 × 4.9 µm; Q 1.27–1.60; Q mean 1.43 (*n* = 20/1)], smooth, broadly amygdaliform to elliptic, mostly with (sub)conical apices, thick-walled, with a distinct apiculus, yellowish brown. *Basidia* 23–27 × 7–9 µm, clavate, hyaline to pale yellowish, often collapsed, 4-sterigmate. *Pleurocystidia* 48–70 × 10–15 µm, fusiform to narrowly fusiform, thick-walled (2.5–4.0 µm), hyaline or pale yellowish; apices obtuse, crystalliferous. *Cheilocystidia* similar to pleurocystidia, mixed with clavate thin-walled paracystidia (poorly reviving in the lectotype). *Caulocystidia* metuloid, fusiform to cylindrical or irregularly cylindrical, also lageniform, somewhat longer, 60–90 × 12–17 µm, at times with distinct necks, descending at least to the centre of the stipe, mixed with few cylindroclavate to clavate thin-walled cells; base of stipe with vestiture of interwoven cylindrical hyphae. *Pileipellis* a cutis, poorly reviving in the lectotype. *Clamp connections* present. Fig. 68.

Occurs in temperate parts of W.A., S.A., N.S.W. and in Tas.; on the ground in eucalypt forest. Fruiting June and Aug.

Additional specimens examined: see under *Inocybe serrata*.

Diagnostic features: Fruit bodies small, slender, not tough. Pileus convex, fibrillose, yellowish brown. Stipe pale brown, pinkish above, pruinose apically, with white villous vesture below. Spores smooth, variable in shape, <9 µm. Caulocystidia larger than hymenial cystidia, numerous on the upper half of the stipe and mixed with caulocystidioid hairs. Vesture hyphae cylindrical and thick-walled.

The gross morphological description is based on the protologue in Cleland (1933), the description in Cleland (1934), and Cleland's notes of AD 40427 (the lectotype). The microscopic description originates from examination of the lectotype as designated in Matheny & Bougher (2010).

The suite of diagnostic features listed above for *I. fulvo-olivacea* is exemplified by PERTH 08511845 (*PBM3432*) in clade III of the *Inocybe serrata* complex (see discussion under *I. serrata*). However, note that other collections represented in clade III (Fig. 66) vary in morphology. For example, PERTH 08242208 (*PBM3457*) has a pale coarsely fibrillose pileus and no pink flush on the stipe (see Pl. 135). Furthermore, the name *I. fulvo-olivacea* has been applied tentatively to collections in all three clades of the *I. serrata* complex and other members of the *I. serrata* group. For further discussion, see under *I. serrata* (species 63).

Plate 122. *Inocybe fulvo-olivacea*. PERTH 08511845 (*PBM3432*, *I. serrata* clade III). Photograph — P.B.Matheny.

Plate 123. *Inocybe granulosipes*. PERTH 08561842 (*PBM3234*, *I. serrata* clade I). Photograph — P.B.Matheny.

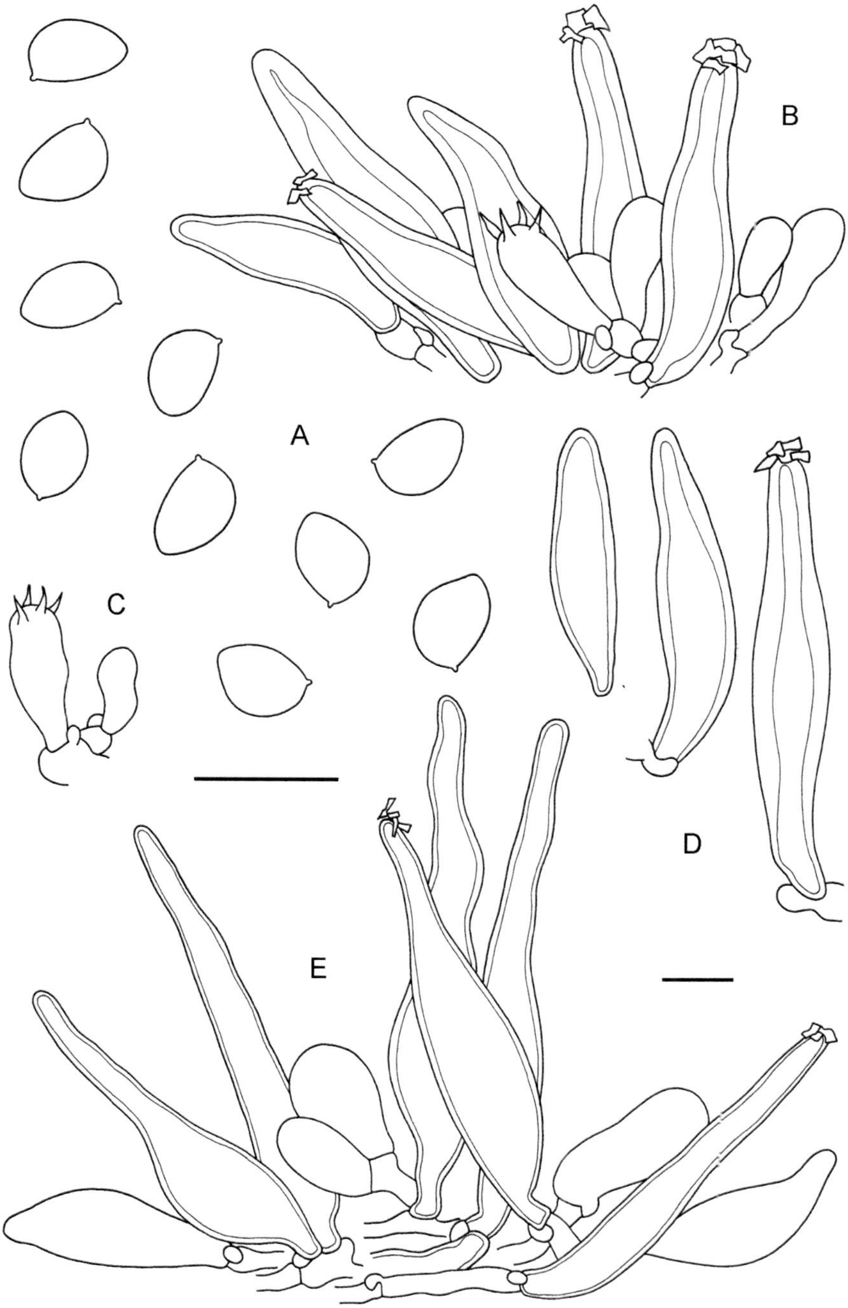

Figure 68. *Inocybe fulvo-olivacea.* A–C, lectotype (AD 40427); D–G, PERTH 08511845 (*PBM3-32*, *I. serrata* clade III). **A,** Basidiospores; **B,** Cheilocystidia (paracystidia not reviving); **C,** Pleurocystidia. **D,** Basidium; **E,** Cheilocystidia and paracystidia; **F,** Basidiospores; **G,** Caulocystidia and cauloparacystidia (upper stipe). Scale bars = 10 µm (longer bar for spores only).

59. Inocybe granulosipes Cleland, *Trans. Roy. Soc. S. Australia* 57: 192 (1933)

Type: Stirling West, Mount Lofty, S.A, gregarious, 23 July 1927, *J.B.Cleland*; lecto: AD 5450. *fide* P.B.Matheny & N.L.Bougher, *Muelleria* 28: 96 (2010)

Pileus 12–18 mm wide, convex, umbonate, fibrillose, brown to dark brown (Snuff Brown to bistre). *Lamellae* adnate, moderately close, subventricose, brown (Snuff Brown). *Stipe* 18–25 × 1–2 mm, rather slender, generally becoming hollow, equal, the buried base slightly bulbous, dark brown (bistre) with abundant whitish mealy to fibrillose granules; context very thin. Pl. 123.

Basidiospores 7.0–8.5 × 4.5–5.5 µm [mean 7.9 × 5.2 µm; Q 1.36–1.70; Q mean 1.53 (*n* =20/1)], smooth, subamygdaliform or broadly so to elliptic with pointed or rounded apices, with a slightly thickened wall and a small but distinct apiculus, yellowish brown. *Basidia* 26–33 × 7–8 µm, clavate, hyaline, 4-sterigmate. *Pleurocystidia* 55–75 × 10–15 µm, narrowly fusiform, less often or rarely lageniform, thick-walled, the walls 2.5–4.0 µm thick, pale yellow or hyaline; apices obtuse, crystalliferous. *Cheilocystidia* similar to pleurocystidia, mixed with clavate thin-walled paracystidia that revive poorly in the lectotype. *Caulocystidia* in dense clusters, metuloid, 83–125 × 9–13 µm at the stipe apex, with clavate cauloparacystidia and cells of intermediate shape and size (mostly fusiform to subcylindrical or irregularly cylindrical, often thick-walled), descending to the centre of the stipe, flexuous; vesture hyphae interwoven and more conspicuous at the base of the stipe, with infrequent caulocystidioid terminal cells. *Pileipellis* a cutis of repent hyphae, these cylindrical to inflated, to 23 µm wide, faintly encrusted and with thin to slightly thickened walls, surmounted by broad smooth thin-walled hyphae 4–16 µm wide. *Clamp connections* present. Fig. 69.

Occurs on the ground in eucalypt forest in south-western W.A., southern S.A. and Tas. Fruiting June and July.

Additional specimens examined: See under *Inocybe serrata*.

Diagnostic features: Fruit bodies small, slender, not tough. Pileus umbonate, fibrillose, brown to dark brown. Stipe dark brown, pruinose apically, with white villous vesture below. Spores smooth, shape variable, <9 µm. Caulocystidia larger than hymenial cystidia, numerous on the upper half of the stipe and mixed with caulocystidioid hairs. Vesture hyphae filiform and thick-walled.

The foregoing morphological description is based on Cleland (1933), the description in Cleland (1934), Cleland's notes, and examination of the lectotype. The microscopic data come from our examination of the lectotype.

Based on the diagnostic features, *I. granulosipes* is best represented by fruit bodies that resemble PERTH 08561842 (PBM3234, *I. serrata* clade I). Indeed, all collections with a dark brown or bistre pileus and stipe coalesce in clade I of the *I. serrata* complex (Fig. 66). However, the external morphology of other collections in clade I varies (some with paler brown pileus, dull brown stipe, or stipe with pinkish tinges). *Inocybe australiensis* also has a dark brown pileus and similar spores, and it was considered synonymous with *I. granulosipes* by Horak (1980). However, it differs by having a squarrose to squamulose or scaly pileus rather than a merely fibrillose pileus as in *I. granulosipes* (Matheny & Bougher, 2010). Further collections and phylogenetic analysis are required to determine whether *I. granulosipes* might be referable to clade I of the *I. serrata* complex. See further discussion under *I. serrata* (species 63).

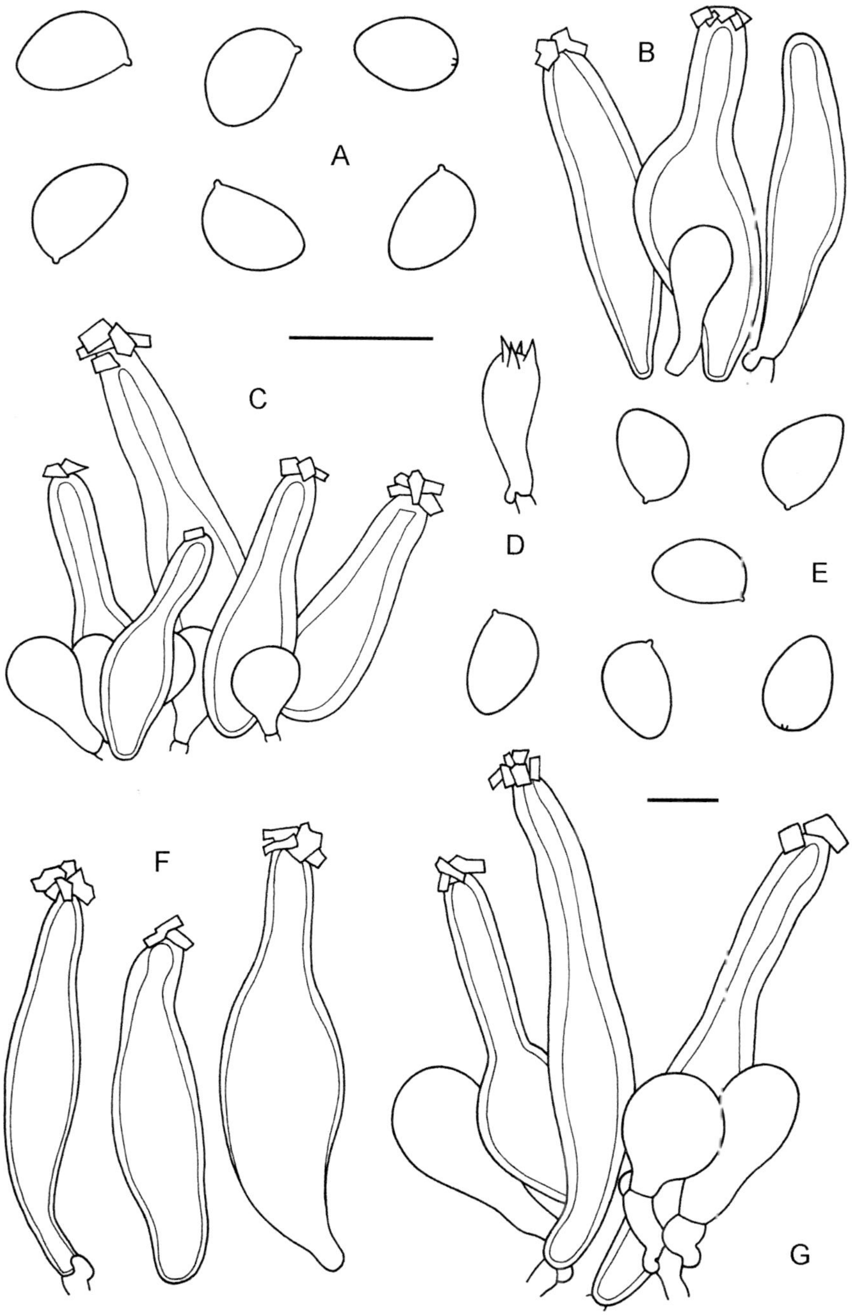

Figure 69. *Inocybe granulosipes.* A, B, F, lectotype (AD 5450); C, D, E, G, PERTH 08561842; *PBM3234*, *I. serrata* clade I). **A,** Basidiospores; **B,** Cheilocystidia and paracystidium; **C,** Cheilocystidia and paracystidia; **D,** Basidium; **E,** Basidiospores; **F,** Pleurocystidia; **G,** Caulocystidia and cauloparacystidia (upper stipe). Scale bars = 10 µm (longer bar for spores only).

60. **Inocybe pallidiflavipes** Matheny, Bougher & G.M.Gates, *sp. nov.*

MycoBank: MB 809426

Type: track to Lady Barron Falls, Mount Field Natl Park, Tas., 42°40'S, 146°43'E, gregarious but solitary or in pairs on well-decayed wood mixed with soil in sclerophyll forest under *Eucalyptus obliqua* and *Acacia*, 1 June 2010, *P.B.Matheny PBM3434*; holo: PERTH 08517096; iso: TENN 067778 (JQ313574-nLSU, KP308799-ITS, KM406207-*rpb2*).

Pileus 10–17 mm wide, conical, soon expanding to plane, with a small nipple-like umbo; margin decurved; velipellis not conspicuous; surface dry, smooth and unbroken in the centre, radially fibrillose and entire or rimulose towards the margin; centre or umbo dark yellowish brown or bay brown (10YR 3/4) or bistre, generally strong brown or umber (7.5YR 4/4–5/6) towards the margin; context very thin-fleshed except under the umbo, pallid with a pale brown tone, not changing colour upon exposure. *Lamellae* ascending-adnate to adnexed, subdistant, 20–24 L, with several tiers of lamellulae, ventricose, light yellowish brown to brown (10YR 6/4–4/3); edges pallid and indistinctly fimbriate. *Stipe* 21–33 × 1–2 mm at the apex, base swollen and up to 3 mm wide, cortina not directly observed but presumably fugacious; surface dry, initially densely white-fibrillose ±throughout, upper half becoming pruinose-granulose; apex light reddish brown (5YR 6/4) to reddish brown (5YR 5/4), darker towards the base, i.e. dark reddish brown (5YR 3/4) to bistre, particularly with age, base pale yellow to yellow (2.5Y 8/6), this noticeable on almost all specimens in the field, rarely whitish; context solid. *Odour* not remarkable to weakly acidulous. Pl. 124, 125.

Basidiospores 8.0–10.0 × 5.0–6.0 µm [mean 9.1 × 5.4 µm; Q 1.42–1.92 (–2.00); Q mean 1.68 (n = 53/3)], smooth, amygdaliform with bluntly pointed apices, yellowish brown; apiculus small but usually distinct. *Basidia* 25–29 × 7–9 µm, clavate, hyaline, 4-sterigmate. *Pleurocystidia* 55–70 × 12–18 µm, fusiform to fusiform-ventricose, apices obtuse and at times bearing crystals, very thick-walled (mostly 4–5 µm thick), pale yellow to yellow, occasionally hyaline. *Cheilocystidia* similar to pleurocystidia but often shorter, at times ventricose or utriform, hyaline, yellow, or with ochraceous contents, mixed with paracystidia. *Caulocystidia* 41–170 × 11–17 µm, fusiform (like hymenial cystidia) but also irregularly cylindrical with tapered apices, in sparse or dense clusters mixed with cauloparacystidia and cells of intermediate size and shape at the stipe apex, mixed with or replaced by caulocystidioid cells, short-clavate cauloparacystidia not observed at the stipe centre and above the stipe base; caulocystidioid cells often elongate, irregularly cylindrical, hyaline, with walls mostly 1.5–2.0 µm thick but some thinner-walled; superficial hyphae densely interwoven above the stipe base. *Pileipellis* composed of an ochraceous-tawny to dark yellowish brown layer of cylindrical to inflated hyphae mostly 10–25 µm wide and strongly encrusted with pigment, septa frequent, walls thin or thickened; in some sections the terminal hyphae pale yellow, smooth, narrower (4–8 µm wide), elongate, the walls 1.0–1.5 µm thick, with tapered apices, at times with encrusted subtending cells. *Clamp connections* present. Fig. 70.

Occurs on soil and on moss-covered logs in dry or wet sclerophyll forest in Tas., with mixed *Eucalyptus*, *Pomaderris* and, sometimes, *Nothofagus* or *Doryphora*. Fruiting May and June.

Additional specimens examined: Tas.: Lower Kermandie, scattered singly on soil in wet sclerophyll forest under *Eucalyptus* and *Pomaderris*, 19 May 2010, *D.Ratkowsky, G.Gates & P.B.Matheny PBM3357* (PERTH 08517118, TENN 067779; KP170967-nLSU, KP308800-ITS, KM406208-*rpb2*); Lower Kermandie, scattered singly on moss-covered log in wet sclerophyll forest/rainforest mixture with *Eucalyptus*, *Nothofagus* and *Pomaderris*, 19 May 2010, *G.Gates, D.Ratkowsky & P.B.Matheny PBM3363* (TENN 065747); track to Lady Barron Falls, Mount Field Natl Park, on soil in wet sclerophyll forest under *E. obliqua*, *Acacia*, *Pomaderris* and *Doryphora*, 1 June 2010, *G.Gates & P.B.Matheny PBM3453* (PERTH 08517126, TENN 067777; KP170968-nLSU, KP308801-ITS, KM406209-*rpb2*).

Etymology: From the Latin *pallidus* (pale) and *flavipes* (yellow foot), in reference to the pale yellow stipe base.

Diagnostic features: Fruit bodies small. Pileus dark yellowish brown to dark brown, with a nipple-like umbo, smooth at the centre, fibrillose towards the margin; upper half of stipe becoming pruinose-granulose; base swollen and pale yellow. Spores smooth. Closely resembles *Inocybe granulosipes* but differs by the pale yellow colouration at the stipe base and somewhat larger spores.

Inocybe pallidiflavipes is a species from Tasmania in the broader *I. serrata* group (Fig. 66) which typically have small fruit bodies with a well-developed superficial layer of fibrils on the lower part of the stipe. The terminal cells of these superficial hyphae are often thick-walled, elongate, and may bear crystals at the apices. The broader *I. serrata* group includes species that are difficult to separate in the field such as *I. granulosipes*, *I. serrata*, *I. fulvo-olivacea*, and *I. albofloccosa*. Two species from *Nothofagus* forests in New Zealand, *I. scabriuscula* and *I. destruens* (Horak 1977), have also been placed in this group based on phylogenetic analyses (Fig. 66). *Inocybe pallidiflavipes* can best be distinguished by the combination of the dark brown pileus, yellowish stipe base, and somewhat larger spores. *Inocybe albofloccosa* forms a distinct monophyletic group with *I. pallidiflavipes*. This relationship is supported by rRNA and *rpb2* gene trees (Fig. 66).

Inocybe pallidiflavipes was previously misidentified as *I. granulosipes* Cleland by Kosentka *et al.* (2013), who reported the absence of muscarine in *PBM3363*.

61. **Inocybe pluviosilvestris** Matheny & Bougher, *sp. nov.*

MycoBank: MB 809427

Type: Mount Michael Track, Blue Tier, Tas., Australia, 41°10'S, 148°01'E, gregarious singly or in small clusters on soil and woody debris in rainforest under *Nothofagus cunninghamii*, alt. 800 m, 22 May 2010, *G.Gates PBM3389*; holo: PERTH 08517134; iso: TENN 068283 (KP170969-nLSU, KP308802-ITS, KM406210-*rpb2*).

Pileus 5–14 mm wide, conical to broadly so, expanding to almost plane, often with a small obtuse umbo; margin decurved; velipellis absent or inconspicuous; surface generally smooth and unbroken on the disc, fibrillose outside, at times rimulose, with margin occasionally denticulate, otherwise entire, dry; colour very dark brown to dark yellowish brown (Snuff Brown or Prout's Brown) or shading to brown towards the margin; umbo at times greyish brown or lighter in colour than the rest of the pileus centre; context thin. *Lamellae* adnexed to subsinuate, subdistant to moderately close, c. 20 L, with several tiers of lamellulae, pale brown to yellowish brown; edges pallid, fimbriate under a lens (or indistinctly so). *Stipe* 15–25 × 1–2 mm, terete, often flexuous, slightly swollen at the base, upper half or apex pruinose, with silky-fibrillose white vesture below; reddish brown (5YR 5/4 to 4/4) to strong brown. *Odour* absent. Pl. 126.

Basidiospores 6.5–8.0 × 4.0–5.0 µm [mean 7.0 × 4.6 µm; Q 1.40–1.75; Q mean 1.55 (n = 46/2)], smooth, amygdaliform to broadly amygdaliform, with pointed to bluntly pointed apices, yellowish brown; apiculus small. *Basidia* 25–27 × 8–9 µm, clavate, hyaline, 4-sterigmate. *Pleurocystidia* 30–60 × 12–15 µm, fusiform, apices obtuse to subacute, at times bearing crystals, thick-walled, the walls 2–4 µm thick, yellow to bright yellow. *Cheilocystidia* similar to pleurocystidia, occasionally shorter, mixed with clavate paracystidia and some basidia. *Caulocystidia* frequent at the apex, mixed with cells of various shapes and sizes, including clavate cauloparacystidia, longer than hymenial cystidia (to 100 µm long), slightly thick-walled (c. 1.5 µm), hyaline to yellow, apices obtuse to subacute, necks long and sometimes flexuous; caulocystidioid hairs arising as terminal cells of vesture hyphae down to the stipe centre, these thick-walled, irregularly cylindrical and often flexuous; lower part of stipe with only cylindrical and flexuous end cells. *Pileipellis* a cutis of cylindrical hyphae mostly 6–12 µm wide, dark yellowish brown to yellowish brown in mass, mostly smooth (at least the terminal cells) otherwise weakly encrusted, with walls 1.0–1.5 µm thick, terminal cells tapered towards the apices or not differentiated. *Clamp connections* present. Fig. 71.

Known only from northern Tas. where it grows on soil or woody debris in temperate rainforest under *Nothofagus cunninghamii*. Fruiting May.

Additional specimen examined: Tas.: Vale of Belvoir, 30 km W of Gowrie Park, on soil along track in rainforest under *Nothofagus cunninghamii*, 25 May 2010, *F.Karstedt PBM3418* (PERTH 08517142, TENN 068284).

Etymology: From the Latin *pluvius* (rainy) and *silvestris* (of the forest), in reference to the habitat of the species.

Diagnostic features: Fruit bodies small; pileus to 14 mm wide, smooth to fibrillose. Spores small, 6.5–8.0 × 4.0–5.0 µm; hymenial cystidia yellow to bright yellow. On soil or woody debris in temperate rainforest under *Nothofagus*.

Plate 124. *Inocybe pallidiflavipes*, holotype (PERTH 08517096). Photograph — P.B.Matheny.

Plate 125. *Inocybe pallidiflavipes*. PERTH 08517126. Photograph — P.B.Matheny.

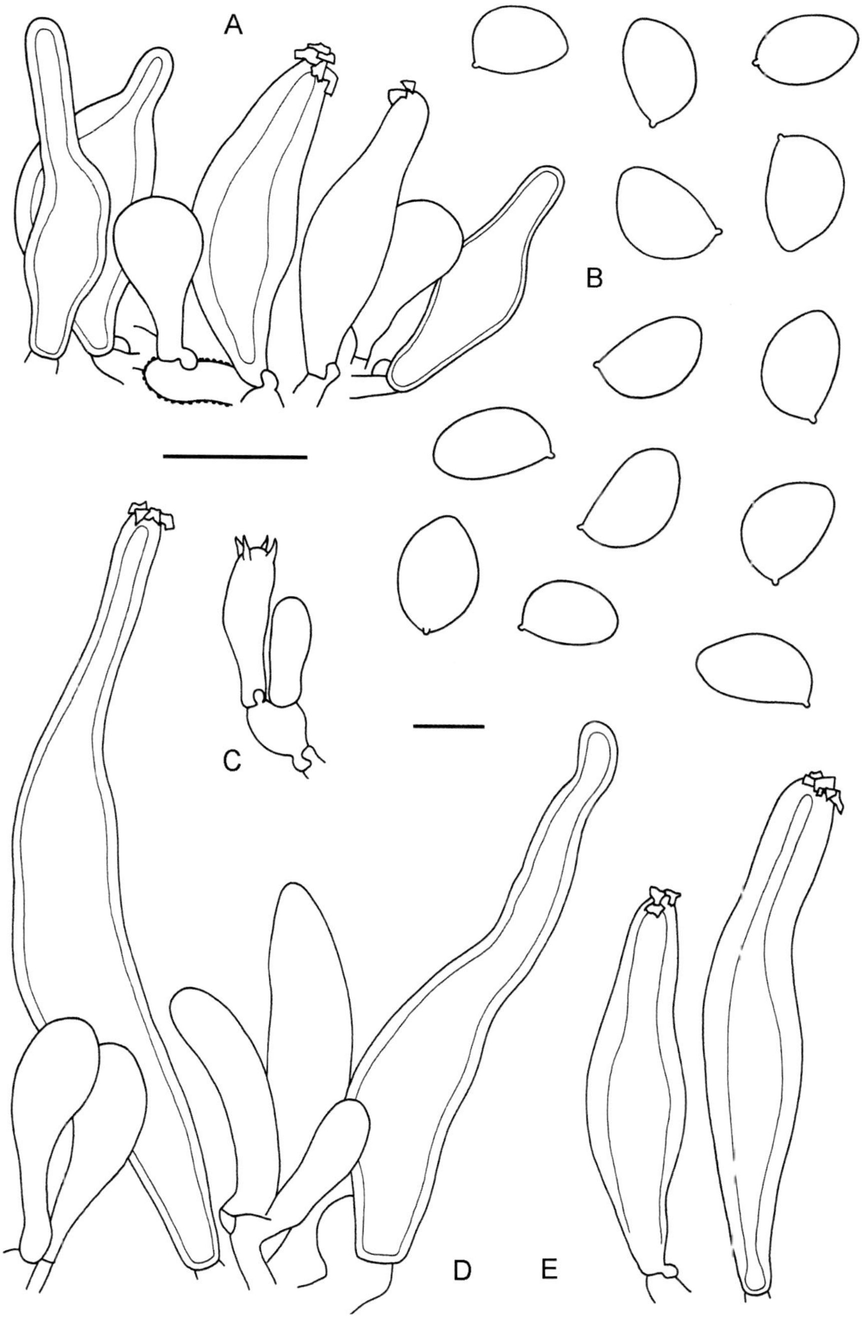

Figure 70. *Inocybe pallidiflavipes.* All holotype (PERTH 08517096). **A,** Cheilocystidia and paracystidia; **B,** Basidiospores; **C,** Basidia; **D,** Caulocystidia and cauloparacystidia (from upper stipe); **E,** Pleurocystidia. Scale bars = 10 µm (longer bar for spores only).

Inocybe pluviosilvestris is the only member of the *I. serrata* group (Fig. 66) known to associate with *Nothofagus* in Australia. Group characteristics include the small size, an often conical pileus, the stipe with a vesture at least over the lower part, and small smooth spores. All Australian species have a smooth to fibrillose pileus.

Inocybe mendica E.Horak (*PBM3101*, TENN 063678) is superficially similar given its very small size, but it differs in its light brown pileus, longer spores (8.0–9.5 × 5.0–5.5 µm), and occurrence in New Zealand. *Inocybe tasmanica* is also very similar to *I. pluviosilvestris*, but it has a tomentose-fibrillose pileus, avellaneous-coloured lamellae, larger spores (9.0–10 × 5.0–6.0 µm), and hyaline hymenial cystidia. Like *I. pluviosilvestris*, *I. tasmanica* is known only from Tasmanian rainforest under *Nothofagus cunninghamii*, but is phylogenetically distantly related to the *I. serrata* group and to *I. mendica* (Fig. 66).

Plate 126. *Inocybe pluviosilvestris*, holotype (PERTH 08517134). Photograph — P.B.Matheny.

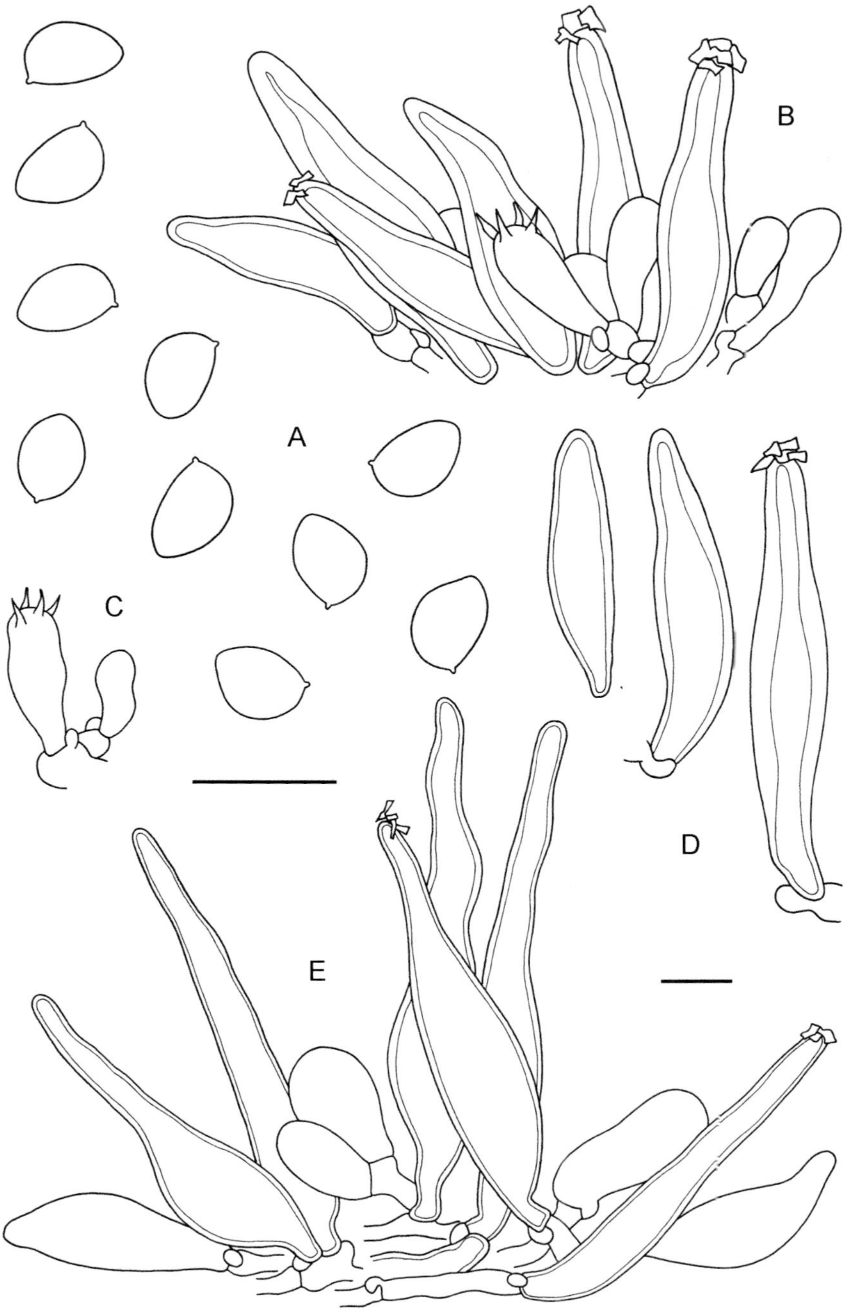

Figure 71. *Inocybe pluviosilvestris.* All holotype (PERTH 08517134). **A,** Basidiospores; **B,** Cheilocystidia and paracystidia; **C,** Basidia; **D,** Pleurocystidia; **E,** Caulocystidia and cauloparacystidia from upper stipe. Scale bars = 10 μm (longer bar for spores only).

62. Inocybe serratoides Matheny & Bougher, *sp. nov.*

MycoBank: MB 809458

Type: O'Neils Creek, Gowrie Park, Tas., 41°28'S, 146°12'E, on soil under *Eucalyptus*, 25 May 2010, *F.Karstedt & P.B.Matheny PBM3422*; holo: PERTH 08557748; iso: TENN 069521 (KP171033-nLSU, KP636829-ITS, KM555129-*rpb2*).

Pileus 6–17 mm wide, parabolic, conical to broadly conical, becoming plano-convex or eventually plane, often with a low obtuse umbo; the margin weakly decurved to almost straight, becoming uplifted with age; surface dry; velipellis often present, centre smooth and unbroken or fibrillose to matted-fibrillose, margin finely fibrillose to fibrillose or occasionally rimulose; edge of margin sometimes with pale veil remnants; disc dark brown (10YR 3/5; Mummy Brown) (where velipellis is obscure), shading to dark yellowish brown (10YR 3/4 to 4/4; Saccardo's Umber) or yellowish brown to brown (10YR 6/3, 5/3, 5/4; Tawny-Olive, Saccardo's Umber) towards the margin; disc very pale brown (10YR 8/3 to 7/3) (where velipellis is dense); context thin-fleshed, up to 1 mm thick under the disc, very thin elsewhere, concolorous with the surface or paler, not changing colour where cut or bruised. *Lamellae* adnexed to uncinate, subdistant or at most moderately close, 20–22 L, with several tiers of lamellulae, brown, strong brown or yellowish brown (10YR 5/3 to 4/3, 7.5YR 4/4); edges pallid, indistinctly fimbriate, ventricose, narrow, to 2 mm deep. *Stipe* 12–35 × 1–2 mm, even, at times flexuous; surface dry, extreme apex (the uppermost 12–17%) densely pruinose; lower levels with a thin pallid or white silky-fibrillose vesture (not mealy); colour light brown (7.5YR 6/4) at the apex, shading to brown (7.5YR 5/4 to 4/4) below or with a reddish yellow or dull pink ground colour (5YR 5/4, 7.5YR 7/4, 7/6, 6.6); context solid, pallid in the centre, cortex concolorous with the surface. *Odour* acidulous or not remarkable. Pl. 127, 128.

Basidiospores 7.0–8.5 × 4.5–5.0 µm [mean 7.6 × 4.8 µm; Q 1.40–1.78; Q mean 1.57 (*n* = 37/2)], smooth, mostly amygdaliform to subamygdaliform, with conical apices, occasionally subelliptic, pale yellowish brown to yellowish brown; apiculus small and not distinctive. *Basidia* 28–33 × 7–8 µm, narrowly clavate, hyaline, 4-sterigmate. *Pleurocystidia* 46–64 × 12–15 µm, fusiform, the apices obtuse and crystalliferous, thick-walled (3–5 µm), the walls pale yellow to yellow. *Cheilocystidia* similar to pleurocystidia, some shorter, mixed with hyaline clavate paracystidia. *Caulocystidia* and cauloparacystidia absent; extreme apex of stipe with numerous scattered clusters of caulocystidioid hairs, these often flexuous or serpentine, narrowly cylindrical to fusiform, very long (86–145 × 10–14 µm), thick-walled (1.0–2.5 µm), hyaline, mixed with short cylindrical thin-walled cells (these not apparently homologous with paracystidia), mostly 17–31 × 8–9 µm; superficial hyphae below and towards the stipe centre and above the base part of a dense interwoven vesture of hyaline cylindrical flexuous hyphae, the terminal cells (and subterminal cells at times) with walls 1.0–2.5 µm thick, terminal cells cylindrical and not differentiated in shape. *Pileipellis* a repent cutis, yellowish brown in mass, the hyphae cylindrical, mostly 6–15 µm wide, encrusted, thin-walled; velipellis hyphae narrowly cylindrical, often 5–7 µm wide, hyaline, smooth, unbranched, with walls 1.0–2.5 µm thick, only occasionally septate. *Clamp connections* present. Fig. 72.

Occurs in N.S.W. and Tas.; grows on soil under *Eucalyptus* and *Pomaderris*. Also under *Kunzea* in New Zealand. Fruiting in May.

Additional specimens examined: N.S.W.: above Kangaroo Valley, 1 km E of Fitzroy Falls, Southern Tablelands, 34°39'S, 150°28'E, scattered singly or in pairs on soil along track under *Eucalyptus*, 18 May 2009, *P.B.Matheny PBM3175* (PERTH 08557713, TENN 063978; KP171035-nLSU, KP636831-ITS). Tas.: Kermandie Falls, lower track, 43°12'S, 146°52'E, scattered singly on soil in wet sclerophyll forest under *Eucalyptus* and *Pomaderris*, 19 May 2010, *P.B.Matheny PBM3356* (PERTH 08557721, TENN 068701; KP171036-nLSU, KP636832-ITS, KM555131-*rpb2*).

Extralimital specimen examined: New Zealand: North Island, Arataki Visitors Centre, Waitakere Ra., W of Auckland, 36°56'49.2"S, 174°36'23.04"E, scattered singly on soil in gully under *Kunzea ericoides* along track, 6 May 2009, *P.B.Matheny PBM3072b* (PDD 97910, TENN 063651; KP171034-nLSU, KP636830-ITS, KM555130-*rpb2*).

Etymology: The epithet *serratoides* is in reference to its similarity to *Inocybe serrata*.

Diagnostic features: Fruit bodies small. Pileus fibrillose; densely pruinose only at the stipe apex (the uppermost 12–17%), with fine white silky vesture elsewhere (not mealy). Odour acidulous or not remarkable. Spores mostly amygdaliform to subamygdaliform, 7.0–8.5 × 4.5–5.0 µm; hymenial cystidia fusiform, thick-walled, often pale yellow or yellow; apex of stipe with clusters of long caulocystidioid hairs, elsewhere with thick-walled cylindrical vesture hyphae.

Spores of the New Zealand specimen can be a little longer (7.5–10.0 × 4.5–5.0 µm) than those of the Australian material. Broadly similar species described by Horak (1977) from New Zealand differ in their much smaller size (*I. mendica* E.Horak), in having a scaly pileus surface (*I. albovestita* E.Horak, *I. destruens* E.Horak and *I. scabriuscula* E.Horak) or in other ways. Molecular analyses confirm the separation from *I. serratoides* from the four other taxa (Fig. 66).

Whereas *I. serratoides* is difficult to separate morphologically from members of the *I. serrata* complex (clades I–III), as well as *I. fulvo-olivacea* and *I. granulosipes*, molecular results are clear (Fig. 66). A clue to its identity might be the absence of cauloparacystidia and true caulocystidia, which are replaced on the stipe by caulocystidioid hairs. In *I. serratoides*, the stipe is not as mealy-fibrillose as in some collections of the *I. serrata* complex (e.g. *I. serrata* clade III, *PBM3354*), and mixed among the caulocystidioid hairs are short, thin- to slightly thick-walled, cylindrical to irregularly cylindrical cells. A more detailed morphological assessment than we can provide here will be necessary to substantiate these differences.

Several sequences from environmental (soil) samples in Queensland form a monophyletic group with *I. serratoides* (Fig. 66). These sequences were produced from hyphal ingrowth bags in field sites dominated by *Eucalyptus pilularis* in wet sclerophyll at Peachester State Forest (26°50'S, 152°53'E; Bastias *et al.*, 2006; see DQ388808, DQ388832, DQ388844). They differ from sequences of *I. serratoides* by 4–5% and so could represent a distinct species. The sequence quality of the environmental samples should be confirmed, but in the interim we refer to them as *I. serratoides*.

The phylogenetic position of *I. serratoides* differs by gene region with significantly supported conflict between nLSU+ITS and *rpb2* gene trees (data not shown). In the former, both *I. albofloccosa* and *I. pallidiflavipes* form a strongly supported monophyletic group with the *I. serrata* complex. In the *rpb2* gene tree, *I. serratoides* is sister to the *I. serrata* complex with strong support. Concatenated data place *I. serratoides*, *I. scabriuscula* and *I. destruens* as sister to the *I. serrata* complex, but, of course, with weak support (Fig. 66). Sequences from additional unlinked markers will be necessary to unravel the evolutionary processes within this group.

Plate 127. *Inocybe serratoides*. PERTH 08557713. Photograph — P.B.Matheny.

Plate 128. *Inocybe serratoides*. PERTH 08557721. Photograph — P.B.Matheny.

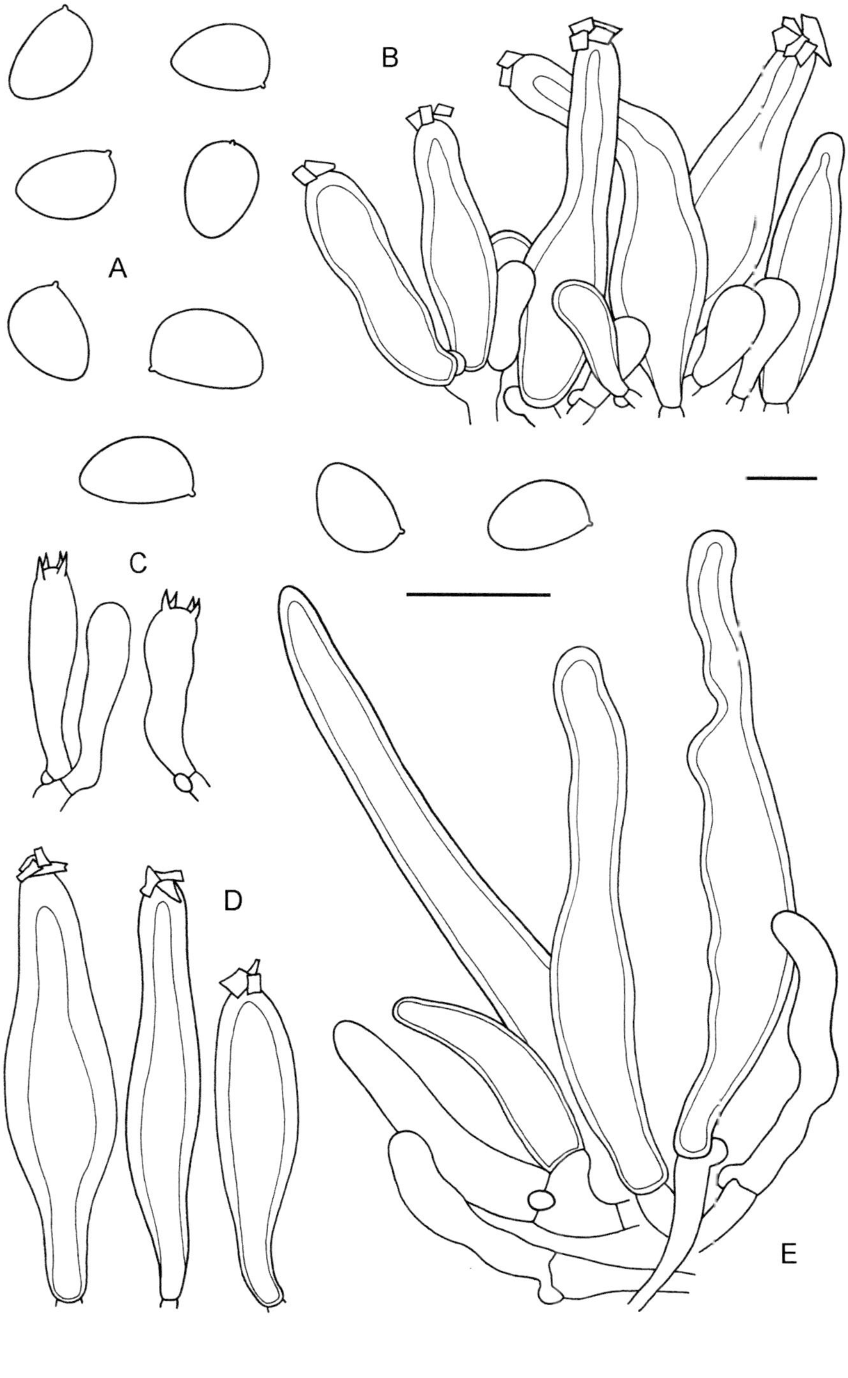

Figure 72. *Inocybe serratoides.* All holotype (PERTH 08557748). **A,** Basidiospores; **B,** Cheilocystidia and paracystidia; **C,** Basidia; **D,** Pleurocystidia; **E,** Caulocystidioid cells from upper stipe. Scale bars = 10 µm (longer bar for spores only).

63. **Inocybe serrata** Cleland, *Trans. Roy. Soc. S. Australia* 57: 192 (1933)

Type: Mount Lofty, S.A., 23 June 1928, *J.B.Cleland*; lecto: AD 5444 (ADW 12669), *fide* P.B.Matheny & N.L.Bougher, *Muelleria* 28: 99 (2010).

[*Inocybe australiensis auct. non* Cleland & Cheel: N.L.Bougher & K.Syme, *Fungi of Southern Australia* 278 (1998)].

Pileus 12–13 mm wide, convex, fibrillose, pallid greyish brown, pallid brown, pale yellow-brown (the colour of dead grass) or brown (Verona Brown). *Lamellae* adnexed to adnate, the edges serrate and sometimes paler, pale cinnamon (near Snuff Brown or Wood Brown). *Stipe* 18–30 × 0.5–2.0 mm, slender, mealy to mealy-striate, pale brown or pallid brownish, with a whitish villous lower part; context firm or hollow, pale brown. *Odour* unknown or in one collection (note by Cleland) "farinaceous". Pl. 129.

Basidiospores 6.5–8.5 (–9.0) × 4.5–5.5 µm [mean 7.6 × 5.0 µm; Q 1.27–1.89 (–2.00); Q mean 1.51 (n = 45/4)], smooth, yellowish brown, variable in shape (amygdaliform, broadly amygdaliform to subtriangular in profile, or elliptic), without a suprahilar depression; apices bluntly pointed or obtuse, with a small apiculus. *Basidia* 23–30 × 7–8 µm, clavate, hyaline, 4-sterigmate. *Pleurocystidia* 53–75 × 11–15 µm, fusiform, thick-walled (walls to 4 µm), hyaline or pale yellowish; apices obtuse and crystalliferous. *Cheilocystidia* similar to pleurocystidia, mixed with clavate thin-walled paracystidia. *Caulocystidia* often lageniform, also fusiform, thick-walled and with crystalliferous apices, 83–110 × 14–18 µm, descending to at least the centre of the stipe, mixed with cauloparacystidia and intermediate cells of varying shape and size, intergrading with caulocystidioid hairs 100–160 µm long, filiform, thick-walled, and with crystalliferous apices; lower part of stipe composed of interwoven vesture hyphae, their terminal cells usually filiform, occasionally flexuous and thick-walled, with few cystidioid terminal cells. *Pileipellis* a cutis of lightly encrusted hyphae to 20 µm wide, strong brown (Ochraceous-Tawny) in mass, thin- to slightly thick-walled, surmounted by an interwoven layer of smooth thin-walled cylindrical hyaline superficial hyphae of the velipellis. *Clamp connections* present. Fig. 73.

Occurs in W.A., S.A., southern N.S.W. and Tas.; on soil in eucalypt forests. Fruiting Apr.–Aug. and Nov.

Additional specimens examined (Cleland material): S.A.: Mount Lofty, 21 June 1924, *J.B.Cleland* (AD 5438) [ADW 12672]; Mount Lofty, 7 Apr. 1917, on ground under log, *J.B.Cleland* (AD 5436) [ADW 12670]; Mount Lofty, 7 Apr. 1917, *J.B.Cleland* (AD 5437) [ADW 12668]; Upper Tunkalilla Ck, subcaespitose on ground, 4 June 1930, *J.B.Cleland* (AD 5446) [ADW 12667], determination questionable (Matheny & Bougher, 2010).

Additional specimens examined (*I. serrata* clade I): W.A.: Acourt Road Bushland, Jandakot Regional Park, Perth, 32°06'S, 115°52' E, on soil under *Kunzea* and *Astartea* shrubs, 5 July 2005, *N.L.Bougher & P.Davison E8260* (PERTH 07681216 "*serrata*"; JN974917-nLSU, KP636818-ITS, KM555119-*rpb2*), see image in Bougher *et al.* (2005) as *Inocybe* sp. specimen ID: 994; Bungendore, Bedfordale, 32°10'52"S, 116°02'26"E, in litter in mixed Jarrah-Marri forest under *Eucalyptus marginata*, *Corymbia calophylla*, *Allocasuarina* and *Dryandra*, 1 July 2007, *J.Froudist & N.L.Bougher NLB338* (PERTH 08073627 "*fulvo-olivacea*"; KP171019-nLSU, KP636816-ITS, KM555117-*rpb2*), see image in Bougher *et al.* (2007c) as *Inocybe* sp. specimen ID: 3112; Bald Island Nature Reserve, 34°55'29"S, 118°27'57"E, on bare soil near granite outcrop under shrubs of *Melaleuca microphylla* and *Andersonia sprengeloides*, 14 Aug. 2007, *N.L.Bougher, T.Friend & L.Bell NLB364* (PERTH 08096198 "*fulvo-olivacea*"; KP171014-nLSU, KP636812-ITS); Heartbreak Trail, Warren Natl Park, 34°29'46"S, 115°57'08.8"E, on soil in Karri-*Allocasuarina* forest, 8 June 2010, *P.B.Matheny PBM3479* (PERTH 08241961, TENN 065781 "*granulosipes*"; KP171018-nLSU, KP636815-ITS, KM555116-*rpb2*); *loc. id.*, 8 June 2010, *P.B.Matheny PBM3490* (PERTH 08241759, TENN 065770 "*fulvo-olivacea*"; KP171020-nLSU, KP636817-ITS, KM555118-*rpb2*); Beedelup Natl Park, 34°25'23"S, 115°50'42"E, on red loam, in Karri-Marri forest including *Bossiaea*, 8 June 2010, *P.B.Matheny PBM3477* (PERTH 08241988, TENN 069658 "*flavopus* ined."; KP171013-nLSU, KP636811-ITS, KM555112-*rpb2*); *loc. id.*, on red loam in Karri-Marri forest, 8 June 2010, *P.B.Matheny PBM3491* (PERTH 08241740, TENN 069650 "*fulvo-olivacea*"; KP171015-nLSU, KP636813-ITS, KM555113-*rpb2*); Bibbulmun Track and Shelley Beach Rd, West Cape Howe Natl Park, 35°05'56.8"S, 117°37'44.1"E, on sandy soil under *Melaleuca*, *Taxandria* and *Agonis flexuosa* in closed canopy *Melaleuca* thicket, 18 July 2011, *P.B.Matheny PBM3643* (PERTH 08319448, TENN 066579 "*serrata*"; KP171016-nLSU, KP636814-ITS, KM555114-*rpb2*). N.S.W.: Jervis Bay Territory, Murray's Beach Rd, 32 km SSE of Shoalhaven River Bridge, Nowra, 35°08'S, 150°43'E, on soil in *Eucalyptus*-dominated forest, alt. 40 m, 30 June 1991, *H.Lepp 577* (CANB 574547; JN974966-nLSU); above Kangaroo Valley, 7 km E of Fitzroy Falls, on soil bank under *Eucalyptus*, 18 May 2009, *P.B.Matheny PBM3176* (TENN 069657 "*fulvo-olivacea*"; KP171017-nLSU, KM555115-*rpb2*); *loc. id.*, on soil bank under *Eucalyptus* and *Allocasuarina*, 18 May 2009, *P.B.Matheny PBM3177* (TENN 065355 "*fulvo-olivacea*"; KP171021-nLSU); *loc. id.*, on soil bank under *Allocasuarina* and *Eucalyptus*, 18 May 2009, *P.B.Matheny PBM3178* (TENN 069656 "*fulvo-olivacea*"; KP171007-nLSU, KM555107-*rpb2*). A.C.T.: Orroral–Cotter Hut road, 38.5 km SW of Capital Hill, Canberra, 35°38'S, 148°57'E, on soil in *Eucalyptus* woodland, 12 June 1993, *H.Lepp 967* (CANB 574596; KP171008-nLSU, KP636805-ITS). Tas.: Mount Field Natl Park, in wet sclerophyll forest with *Eucalyptus regnans*, *Acacia* and *Pomaderris*, 8 June 2009, *P.B.Matheny, G.M.Gates & D.Ratkowsky PBM3221* (TENN 069651 "*granulosipes*"; KP171005-nLSU, KP636803-ITS, KM555105-*rpb2*); *loc. id.*, 8 June 2009, *D.Ratkowsky, G.M.Gates & P.B.Matheny PBM3222* (TENN 069652 "*granulosipes*"; KP171004-nLSU, KP636802-ITS); *loc. id.*, 8 June 2009, *P.B.Matheny, G.M.Gates & D.Ratkowsky PBM3223* (TENN 069653 "*granulosipes*"; KP171006-nLSU, KP636804-ITS, KM555106-*rpb2*); *loc. id.*, 8 June 2009, *G.M.Gates & P.B.Matheny PBM3231* (TENN 069660 "*granulosipes*"; KP171011-nLSU, KP636809-ITS, KM555110-*rpb2*); *loc. id.*, 8 June 2009, *G.M.Gates, D.Ratkowsky & P.B.Matheny PBM3234* (TENN 069654 "*granulosipes*"; KP171009-nLSU, KP636807-ITS); *loc. id.*, 8 June 2009, *P.B.Matheny, D.Ratkowsky & G.M.Gates PBM3235* (TENN 069659 "*granulosipes*"; KP171012-nLSU KP636810-ITS, KM555111-*rpb2*); Chauncy Vale Reserve, on soil near creek in dry sclerophyll forest under *Eucalyptus*, *Acacia* and *Pomaderris*, 11 June 2009, *G.M.Gates, D.Ratkowsky, P.B.Matheny PBM3240* (TENN 065744 "*granulosipes*"; JQ313563-nLSU, KP636806-ITS, KM555108-*rpb2*); Lower Kermandie Falls, on soil in wet sclerophyll forest under *Eucalyptus* and *Pomaderris*, 19 May 2010, *P.B.Matheny PBM3361* (PERTH 08561834, TENN 069655 "yellow gilled *granulosipes*"; KP171010-nLSU, KP636808-ITS, KM555109-*rpb2*).

Additional specimens examined (*I. serrata* clade II): W.A.: Valema Farm, Paddock 51 (soil site 4), W end of Grylls Rd, Corrigin Shire, 32°28'06"S, 117°49'27"E, on soil in remnant vegetation and woodland, under *Gastrolobium* and *Allocasuarina*, 13 Aug. 1999, *S.Q.Bolsenbroek E6320* (PERTH 07612052; EU600856-nLSU, EU600855-*rpb2*); Ryedene property, near Two Peoples Bay, 34°56'24"S, 118°05'08"E, in litter in woodland of *Eucalyptus marginata*, *Corymbia calophylla* and *Xanthorrhoea*, 22 June 2007, *N.L.Bougher & L.Bell NLB331* (PERTH 08073554 "*serrata*"; KP171026-nLSU, KP636823-ITS, KM555124-*rpb2*); Dalgarup Natural Reserve, Wetherley Rd, 25 km W of Bridgetown, 34°00'15.5"S, 116°58'15.5"E, scattered singly or in pairs on gravelly soil alongside road in long unburnt Jarrah-Marri forest, 13 July 2011, *P.B.Matheny PBM3615* (PERTH 08319707, TENN 066590 "*serrata*"; KP171027-nLSU); South Western Hwy, c. 1 km N of Weld River Bridge, among litter in Jarrah forest under *Eucalyptus marginata*, *Corymbia calophylla*, 14 July 2011, *N.L.Bougher NLB814* (PERTH 08318689 "*serrata*"; KP171025-nLSU, KP636822-ITS, KM555123-*rpb2*); Jarrahdale Rd, 8 km W of Albany Hwy, 32°19'44.4"S, 116°07'42.9"E, on moss-covered sandy soil at edge of clearing under *Corymbia calophylla*, *Eucalyptus marginata* and planted non-native *Acacia*, 25 July 2011, *P.B.Matheny PBM3671* (PERTH 08319162, TENN 066455 "*serrata*"; KP171024-nLSU, KP636821-ITS, KM555122-*rpb2*). N.S.W.: Bumberry L., alt. 400 m, on ground under *Eucalyptus sideroxylon* and *E. microcarpa* with a grass ground cover and some *Callitris endlicheri*, *Acacia pravifolia*, *Dillwynia juniperina* and other shrubs, 11 Nov. 1999, *J.Trappe 24894* (OSC, TENN 069664); Genaren, Genaren Hill Sanctuary, near N boundary fence, alt. 265 m, on non-grazed woodland of regrowth *Eucalyptus dealbata*, *Acacia doratoxylon*, *Micromyrtus ciliata* and *Callitris glaucophylla*, 22 Apr. 2000, *J.Trappe 25079* (OSC, TENN 069663; AY380392-nLSU, AY337399-*rpb2*, AY351816-*rpb1* "*crassocystis* ined."); Blackfellas Hand Rd, c. 7 km from Wallerawong, Central Tablelands, on sandy soil under *Eucalyptus* spp., 24 May 2009, *N.Fechner PBM3191* (TENN 063644 "*fulvo-olivacea*"; KP171022-nLSU, KP636819-ITS, KM555120-*rpb2*). Tas.: Chauncy Vale Reserve, on soil in dry sclerophyll forest under *Eucalyptus*, 11 June 2009, *G.M.Gates, D.Ratkowsky & P.B.Matheny PBM3244* (TENN 069665 "*fulvo-olivacea*"; KP171023-nLSU, KP636820-ITS, KM555121-*rpb2*).

Additional specimens examined (*I. serrata* clade III): W.A.: Cane Break Pool Recreation Site, Rapids Conservation Park, 33°52'00"S, 115°17'00"E, on soil in Jarrah-Marri forest under *Eucalyptus marginata* and *Corymbia calophylla*, 6 June 2010, *P.B.Matheny PBM3457* (PERTH 08242208, TENN 069630 "*serrata*"; KP171031-nLSU, KP636827-ITS). Tas.: Lower Kermandie Falls, on soil in wet sclerophyll forest under *Eucalyptus*, *Pomaderris* and *Acacia dealbata*, 19 May 2010, *D.Ratkowsky & P.B.Matheny PBM3354* (TENN 069628 "*fulvo-olivacea*"; KP171029-nLSU, KP636825-ITS, KM555126-*rpb2*); *loc. id.*, 19 May 2010, *P.B.Matheny & G.M.Gates PBM3362* (TENN 069626 "*serrata*"; KP171030-nLSU, KP636826-ITS, KM555127-*rpb2*); Duckhole Lake Track, near Dover, on soil in wet sclerophyll forest under *Eucalyptus* and *Acacia*, 29 May 2010, *P.B.Matheny PBM3430* (TENN 069627 "*serrata*"; KP171032-nLSU, KP636828-ITS, KM555128-*rpb2*); Mount Field Natl Park, on soil in dry sclerophyll forest under *Eucalyptus obliqua* and *Acacia*, 1 June 2010, *P.B.Matheny & G.M.Gates PBM3432* (PERTH 08511845, TENN 069629 "*fulvo-olivacea*"; KP171028-nLSU, KP636824-ITS, KM555125-*rpb2*).

Diagnostic features: Fruit bodies small, slender, not tough. Pileus fibrillose, at times with a pallid velipellis, brownish to pale yellowish brown or brown; stipe pruinose above, white-villous below, even, pale brown elsewhere. Spores smooth, variable in outline, <9.0 µm long; caulocystidia larger than hymenial cystidia, numerous on the upper half of the stipe, mixed with caulocystidioid hairs and caulocystidioid cells of intermediate size and shape; vesture hyphae filiform and thick-walled.

The gross morphological description above is based on notes compiled by Cleland from collections cited in his original works (Cleland, 1933, 1934) and examined by us. The description of microscopic characters is based on the Cleland collections cited in Matheny & Bougher (2010).

Inocybe serrata and its allies are among the most commonly collected species of Inocybaceae in Australia. Collectively, and as a complex, they are easy to identify in the field, yet morphological recognition of the numerous phylogenetic species in the complex is fraught with challenges. Historically, *I. serrata* is one of three similar small *Inocybe* species described by J.B.Cleland (1933). Cleland's concepts of *I. serrata* and two other species, *I. fulvo-olivacea* and *I. granulosipes*, are rather nebulous because of the brief original diagnoses, mixed-species collections, and morphological variation likely to have been influenced by environmental factors. Moreover, collections made by Cleland many years apart and determined by him as *I. serrata* do not appear to be conspecific (Matheny & Bougher, 2010).

Cleland distinguished his three species on the basis of morphological characters such as the serrated edges of the lamellae in *I. serrata*, an umbonate pileus in *I. granulosipes*, a tawny-olive pileus in *I. fulvo-olivacea*, the pallid stipe of *I. serrata* and *I. fulvo-olivacea* in contrast to the bistre stipe of *I. granulosipes*. He separated *I. serrata* from *I. fulvo-olivacea* and *I. granulosipes* by its serrate gill edges. However, we have not observed any Australian collections in this group with distinctive and consistently serrate gill edges (a hand lens is usually needed to observe the cheilocystidia). Cleland ascribed a rosy upper stipe and convex pileus to *I. fulvo-olivacea*, and although we have observed some recent collections with a pinkish upper stipe, some have a convex pileus (e.g. *PBM3362*, PERTH 0811829) while others do not (e.g. *PBM3432*, PERTH 08511845). Such features can be highly variable and influenced by environmental conditions. The collections on which Cleland based the three species were geographically limited, mostly from locations in south-eastern South Australia close to Adelaide. He could not have been aware of the wide variation in morphology now evident among collections of *I. serrata* and its allies from other parts of Australia. For example, it is clear from recent collections that the strength and persistence of the velipellis can greatly influence the shape and colour of the pileus among phylogenetic species in the complex and in the larger group, as suggested for *Inocybe* species in general by Kuyper (1986).

Various interpretations and synonymies have been proposed for *I. serrata*, *I. fulvo-olivacea* and *I. granulosipes* (Horak, 1980; Grgurinovic, 1997; Matheny & Bougher, 2010). Unfortunately, we were unable to obtain DNA sequences from Cleland's type collections. However, in line with the designation of AD 5444 as lectotype of *I. serrata* (Matheny & Bougher, 2010), we suggest that Cleland's *I. serrata*, along with *I. fulvo-olivacea* and *I. granulosipes*, are likely to be aligned with recent collections of *I. serrata*-like fungi from around Australia which aggregate in a strongly supported clade (Fig. 66). We refer to this clade as the "*I. serrata* complex".

DNA sequences from the numerous recent collections ascribed by us to the *I. serrata* complex segregate these collections into three separate clades, labelled as *I. serrata* clade I, *I. serrata* clade II, and *I. serrata* clade III (Fig. 66). The divergence indicated by these results indicates that the collections include three phylogenetic species. Therefore, in view of the potential of the *I. serrata* complex to consist of multiple species, it would not be prudent at this stage to synonymise Cleland's *I. serrata, I. fulvo-olivacea* and *I. granulosipes*.

We also explored the possibility of assigning each of the three clades to one of the three Cleland species and assigning an epitype selected from each of the clades to the corresponding taxon name. However, this option is problematic because collections with key morphological features used by Cleland to recognise each of the three species are not exclusive to individual clades, e.g. collections possibly macroscopically corresponding with Cleland's concept of *I. fulvo-olivacea* include *PBM3178* (PERTH 08561877) in *I. serrata* clade I, *PBM3191* (PERTH 08511810) in *I. serrata* clade II and *PBM3432* (PERTH 088511845) in *I. serrata* clade III.

We also have observed some collections that feature idiosyncratic traits. For example, Trappe 25079 (*I. serrata* clade II) has uniquely thick-walled cystidia (up to 12 μm thick). Similarly, some other collections represent distinct macroscopic variants, e.g. *PBM3361* (PERTH 08561834) with yellow lamellae (Pl. 130), *PBM3477* (PERTH 08241988) with a yellowish stipe base (Pl. 131), *PBM3671* (PERTH 08319162) and *NLB860* (PERTH 08320713) with diminutive fruit bodies (Pl. 132, 133), *PBM3643* (PERTH 08319448) with pale pilei and a thick, persistent velipellis (Pl. 134; see also Pl. 4C), and *PBM3457* (PERTH 08242208) with pale, coarsely fibrillose pilei (Pl. 135).

Further research is required to provide an authoritative interpretation of the *I. serrata* complex and the status of Cleland's three species. Here, we provide descriptions of the three Cleland taxa (*I. serrata*, *I. fulvo-olivacea* and *I. granulosipes*) based on their protologues and type specimens. Further collections should help to assess the consistency of morphological differences, although our data suggest that the use of gross morphology alone is unreliable. Moreover, future work should also aim to distinguish the *I. serrata* complex from other species in the broader *I. serrata* group (Fig. 66), e.g. *I. albofloccosa*, *I. pallidiflavipes*, *I. tasmanica*, *I. pluviosilvestris* and *I. serratoides*.

Plate 129. *Inocybe serrata*. PERTH 08319707 (*PBM3615*, *I. serrata* clade II). Photograph — P.B.Matheny.

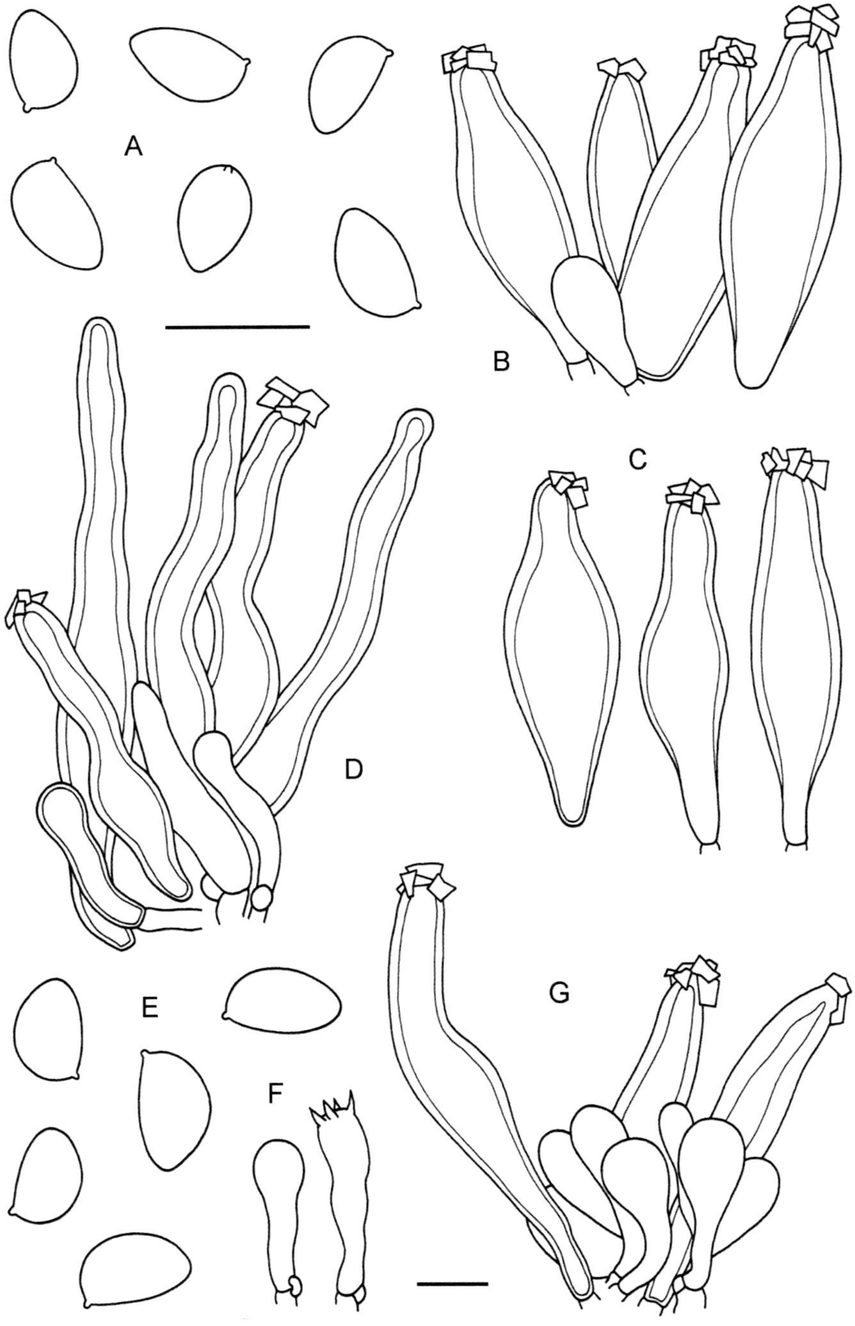

Figure 73. *Inocybe serrata.* A–C, lectotype (AD 5444); D–G, PERTH 08319707 (*PBM3615*, *I. serrata* clade II). **A,** Basidiospores; **B,** Cheilocystidia and paracystidium; **C,** Pleurocystidia; **D,** Caulocystidia and cauloparacystidia; **E,** Basidiospores; **F,** Basidia; **G,** Cheilocystidia and paracystidia. Scale bars = 10 μm (longer bar for spores only).

Some examples in the *Inocybe serrata* complex of variations in macroscopic morphology, some of which may be representative of undescribed species:

Plate 130. PERTH 08561834 (*PBM3361*, *I. serrata* clade I) featuring yellow lamellae. Photograph — P.B.Matheny.

Plate 131. PERTH 08241988 (*PBM3477*, *I. serrata* clade I) featuring yellow lamellae and stipe bases. Photograph — P.B.Matheny.

Plate 132. PERTH 08320713 (*NLB860*) featuring diminutive fruit bodies with pilei mostly <10 mm diam. Photograph — N.L.Bougher.

Plate 133. PERTH 08319162 (*PBM3671*, *I. serrata* clade II) featuring diminutive fruit bodies with pilei mostly <10 mm diam. Photograph — P.B.Matheny.

Plate 134. PERTH 08319448 (*PBM3643*, *I. serrata* clade I) featuring pale pilei with thick, persistent velipellis. Photograph — P.B.Matheny.

Plate 135. PERTH 08242208 (*PBM3457*, *I. serrata* clade III) featuring pale, coarsely fibrillose pilei. Photograph — P.B.Matheny.

OTHER SMOOTH-SPORED SPECIES OF THE *INOCYBE* CLADE

Nine Australian species of the *Inocybe* clade with smooth-spores, the phylogenetic positions of which are outside the STAC group or unknown, are documented below. Three may be allied with the STAC group (*I. spadicea*, *I. murrayana*, *I. paucicolor*), but phylogenetic results are ambiguous or lacking. A tree for *I. flavoalbida* and its closely related north-temperate allies is provided under that species.

64. **Inocybe austrosquarrosa** Matheny, Bougher & G.M.Gates, *sp. nov.*

MycoBank: MB 809264

Type: Australian National Botanic Gardens, 4 km NNW of Capital Hill, Canberra, A.C.T., 35°18'S, 149°06'E, scattered or in small clusters in lawn under scattered *Eucalyptus*, 26 May 1991, *H.Lepp 546*; holo: CANB 574568; iso: PERTH 08516944, TENN 069639 (KJ756438-nLSU).

Pileus 18–45 mm wide, initially convex, becoming conical to plano-convex or plane with a low obtuse umbo; margin incurved at first, later decurved and eventually uplifted; surface dry, squarrose-scaly with pyramidal tufts of squarrose scales at the centre to imbricate-scaly, fibrillose and radially split towards the margin; sepia or dark brown overall but a lighter shade of brown beneath or between the scales and darker fibrils; context dirty white and thin. *Lamellae* adnate, close, with several tiers of lamellulae, up to 4 mm wide, light grey to whitish when very young, becoming dark yellowish brown, drying rusty brown; edges pallid-fimbriate. *Stipe* 18–35 × 4–10 mm wide at the apex, even, central, terete, light brown to reddish brown; surface fibrillose, with fibrils more dense and sepia-coloured or dark greyish brown above the base, nowhere pruinose; cortina present when young but evanescent; context solid, concolorous with the pileus. *Odour* none. Pl. 136.

Basidiospores 7.8–9.7 × 4.5–6.5 µm [mean 8.6 × 5.5 µm; Q (1.29–) 1.36–1.83; Q mean 1.58 (n = 26/2)], smooth, subamygdaliform to amygdaliform, with bluntly pointed or sometimes mucronate apices, yellowish brown; apiculus small but distinct. *Basidia* 25–29 × 9–10 µm, clavate, hyaline, 4-sterigmate. *Pleurocystidia* 50–70 × 10–17 µm, fusiform, thin- to thick-walled (1.0–2.5 µm thick), hyaline; apices obtuse, crystalliferous or sparsely so, at times subcapitate. *Cheilocystidia* similar to pleurocystidia but shorter, mixed with hyaline paracystidia. *Caulocystidia* lacking, but the stipe surface with clusters of brown encrusted terminal hyphae, these generally cylindrical or somewhat fusiform (largely undifferentiated end cells), lacking crystals, with frequent septa, more prevalent and dense on lower part of the stipe. *Pileipellis* with trichodermal tufts of cylindrical hyphae, dark yellowish brown in mass, mostly 5–12 µm wide, encrusted. *Clamp connections* present. Fig. 74.

Known only from two localities in the A.C.T. and Tas.; grows on soil in wet sclerophyll forest under *Pomaderris apetala* seedlings and in a lawn under *Eucalyptus*. Fruiting May and Dec.

Additional specimen examined: Tas.: Pipeline Track, Mt Wellington, 42°55'S, 147°15'E, 26 Dec. 2011, *G.M.Gates & D.Ratkowsky 135* (HO 567551; KJ756439-nLSU, KJ756474-ITS).

Etymology: From the Latin *austro-* (south) and *squarrosus* (scaly), in reference to the scaly pileus and its occurrence in southern Australia.

Diagnostic features: Fruit bodies medium or robust. Pileus dark brown, squarrose to imbricate-scaly; stipe not pruinose, lower part coloured like the pileus and densely fibrillose; odour lacking. Spores smooth; caulocystidia absent.

Inocybe austrosquarrosa, a smooth-spored species with a squarrose pileus, is known from two localities in south-eastern Australia. The fungus is somewhat unusual in that caulocystidia are absent from the stipe surface. While several other species also lack caulocystidia (or these are restricted to the extreme stipe apex), *I. auritelloides* has shorter spores, at times a distinct odour of *Pelargonium* and a habitat in *Nothofagus* rainforest, while *I. lucifera* differs in its tawny or rusty pileus, pungent odour and an association with *Nothofagus* in rainforest. Among other comparable taxa, *I. leucotaenia* lacks the scaly pileus, and its context slowly turns pink, *I. hymenocarpa* has a more slender habit, a hemispherical pileus, a cortinate ring-zone and more slender pleurocystidia, and *I. tomentipes* shares the more slender habit with *I. hymenocarpa* and has a brownish streaked pileus completely lacking scales.

Inocybe austrosquarrosa can be confused with *I. australiensis*, but the latter has a pruinose stipe (at least the upper half), the pileus is much smaller, and the spores are only 6.5–7.5 × 4.0–5.0 µm. *Inocybe phaeosquarrosa* E.Horak, described from rotting wood in *Nothofagus*-dominated rainforest in New Zealand, also differs in having a scaly stipe and thin-walled cystidia (Horak, 1977). *Inocybe umbrosa* E.Horak, a New Zealand endemic, has a dark brown scaly pileus (Horak, 1977), but it differs in its entirely pruinose, reddish brown stipe. It grows on soil or rotten wood in *Nothofagus* forest mixed with *Leptospermum* or *Kunzea*. *Inocybe strobilomyces* E.Horak, one of the most striking species of *Inocybe* in New Zealand (Horak, 1977), is superficially similar to *I. austrosquarrosa*, having similar colours, dimensions and a squarrose pileus. However the spores of *I. strobilomyces* are somewhat smaller (Horak, 1977; 7.0–9.0 × 4.0–5.0 µm), and it grows on rotten wood and bark of *Dacrydium cupressinum* in *Nothofagus* forests.

The phylogenetic placement of *I. austrosquarrosa* is not known with confidence Blast analyses and phylogenetic analyses suggest that it may be nested in the STAC group, but the species clusters with outgroup sequences in our phylogenetic analysis of the STAC group (data not shown). Consequently, it is considered *incertae sedis* within *Inocybe*.

The holotype is a mixed collection including, in addition to *I. austrosquarrosa*, some fruit bodies representing two unidentified taxa: a nodulose-spored species and a smooth-spored species with slender, amygdaliform spores (9.0–11.5 × 4.5–5.5 µm).

Plate 136. *Inocybe austrosquarrosa*. HO 567551. Photograph — G.M.Gates.

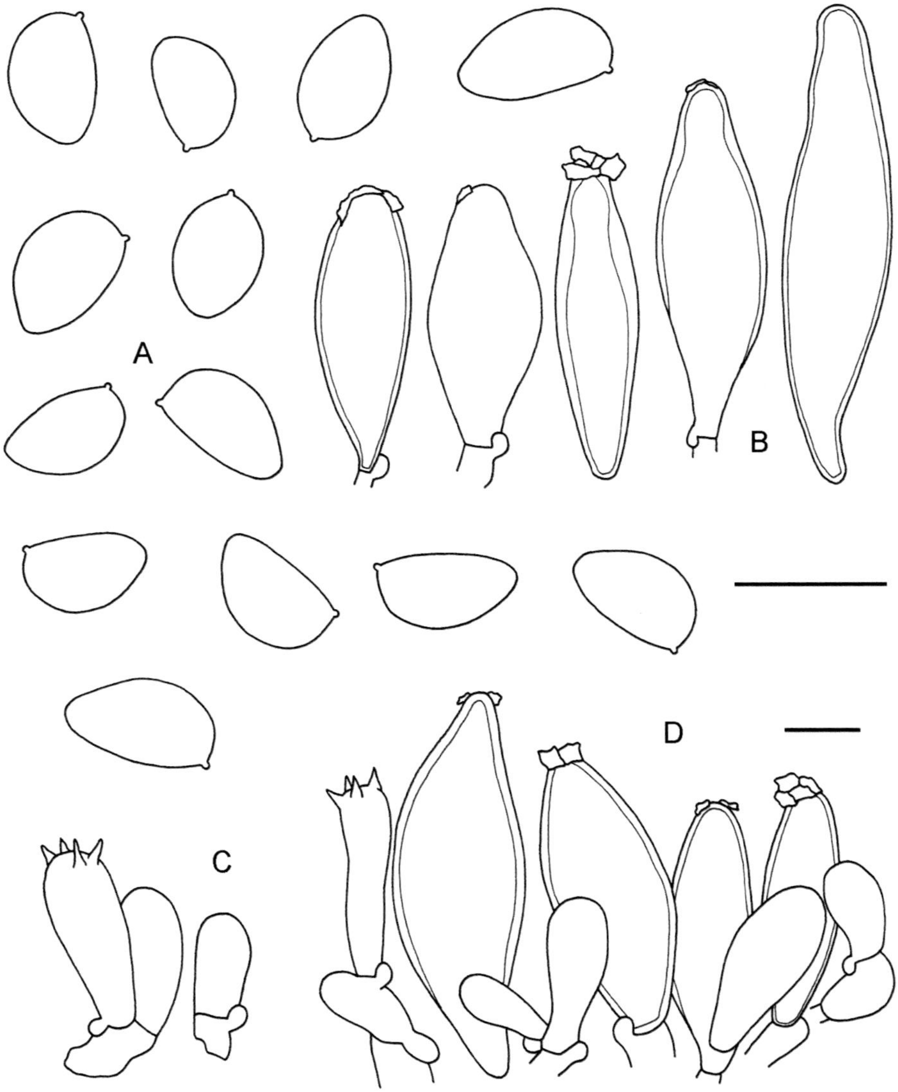

Figure 74. *Inocybe austrosquarrosa.* All holotype (CANB 574568). **A,** Basidiospores; **B,** Pleurocystidia; **C,** Basidia; **D,** Cheilocystidia and paracystidia. Scale bars = 10 µm (longer bar for spores only).

65. **Inocybe bella** Matheny, Bougher & G.M.Gates, *sp. nov.*

MycoBank: MB 809266

Type: Truganini Track, Hobart, Tas., 42°55'S, 147°20'E, with eucalypts, *Pomaderris* and rainforest understory shrubs, 10 Nov. 2004, *G.M.Gates 198*; holo: HO 571334; iso: TENN 065710 (KJ756441-nLSU, KJ756476-ITS).

Pileus 11–26 mm wide, initially hemispherical, becoming broadly convex or developing a slight obtuse umbo; margin decurved; surface dry, radially fibrillose, but the centre becoming cracked and developing fine appressed scales; velipellis absent; reddish brown throughout. *Lamellae* adnexed or sinuate, moderately close, 32–38 L, with several tiers of lamellulae, ventricose, 4-5 mm deep, pale grey with a lilac tint, with edges pallid and fimbriate. *Stipe* 23–32 × 2–5 mm at the apex, 3–6 mm wide at the slightly enlarged base, terete; surface dry, pruinose at the apex (at least the upper third), silky-fibrillose elsewhere; partial veil presumably fugacious; white with lilac hues near the apex; context solid, pale lilac, with faint ochraceous tones at the base. *Odour* acidulous. Pl. 137.

Basidiospores 10.0–11.5 × 5.5–6.0 µm [mean 10.8 × 5.8 µm; Q 1.75–2.00; Q mean 1.87 (n = 17/1)], smooth, amygdaliform to fusiform with pointed apices, yellowish brown, with apiculus prominent. *Basidia* 25–31 × 8–11 µm, clavate to cylindro-clavate, hyaline, 4-sterigmate. *Pleurocystidia* 45–70 × 13–18 µm, fusiform to ventricose with some smaller clavate forms, with obtuse or sometimes subcapitate apices often crystalliferous, thick-walled (to 5 µm), hyaline or pale greenish yellow. *Cheilocystidia* similar to pleurocystidia, occasionally clavate, mixed with crowded thin-walled clavate paracystidia. *Caulocystidia* present at the (extreme) apex in dense scattered clusters, fusoid, often flexuous, some longer than hymenial cystidia, thick-walled (to 5 µm), mixed with caulocystidioid cells of various sizes and shapes, and small clavate cauloparacystidia; superficial hyphae at the stipe centre and below with undifferentiated terminal cells or, occasionally, caulocystidioid terminals. *Pileipellis* a cutis of cylindrical hyphae mostly 7–17 µm wide, encrusted, thin-walled, cinnamon-brown in mass. *Clamp connections* present. Fig. 75.

Known only from the type collection from southern Tas. where it grows on soil under eucalypts, *Pomaderris* and rainforest understory shrubs. Fruiting Nov.

Etymology: From the Latin *bella* (pretty), in reference to that field name applied by the collector, Genevieve Gates.

Diagnostic features: Fruit bodies medium-sized. Pileus reddish brown; lamellae and stipe apex tinged lilac, the stipe white elsewhere. Spores smooth and large, at least 10 µm long, with a distinct apiculus; caulocystidia present at the stipe apex only.

Blastn results of ITS and LSU suggest that *I. bella* might be most closely related to smooth-spored species from the Northern Hemisphere including *I. pseudoreducta* Stangl & Glowinski, *I. griseovelata* Kühner, *I. sindonia* (Fr.) P.Karst. and *I. flocculosa* Sacc. In terms of morphology, it appears most similar to *I. amethystina* Kuyper, which is associated with coniferous and hardwood trees in Europe (Kuyper, 1986). The latter is characterised by its isabella-ochraceous stipe (not white) that is tinged lilac. The widespread temperate-Australian species *I. violaceocaulis* can be uniformly violet when immature, but it differs from *I. bella* in its non-pruinose stipe, spermatic odour and elliptic spores with obtuse apices. *Inocybe caerulata*, from Tasmania and New Zealand, has fruit bodies (pileus and/or stipe) that have bluish areas and somewhat smaller spores that lack a distinct apiculus.

McMullan-Fisher *et al.* (2014) documented a fungus associated with *Lophostemon* in eucalypt forest and heathland in subtropical Australia. They identified it as the northern temperate *I. griseolilacina* J.E.Lange, which is probably a misapplication of the name. While this material shares with *I. bella* a white smooth stipe with lilac colouring and lilac-tinted young lamellae, it differs in its paler pileus, the reported spermatic odour, and smaller spores (8–10 × 5–6 µm).

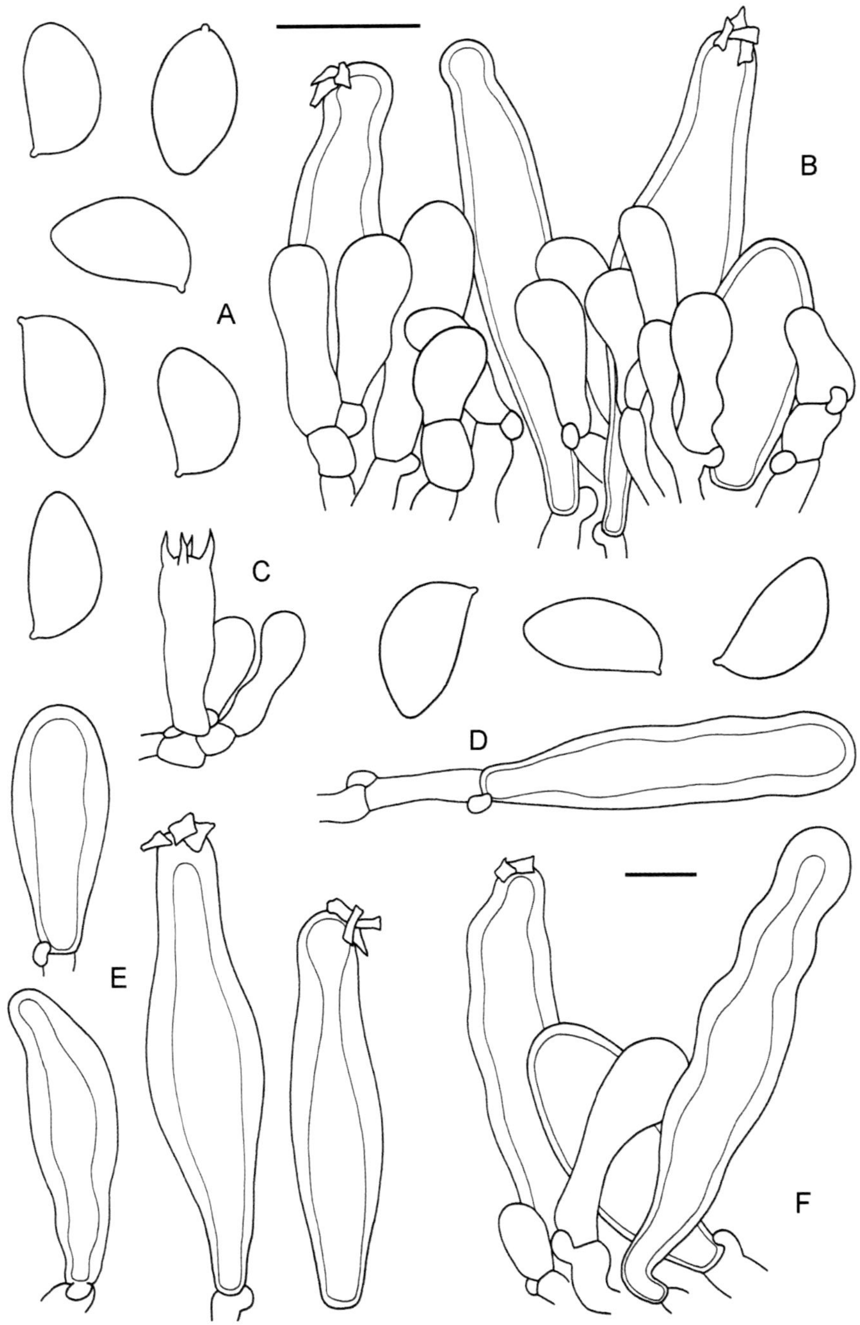

Figure 75. *Inocybe bella.* All holotype (HO 571334). **A,** Basidiospores; **B,** Cheilocystidia and paracystidia; **C,** Basidia; **D,** Caulocystidioid terminal on superficial hyphae from mid-stipe region; **E,** Pleurocystidia; **F,** Caulocystidia and cauloparacystidia. Scale bars = 10 μm (longer bar for spores only).

Plate 137. *Inocybe bella*, holotype (HO 571334). Photograph — G.M.Gates.

66. Inocybe flavoalbida Matheny & Bougher, *sp. nov.*

MycoBank: MB 809326

Type: Mapleton Falls Natl Park, near Mapleton, Qld, 26°37'S, 152°52'E, solitary and scattered on soil in leaf litter under *Eucalyptus* in mixed subtropical rainforest, 29 Feb. 2012, *P.B.Matheny PBM3768*; holo: PERTH 08363196; iso: TENN 067000 (KJ729901-nLSU, KJ729873-ITS, KJ729932-*rpb2*).

Pileus 17–20 mm wide, obtusely conical to campanulate, becoming plane, the centre with a broad low obtuse umbo; margin inrolled, decurved, becoming ±straight with age; surface dry; velipellis absent or at least not noticeable; disc smooth and unbroken, fibrillose towards the margin, rimulose but not rimose; nearest to yellow or very pale brown (2.5Y 7/4 to 10YR 7/4; Honey Yellow with mixtures of Chamois or Cream-Buff); context white, not changing upon exposure, up to 3 mm thick under the disc. *Lamellae* adnexed to ascending-adnate, close, with c. 40 L, with few tiers of lamellulae, light grey (10YR 7/2) to brown (10YR 5/3) with age, narrow, edges indistinctly pallid, not noticeably fimbriate without a hand lens. *Stipe* 20–22 × 3–4 mm, terete, with a 5–6 mm wide marginate bulb; surface dry, pruinose at least to the centre of the stipe, otherwise smooth, white, shiny, developing a dull yellow tinge with age; context solid, uniformly white. *Odour* absent. Pl. 138.

Basidiospores 6.5–8.5 × 4.0–5.0 µm [mean 7.4 × 4.5 µm; Q 1.44–1.78; Q mean 1.64 (*n* = 20/1)], smooth, mostly subamygdaliform, with bluntly pointed apices (mostly elliptic to subamygdaliform with rounded to subconical apices in TR44-06), light yellowish brown; apiculus small but distinct. *Basidia* 23–27 × 7–9 µm, clavate or pedicellate, hyaline, (2–) 4-sterigmate. *Pleurocystidia* 45–75 × 13–17 µm, fusiform to subcylindrical; apices obtuse, crystalliferous, slightly thick-walled (to 1.5 µm) in the type, thicker in the extralimital collections (to 2.5 µm), hyaline to pale yellow. *Cheilocystidia* similar to pleurocystidia, not crowded, mixed with basidia-like paracystidia and some basidia. *Caulocystidia* similar to pleurocystidia, at times shorter, in scattered but often dense clusters with a mixture of shorter cells and cauloparacystidia, descending the entire length of the stipe, less frequent above the stipe base. *Pileipellis* a cutis of cylindrical hyphae, ochre-tawny in mass, mostly 5–10 µm wide, weakly encrusted or smooth, thin-walled. *Clamp connections* present. Fig. 77.

Occurs on soil under *Eucalyptus* in subtropical rainforest in south-eastern Qld and in *Castanopsis* forest in Papua New Guinea. Fruiting Feb. in Qld and Nov.–Jan. in Papua New Guinea.

Extralimital specimens examined: Papua New Guinea: Mt Gahavisuka, landslide area, in mixed *Castanopsis* forest, 19 Nov. 2002, *R.Treu TR104-02* (M 0138262; KJ729874-ITS); Mt Gahavisuka, former coffee garden, in *Castanopsis* forest, 17 Jan. 2006, *R.Treu TR44-06* (M 0138283; JN974960-nLSU, KJ729875-ITS).

Etymology: From the Latin *flavus* (yellow) and *albidus* (whitish), in reference to the combination of the honey yellow to very pale brown pileus and the white stipe.

Diagnostic features: Fruit bodies small to medium. Pileus honey-yellow to very pale brown, smooth to fibrillose, rimulose; stipe white and with a marginate basal bulb, pruinose at least down to the middle of the stipe; spores small, 6.5–8.5 × 4.0–5.0 µm, subamygdaliform to elliptic.

Inocybe flavoalbida has been recorded from Papua New Guinea and eastern subtropical Australia. The two collections sequenced from Papua New Guinea demonstrate little genetic divergence from the Australian collection (Fig. 76). However, we note several morphological differences between collections from the two areas: those from Papua New Guinea have fusiform-ventricose hymenial cystidia with slightly thicker walls and appear somewhat darker (more yellowish brown) in their dried state, including the stipe. The Papua New Guinea samples are recorded with *Castanopsis* and the Queensland collection with *Eucalyptus*. Further collections with careful habitat notes should determine if associated vegetation/habitat is a significant factor in this species.

Inocybe flavoalbida is most closely related to the northern temperate *I. amblyspora* Kühner and *I. praecox* Kropp, Matheny & Nanagy. (Fig. 76), with which it shares small smooth spores, a bulbous or marginate stipe base, and caulocystidia on the lower part of the stipe. *Inocybe flavoalbida* was not documented in Horak (1980), nor in the east Asian flora of Kobayashi (2002), where it keys out to sect. *Splendentes* Singer.

Plate 138. *Inocybe flavoalbida*, holotype (PERTH 08363196). Photograph — P.B.Matheny.

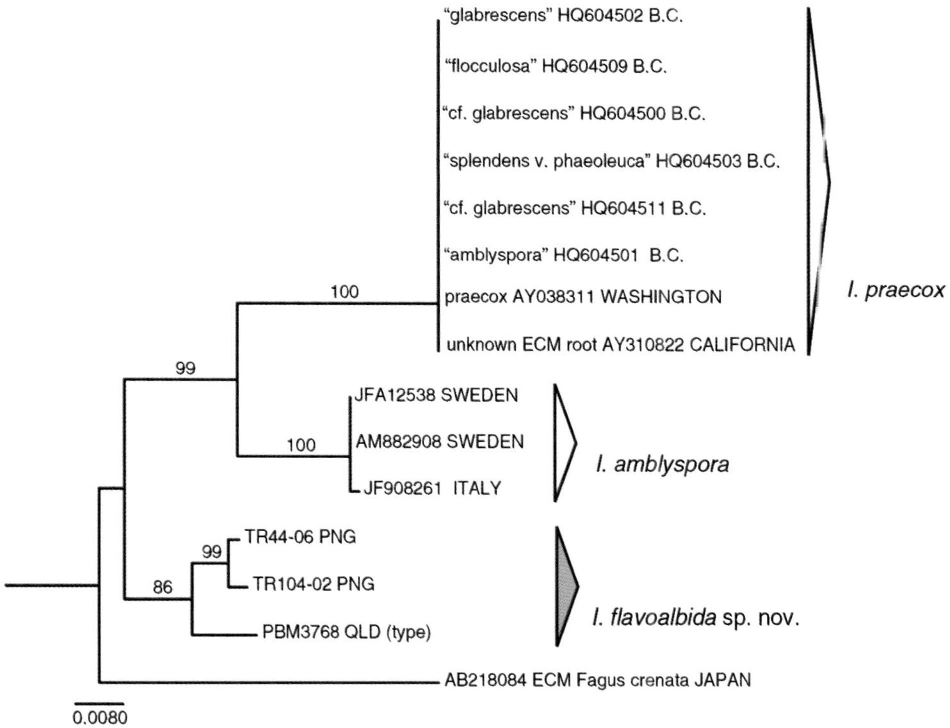

Figure 76. Maximum Likelihood phylogram of *Inocybe flavoalbida* and its north temperate allies. Bootstrap values are shown above branches. The tree was inferred from combined ITS and nLSU gene regions and includes 2176 total sites. The Australasian distribution of *I. flavoalbida* is indicated by the grey-filled triangle. *Inocybe amblyspora* is a European species, and *I. praecox* is known only from the west coast of North America.

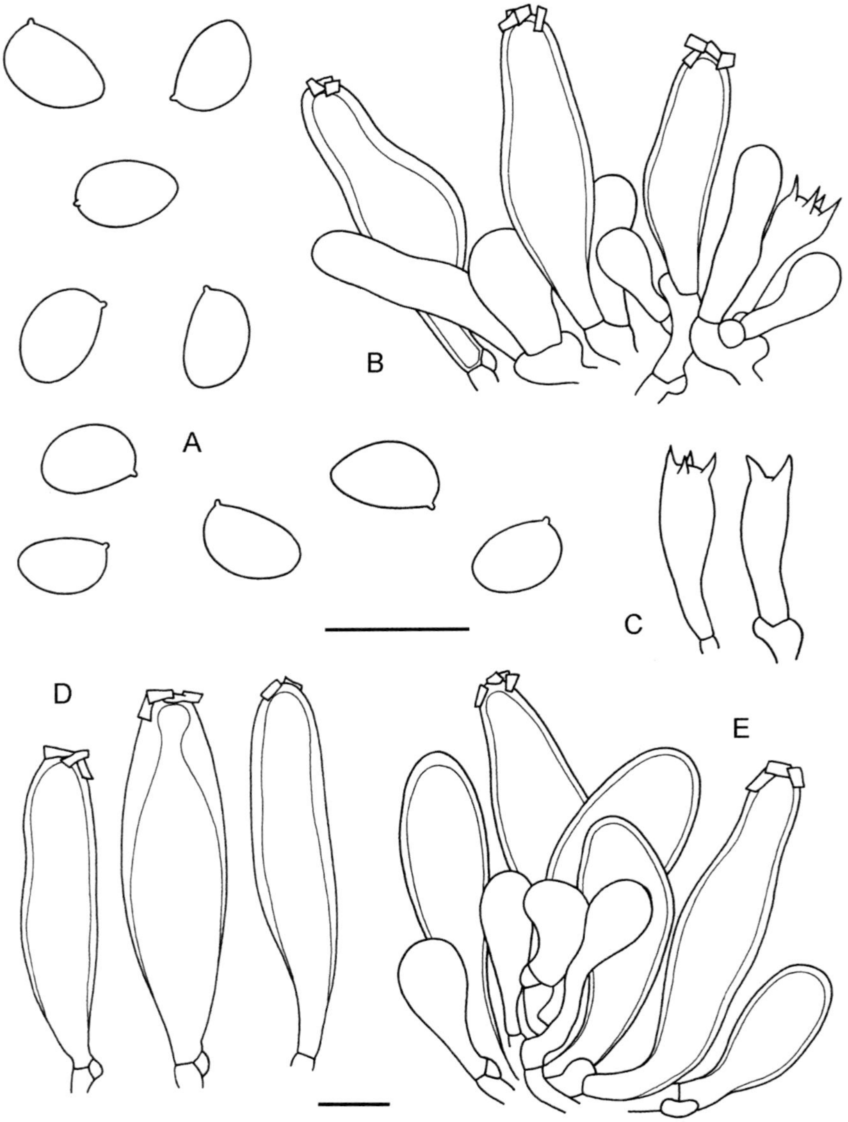

Figure 77. *Inocybe flavoalbida.* All holotype (PERTH 08363196). **A,** Basidiospores; **B,** Cheilocystidia and paracystidia; **C,** Basidia; **D,** Pleurocystidia; **E,** Caulocystidia and cauloparacystidia. Scale bars = 10 μm (longer bar for spores only).

67. Inocybe luteo-olivacea Matheny, Bougher & Halling, *sp. nov.*

MycoBank: MB 809425

Type: Mobo Creek Crater, Danbulla Forest Drive, Danbulla S.F., L. Tinaroo, Atherton Tableland, Qld, 17°08'23"S, 145°37'47"E, on soil under *Acacia* in rainforest, 12 Mar. 2007, *R.E.Halling & N.Fechner REH8914*; holo: BRI 794189, iso: NY 00817398 (JN974991-nLSU, KP308771-ITS, KJ811602-*rpb2*).

Pileus 20–40 mm wide, broadly conical at first, becoming convex to irregularly convex with a low broad umbo; velipellis absent; surface dry, matted on the disc, radially matted-fibrillose and rimulose, especially towards the margin; pale brown to yellowish brown (5C5-5D5) on the disc, initially more yellow towards the margin, that area becoming pale orange-brown, bright yellow showing between the fibrils at the margin; context very pale yellow to white with a very pale yellow tint. *Lamellae* adnexed, close, pale yellow at first, becoming greenish yellow to yellowish green or developing olive tones with age, edges even and not noticeably fimbriate, not changing colour where bruised. *Stipe* 25–40 × 8–10 mm, curved, equal, slightly wider at the base but not bulbous; surface dry, heavily fibrillose to matted-fibrillose; (light) greenish yellow to yellowish green, dull to subshiny; context solid, pale greenish yellow to pale dirty yellow. *Odour* mild. *Taste* mild to somewhat unpleasant. Pl. 139, 140.

Basidiospores 7.5–9.5 (–10.0) × 4.5–6.0 µm [mean 8.4 × 5.2 µm; Q 1.45–1.82; Q mean 1.63 (n = 41/2)], smooth, amygdaliform with pointed apices to citriniform with mucronate apices (mucro sometimes asymmetrical), yellowish brown; apiculus small but distinct. *Basidia* 30–34 × 8–10 µm, clavate, hyaline, 4-sterigmate. *Pleurocystidia* 41–61 × 14–18 µm, utriform, ventricose, broadly fusiform or saccate with obtuse to rounded apices, these crystalliferous or bare, bases tapered or truncate; thick-walled (walls 2–3 µm), yellow to pale yellow. *Cheilocystidia* similar to pleurocystidia but occasionally cylindrical, variable in abundance along the lamella edge, in parts infrequent among a crowded palisade of narrowly clavate paracystidia, elsewhere crowded, with fewer less organised paracystidia. *Caulocystidia* absent. *Pileipellis* a cutis of repent hyphae, these weakly encrusted, thin-walled, cylindrical, 5–10 µm wide, brownish yellow or fulvous in mass. *Clamp connections* present. Fig. 78 (see also Pl. 6J).

Occurs on soil under *Acacia* in rainforest in north-eastern Qld. Fruiting Feb. and Mar.

Additional specimen examined: Qld: Mount Lewis road, W of Julatten, alt. 745 m, 16°35'17.8"S, 145°18'28.3"E, gregarious on soil under *Acacia* in rainforest, 16 Feb. 2012, *R.E.Halling, N.Fechner & K.Bransgrove REH9627* (NY 01491109; KP170946-nLSU, KP308772-ITS, KJ811603-*rpb2*).

Etymology: From the Latin *luteus* (deep yellow) and *olivaceus* (olive-green), in reference to the colours of the fruit bodies.

Diagnostic features: Fruit bodies medium-sized to robust. Pileus yellowish brown, yellow towards the margin or between the spreading fibrils; lamellae pale yellow to olivaceous; stipe not bulbous, surface densely fibrillose. Spores smooth, amygdaliform with pointed to mucronate apices; caulocystidia absent. On soil under *Acacia* in tropical rainforest.

Inocybe luteo-olivacea is a species of tropical Australian rainforest with a unique suite of morphological characters. The pileus is yellowish brown to orange-brown with underlying yellow where the fibrils spread apart. The context of the pileus is tinged yellow, and the lamellae are pale yellow to yellowish green or with clear olive tones with age. The stipe is light greenish yellow to yellowish green and heavily fibrillose.

Distinctly citriniform spores are more prevalent in the holotype than in NY 01491109. The abundance of cheilocystidia varies along the lamellae of both collections, but they are mainly less crowded and seated among a crowded palisade of paracystidia in NY 01491109 (compare Fig. 78A and Fig. 78D).

Inocybe luteo-olivacea is most closely related to specimens from Thailand identified as *I. sphaerospora* Kobayasi (see Horak *et al.* 2015) based on inspection of currently unpublished family level nLSU and *rpb2* gene trees. Their placement within the *Inocybe* clade is not resolved, and the species merit a section of their own, likely with the addition of *I. hinoana* Y.Yukawa & K.Katumoto (which has not been sequenced). Kobayasi (2002) established section *Hinoanae* to accommodate *I. hinoana* and *I. sphaerospora*, and it will require some emendation in order to incorporate *I. luteo-olivacea*. All three species have yellowish colours, smooth spores, and lack caulocystidia, or these restricted to the extreme apex of the stipe. *Inocybe sphaerospora*, originally described from Japan (Kobayasi, 1952), is distinguished by its strong yellow fruit bodies and subglobose spores. *Inocybe hinoana*, another Japanese species, differs from *I. sphaerospora* in its longer and broadly elliptic to subglobose spores.

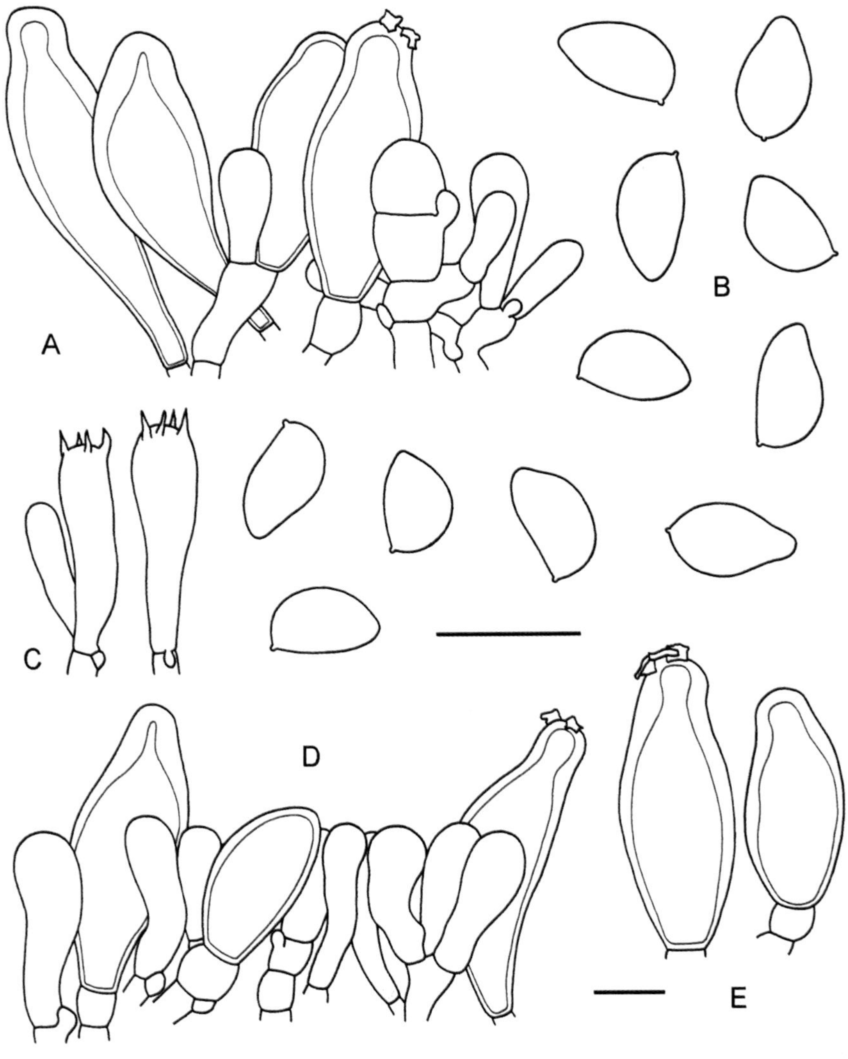

Figure 78. *Inocybe luteo-olivacea.* All holotype (BRI 794189) except D (NY 01491109). **A,** Crowded cheilocystidia and disorganised paracystidia; **B,** Basidiospores; **C,** Basidia; **D,** Less crowded cheilocystidia and palisade of paracystidia; **E,** Pleurocystidia. Scale bars = 10 μm (longer bar for spores only).

Plate 139. *Inocybe luteo-olivacea*, holotype (BRI 794189). Photograph — R.E.Halling.

Plate 140. *Inocybe luteo-olivacea*. NY 01491109. Photograph — R.E.Halling.

68. Inocybe murrayana Cleland, *Trans. Roy. Soc. S. Australia* 57: 192 (1933)

Type: Kinchina, S.A., 1 Aug. 1925, *J.B.Cleland*, lecto: AD 5441, *fide* C.A.Grgurinovic, *Larger Fungi S. Australia* 183 (1997).

Pileus to 16 mm wide, thereafter expanding, ±umbonate, surface finely fibrillose to fibrillose-squamose, silky-sleek, cinnamon-brown to russet-brown (Buckthorn Brown to Dresden Brown). *Lamellae* adnate to adnexed, nearly crowded, the edges subserrate, tobacco-brown to avellaneous. *Stipe* 25 mm long, slender, without a bulb, finely striate to fibrillose or mealy, tobacco-brown to pale brown; context solid.

Basidiospores 9.0–12.0 × 4.5–6.0 (–6.5) µm [mean 10.4 × 5.3 µm; Q 1.73–2.20; Q mean 1.97 (*n* = 20/1)], smooth, amygdaliform to oblong-amygdaliform, with conical apices, slightly thick-walled, with a small but distinct apiculus, yellowish brown. *Basidia* c. 28 × 9 µm, 4-sterigmate. *Pleurocystidia* 65–77 × 11–18 µm, narrowly fusiform, fusiform or lageniform, with elongate necks and a short tapered pedicel; apices obtuse, often crystalliferous, thick-walled (2–3 µm), hyaline. *Cheilocystidia* similar to pleurocystidia, at times yellowish, scattered among hyaline thin-walled clavate paracystidia. *Caulocystidia* similar to pleurocystidia but often lageniform, with lanceolate necks and subacute apices, mixed with cauloparacystidia, together forming scattered clusters at the apex and below the mid-region of the stipe; base with metuloids (caulocystidioid hairs) but cauloparacystidia not observed. *Pileipellis* a cutis of russet-brown to reddish brown hyphae in mass, these cylindrical, 6–15 µm wide, thin-walled or slightly thickened, encrusted. *Clamp connections* present. Fig. 79.

Known only from Kinchina, S.A. and not collected since the 1920s; gregarious on the ground. Fruiting June–Aug.

Additional specimens examined: S.A.: Kinchina, 8 June 1925, *J.B.Cleland s.n.* (AD 5440); Monarto South, Kinchina, 9 July 1922, *J.B.Cleland s.n.* (AD 5442, doubtfully *murrayana*).

Diagnostic features: Fruit bodies small. Pileus fibrillose to fibrillose-scaly, cinnamon-brown to russet-brown; stipe brown to pale brown. Spores amygdaliform to oblong-amygdaliform, 9.0-12.0 × 4.5–6.0 µm; metuloids (as caulocystidioid hairs) present on the lower part of the stipe.

The description of gross morphology is a translation of the Latin protologue supplemented with our own observations and measurements of the lectotype (see also Matheny & Bougher, 2010). We have not seen any recent collections of *I. murrayana*, and we were unable to obtain DNA sequences from the lectotype. An image and description incorrectly labelled as *I. murrayana* in Bougher (2009) is of an unidentified species collected in Western Australia (PERTH 08073775).

Based on anatomical similarities, we suggest that *I. murrayana* might be included in an unresolved complex along with *I. persicinipes* and several other poorly clarified and undetermined specimens (Fig. 32). In terms of morphology, *I. persicinipes* seems most similar to *I. murrayana*, differing in its ochraceous to salmon-coloured stipe and smaller spores. We cannot confidently assign the name *I. murrayana* to any of the unclarified species in the *I. persicinipes* group due to their smaller spore sizes.

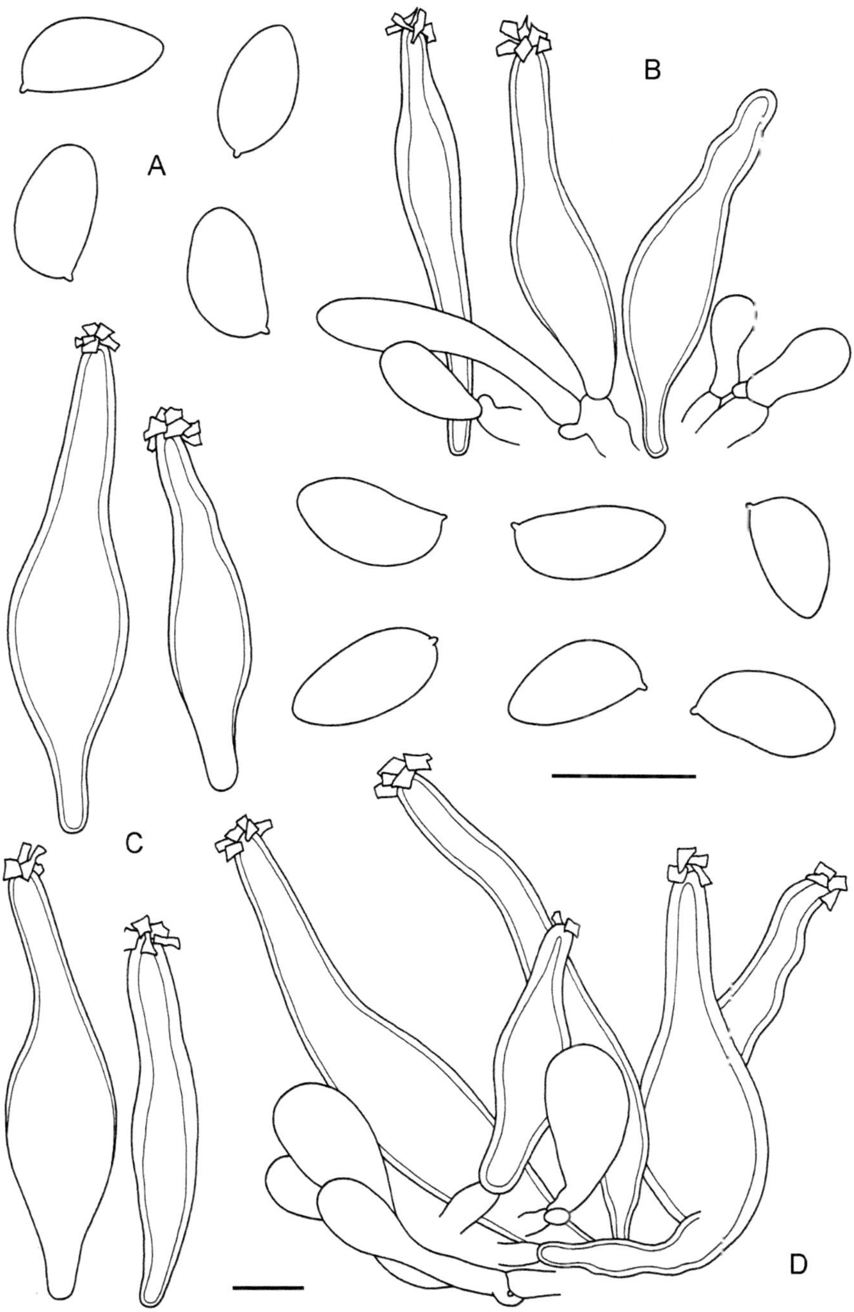

Figure 79. *Inocybe murrayana.* All lectotype (AD 5441). **A,** Basidiospores; **B,** Cheilocystidia and paracystidia; **C,** Pleurocystidia; **D,** Caulocystidia and cauloparacystidia (from upper stipe).
Scale bars = 10 µm (longer bar for spores only).

69. Inocybe paucicolor Matheny & Bougher, *sp. nov.*

MycoBank: MB 809429

Type: firebreak S of Two Peoples Bay Rd, Two Peoples Bay Nature Reserve, Lake Gardner area, Moate's area, W.A., 34°59'04"S, 118°10'59"E, on sandy soil under *Allocasuarina fraseriana* in *Melaleuca* woodland, 11 June 1991, *K.Syme KS71/91*; holo: PERTH 03978494.

Pileus 6–18 mm wide, conical to convex, becoming plane with age; umbo obtuse and persistent; margin decurved; velipellis absent; surface dry, radiating tomentose-fibrillose from the centre; dirty pale ochraceous; context pallid, unchanging where bruised. *Lamellae* adnexed, moderately close, with several tiers of lamellulae, pale buff to ochraceous, becoming somewhat darker with age, edges pallid-fimbriate. *Stipe* 15–24 × 1.5–3.0 mm, terete, even but with a swollen base, not marginate; partial veil present, cortinate on young specimens, fugacious; surface pruinose-fibrillose, pale ochre; context pallid, tinged ochraceous, hollow. *Odour* not distinctive.

Basidiospores 7.0–9.5 (–10) × 4.0–5.0 µm [mean 8.2 × 4.5 µm; Q 1.56–2.25; Q mean 1.84 (*n* = 28/2)], smooth, amygdaliform, cylindrical, at times allantoid with an adaxial depression (minimally angular); apices mostly bluntly pointed or conical, less often obtuse, noticeably pale (light buff to light ochraceous buff); apiculus small. *Basidia* 22–28 × 7–9 µm, clavate, usually hyaline, 4-sterigmate. *Pleurocystidia* 63–83 (–100) × 12–14 µm, fusiform to narrowly fusiform or lageniform with a long neck, apices obtuse, often bearing crystals, thick-walled (walls 1–4 µm), hyaline or pale to bright yellow, abundant. *Cheilocystidia* similar to pleurocystidia, often shorter, mixed with clavate paracystidia and basidia. *Caulocystidia* densely arranged, to 100 µm long, descending the entire length of the stipe, mixed with cells of intermediate size and shape, most often lageniform with lanceolate necks, occasionally flexuous, other cells clavate or subcylindrical, hyaline, thick-walled; cauloparacystidia observed at least at the stipe apex or not at all. *Pileipellis* a cutis of hyaline, mostly thin-walled and non-encrusted hyphae 5–12 µm wide, with some emerging slightly tapering end-cells. *Clamp connections* present. Fig. 80 (see also Pl. 9A, B).

Occurs on soil under *Allocasuarina*, *Melaleuca* or *Eucalyptus marginata*, in *Melaleuca* woodland or Jarrah forest in south-western W.A. between Manjimup and Two Peoples Bay. Fruiting June.

Additional specimen examined: W.A.: Manjimup, Perup region, 34°19'S, 116°25'E, on soil under *Eucalyptus marginata*, 22 June 1985, *N.Malajczuk E0482* (PERTH 07587619, TENN 069517).

Etymology: From the Latin *paucicolor*, weakly pigmented, in reference to the very pale basidiospores.

Diagnostic features: Fruit bodies small; pileus tomentose-fibrillose, pale ochraceous. Spores narrow, pale buff, smooth to minimally angular.

Inocybe paucicolor is characterised by the relatively small and noticeably pale basidiospores. Whereas the basidiospores of most species of *Inocybe* appear yellowish brown in mounts with water, ammonia or KOH, those of *I. paucicolor* are light buff to light ochraceous buff. The cylindrical, narrow, or at times minimally angular spore outline and pale spore colour resemble *I. sambucina* (Fr .: Fr.) Quél. Unfortunately, we were unable to produce DNA sequence data from the type for phylogenetic analysis of this relationship. Although caulocystidioid cells can be found throughout the entire length of the stipe, a partial veil is present on younger specimens.

Inocybe roseola, known only from Tasmania, also has pale spores and subhyaline amygdaliform to broadly amygdaliform spores. However, significant differences from *I. paucicolor* include pinkish tinges on the pileus, pinkish gills, and a white stipe.

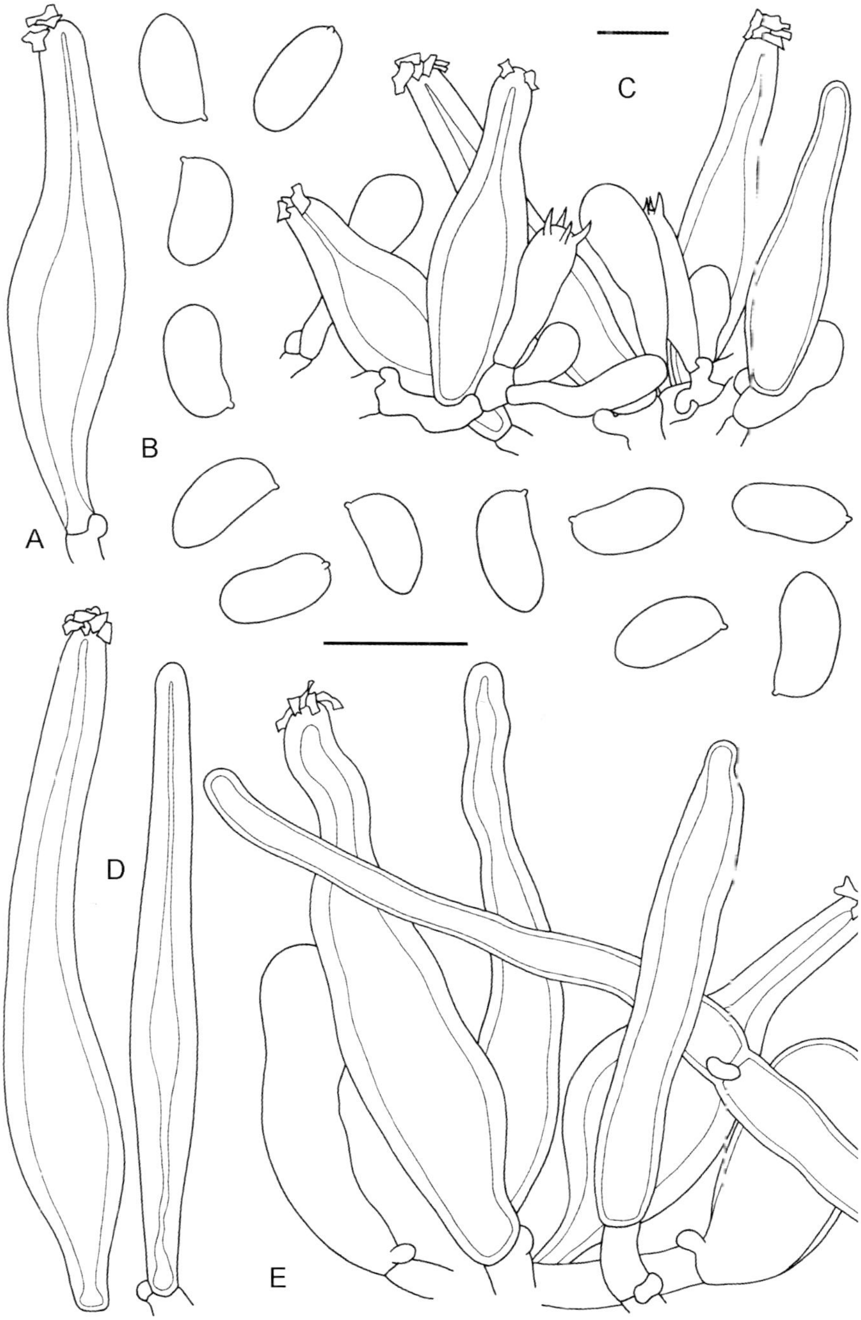

Figure 80. *Inocybe paucicolor.* All holotype (PERTH 03978494), except D (PERTH 07587619). **A,** Pleurocystidium; **B,** Basidiospores; **C,** Cheilocystidia, paracystidia and basidia; **D,** Pleurocystidia; **E,** Caulocystidia and associated cells. Scale bars = 10 µm (longer bar for spores only).

70. **Inocybe rufescens** Matheny & Bougher, *sp. nov.*

MycoBank: MB 809451

Type: Sinker Reef Track, Two Peoples Bay Nature Reserve, W.A., 34°56'54.2"S, 118°09'58.8"E, in litter in heath-scrub under *Agonis flexuosa*, *Spyridium globulosum*, *Acacia* and *Melaleuca*, 19 July 2011, *N.L.Bougher NLB834*; holo: PERTH 08318468; iso: TENN 066758 (KP170991-nLSU, KP308818-ITS, KM406231-*rpb2*).

Pileus 10–35 mm wide, parabolic, with a flattened top when young, becoming campanulate with a low broad umbo, or pulvinate or broadly or irregularly convex; margin decurved; surface uniformly matted-felty when young, later remaining so at the centre, surrounded by fine appressed radial fibrils; squamules absent; uniformly reddish brown (near 2.5YR 3/4 to 5YR 3/3 or 5YR 4/4); margin concolorous with the centre; context white and pearly watery. *Lamellae* broadly adnexed, close, with 2 or 3 tiers of lamellulae, up to 4 mm deep, very pale cream (almost white) or initially light grey, becoming pale brown (10YR 5/3 to 7/3), edges finely fimbriate. *Stipe* 20–30 × 6–8 mm, cylindrical and without a basal bulb, even; surface pruinose at the extreme apex, scattered-furfuraceous with fine white streaked fibrils down to the base, drab whitish with some reddish stains and with a dull pinkish to pinkish cinnamon ground colour; context whitish but staining reddish after being cut, white in the stipe base, red staining persisting for at least 15 minutes. *Odour* strongly spermatic. Pl. 141, 142 (see also Pl. 3C).

Basidiospores 8.5–12.0 × 5.0–6.5 µm [mean 10.3 × 5.9 µm; Q 1.50–2.18; Q mean 1.75 (*n* = 50/2)], smooth, amygdaliform to elliptic; apices usually bluntly pointed, yellowish brown; apiculus not distinct. *Basidia* 34–45 × 10–12 µm, clavate, hyaline, 4-sterigmate, rarely 2-sterigmate. *Pleurocystidia* 67–91 × 13–19 µm, narrowly fusiform, cylindrical or skittle-shaped, usually with a short slender basal pedicel; apices obtuse, occasionally subcapitate, some bearing crystals or with few small granules; thin-walled or less often slightly thickened and up to 1.0-1.5 µm thick apically. *Cheilocystidia* similar to pleurocystidia, clustered with basidiole-like paracystidia and fertile basidia on the gill edge. *Caulocystidia* not observed; stipitipellis with an interwoven shallow vesture of cylindrical hyaline hyphae with undifferentiated end cells; end cells at the stipe base twisted or with gnarled apices. *Pileipellis* a cutis of repent to interwoven cylindrical encrusted hyphae mostly 7–17 µm wide. *Clamp connections* present. Fig. 81.

Known only from south-western W.A.; forming caespitose clusters on soil in low heath-scrub near species of *Acacia*, *Agonis*, *Eucalyptus*, *Melaleuca* and *Spyridium* (type collection) and tall eucalypt forest. Fruiting June, July.

Additional specimen examined: W.A.: Nanga Bridge, Nanga Rd, c. 8 km S of Dwellingup, in caespitose cluster on soil in Jarrah forest, under *Eucalyptus marginata*, with *E. rudis* nearby, 17 June 2012, *N.L.Bougher NLB1003* (PERTH 08383324; KP170992-nLSU, KP308819-ITS, KM406232-*rpb2*).

Etymology: From the Latin *rufescens* (becoming red), in reference to the reddish staining context.

Diagnostic features: Fruit bodies robust; context staining reddish where bruised; odour spermatic. Spores smooth; hymenial cystidia primarily thin-walled; caulocystidia absent.

Inocybe rufescens is unusual among Australian species of *Inocybe*. It is robust, the odour is strongly spermatic, the context in the stipe stains reddish in areas (Pl. 3C), the pleurocystidia are primarily thin-walled, and no caulocystidioid cells were observed on the stipe.

Phylogenetically, the placement of *I. rufescens* based on nLSU DNA sequence data is not clear, but it is one of few smooth-spored of *Inocybe s. str.* recovered outside the STAC group. Data from the *rpb2* gene indicate a robust relationship with two species from the Americas (data not shown), neither of which exhibits reddening context. According to Kuyper (1986) reddening context is a significant taxonomic trait, and it has evolved on several occasions in distantly related groups of Inocybaceae in the *Inosperma* clade, the *Pseudosperma* clade, and in at least three independent groups in the *Inocybe* clade (the *I. corydalina* group, the *I. geophylla* group and *I. godeyi*). This is the first report of reddening context in native Australian species bearing smooth spores and pleurocystidia, although *I. leucotaenia* (species 35) has context that can slowly become pink after cutting.

Some morphological and genetic variation was observed among the two collections of *I. rufescens*. Thus, the spores of the holotype are on average more than 10 μm long, while PERTH 08383324) has slightly shorter spores [7.4–10.3 × 5.0–5.9 μm; mean 9.2 × 5.5 μm; Q 1.42–1.88; Q mean 1.69 (n = 30), and occurs in Jarrah forest rather than scrub heath. The spore size range overlaps between the two collections, and both have variously shaped hymenial cystidia, and cylindrical cystidia are more common in PERTH 08383324. ITS sequence data indicate the two collections are genetically divergent but c. 92% similar to each other. More collections are needed to accurately assess their taxonomic boundaries.

Plate 141. *Inocybe rufescens*. PERTH 08383324. Photograph — N.L.Bougher.

Plate 142. *Inocybe rufescens*, holotype (PERTH 08318468). Photograph — N.L.Bougher.

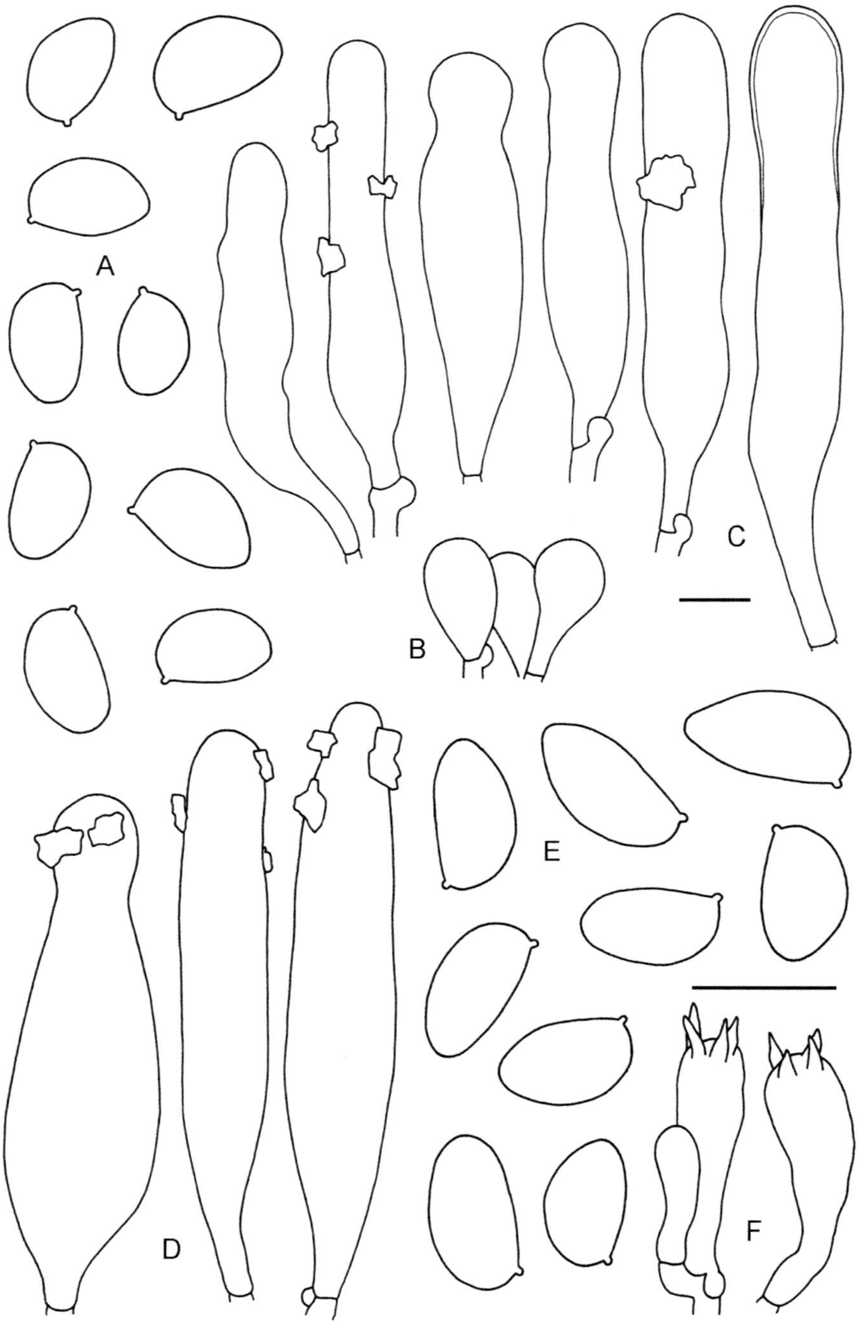

Figure 81. *Inocybe rufescens.* All holotype (PERTH 08318468), except A and C (PERTH 08383324). **A,** Basidiospores; **B,** Paracystidia; **C,** Cheilocystidia; **D,** Pleurocystidia; **E,** Basidiospores; **F,** Basidia. Scale bars = 10 µm (longer bar for spores only).

71. Inocybe spadicea Matheny & Bougher, *sp. nov.*

MycoBank: MB 809461

Type: Shenton Bushland, Shenton Park, Perth, W.A., in disturbed urban Jarrah/Marri woodland on sandy soil under *Eucalyptus marginata*, *Banksia* and *Jacksonia* (Fabaceae), 8 July 2001, *P.B.Matheny & R.Hart PBM2155/E6976*; holo: PERTH 07678444; iso: TENN 066789.

Pileus 15–30 mm wide, convex to plano-convex, becoming plane with age; umbo absent; margin decurved, becoming slightly arched; surface dry; velipellis absent, smooth at the centre, fibrillose to coarsely fibrillose or tomentose-fibrillose towards the margin, splitting or frayed occasionally but not rimose; reddish brown or bright brown, shading brown or yellowish brown towards the margin (5YR 4/4 to 4/6, 7.5YR 4/4; Argus Brown or Amber Brown mixed with Brussels Brown to Sudan Brown or Russet mixed with shades of Cinnamon-Brown and Buckthorn Brown or Verona Brown mixed with Prout's Brown); context pale yellowish to whitish or pallid, not changing colour where bruised. *Lamellae* adnexed, close, 36–46 L, with several tiers of lamellulae; pale yellow to light yellowish brown (2.5Y 7/4 to 6/4; Chamois), becoming light olive-brown (2.5Y 5/6; Isabella Color); ventricose, broad, up to 4 mm deep, edges pallid-fimbriate. *Stipe* 20–35 × 3–6 mm, even or swollen at the base but not bulbous, terete; partial veil not observed, surface furfuraceous-pruinose over upper half, indistinctly fibrillose below, more densely fibrillose at the extreme base; dull yellowish (Colonial Buff) to off-white; context whitish, solid becoming hollow. *Odour* not remarkable or faintly acidulous. Pl. 143.

Basidiospores 10.0–14.0 (–15.0) × 5.0–6.5 µm [mean 11.7 × 5.8 µm; Q 1.57–2.45 (–2.55); Q mean 2.02 (*n* = 43/2)], smooth, oblong-amygdaliform to amygdaliform, occasionally cylindrical or fusiform ("laceroid"), with conical to bluntly pointed apices, often adaxially depressed, yellowish brown; apiculus small but distinct. *Basidia* 33–55 × 11–14 µm, clavate, hyaline, becoming ochraceous (especially in lower part, then throughout), 4-sterigmate. *Pleurocystidia* 64–102 × 15–18 µm, fusiform to subcylindrical, the apices obtuse, often with crystals, at times with a long slender basal pedicel, the walls 1–2 µm thick, hyaline or becoming yellow. *Cheilocystidia* similar to pleurocystidia, mixed with versiform paracystidia (clavate to cylindrical) and some basidia, variably abundant. *Caulocystidia* 30–73 × 12–35 µm, descending below the stipe centre, dominated by clavate to sphaeropedunculate cells mixed with fusiform or irregularly cylindrical cells; mostly thin-walled but occasionally thick-walled, hyaline to ochraceous. *Pileipellis* a cutis of cylindrical bright brown encrusted hyphae 4–15 µm wide. *Clamp connections* present. Pl. 82 (see also Pl. 6H, 11F).

Occurs on soil under eucalypts between Perth and Cervantes in south-western W.A. Fruiting July and Aug.

Additional specimen examined: W.A.: Tuart Reserve, 11 km NE of Cervantes, 30°25'03"S, 115°08'20"E, solitary in sandy soil under *Eucalyptus gomphocephala* and *Agonis* sp., 27 Aug. 2001, *P.B.Matheny PBM2203* (*E7051*) (PERTH 07676794, TENN 066791; EU600865-nLSU, KP636866-ITS, EU600864-*rpb1*).

Etymology: From the Latin *spadicea* (date-coloured or deep reddish brown), in reference to the pileus colour.

Diagnostic features: Fruit bodies medium. Pileus fibrillose to coarsely fibrillose, reddish brown or bright brown, shading brown or yellowish brown towards the margin; lamellae pale yellow to light yellowish brown, becoming olivaceous; stipe furfuraceous-pruinose at least over the upper half, dull yellowish to off-white; odour not remarkable. Spores oblong-amygdaliform, large, 10.0–14.0 µm long; basidia large; caulocystidia descending below the mid-region of the stipe.

Inocybe spadicea is a distinctive new species from Western Australia with brightly coloured fruit bodies, elongate spores and large basidia. *Inocybe aureorimosa* also has brightly coloured fruit bodies, but it differs most markedly in its shorter basidiospores. Phylogenetically, the two taxa are quite distantly related, and while *I. aureorimosa* is supported in the STAC lineage (Fig. 8), *I. spadicea* is at best the poorly supported sister lineage to STAC (Matheny *et al.*, 2009); the species clustered among outgroups in our phylogenetic analysis of the STAC group (results not shown).

Inocybe rufobadia is another species with brightly coloured fruit bodies (within the greater *I. australiensis* group, see Fig. 8) that might be confused with *I. spadicea*. However, it has smaller, often cylindrical spores (mean 8.9 × 4.6 µm) and a bright rusty brown to vivid reddish brown pileus.

Two collections somewhat similar to *I. spadicea* but with broader spores (9.0–12.0 × 5.5–7.0 µm; Q mean 1.65) are compared here with some reservation due to lack of adequate notes, photos or DNA sequences and possible confusion with *I. aureorimosa*: PERTH 00944165, from near Kalamunda, Western Australia, under *Allocasuarina*; and PERTH 03978532, from Two Peoples Bay, Little Bay, Western Australia in coastal heath under *Melaleuca*.

Plate 143. *Inocybe spadicea*, holotype (PERTH 07678444). Photograph — P.B.Matheny.

Figure 82. *Inocybe spadicea.* All holotype (PERTH 07678444). **A,** Basidiospores; **B,** Cheilocystidia and paracystidia; **C,** Basidia; **D,** Pleurocystidia **E,** Caulocystidia and cauloparacystidia. Scale bars = 10 µm (longer bar for spores only).

72. Inocybe violaceocaulis Matheny & Bougher, *Australas. Mycol.* 24: 7 (2005)

Type: Floreat Park, Perth, W.A., on lawn under *Eucalyptus gomphocephala*, 6 Aug. 2001, *P.B.Matheny PBM2164 (E7013)*; holo: PERTH 07671857 (AY732208-LSU).

Pileus 15–35 mm wide, initially conical, expanding with age to obtusely conical or campanulate to planoconvex, with a large but low obtuse umbo; margin decurved, becoming undulate; surface dry but somewhat lubricous when moist, smooth at the centre but it can become diffract-scaly with age, silky-fibrillose to fibrillose towards the margin, with fibrils diverging around the centre with age; margin occasionally split; ground colour brown or umbrinous (Cinnamon Brown, Dresden Brown, Snuff Brown), generally a lighter shade towards the margin (10YR 4/3, 4/4 to 5/3, 5/4; Tawny-Olive), with a violet superficial layer at least when young, this wearing away with age to reveal the umbrinous ground colour; context whitish, unchanging where bruised, to 5 mm thick under the disc. *Lamellae* narrowly adnate, seceding, close, up to 40 L reaching the stipe, with a few tiers of lamellulae, light grey (10YR 6/1) or with a lilac or violet tinge when young, becoming yellowish brown (10YR 5/4), the edges white-fimbriate, ventricose, up to 5 mm deep. *Stipe* 20–40 × 4–6 mm at the apex, enlarged to swollen at the base, at times rounded-bulbous, there up to 9 mm wide; basal mycelium white but sparse and not conspicuous; cortina initially thick but fugacious, pale violet; surface fibrillose, never agglutinated, not pruinose or pruinose only at the extreme apex, violet, lilac or greyish lilac (15B4 to 16B4; Light Wistaria Violet to Plumbago Grey) throughout, except for the pallid to cream-tinged base; dried specimens retaining their violaceous tinges; cortex coloured like the surface, with central and lower context pallid; stipe (and pileus) surface orange with 15% KOH. *Odour* spermatic. Pl. 144–146.

Basidiospores 7.5–9.5 (–10.0) × 5.0–5.5 µm [mean 8.6 × 5.2 µm; Q 1.36–1.90; Q mean 1.66 (*n* = 42/3)], smooth, mostly elliptic but occasionally subreniform to subamygdaliform, sometimes with a suprahilar depression; apices rounded, yellowish brown in KOH with thickened walls; apiculus distinct, dark brown (6F8) in mass. *Basidia* 25–32 × 7–8 µm, 4-sterigmate, clavate, hyaline. *Pleurocystidia* 52–63 × 11–15 µm, fusiform, usually without a distinct neck, thin-walled to only slightly thick-walled, the walls 0.5–1.5 µm thick, hyaline; apices obtuse, sparsely crystalliferous, with a basal pedicel. *Cheilocystidia* similar to pleurocystidia, infrequently vesiculose-ovoid, mostly thin-walled; paracystidia clavate, thin-walled, hyaline. *Caulocystidia* not observed; caulocystidioid hairs infrequent, restricted to the extreme apex, 55–78 × 13–16 µm, fusiform to subcylindrical, thin-walled, hyaline, the apices mostly bare; cauloparacystidia not observed; superficial hyphae interwoven, hyaline, occasionally faintly violet in mass in KOH, but pigments cytoplasmic and not encrusted, the hyphae cylindrical, 4–11 µm diam. *Pileipellis* a cutis of smooth cylindrical hyphae 5–12 µm wide, parallel to somewhat interwoven, no conspicuous velipellis hyphae on mature pilei, light or faint cinnamon in mass. *Clamp connections* present. Fig. 83.

Occurs in W.A., S.A., Qld, Vic. and Tas.; solitary or in clusters on grassy lawns in parks and landscaped areas under various woody myrtaceous plants, including *Eucalyptus gomphocephala*, *E. marginata* and *Corymbia calophylla*. Also under myrtaceous plants in natural ecosystems such as Karri forest under *E. diversicolor*, *Allocasuarina decussata* and *Agonis flexuosa*; in wet sclerophyll forest under *E. regnans*, in moist old sand dunes under *Agonis flexuosa* or *Allocasuarina* and *Eucalyptus*, and in forests or woodlands of *E. obliqua*. In W.A. it occurs under introduced *E. maculata* and *Lophostemon confertus*, and in *E. globulus* plantations. Fruiting May–Aug. and October; also Dec.–June in Tas.

Selected specimens examined: W.A.: Cannington, in sandy soil, *s. dat.*, *R.N.Hilton s.n.* (PERTH 00755702); Kings Park, Perth, in bushland under *Corymbia calophylla* and *Eucalyptus marginata*, 26 June 1985, *E.Horak ZT2760* (PERTH 07712812, as *Inocybe* sp.); Floreat Park, in bushland under eucalypts, 21 June 1990, *N.L.Bougher E0747* (PERTH 07598718); Carpenters block off 7 Day Rd near Manjimup, on red earth soil in 1986 plantings of *Eucalyptus globulus*, 11 October 1991, *N.Malajczuk and G.Hardy E0801* (PERTH 07564732); near Limpoo R., Juniperina Ck area, in moist old sand dunes in moss below *Agonis flexuosa*, 14 Aug. 1991, *K.Syme KS 333/91* (PERTH 05302331); Nedlands, at Lurie Hall, University of W.A. campus, Perth, on the ground under *Agonis flexuosa*, 22 May 1991, *O.K. & H.Miller & L. & M.Bailey OKM24599* (PERTH 08561192); Mount Shadforth Reserve, adjoining NW corner lot 406, in deep litter in Karri forest under *Eucalyptus diversicolor*, *Corymbia calophylla*, *Allocasuarina decussata* and *Agonis flexuosa*, 22 May 1993, *A. & K. Syme KS648/93* (PERTH 07572514); Herdsman L., Perth, N side of lake on edge of mowed grass beside park under introduced

Eucalyptus, 12 July 1997, *N.L. & M.E.Bougher E5916* (PERTH 07608748); Bold Park, E of Perry Lakes Stadium, Perth, on lawn under *Eucalyptus gomphocephala*, 7 Aug. 2001, *P.B.Matheny PBM2167* (PERTH 07671881); *loc. id.*, 7 Aug. 2001, *P.B.Matheny PBM2168* (PERTH 07671903, TENN 066745); Bold Park, along Brookdale Ave., Perth, under *E. gomphocephala*, 8 Aug. 2001, *P.B.Matheny PBM2170* (PERTH 07671938); L. Beedelup, 22 km from Pemberton, walk-through Tree Trail, Karri Valley Resort, near *Eucalyptus diversicolor*, 19 Aug. 2001, *P.B.Matheny PBM 2198* (PERTH 07676735; AY380404-LSU); Kings Park (Subiaco entry), Perth, in lawn on ground under *Eucalyptus marginata*, 27 Aug. 2001, *P.B. & S.D.Matheny PBM 2205* (PERTH 07676816); Cunningham Terrace Park, Daglish, Perth, among mulch and woodchips under *Eucalyptus marginata* and *E. gomphocephala*, 9 July 2004, *N.L.Bougher & R.Hart E8053* (PERTH 07678711); Wembley Downs, corner of Morden Rd and Ednah St, Perth, in grassy lawn under planted *Eucalyptus maculata* and *Lophostemon* (*Tristania*) *confertus*, 16 July 2004, *N.L.Bougher E8055* (PERTH 07678738); Denmark (town), in grassy area under *Eucalyptus diversicolor*, 16 July 2011, *T. Lebel & N.Bougher NLB817* (PERTH 08318654, TENN 066663). S.A. (examined by P.S.Catcheside): Stringybark Walking Trail, Deep Creek Conservation Park, on the ground in *Eucalyptus obliqua* forest, June 2008, *T.W.May 1752* (AD-C 54960, MEL); Coffin Bay Natl Park, Eyre Penin., in sandy soil in sand dunes, heath, under bush and under *Allocasuarina verticillata*, also with *A. verticillata*, *Leucopogon*, *Adriana klotzschii* and daisybush, 15 Aug. 2011, *P.S.Catcheside PSC3671* (AD-C 58138); Kuitpo Forest, near Meadows, scattered to gregarious in soil in open eucalypt woodland, 24 Aug. 2013, *P.S.Catcheside PSC3888* (AD-C 58361). Qld: adjacent to Lake McKenzie car park, Fraser Is., alt. 99 m, 25°26'42.4"S, 153°02'58.4"E, on sand under *Allocasuarina* and *Eucalyptus*, 8 June 2009, *N.Fechner & K.Querengasser REH9202* (NY, TENN 066848; KC305367-nLSU, KP641643-ITS, KC305417-*rpb2*). Vic.: Mortimer Picnic Ground and Nature Trail, 10 May 2009, *P.George* (MEL). Tas.: Keoghs Ck, Arve Rd, Geeveston, 20 Jan. 2004, *G.Gates & D.Ratkowsky I86* (TENN 066846; KC305368-nLSU); Mount Field Natl Park, on disturbed soil along roadside in wet sclerophyll forest under *Eucalyptus regnans*, *Acacia* and *Pomaderris*, 8 June 2009, *P.B.Matheny PBM3215* (TENN 066847; KC305366-nLSU, KP641644-ITS, KC305418-*rpb2*).

Diagnostic features: Fruit bodies medium. Pileus with violet tint at first but becoming cinnamon-brown to umbrinous; stipe violet and pruinose at the extreme apex or nowhere at all; odour strongly spermatic. Spores elliptic with obtuse apices. Metuloids present. Widely distributed in temperate regions.

Inocybe violaceocaulis is readily distinguished from the other Australian species by its violet stipe, a pileus that is violet-tinted at first but becoming cinnamon-brown to umbrinous, and its strongly spermatic odour. It occurs in highly disturbed sites such as under planted trees on suburban street verges and in native forests, but generally in disturbed microhabitats. The name *Inocybe geophylla* var. *lilacina* (Peck) Gillet was misapplied for many years to this conspicuous species (Matheny & Bougher, 2005). Like members of the *I. geophylla* group, the spores of *I. violaceocaulis* are elliptic and the cystidia are at most slightly thick-walled. However, members of the *I. geophylla* group, which occur in north temperate regions, are characterised by numerous caulocystidia at the stipe apex, and agglutinated fibrils at the base of the stipe and on the pileal disc. *Inocybe violaceocaulis* is characterised by infrequent caulocystidioid hairs at the extreme apex of the stipe, the absence of agglutinated fibrils, and an association with myrtaceous plants in Australia.

Since its first collection in 1985 (cited above, and also as "*I. geophylla*"; see Bougher, 2010) and subsequent description (Matheny & Bougher, 2005), *I. violaceocaulis* has been recorded from Victoria, Tasmania, Queensland (see collections cited here) and South Australia (Catcheside & May, 2010). Generally it has been found in association with a wide range of eucalypts, and it has been recorded fruiting between December and June in Tasmania, based on collections in the personal herbarium of G.Gates (I32, I61, I77, I89) and some duplicates in TENN. The species is ilustrated from Tasmania with a colour photograph in Gates & Ratkowsky (2014).

Phylogenetically, *I. violaceocaulis* is rather unusual among Australian smooth-spored Inocybaceae in that it is nested within a clade of otherwise exclusively Northern Hemisphere taxa (Matheny *et al.*, 2009; Braaten *et al.*, 2014). It is most closely related to *I. hainanensis* T.Bau & Y.G.Fan that occurs under Fagaceae in highland tropical forest on Hainan Island, China (Fan & Bau, 2014). Several smooth-spored species of *Inocybe* with violet colouration from Malaysia and Australasia have been documented by Horak (1980). *Inocybe ionides* Corner & E.Horak, described from Sabah, Malaysia, appears similar, but it lacks a cortina and pleurocystidia. *Inocybe subgeophylla* P.Henn. is not conspicuously violet and occurs with fagaceous hosts in Indonesia, while *I. violeipes* E.Horak and *I. violaceovelata* E.Horak, from Papua New Guinea, are squamulose and have stipes with belts of veil tissue; the latter does not have elliptic spores. No violet species with smooth basidiospores have been reported from New Zealand (Horak, 1977). An examination of Cleland's type collections of *Inocybe* at the State Herbarium of South Australia (AD) indicates that Cleland did not previously describe *I. violaceocaulis*; nor was this species described in Grgurinovic (1997).

Plate 144. *Inocybe violaceocaulis* (PERTH 08318654). Immature fruit body showing a thick cortina. Photograph — N.L.Bougher.

Plate 145. *Inocybe violaceocaulis* (PERTH 07678711). Photograph — N.L.Bougher.

Plate 146. *Inocybe violaceocaulis* (PERTH 08318654). Note the paucity of violaceous tinge on the mature pilei, but tinges persist on the stipe. Photograph — N.L.Bougher.

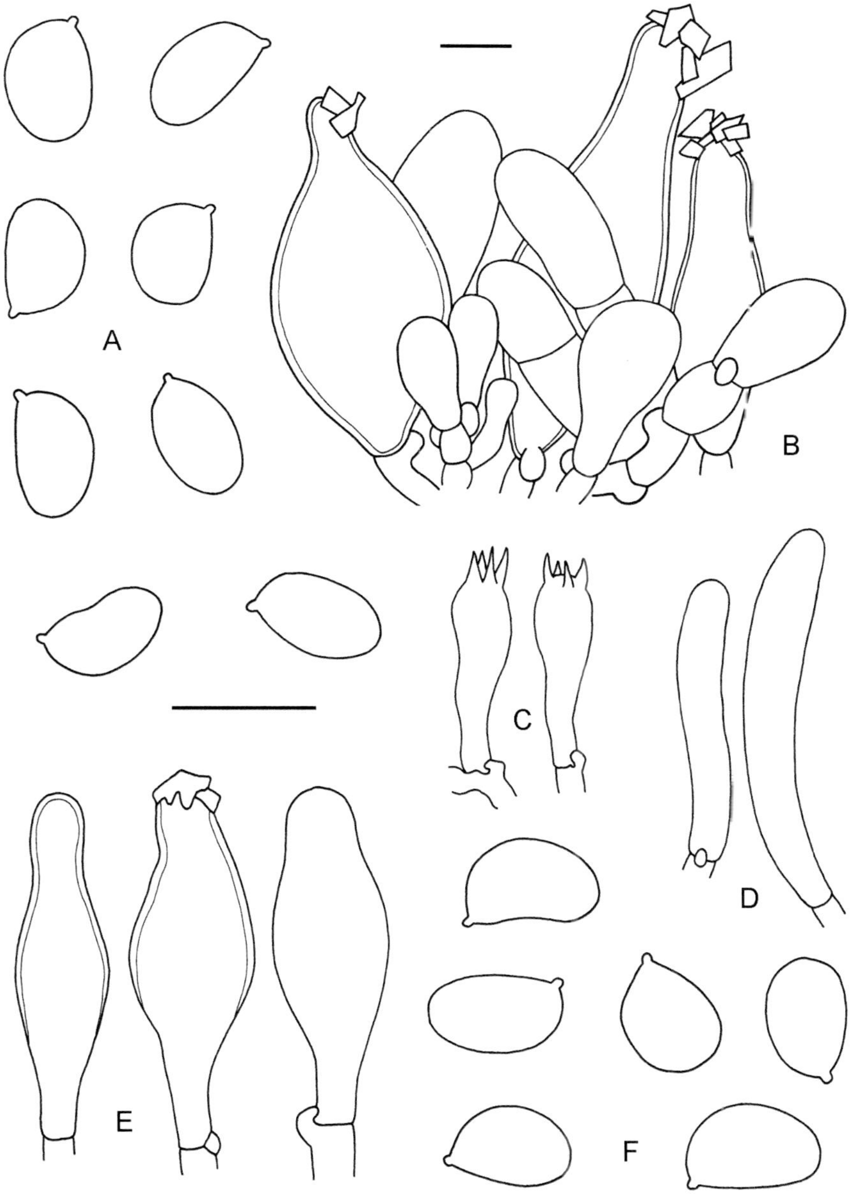

Figure 83. *Inocybe violaceocaulis*. All holotype (PERTH 07671857), except C and F (PERTH 07678738). **A,** Basidiospores; **B,** Cheilocystidia and paracystidia; **C,** Basidia; **D,** Caulocystidioid hairs; **E,** Pleurocystidia; **F,** Basidiospores. Scale bars: 10 µm (longer bar for spores only).

NODULOSE-SPORED SPECIES OF THE *INOCYBE* CLADE

Descriptions for 42 Australian species of the *Inocybe* clade with nodulose spores are documented below. Overall ten groups are presented based on phylogenetic analysis and four species are considered *incertae sedis* (uncertain position). Species of *Inocybe* with nodulose spores, treated by Horak (1977; 1979a) under *Astrosporina*, are not monophyletic (Matheny *et al.*, 2002; Matheny, 2005; Matheny *et al.*, 2009; Ryberg *et al.*, 2010).

INOCYBE ALIENOSPORA GROUP

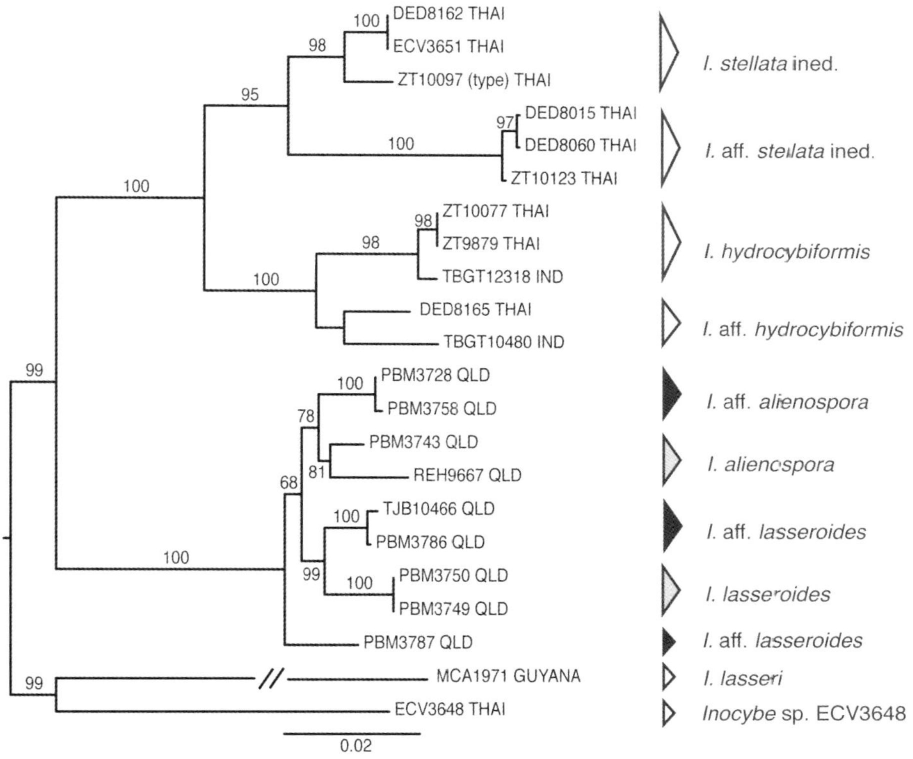

Figure 84. Maximum Likelihood phylogram of Australian samples of *Inocybe alienospora* and *Inocybe lasseroides*. Bootstrap values are shown above branches. The tree was inferred from combined ITS, nLSU, and *rpb2* gene regions and includes 2808 total sites (21 sites too ambiguously aligned were removed from the ITS region before phylogenetic analysis). Species from Australia are indicated by black-filled triangles; grey triangles denote species shared between Australia and other countries; white triangles denote species not found in Australia. The tree is rooted with *Inocybe lasseri* and *Inocybe* sp. ECV3648 based on family-level *rpb2* gene tree analysis (data not shown). Of the 22 branch tips shown, 21 are represented by ITS, all by nLSU, and 20 by *rpb2*.

This remarkable nodulose-spored group is composed of nine phylogenetic species. Five occur in Australasia (Queensland, Papua New Guinea) and form the sister group to a suite of south-east Asian species focused around *I. hydrocybiformis*. More detailed studies are needed in this group to resolve the phylogenetic depth observed in *I. alienospora*, *I. lasseroides*, *I. hydrocybiformis*, and *I. stellata*. As such the treatment here is a preliminary one. Both *I. alienospora* and *I. lasseroides* feature coarsely nodulose spores, the nodules of which may appear crested or saddle-shaped. The lamellae edges are unusual in that both species produce thin-walled cylindric or clavate cheilocystidia with yellow plasmatic inclusions and lack paracystidia typical of species of the *Inocybe* clade.

73. **Inocybe alienospora** (Corner & E.Horak) Garrido, *Biblioth. Mycol.* 120: 176 (1988)

Astrosporina alienospora Corner & E.Horak, *Persoonia* 10: 173 (1979). Type: Botanical Gardens, Singapore, 7 Nov. 1934, *E.J.H.Corner ZT 78/51*; holo: *n.v.*

Pileus 6–25 mm wide, conical with an obtuse to subacute umbo, or at times merely conical, becoming plane with a distinct umbo; margin decurved; surface dry; velipellis absent; disc and outer surface recurved-squamulose to squarrose, fibrillose at the margin edge; scales and overall colour brown (10YR 4/3; Snuff Brown); scales initially lighter in colour than the ground, eventually the colour paler between scales; context very thin except under the disc, pallid with a dull brown tinge, flesh not changing colour where exposed. *Lamellae* adnexed, moderately close, 26–30 L, with several tiers of lamellulae; pale brown to light yellowish brown, becoming brown with age; edges indistinctly pallid-fimbriate, medium. *Stipe* 15–45 × 1–3 mm, even or with a slightly wider base, terete; cortina not observed but ostensibly fugacious; nowhere pruinose, coarsely fibrillose to silky-fibrillose, not at all scaly and lacking girdles of veil tissue; pale brown to brown (10YR 6/3 to 5/3), lighter in colour than the pileus, white at the base; context solid; interior very pale brown. *Odour* weakly of *Pelargonium* or not remarkable. Pl. 147, 148.

Basidiospores 8.0–10.5 × 6.5–8.5 µm [mean 9.5 × 7.6 µm; Q 1.12–1.46; Q mean 1.25 (n = 35/3)], coarsely nodulose with several wedge- or saddle-shaped nodules. *Basidia* 24–32 × 8–11 µm, clavate, at times with a median constriction, hyaline, 4-sterigmate. *Pleurocystidia* 45–80 × 13–19 µm, fusiform to ventricose; apices obtuse to indistinctly subcapitate or swollen, smooth or crystalliferous; thick-walled apically (1–3 µm thick), hyaline, thinner-walled below. *Cheilocystidia* 35–70 × 7–10 µm, cylindrical to subclavate, sometimes curved, thin-walled, with dull yellowish plasmatic pigment or yellow occlusions; apices obtuse, not crystalliferous; numerous, densely arranged; paracystidia not observed, occasional swollen or infrequent metuloid cells observed (similar to pleurocystidia). *Caulocystidia* not seen. *Pileipellis* dark yellowish brown in mass, the hyphae cylindrical, mostly 6–12 µm wide, encrusted, ±thin-walled. *Clamp connections* present. Fig. 85.

Known from north-eastern and south-eastern Qld; grows in mixed rainforest, in sclerophyll vegetation under *Eucalyptus*, *Allocasuarina*, *Acacia* and *Corymbia*, in lowland *Melaleuca* swamp and in paddocks under *Acacia*. Also in Singapore. Fruiting Feb. to Mar. in Qld (Nov. in Singapore).

Additional specimens examined: Qld: Davies Creek Rd, Dinden Natl Park, between Kuranda and Mareeba, 17°01'36.1"S, 145°36'4.3'E, scattered singly on roadbank at woodland-rainforest transition under *Allocasuarina torulosa*, *Eucalyptus grandis* and other *Eucalyptus* spp., 24 Feb. 2012, *P.B.Matheny PBM3743* (PERTH 08362947, TENN 066975; KM197209-nLSU, KP171104-ITS, KM245970-*rpb2*); Ipswich, Harding's Paddock, alt. 85 m, 27°44'06.1"S, 152°46'01.6"E, gregarious on soil under *Acacia*, 6 Mar. 2012, *R.E.Halling*, *N.Fechner REH9667* (BRI, NY; KM197210-nLSU, KP171105-ITS, KM245971-*rpb2*).

Other specimens examined (genetic or morphological variants; *I. aff. alienospora*): Qld: Platypus Ck, Tinaroo area, 2 May 1988, *N.L.Bougher E4001* (PERTH 7709692, TENN 069569); Mt Beerburum, 26°57'29.7"S, 152°57'7.3"E, on soil in litter in woodland of tall eucalypts (including blackbutt) and *Allocasuarina* sp., 28 Feb. 2012, *N.L.Bougher NLB951* (PERTH 08364613, TENN 070170); *loc. id.*, *N.L.Bougher NLB952* (PERTH 08364621); Black Mountain Rd, near Kuranda, 16°47'12.4"S, 145°38'21.9"E, scattered to gregarious on soil among leaf litter under *Corymbia*, *Eucalyptus pellita* and *Acacia celsa*, 23 Feb. 2012, *M.D.Barrett & N.L.Bougher PBM3728* (PERTH 08362866, TENN 066964; KM197211-nLSU, KP171106-ITS, KM245972-*rpb2*); Davies Creek Rd, Dinden Natl Park, between Kuranda and Mareeba, 17°01'36.1"S, 145°36'4.3"E, emerging from litter in woodland-rainforest transition under *Allocasuarina torulosa*, *Eucalyptus grandis* and other *Eucalyptus* spp.,

24 Feb. 2012, *N.L.Bougher NLB938* (PERTH 08373124, TENN 069649); corner of Armstrong Rd and Jeffrey Rd, just N of the Barren R., N of Kuranda, 16°47'54.9"S, 145°33'47.8"E, scattered on wet soil in drainage in *Melaleuca* swamp under *Melaleuca viridiflora*, *Corymbia intermedia* and *Eucalyptus pellita*, 25 Feb. 2012, *M.D.Barrett & P.B.Matheny PBM3758* (PERTH 08363269, TENN 066986; KM197212-nLSU, KP171107-ITS, KM245973-*rpb2*).

Diagnostic features: Fruit bodies small to medium. Pileus conical, often dark brown to dark yellowish brown, coarsely fibrillose to scaly; stipe nowhere pruinose, coarsely fibrillose to silky-fibrillose. Spores less than 10.5 µm long, each with several saddle-shaped nodules; cheilocystidia cylindrical to subclavate, thin-walled, yellow or with yellow occlusions.

Inocybe alienospora is reported here for the first time from Australia. The Australian collections match the description of the type collection from Singapore (Horak, 1979a), including the unusual spores and cheilocystidia. *Inocybe lasseroides* can be confused with *I. alienospora*, but it has larger spores and exhibits phylogenetic divergence (Fig. 84). The lamellar edges of Australian collections of these two species have unusual thin-walled cylindrical to subclavate or clavate cells with yellowish plasmatic pigment. In addition, we have observed infrequent swollen or metuloid cells similar to pleurocystidia on some lamellar edges of *I. alienospora*; in *I. lasseroides* metuloids can be absent, or sparse to abundant on the lamellar edges.

Numerous genetic variants of *I. alienospora* have been detected in Australia (e.g. as listed above, and in Fig. 84), and while some might represent cryptic species, some of these have distinct macromorphologies. For example, *PBM3728* (PERTH 08362866) has fruit bodies resembling *Cortinarius* (*Telamonia*) and, therefore, undoubtedly represents a distinct species (see Pl. 1E). Another collection from north-eastern Queensland [Black Mountain Rd, near Kuranda, 16°47'7.8"S, 145°37'32.1"E, scattered singly on soil along roadside in mixed rainforest under *Allocasuarina littoralis*, *Acacia mangium* and *Eucalyptus pellita*, 23 Feb. 2012, *M.D.Barrett & P.B.Matheny PBM3729* (PERTH 08362874, TENN 066965; PERTH 08362874)] resembles *I. alienospora* but it has globose, echinulate spores. It may be referable to *I. hydrocybiformis* (Corner & E.Horak) Garrido, recently documented from Thailand (Horak *et al.* 2015) and India (Pradeep *et al.* 2016) but further collections are required to determine its status and that of the other anomalous specimens mentioned above.

74. **Inocybe lasseroides** (E.Horak) Garrido, *Biblioth. Mycol.* 120: 177 (1988)

Astrosporina lasseroides E.Horak, *Persoonia* 10: 173 (1979). Type: Bulolo, Watut, Morobe District, Papua New Guinea, on soil under *Castanopsis* and *Lithocarpus* (Fagaceae), 28 Apr. 1972, *E.Horak*; holo: ZT 72/422 *n.v.*

Pileus 6–11 mm wide and 5–7 mm high, conical to deeply so; umbo absent or at least not distinct; margin long-decurved to decurved, initially slightly incurved; surface dry, usually appressed-squamulose in the centre, fibrillose towards the margin; velipellis absent, occasionally rimulose towards the margin; umbrinous or brown to dark yellowish brown (10YR 3/4 to 4/3; Snuff Brown); context very thin, c. 1 mm thick under the disc, pallid, not changing colour where exposed. *Lamellae* subfree to adnexed, moderately close, 24–28 L, with few tiers of lamellulae, light brown to brown, ventricose, broad for the small pileus size, edges indistinctly pallid-fimbriate. *Stipe* 22–43 × 1 mm, even, terete, strict or very straight, like some species of *Nolanea*, not obviously pruinose, silky-fibrillose, cortina presumably fugacious; colour lighter than the pileus, yellowish brown (10YR 5/4), pallid at the extreme base; context solid. *Odour* at first acidulous, becoming like *Pelargonium*. Pl. 149, 150.

Basidiospores 9.5–12.0 × 7.5–10.0 µm [mean 10.7 × 9.0 µm; Q: 1.06–1.33; Q mean 1.18 (n = 23/2)], subglobose to ovate in outline with 10–12 conical, wedge-shaped or saddle-shaped nodules that typically project c. 2 (–3) µm. *Basidia* 24–31 × 9–11 µm, cylindroclavate or subventricose, hyaline, 4-sterigmate. *Pleurocystidia* 49–58 × 17–23 µm, fusiform-ventricose, utriform or broadly fusiform; apices obtuse and at times bearing crystals; thick-walled, especially apically (1.5–3.0 µm), hyaline. *Cheilocystidia* variable in abundance and form; lamellar edges noticeably yellow in mass when seen at low power under the microscope due to cylindrical, narrowly or broadly clavate, thin-walled cystidia (28–52 × 6–15 µm) with yellow intracellular pigmented occlusions which are crowded and exclusive, or mixed with variable numbers of metuloids similar to pleurocystidia (sparsely scattered to abundant).

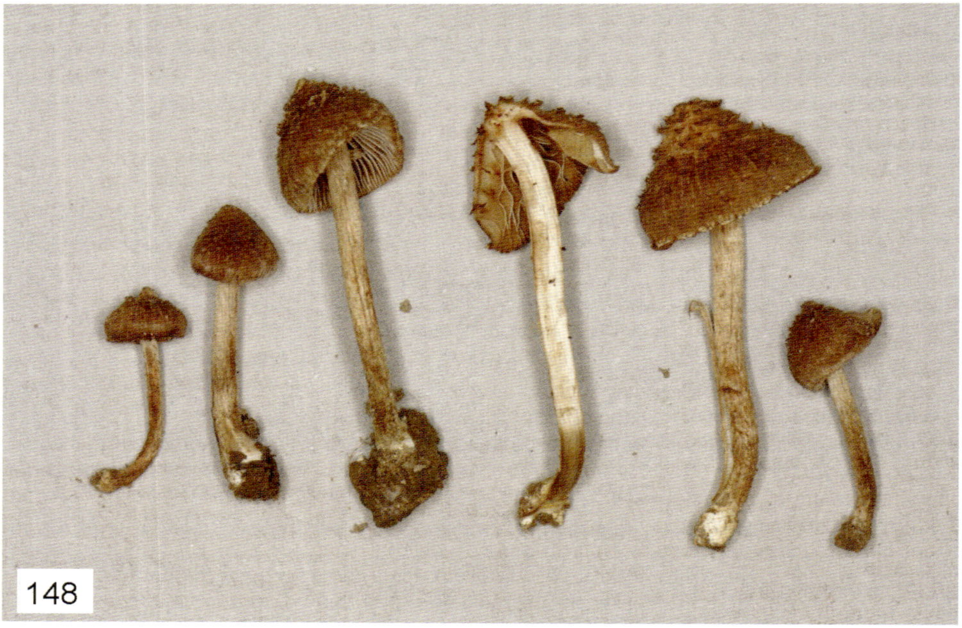

Plate 147. *Inocybe alienospora*. PERTH 08362947. Photograph — P.B.Matheny.

Plate 148. *Inocybe alienospora*. PERTH 08362947. Photograph — P.B.Matheny.

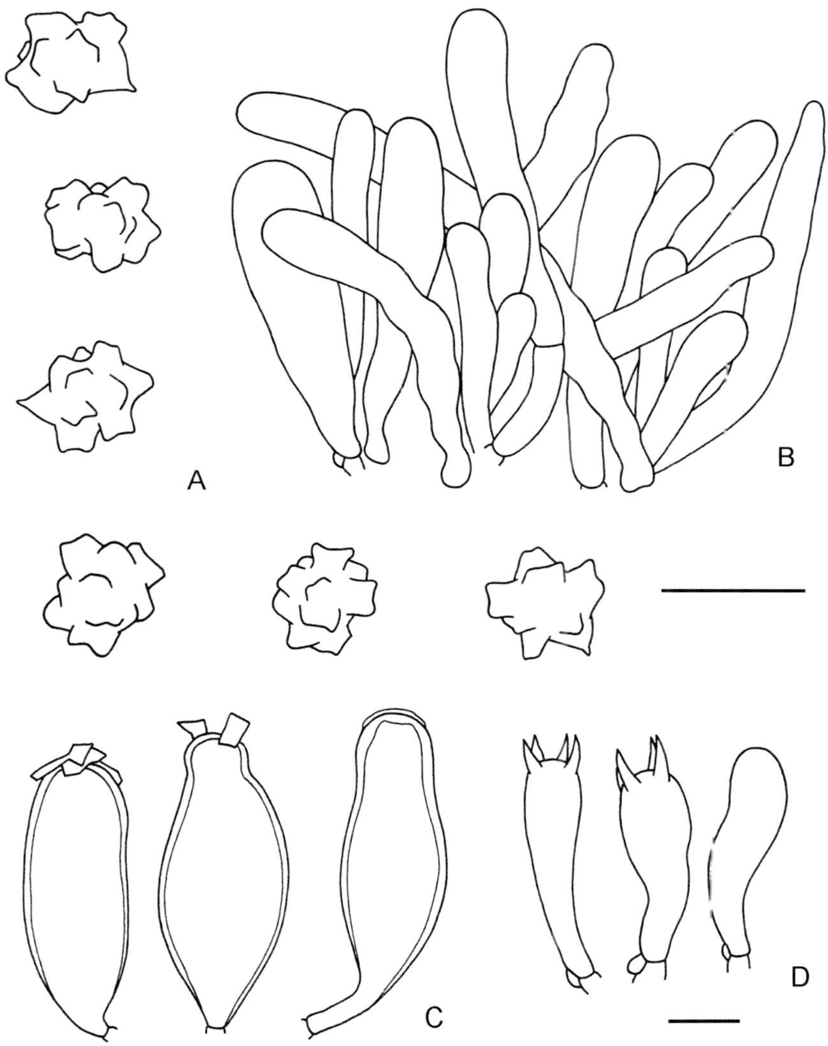

Figure 85. *Inocybe alienospora.* All PERTH 08362947. **A,** Basidiospores; **B,** Cheilocystidia; **C,** Pleurocystidia; **D,** Basidia. Scale bars = 10 μm (longer bar for spores only).

Caulocystidia similar to cheilocystidia, thin-walled, yellow-pigmented, in clusters at the stipe apex, with infrequent metuloids similar to pleurocystidia together with thin-walled, clavate cauloparacystidia. *Pileipellis* composed of cylindrical to inflated hyphae mostly 6–17 µm wide, thin-walled, noticeably encrusted, dark brown in mass. *Clamp connections* present. Fig. 86. (see also Pl. 6D).

Known from north-eastern Qld, under *Allocasuarina* and *Eucalyptus* in sclerophyll forest; fruiting Feb. Also with *Castanopsis* and *Lithocarpus* in Papua New Guinea; fruiting Apr.

Additional specimens examined: Qld: Davies Creek Rd, Dinden Natl Park, between Kuranda and Mareeba, 17°01'36.1"S, 145°36' 04.3"E, scattered singly on soil in leaf litter along roadside at woodland-rainforest transition under *Allocasuarina torulosa*, *Eucalyptus grandis* and other *Eucalyptus* spp., 24 Feb. 2012, *P.B.Matheny PBM3749* (PERTH 08363323, TENN 066979; KP170924-nLSU, KP171145-ITS, KM245993-*rpb2*); *loc. id.*, *P.B.Matheny PBM3750* (PERTH 08363307, TENN 066980; KP170925-nLSU, KP171146-ITS).

Other specimens examined (genetic variants; *I. aff. lasseroides*): Qld: Fraser Is., Great Sandy Natl Park, cross road from Eurong to Pile Valley, on sandy soil on road bank with wet decaying leaves, 18 May 2011, *T.J.Baroni TJB10466* (CORT; KP170928-nLSU, KP171149-ITS, KM245996-*rpb2*); Queen Mary Falls, Main Range Natl Park, 28°20'29.5"S, 152°22'16.9"E, scattered singly on soil in leaf litter under *Allocasuarina*, along bank of track in wet sclerophyll forest, 4 Mar. 2012, *P.B.Matheny PBM3786* (PERTH 08363048, TENN 067018; KP170926-nLSU, KP171147-ITS, KM245994-*rpb2*); *loc. id.*, *P.B.Matheny PBM3787* (PERTH 08363021, TENN 067019; KP170927-nLSU, KP171148-ITS, KM245995-*rpb2*).

Diagnostic features: Fruit bodies small. Pileus conical to deeply so, brown, often scaly but sometimes fibrillose; stipe long and very straight, slender, not obviously pruinose, silky-fibrillose. Cheilocystidia cylindrical to clavate, thin-walled, yellow, mixed with variable numbers of metuloids similar to pleurocystidia; spores larger than *I. alienospora*, with distinctive conical and wedge- or saddle-shaped nodules.

This is the first report of *I. lasseroides* since its description based on a single collection from Papua New Guinea (Horak, 1979a). While it is very similar to the Australasian *I. alienospora* (see above), *I. lasseroides* has larger spores, and differences in DNA sequences also support their separation (Fig. 84).

The lamellar edges of *I. alienospora* are characterised by the dominance of abundant thin-walled cylindrical to subclavate cells with yellowish plasmatic pigment. Although this peculiar cell-type was not reported on the lamellae by Horak for *I. lasseroides*, we observed this feature in both Australian collections. These cystidia render the lamellar edges noticeably yellow in mass when seen at low power under the microscope. In some of the Australian specimens, these elements are crowded and exclusive on the lamellar edges, but in some others they are mixed with variable numbers of metuloids similar to the pleurocystidia (sparsely scattered to abundant). As indicated by Horak (1979) for the type collection of *I. lasseroides*, yellow thin-walled cystidia also occur on the upper stipe of the Australian specimens.

Several genetic variants of *I. alienospora* and *I. lasseroides* have been observed in Queensland (Fig. 84, see discussion under *I. alienospora*, species 73). Further collections and sequence data will be required to detemine whether these variants deserve autonomous status.

Plate 149. *Inocybe lasseroides*. PERTH 08363323. Photograph — P.B.Matheny.

Plate 150. *Inocybe lasseroides*. PERTH 08363307. Photograph — P.B.Matheny.

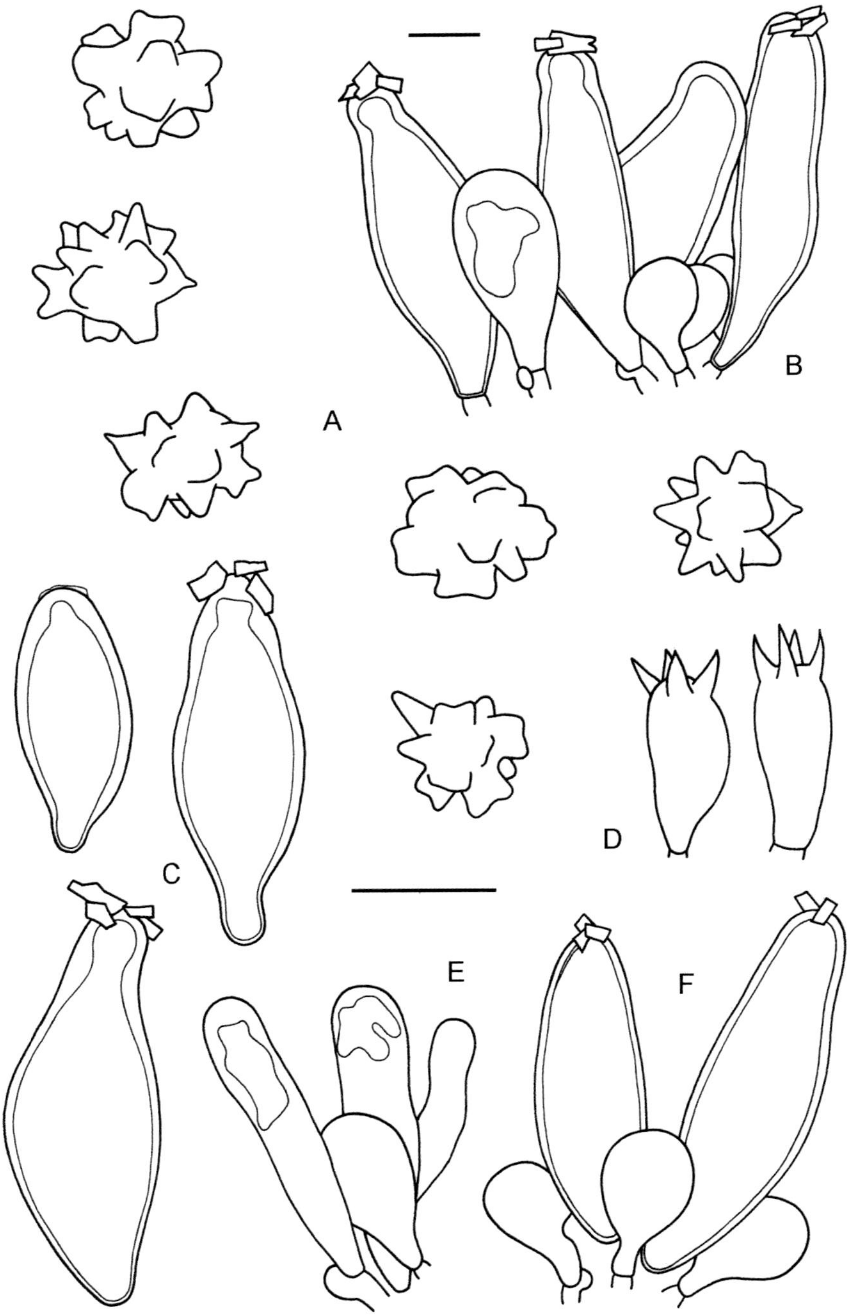

Figure 86. *Inocybe lasseroides.* A, D, PERTH 08363323; B, C, E, F, PERTH 08363307. **A,** Basidiospores; **B,** Metuloid cheilocystidia and thin-walled cystidia often with yellow occlusions; **C,** Pleurocystidia; **D,** Basidia; **E,** Upper stipe - thin-walled yellowish caulocystidia often with yellow occlusions; **F,** Upper stipe - metuloid caulocystidia, and cauloparacystidia. Scale bars = 10 µm (longer bar for spores only).

INOCYBE ASSIMILATA GROUP

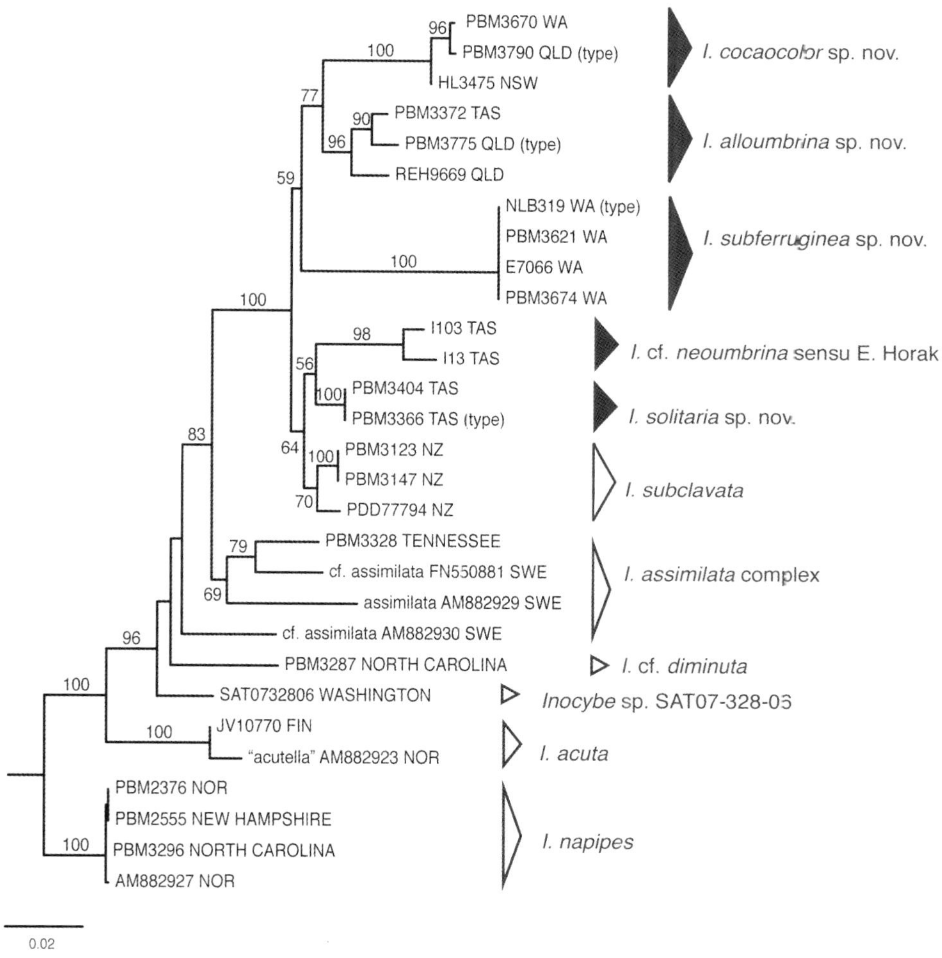

Figure 87. Maximum Likelihood phylogram of the *Inocybe assimilata* group. Bootstrap values are shown above or below branches. The tree was inferred from combined ITS, nLSU, and *rpb2* gene regions and includes 2729 total sites. One hundred and one sites in the ITS gene region were removed due to ambiguous alignment. Species from Australia are indicated by black-filled triangles; white triangles denote extra-Australian species. All Australasian species form a robust monophyletic group. The tree is rooted with *Inocybe napipes* based on inspection of individual family-level trees of nLSU and *rpb2* (data not shown). Of the 29 branch tips included, all are represented by ITS, 24 by nLSU, and 10 by *rpb2*.

This group is well supported by single gene family-level phylogenies and as subclade VII in Ryberg *et al.* (2010) with strong support. The group includes several better known north temperate taxa such as *I. napipes*, *I. acuta*, and *I. assimilata*. Six Australasian species are shown in derived evolutionary positions in this group. Five of the six species are known only from Australia. Four are described here as new. The group is united by the absence of caulocystidia and cauloparacystidia, although some caulocystidioid cells might occasionally be found at the apex of the stipe, and their nodulose spores. The New Zealand species, *I. subclavata*, also occurs in this group.

75. **Inocybe alloumbrina** Matheny & Bougher, *sp. nov.*

MycoBank: MB 809261

Type: Cunninghams Gap, Main Range Natl Park, Qld, 28°02'59"S, 152°23'37.3"E, alt. 776 m, on soil under *Eucalyptus* and *Allocasuarina*, 29 Feb. 2012, *R.E.Halling & N.Fechner PBM3775*; holo: PERTH 08363129; iso: TENN 067007 (KJ801158-nLSU, KJ778848-ITS, KM245974-*rpb2*).

Pileus 14–30 mm wide, almost conical to obtusely so, expanding with age, becoming plano-convex, almost plane, often with an obtuse or subacute umbo; margin decurved; surface dry; velipellis absent or inconspicuous, smooth and unbroken in the centre but becoming recurved-squamulose around the umbo in some collections, fibrillose to coarsely fibrillose towards the margin, rimulose; dark brown (10YR 3/3) or with shades of dark yellowish brown (10YR 4/4) towards the margin; context very thin except under the disc, white or pallid, unchanging upon exposure. *Lamellae* adnexed to sinuate, moderately close, 30–34 L, with several tiers of lamellulae; light yellowish brown or greyish tan, olive-yellow to light olive-brown (2.5Y 6/4, 6/6, 5/4), becoming dark brown with age; edges indistinctly pallid-fimbriate, broad, ventricose. *Stipe* 25–50 × 2–3 mm, terete, swollen or enlarged basally; base up to 5 mm wide; veil (cortina) not observed but presumably fugacious; pruinose at the extreme apex only, elsewhere lightly fibrillose; pale yellow (2.5Y 7/4), light yellowish brown (2.5Y 6/4) to very pale brown (10YR 7/4), generally darker below, mixed with isabelline shades, with age brown in areas, white at the base; context solid, pallid above, interior with an isabelline cortex. *Odour* absent. Pl. 151.

Basidiospores 7.0–8.5 × 5.0–6.0 µm [mean 7.8 × 5.7 µm; Q 1.25–1.60; Q mean 1.39 (n = 20/1), nodulose with mostly 6–9 moderate-sized nodules (projecting 1–2 µm) about a mostly angular outline, yellowish brown; apiculus tapering. *Basidia* 24–28 × 8–10 µm, 4-sterigmate, clavate, hyaline. *Pleurocystidia* 51–70 × 9–20 µm, ventricose or fusiform with a slender basal pedicel up to 20 µm long and an obtuse apex; ±thin-walled with walls at most up to 1.5 µm thick, hyaline; apices crystalliferous to bare. *Cheilocystidia* similar to pleurocystidia but sometimes cylindrical or utriform, abundant, mixed with intermediate shaped elements and small clavate paracystidia. *Caulocystidia* absent, but the upper stipe with scattered areas of caulocystidioid cells, these broadened terminal cells of the stipitipellis hyphae thin-walled to slightly thickened, occasionally similar in shape to hymenial cystidia. *Pileipellis* a cutis of brown cylindrical thin-walled encrusted hyphae 4–12 µm wide; occasionally overlain by a velipellis layer of compact interwoven hyphae, which are thin-walled, cylindrical and mostly 3–7 µm wide. *Clamp connections* present. Fig. 88.

Occurs in north-eastern and south-eastern Qld and Tas.; on soil under *Eucalyptus* and *Allocasuarina*, *Pomaderris* or *Acacia*. Fruiting Feb., Mar. and May.

Additional specimens examined: Qld: Davies Creek Rd, Dinden Natl Park, 17°01'36.1"S, 145°36'04.3"E, emerging from litter on soil at woodland-rainforest interface under *Eucalyptus grandis*, *Allocasuarina torulosa* and other *Eucalyptus*, 24 Feb. 2012, *N.L.Bougher NLB940* (PERTH 08364540, TENN 070162); Mapleton Falls Natl Park, in grassy area cleared of understory, near tall smooth-barked *Eucalyptus* and mixed subtropical rainforest vegetation, 29 Feb. 2012, *N.L.Bougher NLB959* (PERTH 08364699, TENN 070161; cf. *alloumbrina*); Mount Cordeaux Track (near Rainforest Circuit), Cunninghams Gap, Main Range Natl Park, 28°02'59.0"S, 152°23'37.3"E, on soil among litter in tall sclerophyll forest (including eucalypts and *Allocasuarina* sp.), 3 Mar. 2012, *N.L.Bougher NLB970* (PERTH 08364753, TENN 070163); Hardings Paddock, Ipswich, 27°44'06.1"S, 152°46'01.6"E, gregarious on soil under *Acacia*, 6 Mar. 2012, *R.E.Halling & N.Fechner REH9669* (BRI, NY; KJ801159-nLSU, KJ778850-ITS). Tas.: St Columba Falls, near Pyengana, scattered singly on soil along track under *Eucalyptus*, *Pomaderris* and *Acacia*, 21 May 2010, *G.Gates & P.B.Matheny PBM3372* (TENN 068358; KJ778849-ITS).

Etymology: The epithet is derived from the Greek *allos* (other), in reference to a morphological similarity to the northern temperate species *Inocybe umbrina*.

Diagnostic features: Fruit bodies medium. Pileus dark brown, smooth at the centre, fibrillose, coarsely fibrillose to recurved-scaly towards the margin; lamellae with olive tinges, becoming dark brown with age; stipe with an enlarged base, pruinose at the extreme apex, elsewhere lightly fibrillose. Spores nodulose and with more pronounced nodules than *I. solitaria*.

Inocybe alloumbrina is similar morphologically to *I. subclavata* (E.Horak) Garrido from New Zealand (Horak, 1977), *I. neoumbrina* Kobayasi as described from Papua New Guinea (Horak, 1979a) and *I. solitaria* from Tasmania. Indeed, *I. solitaria* and *I. alloumbrina* have been found in the same general locality — St. Columba Falls, near Pyengana. The spores of *I. solitaria* and *I. alloumbrina* are more coarsely nodulose than those of *I. subclavata*, with *I. alloumbrina* having the larger nodules of the Australian species. Phylogenetically, all three taxa are closely related (Fig. 87).

Plate 151. *Inocybe alloumbrina*, holotype (PERTH 08363129). Photograph — P.B.Matheny.

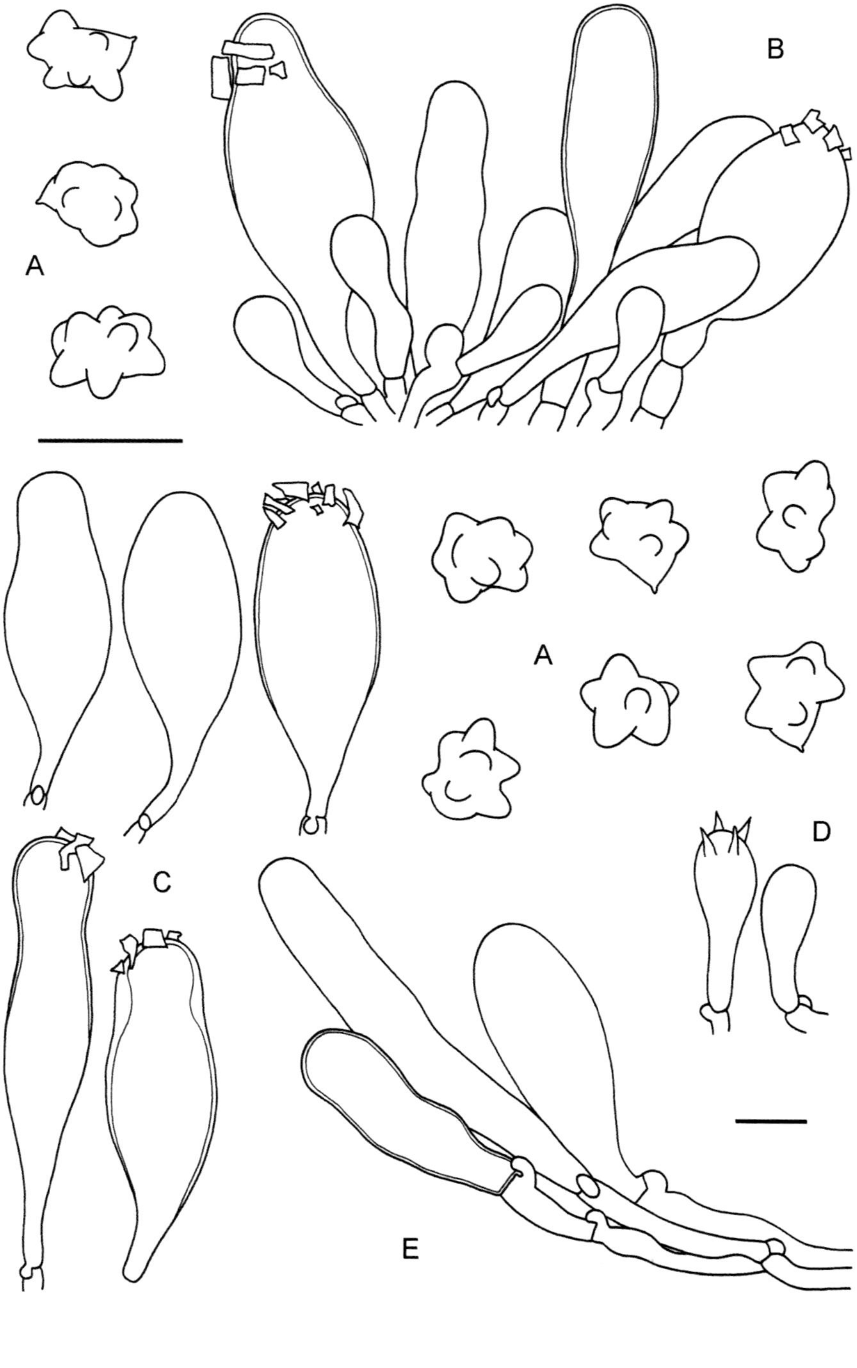

Figure 88. *Inocybe alloumbrina.* All holotype (PERTH 08363129). **A,** Basidiospores; **B,** Cheilocystidia and paracystidia; **C,** Pleurocystidia; **D,** Basidia; **E,** Caulocystidioid hairs (upper stipe). Scale bars = 10 μm (longer bar for spores only).

76. **Inocybe cacaocolor** Matheny & Bougher, *sp. nov.*

MycoBank: MB 809318

Type: Queen Mary Falls, Main Range Natl Park, Qld, 28°20'29.5"S, 152°22'16.9"E, scattered singly on soil along bank of track in wet sclerophyll forest under *Eucalyptus*, 4 Mar. 2012, *P.B.Matheny PBM3790*; holo: PERTH 08363005; iso: TENN 067022 (KJ756464-nLSU, KJ778845-ITS, KJ756422-*rpb2*).

Pileus 5–30 mm wide, obtusely conical to conical or initially pulvinate, later expanding, broadly conical, with a small or low and broad conspicuous obtuse umbo; margin at first incurved, later decurved, becoming straight or torn with age, occasionally with scattered brownish appendiculate veil remnants; velipellis absent, surface dry, matted-fibrillose or coarsely fibrillose to fibrillose when young, developing small appressed scales around the centre, these occasionally recurved, becoming excoriate at the centre; dark yellowish brown (10YR 4/3; Bay Brown) in youth, remaining so at the centre but shading brown or yellowish brown (10YR 4/3; Snuff Brown) towards the margin or maintaining a uniform colour, disc sometimes becoming blackish brown (Mummy Brown); context pallid or sordid yellowish brown to dark yellowish brown, not changing colour upon exposure, thin. *Lamellae* adnexed to sinuate, moderately close to close, 30–48 L, with several tiers of lamellulae; very pale brown, light yellowish brown, to yellowish brown (10YR 7/3 to 6/4 to 4/3), becoming chocolate-brown with age; ventricose, medium, 2–3 mm deep; edges indistinctly pallid-fimbriate. *Stipe* 12–35 × 1–4 mm, terete, even or with a slightly enlarged base up to 5 mm wide, lacking a distinct bulb, partial veil present as a cortina when very young; pruinose at the extreme apex or not at all, elsewhere or entirely fibrillose, at times with scattered loose brown fibrils on the lower part or with a fibrillose streaked appearance; whitish or very pale brown above shading to brown (Snuff Brown) on the lower part, with age becoming brown above and dark yellowish brown below usually mixed with streaks of pallid or greyish fibrils; context solid, coloured like the surface with a pale interior. *Odour* absent or rarely weakly of bruised *Pelargonium* leaves. Pl. 152–154.

Basidiospores 6.5–9.5 (–10.5) × 5.5–7.0 µm [mean 8.1 × 6.1 µm; Q: 1.14–1.40 (–1.58); Q mean 1.30 (n = 60/3)], nodulose, typically with 7–10 moderate or small obtuse nodules, some spores with 5–7 nodules or corners, brownish yellow to yellowish brown; apiculus small, tapering. *Basidia* 22–34 (–40) × 8–10 (–14) µm, clavate, hyaline, rarely ochraceous, 4-sterigmate, with some rather large basidia (see *PBM3670*). *Pleurocystidia* 36–55 × 13–19 µm, ventricose, broadly fusiform, utriform, often with very short necks and broadly rounded or obtuse apices, without crystals, with a short or long tapered basal pedicel; predominantly thin-walled (some up to 1.5 µm thick), hyaline or with brownish cytoplasm in mature specimens (see *BOU863*), infrequent. *Cheilocystidia* similar to pleurocystidia, also oblong-clavate, frequent on lamella edge, mixed with variously-shaped paracystidia (e.g. clavate, pyriform, ventricose). *Caulocystidia* absent, some caulocystidioid end-cells at the stipe apex. *Pileipellis* a cutis of russet-brown hyphae in mass, cylindrical, mostly 5–12 µm wide, encrusted, with slightly thickened walls. *Clamp connections* present. Fig. 89.

Known from single localities in south-western W.A., south-eastern Qld, and eastern N.S.W.; on soil under eucalypts and/or *Acacia*. Fruiting Mar. (Qld), July (W.A.) and Nov. (N.S.W.)

Additional specimens examined: W.A.: Jarrahdale Rd, 7.7 km W of Albany Hwy, 32°19'44.4"S, 116°07'42.9"E, solitary or in small clusters in Jarrah-Marri forest under *Corymbia calophylla*, *Eucalyptus marginata* and planted *Acacia*, 25 July 2011, *M.R.Ryberg MR00273* (PERTH 08256179, TENN 066616); *loc. id.*, on sandy soil at edge of clearing, 25 July 2011, *P.B.Matheny PBM3670* (PERTH 08319170, TENN 066465; KJ756466-nLSU, KJ778847-ITS, KJ756422-*rpb2*); *loc. id.*, *N.L.Bougher NLB863* (PERTH 08320683, TENN 066669). N.S.W.: Spring Creek Trail, 4 km along trail from Baldry–Trewilga road, Goobang Natl Park, 32°51'25"S, 148°23'35"E, on soil among mostly low grasses, in open *Eucalyptus* and *Callitris* woodland, 9 Nov. 2001, *H.Lepp 3475* (CANB 638191; KJ756465-nLSU, KJ778846-ITS).

Etymology: From the Latin *cacaocolor* (chocolate-coloured), in reference to the chocolate-brown colour of the lamellae on mature specimens.

Diagnostic features: Fruit bodies mostly medium sized. Pileus dark yellowish brown, imbricate-scaly to coarsely fibrillose, often umbonate; lamellae pale brown, becoming chocolate-brown with age; partial veil present; stipe even to slightly enlarged at the base, pruinose only at the extreme apex or not at all. Spores nodulose, cystidia thin-walled.

Inocybe cacaocolor has not been collected very often, but it is widespread in Australia, having been found in isolated localities in New South Wales, south-eastern Queensland and Western Australia where it was seen on the edges of forest clearings under eucalypts and planted *Acacia*. Therefore, we cannot reject the possibility the species has been introduced to Western Australia.

Inocybe subferruginea could be most confused, with *I. cacaocolor* differing mainly by rRNA and *rpb2* gene sequences (Fig. 87). The two form a pair of otherwise cryptic species.

Inocybe cacaocolor is most readily distinguished from others in Australia, New Zealand and southern South America by the thin-walled and typically ventricose to utriform hymenial cystidia. The overall fruit body colours are dark and sombre, and the spores are nodulose and mostly <10 µm long. *Inocybe subclavata* (E.Horak) Garrido, described from New Zealand, has a narrowly clavate stipe with a swollen rounded or truncate base that is wider than elsewhere. Its spores are similar to the New South Wales collection of *I. cacaocolor* (7–8 × 5–6 µm), and it has relatively few 5–7 low or obscure nodules. *Inocybe solitaria* Matheny, Bougher & G.M.Gates differs principally by the presence of a napiform to submarginate bulb on the stipe, occurrence under *Nothofagus* or in wet sclerophyll habitat in Tasmania, and its unique phylogenetic position. *Inocybe cerasphora* Singer, from southern South America (Singer and Digilio, 1952), is superficially similar, but it is nested in the *I. lanuginosa* (Bull. : Fr.) P.Kumm. group. It differs by its woolly lower stipe and larger spores (11.0–13.0 × 7.5–10.5 µm), each with 13–16 prominent nodules. *Inocybe gibbosula* (E.Horak) Garrido, from Tierra del Fuego, Argentina (Horak, 1979), differs in having metuloid hymenial cystidia.

Inocybe cacaocolor, together with five other species from Australia and New Zealand, are derived within a temperate Northern Hemisphere group of well known taxa, *I. assimilata* Britzelm. (= *I. umbrina* Bres. *non* Massee) and *I. napipes* J.E.Lange (Fig. 87). The austral clade, in addition to *I. cacaocolor*, includes *I. subclavata*, *I. solitaria* from Tasmania, *I.* cf. *neoumbrina* Kobayasi from Tasmania, *I. subferruginea* from Western Australia and *I. alloumbrina* from south-eastern Queensland and Tasmania. These six species form a robust monophyletic group (Fig. 87) and share a similar overall coloration (sombre and umbrinous), nodulose spores, a partial veil, pruinosity only at the stipe apex or nowhere at all, and an association primarily with Myrtaceae.

The spores of *I. cacaocolor* vary in shape and size. Thus the New South Wales material has spores in the range of 7.0–9.0 × 5.5–6.5 µm with comparatively few (5–7) nodules. The specimen from Western Australia, however, has spores in the range of 8.0–10.5 × 5.5–7.0 µm, typically with 7–10 nodules. The micromorphological characters of Western Australian material are otherwise similar to those of the type, but some features of mature specimens are noteworthy (e.g. in PERTH 08320683): (i) pleurocystidia tend to be wider (–30 µm); (ii) many of the cheilocystidia have pale brown cytoplasm; (iii) some spores have a long, pronounced adaxial depression. DNA sequences (ITS and nLSU) of two collections (CANB 638191 from New South Wales and PERTH 08319170 from Western Australia) are the same. Spores of the type resemble the New South Wales collection in size but typically are more coarsely nodulose (mostly 7–9 nodules) like the Western Australian material.

One distinctive collection (*E6204*, PERTH 07659911), which resembles *I. fibrillosibrunnea* in the field, is similar in outward appearance to *I. cacaocolor*, but it is more robust and has longer, slender, fusiform pleurocystidia (72–88 × 11–14 µm). The spores are 7.5–10.0 × 5.5–7.0 µm with mostly 5–7 low and indistinct nodules.

Plate 152. *Inocybe cacaocolor*, holotype (PERTH 008363005). Photograph — P.B.Matheny.

Plate 153. *Inocybe cacaocolor*. PERTH 008320683. Showing colour of mature lamellae. Photograph — N.L.Bougher.

Plate 154. *Inocybe cacaocolor*. PERTH 08319170. Photograph — P.B.Matheny.

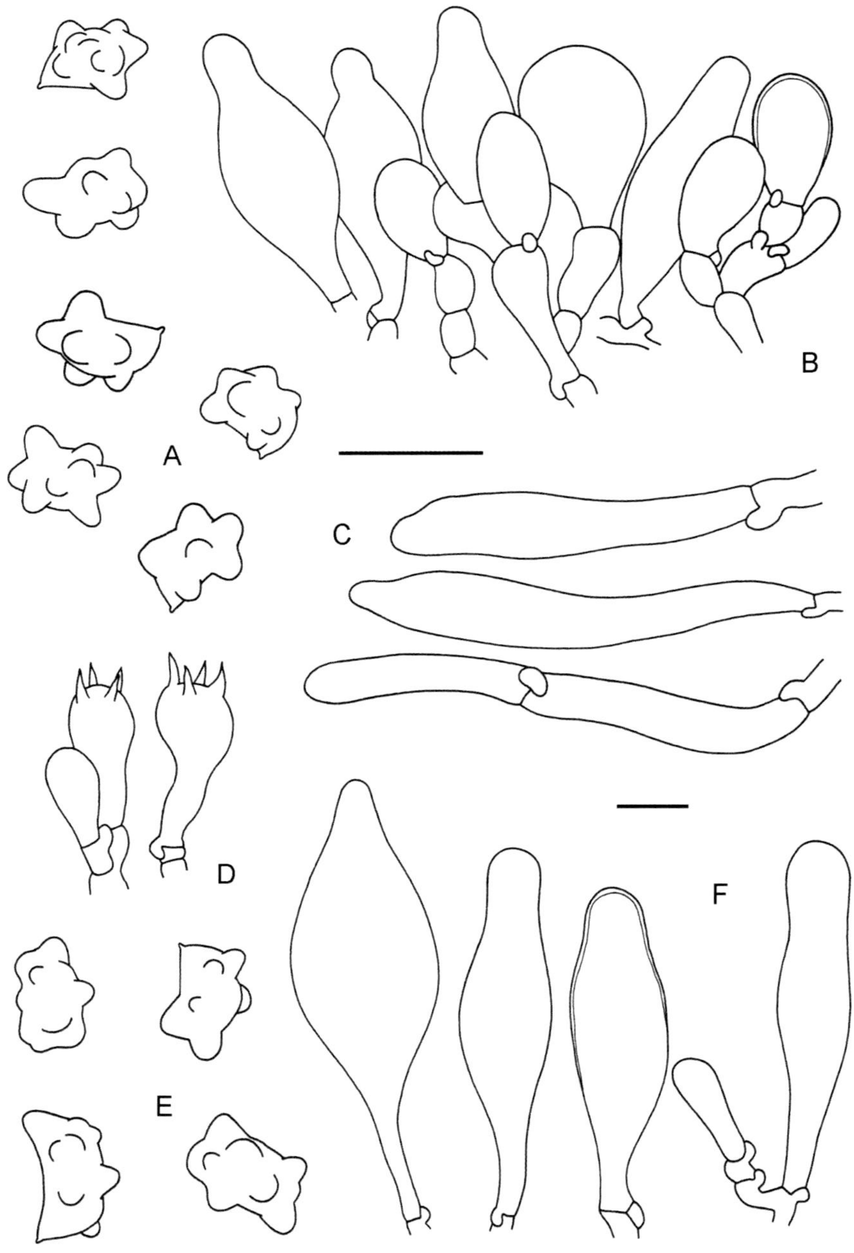

Figure 89. *Inocybe cacaocolor.* All holotype (PERTH 08363005), except E (PERTH 08320683). **A,** Basidiospores; **B,** Cheilocystidia and paracystidia; **C,** Caulocystidioid elements from stipe apex; **D,** Basidia; **E,** Basidiospores; **F,** Pleurocystidia. Scale bars = 10 μm (longer bar for spores only).

77. Inocybe solitaria Matheny, Bougher & G.M.Gates, *sp. nov.*

MycoBank: MB 809460

Type: St. Columba Falls, near Pyengana, Tas., scattered singly on soil under *Eucalyptus*, *Acacia* and *Pomaderris*, 21 May 2010, *P.B.Matheny, G.Gates & D.Ratkowsky PBM3366*; holo: PERTH 08516774; iso: TENN 066002 (KP171038-nLSU, KM555133-*rpb2*).

Pileus 20–25 mm wide, broadly conical to plano-convex with a low broad obtuse umbo; margin decurved but uplifted somewhat with age; surface smooth and unbroken on the disc; velipellis absent or present; dry but shiny, radially fibrillose and rimulose towards the margin, not distinctly rimose; uniformly dark brown or dark yellowish brown (10YR 3/3, 3/4, 4/3) other than when a pallid (whitish) velipellis is present over the disc; context whitish or brownish, thick under the disc but thin elsewhere. *Lamellae* adnexed to sinuate or ascending-adnate, close to crowded, L c. 40, with several tiers of lamellulae; pale brown to light yellowish brown (10YR 6/3 to 6/4); edges pallid and indistinctly fimbriate, ventricose. *Stipe* 30–45 × 4–5 mm at the apex, terete; base bulbous, napiform or submarginate, up to 7 mm wide, surface satiny silky-fibrillose, nowhere pruinose except at the extreme apex, if at all; veil evanescent; ground colour light brown to brown above the base, generally pallid above and darker below; context solid, coloured like the surface, pallid in most areas but can be mixed with brown in age. *Odour* not remarkable, or weakly to strongly acidulous, similar to the odour of *Lycoperdon*. Pl. 155, 156.

Basidiospores 6.0–7.5 (–8.0) × 5.0–6.0 (–6.5) µm [mean 6.9 × 5.4 µm; Q 1.09–1.50; Q mean 1.30 (n = 30/2)], with mostly 5–7 (–8) low to moderate-sized nodules or merely corners about an angular or polygonal outline, yellowish brown; apiculus tapering. *Basidia* 30–31 × 7–9 µm, 4-sterigmate, clavate, hyaline. *Pleurocystidia* 45–63 × 13–20 µm, fusiform or ventricose, with a slender basal pedicel and an obtuse apex, ±thin-walled (walls to c. 1 µm thick), hyaline; apices crystalliferous or not. *Cheilocystidia* similar to pleurocystidia, sparse or absent on many lamellae that have a fertile edge; paracystidia not observed. *Caulocystidia* absent, terminal elements undifferentiated. *Pileipellis* a cutis of lightly encrusted, cylindrical hyphae, mostly 4-10 µm wide, dark yellowish brown in mass, with sometimes an interwoven superimposed velipellar layer comprised of narrow (3–5 µm wide) cylindrical smooth hyaline hyphae. *Clamp connections* present. Fig. 90.

Occurs in Tas.; singly on soil under *Nothofagus cunninghamii* in rainforest or in wet sclerophyll forest under *Eucalyptus*, *Acacia*, and *Pomaderris*. Fruiting May.

Additional specimens examined: Tas.: type locality, 21 May 2010, *P.B.Matheny PBM3368* (TENN 068357); *loc. id.*, *G.Gates & P.B.Matheny PBM3369* (TENN 069572); Weldborough Pass, near Pyengana, solitary on road bank under *Nothofagus cunninghamii* in rainforest, 24 May 2010, *P.B.Matheny PBM3404* (TENN 066001; KP171039-nLSU, KP636834-ITS).

Etymology: From the Latin *solitarius* (alone, solitary), in reference to the solitary nature of the fruit bodies.

Diagnostic features: Fruit bodies medium. Pileus dark brown to dark yellowish brown, smooth in the centre, with a rimulose margin; stipe bulbous with a napiform to submarginate bulb; veil evanescent. Spores nodulose; pleurocystidia ±thin-walled (to 1 µm thick); caulocystidia absent.

Inocybe solitaria is similar to several other species reported from Papua New Guinea, south-eastern Asia and New Zealand. These include *I. neoumbrina* Kobayasi, *I. mediocris* (Corner & E.Horak) Garrido and *I. subclavata* (E.Horak) Garrido. Horak's description of *I. neoumbrina* (Horak, 1979), based on material collected in Papua New Guinea, is most similar to *I. solitaria*, differing only in the smaller size of the stipe. *Inocybe neoumbrina* (type from Japan) might be more appropriate for two other Tasmanian collections (*I103*, *I13*; as *I.* cf. *neoumbrina*, see Fig. 87). Our notes of these latter collections are incomplete, but the fruit bodies are smaller than *I. solitaria*, and they have more slender stipes consistent with what has been ascribed to *I. neoumbrina* by Horak (1979a). However, in one of these collections the stipes are noticeably dark or fuscous or may be fuscescent after drying (but also observed in *PBM3369*; *I. solitaria*), features that might be diagnostic for this species.

Inocybe subclavata, described from New Zealand (Horak, 1977), differs from *I. solitaria* by its lighter (brown) pileus and ITS sequence divergence (5% dissimilarity; GU233360). In addition, the spores of *I. solitaria* are more coarsely nodulose than those of *I. subclavata*. This difference is evident from Horak's (1977) illustration of the holotype, and we confirmed it by examining the holotype of *I. subclavata* (PDD 2711) and two other collections from New Zealand: PDD 89376 (collected and identified by E.Horak in 2000) and PDD 97858 (collected and identified by P.B.Matheny in 2009). *Inocybe mediocris* (Corner & E.Horak) Garrido, described by Horak (1979a) from Singapore, is also similar to *I. solitaria*, but differs in its significantly paler colouration, *viz.* a pale ochraceous pileus covered with ochraceous to fuscous fibrils and a white stipe that becomes pale ochraceous.

Inocybe alloumbrina, another species that could be confused with *I. solitaria*, has more coarsely nodulose spores. DNA sequence analysis clusters it with two species with much darker lamellae and tendency towards more robust fruit bodies (*I. cacaocolor* and *I. subferruginea*; Fig. 87).

Inocybe solitaria and *I. subclavata* may be capable of forming ectomycorrhizal associations with a broad range of host plants. *Inocybe solitaria* occurs in wet sclerophyll forest and *Nothofagus* rainforest communities in Tasmania, while *I. subclavata* has been collected from myrtaceous (*Kunzea*) and *Nothofagus* forests in New Zealand.

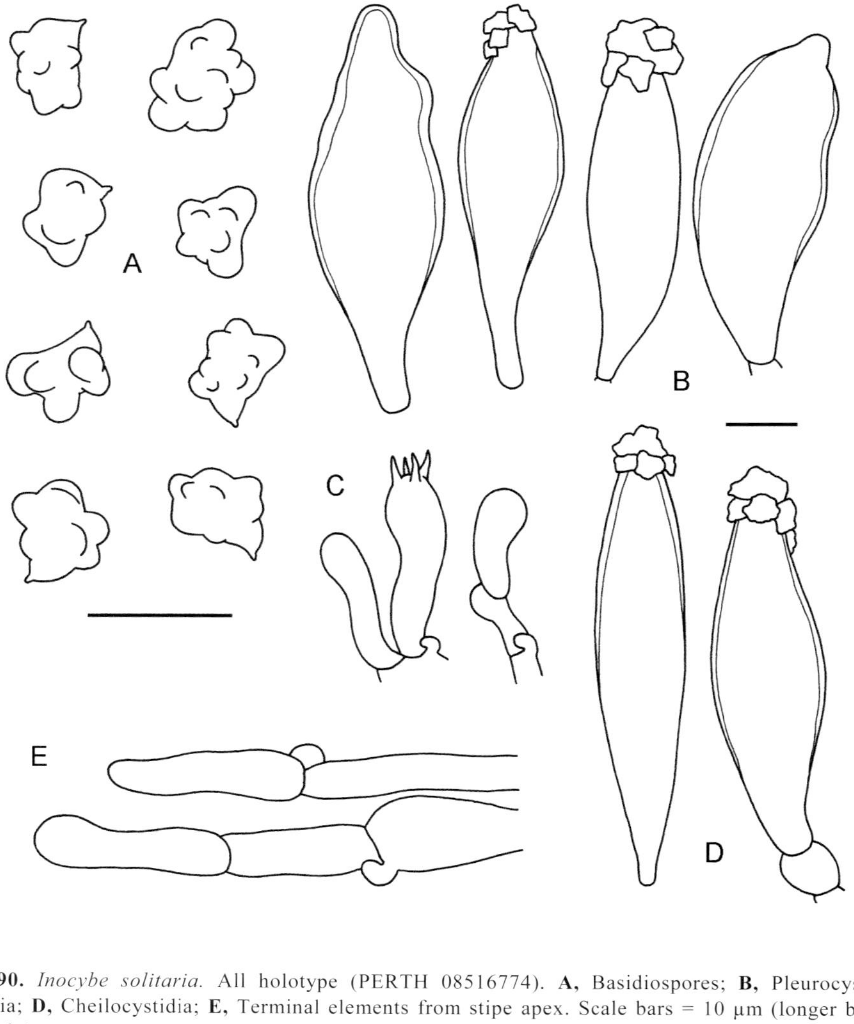

Figure 90. *Inocybe solitaria.* All holotype (PERTH 08516774). **A,** Basidiospores; **B,** Pleurocystidia; **C,** Basidia; **D,** Cheilocystidia; **E,** Terminal elements from stipe apex. Scale bars = 10 μm (longer bar for spores only).

Plate 155. *Inocybe solitaria*, holotype (PERTH 08516774). Photograph — P.B.Matheny.

Plate 156. *Inocybe solitaria*. TENN 066001. Photograph — P.B.Matheny.

78. Inocybe subferruginea Matheny & Bougher, *sp. nov.*

MycoBank: MB 809464

Type: Ryedene property, W of Two Peoples Bay, W.A., in litter in woodland of *Eucalyptus marginata*, *Corymbia calophylla* and *Allocasuarina*, 22 June 2007, *N.L.Bougher & L.Bell NLB319*; holo: PERTH 08074437 (KP171070-nLSU, KP636874-ITS, KM656115-*rpb2*).

Pileus 10–40 mm wide, bluntly conical to obtusely conical when immature, expanding to convex or campanulate, eventually plane, with a depressed centre and an uplifted margin; umbo developing or not; margin initially incurved, soon decurved, with overhanging brown fibrils when young; surface dry; velipellis absent; typically appressed-squamulose at the centre and occasionally towards the margin; squamules recurved at times, radially-fibrillose and rimulose to rimose towards the margin; very dark brown, dark yellowish brown, to dark reddish brown (6F5/6; Mummy Brown, Bistre, or Prout's Brown), paler between the spreading fibrils towards the margin; context pallid, thickest under the disc. *Lamellae* adnexed to sinuate, moderately close to close, 30–40 L, with several tiers of lamellulae, initially pale grey to pinkish grey (near 5B3 or 6C3), becoming light brown (7.5YR 6/4), brown (7.5YR 5/4) and eventually dark yellowish brown (10YR 4/4) and rusty with extreme age, medium to broad, 3–5 mm deep, becoming ragged; edges indistinctly pallid-fimbriate. *Stipe* 30–35 × 3–5 mm at the apex, terete, even or somewhat widened at the base but lacking a distinct bulb; partial veil a cortina, fugacious; surface pruinose at the extreme apex only or nowhere obviously pruinose, tattered to densely fibrillose with brown fibrils; apex initially pallid, elsewhere brown to yellowish brown and at times darker near the base, clods of soil often present at the base; context solid; cortex brown, interior pallid and pithy. *Odour* none. Pl. 157, 158.

Basidiospores 7.0–10 (–10.5) × 5.5–7.0 μm [mean 8.3 × 6.2 μm; Q 1.08–1.57 (–1.64); Q mean 1.33 (n = 49/3)], coarsely nodulose about a polygonal (amygdaliform) outline but shape variable in size and magnitude and frequency of nodulation, mostly with 8–12 moderate-sized nodules, yellowish brown to brownish yellow; apiculus small. *Basidia* 27–30 × 8–10 μm, clavate, hyaline, 4-sterigmate (rarely 2-sterigmate). *Pleurocystidia* 51–65 × 14–22 μm, fusiform to fusiform-ventricose, necks short or well-developed, with a short basal pedicel; apices rounded to obtuse, crystals not observed; thin-walled, hyaline, infrequent. *Cheilocystidia* similar to pleurocystidia but more variable in shape, some shorter, mixed with clavate paracystidia, frequent. *Caulocystidia* present as caulocystidioid cells; these terminal and differentiated cells of vesture hyphae, fusiform to lageniform or more variably shaped, at times with cinnamon pigmentation and with slightly thickened walls; cauloparacystidia not observed. *Pileipellis* a cutis of russet or dark reddish brown hyphae in mass, cylindrical to inflated, 7–18 μm wide, thin-walled or with slightly thickened walls, encrusted; septa rather frequent. *Clamp connections* present. Fig. 91.

Occurs in south-western W.A., between Perth, Denmark and Two Peoples Bay; on soil in Jarrah woodland and forest under various eucalypts and/or planted *Acacia*, or in grass under planted *Corymbia ficifolia*. Fruiting June–Nov.

Additional specimens examined: W.A.: Bold Park, near corner of Rochdale Rd and Stevenson Ave., peg 3173, City Beach, Perth, under *Eucalyptus marginata* and *Banksia menziesii*, 23 June 1999, *N.L.Bougher, J.Catchpole & S.Q.Bolsenbroek E6127* (PERTH 7589255); near L. Byleveld, outside William Bay Natl Park, on ground and on buried rotten wood and sticks, under *Corymbia calophylla*, *E. patens* and introduced *Eucalyptus* sp., 23 June 2001, *N.L.Bougher PBM2123 / E6750* (PERTH 07652003); *loc. id.*, *N.L.Bougher PBM2126 / E6759* (PERTH 07677650); Whiteman Park, Perth greater metropolitan area, under *Eucalyptus rudis*, 16 Sep. 2001, *I.C.Tommerup E7066* (PERTH 07676905; EU600894-nLSU, KP636875-ITS); *loc. id.*, on open ground among *Lomandra* sp. in *Melaleuca raphiophylla* woodland, 13 Nov. 2011, *K.J.Tullis KT11-010* (PERTH 08354642); Walpole, among grass under planted *Corymbia ficifolia*, 24 June 2006, *N.L.Bougher NLB226* (PERTH 07699700; KP636876-ITS); *loc. id.*, 14 July 2011, *N.L.Bougher, M.Ryberg & P.B.Matheny PBM3621* (PERTH 08319650, TENN 066658; KP171071-nLSU, KP636877-ITS, KM656116-*rpb2*); Jarrahdale Rd, 7.7 km W of Albany Hwy, 32°19'44.4"S, 116°07'42.9"E, scattered singly on sandy soil at edge of clearing under *Corymbia calophylla*, *Eucalyptus marginata* and planted *Acacia*, 25 July 2011, *P.B.Matheny PBM3674* (TENN 066497; KP171072-nLSU, KP636878-ITS); Porongorup Natl Park, 34°41'05.3"S, 117°55'53.7"E, in mixed forest of *Corymbia calophylla*, *Eucalyptus diversicolor* and *E. marginata*, 21 July 2011, *N.L.Bougher NLB851* (PERTH 08320810).

Plate 157. *Inocybe subferruginea*, holotype (PERTH 08074437). Photograph — N.L.Bougher.

Plate 158. *Inocybe subferruginea*. PERTH 08319650. Photograph — P.B.Matheny.

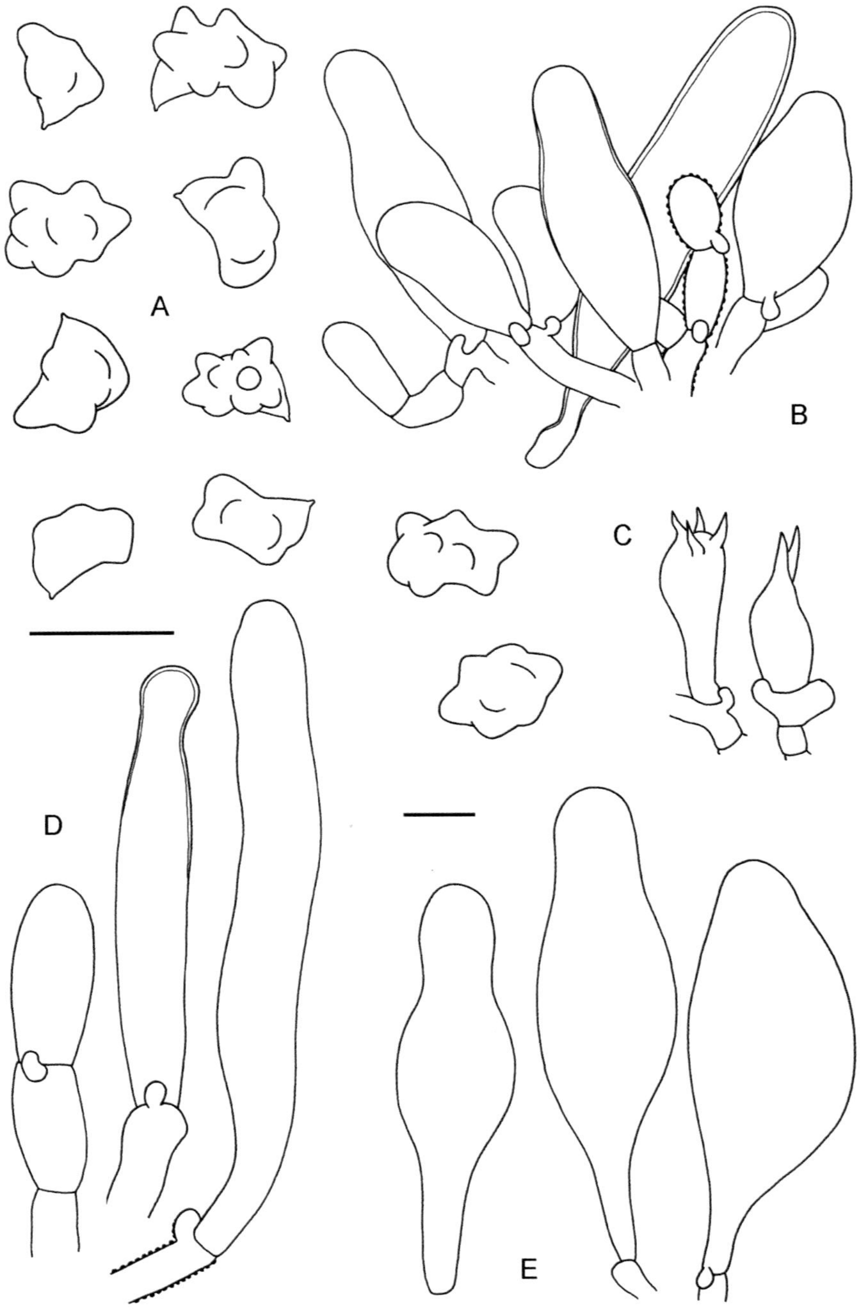

Figure 91. *Inocybe subferruginea.* All holotype (PERTH 08074437). **A,** Basidiospores; **B,** Cheilocystidia and paracystidia; **C,** Basidia; **D,** Caulocystidioid elements from upper stipe; **E,** Pleurocystidia. Scale bars = 10 μm (longer bar for spores only).

Etymology: From the Latin prefix *sub-* (almost) and *ferrugineus* (rust-coloured), in reference to the overall dark brown to dark reddish brown colouration and the rusty-brown lamellae of old specimens.

Diagnostic features: Fruit bodies medium. Pileus typically appressed-squamulose, very dark brown, dark yellowish brown to dark reddish brown; lamellae rusty-brown with age; stipe coarsely fibrillose, pruinose at the extreme apex only.

Inocybe subferruginea is most closely related to *I. cacaocolor* and the two species form a cryptic species pair in the *I. assimilata–I. napipes* group (Fig. 87). For a discussion of related and superficially similar taxa, see comments under *I. cacaocolor*.

79. Inocybe subangustifolia Matheny, Bougher & Halling, *sp. nov.*

MycoBank: MB 809463

Type: Cathedral Beach, Fraser Island, Qld, 25°13'07.8"S, 153°15'22.8"E, alt. 40 m, gregarious on sand under *Eucalyptus*, 18 May 2010, *R.E.Halling, N.Fechner & M.Castellano REH9319*; holo: BRI 794269; iso: PERTH 08517282, NY (KP171067-nLSU, KP636870-ITS, KM656112-*rpb2*).

Pileus 10–24 mm wide, convex to plane, sometimes with a low to acute umbo; margin initially slightly incurved, later decurved or upturned with age; surface dry, densely radially appressed-fibrillose, with a matted fibrillose centre, not rimose; margin entire (not translucent-striate); yellow or dull yellowish (near 5Y 7/6) to greenish yellow or olivaceous yellow, becoming pale orange-brown with age and drying *in situ*; context pastel-yellow to watery greenish yellow, orange where old and damaged but not changing from yellow where cut or bruised. *Lamellae* adnexed, close to crowded, up to 52 L with abundant lamellulae, narrow, shallow (up to 1.5 mm deep), not ventricose; greenish yellow to bright yellow, becoming pale olive-yellow or dull yellow-olivaceous (near 2.5Y 6/6, 3B5) with age; edges concolorous with the faces and minutely fimbriate. *Stipe* 15–40 × 2–3 mm at the apex, equal, without a bulbous base, cylindrical but often bent; surface dry; partial veil absent; pruinose for its entire length, silky-fibrillose to finely fibrillose-striate; shiny pastel-yellow (2A4), pale greenish yellow or pale yellow, developing some pale brownish to pale pinkish brown stains with age; context coloured like the surface, not changing colour where bruised, more orange where damaged or old. *Odour* spermatic. Pl. 159, 160.

Basidiospores 6.5–8.5 (–10.0) × 5.0–6.5 µm [mean 7.4 × 5.4 µm; Q 1.17–1.60; Q mean 1.38 (n = 20/1)], nodulose with mostly 7–10 moderate-sized conical or obtuse or less often flat-topped nodules, yellowish brown. *Basidia* 23–28 × 7–9 µm, clavate, hyaline, 4-sterigmate. *Pleurocystidia* 40–53 × 12–16 µm, mostly fusiform, occasionally subcylindrical or sublageniform, generally tapered at both ends, lacking a distinct basal pedicel, at times with a truncate base; thick-walled (3–4 µm thick), the walls yellow to hyaline; apices obtuse and crystalliferous. *Cheilocystidia* similar to pleurocystidia but at times smaller, mixed with hyaline paracystidia. *Caulocystidia* descending the entire length of the stipe, mixed with hyaline cauloparacystidia, similar to pleurocystidia and cheilocystidia but hyaline. *Pileipellis* with an interwoven layer of velipellis hyphae which are narrow, 2–4 µm wide, smooth, superimposed on top of somewhat broader faintly encrusted hyphae to 12 µm; buff overall in mass. *Clamp connections* present. Fig. 92.

Occurs on sand or soil under *Eucalyptus* spp. or *Acacia* spp. in north-eastern and south-eastern Qld. Fruiting Feb.–May.

Additional specimens examined: Qld: Mount Lewis road, Julatten, c. 8 km W of Mount Molloy–Mossman road, near *Acacia* and occasional *Eucalyptus*, 11 Feb. 2006, *T.Osmundson 1109* (BRI 794596); Queen Mary Falls, Main Range Natl Park, on soil among litter in eucalypt forest (mixed *Eucalyptus* species), 4 Mar. 2012, *N.L.Bougher NLB983* (PERTH 08364850; KP171068-nLSU, KP636871-ITS, KM656113-*rpb2*).

Etymology: From the Latin prefix *sub-* (almost) and *angustifolia* (narrow-leaved), in reference to its overall similarity to *Inocybe angustifolia*.

Diagnostic features: Fruit bodies small. Pileus yellow to olive-yellow, fibrillose to matted-fibrillose; lamellae narrow, close to crowded, yellow to olive-yellow; stipe even (not bulbous), entirely pruinose, yellow to pale olive-yellow; odour spermatic. Spores nodulose; hymenial cystidia thick-walled. On soil in tropical to warm-temperate Queensland.

Inocybe subangustifolia is very similar to *I. angustifolia* (Corner & E.Horak) Garrido (Horak, 1979a) from south-eastern Asia, but differs in its smaller size, ecology, the absence of a marginate basal bulb, and in DNA sequence differences (ITS, nLSU, and *rpb2*; compared to specimens of *I. angustifolia* from Thailand; GQ892990, GQ892989, GQ892988; Horak *et al.* 2015). Turnbull & Watling (1999) suggested *I. angustifolia* is the most widespread species of *Inocybe* in Indomalaya and Australasia, but this species has not yet been confirmed to occur in Australia.

This is the first report of a species of the *I. angustifolia* subgroup from Australia. *Inocybe subangustifolia* is also phylogenetically related to several northern temperate species, *viz. I. paludinella* (Peck) Sacc., *I. alabamensis* Kauffman (a taxonomic synonym of *I. paludinella*), *I. argenteolutea* Vauras and *I. umbratica* Quél. *Inocybe straminea* (E.Horak) Garrido, known only from its type collection in New Zealand, differs in its larger stature, larger spores (8–10 × 5–6 µm), and longer cystidia, but it shares a similar overall colouration (yellow throughout), and occurrence in sand (Horak, 1977).

Blastn searches based on the ITS sequence of *I. subangustifolia* found two other sequences with 98% similarity from roots of *Pinus thunbergii* in South Korea (Obase *et al.*, 2011), suggesting that *I. subangustifolia* might be more widespread than is currently recognised and composed of genetic variants.

Plate 159. *Inocybe subangustifolia*, holotype (BRI 794269). Photograph — R.E.Halling.

Plate 160. *Inocybe subangustifolia*. PERTH 08364850. Photograph — N.L.Bougher.

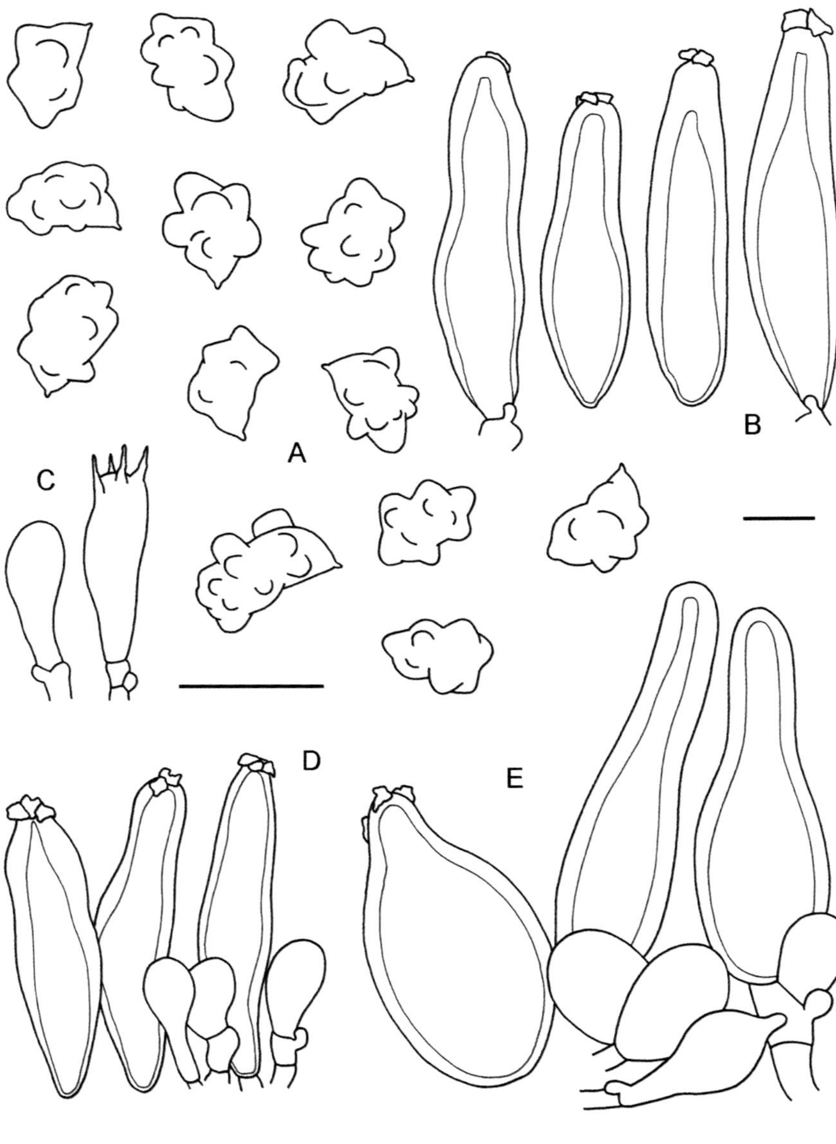

Figure 92. *Inocybe subangustifolia.* All holotype (BRI 794269). **A,** Basidiospores; **B,** Pleurocystidia; **C,** Basidia; **D,** Cheilocystidia and paracystidia. **E,** Caulocystidia and cauloparacystidia. Scale bars = 10 µm (longer bar for spores only).

INOCYBE ASTEROSPORA GROUP

The *I. asterospora* group is largely a north temperate and south-east Asian clade which is characterized by species with an entirely pruinose stipe and marginate bulb, stellate spores, and presence of caulocystidia the entire length of the stipe. We recognize two new Australian species, *I. nobilissima*, and *I. vagata*, the latter also occurring in New Zealand. As in the *I. assimilata* group, the austral species occupy evolutionarily derived phylogenetic positions. Both of the austral species also feature strong spermatic odours.

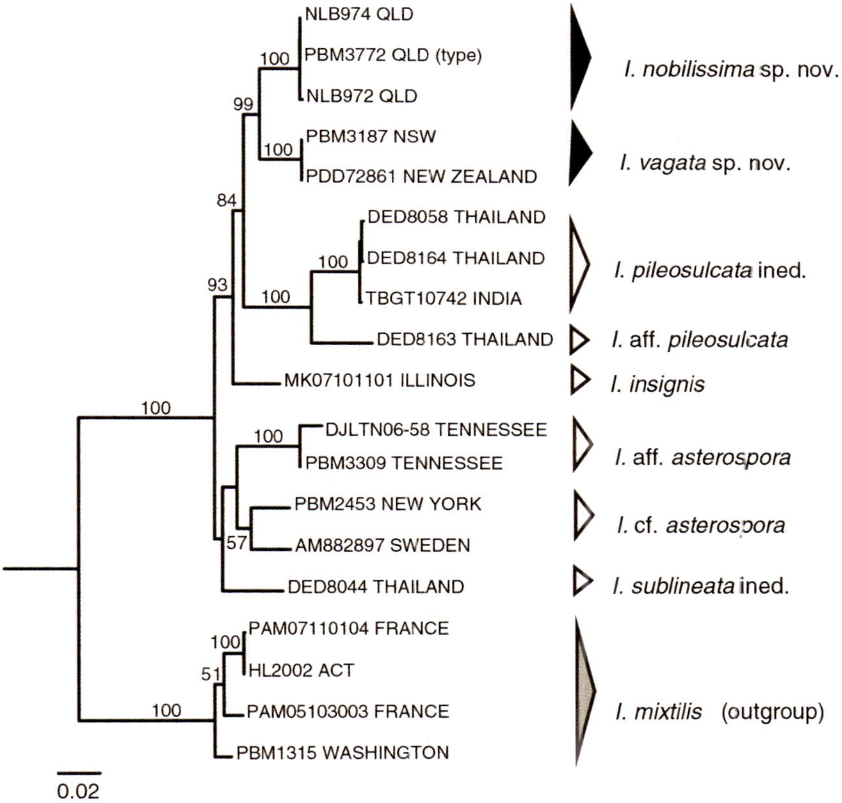

Figure 93. Maximum Likelihood phylogram of the *Inocybe asterospora* group. Bootstrap values are shown above the branches. The tree was inferred from combined ITS, nLSU, and *rpb2* gene regions and includes 2874 total sites. Twenty eight sites from the ITS region too ambiguous to align were removed prior to analysis. Australian species are indicated by black triangles. Species from Australia and New Zealand, or elsewhere, are indicated by grey-filled triangles. White triangles indicate taxa from non-Australasian regions. Of the 19 branch tips included, 13 are represented by ITS sequences, all by nLSU, and 13 by *rpb2*. *Inocybe mixtilis* (introduced to Australia) was used for outgroup purposes.

80. Inocybe nobilissima Matheny & Bougher, *sp. nov.*

MycoBank: MB 809394

Type: Cunninghams Gap, Main Range Natl Park, Qld, 28°02'9"S, 152°23'37.3"E, alt. 776 m, on soil in dry sclerophyll forest under *Eucalyptus* and *Allocasuarina*, 29 Feb. 2012, *R.E.Halling & N.Fechner PBM3772*; holo: PERTH 08363153; iso: TENN 67004 (KP170959-nLSU, KP308789-ITS, KM406202-*rpb2*).

Pileus 13–35 mm wide, deeply conical to bluntly conico-campanulate; margin initially incurved, later decurved; surface dry; velipellis absent or an inconspicuous to conspicuous greyish area over the disc when young, otherwise smooth and unbroken over the disc, shiny, and initially finely fibrillose towards the margin and its edge with scattered white silky fibrils; with age developing a large obtuse umbo; margin long-rimose and radially fibrillose after pileal expansion; uniformly dark brown (7.5YR 2/2) in youth, remaining so at the centre but shading umber or strong brown (7.5YR 4/4) to dark greyish brown (Bay or 10YR 3/4) towards the margin, much paler brown between spreading fibrils; context pallid or almost white, not changing color upon exposure, thin except under the disc. *Lamellae* adnexed to subfree, crowded or closely spaced, c. 50–60 L, with few or several tiers of lamellulae; initially very pale brown or pallid (10YR 8/3), becoming brown to yellowish brown (10YR 5/3–5/4), with edges indistinctly pallid-fimbriate, subventricose or slightly ventricose, narrow to medium, 2-4 mm deep. *Stipe* 45–75 × 3–4 mm, terete to slightly compressed, at times flexuous, partial veil not observed, pruinose the entire length down to just above the distinct marginate bulb, satiny-fibrillose in appearance, the bulb up to 8 mm wide; surface Pinkish Buff to mostly Cinnamon Buff, at times a little paler at the apex (near "pink" 7.5YR 7/4 or pinkish grey) or very pale brown (10YR 8/4), the bulb paler or whitish; context solid, with a pallid interior, at times with pinkish tinges apically, white in the base. *Odour* strongly spermatic when first cut. Pl. 161–163.

Basidiospores 9.0–12.0 × 8.0–10.0 (–10.5) µm [mean 10.5 × 8.8 µm; Q 1.00–1.33; Q mean 1.19 (*n* = 34/3)], stellate with mostly 8–10 large conical projections to 2–3 µm high, yellowish brown, with a small indistinct apiculus. *Basidia* 31–37 × 10-12 µm, 4-sterigmate, clavate with a broad apical region, hyaline. *Pleurocystidia* 47–61 × 14–18 µm, fusiform, thick-walled; walls 2–4 µm thick, hyaline; apices crystalliferous. *Cheilocystidia* similar to pleurocystidia, mixed with crowded clavate paracystidia. *Caulocystidia* similar to cheilocystidia and pleurocystidia but more variable in shape and size, clustered, becoming more scattered below the stipe centre and towards its base, mixed with clavate and variably-shaped cauloparacystidia. *Pileipellis* a cutis of repent cylindrical encrusted hyphae mostly 5–12 µm wide, strong brown in mass. *Clamp connections* present. Fig. 94 (see also Pl. 7I, 8D).

Occurs in south-eastern Qld; on soil among litter under *Acacia*, *Allocasuarina* and *Eucalyptus* spp. in sclerophyll forest. Fruiting Feb.–Apr., and (once) in Oct.

Additional specimens examined: Qld: Mt Glorious, on soil, 1 Apr. 1972, *A.Young 193* (BRI 775475); Bunya Mountains Natl Park, gregarious on soil beside track in eucalypt forest, 15 June 1987, *A.Young 1150* (BRI 808422); *loc. id.*, gregarious on soil under *Acacia* in eucalypt forest, 1 Apr. 1989, *A.Young 1325* (BRI 808409); Banksia Beach, Bribie Is., Wallum, on sandy soil near *Banksia* sp., 27 June 2009, *P.Leonard 25609* (BRI 793924); Balmoral Bridge, 'Dilkusha' Nature Refuge, Maleny, on crusher dust/gravel in plant nursery area, 11 Oct. 2010, *F.E.Guard FEG210002* (BRI 795089); Mount Cordeaux Track (near Rainforest Circuit), Cunninghams Gap, Main Range Natl Park, 28°02'59"S, 152°23'37.3"E, on soil among litter in tall sclerophyll forest including eucalypts and *Allocasuarina*, 3 Mar. 2012, *N.L.Bougher NLB972* (PERTH 08364788, TENN 067749; KP170960-nLSU, KP308790-ITS, KM406203-*rpb2*); *loc. id.*, 3 Mar. 2012, *N.L.Bougher NLB974* (PERTH 8364818, TENN 067751; KP170961-nLSU, KP308791-ITS, KM406204-*rpb2*).

Etymology: From the Latin *nobilissima* (most noble), in reference to the stately appearance and stellate spores.

Diagnostic features: Pileus dark brown to strong brown, conspicuously rimose; lamellae narrow or medium but crowded; stipe long and with a distinct marginate bulb; odour strongly spermatic. Spores large and stellate.

This is one of two Australian species in the *I. asterospora* Quél. group (Fig. 93), the other being *I. vagata* Matheny & Bougher. *Inocybe nobilissima* and *I. vagata* can only be accurately distinguished by DNA sequence data and are sister taxa (Fig. 93). *Inocybe nobilissima* has a more richly coloured brown pileus than *I. vagata* and it can have a more slender, taller aspect with more crowded and narrow lamellae. Additional field collections are needed to determine if these gross morphological differences are reliable. Microscopically, the two species cannot be distinguished.

Inocybe nobilissima and *I. vagata* have a strong spermatic odour that is similar to the New Zealand species *I. graveolens* (E.Horak) Garrido, and *I. nobilissima* shares its stature with *I. horakomyces* Garrido (= *Astrosporina avellana* E.Horak) also described from New Zealand. However, those two taxa have not been recorded in Australia, and neither one has stellate spores. In addition, *I. horakomyces*, *I. calopedes*, *I. scissa* (E.Horak) Garrido and *I. leptospermi* (E.Horak) Garrido form a clade (Fig. 122) distantly related to the *I. asterospora* group (Fig. 93), and they also lack the diagnostic stellate spores of that group.

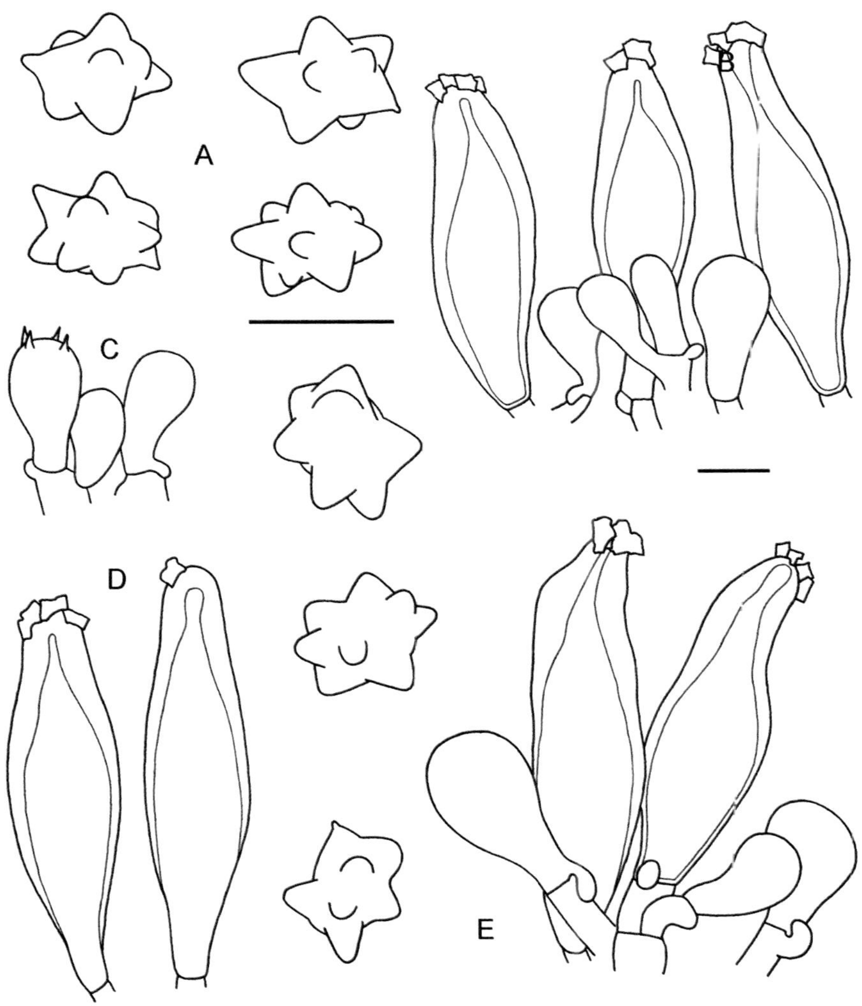

Figure 94. *Inocybe nobilissima.* All holotype (PERTH 08363153). **A,** Basidiospores; **B,** Cheilocystidia and paracystidia; **C,** Basidia; **D,** Pleurocystidia; **E,** Caulocystidia and cauloparacystidia. Scale bars = 10 μm (longer bar for spores only).

Plate 161. *Inocybe nobilissima*, holotype (PERTH 08363153). Photograph — P.B.Matheny.

Plates 162, 163. *Inocybe nobilissima*. PERTH 08364788. Photographs — N.L.Bougher.

Inocybe nobilissima has been confirmed only from mixed sclerophyll forest in south-eastern Queensland, whereas *I. vagata* occurs in New South Wales and New Zealand. While the two taxa do not appear to be sympatric, additional collections should clarify their geographic ranges.

Another specimen (NLB953, PERTH 08364648), from eucalypt woodland mixed with *Allocasuarina* in south-eastern Queensland (Glass House Mountains Natl Park), also has stellate spores, but it differs from *I. nobilissima* in its entirely white stipe, yellowish brown or fulvous pileus, and a lack of any odour. It is currently unidentified. The recently described *I. pileosulcata* E.Horak, Matheny & Desjardin from Thailand and Malaysia (Horak *et al.* 2015) and India (Pradeep *et al.* 2016) is morphologically similar and phylogenetically close to *I. nobilissima* but is distinguished by its lack of any distinctive odour and by ecological association with dipterocarps and *Castanopsis*, these at times mixed with *Pinus*. *Inocybe insignis* A.H.Sm. is another closely related species, known only from North America. It differs most obviously by its greenish blue bruising reaction of the fruit bodies (Kuo & Matheny 2015).

81. Inocybe vagata Matheny & Bougher, *sp. nov.*

MycoBank: MB 809477

Type: Hammerhead Point, Jervis Bay Natl Park, N.S.W., 34°59'E, 150°47'S, solitary on soil under leaf litter beneath *Eucalyptus*, 19 May 2009, *P.B.Matheny PBM3186*; holo: CANB 782346; iso: TENN 066142.

[*Astrosporina asterospora auct. non* (Quél.) Rea: E.Horak, *New Zealand J. Bot.* 15: 739 (1977)]

Pileus to 20 mm wide, conical, lacking an umbo; disc with adhering soil particles but not viscid; velipellis absent; surface finely silky-fibrillose, often rimose or becoming strongly rimose with age; centre brown (10YR4/3; Saccardo's Umber), shading to yellowish brown (10YR 5/3 to 5/4; Tawny Olive) towards the margin; context firm, pallid, not changing colour when cut or bruised. *Lamellae* adnexed, moderately close, pale brown to brown (10YR 6/3 to 5/3), with edges pallid-fimbriate. *Stipe* to 30 × 4 mm at the apex, below with a distinct marginate bulb to 8 mm wide, lacking a veil; surface densely pallid-pruinose with a satiny streaked appearance; ground color light brown (7.5YR 5/4; near Wood Brown or Avellaneous) with a flesh-coloured tint; context pallid with a pinkish cinnamon cortex, white in the bulb. *Odour* strongly spermatic. Pl. 164.

Spores 10.0–13.0 × 7.0–10.0 µm [mean 11.3 × 8.9 µm; Q (1.07–) 1.14–1.48 (–1.57); Q mean 1.27 (n = 40/2)], stellate to coarsely nodulose, with 10–12 conspicuous conical nodules up to 4 µm tall, yellowish brown; apiculus tapered and often obscured. *Basidia* 33–40 × 11–13 µm, clavate, hyaline, 4-sterigmate. *Pleurocystidia* 62–70 × 14–16 µm, sublageniform or fusiform, at times with long narrow necks and subacute crystalliferous apices, thick-walled (2–3 µm), hyaline. *Cheilocystidia* similar to pleurocystidia, numerous, mixed with hyaline paracystidia. *Caulocystidia* numerous, similar to pleurocystidia and cheilocystidia, but some longer and cylindrical or tapering, mixed with hyaline cauloparacystidia, descending the entire length of the stipe. *Pileipellis* a cutis with an outer layer composed of subparallel to interwoven hyphae, these cylindrical and mostly 5–8 µm wide, subhyaline to light yellowish brown in mass; outer layer subtended by a darker brown layer of encrusted hyphae to 15 µm wide. *Clamp connections* present. Fig. 95.

In Australia, known from soil under leaf litter beneath *Eucalyptus* at the type locality in south-eastern N.S.W. Also in New Zealand, on soil under *Nothofagus menziesii* or *Leptospermum scoparium* and *Kunzea ericoides*. Fruiting May in Australia; Mar., May and June in New Zealand.

Additional specimen examined: N.S.W.: type locality, 19 May 2009, *P.B.Matheny PBM3187* (CANB 782347, TENN 063979; KP171091-nLSU, KP641639-ITS, KM656135-*rpb2*).

Extralimital specimens examined: New Zealand: North Island: Clark's Bush, Titirangi, Waitemata Co., Auckland, on soil in association with *Leptospermum*, 13 June 1973, *J.M.Dingley* (PDD 30849); South Head, Hokianga Harbour, Omapere, Northland District, on soil under *Leptospermum scoparium* and *Kunzea ericoides*, 24 May 2001, *E.Horak & A.Horak* (PDD 72861; HM060325-nLSU, HM060326-ITS). South Island: Stewart Falls, Haast Pass, Otago Lakes area, on soil in association with *Nothofagus menziesii*, 6 Mar. 1983, *E.Horak* (PDD 89389); Reefton, Little Grey R. valley, Maimai, Buller area, on soil in association with *Nothofagus menziesii*/*truncata*, 19 Mar. 1983, *E.Horak* (PDD 89390).

Etymology: From the Latin *vagatus* (wandering), in reference to the comparatively broad geographic distribution of this species in eastern Australia and New Zealand.

Diagnostic features: Fruit bodies medium sized; pileus brown, rimose; stipe pruinose with a distinct marginate bulb; odour strongly spermatic. Spores stellate and large, 10.0–13.0 × 7.0–10.0 µm.

Inocybe vagata is described here as a new species to circumscribe *Astrosporina asterospora sensu* Horak. The northern temperate name *Astrosporina asterospora* (Quél.) Rea was applied in a broad sense by Horak to include samples from Papua New Guinea, the Solomon Islands, Malaysia and New Zealand (Horak, 1977). Here, we confirm the conspecificity of Australian (N.S.W.) and New Zealand collections (PDD 72861, identified by Horak as "*A. asterospora*") based on ITS and nLSU sequence comparisons. Furthermore, *I. vagata* shares only an 86–87% similarity with northern European sequences of *I. asterospora* Quél. Note that *Astrosporina* is not monophyletic (Matheny *et al*., 2002; Matheny, 2005), thus *Inocybe* is used.

The hymenial cystidia of the Australian collections of *I. vagata*, are predominantly more slender and attenuated in their upper part than those of the New Zealand material. We confirm that PDD 30849 (illustrated as "*Astrosporina asterospora*" by Horak, 1977) has cystidia that are mostly broader in their distal region. However, in some of the other collections from New Zealand (e.g. PDD 72861), we observed a range of cystidial forms. Therefore, we consider this variation in cystidial form to be acceptable within the concept of this species. It is notable that the illustration of *A. asterospora* provided by Horak (1979a) from Borneo depicts fusoid/lageniform cystidia similar to the dominant form in the holotype of *I. vagata*.

Inocybe vagata is closely related to *I. nobilissima* from Queensland. The two species cluster in the *I. asterospora* group (Fig. 93), but microscopically they cannot be readily distinguished. *Inocybe nobilissima* has a more richly coloured brown pileus than *I. vagata*, and it can have a more slender and taller aspect with more crowded, narrow lamellae.

Plate 164. *Inocybe vagata*, holotype. Photograph — P.B.Matheny.

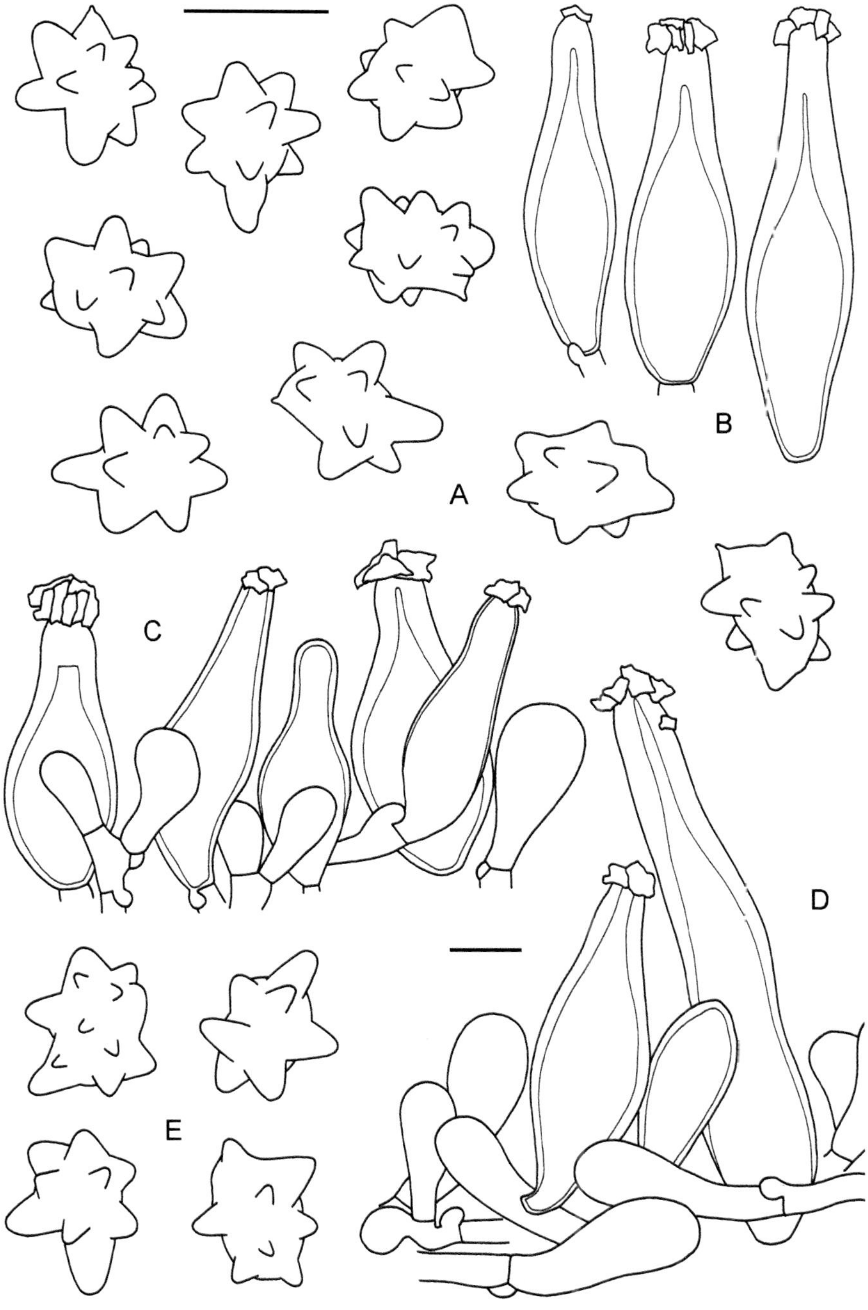

Figure 95. *Inocybe vagata.* All holotype, except E (PDD 72861). **A,** Basidiospores; **B,** Pleurocystidia; **C,** Cheilocystidia and paracystidia; **D,** Caulocystidia and cauloparacystidia; **E,** Basidiospores.
Scale bars = 10 µm (longer bar for spores only).

INOCYBE CALIDA GROUP

Three Australian species from Queensland and the Northern Territory are nested in this otherwise temperate and south-east Asian clade. The group is characterized by a pruinose stipe often with a marginate bulb, presence of caulocystidia and cauloparacystidia below the mid-region of the stipe, and nodulose (non-stellate) spores. The Australian species lack any remarkable odours.

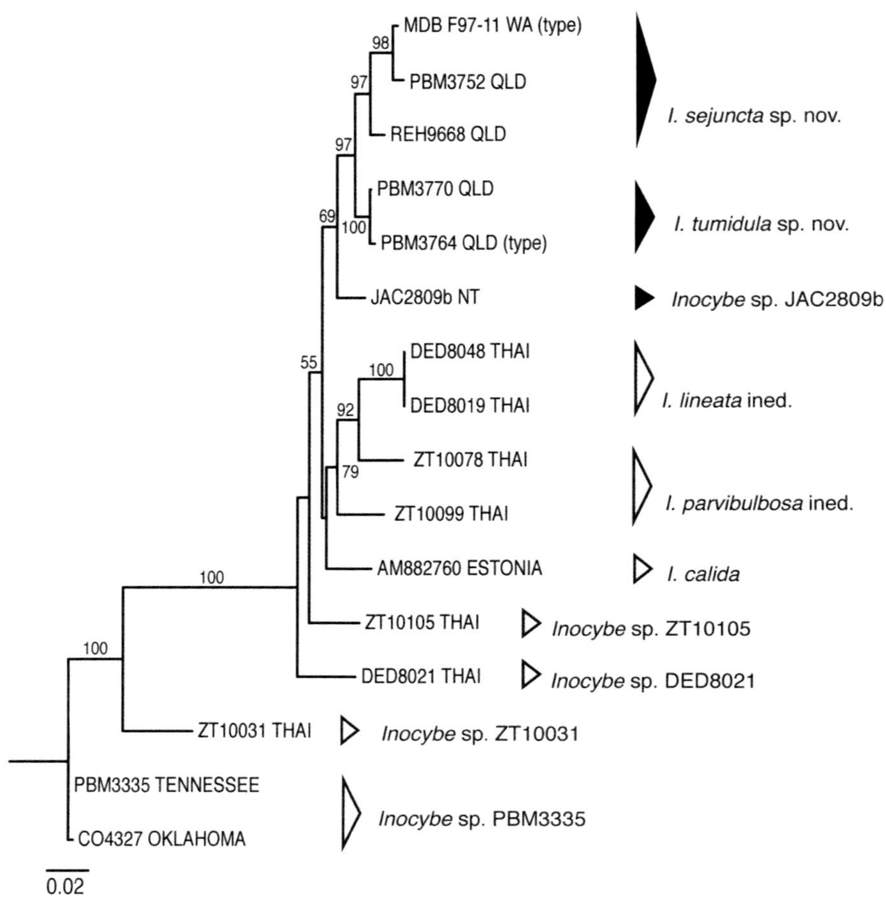

Figure 96. Maximum Likelihood phylogram of the *Inocybe calida* group. Bootstrap values are shown above or below branches (or next to nodes). The tree was inferred from combined ITS, nLSU, and *rpb2* gene regions and includes 2892 total sites. Species from Australia are indicated by black-filled triangles. White triangles denote non-Australian species. The tree is rooted with *Inocybe* sp. *PBM3335* based on inspection of individual family-level trees of nLSU (data not shown). Sixteen branch tips are included, all represented by nLSU, 12 by ITS, and 7 by *rpb2*.

82. **Inocybe sejuncta** Matheny, Bougher & M.D.Barrett, *sp. nov.*

MycoBank: MB 809457

Type: Great Northern Highway, 69 km due ENE of Broome, West Kimberley region, W.A., 17°47'10"S, 122°49'41"E, on sand in swampy, grassy savannah under *Eucalyptus tectifica*, *Corymbia polycarpa* and *Melaleuca* sp., 15 Feb. 2011, *M.D.Barrett MDB F97/11*; holo: PERTH 08383103; iso: TENN 068374 (KP171001-nLSU, KM555102-*rpb2*).

Pileus 13–22 mm wide, broadly conical, becoming almost plane, with a low rounded umbo; margin slightly decurved; surface dry; velipellis absent or present over the disc, smooth and unbroken in the centre, radially fibrillose or typically rimose towards the margin; warm reddish brown or umber to dark brown (6D8 to 6E8; 7.5YR 4/4), paler between the spreading fibrils. *Lamellae* adnexed, close, c. 40 L, with several tiers of lamellulae, ventricose, 2–3 mm deep; initially greyish brown or pale brown (5C4), becoming brown (5C6) with age; edges pale or nearly concolorous with faces, indistinctly fimbriate. *Stipe* 22–34 × 2.5–3.5 mm, central, cylindrical with a weak or distinct marginate to rounded basal bulb 4.0–5.5 mm wide, with tightly bound soil; surface dry, entirely pruinose with a streaked satiny appearance; buff to pale brown or becoming reddish brown apically (6B5, 6C5, 5B5; 10YR 7/3 to 6/3) mixed with white or isabelline to dull olive streaks; context hollow in the centre, fleshy fibrous, ±concolorous with the surface, whitish in the bulb. *Odour* not remarkable. Pl. 165, 166.

Basidiospores 9.0–11.5 × 5.0–6.5 µm [mean 9.9 × 5.8 µm; Q 1.50–1.92; Q mean 1.72 (*n* = 34/2)], angular to oblong-angular (particularly in face view) to trapeziform or subamygdaliform in outline, with 5–7 weakly formed nodules or corners, sometimes with one or more larger nodules; apiculus small but distinctive, yellowish brown. *Basidia* 26–29 × 8–10 µm, clavate to subventricose, occasionally widest below the apex, hyaline, 4-sterigmate. *Pleurocystidia* 55–65 × 15–20 µm, fusiform-ventricose, ventricose or utriform, tapered towards the base, with or without a distinct pedicel, thick-walled (mostly 2–4 µm), hyaline; apices obtuse, crystalliferous or not. *Cheilocystidia* similar to pleurocystidia but often shorter and with short necks, thick-walled especially at the apices, mixed with clavate hyaline thin-walled paracystidia often in a crowded palisade. *Caulocystidia* descending the entire length of the stipe, similar to cheilocystidia but more variable, utriform, subcylindrical or subfusiform, hyaline, thick-walled, in scattered clusters or isolated and accompanying slender clavate to subcylindrical hyaline thin-walled cauloparacystidia often in crowded clusters. *Pileipellis* a cutis of cylindrical hyphae mostly 7–15 µm wide, rusty-tawny in mass, walls thin or slightly thickened, encrusted but also with numerous coarse rusty interhyphal pigment deposits. *Clamp connections* present. Fig. 97 (see also Pl. 6B).

Occurs in the western Kimberley region, W.A. and in north-eastern and south-eastern Qld; on sand in swampy, grassy savannah under *Eucalyptus*, *Corymbia* and *Melaleuca* (W.A.), and on soil in woodland under eucalypts or *Acacia* (Qld). Fruiting Feb. and Mar.

Additional specimens examined: Qld: end of Emerald Creek Rd, Dinden S.F., 17°03'08.5"S, 145°32'24.2"E, in woodland near creek edge under eucalypts, 24 Feb. 2012, *P.B.Matheny PBM3752* (PERTH 08363293, TENN 066981; KP171002-nLSU, KM555103-*rpb2*); Hardings Paddock, Ipswich, 27°44'06.[illegible]"S, 152°46'01.6"E, gregarious on soil under *Acacia*, 6 Mar. 2012, *R.E.Halling & N.Fechner REH9668* (BRI, NY, PERTH 08516677; KP171003-nLSU, KP308828-ITS, KM555104-*rpb2*).

Additional specimen examined (genetic and/or morphological variants, as *Inocybe* sp.): N.T.: Cutta Cutta Caves, 25 km SE of Katherine, 14°35'S, 132°28'E, limestone outcrop in *Eucalyptus* woodland, on damp soil inside cave, 10 Apr. 1989, *J.A.Curnow JAC2809b* (CANB 8916878, in part; KP636838-ITS).

Etymology: From the Latin *sejuncta* (isolated), in reference to its occurrence in remote regions of northern Australia.

Diagnostic features: Similar to *Inocybe calocephala*, *I. calopedes* and *I. leptospermi*, but differs from these species by the angular to oblong-angular basidiospores with few (5–7) low nodules or corners, and a mainly tropical distribution in northern Australia.

Inocybe sejuncta superficially resembles *I. calocephala* and *I. calopedes*; all three have a dark or warm reddish brown pileus and pruinose stipe. The angular to oblong-angular spores with few low nodules or corners separates *I. sejuncta* from these species. *Inocybe leptospermi* (E.Horak) Garrido can be best distinguished in the field by its dark brown pileus, whitish stipe lacking pink tones, and non-oblong spores. We have yet to see *I. leptospermi*, *I. calopedes* or *I. calocephala* from northern regions of Australia where *I. sejuncta* has been found, but their latitudinal ranges overlap in south-eastern Queensland.

Inocybe sejuncta forms a monophyletic group with other taxa from Queensland and Northern Territory, including *I. tumidula* Matheny & Bougher, several undescribed taxa from Thailand, and the European species *I. calida* Velen. (Fig. 96). This topology supports a very recent derivation of the northerly Queensland lineages from south-east Asian lineages.

Inocybe tumidula can be distinguished from *I. sejuncta* by a subtle difference in pileus colour, at least based on the few collections examined: golden brown (7.5YR 7/6; 5D8), fulvous, or yellowish brown compared to the mostly warm reddish brown colour of *I. sejuncta*. Furthermore, the spores of *I. tumidula* are not oblong but rather angular-nodulose and broader (mostly 7.5–9.5 × 5.5–7.5 µm) but with few (6–8) nodules.

An unidentified specimen from the Northern Territory (*Inocybe* sp. *JAC2809b*) was also examined (Fig. 96). At present, our understanding of this specimen is incomplete due to inadequate documentation and insufficient material. Phylogenetically, it is closely related to *I. sejuncta* and *I. tumidula*.

The collection *REH9668* (PERTH 08516677) is marginally different from the holotype of *I. sejuncta* (c. 2% difference across ITS, nLSU, and *rpb2* loci; Fig. 96). It does not feature the warm reddish brown pileus colour, but does include some oblong-angular spores similar to those of the holotype of *I. sejuncta*. Thus, we refer to it as *I. sejuncta*.

83. **Inocybe tumidula** Matheny & Bougher, *sp. nov.*

MycoBank: MB 809476

Type: Mt Beerburrum, Glass House Mountains Natl Park, Qld, 26°57'29.7"S, 152°57'07.3"E, gregarious to scattered on soil in leaf litter in dry sclerophyll woodland under *Eucalyptus pilularis* and other *Eucalyptus* spp., 28 Feb. 2012, *R.E.Halling PBM3764*; holo: PERTH 08557772; iso: TENN 066996 (KP171089-nLSU, KP641637-ITS, KM656133-*rpb2*).

Pileus 10–21 mm wide, conical, expanding with age, becoming plane; often umbonate; umbo obtuse, at times prominent; margin decurved; surface dry; velipellis present or absent; disc smooth and unbroken; smooth to fibrillose towards the margin, not rimose but becoming rimulose with age or when torn; strong brown (7.5YR 5/6; Sayal Brown; near 5D8) or nearly yellowish brown or fulvous (10YR 5/6 to 5/8); context pallid, thin to very thin except under the disc, not changing colour upon exposure, up to 2 mm thick under the disc. *Lamellae* adnexed, close to almost crowded, 45–50 L, with several tiers of lamellulae, very pale brown to brown (10YR 7/3, 6/3, 5/3), becoming dark yellowish brown (10YR 4/4), edges indistinctly pallid-fimbriate, ventricose, narrow to medium, to 3 mm deep. *Stipe* 25–32 × 1–3 mm, terete, flexuous; base with a marginate bulb up to 6 mm wide; cortina absent; pruinose for its entire length or at least beyond the centre of the stipe; surface satiny in appearance, pallid with a very pale brown tinge (10YR 7/4) or pale brown; context solid, coloured like the surface, white in the bulb, at times fragile. *Odour* mild and not remarkable. Pl. 167, 168.

Basidiospores 7.5–9.5 × 5.0–6.5 µm [mean 8.1 × 5.8 µm; Q 1.23–1.60; Q mean 1.39 (n = 20/1)], nodulose with mostly 6–8 moderate sized nodules and angular corners, yellowish brown; apiculus tapering. *Basidia* 23–28 × 8–10 µm, clavate, hyaline, 4-sterigmate. *Pleurocystidia* 45-61 × 18–27 µm, broadly fusiform, lageniform or ventricose; apices obtuse and often crystalliferous; base broadly rounded or narrow; thick-walled; walls mostly 2–3 (–4.5) µm thick, especially thick at the apex, hyaline. *Cheilocystidia* similar to pleurocystidia, mixed with mainly clavate but sometimes subcylindrical paracystidia. *Caulocystidia* similar to cheilocystidia, mixed with crowded clusters of cauloparacystidia, present below the mid-region of the stipe.

Plate 165. *Inocybe sejuncta*, holotype (PERTH 08383103). Photograph — M.D.Barrett.

Plate 166. *Inocybe sejuncta*. PERTH 08516677. Photograph — R.E.Halling.

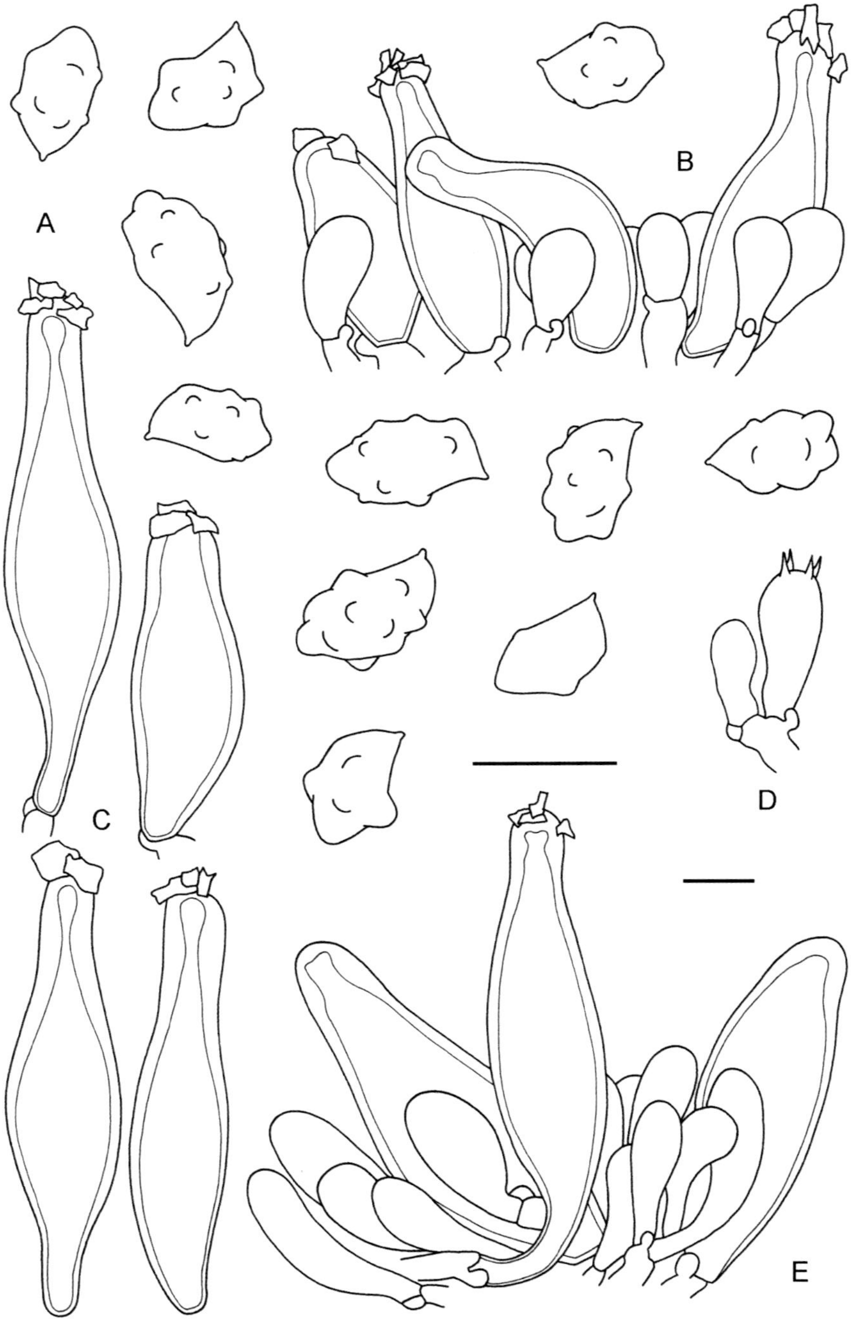

Figure 97. *Inocybe sejuncta.* All holotype (PERTH 08383103). **A,** Basidiospores; **B,** Cheilocystidia and paracystidia; **C,** Pleurocystidia; **D,** Basidia; **E,** Caulocystidia and cauloparacystidia. Scale bars = 10 µm (longer bar for spores only).

Pileipellis a cutis of cylindrical hyphae mostly 7–15 µm wide, ochre-tawny in mass, encrusted with scattered to numerous interhyphal refractive yellow deposits. *Clamp connections* present. Fig. 98.

Known only from south-eastern Qld; grows in soil under *Eucalyptus* and/or *Eucalyptus* mixed with *Allocasuarina*. Fruiting Feb.

Additional specimen examined: Qld: Keith and Ivy Boon Memorial Park, corner of Grigor Rd and Maleny–Kenilworth road, E of Conondale Natl Park, scattered singly on soil in grass in park under *Eucalyptus*, 29 Feb. 2012, *P.B.Matheny PBM3770* (PERTH 08363188, TENN067002; KP171090-nLSU, KP641638-ITS, KM656134-*rpb2*).

Etymology: From the Latin *tumidulus* (somewhat swollen), in reference to the marginate basal bulb of the stipe.

Diagnostic features: Fruit bodies small. Pileus strong brown, yellowish brown or fulvous, smooth to fibrillose, not rimose; lamellae close to crowded, very pale brown to brown or dark yellowish brown; stipe pruinose at least beyond its mid-region, pallid but with a pale brown tinge. Spores nodulose, mostly with 6–8 moderate-sized nodules and angular corners. Pleurocystidia broad, broadly fusiform, lageniform or ventricose, thick-walled.

Inocybe tumidula is one of three Australian species in the *I. calida* Velen. group (Fig. 96) thought to have been derived from Asian ancestors. *Inocybe tumidula* is most closely related to the tropical Australian *I. sejuncta* which has longer spores (9.0–11.5 × 5.0–6.5 µm). The unclarified lineage (*JAC2809b*; Fig. 96) was detected from a heterogeneous assemblage of specimens mostly referable to as *I. gracilissima* in the *Pseudosperma* clade. The lineage forms a moderately well-supported group with *I. tumidula* and *I. sejuncta*, but lacks adequate documentation for a thorough description at present.

Several other more distantly related species are superficially similar to *I. tumidula*, although they are predoninantly southern-temperate in distribution, e.g. *I. dewrangia*, *I. invadens* and *I. fulvilubrica*. Immature fruit bodies of *I. dewrangia* have a universal veil connecting the margin of the pileus to the edge of the marginate bulb. The species also has somewhat larger fruit bodies than *I. tumidula*. *Inocybe invadens* has very thick-walled hymenial cystidia, spores with more numerous nodules, and produces larger fruit bodies. *Inocybe fulvilubrica* has larger fruit bodies and spores with considerably more nodules than *I. tumidula*.

Plate 167. *Inocybe tumidula*, holotype (PERTH 08557772). Photograph — P.B.Matheny.

Plate 168. *Inocybe tumidula*. PERTH 08363188. Photograph — P.B.Matheny.

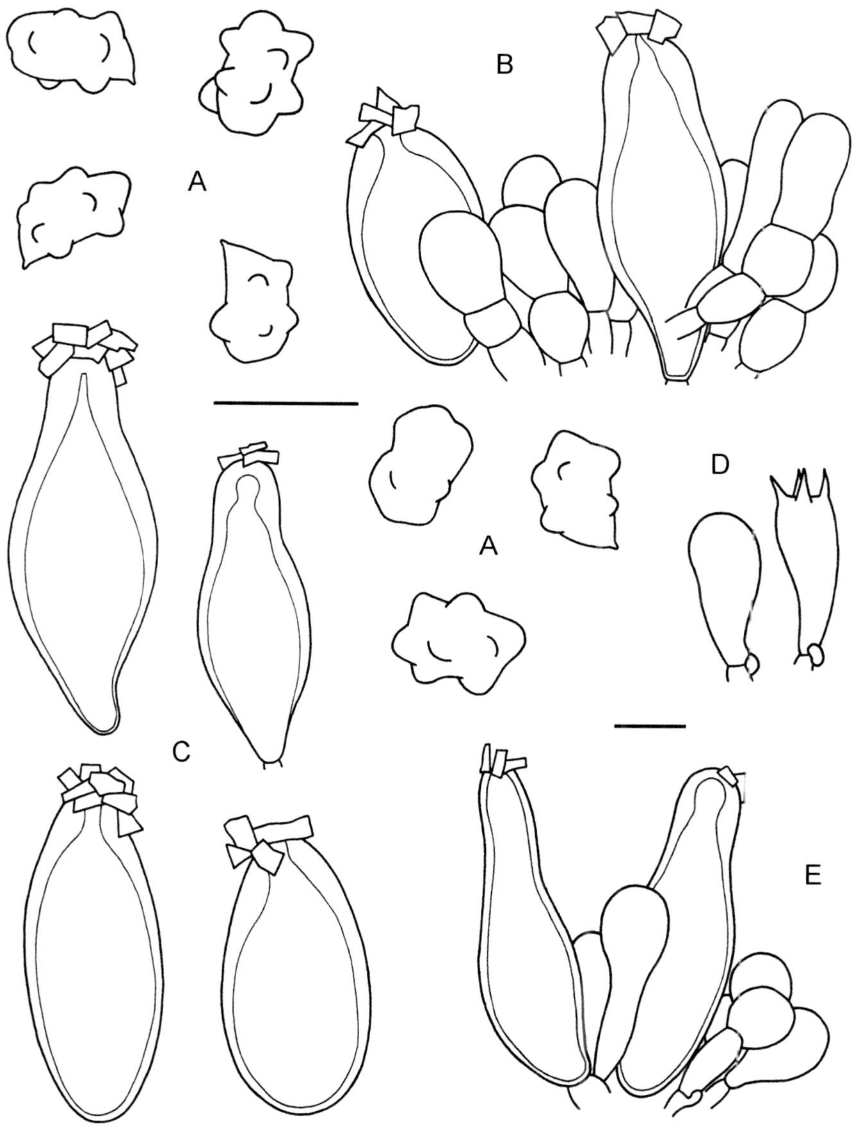

Figure 98. *Inocybe tumidula.* All holotype (PERTH 08557772). **A,** Basidiospores; **B,** Cheilocystidia and paracystidia; **C,** Pleurocystidia; **D,** Basidia; **E,** Caulocystidia and cauloparacystidia. Scale bars = 10 µm (longer bar for spores only).

INOCYBE CALOSPORA AND *INOCYBE PRAETERVISA* GROUPS

These two groups form a large inclusive and moderately supported clade (clade IV in Matheny, 2005; clade VIII sister to *I. calospora* in Ryberg *et al.* (2010) but with weak support; and as monophyletic in Matheny *et al.* (2009) but with weak support. In this analysis species with spinose spores form a paraphyletic group giving rise to a large group of species focused around *I. praetervisa* and its mostly north-temperate allies. Only the four Australian species presented below have been detected in what is otherwise a highly diversified north-temperate group. Three species with spinose spores from northern Australia are described here (*I. spiniformis*, *I. acanthosperma*, *I. tertia*). *Inocybe invadens*, with nodulose spores and entirely pruinose stipe with a marginate bulb, is also described as new and is found most closely related to north-temperate taxa.

Figure 99 (opposite). Maximum Likelihood phylogram of the *Inocybe calospora*/*Inocybe praetervisa* group. Bootstrap values are shown above branches. The tree was inferred from combined nLSU and *rpb2* gene regions and includes 2187 total sites. Species from Australia are indicated by black-filled triangles. White triangles denote non-Australian species. The tree is rooted with *Inocybe lacera*, which is supported outside the ingroup based on earlier studies (e.g., Matheny 2005). Seventy three branch tips are included, all represented by nLSU and 26 by *rpb2*.

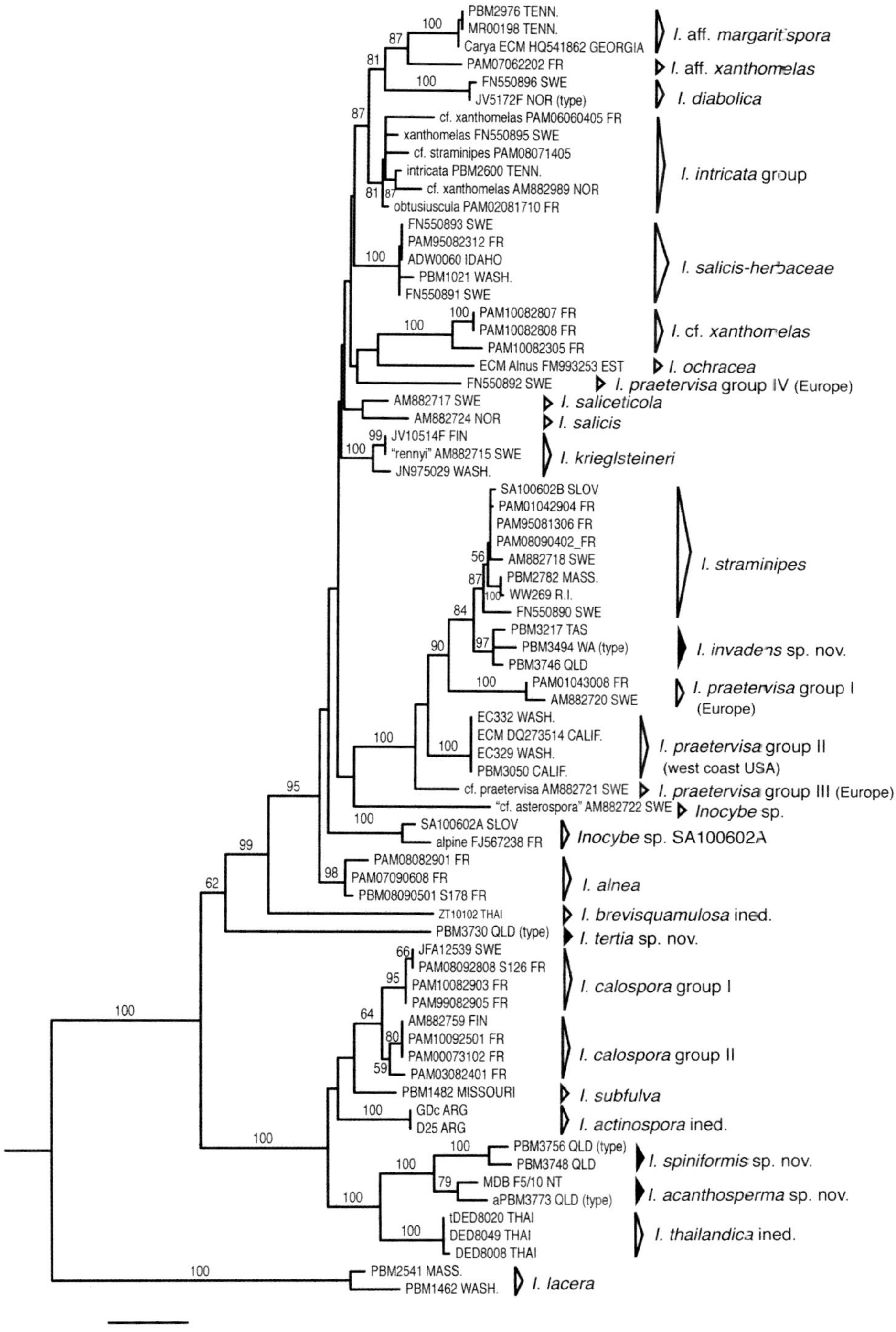
PBM2976 TENN.
MR00198 TENN.
Carya ECM HQ541862 GEORGIA
I. aff. margaritispora
PAM07062202 FR
I. aff. xanthomelas
FN550896 SWE
JV5172F NOR (type)
I. diabolica
cf. xanthomelas PAM06060405 FR
xanthomelas FN550895 SWE
cf. straminipes PAM08071405
intricata PBM2600 TENN.
cf. xanthomelas AM882989 NOR
obtusiuscula PAM02081710 FR
I. intricata group
FN550893 SWE
PAM95082312 FR
ADW0060 IDAHO
PBM1021 WASH.
FN550891 SWE
I. salicis-herbaceae
PAM10082807 FR
PAM10082808 FR
PAM10082305 FR
I. cf. xanthomelas
ECM Alnus FM993253 EST
I. ochracea
FN550892 SWE
I. praetervisa group IV (Europe)
AM882717 SWE
I. saliceticola
AM882724 NOR
I. salicis
JV10514F FIN
"rennyi" AM882715 SWE
JN975029 WASH.
I. krieglsteineri
SA100602B SLOV
PAM01042904 FR
PAM95081306 FR
PAM08090402_FR
AM882718 SWE
PBM2782 MASS.
WW269 R.I.
FN550890 SWE
I. straminipes
PBM3217 TAS
PBM3494 WA (type)
PBM3746 QLD
I. invadens sp. nov.
PAM01043008 FR
AM882720 SWE
I. praetervisa group I (Europe)
EC332 WASH.
ECM DQ273514 CALIF.
EC329 WASH.
PBM3050 CALIF.
I. praetervisa group II (west coast USA)
cf. praetervisa AM882721 SWE
I. praetervisa group III (Europe)
"cf. asterospora" AM882722 SWE
Inocybe sp.
SA100602A SLOV
alpine FJ567238 FR
Inocybe sp. SA100602A
PAM08082901 FR
PAM07090608 FR
PBM08090501 S178 FR
I. alnea
ZT10102 THAI
I. brevisquamulosa ined.
PBM3730 QLD (type)
I. tertia sp. nov.
JFA12539 SWE
PAM08092808 S126 FR
PAM10082903 FR
PAM99082905 FR
I. calospora group I
AM882759 FIN
PAM10092501 FR
PAM00073102 FR
PAM03082401 FR
I. calospora group II
PBM1482 MISSOURI
I. subfulva
GDc ARG
D25 ARG
I. actinospora ined.
PBM3756 QLD (type)
PBM3748 QLD
I. spiniformis sp. nov.
MDB F5/10 NT
aPBM3773 QLD (type)
I. acanthosperma sp. nov.
tDED8020 THAI
DED8049 THAI
DED8008 THAI
I. thailandica ined.
PBM2541 MASS.
PBM1462 WASH.
I. lacera
0.02

84. Inocybe acanthosperma Matheny & Bougher, *sp. nov.*

MycoBank: MB 808680

Type: Cunninghams Gap, Main Range Natl Park, Qld, 28°02'59"S, 152°23'37.3"E, scattered singly on soil in dry sclerophyll forest under *Eucalyptus* and *Allocasuarina*, 25 Feb. 2012, *R.E.Halling & N.Fechner PBM3773*; holo: PERTH 08363145; iso: TENN 067005 (KP171089-nLSU, KP641637-ITS, KM656133-*rpb2*).

Pileus 15–28 mm wide, hemispherical to obtusely conical in youth, expanding with age; umbo absent; margin at first slightly incurved, becoming decurved; surface dry; velipellis absent; recurved squamulose at the centre and outwards, coarsely fibrillose at edge of margin, at most becoming weakly rimulose; dark brown (10RY 3/3 to 3/4; Bay Brown); context pallid, not changing colour where cut or bruised, thin except under the disc. *Lamellae* adnexed, moderately close, 26–36 L, with several tiers of lamellulae, very pale brown to pale brown (10YR 7/3 to 6/3) when young, becoming yellowish brown (10YR 5/4 to 5/6; near Buckthorn Brown) at maturity, ventricose, broad, the edges indistinctly pallid-fimbriate. *Stipe* 25–50 × 3 mm, terete, occasionally split, swollen at the base into a rounded bulb up to 8 mm wide; veil not observed; surface pruinose at the apex, not visibly so below the stipe centre, instead with light superficial pallid fibrils (very thin vesture); base white, whitish above the base, elsewhere reddish brown (5YR 5/4 to 4/4); context solid, concolorous with the surface except for the paler interior, white in the base. *Odour* not remarkable, at most slightly acidulous when first cut. Pl. 169, 170.

Basidiospores 11.5–14.0 × 11.0–13.5 µm [mean 12.9 × 12.3 µm; Q 1.00–1.11; Q mean 1.05 (*n* = 20/1)], spinose, globose, with 17–23 acutely conical projections (c. 2.0–2.5 µm high); yellowish brown; apiculus broad, truncate but not often discernible. *Basidia* 37–50 × 9–13 µm, cylindroclavate, often with an apical constriction, hyaline, 4-sterigmate. *Pleurocystidia* 38–50 × 9–15 µm, utriform to fusiform or ventricose, tapered towards the base; apices obtuse, crystalliferous, at times with a resinous cap; walls 1.5–2.5 µm thick, especially apically, hyaline. *Cheilocystidia* similar to pleurocystidia, sometimes thin-walled, mixed with crowded, hyaline, clavate or pyriform paracystidia and some basidia. *Caulocystidia* mixed with cauloparacystidia in upper ½ of stipe only; shape variable, including long-cylindrical and lageniform individuals, in scattered clusters, some thin-walled, hyaline. *Pileipellis* composed of cylindrical to inflated repent hyphae mostly 5–17 µm wide, tawny or reddish brown in mass, weakly encrusted, with thin or slightly thickened walls. *Clamp connections* present. Fig. 100 (see also see Pl. 6E, 8B).

Occurs in south-eastern Qld; in soil under *Eucalyptus* and/or *Allocasuarina* in warm-temperate sclerophyll forest. Fruiting Feb.

Additional specimen examined: Qld: type locality, 29 Feb. 2012, *R.E.Halling & N.Fechner PBM3778* (PERTH 08363102, TENN 067010).

Etymology: From the Greek *akanthos* (prickly) and *sperma* (a seed), in reference to the globose spores with numerous conical projections.

Diagnostic features: Fruit bodies medium-sized. Pileus scaly; stipe pruinose over upper half, fibrillose below, with a rounded basal bulb; odour not remarkable. Spores large, spinose, globose. In warm-temperate sclerophyll forest under *Eucalyptus* and/or *Allocasuarina*.

Inocybe acanthosperma shares a similar cystidial and spore morphology with *I. spiniformis*, also described from northern Australia. However, the latter has small, slender fruit bodies with a furfuraceous-scaly and striate pileus surface and broadly elliptic spores. Phylogenetic analyses support the separation of the two species but group both with holarctic and temperate South American species related to *I. calospora* Quél. (Fig. 99). Another Australian species, *I. tertia*, has broadly elliptic spores with fewer (13–16) acutely conical projections.

The microscopic morphology of *I. acanthosperma* also resembles that described for *I. echinosimilis* (E.Horak) Garrido from Indonesia and New Guinea (Horak, 1979a). The latter is associated with Fagaceae and has a fruity or pleasant odour and smaller spores. For further comparisons with species described by Horak (1979a) from Indonesia and New Guinea, see the discussion under *I. spiniformis*.

One collection from the Northern Territory has sequences similar to those of the type specimen, but its spores are broadly elliptic and its hymenial cystidia have thicker walls. Its collection details are: Fogg Dam, 52 km ESE of Darwin, N.T., 12°34'08"S, 131°18'29"E, on soil at edge of swampy rainforest near road under *Corymbia*, *Lophostemon lactifluus*, *Pandanus spiralis*, and rainforest trees, 1 Jan. 2010, *M.D.Barrett MDB F5/10* (PERTH 08383286; KJ729890-nLSU, KJ729862-ITS).

85. Inocybe spiniformis Matheny & Bougher, *sp. nov.*

MycoBank: MB 809462.

Type: Mount Whitfield Conservation Park, Cairns, Queensland, Australia, 16°53'53.8"S, 145°45'03.1"E, gregarious or scattered singly on soil in *Melaleuca* swamp in lowland rainforest under *Melaleuca leucadendra*, *Lophostemon suaveolens*, *Corymbia tessularis* and *Acacia mangium*, 25 Feb. 2012, *N.L.Bougher PBM3756*; holo: PERTH 08363277; iso: TENN 066984, CNS 140400 (KP171063-nLSU, KP636867-ITS, KM656102-*rpb2*).

Pileus 3–15 mm wide, conical, expanding to plane, often with a conical or small nipple-like umbo; margin initially decurved, becoming slightly upturned; surface dry; velipellis absent, typically with small appressed scales on the disc and around the centre giving it a furfuraceous appearance, fibrillose to smooth towards the margin, translucent-striate towards margin when fresh; umber (7.5YR 4/4) or almost (dark) reddish brown (2.5 YR 3/4) throughout or somewhat paler towards the margin, a shade lighter between striae, with furfuraceous scales also lighter in colour than the surface; context very thin, dull whitish or pale brown, not changing colour on cutting or bruising. *Lamellae* adnexed to subfree, subdistant, 25–30 L, with several tiers of lamellulae, pale brown, brown, to yellowish brown (7.5 YR 4/4, 10YR 6/3, 5/3, 5/4), ventricose, medium to broad, to 2 mm deep. *Stipe* 12–30 × 0.5–1.5 mm, even, terete, not swollen at the base; surface pruinose at the extreme apex above the pileal insertion or weakly pruinose-furfuraceous above to about the centre of the stipe, lightly fibrillose below or nearly throughout with age; veil (cortina) fugacious; very pale brown to light yellowish brown (10YR 7/4 to 6/4); context solid, becoming hollow, pallid at the extreme base. *Odour* acidulous when first cut or faintly musky, but soon not remarkable or becoming grassy. Pl. 171, 172.

Basidiospores (10.0–) 10.5–13.5 × 9.0–11.5 (–12.0) µm [mean 12.2 × 10.4 µm; Q 1.05–1.32 (-1.37); Q mean 1.18 (n = 38/3)], spinose, broadly elliptic to subglobose, with 16–24 sharp conical projections c. 2–3 µm high, yellowish brown; apiculus broad, truncate but not always apparent. *Basidia* 31–45 × 10–11 µm, cylindroclavate, hyaline, 4-sterigmate, often as long as or longer than the hymenial cystidia. *Pleurocystidia* 36–50 × 11–18 µm, clavate, cylindrical, utriform or less often ventricose or fusiform; apices well-rounded, bearing crystals, sometimes with a resinous cap; thick-walled, 1–3 (–5) µm thick especially apically, pale yellow, tapering towards the base or truncate. *Cheilocystidia* similar to pleurocystidia but shorter, mixed with abundant hyaline clavate paracystidia. *Caulocystidia* similar to cheilocystidia but some lageniform with a long narrow neck, hyaline, mixed with cauloparacystidia, in dense clusters down to the stipe centre but absent below and above the stipe base. *Pileipellis* composed of mostly inflated hyphae 10–26 µm wide, tawny to rusty-brown in mass, weakly encrusted and with thickened walls. *Clamp connections* present. Fig. 101 (see also Pl. 7D).

Occurs in north-eastern Qld; on soil in leaf litter in *Melaleuca* swamp in lowland forest including *Lophostemon*, *Corymbia*, and *Acacia*. Fruiting Feb.

Additional specimen examined: Qld: type locality, 25 Feb. 2012, *N.L.Bougher NLB949* (PERTH 08364591, TENN 068778).

Etymology: From the Latin *spina* (a thorn) and the suffix *-formis* (having the shape of), in reference to the spinose basidiospores.

Diagnostic features: Fruit bodies small and slender; pileus umbonate and often with small scales over the centre, translucent-striate towards the margin; upper half of stipe pruinose, fibrillose below. Spores large, broadly elliptic to subglobose, spinose with 16–24 sharp conical projections; pleurocystidia short, variable in shape but mostly clavate, cylindrical or utriform. On soil under various Myrtaceae (not *Eucalyptus*) or mixture of these with *Acacia*; tropical Queensland.

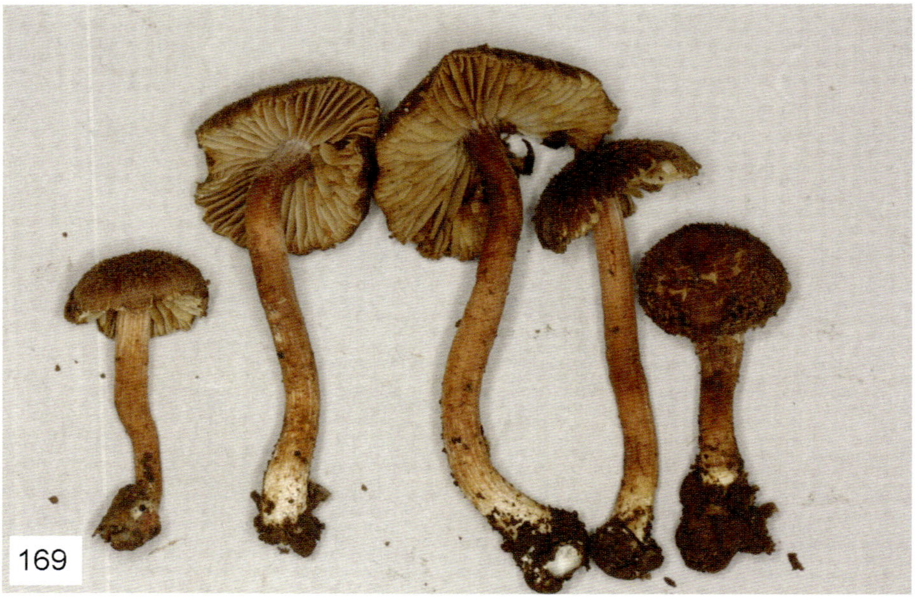

Plate 169. *Inocybe acanthosperma*, holotype (PERTH 08363145). Photograph — P.B.Matheny.

Plate 170. *Inocybe acanthosperma*. PERTH 08363102. Photograph — P.B.Matheny.

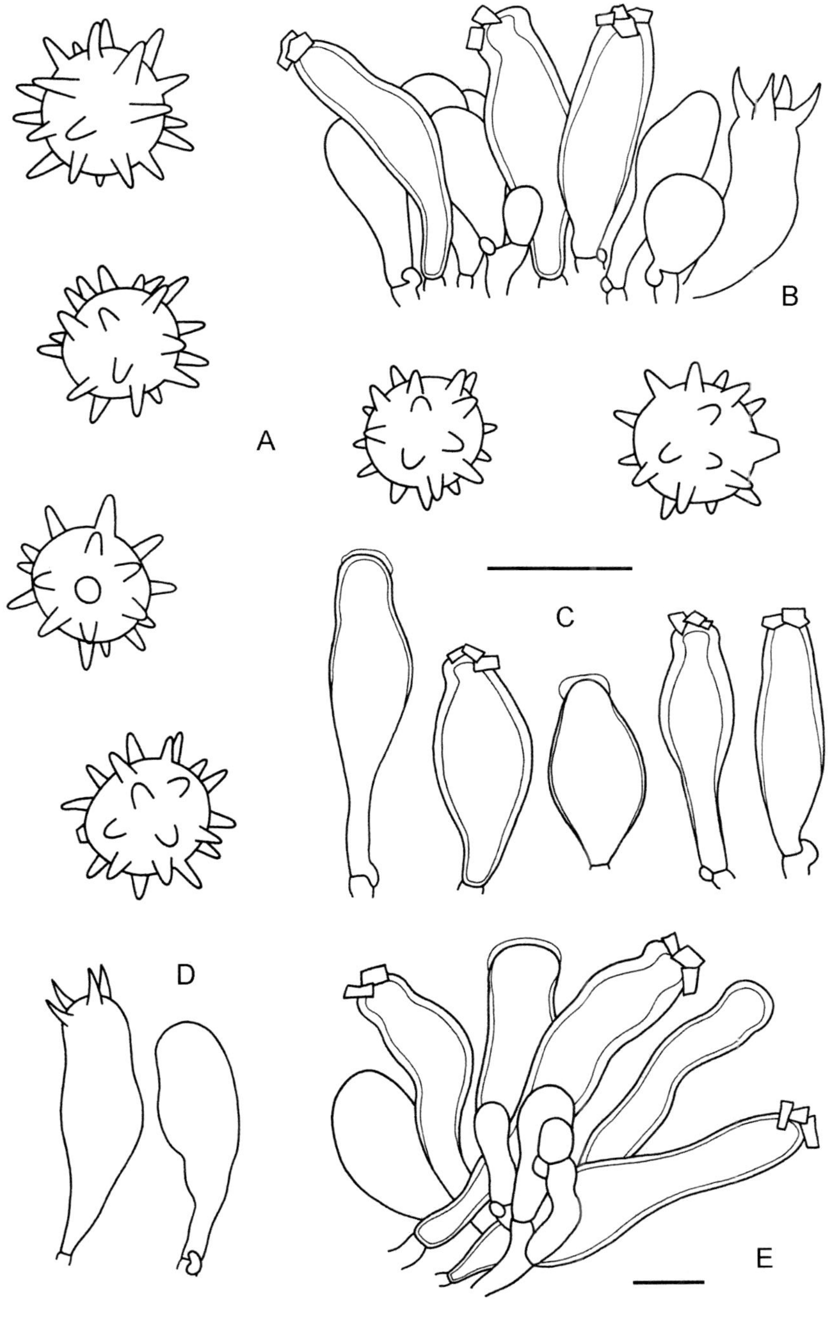

Figure 100. *Inocybe acanthosperma.* All holotype (PERTH 08363145). **A,** Basidiospores; **B,** Cheilocystidia and paracystidia; **C,** Pleurocystidia; **D,** Basidia; **E,** Caulocystidia and cauloparacystidia. Scale bars = 10 µm (longer bar for spores only).

Inocybe spiniformis is one of three newly described Australian spinose-spored species, all allied to the European species *I. calospora* Quél. (Fig. 99). *Inocybe acanthosperma* differs from *I. spiniformis* in its more robust fruit bodies, the stipe base with a rounded bulb and globose spores, while *I. tertia* has broadly elliptic spores with fewer (13–16) sharp conical projections.

Miscroscopic features of these taxa resemble some of those described by Horak (1979a) from Indonesia and New Guinea, but some of the spinose-spored taxa described by Horak (1979a) have entirely pruinose stipes. By contrast, the stipes of *I. spiniformis* and *I. acanthosperma* are pruinose at the apex only, or generally down as far as the mid-region. The spores of *I. spiniformis* most closely resemble those of *I. gemina* (E.Horak) Garrido, and the spores of *I. acanthosperma* resemble those illustrated for *I. echinosimilis* (E.Horak) Garrido. These two species were described by Horak (1979a) under *Lithocarpus* and *Castanopsis* in Papua New Guinea and/or Sabah (Malaysia). Aside from ecological differences (the fagaceous hosts), *I. gemina* differs from *I. spiniformis* in its non-pruinose stipe that is decorated with girdles of veil tissue and white appendiculate veil remnants along the margin of the pileus. The clavate, cylindrical or utriform hymenial cystidia typical of *I. spiniformis* were not reported for *I. gemina* or *I. echinosimilis* by Horak (1979a). The last-named species differs from *I. spiniformis* in its larger size, scaly stipe surface except at the pruinose apex, and smaller globose spores (7.5–10.0 µm). We note that Horak (1979a) described the spores of *I. echinosimilis* as "subglobose to ovoid", but illustrated them as consistently globose.

Inocybe spiniformis can be mistaken in the field for *I. humidicola*, a tropical species found in similar habitats where *Eucalyptus grandis* occurs. However, *I. humidicola* lacks a scaly pileus and is most readily distinguished by its smaller, non-spinose, nodulose spores.

Another similar entity from Queensland, which we refer to as *Inocybe* cf. *spiniformis* pending further studies, is represented by the collections listed below. It differs from *I. spiniformis* by occurring with *Eucalyptus*, often in a transition zone between sclerophyll and rainforest vegetation, having a fibrillose rather than scaly pileus, becoming rimose (not translucent-striate) near the margin, and having smaller spores (9.0–11.0 × 8.0–9.5 µm) with coarser, broader-based spines. The ITS sequence of one specimen, *PBM3748*, is 96% similar to that of the type of *I. spiniformis* (Fig. 101). Additional collections are required to determine the constancy and reliability of these differences, but for completeness, a description of *PBM3748* is provided here.

Pileus 6–21 mm wide, initially conical, expanding with age to broadly conical, with or without a subacute umbo; margin decurved; pallid partial veil remnants observed at edge of margin in youth; surface dry; velipellis absent; centre smooth and unbroken, becoming radially rimose towards the margin; dark brown to dark yellowish brown (10YR 3/3–3/4); context pale brown, very thin. *Lamellae* adnexed, moderately close, 24–30 L, with several tiers of lamellulae, ventricose, broad, pale brown to yellowish brown (10YR 6/3, 5/3, 5/4); edges indistinctly pallid-fimbriate. *Stipe* 18–30 × 1–2 mm, terete, even, with a slightly swollen base; cortina fugacious; pruinose at the apex but not below the mid-region, elsewhere lightly fibrillose; very pale brown to light yellowish brown (10YR 7/3 to 6/4); context not firm, brittle, solid, becoming hollow. *Odour* none.

Basidiospores 9.0–11.0 × 8.0–9.5 µm [mean 10.0 × 8.7 µm; Q 1.05–1.29; Q mean 1.15 (n = 23/1)], spinose, subglobose to broadly elliptic, with 14–20 sharp conical projections c. 2.0–2.5 µm high, yellowish brown; apiculus not discernible. Other microscopic features are similar to *I. spiniformis*.

Specimens examined: Qld: Tully Falls area, Red Rd, site 1, beside track, under *Eucalyptus grandis*, 4 May 1988, *N.L.Bougher E4011* (PERTH 07709714, TENN 068779); Davies Creek Rd, Dinden Natl Park, between Kuranda and Mareeba, 17°01'36.1"S, 145°36'04.3"E, scattered to gregarious on soil along roadside in leaf litter under *Allocasuarina torulosa*, *Eucalyptus grandis* and other eucalypts at woodland-rainforest interface, 24 Feb. 2012, *P.B.Matheny PBM3748* (PERTH 08363331, TENN 066949; KP171064-nLSU, KP636868-ITS, KM656103-*rpb2*); Keith and Ivy Boon Memorial Park, corner of Grigor Rd and Maleny–Kenilworth road, 26°42'14.5"S, 152°41'05.0"E, among grass under mixed *Eucalyptus* (smooth and rough bark), 29 Feb. 2012, *N.L. Bougher NLB957* (PERTH 08364672, TENN 069828).

Plate 171. *Inocybe spiniformis*, holotype (PERTH 08363277). Photograph — P.B.Matheny.

Plate 172. *Inocybe spiniformis*. PERTH 08364591. Photograph — N.L.Bougher.

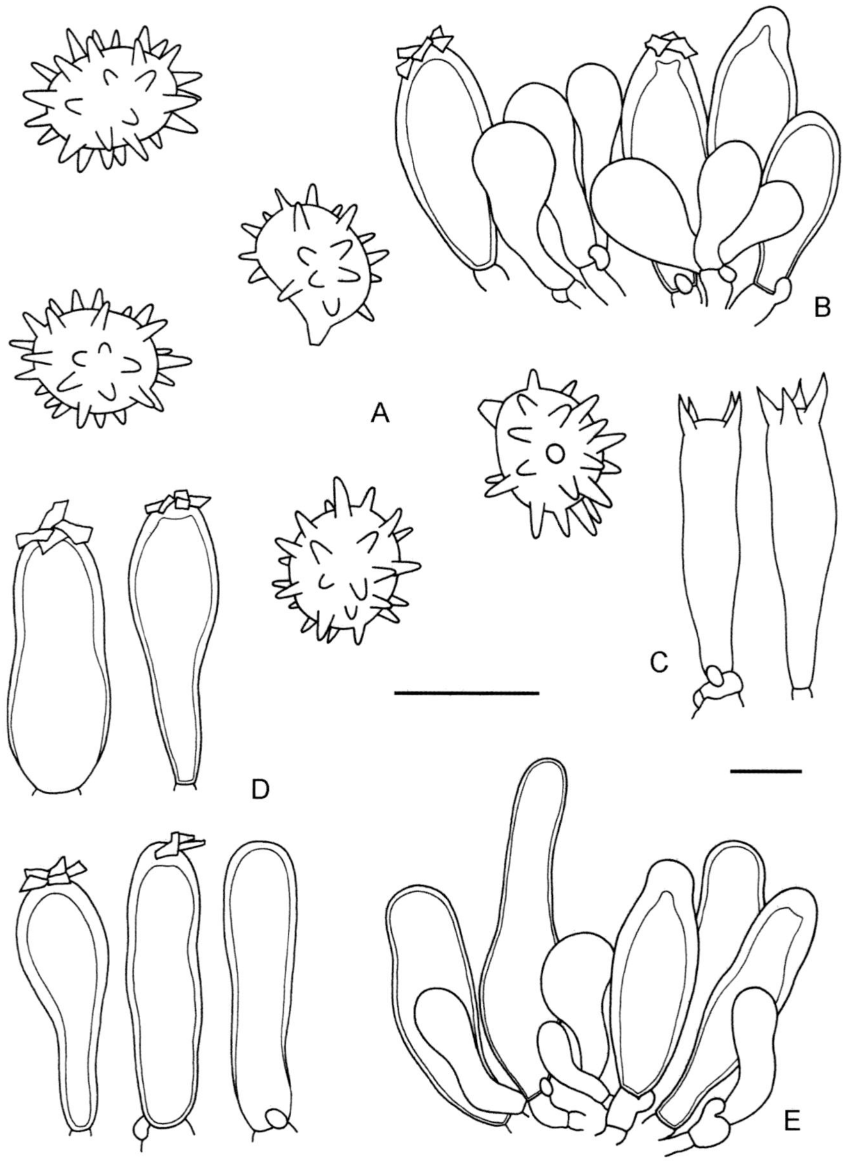

Figure 101. *Inocybe spiniformis.* All holotype (PERTH 08363277). **A,** Basidiospores; **B,** Cheilocystidia and paracystidia; **C,** Basidia; **D,** Pleurocystidia; **E,** Caulocystidia and cauloparacystidia.

Scale bars = 10 μm (longer bar for spores only).

86. Inocybe tertia Matheny & Bougher, *sp. nov.*

MycoBank: MB 809473

Type: Black Mountain Rd, near Kuranda, Qld, 16°47'07.8"S, 145°37'32.1"E, solitary on soil along track under *Allocasuarina littoralis*, *Acacia mangium* and *Eucalyptus pellita* in mixed rainforest, 23 Feb. 2012, *M.D.Barrett & P.B.Matheny PBM3730*; holo: PERTH 08362882; iso: TENN 066951 (KP171086-nLSU KP641633-ITS, KM656128-*rpb2*).

Pileus up to 22 mm wide, plano-convex, lacking a velipellis and umbo; margin decurved; surface dry; disc and centre recurved-squamulose; appressed-squamulose towards the margin; coarsely fibrillose and rimulose at edge of margin; brown or umber (7.5YR 4/4; Snuff Brown) throughout, pale brown between the spreading fibrils; context thin but up to 2 mm thick under the disc, white, no colour changes where cut or bruised, not confluent with the stipe. *Lamellae* adnexed, moderately close, c. 30 L, with several tiers of lamellulae, pale brown to yellowish brown (10YR 6/3 to 5/4), ventricose, medium, up to 3.5 mm deep; edges indistinctly pallid-fimbriate. *Stipe* 32 × 2–3 mm, terete but slightly swollen at the base (to 4 mm wide); veil probably absent; pruinose below the centre of the stipe; indistinctly fibrillose above the base; reddish brown (5YR 5/4); context solid, coloured like the surface, pallid at the extreme base. *Odour* absent. Pl. 173.

Basidiospores 11.5–14.0 (–14.5) × 10.0–12.0 µm [mean 12.6 × 10.8 µm; Q 1.09–1.30; Q mean 1.16 (n = 22/1)], spinose with numerous (13–16) conical projections up to 4.5 (–5.0) µm tall, most spores with at least 1 bifid nodule, broadly elliptic to subglobose, yellowish brown. *Basidia* 34–38 × 11–12 µm, cylindroclavate, at times with an apical constriction, hyaline, 4-sterigmate. *Pleurocystidia* 31–45 × 14–19 µm, utriform, ventricose or fusiform, or less often clavate; apices obtuse to well-rounded, some bearing crystals; thick-walled (4–6 µm), yellow. *Cheilocystidia* similar to pleurocystidia, often shorter, the apices obtuse to well-rounded, some bearing crystals, others with a resinous cap; mixed with hyaline and mostly clavate (to cylindrical) paracystidia and thin-walled utriform or ventricose cells. *Caulocystidia* 26–52 × 10–17 µm, descending at least to the stipe centre, absent above the base, in scattered clusters; utriform, lageniform or ventricose, mixed with cylindrical cauloparacystidia; moderately thick-walled (1.5–2.5 µm), occasionally thinner-walled, hyaline. *Pileipellis* a cutis giving rise to upturned fascicles of hyphae which are dark tawny or rusty-brown in mass, cylindrical to inflated, mostly 5–17 µm wide, weakly encrusted, occasionally with thickened walls. *Clamp connections* present. Fig. 102 (see also Pl. 7E).

Known only from soil in mixed rainforest under *Allocasuarina*, *Acacia* and *Eucalyptus* in north-eastern Qld. Fruiting Feb.

Additional specimen examined: Qld: Davies Creek Rd, Dinden Natl Park, 17°01'58.5'S, 145°36'33.1"E, emerging from embankment near *Eucalyptus grandis* and *Allocasuarina torulosa*, 20 Feb. 2012, *N.L.Bougher NLB920* (PERTH 08373000, TENN 070167).

Etymology: From the Latin *tertius* (the third), in reference to this being the third spinose-spored species of *Inocybe* recorded in Australia.

Diagnostic features: Fruit bodies small to medium-sized. Pileus scaly at the centre, rimulose towards the margin, brown or umbrinous; stipe reddish brown, pruinose below the centre, but fibrillose above the base. Spores with 13–16 conical projections and a bifid nodule, large, broadly elliptic to subglobose; pleurocystidia present.

Of the species described by Horak (1979a) from south-eastern Asia, *I. hydrocybiformis* (Corner & E.Horak) Garrido has spores most similar to those of *I. tertia*. However, the former lacks pleurocystidia and has veil remnants on the stipe. *Inocybe pahangi* (E.Horak & Corner) Garrido has spores that are more stellate in appearance and much longer pleurocystidia, while *I. echinosimilis* has globose-spinose spores and a lower stipe bearing minute floccose scales or fibrils.

Inocybe tertia is phlogenetically closely related to northern temperate species lacking a veil or having an entirely pruinose stipe, often with a marginate basal bulb, *viz. I. straminipes* Romagn., *I. salicis-herbaceae* Kühner, *I. krieglsteineri* Fernández Sas., and members of the *I. xanthomelas* Boursier & Kühner group and *I. praetervisa* Quél. group (Fig. 99). In view of this relationship it may be inferred that *I. tertia* probably lacks a veil, despite the absence of caulocystidia above its stipe base.

In the field, *I. tertia* could be mistaken for a member of the smooth-spored *I. australiensis* group given the umber-coloured, scaly pileus and the reddish brown, partly pruinose stipe.

Plate 173. *Inocybe tertia*, holotype (PERTH 08362882). Photograph — P.B.Matheny.

87. Inocybe invadens Matheny, Bougher & G.M.Gates, *sp. nov.*

MycoBank: MB 809382

Type: intersection of Shedley Drive and Tingle Drive, Walpole-Nornalup Natl Park, W.A., 34°59'24.5"S, 116°39'20.6"E, on soil in tall forest with Karri (*Eucalyptus diversicolor*), Tingle (*E. jacksonii*) and *Allocasuarina*, 9 June 2010, *N.L.Bougher PBM3494*; holo: PERTH 08241716; iso: TENN 069536 (KJ801185-nLSU).

Pileus 13–45 mm wide, conical to broadly so or plano-convex, becoming plano-convex with a low broad obtuse umbo; margin decurved but occasionally becoming uplifted; surface dry; velipellis absent; disc smooth and unbroken, radially fibrillose-rimulose towards the margin; fulvous or yellowish brown, or brown (10YR 4/3 or 5/6 to 5/8) at the centre and shading yellowish brown to strong brown (10YR 5/4 to 7.5YR 5/6) towards the margin, context thin or to to 4 mm thick under the disc, dingy isabelline when moist to pallid or pale brown or light grey, not changing colour where bruised. *Lamellae* adnexed to almost free, moderately close with c. 30 L and several tiers of lamellulae; light grey (10YR 7/2) or very pale brown (10YR 7/3 to 6/3), becoming light yellowish brown (2.5Y 6/4), yellowish brown (10YR 5/4), or olive-brown; medium to broad, up to 3 mm deep, ventricose, the edges pallid and fimbriate. *Stipe* 20–65 × 2–6 mm, even to slightly compressed, occasionally flexuous, with a (weak) marginate bulb to 5–10 mm wide; surface dry, pruinose for its entire length; veil absent; white to whitish with a dull pale yellow or buff undertone (2.5Y 7/4 to 8/4); context whitish, solid, white in the bulb. *Odour* none or not remarkable. Pl. 174, 175 (see also Pl. 2B).

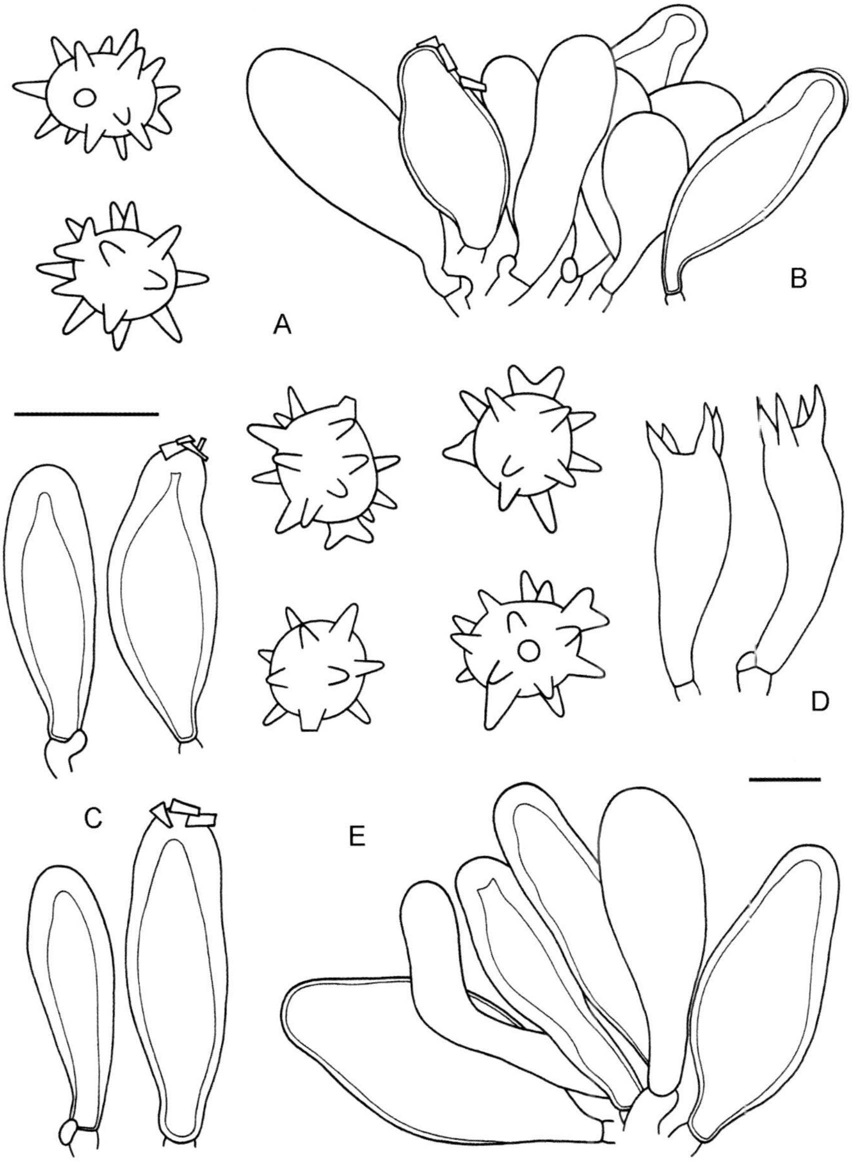

Figure 102. *Inocybe tertia.* All holotype (PERTH 08362882). **A,** Basidiospores; **B,** Cheilocystidia and paracystidia; **C,** Pleurocystidia; **D,** Basidia; **E,** Caulocystidia and cauloparacystidia from upper stipe. Scale bars = 10 µm (longer bar for spores only).

Basidiospores 7.5–10.0 (–12.0) × 5.0–7.5 µm [mean 8.9 × 6.4 µm; Q (1.14–) 1.18–1.62 (–1.67); Q mean 1.40 (n = 68/2)], nodulose, often with 7–11 obtuse nodules up to 1.5 µm high; nodule at spore apex to 3 µm long, yellowish brown; apiculus small. *Basidia* 23–35 × 7.0–9.5 µm, narrowly clavate, hyaline, 4-sterigmate. *Pleurocystidia* 43–63 × 14–21 µm, mostly ventricose, utriform or fusiform-ventricose, occasionally saccate or subcylindrical, usually with obtuse apices, often crystalliferous; thick-walled (4–6 µm thick in the holotype, thinner in others), especially in upper part, dull yellowish. *Cheilocystidia* similar to pleurocystidia, abundant in crowded but isolated clusters, associated with densely crowded clusters of small (to 20 µm long) clavate to subcylindrical hyaline paracystidia. *Caulocystidia* fusoid, lageniform, thick-walled, present along the entire stipe in scattered clusters among clusters of broadly clavate or pyriform thin-walled cauloparacystidia. *Pileipellis* a cutis of encrusted yellowish brown hyphae 5–15 µm wide. *Clamp connections* present. Fig. 103.

Found in W.A., Qld and Tas. Occurs on soil in grassy area near *Acacia* and *Syncarpia* and in woodland-rainforest transition under *Allocasuarina torulosa*, *Eucalyptus grandis* and other *Eucalyptus* spp. in north-eastern Qld; in wet sclerophyll forest under *Eucalyptus regnans*, *Acacia* and *Pomaderris* in Tas.; and in Tingle-Karri-*Allocasuarina* forest in south-western W.A. Fruiting Feb. and May (Qld); June (Tas., W.A.)

Additional specimens examined: W.A.: Bee Farm, off Albany Hwy, 4 July 1981, *R.N.Hilton UWA2557* (PERTH 00908800). Qld: Tully Falls, grassy area with *Acacia* and *Syncarpia*, 4 May 1988, *J.M.Trappe E4019* (PERTH 07709323, TENN 069537); Davies Creek Rd, Dinden Natl Park, between Kuranda and Mareeba, 17°01'36.1"S, 145°36'04.3"E, scattered singly on soil on roadbank at woodland-rainforest transition under *Allocasuarina torulosa*, *Eucalyptus grandis* and other *Eucalyptus* spp., 24 Feb. 2012, *P.B.Matheny PBM3746* (PERTH 08363374, TENN 066978; JX258832-ITS, KJ801187-nLSU, KJ811578-*rpb2*); *loc. id.*, on soil under *Eucalyptus grandis*, *Allocasuarina torulosa* and other eucalypts, 24 Feb. 2012, *M.D.Barrett & N.L.Bougher NLB941* (PERTH 08364559, TENN 070159). Tas.: Mount Field Natl Park, solitary on soil in wet sclerophyll forest under *Eucalyptus regnans*, *Acacia* and *Pomaderris*, 8 June 2009, *G.M.Gates, D.Ratkowsky & P.B.Matheny PBM3217* (PERTH 08516588, TENN 069535; KJ801186-nLSU, KP171133-ITS, KJ811577-*rpb2*).

Etymology: From the Latin *invadens* (attacking, invading), in reference to this species having evolved only recently in Australia from north-temperate ancestors

Diagnostic features: Fruit bodies medium-sized. Pileus smooth on the disc, radially rimulose towards the margin, usually yellowish brown; lamellae light grey to yellowish brown, at times with olivaceous tinges; stipe with a marginate bulb, white to buff. Spores nodulose, typically with 7–11 moderate-sized nodules. Occurs under eucalypts, *Allocasuarina*, *Syncarpia* and *Acacia*.

A member of the *I. praetervisa* Quél. group (Fig. 99), *I. invadens* has a broad Australian distribution including south-western Western Australia, Tasmania and north-eastern Queensland. It shares with northern-temperate species in this group a fulvous to brown rimulose pileus, pruinose whitish stipe with a marginate bulb, the absence of a veil, and nodulose basidiospores.

Inocybe invadens is superficially very similar to several other Australian species, and careful examination of development, stipe morphology, the presence/absence of a velipellis and anatomy is required to recognise these species, apart from DNA sequencing. *Inocybe fulvilubrica* is characterised by its greasy or lubricous pileus that often includes a velipellis, and more coarsely nodulose spores with 12–15 obtusely conical nodules; phylogenetically, the two species are distantly related. *Inocybe dewrangia* differs from *I. invadens* by the presence of a veil connecting the pileus margin to the stipe base, somewhat larger spores, and larger paracystidia on the lamellar edges. *Inocybe sylvicola*, known mainly from *Nothofagus* forests in southern Australia, has a straw-yellow pileus that is long-rimose, while *I. exigua* differs from *I. invadens* in lacking a basal bulb.

Plate 174. *Inocybe invadens*, holotype (PERTH 08241716). Photograph — P.B.Matheny.

Plate 175. *Inocybe invadens*. PERTH 08363374. Photograph — P.B.Matheny.

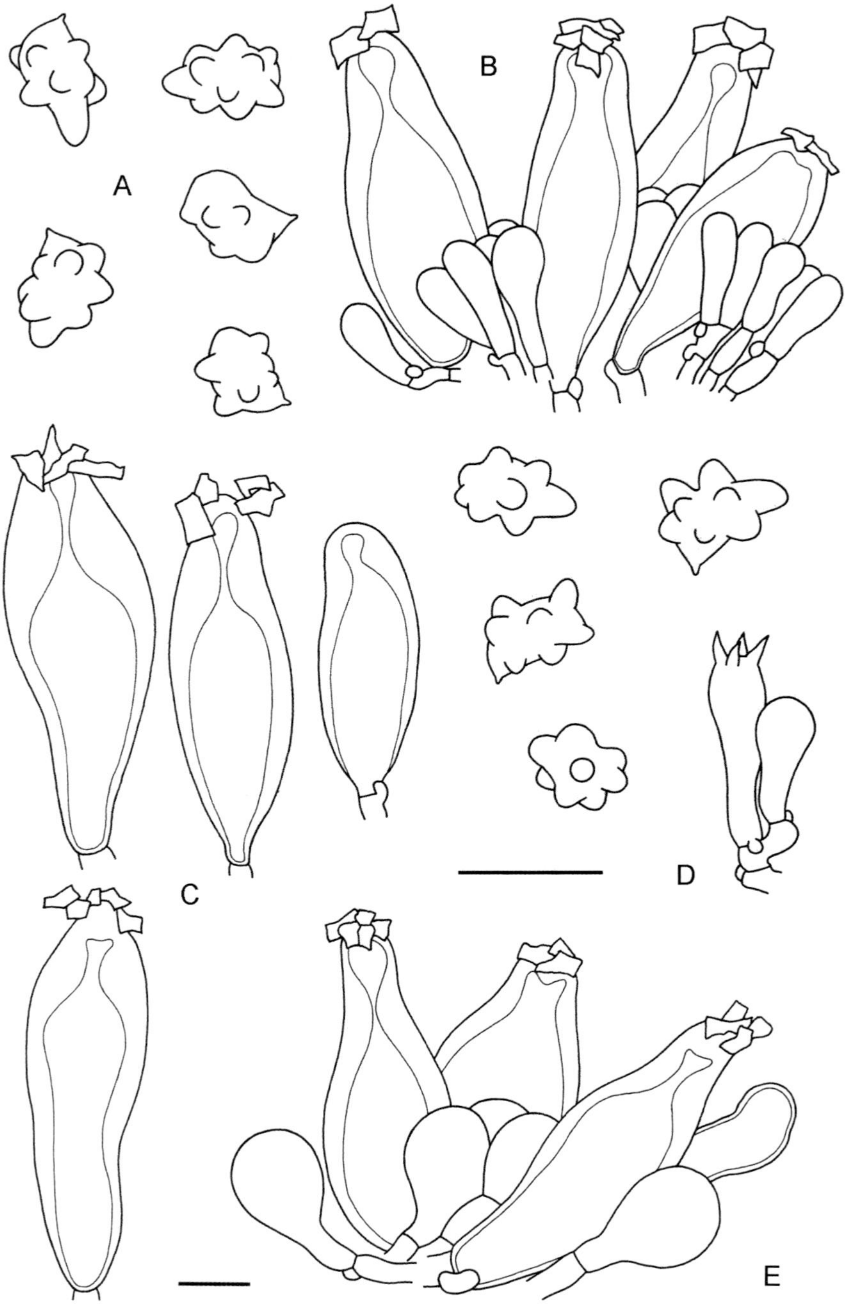

Figure 103. *Inocybe invadens.* All holotype (PERTH 08241716). **A,** Basidiospores; **B,** Cheilocystidia and paracystidia; **C,** Pleurocystidia; **D,** Basidia; **E,** Caulocystidia and cauloparacystidia. Scale bars = 10 μm (longer bar for spores only).

INOCYBE EMERGENS GROUP

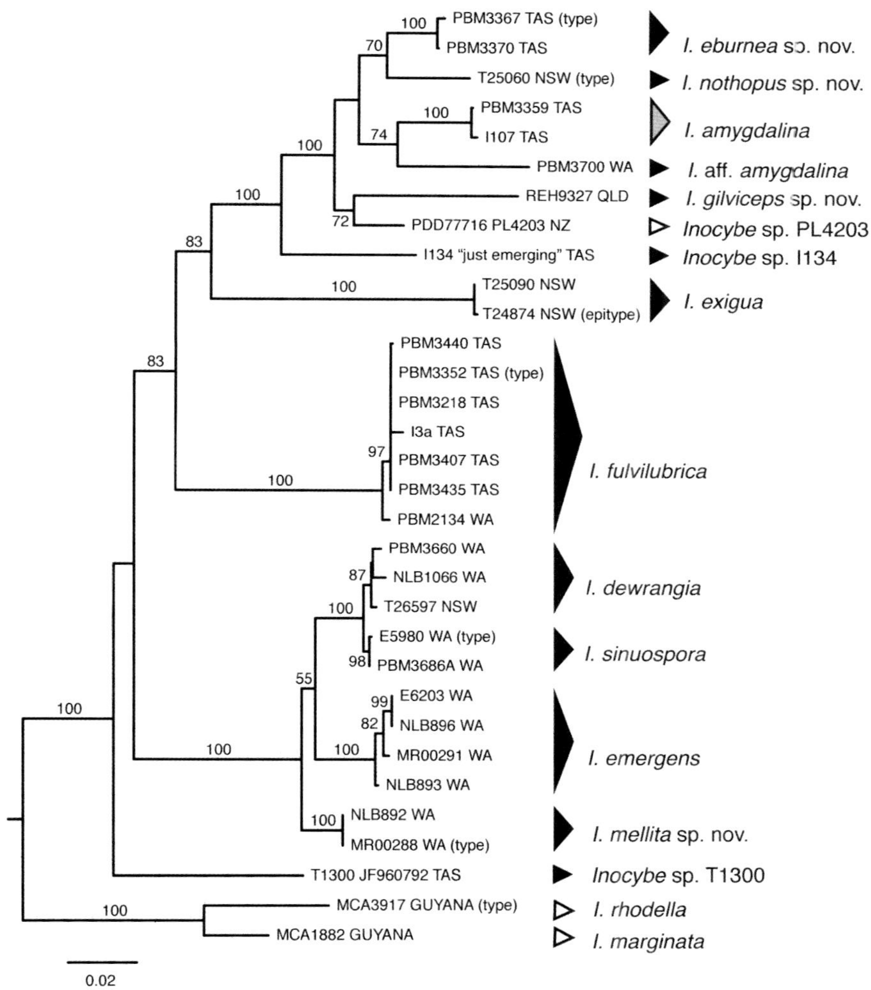

Figure 104. Maximum Likelihood phylogram of the *Inocybe emergens* group. Bootstrap values are shown above or below branches. The tree was inferred from combined ITS, nLSU, and *rpb2* gene regions and includes a total of 2870 sites. Fifty sites too ambiguous to align in the ITS region were removed prior to analysis. Species from Australia are indicated by black-filled triangles. Grey-filled triangles indicate species that occur in both Australia and New Zealand. White triangles indicate non-Australian taxa. Thirty two branch tips are included with 24 represented by ITS, 31 by nLSU, and 22 by *rpb2*. The tree is rooted using *Inocybe marginata* and *Inocybe rhodella*, two neotropical taxa positioned outside the *I. emergens* group (Matheny *et. al.*, 2009).

The *I. emergens* group contains 14 species from temperate regions of Australia and/or New Zealand. Thirteen of the species occur in Australia. Three are described here as new, and ten are documented below. Of the Australian species studied by us, most are clearly pruinose at least to the mid-region of the stipe (*I. nothopus* and *I. mellita* appear pruinose the entire stipe length). However, caulocystidia and cauloparacystidia can be found on the lower part of stipe. The cluster of *I. dewrangia*, *I. sinuospora*, *I. emergens*, and *I. mellita* share the presence of angular-nodulose, oblong-angular, or spores with a sinuose outline. All other species in the group at large that we have studied feature nodulose spores. Only two of the Australian species feature a distinct marginate bulb (*I. fulvilubrica*, *I. dewrangia*). The remainder typically have a weakly bulbous, swollen, or even stipe base.

Matheny *et al.* (2009) indicate the *I. leptospermi* group (see below) is nested within the *I. emergens* group, but support for this result was low. Phylogenetic studies of *Inocybe* as a whole will be necessary to reconcile relationships between the two groups.

88. Inocybe amygdalina (E.Horak) Garrido, *Biblioth. Mycol.* 120: 176 (1988)

Astrosporina amygdalina E.Horak, *New Zealand J. Bot.* 15: 739 ('1978') [1977]. Type: Okarito Trig., Westland, New Zealand, on the ground under *Leptospermum scoparium*, 19 Feb. 1969, *E.Horak 69/97*; holo: PDD 27114.

Pileus 20–45 mm wide, conical to plano-convex, with or without an obtuse umbo; margin initially incurved, soon decurved; surface dry and dull, fibrillose, not rimose; a distinct velipellis not noticeable; colour light yellowish brown or blonde (10YR 6/4) over the disc, fulvous or yellowish brown (10YR 5/6) or brownish orange (5C4) towards the margin; context pallid to watery pale brown, several mm thick under the disc. *Lamellae* adnexed, close, c. 34 L, with several tiers of lamellulae, pale brown (10YR 6/3), edges pallid, indistinctly fimbriate, to 3 mm deep. *Stipe* 30–57 × 4–5 mm at the apex; base slightly bulbous and up to 5–7 mm wide, swollen or indistinctly marginate to submarginate; upper half satiny-pruinose, lower part finely-fibrillose, cortina not observed; white on the upper half, patchily pale brown (10YR 6/3) below; context pallid with pale brown areas below. *Odour* resembling almond extract or marzipan when cut. Pl. 176, 177.

Basidiospores 7.5–9.5 × 5.5–7.0 µm [mean 8.8 × 6.1 µm; Q (1.22–) 1.26–1.54 (–1.59); Q mean 1.44 (n = 30/2)], gibbous with mostly 8–10 moderate sized nodules, yellowish brown; apiculus usually indistinct. *Basidia* 26–34 × 7–9 µm, clavate, hyaline, 4-sterigmate. *Pleurocystidia* 40–56 × 14–20 µm, utriform, ventricose or (broadly) fusiform, necks not well-developed, tapered at the base with a basal pedicel; apices obtuse and crystalliferous; thick-walled (2–4 µm), pale yellow to hyaline. *Cheilocystidia* similar to pleurocystidia, fusiform, ventricose or lageniform, mixed with crowded clavate hyaline paracystidia. *Caulocystidia* descending the entire length of the stipe, generally larger than pleurocystidia (50–70 × 16–18 µm), mostly fusiform to sublageniform or saccate, mixed with cauloparacystidia and cells intermediate in size and shape. *Pileipellis* composed of an epicutis of hyaline smooth or lightly encrusted hyphae 5–10 µm wide, with a distinct pigmented subcutis below, yellowish brown in mass; hyphae distinctly encrusted, to 20 µm wide. *Clamp connections* present. Fig. 105.

Known from Tas., where it grows on soil under myrtaceous plants in wet sclerophyll forest.; also in New Zealand. Fruiting Jan.–Feb., May.

Additional specimens examined: Tas.: Lower Kermandie, on soil in wet sclerophyll forest, 12 Feb. 2005, *G.M.Gates & D.Ratkowsky 1107* (HO 568205, TENN 066004; KJ729902-nLSU, KJ729876-ITS, KJ729933-*rpb2*); Lower Kermandie, singly on soil along track in wet sclerophyll forest under *Eucalyptus*, *Acacia dealbata* and *Pomaderris*, 19 May 2010, *G.M.Gates PBM3359* (PERTH 08516553, TENN 066003; KJ729903-nLSU, KJ729877-ITS, KJ729934-*rpb2*); Lower Kermandie, 21 Jan. 2011, *G.M.Gates 1146* (HO 562676, TENN 067752).

Diagnostic features: Fruit bodies medium-sized. Pileus light yellowish brown to yellowish brown or fulvous to brownish orange, fibrillose, not distinctly rimose; stipe white but pale brown in areas, the base swollen or indistinctly marginate, pruinose at least over the upper half, partial veil remnants not observed; odour of cut flesh resembling almond extract. Spores nodulose, hymenial; cystidia thick-walled, tapered at the base with a pedicel. Growing under myrtaceous plants.

This is the first report of *I. amygdalina* occuring in Australia. Our collections largely match Horak's (1977) description of habit, microscopic features and ecology (association with ECM Myrtaceae), apart from his statements that caulocystidia are not distinctive and that the pileus becomes scaly at the disc. However, we suspect Horak may have meant that *caulocystidia* were not distinct from cheilocystidia and pleurocystidia, and this is supported by his illustration of a thick-walled element (Horak, 1977). Our examination of the holotype revealed abundant caulocystidia in vast carpets on the stipe (not restricted to isolated clusters), including thick-walled metuloids with apical crystals, thin-walled elements of various shapes, and some multi-septate forms as illustrated by Horak (1977). Australian collections have distinctive caulocystidia even on the lower part of the stipe, although none were observed with narrow-beaked apices or a multi-septate form. Furthermore, we have not seen any scales at the centre of the pileus. Despite these comparatively minor disparities, we are prepared to accept that *I. amygdalina* also occurs in Tasmania, although genetic data from New Zealand material (type locality) are lacking. We examined two other collections from New Zealand labelled as "*I. amygdalina*" (PDD 89377 and PDD 89378). These have more coarsely nodulose spores with nodules to 5 µm high and round-based cystidia typical of *I. scissa* (E.Horak) Garrido, a species we have yet to confirm as occuring in Australia.

At least two Western Australian collections are similar in outward appearance to *I. amygdalina* and also have a distinctive odour of almonds. However, DNA sequences of one of these, *PBM3700* (PERTH 08318840, TENN 066585; KJ756467-nLSU, KJ756497-ITS, KJ756403-*rpb2*), collected on 27 July 2011 under *Allocasuarina* near Corrigin in south-western Western Australia, are noticeably different from those of Tasmanian collections, although the sequences do form a robust monophyletic group (Fig. 104). The anatomy of *PBM3700* matches the Tasmanian material and the type of *I. amygdalina*. However, it has cystidia with particularly pronounced pedicels (to 20 × 4 µm), and its basidia are consistently larger (to 45 × 12 µm). Reference to a second similar collection, not yet confirmed as *I. amygdalina*, from Western Australia [Mullering Brook, along Brand Hwy, 206 km N of Perth, in association with *Eucalyptus*, *Hypocalymma angustifolium* and *Chamelaucium uncinatum*, 30 May 1989, *O.K.Miller*] was made by Miller (unpublished data) and labelled *E0565* (OKM 23766, TENN 069546).

89. **Inocybe dewrangia** Grgur., *Larger Fungi of S. Australia* 185 (1997)

Type: Mount Lofty, S.A., 7 Apr. 1917, *J.B.Cleland*; holo: AD 12298.

[*Inocybe asterospora auct. non* Quél.: Cleland & Cheel, *Trans. Roy. Soc. S. Australia* 42: 105 (1918)]

[*Astrosporina asterospora auct. non* (Quél.) Rea: Cleland, *Toadstools & Mushrooms S. Australia* 118 (1934–1935)]

Pileus 9–25 mm wide, initially conical, becoming plano-convex and eventually plane, often with a small obtuse umbo; margin slightly incurved, soon decurved or almost straight with age and/or uplifted; surface dry; velipellis absent; centre smooth and unbroken; finely fibrillose towards the margin, later torn, becoming rimose, when young the edge of the margin with white universal veil remnants appearing as inconspicuous appendiculate fibrils, these soon disappearing; colour brownish yellow to yellowish brown (10YR 6/6 to 5/4), when older developing olivaceous or isabelline tones; context white, not changing colour where cut, up to 3 mm thick under the disc. *Lamellae* adnexed, moderately close, 26–30 L, with several tiers of lamellulae, pallid to light grey when young, becoming light yellowish brown to dark yellowish brown at maturity or later, ventricose, broad, edges indistinctly pallid-fimbriate. *Stipe* 14–30 × 3–5 mm, base with a marginate bulb up to 7 mm wide, at least the upper half pruinose, below lightly fibrillose, glabrescent below, at first white or pallid but becoming tinged buff or pale dingy yellow; context pallid, solid. *Odour* absent. Pl. 178–180.

Basidiospores 8.0–10.0 × 5.5–6.5 µm [mean 9.0 × 5.9 µm; Q 1.36–1.74; Q mean 1.53 (n = 30/1)], angular-nodulose, mostly with 9–11 small nodules, yellowish brown; apiculus small but distinct. *Basidia* 26–33 × 9–11 µm, clavate, hyaline, 4-sterigmate. *Pleurocystidia* 46–67 × 12–19 µm, utriform or fusiform; apices obtuse and crystalliferous; thick-walled especially apically (2–3 µm thick), hyaline to pale yellowish. *Cheilocystidia* similar to pleurocystidia, among abundant clavate and pyriform paracystidia.

Plate 176. *Inocybe amygdalina*. PERTH 08516553. Photograph — P.B.Matheny.

Plate 177. *Inocybe amygdalina*. HO 562676. Photograph — G.M.Gates.

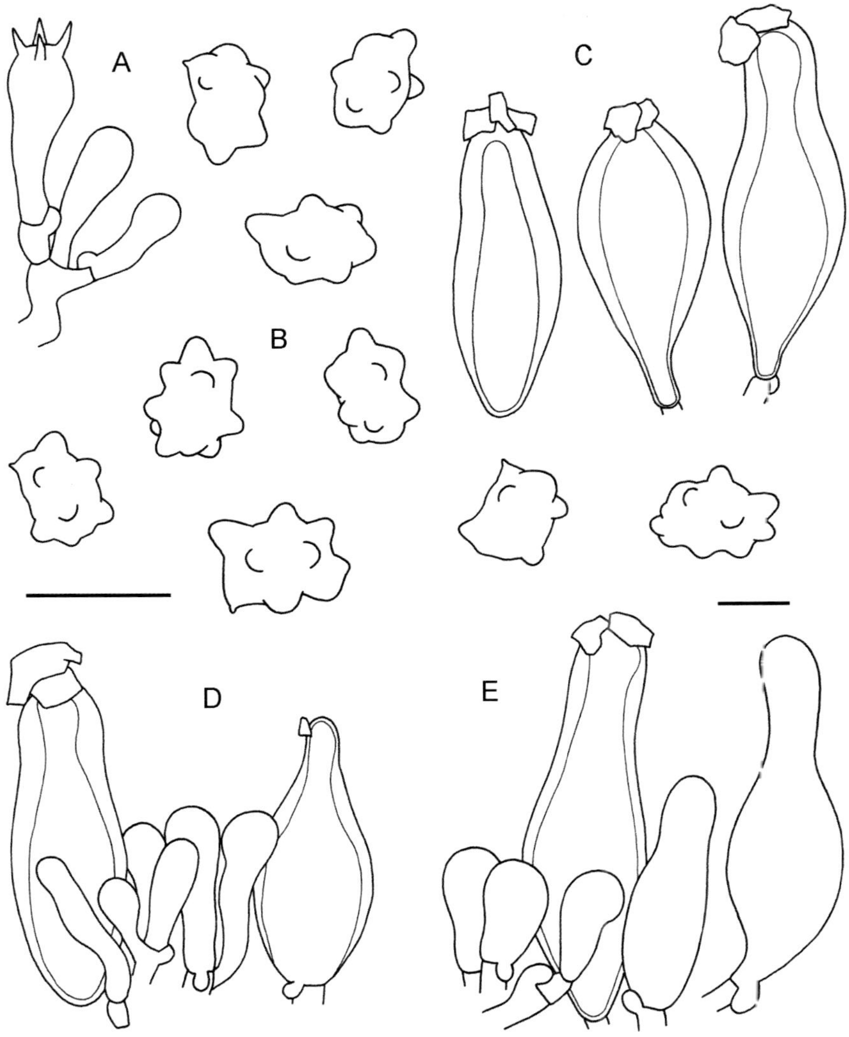

Figure 105. *Inocybe amygdalina.* All PERTH 08516553. **A,** Basidia; **B,** Basidiospores; **C,** Pleurocystidia; **D,** Cheilocystidia and paracystidia; **E,** Caulocystidia and cauloparacystidia Scale bars = 10 µm (longer bar for spores only).

Caulocystidia similar to pleurocystidia, but also fusiform to cylindrical or lageniform, up to 85 µm long, mixed with cauloparacystidia, descending below the mid-region of the stipe. *Pileipellis* a cutis of cylindrical hyphae, light reddish brown in mass, in the type surmounted by a sparse velipellar layer of narrowly cylindrical hyphae. *Clamp connections* present. Fig. 106.

Known from south-western W.A., south-eastern S.A. and south-western N.S.W.; grows on soil under *Eucalyptus*, *Corymbia* and/or *Gastrolobium*. Fruiting Apr.–July, Nov.

Additional specimens examined: W.A.: Woodlands Rd, Porongurup Natl Park, 34°40'06.4"S, 117°52'53.6"E, scattered singly on moss-covered soil in Karri-Marri-Jarrah forest under *Eucalyptus diversicolor* and *Corymbia calophylla*, 21 July 2011, *P.B.Matheny PBM3660* (PERTH 08319286, TENN 066471; KP170902-nLSU, KP171114-ITS); near Perup Ecology Centre, Perup Nature Reserve, just S of Boyup Brook–Cranbrook road, among moss in open forest dominated by *Eucalyptus marginata* with understorey of *Gastrolobium bilobum*, 1 June 2013, *N.L.Bougher NLB1066* (PERTH 08561184; KP170903-nLSU, KP171115-ITS, KM245979-*rpb2*); Tindale Rd, Mount Romance Farm, Denmark, 34°58'S, 117°21'E, in moss under *Corymbia calophylla*, 22 July 2008, *D.Fewings MR9/08* (PERTH 08129851). N.S.W.: sands plot 34, Ringwood State Forest, Riverina, on soil under *Eucalyptus microcarpa*, 6 Nov. 2000, *J.Trappe T26597* (OSC; KP170904-nLSU, KP171116-ITS, KM245980-*rpb2*).

Diagnostic features: Fruit bodies medium. Pileus brownish yellow to yellowish brown or golden brown, not scaly, rimose; stipe pruinose at least on the upper half, base with a marginate bulb; veil remnants present, initially connecting the pileus margin to the edge of the marginate bulb. Spores angular-nodulose, with 9–11 small nodules; cystidia mostly fusiform to utriform. Caulocystidia present below the mid-region of the stipe.

The description provided here is based on an examination of the South Australian type specimen supplemented by recent collections from south-western Western Australia. Previous descriptions (e.g. Cleland & Cheel, 1918, as *I. asterospora*; Grgurinovic, 1997; Matheny & Bougher, 2010) were based solely on Cleland material, while recent collections of *I. dewrangia* have proved to be distinctive due to their golden brown, rimose pileus that becomes quite ragged with age, and the pallid stipe. Indeed Cleland & Cheel (1918), who provided the first descriptive notes of this entity, described the pileus as "somewhat golden brown [rimose]". Cleland & Cheel's (1918) notes were based on AD 12298 (originally labeled *I. asterospora*), which was selected by Grgurinovic (1997) as the holotype of *I. dewrangia*. However, Grgurinovic's protologue was based on two of Cleland's collections (AD 12298 and AD 12143) that, unfortunately, represent different species (Matheny & Bougher, 2010). Thus, Grgurinovic (1997) described the pileus of *I. dewrangia* as "fibrilloso-squamosus" and the odour as "strong, rather seminal"; however, these are attributes of AD 12143, but not of the holotype (AD 12298). The former was subsequently identified as *I. curvipes* P.Karst., a northern temperate species introduced into Australia (Matheny & Bougher, 2011).

Matheny & Bougher (2010) emended the description of *I. dewrangia* to eliminate the attributes of AD 12143 but, like the previous authors, we only had access to the type collection. Some features considered to be distinctive for *I. dewrangia* included the golden brown, rimose pileus, its medium size, a finely fibrillose stipe that is swollen or somewhat enlarged towards the base but not bulbous; caulocystidia restricted to the apex of the stipe; fusiform to utriform, thick-walled pleurocystidia; and nodulose spores. Recent collections of *I. dewrangia* generally confirm these diagnostic features, but it has become evident that the stipe base can bear a distinct marginate bulb, and caulocystidia (metuloids) occur below the mid-region of the stipe. However, the latter are sparsely scattered on the lower stipe, and therefore the lower stipe does not usually appear to be pruinose.

Several other species can be confused with *I. dewrangia*. For example, *I. fulvilubrica* is characterised by its typically lubricous pileus, the lack of veil remnants, and more coarsely nodulose spores; the spores of *I. invadens* have fewer but larger nodules, cystidia with much thicker walls, and the absence of veil remnants; *I. sylvicola* occurs mainly in *Nothofagus* forests, has a honey-yellow and strongly rimose pileus, and considerably thicker-walled cystidia; while *I. exigua* has comparatively small fruit bodies, a stipe lacking a basal bulb, and short cystidia.

DNA sequences of *I. dewrangia* cluster very closely to those of *I. sinuospora* (Fig. 104). The latter is distinguished by its oblong and sinuous spores.

Plates 178, 179. *Inocybe dewrangia*. PERTH 08319286. Note in Plate 179 that the margin of the pileus is adorned with white, appendiculate, veil fibrils - but only when very young. Also note that the upper region of the stipe is more densely pruinose than the lower part. Photographs — P.B.Matheny.

Plate 180. *Inocybe dewrangia*. PERTH 08241716. Photograph — N.L.Bougher.

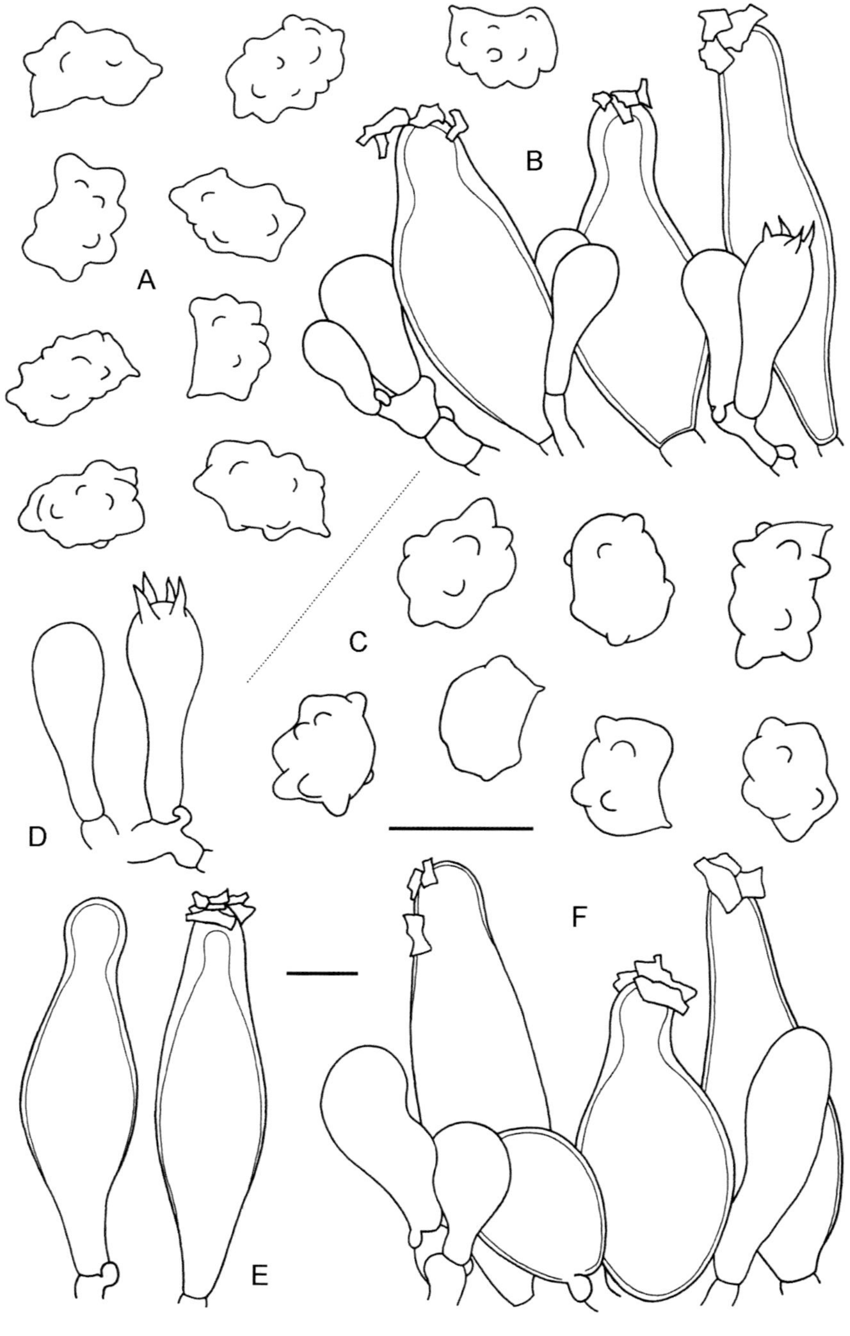

Figure 106. *Inocybe dewrangia.* All PERTH 08319286 except C. **A,** Basidiospores; **B,** Cheilocystidia; **C,** Basidiospores (holotype); **D,** Basidia; **E,** Cheilocystidia; **F.** Caulocystidia and cauloparacystidia from stipe apex. Scale bars = 10 μm (longer bar for spores only).

90. Inocybe eburnea Matheny, Bougher & G.M.Gates, *sp. nov.*

MycoBank: MB 809320

Type: St. Columba Falls Track, near Pyengana, Tas., 41°17'22"S, 148°00'17" E, on soil along track in wet sclerophyll forest under *Eucalyptus*, *Acacia* and *Pomaderris*, 21 May 2010, *G.M.Gates, D.Ratkowsky & P.B.Matheny PBM3367*; holo: PERTH 08516987; iso: TENN 068286 (KJ801162-nLSU, KJ778851-ITS, KJ756424-*rpb2*).

Pileus 20–42 mm wide, conical to convex, expanding with age, lacking an umbo; margin decurved; surface dry, with an expansive thin whitish or pallid velipellis, fibrillose to coarsely fibrillose, not rimose, at most fibrils weakly diverging near the margin that can become torn with age; colour light blonde overall where covered by velipellis, otherwise the ground colour yellow (2.5Y 7/6) to pale yellow (2.5Y 7/4), becoming light yellowish brown (2.5Y 6/4) with age; context pallid, up to 5 mm thick under the disc, not changing colour where bruised. *Lamellae* adnexed to ascending-adnate or sinuate, very close, 50–60 L, with several tiers of lamellulae, pale brown to fawn (Avellaneous), becoming light yellowish brown to dark yellowish brown with age, up to 6 mm deep; edges pallid but not appearing fimbriate. *Stipe* 25–45 × 3–6 mm, terete, even above the weakly bulbous base, with point of attachment embedded in a dense and compact mycelium aggregated with soil (most unusual in *Inocybe*); surface densely and uniformly pruinose, the pruinosity imparting a whitish colour, but the underlying ground colour is yellowish to ochraceous; context solid. *Odour* mild. Pl. 181.

Basidiospores 8.0–10.5 (–11.0) × 6.5–9.0 µm [mean 9.3 × 7.4 µm; Q 1.12–1.43; Q mean 1.27 (*n* = 24/2)], nodulose about an elliptic outline, with mostly 10–12 small to large nodules, yellowish brown. *Basidia* 29–36 × 8–11 µm, clavate, hyaline, 4-sterigmate, occasionally 1-spored (or then at least with asynchronous development of the sterigmata). *Pleurocystidia* 38–52 × 13–20 µm, short, mostly utriform or ventricose to subfusiform or subcylindrical, tapered at the base; apices well rounded or obtuse and bearing crystals, scattered singly; thick-walled (walls mostly 3–5 µm thick), hyaline. *Cheilocystidia* similar to pleurocystidia, short (<50 µm long) and not projecting far beyond the hymenium, mixed with basidia and clavate or pyriform hyaline crowded paracystidia. *Caulocystidia* to 70 µm long, fusiform to ventricose, with an obtuse or sometimes subcapitate apex; descending the entire length of the stipe, in scattered but dense clusters mixed with cells of intermediate size and shape and clavate to subcylindrical cauloparacystidia, these thin- to thick-walled. *Pileipellis* a cutis of cylindrical hyphae, ochraceous in mass, mostly 5–15 µm wide, thin-walled, with faint encrustations or sometimes appearing smooth. *Clamp connections* present. Fig. 107.

Known only from north-eastern Tas. at the type locality in wet sclerophyll forest. Fruiting in May.

Additional specimen examined: Tas.: type locality, *P.B.Matheny PBM3370* (PERTH 08516995, TENN 065786; KJ801163-nLSU, KJ778852-ITS, KJ756425-*rpb2*).

Etymology: From the Latin *eburnea* (ivory-white, white verging on yellow), in reference to the pale fruit bodies.

Diagnostic features: Fruit bodies medium. Pileus dry and ivory-white with a pale yellow ground colour; lamellae avellaneous, lacking distinct fimbriate edges; stipe entirely pruinose and with a weak bulbous base. Spores nodulose; pleurocystidia short (38–52 µm long), mostly utriform or ventricose.

Inocybe eburnea is somewhat similar to *I. viscata* (E.Horak) Garrido, both having short hymenial cystidia. However, the latter has spores with fewer nodules and a viscid pileus with a broad gelatinised epicutis. Phylogenetically, it is most closely related to *I. nothopus*, *I. amygdalina* and *I. gilviceps* and a few other poorly known and unresolved lineages from Australia and New Zealand (Fig. 104).

The second collection of *I. eburnea*, *PBM3370*, is morphologically very similar to the type (Fig. 104). However, *PBM3370* has a greater proportion of coarsely nodulose spores bearing multiple nodules exceeding 3 µm in height, while the hymenial cystidia of the type have larger, more block-like, apical crystals.

Plate 181. *Inocybe eburnea*. Holotype (PERTH 08516987). Photograph — P.B.Matheny.

91. Inocybe emergens (Cleland) Grgur., *Larger Fungi S. Australia* 192 (1997)

Astrosporina emergens Cleland, *Trans. Roy. Soc. S. Australia* 57: 192 (1933). Type: Kinchina, S.A., 8 June 1925, *J.B.Cleland*; holo: AD 40426.

Pileus 10–30 mm wide, plano-convex to plane or applanate, lacking an umbo; margin inrolled when young, becoming decurved, arching slightly upwards with age; surface dry but sometimes with much adhering soil, finely fibrillose to felty, especially the inrolled margin when young, occasionally with scattered whitish patches of velipellis material, foxy-brown (6D7 to 6E7; Amber Brown, Hazel, or Cinnamon-Brown) to pale brown (5B4) or various intermediate shades, lighter or almost whitish where densely covered with veil material, with inrolled margin almost white when immature; context pallid, not changing colour where bruised. *Lamellae* adnexed, close, with numerous tiers of lamellulae, white when young, becoming grey-brown (5C4) to pale brown (near 6D6); edges pallid or concolorous, fimbriate (at times indistinctly so), up to 5 mm deep. *Stipe* 21–25 × 5–6 mm, even, terete, with base swollen or with a weak marginate bulb that can be covered with soil or sand; relatively robust compared to the pileus; upper part noticeably pruinose, overall with a silky-fibrillose appearance, glabrescent, less obviously pruinose with age, white to pale yellow but becoming duller with age; cortina not seen; context solid, white. *Odour* none. Pl. 182, 183 (see also Pl. 5A).

Basidiospores 8.0–11.0 (–13.0) × 5.5–7.0 (–7.5) µm [mean 9.9 × 6.3 µm; Q 1.33–1.75 (–1.96); Q mean 1.57 (*n* = 42/4)], oblong-angular, occasionally angular-subphaseoliform to subtrapeziform, yellowish brown, often with a ventral concave depression, usually with 8–11 low obtusely conical nodules or weak corners, often with a distinct apical nodule. *Basidia* 32–40 × 9–10 µm, clavate, 4-sterigmate, occasionally 2-sterigmate, hyaline. *Pleurocystidia* 40–75 × 11–21 µm, utriform to fusiform, at times subclavate, subcylindrical or broadly fusiform, less often saccate, with apices well-rounded to obtuse and crystalliferous and a basal pedicel, mostly thick-walled (walls 1.5–4.0 µm), occasionally up to 1.5 µm thick apically, hyaline to pale yellow in KOH, frequent but less so in some collections (as in *E5551* and the type).

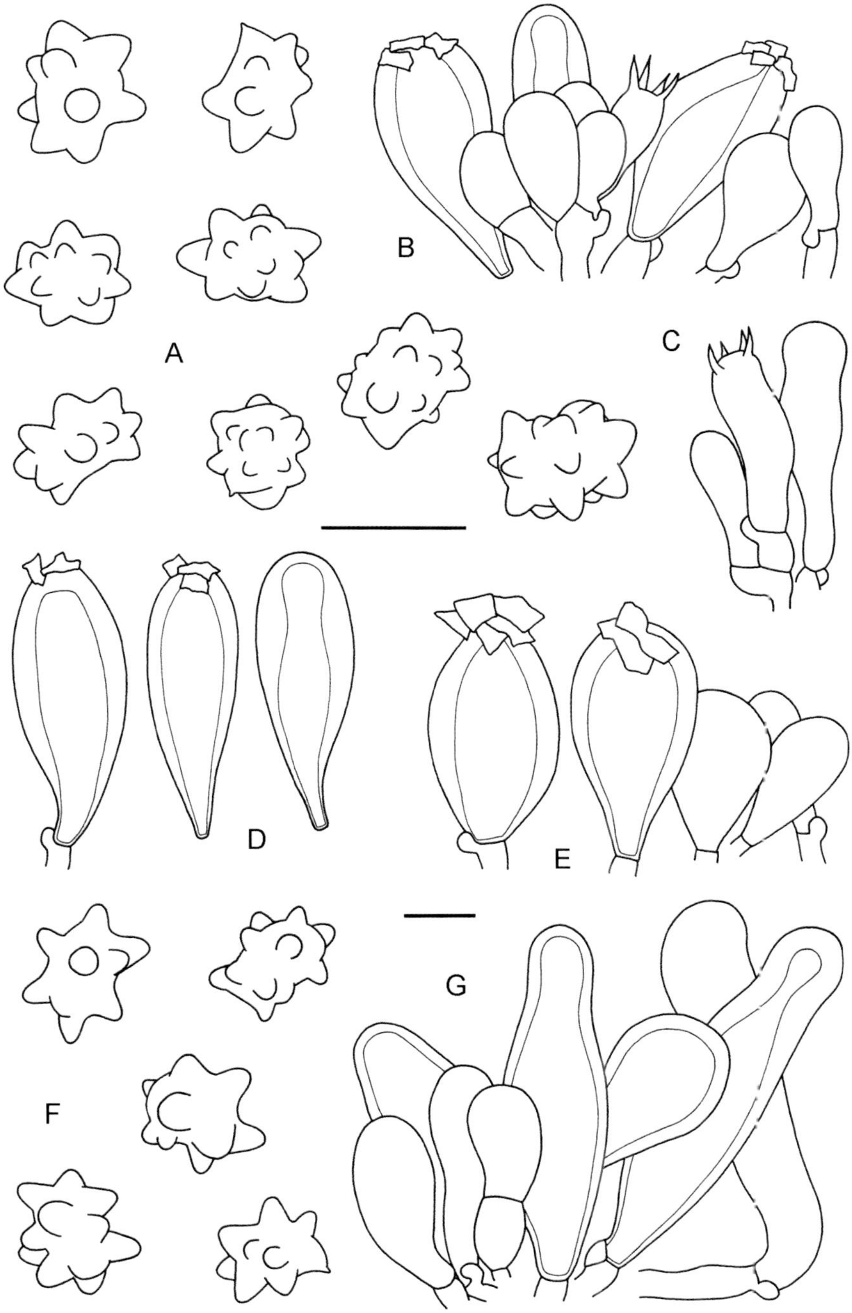

Figure 107. *Inocybe eburnea.* All holotype (PERTH 08516987), except E, F (PERTH 08516995). **A,** Basidiospores; **B,** Cheilocystidia and paracystidia; **C,** Basidia; **D,** Pleurocystidia; **E,** Cheilocystidia and paracystidia; **F,** Basidiospores; **G,** Caulocystidia and cauloparacystidia. Scale bars = 10 μm (longer bar for spores only).

Cheilocystidia 27–75 × 8–20 µm, similar to pleurocystidia, at times with swollen or indistinctly subcapitate apices, thick- or thin-walled, occasionally partly or entirely with brown cytoplasm, mixed with clavate thin-walled hyaline paracystidia. *Caulocystidia* similar to cheilocystidia, at times with swollen or subcapitate apices (typically lacking crystals), occasionally irregular in shape, thick-walled, rarely thin-walled (as in E6631), descending the entire length of the stipe; caulocystidia mixed with clusters of cauloparacystidia. *Pileipellis* with a superficial interwoven layer of hyaline smooth narrowly cylindrical velipellis hyphae 3–8 µm wide; subcutis of repent hyphae, ochraceous brown in mass; hyphae cylindrical, to 15 µm wide, encrusted. *Clamp connections* present in all tissues. Fig. 108.

Common in the wheatbelt region of W.A. and known along the southern coast east of Albany. The type is from South Australia, but *I. emergens* has not been reported from south-eastern Australia. Occurs in diverse vegetation types such as those containing *Allocasuarina* (Casuarinaceae), *Acacia* and *Gastrolobium* (Fabaceae), species of *Eucalyptus*, *Corymbia*, *Melaleuca*, *Leptospermum*, *Hypocalymma* and *Agonis* (Myrtaceae) and *Spyridium* (Rhamnaceae). Fruiting May–July in W.A., June in S.A.

Additional specimens examined: W.A.: 39.7 km S of Dongara, on sandy soil, 15 June 1989, *B.Dell E645*, *OKM23943* (PERTH 07548257); site 17, Higginson Rd, 20 km N of Kellerberrin, on ungrazed ground under *Gastrolobium* and *Eucalyptus salmonophloia*, 15 June 1995, *N.L.Bougher E5458* (PERTH 07709943, TENN 068368); *loc. id.*, on ground under *Allocasuarina* and *Acacia*, 15 June 1995, *W.Dunstan & I.Tommerup E5462* (PERTH 07710062, TENN 068369); hilltop site 17B, Higginson Rd, 16 km N of Kellerberrin, under *Allocasuarina campestris* and *Acacia* sp., 5 July 1995, *I.Tommerup E5531* (PERTH 07605323, TENN 068370); *loc. id.*, under *Eucalyptus wandoo* and *Allocasuarina campestris*, 19 July 1995, *W.Dunstan E5551* (PERTH 07608144, TENN 068371); site 18, 31 km N of Kellerberrin, under *Eucalyptus salmonophloia*, 20 July 1995, *I.Tommerup E5555* (PERTH 07608136, TENN 068372); Vinegar Well, off Barber Rd, Kunjin, 20 km W of Corrigin, in litter under *Eucalyptus loxophleba* and *Acacia acuminata* in open woodland, 1 July 1999, *J.Catchpole, I.Tommerup & N.L.Bougher E6203* (PERTH 07659903, TENN 068366; EU569838–nLSU, JQ085935–ITS, EU569837–*rpb1*); *loc. id.*, 19 July 2000, *N.L.Bougher & I.Tommerup E6631* (PERTH 07674325); *loc. id.*, scattered on ground under *Acacia acuminata*, 19 July 2000, *N.L.Bougher & I.Tommerup E6631* (PERTH 07674325); *loc. id.*, in sand, solitary, often buried, under *Allocasuarina* sp., 27 July 2011, *P.B.Matheny PBM3687* (PERTH 08318980, TENN066554); *loc. id.*, scattered singly in open woodland of *Eucalyptus kondinensis* and *Acacia acuminata*, 27 July 2011, *N.L.Bougher NLB893* (PERTH 08320365, TENN 066831; KP170905-nLSU, KM245982-*rpb2*); *loc. id.*, scattered and solitary near *Acacia acuminata* and *Allocasuarina huegeliana*, 27 July 2011, *N.L.Bougher NLB896* (PERTH 08320349, TENN 066778; KJ729915-nLSU, JX258837-ITS, KJ729943-*rpb2*); *loc. id.*, in sandy soil under *Allocasuarina*, 27 July 2011, *M.Ryberg MR00291* (TENN 066564; KJ729916-nLSU, KJ729944-*rpb2*); paddock 51, Valema Farm, Grylls Rd, Corrigin Shire, on ground with moss under *Melaleuca spicigera*, *M. uncinata* and *Leptospermum erubescens*, 12 July 1999, *J.Catchpole, I.Tommerup, S.Bolsenbroek & N.L.Bougher E6250* (PERTH 07658664, TENN 068373); fungi survey site 5, Dryandra S.F., under *Corymbia calophylla* in a low woodland over low heath including *Gastrolobium microcarpa* and *Hypocalymma angustifolium*, 30 May 1992, *Australian Plant Pathology Society Dryandra 7* (PERTH 02342626, TENN 068367); Rocky Point Track, Two Peoples Bay Nature Reserve, under *Melaleuca thymoides*, *Agonis flexuosa*, *Spyridium globulosum* and *Westringia*, 28 May 1992, *K.Syme KS539/92A* (PERTH 05429684; heterogeneous); old dismantled Kunjin railway, c. 14 km W of Corrigin, solitary near planted eucalypts and *Allocasuarina* bushes, 26 July 2011, *N.L.Bougher NLB878* (PERTH 08320527, TENN 066729); Corrigin Wildflower Drive, 5 km W of Corrigin, scattered singly in *Allocasuarina campestris* thicket, 27 July 2011, *N.L.Bougher NLB884* (PERTH 08320462, TENN 066798).

Diagnostic features: Fruit bodies medium-sized, at times robust. Pileus orange-brown, brownish yellow to pale brown, at times with white velipellis; stipe pruinose at least on the upper half, with a swollen or indistinctly bulbous base, usually not longer than pileus width. Spores elongate and angular, with 7–11 indistinct nodules and a ventral concave depression; caulocystidia present along the entire length of the stipe, some with swollen or subcapitate apices.

Inocybe emergens is characterised by the predominantly angular spore outline with 7–11 indistinct nodules and a ventral concave depression; some collections have a minority of spores with more prominent obtuse nodules. Caulocystidia can be found the entire length of the stipe and characteristically include some with swollen or subcapitate apices. This latter feature also occurs in the closely related and sometimes co-occurring *I. sinuospora* (Bougher *et al.*, 2012). *Inocybe mellita* also occurs sympatrically and is rather similar to *I. emergens*, but it has a longer stipe (giving it a more stately appearance), it lacks a distinct bulb, and has an odour of honey.

In the field, the pileus colour of *I. emergens* varies from orange-brown to brownish yellow and pale brown, at times with conspicuous patches of an appressed, thin, white velipellis. The extent of velipellar development influences the overall pileus colour. The stipe is conspicuously pruinose (at least in immature specimens) along its entire length, and it is generally robust with a swollen or indistinctly marginate basal bulb. Many collections of *I. emergens* have squat fruit bodies where the stipe length rarely exceeds the pileus diameter.

Western Australian material corresponds well microscopically with the South Australian type specimen (Matheny & Bougher, 2010), and while spore dimensions of most western collections tend to be somewhat larger than the type, they do overlap in size. Cleland (1933) originally described the pileus as "pallido-luteus", which could be construed as being from material with a prominent velipellis.

In addition to gross morphological similarities with *I. sinuospora* and *I. mellita*, fruit bodies of *I. emergens* resemble at least one undescribed, smooth-spored species (Bougher *et al.*, 2012). In the field care should be taken to avoid mixed collections. For example, PERTH 05429864 is a heterogeneous mixture of a smooth-spored species and specimens of *I. emergens*.

Phylogenetic data suggests that *I. emergens* is most closely related to *I. sinuospora*, *I. mellita* and *I. dewrangia* (Fig. 104).

92. **Inocybe exigua** (Cleland) Grgur., *Larger Fungi S. Australia* 184 (1997)

Astrosporina exigua Cleland, *Trans. Roy. Soc. S. Australia* 57: 192 (1933). Type: Hope Valley, near Adelaide, S.A., on sandy soil, 29 Apr. 1923, *J.B.Cleland s.n.*; holo: AD 12226; epi: PERTH 08517290 (selected here): near south-central gate, Genaren Hill Sanctuary, Parkes Shire, N.S.W., on rocky soil in ungrazed woodland of regrowth *Eucalyptus dealbata*, *Acacia doratoxylon*, *Callitris glaucophylla*, *Micromyrtus ciliata* and *Helichrysum viscosum*, alt. 265 m, 9 Nov. 1999, *J.Trappe T24874*; isoepi: TENN 068378 (AY380389-nLSU, KJ778856-ITS, AY351809-*rpb2*, AY337393-*rpb1*).

Pileus 12–25 mm wide, initially subconical, expanding to broadly conical, broadly convex, campanulate or almost plane, at first subumbonate, becoming obtusely umbonate with age, or an umbo lacking; margin incurved when young, becoming decurved; surface dry, minutely squamulose on the disc, fibrillose towards the sparingly lacerate margin; light brown to cinnamon-brown, greyish brown to light yellowish brown, generally darker in the centre than at the margin; context to 1 mm thick, brownish white. *Lamellae* adnexed to adnate, moderately close, pallid to pale brown when young, becoming pale brown with age. *Stipe* 12–27 × 2–4 mm, even, pruinose at least over the upper half, otherwise silky in appearance, or appearing glabrous, white or brownish white; veil absent. *Odour* not distinctive.

Basidiospores 8.0–10.5 × 6.0–7.0 (–7.5) µm [mean 9.0 × 6.6 µm; Q 1.21–1.58; Q mean 1.36 (n = 51/3)], nodulose and usually with 9–13 distinct subconical (seldom blunt) nodules around a subelliptic to subangular outline, yellowish brown to brownish yellow; apiculus small. *Basidia* 25–38 × 8–11 µm, clavate to utriform, hyaline, 4-sterigmate. *Pleurocystidia* 31–65 × 11–27 µm, clavate, (short-) saccate, utriform, cylindrical, or elliptic, necks not differentiated; apices well-rounded and often crystalliferous; thick-walled, walls 1.5–3.0 µm thick, some (holotype) nearly thin-walled, hyaline, not frequent. *Cheilocystidia* similar to pleurocystidia but generally smaller and more often elliptic in outline, not crowded among paracystidia and basidia; paracystidia pyriform to subglobose, thin-walled, hyaline. *Caulocystidia* similar to cheilocystidia, mixed with thin- and thick-walled subcylindrical clavate and lageniform cells, frequent on the upper half of the stipe, less frequent below the centre of the stipe and near the base, these caulocystidia at times thin-walled but mixed with cauloparacystidia. *Pileipellis* an irregular to subregular cutis of dark brown hyphae (Tawny to Ochraceous-Tawny in mass) that are faintly encrusted, 5–8 (–20) µm wide; above with an irregular subhyaline layer of narrow smooth velipellis hyphae 4–6 µm wide. *Clamp connections* present. Fig. 109.

Occurs in S.A. and N.S.W.; on soil in *Eucalyptus* woodland. Fruiting Apr. and Nov.

Additional specimen examined: N.S.W.: epitype locality, 21 Apr. 2000, *J.Trappe T25090* (PERTH 08517304, TENN 068379; KP170908-nLSU, KM245985-*rpb2*).

Plate 182. *Inocybe emergens*. PERTH 07659903. Photograph — N.L.Bougher.

Plate 183. *Inocybe emergens*. PERTH 08320462. Photograph — N.L.Bougher.

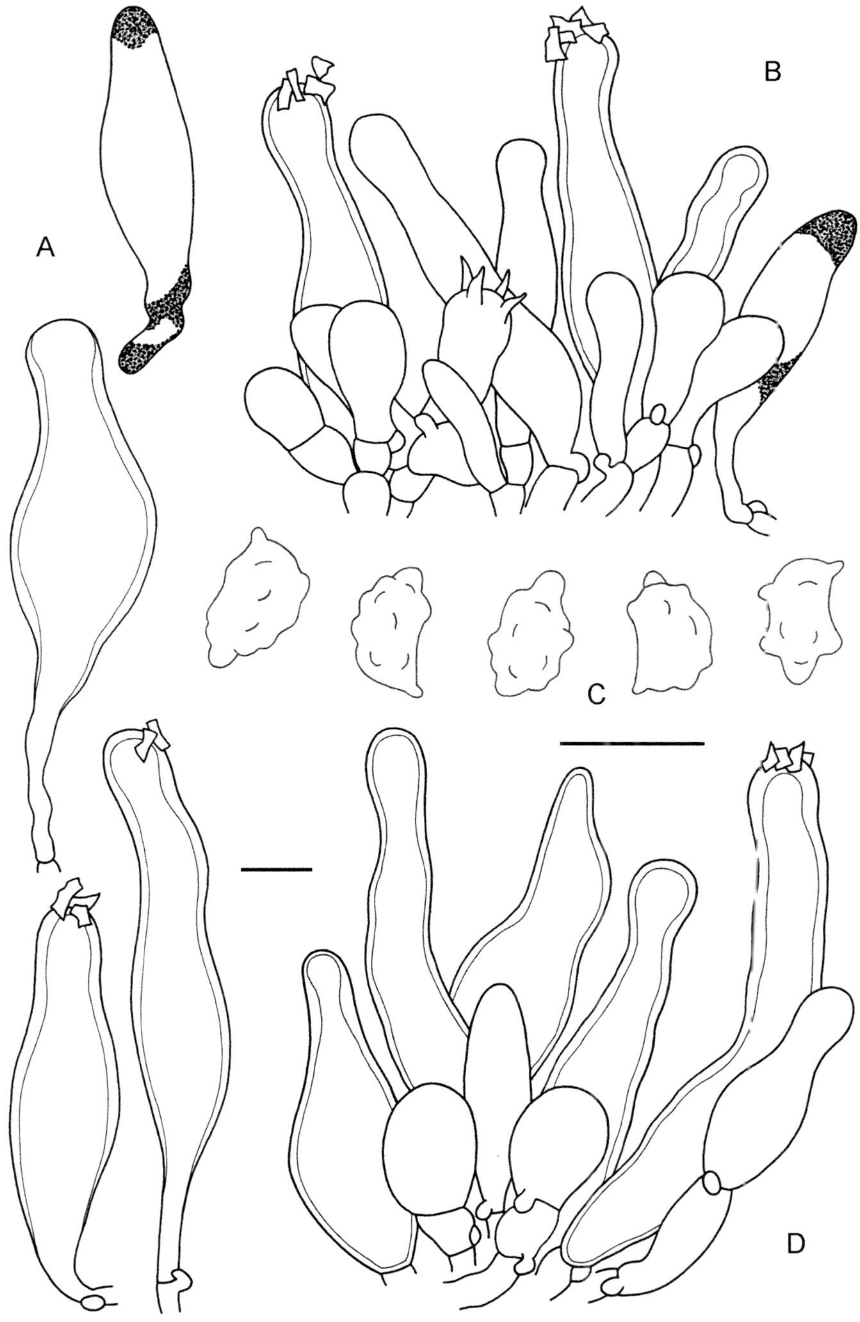

Figure 108. *Inocybe emergens*. All PERTH 07659903. **A,** Pleurocystidia; **B,** Cheilocystidia, paracystidia and basidia; **C,** Basidiospores; **D,** Caulocystidia and cauloparacystidia from upper stipe. Scale bars = 10 μm (longer bar for spores only).

Diagnostic features: Fruit bodies small. Pileus light brown to cinnamon-brown; stipe pallid and pruinose but lacking a distinctive marginate bulb. Pleurocystidia short, clavate, (short-) saccate or utriform, thick-walled to slightly so; spores coarsely nodulose, mostly with 9–13 distinct nodules about a subelliptic outline; caulocystidia descending below the stipe centre.

The foregoing description expands on previous accounts of *I. exigua* (Cleland, 1933; Horak, 1979a; Grgurinovic, 1997; Matheny & Bougher, 2010) by incorporating the attributes of two recently acquired collections. As mentioned by Horak (1979a), the holotype is in poor condition and very small and fragmented. Critical macroscopic attributes, such as presence or absence of a veil, were not recorded, microscopic features do not revive well, and our attempts at PCR amplification from the holotype have been unsuccessful. Therefore, we designate PERTH 08517290 as a supporting epitype of *I. exigua*, and we have captured images of microscopic features and obtained DNA sequence data from that specimen (Fig. 104).

Inocybe exigua has pleurocystidia that are noticeably short, and this trait combined with the coarsely nodulose spores, small size, and pruinose, even stipe forms a distinctive suite of diagnostic characteristics. Similar Australian species include *I. emergens* (Cleland) Grgur., which differs by its oblong spores with indistinct nodules, and *I. dewrangia* Grgur., which has larger spores and a stout stipe. Phylogenetically, *I. exigua* is sister to a number of taxa (Fig. 104), but it is most closely related to *I. nothopus* from New South Wales [as "*I. nothopedes* nom. prov." in Matheny *et al.* (2009)]. The latter can be distinguished from *I. exigua* by its tapered stipe, smaller spores with fewer (7–11) nodules, and cylindrical to subfusiform pleurocystidia. Matheny *et al.* (2009) referred to the newly selected epitype of *I. exigua* as "*Inocybe pseudocystis* nom. prov.".

93. Inocybe fulvilubrica Matheny, Bougher & G.M.Gates, *Nuytsia* 22: 62 (2012)

Type: Lower Track, Kermandie Falls, Tas., gregarious to scattered singly on soil along track in wet sclerophyll forest under *Eucalyptus*, *Acacia dealbata* and *Pomaderris apetala*, 9 May 2010, *G.M.Gates & P.B.Matheny PBM3352*; holo: TENN 065756; iso: PERTH 08309051 (JQ085922-nLSU).

Pileus 15–40 mm wide, parabolic or conical when young, expanding to plano-convex, at times campanulate, when expanded often with an obtuse umbo; margin at first incurved, later decurved or uplifted, becoming undulate; surface soapy-lubricous when wet, otherwise with a dry texture; velipellis present as scattered white mouldy patches, or becoming absent with age, or absent throughout development; disc smooth and unbroken, radially appressed-fibrillose or at most weakly rimulose towards the margin, at times torn, yellowish brown (10YR 5/6 to 5/4) or brownish yellow (10YR 6/6 to 5/6), later with shades of brown (10YR 5/3), at times light olive-brown or isabelline (2.5Y 5/6) towards the margin; context pallid, not changing colour where bruised. *Lamellae* adnexed or sinuate to subfree, moderately close to close, c. 30–40 L, with several tiers of lamellulae; pale yellow (5Y 7/4), pale olive (5Y 6/4) or very pale brown (10YR 7/3), becoming pale brown and eventually brown (10YR 6/3 to 5/3) or with olivaceous tints (5Y 5/6) in age; narrow to medium, to 4 mm deep; edges pallid and indistinctly fimbriate. *Stipe* 20–45 × 3–7 mm at the apex, terete or compressed, even down to the marginately bulbous 5–9 mm wide base; white but developing a pale buff or yellow undertone; surface pruinose and densely so (Tas. collections) or less dense (W.A. collections) along the entire length with a silky-satiny appearance; context solid, white or pallid. *Odour* not remarkable (Tas. collections) or weakly spermatic (W.A. collections). Pl. 184–186.

Basidiospores (7.0–) 7.5–10 × (5.0–) 5.5–7.0 (–7.5) µm [mean 8.3 × 5.8 µm; Q 1.24–1.72; Q mean 1.46 (n = 30/1)], coarsely gibbous, not stellate, with numerous (12–15) distinct conical nodules, at times with a single wedge-shaped nodule up to 3.5 µm high, yellowish brown in KOH. *Basidia* 22–31 × 7–10 µm, clavate, hyaline, 4-sterigmate. *Pleurocystidia* 54–72 × 14–20 µm, mostly narrowly fusiform, occasionally lageniform, (very) thick-walled (walls 4–6 µm thick), pale yellow; apices obtuse, crystalliferous. *Cheilocystidia* 40–60 × 13–18 µm, cylindrical, fusiform or narrowly lageniform; apices obtuse; walls to 5 µm thick. *Caulocystidia* descending the entire length of the stipe, occasionally less abundant below the centre, similar to pleurocystidia but often lageniform, scattered in association with dense clusters of broadly clavate or pyriform thin-walled and hyaline cauloparacystidia.

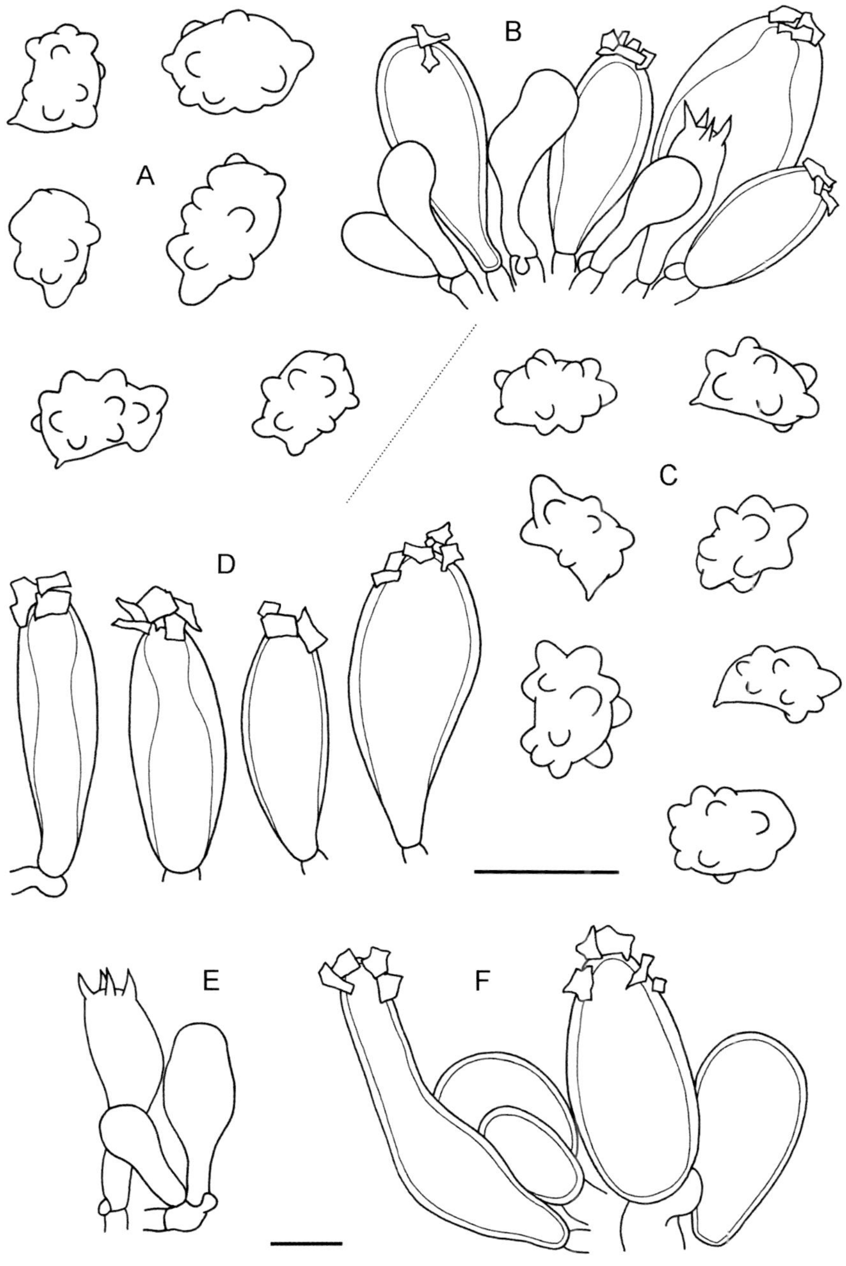

Figure 109. *Inocybe exigua.* All epitype (PERTH 08517290), except C, holotype (AD 12226). **A,** Basidiospores; **B,** Cheilocystidia and paracystidia; **C,** Basidiospores; **D,** Pleurocystidia; **E,** Basidia; **F,** Caulocystidia and associated elements from upper stipe. Scale bars = 10 µm (longer bar for spores only).

Pileipellis a cutis of yellowish brown pigmented hyphae, these cylindrical, encrusted and up to 18 µm wide, overlain by a velipellis layer of mostly repent cylindrical and smooth hyphae 3–8 µm wide. *Clamp connections* present. Fig. 110.

Occurs in south-western W.A. and Tas.; on soil in wet and dry sclerophyll forest in association with *Eucalyptus* and *Corymbia* or *Eucalyptus*, *Acacia* and/or *Pomaderris*. Fruiting Feb.–July.

Additional specimens examined: W.A.: Walpole–Nornalup Natl Park, scattered to gregarious, solitary or in pairs, on sandy soil under *Eucalyptus patens* near track in disturbed Karri–Marri forest, 25 June 2001, *N.L.Bougher E6776 (PBM2134)* (PERTH 07651708, TENN 066715; JQ085932-ITS, EU569849-*rpb2*, EU569848-*rpb1*, EU569850-nLSU); Jarrahdale Rd, 8 km W of Albany Hwy, solitary among litter under *Eucalyptus marginata* and *Corymbia calophylla* in Jarrah–Marri forest, 25 July 2011, *N.L.Bougher NLB867* (PERTH 08320640). Tas.: Mount Field Natl Park, on soil in wet sclerophyll forest under *Eucalyptus regnans*, *Acacia* and *Pomaderris*, 8 June 2009, *G.M.Gates, D.Ratkowsky & P.B.Matheny PBM3218* (TENN 065760; JQ085923-nLSU); Cuckoo Falls, near Scottsdale, on soil in wet sclerophyll under *Acacia*, *Pomaderris* and *Eucalyptus*, 24 May 2010, *P.B.Matheny PBM3407* (TENN 065755; JQ085924-ITS, JQ085925-nLSU); Mount Field Natl Park, track to Lady Barron Falls, scattered singly on soil in dry sclerophyll forest under *Eucalyptus obliqua* and *Acacia*, 1 June 2010, *G.M.Gates & P.B.Matheny PBM3435* (TENN 065758; JQ085926-ITS, JQ085927-nLSU); *loc. id.*, 1 June 2010, *G.M.Gates & P.B.Matheny PBM3436*; TENN 065757); *loc. id.*, scattered singly in wet sclerophyll under *Nothofagus*, *Acacia*, *Pomaderris* and *Eucalyptus regnans*, 1 June 2010, *G.M.Gates & P.B.Matheny PBM3440* (TENN 065759; JQ085928-ITS, JQ085929-nLSU, JQ085930-*rpb2*); Myrtle Gully, 9 Nov. 2000, *G.M.Gates (I 3a) & D.Ratkowsky* (TENN 065761, HO 563719; JQ085931-nLSU); Leap Loop, 5 Feb. 2011, *G.M.Gates I150(a, b)* (TENN 068375, HO 562692).

Diagnostic features: Fruit bodies medium-sized. Pileus yellowish brown, soapy-lubricous when wet, often with patches of white velipellis tissue; stipe white, pruinose throughout or at least below the mid-region; stipe base with a marginate bulb. Spores coarsely nodulose.

184

185

186

Plate 184. *Inocybe fulvilubrica*. PERTH 07651708. Photograph — N.L.Bougher.

Plate 185. *Inocybe fulvilubrica*. PERTH 08320640. Photograph — N.L.Bougher.

Plate 186. *Inocybe fulvilubrica*. PERTH 08320640. Photograph — N.L.Bougher.

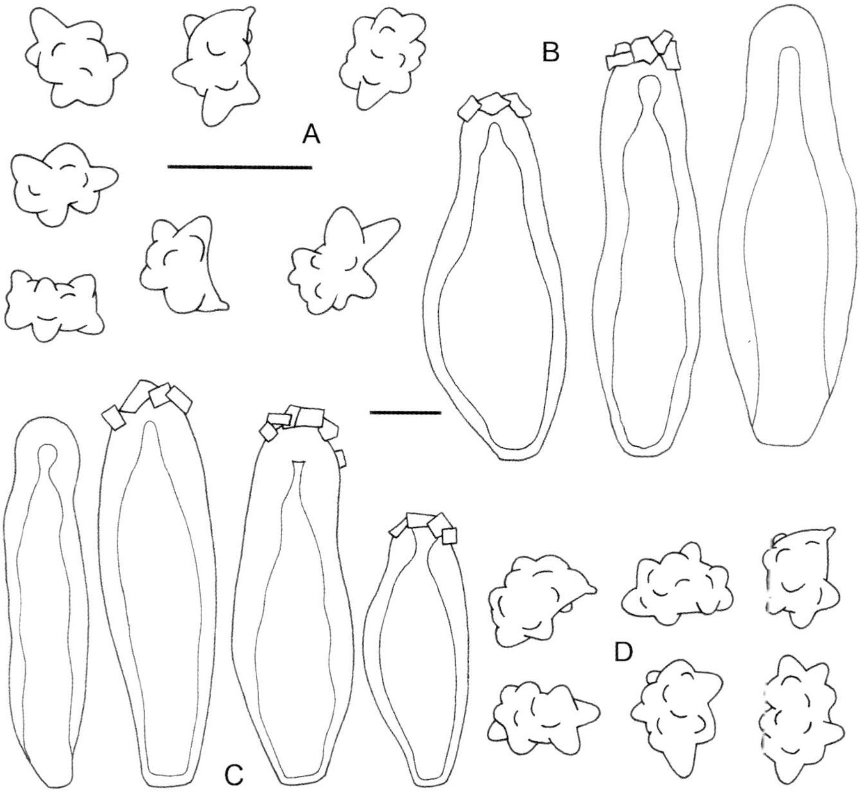

Figure 110. *Inocybe fulvilubrica.* All PERTH 07651708, except A, holotype (TENN 065756). **A,** Basidiospores; **B,** Pleurocystidia; **C,** Cheilocystidia; **D,** Basidiospores. Scale bars = 10 μm (longer bar for spores only).

The foregoing description is based on numerous collections confirmed as being conspecific by DNA sequence analysis (Fig. 104). These collections included two distinct morphological variants: one with a distinct lubricous pileus lacking a velipellis; and a second taxon with a drier pileus but with conspicuous white patches of veil material. However, DNA sequences of these collections are nearly identical, with minor divergence in Western Australian material (lubricous, smooth pileus). One Tasmanian collection (TENN 065759) differs slightly (3–4 bp at nLSU and *rpb2* loci, respectively) from one of the two Western Australian specimens (PERTH 07651708). However, microscopic examination of the pileipellis of the latter collection reveals the presence of superficial velipellis hyphae. Among the Tasmanian collections, the presence/absence of the velar remnants on the pileus is highly variable, as also appears to be the case in some other Australian species such as *I. emergens*.

Inocybe fulvilubrica is allied phylogenetically with several other nodulose-spored taxa from eastern Australia and New Zealand, all of which are characterised by an entirely or almost entirely pruinose stipe; these include *I. amygdalina* and *I. gilviceps* (Fig. 104). While the remaining species are not well known, *I. invadens* could be most easily confused with *I. fulvilubrica*, but it differs in having a dry pileus without a velipellis and spores with fewer nodules.

94. **Inocybe gilviceps** Matheny, Bougher & Halling, *sp. nov.*

MycoBank: MB 809333

Type: Awinya Rd, Fraser Island, Qld, 25°05'35.7"S, 153°10'28.1"E, gregarious on sand under *Eucalyptus racemosa*, 19 May 2010, *R.E.Halling, N.Fechner & M.Castellano REH9327*; holo: BRI 794277; iso: PERTH 08517274, NY (KJ801177-nLSU, KP171121-ITS, KJ811573-*rpb2*).

Pileus 15–35 mm wide, obtusely conical to convex-umbonate; margin decurved; surface dry, smooth and unbroken at the centre; velipellis absent, shaggy appressed-fibrillose to nearly appressed-squamulose towards the margin, rimulose to rimose towards the margin; yellowish brown to golden brown (5C7–5D8–5D7); context white, unchanging where bruised or cut. *Lamellae* adnexed, moderately close with several tiers of lamellulae; tan, pale greyish brown to yellowish brown; edges with an even appearance. *Stipe* 10–40 × 3–4 mm, strict, equal with a slightly bulbous base; surface lightly pruinose above, slightly subfibrillose below, above the base with a fine obscure pale orange-brown zone at the top of the bulb; nearly white overall when young, greyish yellow (4B4) with age; context with a pithy interior, becoming hollow. *Odour* mild. Pl. 187.

Plate 187. *Inocybe gilviceps*. Holotype (BRI 794277). Photograph — R.E.Halling.

Basidiospores 8.5–10.5 × 6.0–7.5 µm [mean 9.5 × 6.8 µm; Q 1.21–1.54; Q mean 1.40 (*n* = 20/1)], gibbous, often with 10 or 11 small to moderate-sized conical nodules, at times with a pronounced apical nodule, yellowish brown. *Basidia* 29–33 × 7–10 µm, clavate, hyaline, often 4-sterigmate, but 2-sterigmate basidia also observed. *Pleurocystidia* 55–68 × 14–17 µm, mostly fusiform and tapered at both ends, occasionally lageniform, deeply embedded within the subhymenium; apices obtuse with crystalliferous apices; thick-walled, the walls 4–5 µm thick, yellow in KOH, hyaline in water. *Cheilocystidia* similar to pleurocystidia but shorter, scattered among clavate paracystidia and/or basidia. *Caulocystidia* descending the entire length of the stipe, similar to pleurocystidia but hyaline, in clusters and mixed with cauloparacystidia and cells of intermediate size and shape. *Pileipellis* a cutis, ochraceous-tawny in mass, composed of cylindrical encrusted hyphae up to 15 µm wide. *Clamp connections* present. Fig. 111.

Occurs in south-eastern Qld. (Fraser Island); grows on sand under *Eucalyptus racemosa*. Fruiting May.

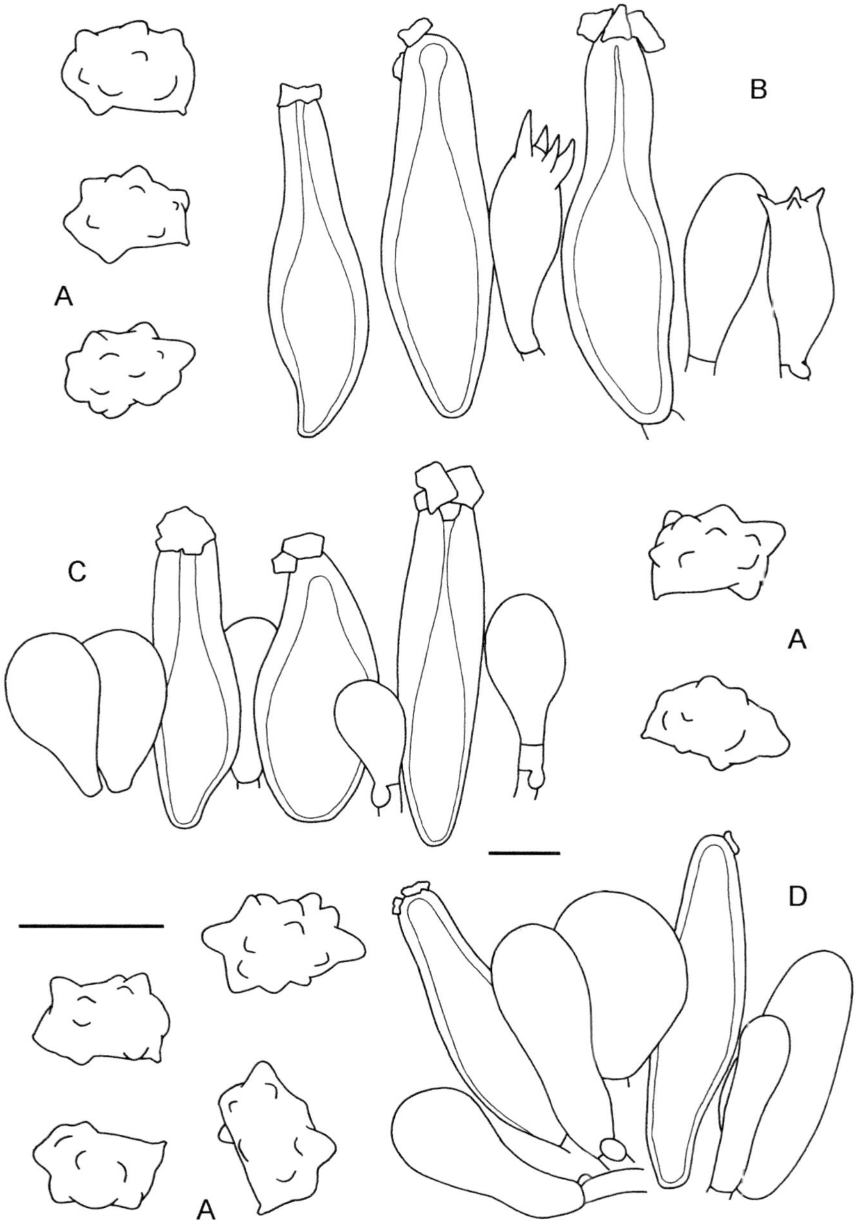

Figure 111. *Inocybe gilviceps.* All holotype (BRI 794277). **A,** Basidiospores; **B,** Pleurocystidia and basidia; **C,** Cheilocystidia and paracystidia; **D,** Caulocystidia and cauloparacystidia. Scale bars = 10 μm (longer bar for spores only).

Etymology: From the Latin *gilvus* (yellowish tan) and the suffix *-ceps* (referring to a head); indicating the distinctively coloured pileus of this species.

Diagnostic features: Pileus yellowish brown to golden brown and coarsely fibrillose to subsquamulose; stipe whitish-pruinose below the centre; base slightly bulbous, with a fine obscure pale orange-brown zone at the top of the bulb; odour mild.

A subtle but unique feature of *I. gilviceps* is the fine obscure pale orange zone at the top of the stipe bulb. Otherwise this species is similar to other temperate Southern Hemisphere marginate or bulbous, nodulose-spored species with a rimulose pileus. Thus, it shares morphological and phylogenetic affinities with *I. amygdalina*, *I. nothopus*, *I. eburnea* and one unclarified lineage from New Zealand (Fig. 104). *Inocybe amygdalina* is readily characterised by its almond-like odour, whereas no remarkable odour was noted for *I. gilviceps*. *Inocybe nothopus* has a tapered stipe base and short pleurocystidia, while *I. eburnea* differs in its whitish to ivory-yellow colours.

95. **Inocybe mellita** Matheny, Bougher & Ryberg, *sp. nov.*

MycoBank: MB 809390

Type: Vinegar Well, off Barber Rd, Kunjin, W of Corrigin, W.A., 32°23'31.7"S, 117°40'06.5"E, solitary or in pairs in sandy soil, in woodland under a mixture of *Eucalyptus kondinensis* and *Acacia acuminata*, 27 July 2011, *M.Ryberg MR00288*; holo: PERTH 08306621; iso: TENN 066622 (KP170949-nLSU, KP308776-ITS, KM406190-*rpb2*).

Pileus 17–22 mm wide, broadly conical to plane, occasionally undulate, umbonate; margin decurved, becoming straight; velipellis present over the centre; surface with much adhering sand and soil debris, smooth and unbroken at the centre, finely to densely fibrillose towards the margin; brownish yellow (10YR 6/6) to yellow (10YR 7/8), becoming yellowish brown (10YR 5/6); context white. *Lamellae* adnexed, close, 40–44 L, with several tiers of lamellulae; medium, 1–2 mm deep; light grey to pale brown (10YR 6/3) or yellowish brown (5C5), with edges pallid but not conspicuously fimbriate. *Stipe* 20–28 × 3–5 mm, even, at times tapering towards the middle, central in attachment or less often eccentric, terete, base at most barely swollen but lacking a conspicuous bulb; surface lightly pruinose for its entire length with a satiny or finely streaked appearance; white to pale yellow (2.5Y 8/4 or 7/4) or yellow (10YR 7/6); context white or off-white. *Odour* similar to honey or sweet and nutty, sometimes not evident immediately in the field. Pl. 188, 189.

Basidiospores 8.0–11.0 × 5.5–7.0 µm [mean 9.2 × 6.1 µm; Q 1.23–1.75; Q mean 1.50 (n = 38/2)], typically angular-nodulose (less often oblong or elliptic), with mostly 6–9 small nodules or corners; nodules obtuse, sometimes up to 2 µm tall, yellowish brown; apiculus small and inconspicuous. *Basidia* 23–34 × 7–10 µm, clavate, hyaline, 4-sterigmate. *Pleurocystidia* 40–62 × 12–18 µm, utriform or broadly ventricose or fusiform but often with a long narrow slender pedicel (to 20 µm long), often with obtuse broadly rounded or truncate apices; walls to 1.5 (-2.0) µm thick, especially in the upper part, hyaline or sometimes with ochraceous contents. *Cheilocystidia* similar to pleurocystidia but more often shorter and wider (to 20 µm), abundant but sometimes patchy on the gill edge, mixed with small clavate paracystidia and basidia. *Caulocystidia* present for the entire length of the stipe, in scattered clusters, ventricose, fusoid, cylindrical, many longer than hymenial cystidia, walls to 1.5 (–2.0) µm thick, hyaline or pale yellowish and sometimes with ochraceous contents, mixed with clavate cauloparacystidia and variously-shaped caulocystidioid cells. *Pileipellis* a compact cutis of subregular hyphae that are cylindrical and mostly 6–15 µm wide, ochraceous in mass, thin-walled, smooth or faintly encrusted. *Clamp connections* present. Fig. 112 (see also Pl. 6F).

Known only from south-western W.A.; grows in soil in open woodland dominated by *Eucalyptus kondinensis* and *Acacia acuminata* in the wheatbelt. Fruiting July.

Additional specimens examined: W.A.: type locality, in soil but emerging well above the soil level in a creek bed near *Eucalyptus kondinensis*, 27 July 2011, *N.L.Bougher NLB892* (PERTH 08320373, TENN 066720; KP170950-nLSU, KP308777-ITS, KM406191-*rpb2*).

Plate 188. *Inocybe mellita*, holotype (PERTH 08306621). Photograph — M.Ryberg.

Plate 189. *Inocybe mellita*. PERTH 08320373. Photograph — N.L.Bougher.

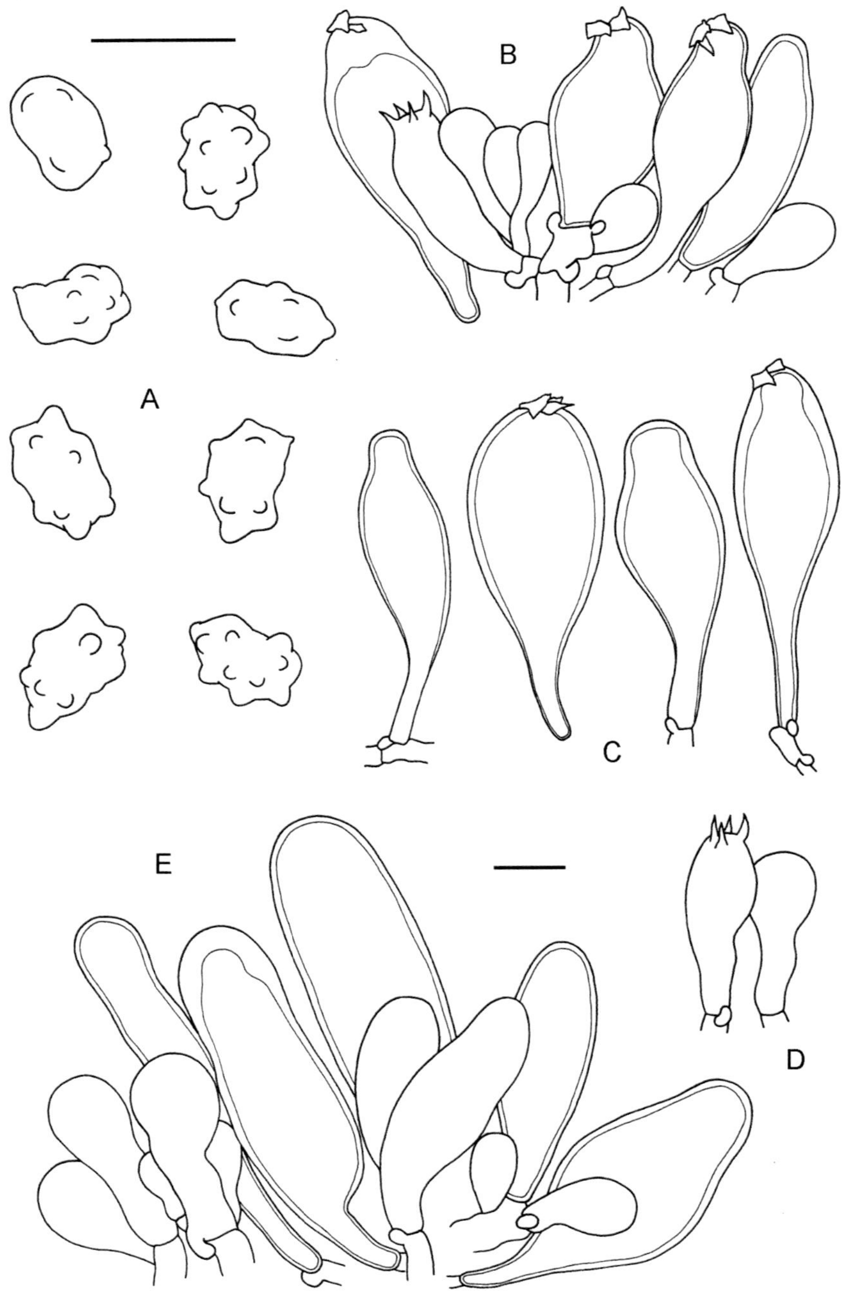

Figure 112. *Inocybe mellita.* All holotype (PERTH 08306621). **A,** Basidiospores; **B,** Cheilocystidia and paracystidia; **C,** Pleurocystidia; **D,** Basidia; **E,** Caulocystidia and cauloparacystidia. Scale bars = 10 μm (longer bar for spores only).

Etymology: From the Latin *mellita* (sweet, pertaining to honey), in reference to the honey odour.

Diagnostic features: Pileus brownish yellow, smooth; stipe white to yellowish, pruinose, even without a distinct bulb; odour of honey which intensifies after collecting.

Inocybe mellita is sympatric with two closely related but distinct species, *I. emergens* and *I. sinuospora*, both of which lack the honey odour and possess a marginate stipe bulb. All three species can be found fruiting together at the same locality in the wheatbelt region in south-western W.A. Phylogenetic analysis supports a grouping of the three species, in which *I. mellita* is the sister lineage to *I. emergens* and *I. sinuospora* (Fig. 104).

Inocybe fulvilubrica is most readily distinguished from *I. mellita* by its marginate stipe base, thick-walled (3–6 µm) pleurocystidia that lack the long, slender pedicel typical of *I. mellita*, and more coarsely nodulose spores (with 12–15 nodules). *Inocybe dewrangia* possesses a marginate stipe base, it lacks the honey odour of *I. mellita*, but has fibrillose universal veil remnants (when immature) along the margin of the pileus and at the edge of the basal bulb.

96. **Inocybe nothopus** Matheny, Bougher & Trappe, *sp. nov.*

MycoBank: MB 809395

Type: Wilga Park, Parkes Shire, New South Wales, alt. 260 m, on rocky soil in lightly grazed woodland of regrowth *Eucalyptus dealbata*, *Acacia doratoxylon* and *Micromyrtus ciliata*, 21 Apr. 2000, *J.Trappe T25060*; holo: PERTH 08517088; iso: TENN 068377 (AY380388-nLSU, KP308792-ITS, AY337391-*rpb2*, AY351815-*rpb1*).

Pileus 10–20 mm wide, convex to plane; surface dry, fibrillose, light yellowish brown; context white, c. 2 mm thick. *Lamellae* moderately close, adnexed to adnate, brown, edges pallid but indistinctly fimbriate. *Stipe* 10–20 × 2.5–4.0 mm, terete but tapering towards the cylindrical (not bulbous) base, slightly pruinose at least at the apex, indistinctly so on lower part; partial veil not observed; brownish white to light brown throughout. *Odour* not distinctive.

Basidiospores 7.0–9.0 × 5.0–6.0 µm [mean 8.0 × 5.7; Q 1.17–1.64 (–1.70); Q mean 1.40 (n = 20/1)], with 7–11 small to moderate-sized nodules, often with a distinct apical nodule ochraceous-buff; apiculus small and tapering. *Basidia* 25–32 × 7–10 µm, subclavate to subcylindrical, hyaline, 4-sterigmate. *Pleurocystidia* 36–52 × 12–16 µm, subcylindrical to subfusiform, sessile; apices obtuse, crystalliferous, sometimes subcapitate, thick-walled (3-4 µm), pale yellow to hyaline. *Cheilocystidia* similar to pleurocystidia but some shorter, at times broadly utriform or narrowly cylindrical; paracystidia clavate to elliptic, hyaline. *Caulocystidia* similar to hymenial cystidia but occasionally longer, descending the entire length of the stipe, mixed with cauloparacystidia. *Pileipellis* an ochraceous-tawny pigmented cutis of encrusted cylindrical to inflated hyphae mostly 5–15 µm wide; velipellis hyphae narrow, compact, interwoven, thin-walled, hyaline, mostly 3–5 µm wide. *Clamp connections* present. Fig. 113.

Occurs in the central-west of N.S.W.; grows on soil under *Eucalyptus*, *Acacia* and *Micromyrtus*. Fruiting Apr.

Etymology: From the Greek *nothos* (false) and *-pus* (footed), in reference to the tapered (non-marginate), pruinose stipe.

Diagnostic features: Fruit bodies small. Stipe entirely pruinose, tapering towards the base. Spores nodulose, typically with a distinct apical nodule; pleurocystidia subcylindrical to subfusiform, short.

Inocybe nothopus is unusual among nodulose-spored species in Australia in having a tapered stipe and lacking a marginate bulb. Phylogenetically, it is most closely related to a group of taxa with entirely or predominantly pruinose stipes (Fig. 104). These species vary in colour and fruit body size, and typically have a marginate basal bulb. *Inocybe nothopus* was previously given the informal name "*Inocybe nothopedes nom. prov.*" by Matheny *et al.* (2009).

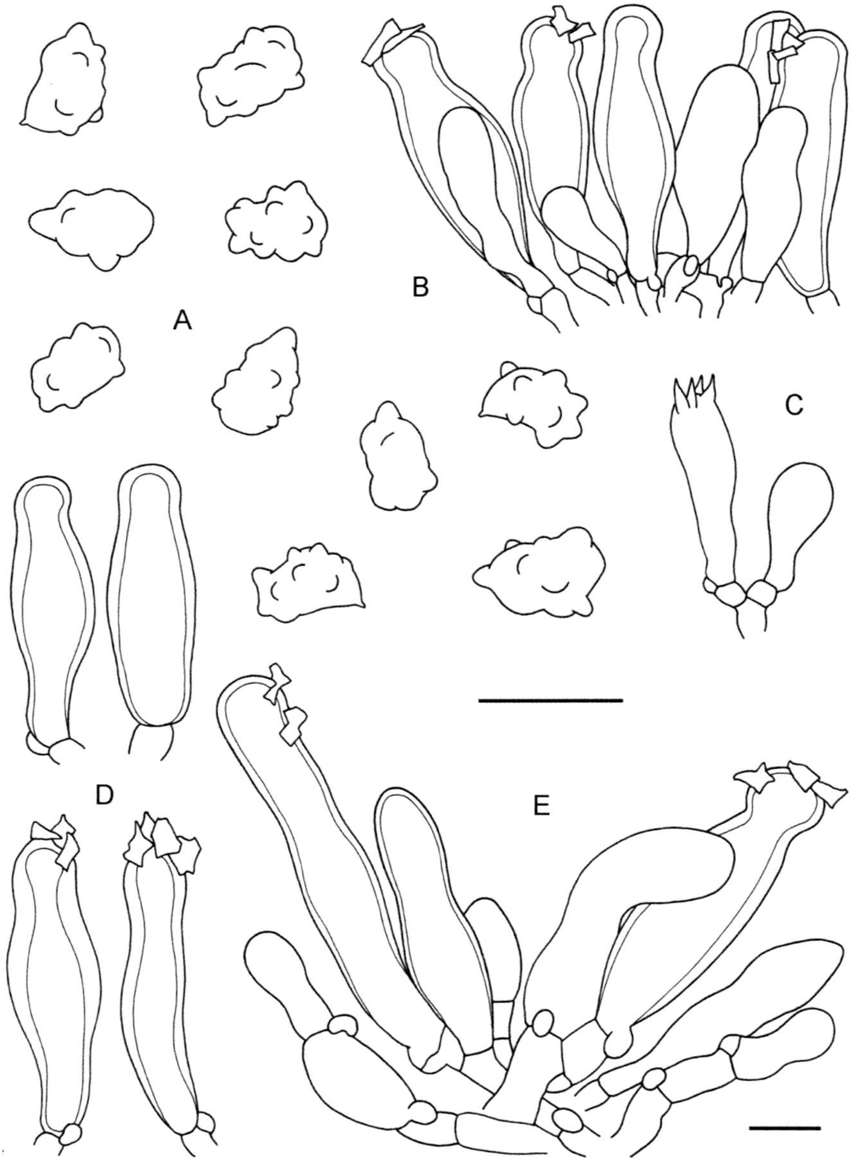

Figure 113. *Inocybe nothopus.* All holotype (PERTH 08517088). **A,** Basidiospores; **B,** Cheilocystidia and paracystidia; **C,** Basidia; **D,** Pleurocystidia; **E,** Caulocystidia and cauloparacystidia. Scale bars = 10 µm (longer bar for spores only).

97. **Inocybe sinuospora** Matheny & Bougher, *Nuytsia* 22: 67 (2012)

Type: Hilltop site (17A/B), Higginson Rd, 16 km N of Kellerberrin, along Bencubbin–Kellerberrin road, W.A, 31°30'S, 117°43'E, scattered on ground under *Allocasuarina*, 29 June 1998, *I.Tommerup E5980*; holo: PERTH 07606532; iso: TENN 068365 (JQ085936-ITS, EU600860-*rpb1*, EU600861-nLSU).

Pileus 10–30 mm wide, broadly convex to plano-convex even when relatively young, soon plane; margin decurved to straight; disc at times with a small obtuse umbo; surface dry but occasionally with adhering grains of sand; velipellis sometimes present over the disc; surface fibrillose but not coarsely so; fulvous or brownish yellow to dull yellow-brown or cinnamon-brown (near 6D6 to 6D5; Buckthorn Brown) or a mixture of these shades, with age dark yellowish brown (Dresden Brown); context pallid with a yellowish tinge, not changing colour upon exposure, to 4 mm thick. *Lamellae* adnexed to adnate, ascending-adnate or uncinate, moderately close to close (44–54 L) with several tiers of lamellulae, ventricose, to 5 mm broad, very pale yellowish brown to dark yellowish brown (6E7; 10YR 4/4); edges pallid and indistinctly fimbriate. *Stipe* 11–30 × 3–10 mm, terete, even, with a small non-marginate basal bulb or appearing even throughout; upper surface pruinose, fibrillose to slightly furfuraceous below, glabrescent with age, pale honey or buff; context initially solid, becoming hollow, pallid to pale brownish or at times with dull yellow tinge, white near the base. *Odour* none. Pl. 190, 191.

Basidiospores (9.5–) 10.0–12.5 (–13.0) × 4.5–6.0 µm [mean 10 × 5.3 µm; Q (1.82–) 1.90–2.33 (–2.56); Q mean 2.10 (*n* = 30/1)], oblong-angular, subtrapeziform, oblong-amygdaliform, often with a few shallow sinuses, sometimes with a ventral depression; apices variable, attenuated and narrower than the base, or bluntly pointed to rounded and broader than the base, gently undulate, rarely with 1 or 2 indistinct low nodules or corners, rarely entirely smooth; colour yellowish brown in KOH; apiculus small but prominent; walls thin or occasionally thick. *Basidia* 29–37 × 8–12 µm, clavate, 4-sterigmate, hyaline or collapsed and filled with yellow granular content. *Pleurocystidia* rather frequent, 51–70 × 14–18 µm, utriform to subfusiform with short necks, with a short or slender basal pedicel; apices obtuse, crystalliferous or not; walls thick apically (to 3 µm), hyaline. *Cheilocystidia* 50–70 × 12–20 µm, similar to pleurocystidia, occasionally ochraceous, mixed with clavate to subelliptic paracystidia that are hyaline and thin-walled; occasional basidia observed. *Caulocystidia* descending to near the stipe base, similar to cheilocystidia and pleurocystidia but also lageniform and at times with subcapitate apices (as in *I. emergens*) or thin-walled, less frequent on the lower half of the stipe, but mixed with cauloparacystidia; refractive hyphae and pigment deposits frequent in the stipe trama. *Pileipellis* a cutis of regular hyphae, cylindrical, yellowish brown (Ochraceous-Tawny) in mass, mostly 7–13 µm wide, lightly encrusted, thin-walled. *Clamp connections* present in all tissues. Fig. 114 (see also Pl. 6A).

Occurs in the central wheatbelt region of south-western W.A.; on the ground under *Allocasuarina*. Fruiting June and July.

Additional specimens examined: W.A.: Old dismantled Kunjin railway, c. 14 km W of Corrigin, 32°21'25.9"S, 117°44'17.1"E, solitary near planted eucalypts and *Allocasuarina* bushes, 26 July 2011, *N.L.Bougher NLB874* (PERTH 08320578); Vinegar Well, off Barber Rd, Kunjin, 20 km W of Corrigin, 32°23'31.7"S, 117°40'6.5"E, scattered and often deeply buried in sand, under *Allocasuarina*, 27 July 2011, *P.B.Matheny PBM3686A* (TENN 066548, PERTH 08318999; KP171037-nLSU, JX258838-ITS, KM555132-*rpb2*).

Diagnostic features: Stipe pale honey-coloured or buff, without a distinct marginate bulb, at least the upper surface visibly pruinose; odour absent. Spores oblong-amygdaliform with a sinuous outline; caulocystidia descending below the mid-region of the stipe.

Inocybe sinuospora and *I. emergens* have fruit bodies of similar general appearance (Bougher *et al.*, 2012), and these species are sympatric at least in some locations in the central wheatbelt of Western Australia, in fragmented remnants of native woodland highly modified by agriculture and increasing salinity. Phylogenetic analyses support a close but distinct relationship between the species (Fig. 104; Matheny *et al.*, 2009). An ITS sequence from the holotype of *I. sinuospora* differs at a minimum of 40 nucleotide positions (94% similarity, not including gaps) from *I. emergens* (PERTH 07659903) reinforcing their taxonomic separation. Other differences include the stipe of *I. sinuospora* lacking the distinct marginate bulb of *I. emergens*.

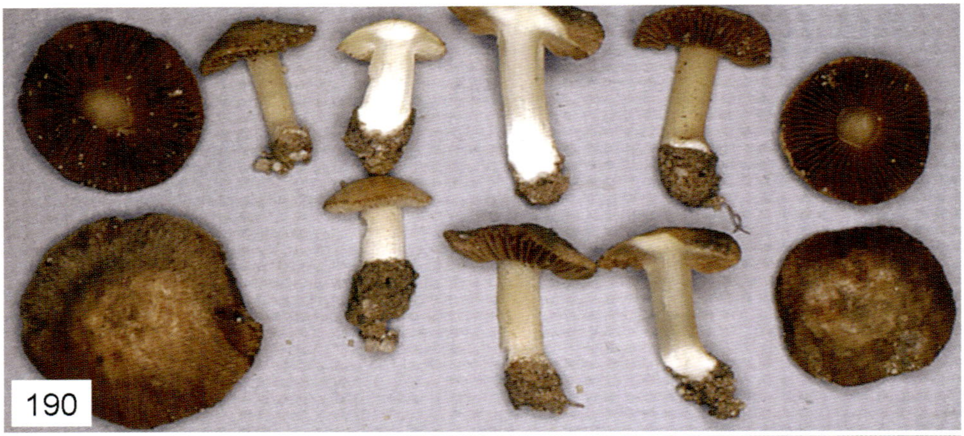

Plate 190. *Inocybe sinuospora,* holotype (PERTH 0706532). Photograph — N.L.Bougher.

Plate 191. *Inocybe sinuospora.* PERTH 08318999. Photograph — P.B.Matheny.

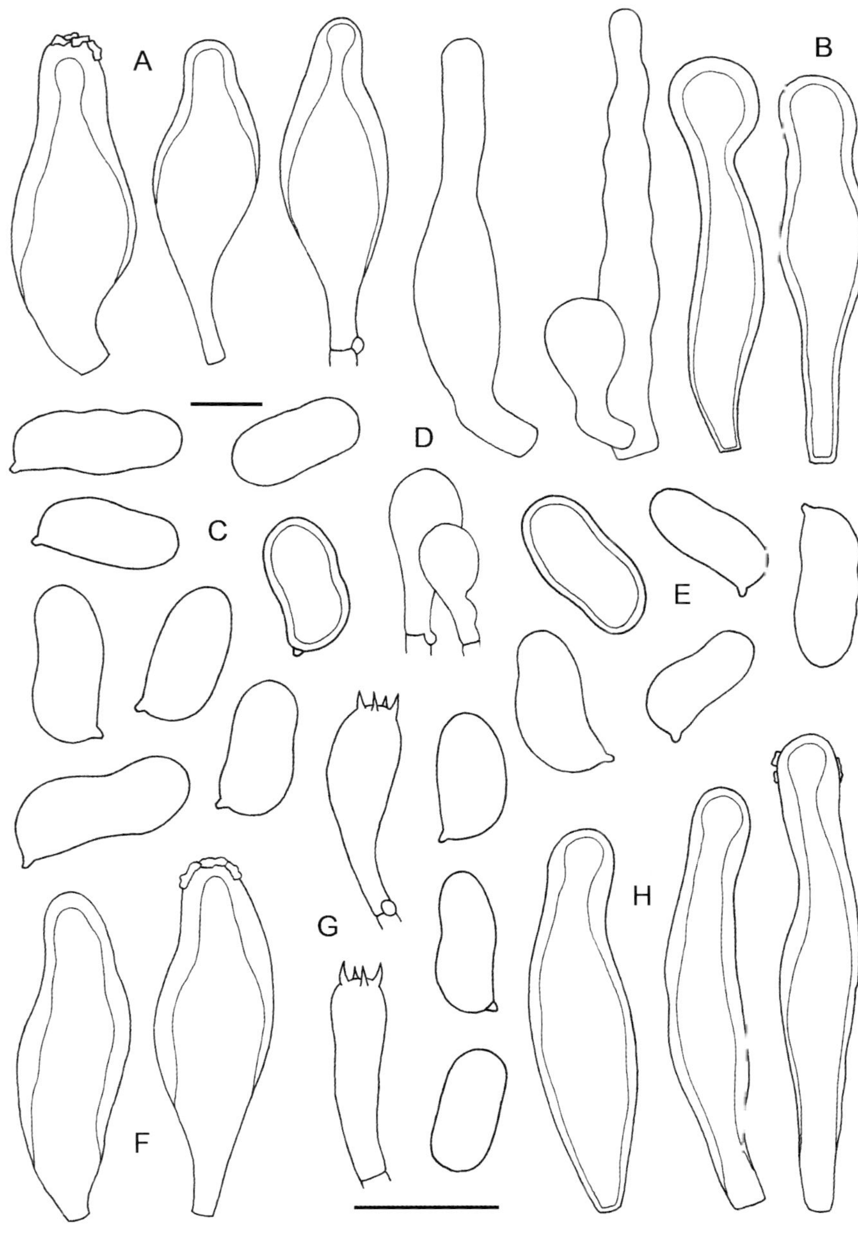

Figure 114. *Inocybe sinuospora* (A, D – G PERTH 0706532; B, C, H TENN 066548). **A,** Cheilocystidia; **B,** Caulocystidia, caulocystidioid hairs and cauloparacystidium; **C,** Basidiospores; **D,** Paracystidia; **E,** Basidiospores; **F,** Pleurocystidia; **G,** Basidia; **H,** Pleurocystidia. Scale bars = 10 μm (longer bar for spores only).

Furthermore, the two species are most readily distinguished microscopically, as the spores of *I. sinuospora* have a sinuous outline whereas those of *I. emergens* are angular to nodulose.

Inocybe sinuospora shares some similarities with *I. casuarinae* Corner & E.Horak from Malaysia (Horal, 1980). Both fungi are associated with Casuarinaceae (*Allocasuarina* and *Casuarina*, respectively), and they share a similar spore outline. However, the spores of *I. casuarinae* are much shorter (6.5–9.0 × 4–5 µm vs 9.5–13.0 × 4.5–6.0 µm).

An isotype was not cited in the protologue of *I. sinuospora* (Bougher *et al.*, 2012), but this has since been identified and accessioned as TENN 068365.

INOCYBE LANUGINOSA GROUP

The *I. lanuginosa* group is a pan-temperate clade containing at least 14 species (only 12 are indicated in the tree figure). Unlike other groups (e.g. the *I. assimilata* group), north-temperate lineages are indicated as derived from those in the southern hemisphere. Eight Australian species (six in the tree figure; *I. trachysperma* and *I. imbricata* were not included) have been detected in the group, six of which are documented below (four as new). *Inocybe subpulverulenta*, which forms the sister lineage to the remainder of the *I. lanuginosa* group, is unusual in that it is the only species to feature a pruinose stipe (likely the entire length) as caulocystidia and cauloparacystidia can be found above the rounded basal bulb, and no partial veil was observed. The remainder possess a partial veil, lack caulocystidia, and have an even stipe or at most a stipe with a slightly swollen base. Most of the species feature a coarsely scaly pileus. Pleurocystidia are absent or rare in *I. nigricans*, a trait also shared with the north temperate *I. leptophylla*.

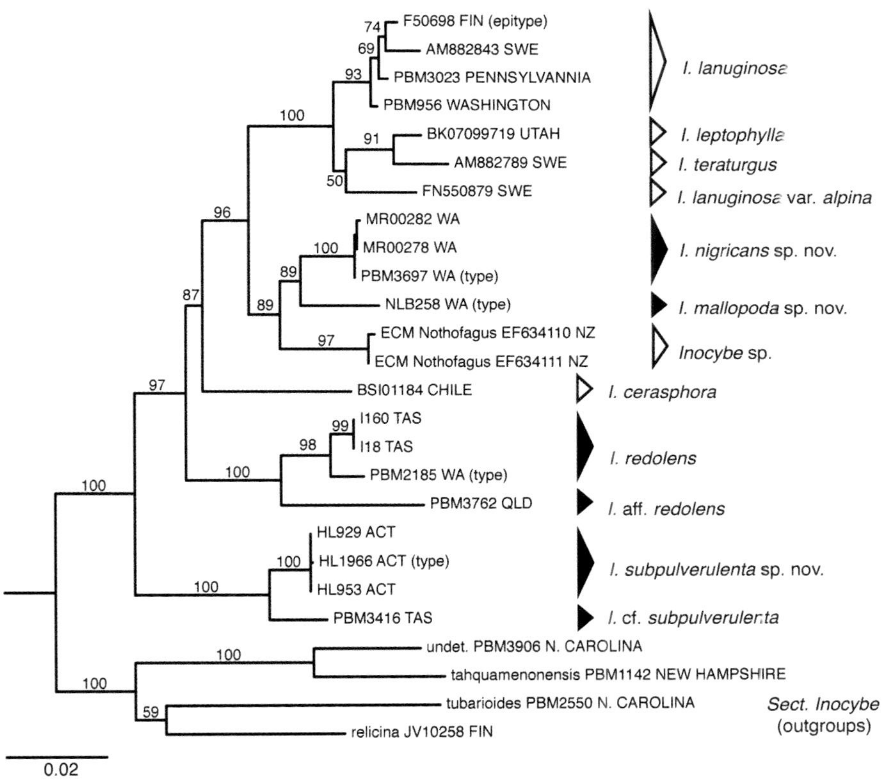

Figure 115. Maximum Likelihood phylogram of the *Inocybe lanuginosa* group. Bootstrap values are shown above and below branches. The tree was inferred from combined ITS, nLSU, and *rpb2* gene regions and includes a total of 2864 sites. Thirty nine ambiguously aligned positions in the ITS region were removed before analysis. Species from Australia are indicated by black-filled triangles. Species from New Zealand, Chile, and the northern hemisphere are indicated by white triangles. Twenty six branch tips are included with 16 represented by ITS, 24 by nLSU, and 17 by *rpb2*. Outgroup choice is based on Matheny *et al.* (2009).

98. Inocybe mallopoda Matheny & Bougher, *sp. nov.*

MycoBank: MB 809389.

Type: West Bay, Augusta, Leeuwin-Naturalist Natl Park, W.A., among moss on white-grey sand in woodland of *Eucalyptus marginata*, *Corymbia calophylla* and myrtaceous shrubs (*Leptospermum*), 17 June 2007, *N.L.Bougher NLB258*; holo: PERTH 08092214 (JN974995-nLSU, KP308774-ITS, KM246000-*rpb2*).

Pileus 12–22 mm wide, convex then pulvinate or applanate, with a non-striate fibrillose margin with tufts of fibrils forming an uneven appendiculate margin; margin at first slightly incurved, soon decurved; surface densely radially fibrillose with the fibrila forming upright sqamules near the pileus centre, uniformly dark yellowish brown. *Lamellae* adnexed, closely spaced but not crowded, 26–32 L, with several tiers of lamellae, dull livid brown or pale brown to strong brown with age, up to 5 mm deep; edges pallid and indistinctly fimbriate. *Stipe* 15–25 × 3–4 mm, cylindrical, without a swollen base; surface longitudinally appressed silky-fibrillose overlain by very dark brown scattered and conspicuous fibrillose scales, not pruinose; partial veil not evident but presumably fugacious; context dull cream, dulling further gradually after cut. *Odour* none. Pl. 192, 193.

Basidiospores 10.5–12.5 × 7.0–8.0 µm [mean 11.4 × 7.7 µm; Q 1.38–1.67; Q mean 1.48 (n = 21/1)], angular-nodulose, with a concave ventral depression, typically with 9–14 low or small nodules, at times coarsely nodulose; nodules generally 1.0–1.5 µm tall; some spores with fewer nodules and corners, pale brown to yellowish brown; apiculus small and not very distinctive. *Basidia* 30–43 × 10–13 µm, clavate, hyaline, 4-sterigmate. *Pleurocystidia* 60–74 × 13–17 µm, fusiform to skittle-shaped; apices obtuse and occasionally subcapitate, typically lacking crystals, often with a slender basal pedicel; thin-walled or at most slightly thickened (to 1 µm thick), hyaline, scattered singly, projecting well beyond the hymenium, easy to locate. *Cheilocystidia* similar to pleurocystidia but sometimes shorter, fusiform, cylindrical to skittle-shaped; sometimes emerging from branched multiseptate elements, mixed with young basidia-like or clavate-vesiculose cells. *Caulocystidia* absent, terminal cells of stipitipellis clavate or cylindrical, brown, thick-walled, some with encrusted pigment. *Pileipellis* a trichoderm composed of tufts of erect hyphae, these dark yellowish brown in mass, mostly 7–17 µm wide and slightly thick-walled (1–2 µm); pigment cytoplasmic; encrusting pigments weak or not observed. *Clamp connections* present. Fig. 116.

Occurs in south-western W.A.; grows among moss on soil in woodland near *Eucalyptus*, *Corymbia*, *Leptospermum*, *Agonis* and/or *Allocasuarina*. Fruiting Apr.–June.

Additional specimen examined: W.A.: Moates Paddock Rd, Two Peoples Bay Nature Reserve, 34°59'04"S, 118°10'59"E, gregarious or at times caespitose under *Eucalyptus marginata*, *Allocasuarina fraseriana* and *Agonis hypericifolia* [*A. theiformis*], 16 Apr. 1992, *K.Syme, M.Hart & R.Hart KS 399/92* (PERTH 05254396).

Etymology: From the Greek *mallos* (wool) and *podos* (a stalk or foot), in reference to the densely fibrillose scales on the stipe, more coarsely so on the lower part.

Diagnostic features: Pileus dark yellowish brown, with recurved fibrillose scales; stipe with dark brown woolly scales contrasting with the pale background. Spores large and nodulose; pleurocystidia primarily thin-walled.

Inocybe mallopoda is a sister species of *I. nigricans*, also from Western Australia, but which occurs in the wheatbelt region about 350 km north-east of the type locality of *I. mallopoda* (Fig. 115). The two taxa share a similar spore morphology (nodulose to simply angular), although the spores of *I. mallopoda* are generally more coarsely nodulose. The most conspicuous morphological difference is that the fruit bodies of *I. mallopoda* lack the dark blackish brown or sombre colouration typical of *I. nigricans*. Microscopic differences include the presence of pleurocystidia in *I. mallopoda* and their absence or rarity in *I. nigricans*. However, material from Two Peoples Bay (PERTH 05254396) that is best ascribed to *I. mallopoda*, based on colouration and association with eucalypts, has rare pleurocystidia. In both species some of the cheilocystidia arise from broad swollen elements, with multibranched forms more abundant in *I. mallopoda*. The two taxa may have different putative mycorrhizal associates, i.e. *I. mallopoda* occurs near eucalypts, whereas *I. nigricans* has been collected in vegetation lacking eucalypts.

Inocybe mallopoda and *I. nigricans*, along with *Nothofagus* ectomycorrhizal isolates from New Zealand, form a sister lineage to a north temperate species clade containing *I. lanuginosa* (Bull. : Fr.) P.Kumm., *I. leptophylla* G.F.Atk., *I. teraturgus* M.M.Moser and *I. lanuginosa* var. *alpina* Schmid-Heckel, an ensemble of species we refer to as the *I. lanuginosa* group (Fig. 115; Matheny & Kropp, 2001). Pleurocystidia are absent or rare in *I. leptophylla* and *I. nigricans*.

Inocybe cerasphora Singer, an associate of *Nothofagus* in southern South America, has large similarly pigmented spores and an overall fruit body colouration not unlike that of *I. mallopoda*. Its cystidia are thick-walled, and it is phylogenetically distinct from Australian species of the *I. lanuginosa* group. *Inocybe paracerasphora* (E.Horak) Garrido, from New Zealand (Horak, 1977) and reported at high elevation in Sabah, Malaysia (Horak, 1979a), differs, according to the protologue, by its occurrence on rotten *Nothofagus* wood and more prominent nodules on the spores.

192

193

Plate 192. *Inocybe mallopoda*, holotype (PERTH 08092214). Photograph — N.L.Bougher.

Plate 193. *Inocybe mallopoda*, holotype (PERTH 08092214). Photograph — N.L.Bougher.

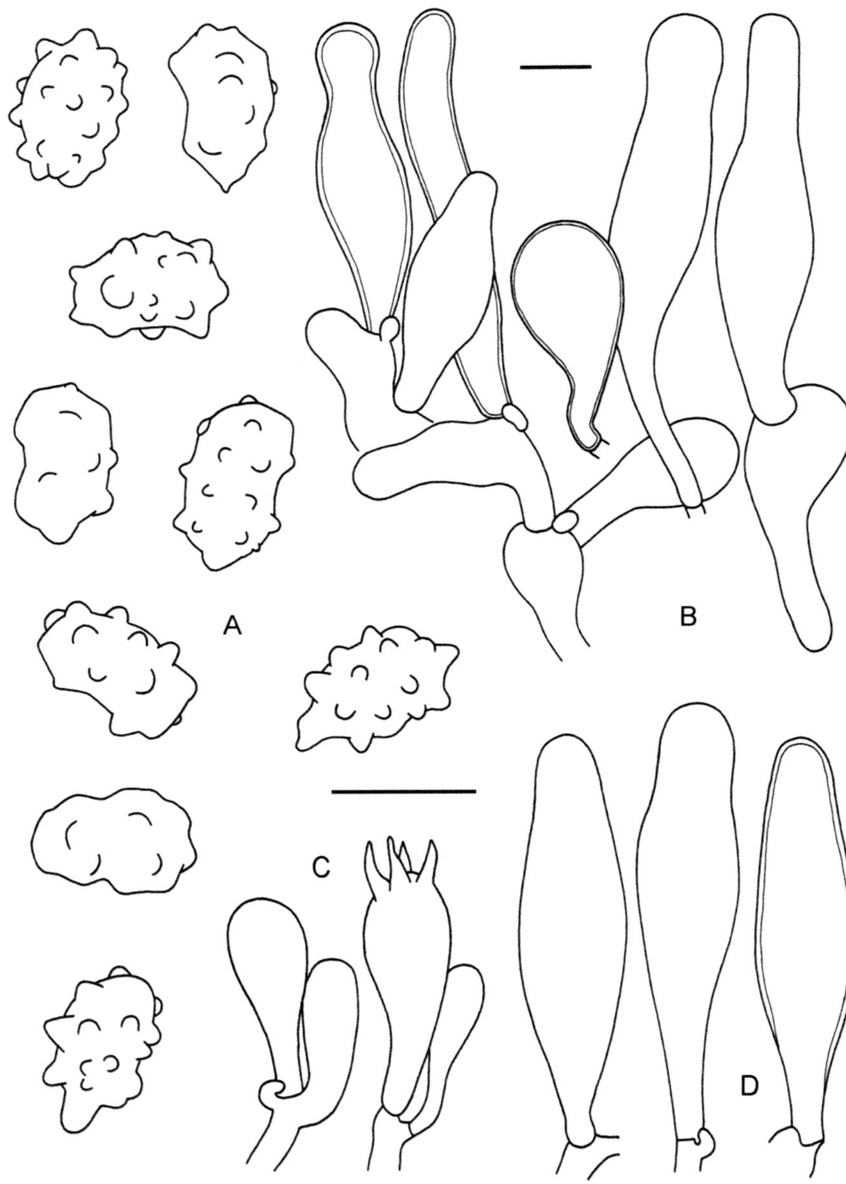

Figure 116. *Inocybe mallopoda.* All holotype (PERTH 08092214). **A,** Basidiospores; **B,** Cheilocystidia; **C,** Basidia; **D,** Pleurocystidia. Scale bars = 10 µm (longer bar for spores only).

99. **Inocybe nigricans** Matheny, Bougher & Ryberg, *sp. nov.*

MycoBank: MB 809393

Type: Corrigin Wildflower Drive, 5 km W of Corrigin, W.A., 32°20'19.1"S, 117°49'20.0"E, gregarious on sandy soil under *Allocasuarina obesa* (Black Tamma), 27 July 2011, *P.B.Matheny PBM3697*; holo: PERTH 08318875; iso: TENN 066635 (KP170956-nLSU, KP308787-ITS, KM406199-*rpb2*).

Pileus 8–35 mm wide, pulvinate to convex, plano-convex to plane with age and with an upturned margin; umbo absent; centre depressed at maturity; margin initially incurved, soon decurved; surface dry; velipellis absent; coarsely squarrose to fibrillose-squamulose; scales with recurved tips or appressed, these progressing outwards to the margin, sometimes woolly-fibrillose and breaking up into fine appressed scales; scales concolorous with the surface or becoming blackish brown against the dark yellowish brown ground color, almost blackish brown throughout with age or when excessively moist; context brownish, with a pallid area beneath the disc, not changing colour upon exposure, up to 5 mm thick. *Lamellae* broadly attached, initially arcuate, later adnate to triangular and running slightly down the stipe, subdistant, 20–30 L, with few tiers of lamellulae; initially very pale brown to light yellowish brown (10YR 7/3 to 6/4), becoming umbrinous (7.5YR 4/4; Snuff Brown) or reddish brown to dark reddish brown (5YR 4/4 to 3/3), eventually dark brown to dark yellowish brown (5YR 3/3, 10YR 3/4 to 3/3; Bay Brown), overall very dark and sombre, blackish brown when overly moist, broad, up to 5 mm deep, somewhat thick, edges not distinctly pallid-fimbriate. *Stipe* 10–26 × 2–5 mm, usually terete, occasionally compressed, central but sometimes excentric, generally squat or the length not greater than the pileus diameter, even; veil (cortina) presumably present but not directly observed; surface not pruinose, with whitish or pale furfuraceous fibrils or with brownish fibrils on the lower part, ground surface lighter in colour than the pileus (grey, greyish brown, brown or vinaceous brown to yellowish brown), becoming dark brown below with age or blackish throughout when overly moist; context solid, becoming hollow, coloured like the surface. *Odour* none. Pl. 194, 195.

Basidiospores 10.0–14.0 × 5.5–8.0 µm [mean 11.8 × 6.6 µm; Q 1.44–2.10 (–2.[illegible]7); Q mean 1.78 (*n* = 44/3)], angular to oblong-angular, with 6–9 small nodules mostly < 1.2 µm tall (c. 5% of spores with nodules > 2 µm tall), or corners, occasionally with a ventral depression; apex bluntly rounded or pointed, pale yellowish brown, the apiculus small. *Basidia* 29–45 × 10–13 µm, clavate to broadly clavate, hyaline, becoming ochraceous to rusty-ochraceous and collapsing with age, mostly 4-sterigmate, occasionally 2- and 1-sterigmate. *Pleurocystidia* absent, rare or infrequent, similar to cheilocystidia. *Cheilocystidia* 43–80 × 11–17 µm, very variable in shape (mostly narrowly fusiform or irregularly cylindrical, also utriform, skittle-shaped or clavate), mostly thin-walled but occasionally with walls up to [illegible].5 µm thick and hyaline, sometimes rusty-ochraceous with age; apices obtuse or rounded when clavate; crystals usually absent, occasionally numerous, mixed with mostly clavate to vesiculose paracystidia, sometimes arising from swollen subtending elements. *Caulocystidia* not observed, but the stipe surface with tangled aggregations of clavate ventricose and cylindrical terminal elements of superficial fibrils, some with brownish and finely encrusted walls. *Pileipellis* a trichoderm composed of scattered tufts of ascending or vertical hyphae, which are very dark brown in mass, cylindrical, mostly 6–15 µm wide; walls encrusted and slightly thickened. *Clamp connections* present. Fig. 117.

Known only from the wheatbelt region of south-western W.A.; often in troops on sandy or laterite soil in shrubland under *Casuarina* or *Allocasuarina* or in open woodland of *Acacia acuminata* and *Allocasuarina huegeliana*. Fruiting July.

Additional specimens examined: W.A.: Old Kunjin railway, 14 km W of Corrigin, 32°21'25.4"S, 117°44'17.1"E, on sandy soil in open shrubland under *Allocasuarina*, 26 July 2011, *M.Ryberg MR00277* (PERTH 08306893, TENN 066621); *loc. id.*, on sandy soil in shrubland under *Allocasuarina*, 26 July 2011, *M.Ryberg MR00278* (TENN 066463; KP170957-nLSU, KP308788-ITS, KM406200-*rpb2*); *loc. id.* on sandy soil under planted eucalypts and *Allocasuarina* bushes, along main road, 26 July 2011, *N.L.Bougher NLB871* (PERTH 08320608, TENN 066739); Corrigin Wildflower Drive, by airfield, 5 km W of Corrigin, 32°20'19.1"S, 117°49'20.0"E, in sandy to laterite soil in open space under *Casuarina obesa*, 27 July 2011, *M.Ryberg MR00282* (PERTH 08306850, TENN 066466; (KP170958-nLSU, KM406201-*rpb2*); *loc. id.*, under *Casuarina obesa*, 27 July 2011, *M.Ryberg MR00284* (TENN 066540); *loc. id.*, on sandy soil under *Casuarina obesa*, 27 July 2011, *P.B.Matheny PBM3701* (PERTH 08319332, TENN 066549); *loc. id.*, on sandy soil under *Casuarina obesa*,

27 July 2011, *P.B.Matheny PBM3702* (PERTH 08319324, TENN 066613); *loc. id.*, on laterite soil in thicket of *Allocasuarina campestris*, 27 July 2011, *N.L.Bougher NLB886* (PERTH 08320446, TENN 066718); *loc. id.*, on laterite soil, in *Allocasuarina campestris* thicket, 27 July 2011, *N.L.Bougher NLB887* (PERTH 08320438).

Etymology: From the Latin *nigricans* (blackish, swarthy), in reference to the fruit body colour.

Diagnostic features: Fruit bodies small to medium-sized. Pileus dark blackish brown or dark brown, reddish brown, sombre coloured; lamellae subdistant. Spores large, 10.0–14.0 × 5.5–8.0 µm, angular to oblong-angular in outline with small nodules or corners; pleurocystidia rarely observed or absent. Putative ectomycorrhizal association with *Casuarina*/*Allocasuarina* and/or *Acacia*.

Inocybe nigricans is variable in its outward appearance, having a woolly-fibrillose to appressed-squamulose or squarrose pileus surface. Most collections are typically blackish brown, small, squat or stubby in stature, and have reddish brown and subdistant lamellae. It is most distinctive under the microscope given the large spores with small nodules or corners, basidia that turn rusty-ochraceous and collapse with age, and rare or absent pleurocystidia. The last trait may be noteworthy because *I. nigricans* is closely related (Fig. 115) to the northern-temperate *I. leptophylla* G.F.Atk., which lacks pleurocystidia.

The spores of *I. nigricans* are very similar to those of *I. imbricata*, first described from South Ausralia; the latter differs in its distinct, thick-walled pleuro- and cheilocystidia (Matheny & Bougher, 2010). Genetically, *I. nigricans* is most closely related to *I. mallopoda*, also from the south-west of Western Australia. (Fig. 115); *I. mallopoda* has a dark yellowish brown pileus, abundant pleurocystidia, a proportion of its spores having larger and numerous nodules, and it has an association with eucalypts.

All collections of *I. nigricans* have been made in shrubland and thickets of *Casuarina*/*Allocasuarina* or in woodland with a mixture of *Allocasuarina* and *Acacia*. No eucalypts were found nearby, so a putative ectomycorrhizal association with *Casuarina*/*Allocasuarina* and/or *Acacia* is plausible.

100. Inocybe redolens Matheny, Bougher & G.M.Gates, *Nuytsia* 22: 65 (2012)

Type: Lake Beedelup, 22 km W of Pemberton, W.A., in soil in forest dominated by *Eucalyptus diversicolor*, 18 Aug. 2001, *P.B.Matheny E7034, PBM2185*; holo: PERTH 07676654 (JQ085933-ITS, EU600843-*rpb1*, EU600844-nLSU).

Pileus 7–12 mm wide, conical to obtusely so; margin decurved; umbo absent; surface dry, with small appressed fibrillose squamules around the centre and towards the margin, with the edge of the margin fibrillose; colour dark brown (10YR 3/3; Bistre to Mummy Brown) at the centre, shading to (dark) yellowish brown or umbrinous (Snuff Brown to Prout's Brown) towards the margin; context pallid, thin, c. 1 mm thick. *Lamellae* narrowly adnate to adnexed, ventricose, moderately close, c. 30 L, with several tiers of lamellulae, light grey to pale brown; edges pallid, fimbriate. *Stipe* 15–20 × 1.5–3.0 mm at the apex, terete, even, with a slightly swollen base to 3 mm wide; veil cortinate, fugacious; entire surface overlain with a pallid silky-fibrillose superficial layer; apex faintly pruinose; base densely covered with pallid fibrils; ground colour light brown to pink or reddish yellow (7.5 YR 6/4–7/4–7/6; Avellaneous to Light Pinkish Cinnamon); context with light dull pinkish tinge to the cortex, and interior pallid, solid. *Odour* of *Pelargonium*. Pl. 196, 197.

Basidiospores (7.0–) 7.5–9.0 × 5.5–7.5 (–8.5) µm [mean 8.0 × 5.8 µm; Q: 1.12–1.65; Q mean 1.39 (n = 30/1)], nodulose with mostly 12–14 moderately-sized hemispherical or obtusely conical nodules around an elliptic outline, yellowish brown in KOH. *Basidia* 30–32 × 8–10 µm, clavate, 4-sterigmate, hyaline. *Pleurocystidia* 63–73 × 14–15 µm, ±cylindrical to narrowly utriform; pedicel long and slender; apices at times swollen or indistinctly subcapitate, crystalliferous; walls 1.5–2.5 µm thick. *Cheilocystidia* 55–70 × 12–14 µm, similar to pleurocystidia, at times intermediate in size and shape, crowded and mixed with abundant paracystidia. *Stipitipellis* lacking caulocystidia but caulocystidioid cells present and restricted to the apex, these similar to cheilocystidia but cauloparacystidia not observed; vesture hyphae present below, these thin-walled and denser towards the lower part of the stipe. *Pileipellis* a cutis of brown (in KOH) hyphae 6–9 µm wide; hyphae mainly thin-walled but some with walls

Plate 194. *Inocybe nigricans*, holotype (PERTH 08318875). Photograph — P.B.Matheny.

Plate 195. *Inocybe nigricans*. PERTH 08319332. Photograph — P.B.Matheny.

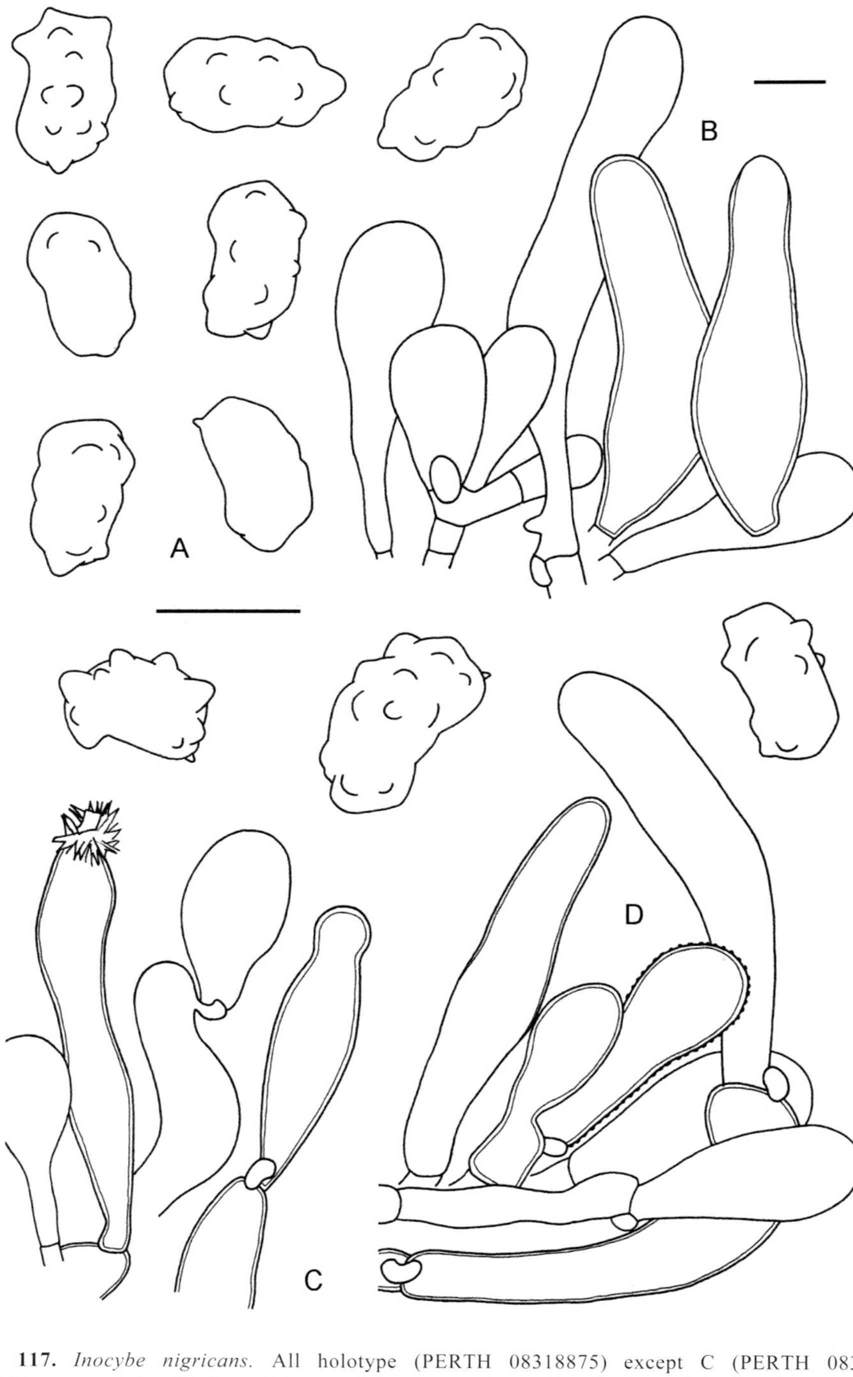

Figure 117. *Inocybe nigricans.* All holotype (PERTH 08318875) except C (PERTH 08306850). **A,** Basidiospores; **B,** Cheilocystidia and paracystidia; **C,** Cheilocystidia and paracystidia; **D,** Caulocystidioid hairs from stipe apex. Scale bars = 10 μm (longer bar for spores only).

up to 1 μm thick and minutely encrusted. *Clamp connections* present in all tissues. Fig. 118 (see also Pl. 7C, 8C).

Occurs in south-western W.A. on soil in Karri forest, under *Eucalyptus diversicolor* and *Corymbia calophylla*, and in Tas. in mixed forest of *Eucalyptus*, *Melaleuca*, *Nothofagus* and *Phyllocladus*. Fruiting Aug.–Dec.

Additional specimens examined: Tas.: Mount Mangana, Bruny Island, on soil in mixed forest of *Eucalyptus*, *Melaleuca*, *Nothofagus* and *Phyllocladus*, 15 Nov. 2001, *G.M.Gates & D.Ratkowsky 118* (HO 563720; JQ085934-nLSU); Duckhole Lake, 22 Dec. 2012, *G.M.Gates 1160* (TENN067764); *loc. id.*, 22 Dec. 2012, *G.M.Gates 1160* (TENN 067763; KP170987-nLSU, KM406227-*rpb2*).

Diagnostic features: Fruit bodies small. Pileus appressed-scaly, dark brown to brown; stipe not pruinose; odour of *Pelargonium*. Spores nodulose with numerous, small, rounded nodules about an ellipsoid outline.

The spores of *I. redolens* are similar to those of *I. exigua*, but that species differs from *I. redolens* in the shape of the hymenial cystidia, the absence of a *Pelargonium*-like odour, and the descent of caulocystidia along the entire length of the stipe.

Inocybe redolens is phylogenetically related to the *I. lanuginosa* group, along with several other newly described Australian species, *viz. I. mallopoda*, *I. nigricans* and *I. subpulverulenta*. Seven base pair positions, three of which are polymorphic, differ at the nLSU locus between Tasmanian and Western Australian collections. However, we do not consider this difference to warrant separation of the collections at the species level at this time given the insufficient taxon sampling. *Inocybe redolens*, together with other austral species of the *I. lanuginosa* group (Fig. 115), form a grade (a paraphyletic group) from which *I. lanuginosa* and its allied north-temperate species are derived.

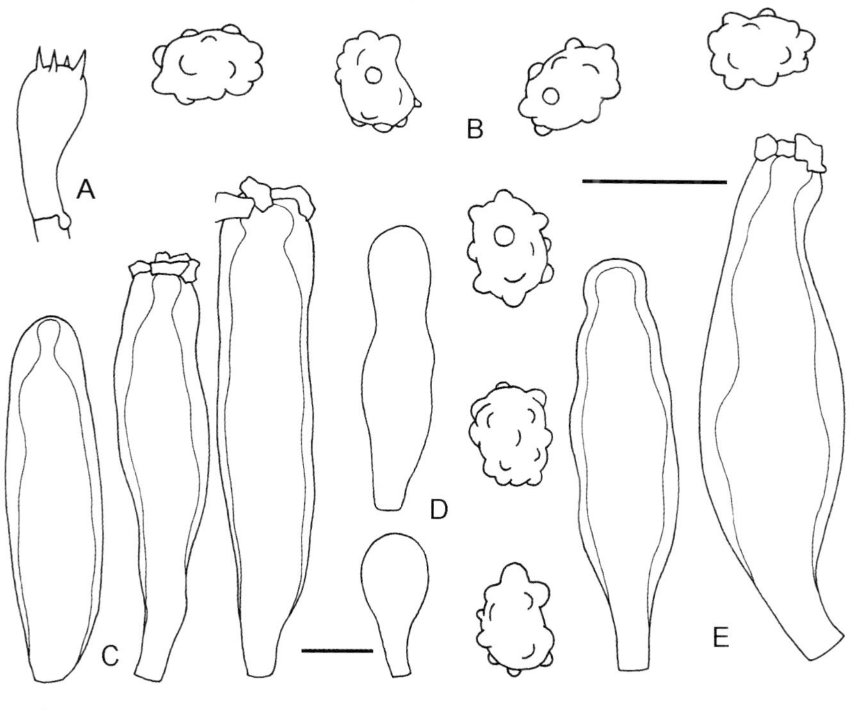

Figure 118. *Inocybe redolens,* holotype (PERTH 07676654). **A,** Basidium; **B,** Basidiospores; **C,** Cheilocystidia; **D,** Paracystidia from lamella margin; **E,** Pleurocystidia. Scale bars = 10 μm (longer bar for spores only).

Plate 196. *Inocybe redolens*, holotype (PERTH 07676654). Photograph — N.L.Bougher.

Plate 197. *Inocybe redolens* TENN067764. Photograph — G.M.Gates

An unclarified taxon (*I. aff. redolens*; Fig. 115) closely resembles *I. redolens* in outward appearance (and including the strong *Pelargonium* odour) but it awaits formal description. It has been collected only once in dry sclerophyll woodland under eucalypts in the Glasshouse Mountains Natl Park in south-eastern Queensland (PBM3762; PERTH 08363234, TENN 066994, KP170988-LSU, KP308816-ITS, KM406228-*rpb2*).

Inocybe trachysperma could be confused with *I. redolens*, and indeed both share similar microscopic features. However, the former has a fibrillose pileus (no scales) and it lacks the *Pelargonium* odour.

101. Inocybe subpulverulenta Matheny, Bougher & Lepp, *sp. nov.*

MycoBank: MB 809466

Type: Australian National Botanic Gardens, 4 km NNW of Capital Hill, Canberra, A.C.T., 35°17'S, 149°06'E, alt. c. 600 m, solitary or in small clusters on soil in garden bed with mixed plantings, *Eucalyptus* dominant, 28 Apr. 1998, *H.Lepp 1966*; holo: CANB 574535; iso: PERTH 08517177, TENN 067774 (KP171078-nLSU, KP641624-ITS, KM656122-*rpb2*).

Pileus 20–60 mm wide, parabolic, becoming conical and eventually applanate, often with an obtusely conical umbo; margin decurved; surface dry; velipellis absent; imbricate-scaly at least around the centre and towards the margin; scales appressed; margin splitting with age, with edge at times with whitish fibrillose veil remnants giving it a slightly shaggy appearance; initially dark brown to blackish brown, later with dark brown to blackish brown (6F8 or darker) fibrillose scales over a pale yellowish brown (5D8 to 5D7) ground colour; context white. *Lamellae* adnexed to uncinate, very close to crowded, c. 40–46 L, with several tiers of lamellulae, initially very pale brown to dirty white, becoming pale fawn to dark yellowish brown, broad; edges not distinctly fimbriate. *Stipe* 25–60 × 3–6 mm, even but with a rounded basal bulb up to 12 mm wide, terete, central; surface with a satiny sheen, almost glossy; partial veil fragments not observed; lower part at top of the basal bulb with aggregated brown tufts of fibrils; colour contrasting with that of the pileus, initially off-white to buff or pale yellow (near 3A3) but becoming light yellowish brown (5B5); base white with a ±tomentose mycelium around the bulb; context pale yellowish brown, hollow. *Odour* not remarkable. Pl. 198, 199.

Basidiospores (8.5–) 9.0–13.0 (–14.5) × (6.5–) 7.0–9.5 (–11.0) µm [mean 11.1 × 8.2 µm; Q 1.18–1.55 (–1.71); Q mean 1.35 (n = 43/2)], coarsely nodulose with mostly 10–14 moderate-sized to conspicuous nodules 1–3 (–4) µm high about a mostly elliptical outline; apiculus small, tapering and not distinct; yellowish brown. *Basidia* 29–34 × 9–11 µm, clavate, hyaline, 4-sterigmate. *Pleurocystidia* 47–67 × 12–17 µm, utriform to fusiform-ventricose, less often fusiform or subcylindrical, lacking well-developed necks; apices rounded to obtuse, often bearing crystals; thick-walled (2.5–4.0 µm), especially in the apical portion, hyaline to pale yellowish. *Cheilocystidia* similar to pleurocystidia, not crowded, scattered among crowded palisade of small clavate to subcylindrical paracystidia, hyaline or some becoming ochraceous with age and with thickened walls. *Caulocystidia* similar to cheilocystidia, mixed with clavate cauloparacystidia in crowded clusters, descending to just above the stipe base. *Pileipellis* a cutis of cylindrical to inflated hyphae, these mostly 7–22 µm wide, encrusted but not coarsely so, thin-walled, yellowish brown in mass. *Clamp connections* present. Fig. 119.

Occurs only in the A.C.T.; on soil in mulched landscape areas under *Eucalyptus* and in dry sclerophyll woodland under *Acacia*. Fruiting Apr.–June.

Additional specimens examined: A.C.T.: Australian National Botanic Gardens, Canberra, section 46, 4 km NNW of Capital Hill, 35°17'S, 148°46'E, alt. 620 m, on soil in a somewhat sheltered location in an *Acacia* dry sclerophyll community, 17 May 1993, *G.Browning & H.Lepp 929* (CANB 574588, PERTH 08517193, TENN 067772; KP171079-nLSU, KP641625-ITS); *loc. id.*, singly or in pairs in a garden bed among leaf litter and woodchip mulch, 7 June 1993, *H.Lepp 953* (CANB 574592, PERTH 08517185, TENN 067773; KP171080-nLSU, KP641626-ITS, KM656123-*rpb2*).

Etymology: From the Latin *subpulverulenta* (almost powdery or dusty), in reference to the satiny or almost glossy appearance to the stipe above the basal bulb.

Diagnostic features: Fruit bodies large or medium-sized. Pileus imbricate-scaly, dark brown to blackish brown; stipe with a satiny surface, bulb rounded. Spores large, coarsely nodulose; caulocystidia present the entire length of the stipe.

Inocybe subpulverulenta, part of the greater *I. lanuginosa* group (Fig. 115), is characterised by its appressed-scaly pileus, dark brown to blackish brown pileal colours, and large, coarsely nodulose spores. It has a satiny stipe covering composed of caulocystidia descending the entire length to just above the rounded bulbous base. However, field notes associated with specimens documented here do not describe the stipe as "pruinose". Terms such as satiny and almost glossy were noted by the collectors. Given the presence of caulocystidia throughout the stipe, we judge that the stipe would appear pruinose when viewed with a hand lens. Whitish fibrillose remnants impart a shaggy-fibrillose texture to the edge of the pileal margin, and aggregated brown fibrils have been observed just above the base. Consequently, we cannot rule out the presence of a veil.

Although we have not seen fresh material of this species, its does have a unique combination of morphological characters and phylogenetic position (Fig. 115), and probable association with native plants (*Eucalyptus*, *Acacia*).

Inocybe imbricata is much smaller (pileus to 16 mm wide), but its pileal morphology is similar. That species has more slender spores (9.5–11.0 × 5.5–6.5 µm) than *I. subpulverulenta* with only 5–8 inconspicuous nodules, and it bears only few thin-walled caulocystidia-like cells at the stipe apex. *Inocybe cerasphora* Singer, described in association with *Nothofagus* from southern South America (Singer, 1953), is closely related (Fig. 115) and shares a similar spore morphology (large with numerous nodules), but it differs in having a cortina and a stipe that lacks caulocystidia. *Inocybe paracerasphora* (E.Horak) Garrido also occurs in association with *Nothofagus* but on rotten wood in New Zealand, is considerably smaller than *I. subpulverulenta* and also lacks caulocystidia on the stipe (Horak, 1977; as *Astrosporina paracerasphora*).

As regards other Australian species in the *I. lanuginosa* group, *I. redolens* has an odour of *Pelargonium* and spores with numerous small nodules about an elliptic outline. *Inocybe nigricans* has a distinctive squat habit, subdistant lamellae, pleurocystidia are absent or rare, and it has more slender spores with fewer smaller nodules. Both *I. nigricans* and *I. mallopoda* lack the readily observed caulocystidia of *I. subpulverulenta*.

An additional unclarified taxon (*PBM3416*, PERTH 08517355, TENN 067775) is outwardly similar to *I. subpulverulenta*, but it has distinct brownish fibrils on the lower part of the stipe above its swollen base (Pl. 199). Caulocystidia are easily observed descending at least to the stipe centre, but the sometimes distinctly stellate spores have fewer but prominent nodules projecting up to 5 µm (Fig. 119F); 2-sterigmate and 4-sterigmate basidia are present, and it occurs in *Nothofagus* rainforest in Tasmania. DNA sequence data (KP171081-nLSU, KP641627-ITS, KM656124-*rpb2*) indicates a close affinity with *I. subpulverulenta* (Fig. 115).

102. Inocybe imbricata (Cleland) Garrido, *Biblioth. Mycol.* 120: 176 (1988)

Astrosporina imbricata Cleland, *Trans. Roy. Soc. S. Australia* 57: 192 (1933). Type: Kinchina, S.A., 8 June 1925, *J.B.Cleland s.n.*; holo: AD 12227.

Pileus to 16 mm wide, convex to subconico-campanulate, with appressed-imbricate fibrils, fuscous brown; context pale. *Lamellae* adnate, seceding, almost crowded, subventricose, fuscous brown. *Stipe* 16 mm long, almost thin, equal, hollow within, fibrillose; context somewhat brown.

Basidiospores 9.0–11.5 × 5.5–6.5 µm [mean 10.4 × 5.9 µm; Q 1.54–2.09; Q mean 1.77 (*n* = 20/1)], oblong-angular with 5–8 (–10) inconspicuous nodules, at times these merely corners, around a rectangular to trapeziform outline, usually with a conical apical nodule, occasionally one side with a concave depression, yellowish brown. *Basidia* 30–42 × 7–10 µm, 4-sterigmate, clavate, hyaline to pale ochraceous. *Pleurocystidia* 58–76 × 11–17 µm, fusiform or narrowly so to subcylindrical, thick-walled (2–3 µm), hyaline, usually with a distinct neck and a slender basal pedicel; apices obtuse and lacking crystals. *Cheilocystidia* similar to pleurocystidia, often shorter, at times ochraceous. *Caulocystidia* not observed, but with scattered cystidioid end cells at the apex, these generally cylindrical or fusiform and thin-walled; cauloparacystidia not observed; stipe centre with interwoven vestiture hyphae terminating in caulocystidioid hairs, vestiture hyphae descending to base of stipe. *Pileipellis* a cutis giving rise to broad fascicles of somewhat ascending hyphae that are often inflated, to 25 µm wide, the end cells not differentiated, light reddish brown or tawny pigmented in mass, walls thin or slightly thickened, encrustations not evident. *Clamp connections* present. Fig. 120 (see also Pl. 7H).

Known with certainty only from S.A.; on the ground, presumably in native vegetation. Fruiting June.

Additional specimen examined: S.A.: Kinchina, on ground, 8 June 1925, *J.B.Cleland s.n.* (AD 5439) [AD 12673B, originally mixed with *I. murrayana* AD 5440].

Plate 198. *Inocybe subpulverulenta.* CANB 574588. Photograph — H.Lepp.

Plate 199. *Inocybe* cf. *subpulverulenta*. PERTH 08517355. Photograph — P.B.Matheny.

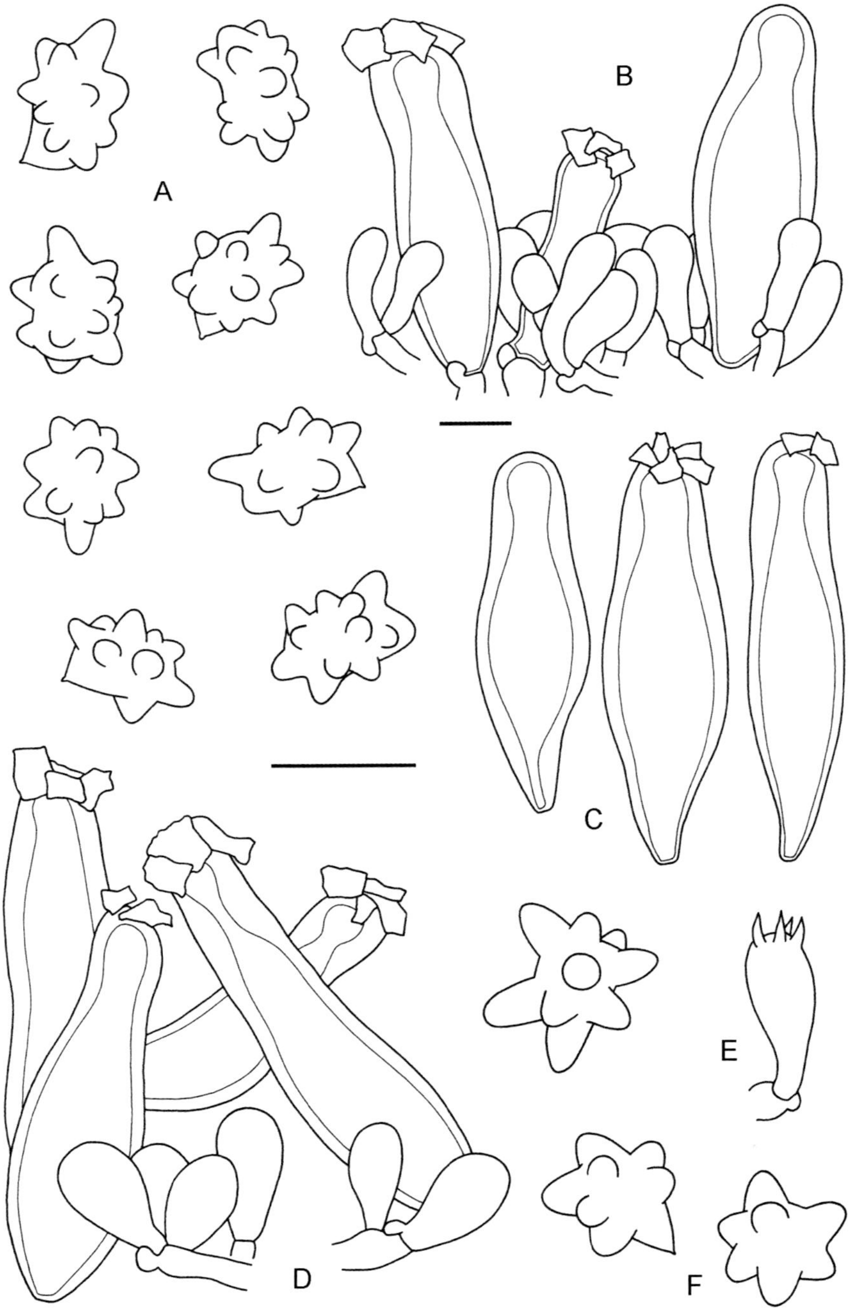

Figure 119. *Inocybe subpulverulenta.* All holotype (CANB 574535) except F (*I.* cf. *subpulverulenta* PERTH 08517355). **A,** Basidiospores; **B,** Cheilocystidia and paracystidia; **C,** Pleurocystidia; **D,** Caulocystidia and cauloparacystidia; **E,** Basidium; **F,** Basidiospores. Scale bars = 10 μm (longer bar for spores only).

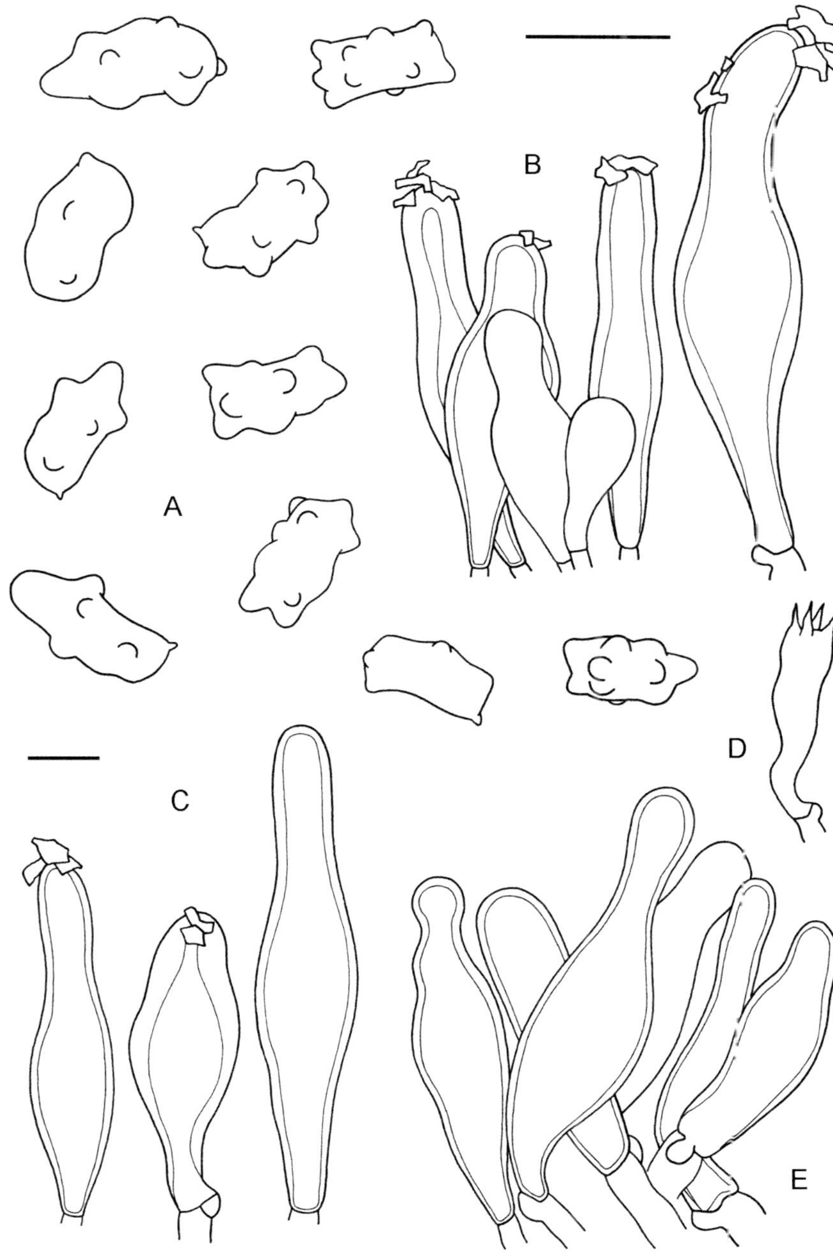

Figure 120. *Inocybe imbricata.* All holotype (AD 12227). **A,** Basidiospores; **B,** Cheilocystidia and paracystidia; **C,** Pleurocystidia; **D,** Basidium; **E,** Caulocystidia. Scale bars = 10 μm (longer bar for spores only).

Diagnostic features: Fruit bodies small; pileus dark brown and shaggy-squamulose; stipe fibrillose, pale brownish, even. Spores oblong-angular around a rectangular to trapeziform outline with few nodules and corners; pleurocystidia thick-walled.

The gross morphological description is a translation of Cleland's original Latin diagnosis, and our study here is an extension of that presented in Matheny & Bougher (2010). *Inocybe imbricata* appears to be most phenotypically similar to a suite of Southern Hemisphere taxa related to the northern-temperate *I. lanuginosa* group, including *I. mallopoda* and *I. nigricans* (Fig. 115), both of which have comparatively long spores (>10 µm) similar to *I. imbricata*. However, *I. mallopoda* has thin-walled pleurocystidia and more coarsely nodulose spores (typically with 9–14 nodules), while *I. nigricans* tends to lack pleurocystidia. *Inocybe cacaocolor* and *I. subferruginea* have shorter spores (<10 µm long) and thin-walled pleurocystidia.

Inocybe paracerasphora, from New Zealand, grows on rotten wood of *Nothofagus* and has broader spores (10.0–11.5 × 7.0–9.0 µm) with 9–11 more conspicuous nodules (Horak, 1977). *Inocybe cerasphora* Singer, occupies a similar habitat in southern South America and has larger spores (11.0–13.0 × 7.5–10.5 µm), each with 13–16 more conspicuously nodules.

The spores of *I. imbricata* resemble those of *I. curvipes* in their trapeziform outline, but the latter has broader hymenial cystidia and is associated with exotic trees.

We have not located any modern collections of *I. imbricata*, although the following collections have similar cystidia and angular, sparsely nodulose spores, and may be affiliated with *I. imbricata*: (1) small fruit bodies from a pot with soil from Jabiru, N.T. and a seedling of *Eucalyptus bleesari*, in a glasshouse at CSIRO Floreat Park, W.A., 6 Apr. 1992, *S.Snelling, N.L.Bougher & I.C.Tommerup E815* (PERTH 07564791); (2) Salt River Rd, Camel Lake Nature Reserve, W.A., gregarious in mallee with *Melaleuca* sp. and *Gastrolobium* sp., 17 July 2007, *K.Syme & S.Ellis KS2029/07* (PERTH 08105960). A colour image of fresh fruit bodies of PERTH 08105960 shows that they resemble *I. nigricans*.

103. **Inocybe trachysperma** Matheny & Bougher, *sp. nov.*

MycoBank: MB 809475

Type: Property 3298, near the junction of Mount McLeod Rd and Bell Rd, along old railway line road (Denmark-Nornalup Heritage Trail), 12 km W of Denmark, W.A, on soil at edge of woodland under *Eucalyptus diversicolor* and *Corymbia calophylla*, 24 June 2001, *P.B.Matheny PBM2127/E6760*; holo: PERTH 07677669; iso: TENN 066787.

Pileus 13–22 mm wide, initially conical, becoming plane with a nipple-like umbo; margin decurved and uplifted with age; velipellis absent; surface dry, somewhat glossy, fibrillose, not rimose, not scaly; brown (Snuff Brown) over the centre, shading to yellowish brown (10YR 5/4; Tawny-Olive) towards the margin; context pallid, not changing colour where bruised, up to 2 mm thick under the disc, thin elsewhere. *Lamellae* narrowly adnate to uncinate, moderately close, 30–34 L, with several tiers of lamellulae, pale brown to brown (10YR 6/3 to 5/3), subventricose, 2–3 mm deep; edges pallid-fimbriate. *Stipe* 25–35 × 2–3 mm, even or at most slightly swollen at the base (not bulbous), terete, often flexuous, rather fragile; partial veil cortinate, fugacious; surface dry, tomentose-fibrillose or silky-fibrillose with a thin pallid vesture, nowhere pruinose; pallid with a dull ochraceous ground colour (2.5Y 7/4 or 6/6); context solid, very pale brown (10YR 7/4), white at the base. *Odour* absent. Pl. 200.

Basidiospores 8.0–9.5 × 6.0–6.5 µm [mean 8.4 × 6.2 µm; Q 1.23–1.58; Q mean 1.37 (n = 21/1)], nodulose with numerous (12–16) small nodules (to 1 µm high), typically with an apical nodule, yellowish brown; apiculus indistinct. *Basidia* 25–32 × 7–10 µm, clavate, hyaline, 4-sterigmate. *Pleurocystidia* 47–60 × 12–15 µm, fusiform to subutriform, apices obtuse, bearing crystals, necks often poorly differentiated, with a tapered basal pedicel, thick-walled (2–3 µm), hyaline. *Cheilocystidia* similar to pleurocystidia, also utriform or sublageniform, hyaline, yellowish or brownish, mixed with paracystidia. *Caulocystidia* absent. *Pileipellis* a cutis of cylindrical hyphae up to 15 µm wide, brown in mass, encrusted. *Clamp connections* present. Fig. 121.

Occurs in south-western W.A., on soil at edge of forest with *Eucalyptus diversicolor* and *Corymbia calophylla*. Fruiting June.

Etymology: From the Greek *trachys* (rough) and *sperma* (a seed), in reference to the basidiospores with numerous small nodules.

Diagnostic features: Differs from *I. redolens* by the fibrillose and non-scaly pileus and the lack of a *Pelargonium* odour.

Inocybe trachysperma is most similar to *I. redolens*, but the latter has a strong odour of *Pelargonium* that is detectable even from unbroken flesh, and by the presence of scales on the pileus surface. *Inocybe trachysperma* lacks an odour and has a fibrillose pileus.

Other species that might be confused with *I. trachysperma* include *I. subferruginea* and *I. cacaocolor*, but these differ in their thin-walled hymenial cystidia, coarsely fibrillose to appressed-squamulose pilei, and rusty or dark brown lamellae at maturity. *Inocybe humidicola* differs in its thin-walled hymenial cystidia and tropical distribution in north-eastern Queensland.

Plate 200. *Inocybe trachysperma*, holotype (PERTH 07677669). Photograph — P.B.Matheny.

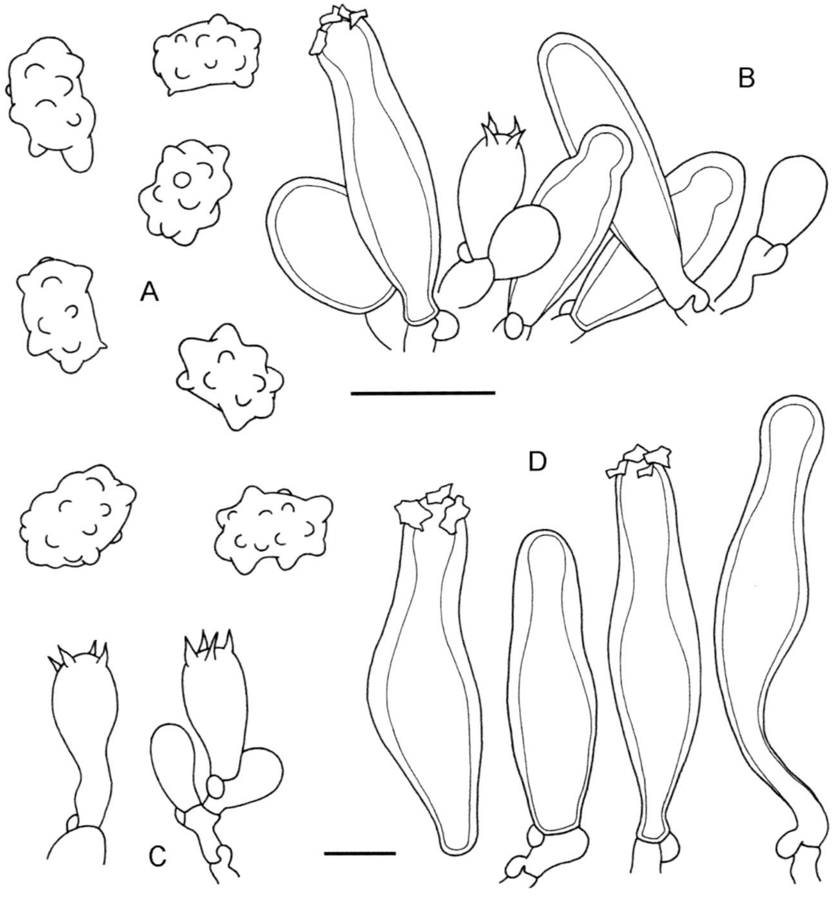

Figure 121. *Inocybe trachysperma.* All holotype (PERTH 07677669). **A,** Basidiospores; **B,** Cheilocystidia and paracystidia; **C,** Basidia; **D,** Pleurocystidia. Scale bars = 10 µm (longer bar for spores only).

INOCYBE LEPTOSPERMI GROUP

The *I. leptospermi* group is a south-temperate clade of species from Australia, New Zealand, and southern South America with nodulose spores and entirely pruinose stipes. The stipe base is typically distinctly marginate, unlike most species of the *I. emergens* group. Of the ten species in the group, several associate with *Nothofagus*: *I. sylvicola* in Australia and *I. horakomyces* and *I. scissa* in New Zealand. Six species in the *I. leptospermi* group occur in Australia, five of which are documented below (three as new). *Inocybe leptospermi* is one of the most widespread species of Inocybaceae in Australasia, having been recorded in N.Z., Tasmania, New South Wales, Australian Capital Territory, and Western Australia.

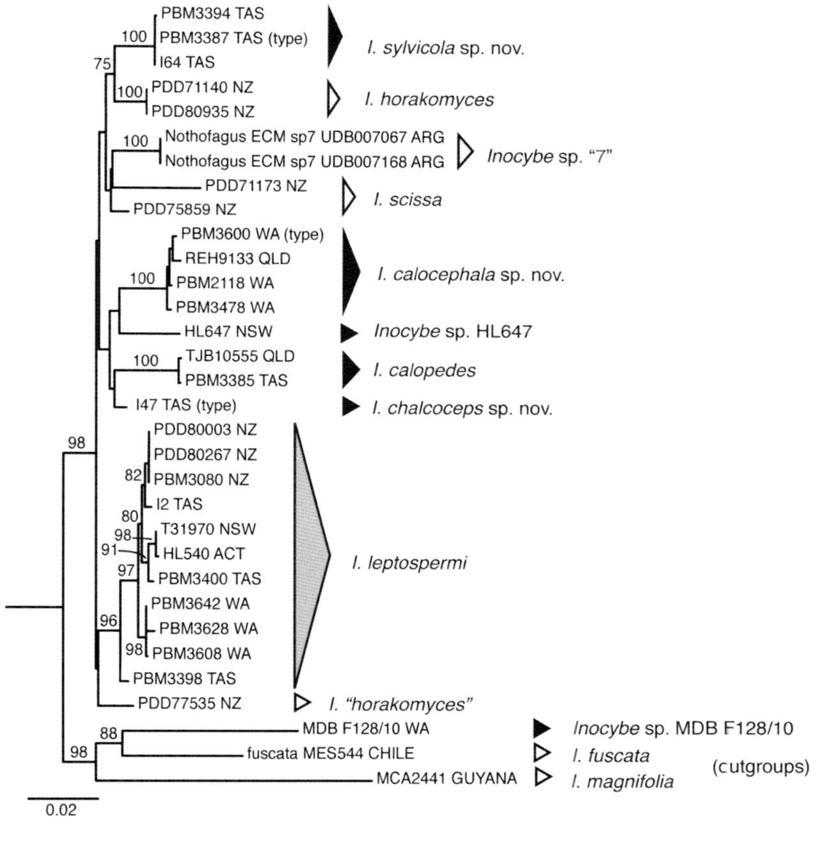

Figure 122. Maximum Likelihood phylogram of the *Inocybe leptospermi* group. Bootstrap values are shown above and below branches. The tree was inferred from combined ITS, nLSU, and *rpb2* gene regions and includes a total of 2871 sites. Forty five ambiguously aligned positions in the ITS region were removed prior to analysis. Species from Australia are indicated by black-filled triangles. Species from Australia and New Zealand are indicated by grey-filled triangles. Branch tips labeled with white triangles are from New Zealand or South America. Thirty two tips are included, with 28 represented by ITS and nLSU, and 18 by *rpb2*. The tree is rooted by *Inocybe magnifolia*, *Inocybe fuscata*, and an unclarified species from the Kimberley region based on Matheny *et al.* (2009).

104. **Inocybe calocephala** Matheny & Bougher, *sp. nov.*

MycoBank: MB 809273

Type: near Greenbushes, 1 km N of Catterick Rd, W.A., 33°50'50.6"S, 116°04'50.3"E, scattered singly or in pairs on unburnt soil in Jarrah forest under *Eucalyptus marginata* and *Corymbia calophylla*, 12 July 2011, *P.B.Matheny PBM3600*; holo: PERTH 08319855; iso: TENN 066460; KJ756455-nLSU, KJ756413-*rpb2*).

Pileus 10–32 mm wide, broadly convex to plano-convex, eventually plane with an uplifted and undulate margin; margin otherwise decurved, at times with a low obtuse umbo; surface dry, moist or subviscid, glossy, at times with adhering debris over the centre; velipellis absent; centre smooth to fibrillose or matted-fibrillose, radially fibrillose and rimose towards the margin; centre dark brown (7.5YR 3/2; Bistre) shading to dark yellowish brown (10YR 3/4; Warm Sepia, Bay Brown, Prout's Brown; 6E8–F8) towards the margin, or dark reddish brown (5YR 2/2, 3/2, 3/3; near Warm Sepia to almost Verona Brown) throughout; edge of margin with sparse veil remnants when very young; context watery grey or brownish when moist, thin, not changing colour where exposed, not confluent with the stipe. *Lamellae* adnexed to almost free, close to crowded, 36–52 L with several tiers of lamellulae; pallid when young, becoming light grey, light brownish grey (10YR 6/2), pale brown, clay brown or eventually brown (10YR 6/3 to 5/3); edges pallid but indistinctly fimbriate, up to 4 mm deep. *Stipe* 15–40 × 2–4 mm, terete to subcompressed; base with a weakly marginate bulb up to 6 mm wide; surface pruinose for its entire length with a fine silky fibrillose appearance, pallid with distinct pinkish tinges especially above (7.5YR 8/4 to 7/4), or with a light ochre buff tinge; context pinkish ochraceous to pale pinkish (or darker) at the apex, white in the bulb, solid but becoming hollow. *Odour* none to weakly acidulous. Pl. 201, 202 (see also Pl. 2B).

Basidiospores 7.0–10.0 × 5.0–7.0 µm [mean 8.4 × 6.0 µm; Q (1.15–) 1.18–1.66 (–1.70); Q mean 1.42 (n = 26/3)], nodulose, with 7–10 conspicuous conical nodules, at times with a pronounced apical nodule, yellowish brown. *Basidia* 22–30 × 8–10 µm, clavate, hyaline, 4-sterigmate. *Pleurocystidia* 58–75 × 16–22 µm, fusiform to lageniform; apices obtuse and often crystalliferous; base with a short pedicel or rounded; thick-walled (2–4 µm), hyaline or pale yellow. *Cheilocystidia* similar to pleurocystidia, crowded and mixed with small clavate to sphaeropedunculate paracystidia. *Caulocystidia* similar to pleurocystidia but sometimes longer, descending the entire length of the stipe, in isolated but dense clusters mixed with elliptic to pyriform cauloparacystidia. *Pileipellis* a cutis of cylindrical to inflated hyphae to 18 µm wide, encrusted with russet-brown pigment, thin-walled. *Clamp connections* present. Fig. 123.

Occurs in south-western W.A. and south-eastern Qld; on soil or sand under *Eucalyptus*, *Corymbia*, *Syncarpia* or these mixed with *Leptospermum* or *Agonis*. Fruiting June and July.

Additional specimens examined: W.A.: Lot 3088, Hugh Wilson's N boundary, Denmark, in litter below *Eucalyptus megacarpa*, *Agonis flexuosa* and *Trymalium floribundum*, 21 July 1994, *K.Syme KS768/94* (PERTH 05306272, TENN 066772); Warren River Loop Trail, Warren River Natl Park, in Karri (*Eucalyptus diversicolor*) forest with *Allocasuarina* sp., 8 June 2010, *P.B.Matheny PBM3478* (PERTH 08241708, TENN 069507; KJ801161-nLSU); near L. Byleveld, beside William Bay Natl Park, 15 km from Denmark, under *Corymbia calophylla* and *Agonis* sp., 23 June 2001, *N.L.Bougher E6735 (PBM2118)* (PERTH 07671393, TENN 068509; KJ756496-ITS); *loc. id.*, *P.B.Matheny PBM2125 (E6758)* (PERTH 07652097; EU555456-nLSU, EU555455-*rpb1*); intersection of Shelley Beach Rd and Bibbulmun Track, West Cape Howe Natl Park, in paperbark-peppermint woodland near *Melaleuca* sp., *Eucalyptus marginata* and fabaceous shrub (*Mirbelia?*), 18 July 2011, *N.L.Bougher NLB827* (PERTH 08318530, TENN 066757). Qld: Central Station, Fraser Island, 25°28'34"S, 153°03'20"E, alt. 100 m, gregarious on sand under *Eucalyptus*, *Syncarpia* and *Leptospermum*, 2 June 2009, *R.E.Halling*, *N.Fechner & K.Querengasser REH9133* (BRI, NY, TENN 069508; KJ756456-nLSU, KJ756495-ITS).

Etymology: From the Greek *kalos* (beautiful) and *kephale* (a head); in reference to the attractive fruit bodies.

Diagnostic features: Fruit bodies medium, somewhat fragile overall. Pileus dark reddish brown, dark brown to brown, not scaly, rimose, dry to subviscid; stipe pallid but often with pink tinges, pruinose for its entire length, with a weak marginate basal bulb. Spores not oblong. On soil or sand with Myrtaceae.

Inocybe calocephala is described here to recognise an independent phylogenetic lineage represented by collections from the south-west of Western Australia and south-eastern

Queensland, with morphological similarities to *I. calopedes* Matheny & Bougher, *I. leptospermi* (E.Horak) Garrido, *I. graveolens* (E.Horak) Garrido and *I. horakomyces* Garrido. The group is difficult to separate on morphological characters, and this causes considerable taxonomic confusion, often with the same name being applied to different phylogenetic lineages.

Inocybe calocephala differs from *I. calopedes* in having an often pink-tinged stipe and somewhat smaller spores, while DNA sequences clearly separate the two taxa (Fig. 122). We have referred to collections of *I. calocephala* as *I. calopedes* in the past (Matheny & Bougher, 2010). *Inocybe sejuncta* is superficially similar to *I. calocephala*, but differs readily by its oblong and weakly nodulose spores.

Inocybe leptospermi, described from New Zealand (Horak, 1977), is also widespread in Australia, and while some minor genetic variation separates New Zealand from Australian populations, the latter tend to lack the pinkish stipes of New Zealand material, and are dark brown, brown or yellowish brown, lacking the often dark red-brown pileal colour of *I. calocephala* and *I. calopedes*. Once again, DNA sequences separate *I. leptospermi* from other similar species (Fig. 122).

The New Zealand species, *I. horakomyces* Garrido (syn: *Astrosporina avellana* E.Horak) can be distinguished from *I. calocephala* by having longer pleurocystidia (Horak, 1977), somewhat larger spores (9–11 × 7–10 µm) with 9–12 nodules, and an association with *Nothofagus*. *Inocybe graveolens*, another species from New Zealand (Horak, 1977), differs from *I. calocephala* in having a spermatic odour and larger spores (10.0–13.0 × 6.5–9.0 µm; Horak, 1977).

105. Inocybe calopedes Matheny & Bougher, *Muelleria* 28: 90 (2010)

Astrosporina discissa Cleland, *Trans. Roy. Soc. S. Australia* 57: 192 (1933); *Inocybe discissa* (Cleland) Grgur., *Larger Fungi S. Australia* 182 (1997), *non Inocybe descissa* (Fr.) Quél. (1872). Type: Upper Tunkalilla Ck, S.A., 4 June 1930, *J.B.Cleland*; holo: AD 12230.

Pileus 18–35 mm wide, initially conical, expanding with age, eventually becoming plane, with a slight or low obtuse umbo; margin decurved; surface dry, glossy, with a smooth and unbroken disc that can break up into small appressed scales; velipellis absent or present (this can influence overall pileus colour); fibrillose and rimose towards the sometimes torn margin; centre dark reddish brown or very dark brown (without velipellis), reddish brown towards the margin (Chestnut Brown), or strong brown (6D3 to 6D4; Sayal Brown) when covered by a greyish white velipellis; context thin-fleshed, 1–2 mm thick, fragile, resembling a *Psathyrella*. *Lamellae* adnexed to subfree, moderately close, with c. 30 L and several tiers of lamellulae, grey (c. 6C2), greyish brown (6D3), becoming brown to yellowish brown (10YR 5/3 to 6/3) with age; edges indistinctly pallid-fimbriate; soft, broad, to 4 mm deep. *Stipe* 25–40 × 2–4 mm at the apex, terete, gradually enlarged towards the marginate 6 mm wide base; surface pruinose for its entire length, pallid overall to ±white, with a dull pale yellow tinge throughout, contrasting with the white bulb; context solid, concolorous with the surface, whitish in the bulb. *Odour* absent or mild. Pl. 203, 204.

Spores 8.5–10.5 (–11.0) × (5.5–) 6.0–7.5 (–8.0) µm [mean 9.4–6.8; Q 1.20–1.62; Q mean 1.39 (n = 41/2)], usually with 7–9 prominent conical nodules, yellowish brown; apiculus small. *Basidia* 23–30 × 9–10 µm, 4-sterigmate, clavate to cylindroclavate, hyaline. *Pleurocystidia* 50–60 × 13–25 µm, utriform, ventricose or fusiform; apices rounded to obtuse, crystalliferous; generally tapered at both ends but lacking distinct necks and basal pedicel, often sessile with rounded-truncate bases; thick-walled (mostly 2.5–4.0 µm) especially apically, hyaline or pale yellow. *Cheilocystidia* similar to pleurocystidia but often shorter and more flask-shaped, with walls hyaline to pale yellow, mixed with paracystidia. *Caulocystidia* descending below the stipe centre, in scattered dense clusters, thick-walled (mostly 2–3 µm) apically, hyaline, generally ventricose to broadly fusiform, with necks not well developed, mixed with cauloparacystidia. Fig. 124.

Occurs in S.A., eastern Qld and Tas.; grows on soil under *Leptospermum* with *Nothofagus cunninghamii* in the vicinity (in Tas.), elsewhere under *Eucalyptus*. Fruiting May and June.

Plate 201. *Inocybe calocephala*, holotype (PERTH 08319855). Photograph — P.B.Matheny.

Plate 202. *Inocybe calocephala*. PERTH 07671393. Photograph — N.L.Bougher.

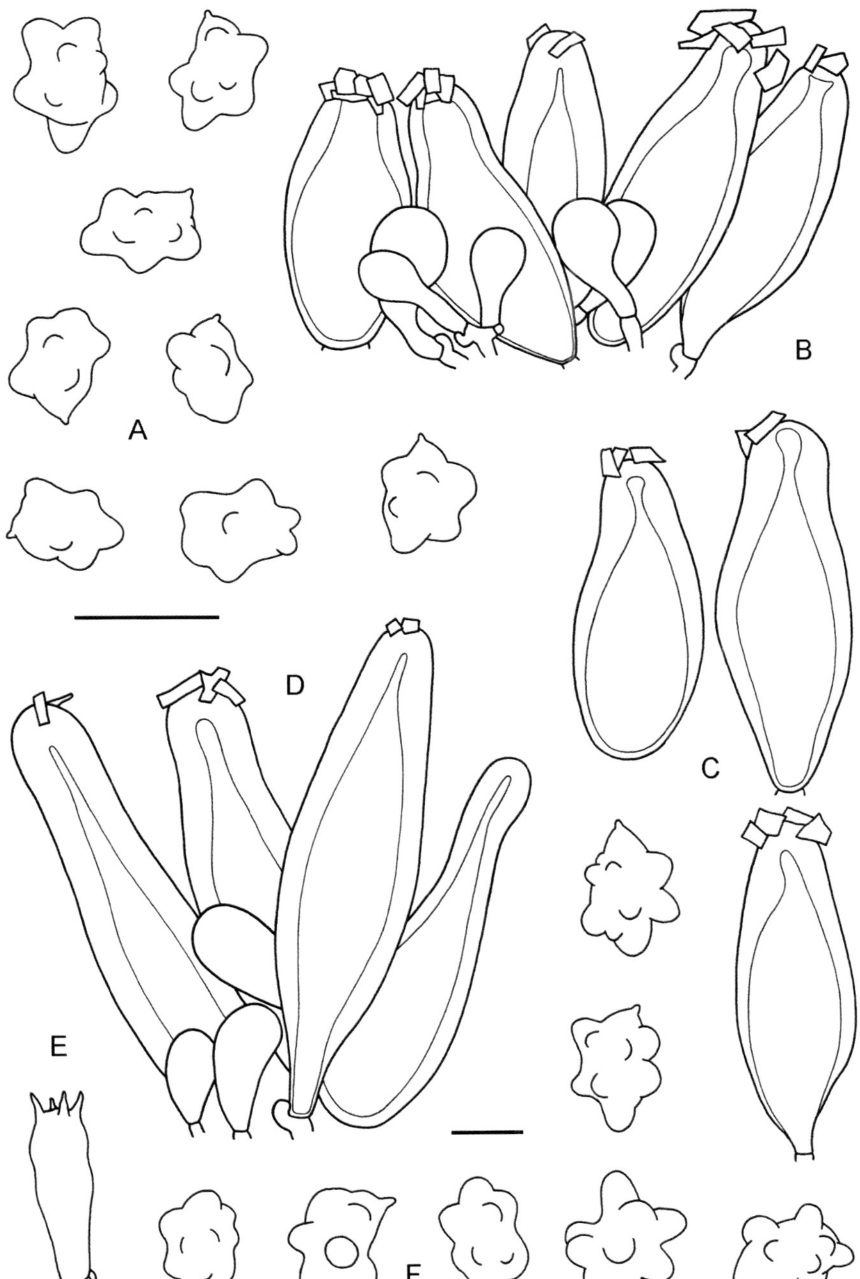

Figure 123. *Inocybe calocephala.* All holotype (PERTH 08319855) except E, F (PERTH 07671393). **A,** Basidiospores; **B,** Cheilocystidia and paracystidia; **C,** Pleurocystidia; **D,** Caulocystidia and cauloparacystidia from stipe apex; **E,** Basidium; **F,** Basidiospores. Scale bars = 10 µm (longer bar for spores only).

Plate 203. *Inocybe calopedes*. PERTH 08516626. Photograph — P.B.Matheny.

Plate 204. *Inocybe calopedes*. TJB 10555. Photograph — T.J.Baroni.

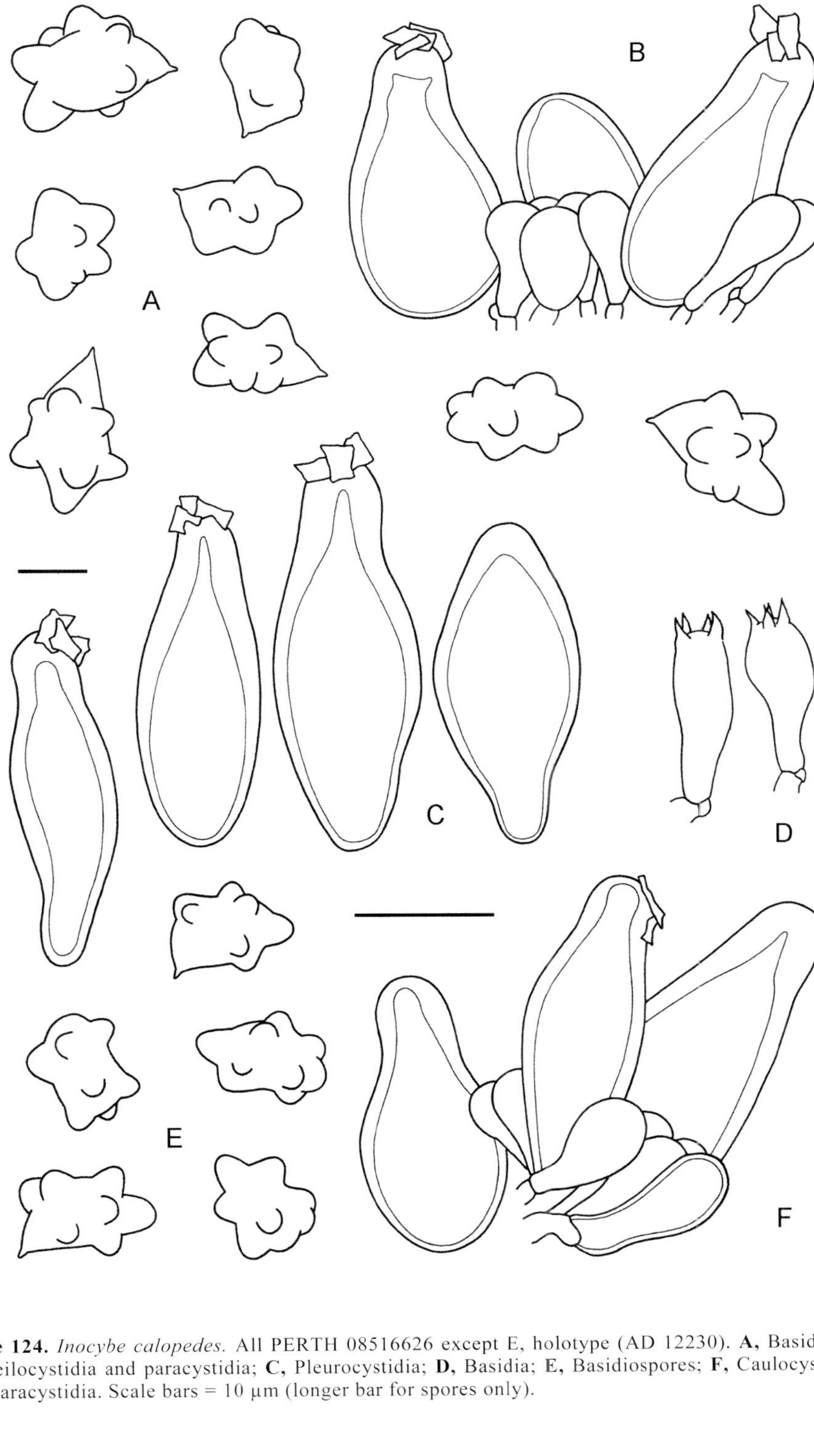

Figure 124. *Inocybe calopedes.* All PERTH 08516626 except E, holotype (AD 12230). **A,** Basidiospores; **B,** Cheilocystidia and paracystidia; **C,** Pleurocystidia; **D,** Basidia; **E,** Basidiospores; **F,** Caulocystidia and cauloparacystidia. Scale bars = 10 µm (longer bar for spores only).

Additional specimens examined: Qld: Apple Tree Park, Springbrook Natl Park, gregarious on sandy soil in wet sclerophyll forest, 2 June 2011, *N.Fechner & T.J.Baroni TJB10555* (BRI, CORT; KJ756457-nLSU, KJ778838-ITS, KJ756414-*rpb2*). Tas.: Mount Michael Track, Blue Tier, alt. 800 m, scattered singly on soil under *Leptospermum* with *Nothofagus cunninghamii* nearby, 22 May 2010, *P.B.Matheny PBM3385* (PERTH 08516626, TENN 068947; JQ313572-nLSU, KJ778839-ITS, KJ756415-*rpb2*).

Diagnostic features: Medium-sized, somewhat fragile. Pileus reddish brown to dark reddish brown, or brown when velipellis is present and heavy; centre smooth but rimose towards the margin; stipe pallid with pale yellow tinges, uniformly pruinose, with a marginate basal bulb. Spores nodulose, mostly 9.0–10.5 × 6.0–7.5 µm with 7–9 conspicuous nodules; pleurocystidia short-fusiform to utriform, thick-walled (2.5–3.0 µm).

Inocybe calopedes was introduced by Matheny & Bougher (2010) as a new name for the illegitimate *I. discissa* (Cleland) Grgur. (*non I. descissa* (Fr.) Quél.). In general the pileus of *I. calopedes* is dark reddish brown to brown and the stipe nearly white but with a dull yellow tinge. The collections cited above best correspond to the type (Matheny & Bougher, 2010). However, some morphologically similar but phylogenetically distinct species can be confused with *I. calopedes*:

(1) We now recognise several collections from Western Australia and south-eastern Queensland as *I. calocephala* (Fig. 122). The latter is almost identical morphologically to *I. calopedes*, but specimens have smaller spores and pink tinges on the stipe.

(2) Collections not having a decidedly dark reddish brown pileus could be confused with *I. leptospermi*, which lacks the warm reddish tinges to the pileus. The pleurocystidia of *I. leptospermi* are thicker-walled than those in *I. calopedes* (3.0–7.0 µm *vs* 2.5–4.0 µm).

(3) *Inocybe chalcoceps* has smaller spores with low nodules and pinkish-tinged stipe. Its fruit bodies are more robust, and the species has a weak spermatic odour.

(4) *Inocybe graveolens* has a strong spermatic odour, unlike *I. calopedes* and *I. calocephala*.

(5) *Inocybe sylvicola* from Tasmania and *I. scissa* (E.Horak) Garrido from New Zealand have straw-yellow to yellow pilei, and both are usually found in association with *Nothofagus*.

Matheny & Bougher (2010) identified two Tasmanian collections, "*I2*" (TENN 067033) and "*I100*" (herb. G.M.Gates), as *I. calopedes*. However, DNA sequences of *I2* (KP170932-nLSU, KP308754-ITS) have since confirmed this collection as *I. leptospermi* (Fig. 122). Collection *I100* has not been sequenced, but it was collected at the same location as *I2* during November in Tasmania, and appears to be conspecific.

A photograph of *I. calopedes* from Tas. is reproduced in Gates & Ratkowsky (2014).

106. Inocybe chalcoceps Matheny, Bougher & G.M.Gates, *sp. nov.*

MycoBank: MB 809275

Type: Lower Kermandie, Tas., on soil in wet sclerophyll forest, 5 Jan. 2002, *G.M.Gates & D.Ratkowsky I47*; holo: PERTH 08556474; iso: TENN 068946 (KJ729914-nLSU).

Pileus 21–35 mm wide, conical to conico-convex with a rounded apex, becoming plane with a low obtuse umbo; margin decurved; surface glossy, moist to dry with some adhering debris; centre smooth and unbroken or with an innate velipellis; margin rimulose and finely fibrillose; bronze brown or light reddish brown (5E7 to 6D7; 5YR 5/4 to 4/4), with very pale or yellow buff disc (velipellis) evident on some specimens; context 2–3 mm thick under the centre, sordid watery grey to pallid, not changing colour upon exposure. *Lamellae* adnexed, close, with several tiers of lamellulae; light grey, sordid cream, becoming light brown; slightly ventricose, up to 3.5 mm deep, thin. *Stipe* 26–42 mm × 3.5–6.0 mm, 5–8 mm across the marginate bulb, central, terete; surface dry, uniformly pruinose, with a satiny or longitudinally striate appearance; pale shell pink with a white base; context fibrous, solid with a pale pink to pallid interior, white in the bulb. *Odour* faintly spermatic. Pl. 205.

Basidiospores 6.0–8.5 (–9.0) × 5.0–6.5 µm [mean 7.3 × 5.7 µm; Q 1.15–1.54 (–1.70); Q mean 1.28 (*n* = 20/1)], angular-nodulose with mostly 7–9 low obtusely conical nodules, an apical nodule projecting to 2.0 (–2.5) µm, yellowish brown; apiculus small. *Basidia* 21–26 × 7–8 µm, clavate, hyaline, 4-sterigmate. *Pleurocystidia* 39–50 × 14–18 µm, mostly utriform or ventricose, with rounded or narrowly truncate bases; apices rounded to obtuse, crystalliferous; thick-walled, especially apically (2.5–4.0 µm thick), hyaline. *Cheilocystidia* similar to pleurocystidia but shorter and more ventricose, mixed with crowded narrowly clavate paracystidia. *Caulocystidia* similar to cheilocystidia, descending the entire length of stipe, mixed with clavate and pyriform cauloparacystidia. *Pileipellis* a cutis of cylindrical hyphae, these brownish yellow or light brown in mass, cylindrical, mostly 5–12 µm wide, smooth; some sparse narrowly cylindrical velipellis hyphae observed above the cutis, these interwoven, narrow, hyaline, thin-walled. *Clamp connections* present. Fig. 125.

Occurs on soil in wet sclerophyll vegetation in Tas., under *Eucalyptus* and *Pomaderris*. Known only from the type specimen. Fruiting Jan.

Plate 205. *Inocybe chalcoceps*, holotype (PERTH 08556474). Photograph — G.M.Gates.

Etymology: From the Greek *chalcos* (copper), and the suffix *-ceps* (referring to a head); in reference to the bronze-brown to light reddish brown pileus.

Diagnostic features: Fruit bodies medium. Pileus bronze-brown to light reddish brown, often with a pallid velipellis over the disc; stipe pruinose for its entire length, with a pale pink cast and a white marginate bulb. Odour weakly spermatic. Spores mostly 6.0–8.5 × 5.0–6.5 µm, angular-nodulose, with 7–9 low obtusely conical nodules. Pleurocystidia often with rounded bases. Associated with *Eucalyptus* and *Pomaderris*.

Inocybe calopedes (*q.v.*) differs from *I. chalcoceps* in its larger spores (8.5–10.5 × 6.0–8.0 µm) with more pronounced nodules and the stipe lacking pink tinges. *Inocybe calocephala* shares the pinkish tinged stipe, and the spores overlap in size (7.0–10.0 × 5.0–7.0 µm) with *I. chalcoceps*; however, the spores of *I. calocephala* have more conspicuous conical nodules, and DNA sequence analysis readily separates them (Fig. 122). The odour of *I. calopedes* is acidulous, whereas in *I. chalcoceps* it is faintly spermatic, although more field collections of the latter are needed to determine the constancy of difference in smell. Finally, *I. leptospermi* has a weak acidulous odour, somewhat larger spores (mostly 8.0–10.5 × 5.5–7.5 µm), a darker brown pileus, and it rarely has pink on the stipe. *Inocybe leptospermi* appears to be the most common and widespread of this species-group in Australia.

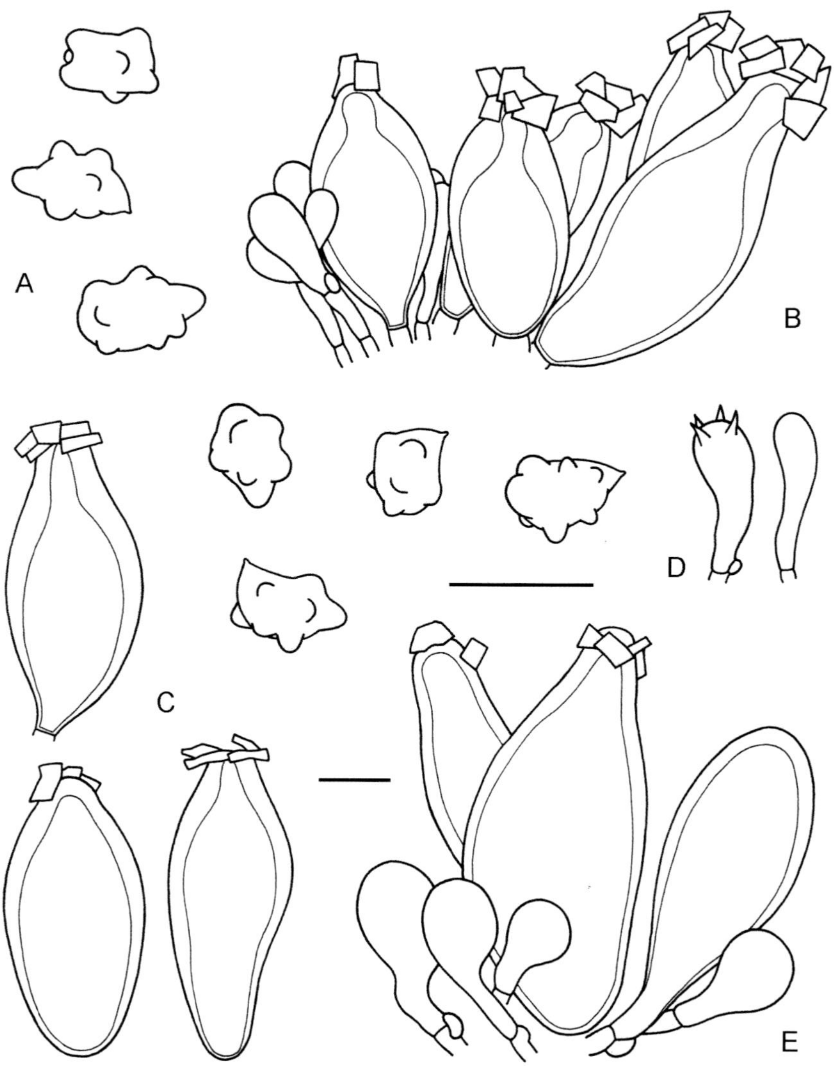

Figure 125. *Inocybe chalcoceps.* All holotype (PERTH 08556474). **A,** Basidiospores; **B,** Cheilocystidia and paracystidia; **C,** Pleurocystidia; **D,** Basidia; **E,** Caulocystidia and cauloparacystidia.

Scale bars = 10 µm (longer bar for spores only).

107. **Inocybe leptospermi** (E.Horak) Garrido, *Biblioth. Mycol.* 120: 177 (1988)

Astrosporina leptospermi E.Horak, *New Zealand J. Bot.* 15: 744 ('1978') [1977]. Type: Long Harry Bay, Stewart Island, New Zealand, on the ground under *Leptospermum scoparium*, 18 Mar. 1969, *E.Horak*; holo: PDD 27117.

Pileus 13–35 mm, obtusely conical, plano-convex to plane with a decurved margin, at times with a low obtuse umbo; margin becoming slightly raised; surface smooth and unbroken on the disc but can become become excoriate or weakly scaly, rimulose to rimose towards the margin, dry but at times with adhering soil particles as though subviscid, dark brown or brown shading to brown or yellowish brown towards the margin (10YR 3/3, 4/3, 5/4; Bistre, Snuff Brown, to Tawny-Olive), with age brown (7.5YR 4/4; Cinnamon-Brown); context thin, white, colour unchanging upon exposure. *Lamellae* adnexed to subfree, close, 34–50 L, with several tiers of lamellulae; grey, greyish brown to yellowish brown, ventricose, medium; edges indistinctly pallid-fimbriate. *Stipe* 20–42 × 2–5 mm at the apex, tapered slightly just below the apex, terete or slightly compressed; base with a marginate bulb to 7 mm wide; pruinose its entire length, striatulate under a hand lens or glabrous, pallid or white, pinkish tones infrequently observed, less often with a very pale brown tone; context solid, becoming hollow, whitish or less often with pinkish tinges, white in the bulb. *Odour* none, not remarkable, or slightly acidulous. Pl. 206, 207.

Basidiospores (7.5–) 8.0–10.5 (–11.0) × 5.5–7.5 µm [mean 9.2 × 6.4 µm; Q 1.21–1.63 (–1.73); Q mean 1.43 (n = 70/5)], usually with 7–10 obtuse to subconical nodules, yellowish brown; apiculus small and tapering. *Basidia* 26–30 × 9–10 µm, clavate, hyaline, 4-sterigmate. *Pleurocystidia* 53–62 × 14–20 (–28) µm, broadly fusiform, occasionally fusiform or utriform, thick-walled (3–7 µm thick), hyaline. *Cheilocystidia* similar to pleurocystidia but typically shorter and mixed with abundant small, clavate, hyaline paracystidia. *Caulocystidia* similar to cheilo- and pleurocystidia, numerous along the entire stipe surface, mixed with thin-walled and thick-walled cauloparacystidia. *Pileipellis* composed of a shallow superficial layer of subparallel to interwoven hyphae mostly 4–8 µm wide, cylindrical and smooth in appearance, lacking distinct pigment; subtending layer distinct, pigmented ochre-tawny. *Clamp connections* present. Fig. 126.

Widespread in W.A., Qld, N.S.W., A.C.T. and Tas.; on soil under *Agonis*, *Corymbia*, *Eucalyptus*, *Melaleuca* and (rarely) *Nothofagus*. Recorded in New Zealand under *Kunzea ericoides* and *Leptospermum scoparium*. Fruiting Feb., Apr., May, July and Nov.

Additional specimens examined: W.A.: Cascades Rd, near Greenbushes, N of Bridgetown, 33°53'24.4"S, 116°08'9.7"E, on soil in two year-old burn site, Jarrah-Marri forest, under *Corymbia calophylla* and *Eucalyptus marginata*, 13 July 2011, *P.B.Matheny PBM3608* (PERTH 08319782, TENN 066567; KP170934-nLSU, KP308757-ITS, KJ811591-*rpb2*); S of Denmark, Reserve 41815, on sandy soil in Jarrah-Marri forest under *Eucalyptus todtiana*, *E. patens* and *Melaleuca*, *E.Davison & P.Davison PBM3628* (TENN 066641; KP170935-nLSU, KP308757-ITS, KJ811592-*rpb2*); Bibbulmun Track, Shelley Beach Rd, West Cape Howe Natl Park, between Denmark and Albany, on soil under *Melaleuca* and *Agonis flexuosa* in closed canopy *Melaleuca* thicket, 35°05'56.8"S, 117°37'44.1"E, 18 July 2011, *P.B.Matheny PBM3642* (PERTH 08319456, TENN 066610) (KP170936-nLSU, KP308759-ITS, KJ811593-*rpb2*). Qld: Queen Mary Falls, Main Range Natl Park, 28°20'29.5"S, 152°22'16.9"E, on soil in wet sclerophyll forest under *Eucalyptus*, 4 Mar. 2012, *N.Fechner PBM3788* (PERTH 08363013, TENN 067020). N.S.W.: Weemala Lookout, Batlow, under *Eucalyptus*, 1 July 2007, *J.Trappe T31970* (OSC, PERTH 08516693). A.C.T.: Australian National Botanic Gardens, Canberra, in lawn under *Eucalyptus*, 26 May 1991, *H.Lepp 540* (CANB 574560; KP170931-nLSU, KP308753-ITS, KJ811588-*rpb2*). Tas.: Myrtle Gully, 9 Nov. 2000, *G.Gates & D.Ratkowsky 12* (TENN067033; KP170932-nLSU, KP308754-ITS); Ralph Falls Track, West Pyengana, 800 m alt., on soil in rainforest under *Nothofagus cunninghami*, 23 May 2010, *P.B.Matheny, G.Gates & D.Ratkowsky PBM3398* (TENN 065746; JQ313561-nLSU, KP308755-ITS, KJ811589-*rpb2*); *loc. id.*, on soil in Tea Tree forest under *Leptospermum*, 23 May 2010, *P.B.Matheny & D.Ratkowsky PBM3399* (TENN 067031); *loc. id.*, *P.B.Matheny & D.Ratkowsky PBM3400* (TENN 067032; KP170933-nLSU, KP308756-ITS, KJ811590-*rpb2*); Leap Loop, 5 Feb. 2011, *G.Gates 1148* (TENN 068376, HO 562689).

Extralimital specimens examined: New Zealand. North Island, Waitakere Ra., W of Auckland, Arataki Visitors Centre, on soil along trail under *Kunzea ericoides*, 6 May 2009, *P.B.Matheny PBM3080* (TENN 063625; JN974983-nLSU, KP308760-ITS, KJ811594-*rpb2*). South Island, Maruia, in *Leptospermum* scrub under *Kunzea ericoides*, 25 May 2003, *J.A.Cooper* (PDD 80003, PERTH 08516596, as *graveolens*; KP170937-nLSU, KP308761-ITS); Hinewai Reserve near Akaroa, under *Kunzea ericoides* in broadleaved forest, 10 Apr. 2003, *J.A.Cooper* (PDD 80267 as *graveolens*; KP170938-nLSU, KP308762-ITS).

Plate 206. *Inocybe leptospermi*. PERTH 08319782. Photograph — P.B.Matheny.

Plate 207. *Inocybe leptospermi*. PERTH 08363013. Photograph — P.B.Matheny.

Diagnostic features: Fruit bodies medium-sized. Pileus dark brown, brown or umbrinous; disc often smooth and unbroken; margin rimulose to rimose; stipe white to pallid, less often with pink tinges; base with a marginate bulb; odour not remarkable to weakly acidulous. Spores nodulose, (7.5–) 8.0–10.5 (–11.0) µm long; hymenial cystidia very thick-walled. On soil under diverse myrtaceous trees or shrubs, rarely under *Nothofagus*.

We apply the name *I. leptospermi* to a distinct clade of numerous collections that were widely sampled from New Zealand and temperate Australia (Fig. 122). This species may have the widest distribution of any *Inocybe* in the South Pacific. We recognise it morphologically by the smooth, brown, umbrinous or dark yellowish brown pileus, rimose pileal margin, pallid (rarely with pink tinges) and entirely pruinose stipe, the presence of a distinct marginate bulb, and the lack of any distinctive odour. It most closely resembles *I. calopedes* Matheny & Bougher, which has a dark red-brown pileus but is genetically distinct.

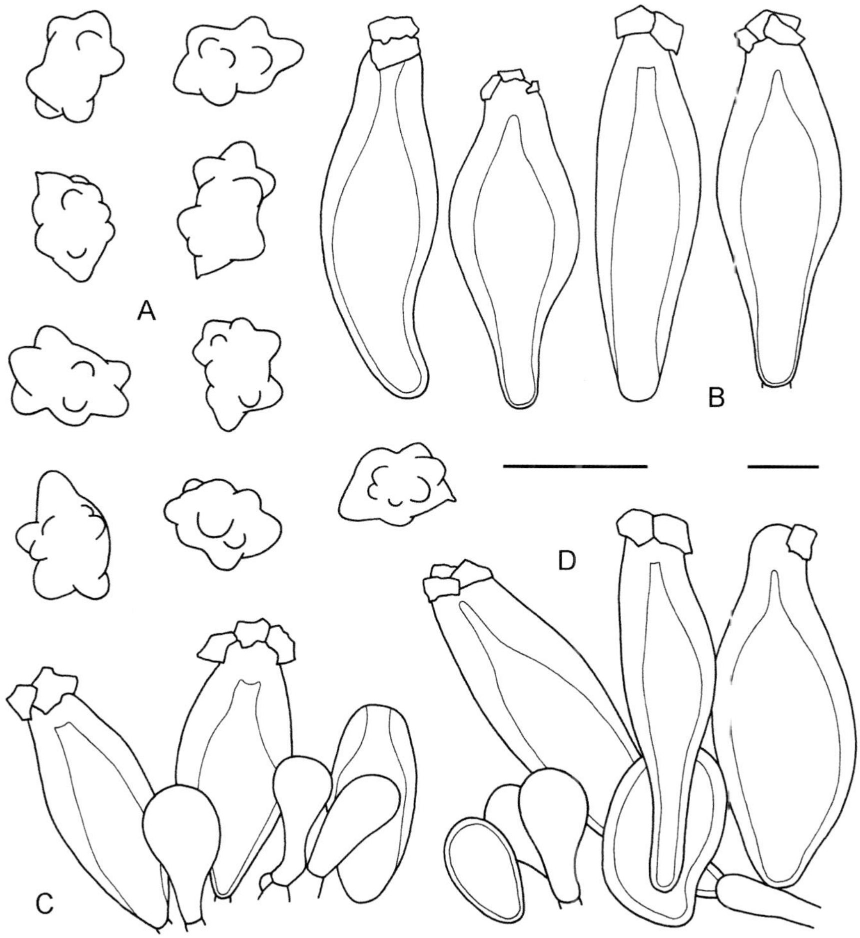

Figure 126. *Inocybe leptospermi.* All PERTH 08319782. **A,** Basidiospores; **B,** Pleurocystidia; **C,** Cheilocystidia and paracystidia; **D,** Caulocystidia and cauloparacystidia. Scale bars = 10 µm (longer bar for spores only).

Inocybe graveolens (E.Horak) Garrido, from New Zealand, is also similar to *I. leptospermi* but differs most noticeably by its strong spermatic odour and larger spores (10–13 × 7–10 µm; Horak, 1977). *Inocybe scissa* (E.Horak) Garrido and *I. sylvicola* Matheny, Bougher & G.M.Gates, like *I. calopedes* and *I. calocephala* Matheny & Bougher, are closely related to *I. leptospermi* based on molecular phylogenetic analysis (Fig. 122). However, they are distinguished primarily by their yellowish to brownish yellow pilei and broader cystidia with broad rounded bases and their usual occurrence in *Nothofagus* forests. According to Horak (1977), *I. horakomyces* Garrido (as *Astrosporina avellana* E.Horak), which is not known from Australia, differs from *I. leptospermi* by its longer pleurocystidia and the more brownish yellow or yellowish brown pileus. Horak (1977) also distinguished these two species ecologically, *I. leptospermi* being associated with *Leptospermum*, and *I. horakomyces* with *Nothofagus*.

However, while we found *I. leptospermi* mainly in association with *Kunzea*, *Leptospermum*, *Melaleuca* or *Eucalyptus*, we observed one collection under *Nothofagus* in Tasmanian rainforest. *Inocybe leptospermi* appears to be composed of several distinct genetic variants in Australia, with at least three different genotypes in Tasmania from which the Western Australian and New Zealand genotypes (possibly cryptic species) are derived (Fig. 122).

Muscarine has been detected in *I. leptospermi* (as *I. cf. graveolens*) by Kosentka *et al.* (2013) (coll. *PBM3398* from Tasmania).

108. Inocybe sylvicola Matheny, Bougher & G.M.Gates, *sp. nov.*

MycoBank: MB 809469

Type: Mount Michael Track, Blue Tier, Tas., alt. 800 m, on soil in rainforest under *Nothofagus cunninghamii*, 22 May 2010, *G.M.Gates & P.B.Matheny PBM3387*; holo: PERTH 08517215; iso: TENN 065735 (JQ313576-nLSU originally as "*Inocybe scissa*", KP641630-ITS, KM656126-*rpb2*).

Pileus 15–42 mm wide, conical to broadly conical, with a slight broad obtuse umbo developing with age or upon expansion; margin initially incurved, later decurved; velipellis absent; disc smooth and unbroken; surface dry, not lubricous, radially rimulose to rimose but mainly striolate or sulcate with age, fibrillose towards the margin which is occasionally torn; pale yellow to olive-yellow (Buffy Yellow) or straw-yellow almost throughout (2.5Y 8/6–7/4–6/6) or with an isabelline or light olive-brown margin (2.5Y 5/6–6/6), at times isabelline almost throughout or bicolorous with a fulvous disc and isabelline margin; thin-fleshed, up to 2 mm thick under the disc, pallid or watery grey. *Lamellae* adnexed to subfree, very close, c. 40–46 L, with several tiers of lamellulae, light grey to very pale brown (10YR 7/2–7/3), pale brown to yellowish brown or brown (10YR 6/3–5/3), (sub)ventricose; edges pallid, indistinctly fimbriate. *Stipe* 25–53 (–70) × 2–5 mm at the apex, terete, even to the slightly to abruptly marginate basal bulb which is up to 10 mm wide; surface pruinose for its entire length with a silky-satiny appearance, pallid or almost white, rarely with any pink tinge; context solid, brittle, pallid. *Odour* none or slightly acidulous. Pl. 208, 209.

Basidiospores (7.5–) 8.0–11.0 × 5.0–8.0 (–8.8) µm [mean 9.5 × 6.6 µm; Q 1.15–1.79 (–1.85); Q mean 1.45 (n = 42/3)], usually with 7–10 prominent nodules, yellowish brown. *Basidia* 26–30 × 9–11 µm, 4-sterigmate, occasionally 2-sterigmate, clavate, hyaline. *Pleurocystidia* 52–65 × 18–28 µm, ventricose to broadly lageniform or utriform, generally tapered at both ends but often lacking a distinct neck and pedicel, with a narrow often truncate base; very thick-walled (to 5 µm thick), hyaline; apices crystalliferous. *Cheilocystidia* similar to pleurocystidia but smaller, mixed with abundant densely crowded small hyaline paracystidia up to 20 × 10 µm. *Caulocystidia* similar to pleurocystidia, descending the entire length of the stipe in scattered dense clusters, often mixed with cells of intermediate size and shape. *Pileipellis* a cutis of cylindrical hyphae 5–10 µm wide, smooth or weakly encrusted, light yellowish brown in mass; tramal hyphae with conspicuous ochraceous refractive hyphae also present. *Clamp connections* present. Fig. 127.

Occurs in Tas., on soil in rainforest under *Nothofagus cunninghamii*; also reported from Vic. under *Nothofagus* and in wet *Eucalyptus* forest. Fruiting Mar.–May.

Additional specimens examined: Tas.: Creepy Crawly Nature Trail, Southwest Natl Park, in rainforest, 16 Mar. 2002, *G.M.Gates & D.Ratkowsky I64* (TENN 065737; KP171085-nLSU, KP641631-ITS); Mount Michael Track, 800 m alt., Blue Tier, on soil in rainforest under *Nothofagus cunninghamii*, 22 May 2010, *G.M.Gates & P.B.Matheny PBM3388* (TENN 065736); Ralphs Falls, West Pyengana, 800 m alt., on soil under *Nothofagus cunninghamii* in rainforest, 23 May 2010, *G.M.Gates, D.Ratkowsky & P.B.Matheny PBM3392* (TENN 065738); *loc. id.*, 23 May 2010, *G.M.Gates, D.Ratkowsky & P.B.Matheny PBM3393* (TENN 065739); *loc. id.*, 23 May 2010, *M.Pilkington & P.B.Matheny PBM3394* (PERTH 08517223, TENN 065745) (JQ313568-nLSU, KP641632-ITS, KM656127-*rpb2*); Weldborough Pass, near Pyengana, on soil in rainforest under *Nothofagus cunninghamii*, 24 May 2010, *G.M.Gates & P.B.Matheny PBM3405* (TENN 065740); Vale of Belvoir, 30 km outside Gowrie Park, on soil along track in rainforest under *Nothofagus cunninghamii*, 25 May 2010, *P.B.Matheny PBM3412* (TENN 065741). Vic.: Melba Gully State Park, W of Apollo Bay, under *Nothofagus cunninghamii*, 3 May 1995, *N.L.Bougher, R.Vilgalys, S.Bolsenbroek & M.Brundrett E5353* (PERTH 07549687).

Etymology: From the Latin *sylva* (forest) and the suffix *-cola* (inhabitant), in reference to the primary location of this species in *Nothofagus* rainforest.

Diagnostic features: Fruit bodies medium-sized. Pileus buff-yellow to straw-yellow, smooth on the disc, striolate or sulcate towards the margin; stipe pallid or almost white, rarely with any pink tones, entirely pruinose, base a marginate bulb; odour none or weakly acidulous. Spores nodulose, on average <10 µm long; pleurocystidia ventricose or broadly lageniform to utriform, thick-walled. On soil in rainforest under *Nothofagus*.

Inocybe sylvicola can be easily confused morphologically with several closely related species that occur in New Zealand and/or temperate Australia. *Inocybe horakomyces* Garrido, from New Zealand but not known from Australia, has a beige to light brown pileus, and the stipe becomes light brown with reddish or pinkish tints (Horak, 1977, as *Astrosporina avellana* E.Horak); it is microscopically very similar to *I. sylvicola*. Horak (1979a) also reported this species from Papua New Guinea but with stellate spores. Both *I. sylvicola* and New Zealand specimens of *I. horakomyces*, including material examined by E.Horak, are closely related phylogenetically (Fig. 122).

Inocybe scissa (E.Horak) Garrido, from New Zealand, can be separated phylogenetically (Fig. 122) and morphologically from *I. sylvicola*. While both taxa have a straw-yellow pileus, pruinose stipe and nodulose spores, *I. sylvicola* differs subtly in the almost invariable absence of pink or reddish tones on the stipe and a less noticeably rimose pileus. *Inocybe sylvicola* also lacks the pyriform cystidia with a broad, rounded base typical of *I. scissa*. Initially, we accepted specimens from Tasmania and Victoria as being *I. scissa*, but after an examination of the holotype and phylogenetic comparison with two collections of *I. scissa* from New Zealand (PDD 71173, PDD 75859), we conclude the Australian specimens represent a distinct taxon. We have yet to confirm the presence of *I. scissa* and *I. horakomyces* in Australia.

Inocybe sylvicola may have been reported for the first time from Australia as *I. scissa* (May, 1989; Ford & May, 1998) in wet forests in Victoria; but these specimens have not been re-examined. We cite an additional collection here from a *Nothofagus* rainforest habitat in Victoria (PERTH 07549687).

Another New Zealand species, *I. graveolens* (E.Horak) Garrido, differs from *I. sylvicola* in having a reddish brown pileus, strong spermatic odour, and an association with *Leptospermum* (Horak, 1977). *Inocybe straminea* (E.Horak) Garrido is yellow throughout (pileus, lamellae and stipe), and is known only from sand dunes under *Leptospermum* at its type locality in New Zealand (Horak, 1977), while *I. calopedes*, from south-western Western Australia and south-eastern Queensland, differs in its reddish brown pileus and the absence of an odour. *Inocybe leptospermi*, a complex of geographical variants found in New Zealand and throughout much of Australia (Fig. 122), has a brown to dark brown, fibrillose pileus, while *I. aequalis* (E.Horak) Garrido differs from *I. sylvicola* in its beige to cinnamon or avellaneous pileus and concolorous or paler stipe.

Inocybe sylvicola is potentially highly toxic. A sample (*PBM3394*, PERTH 08517223, TENN 065745) was analysed for presence of muscarine using high-performance liquid chromatography and was found to have large concentrations of muscarine compared to other species of *Inocybe* examined (Kosentka *et al.*, 2013).

Plate 208. *Inocybe sylvicola*, holotype (PERTH 08517215). Photograph — P.B.Matheny.

Plate 209. *Inocybe sylvicola.* TENN 065740. Photograph — P.B.Matheny.

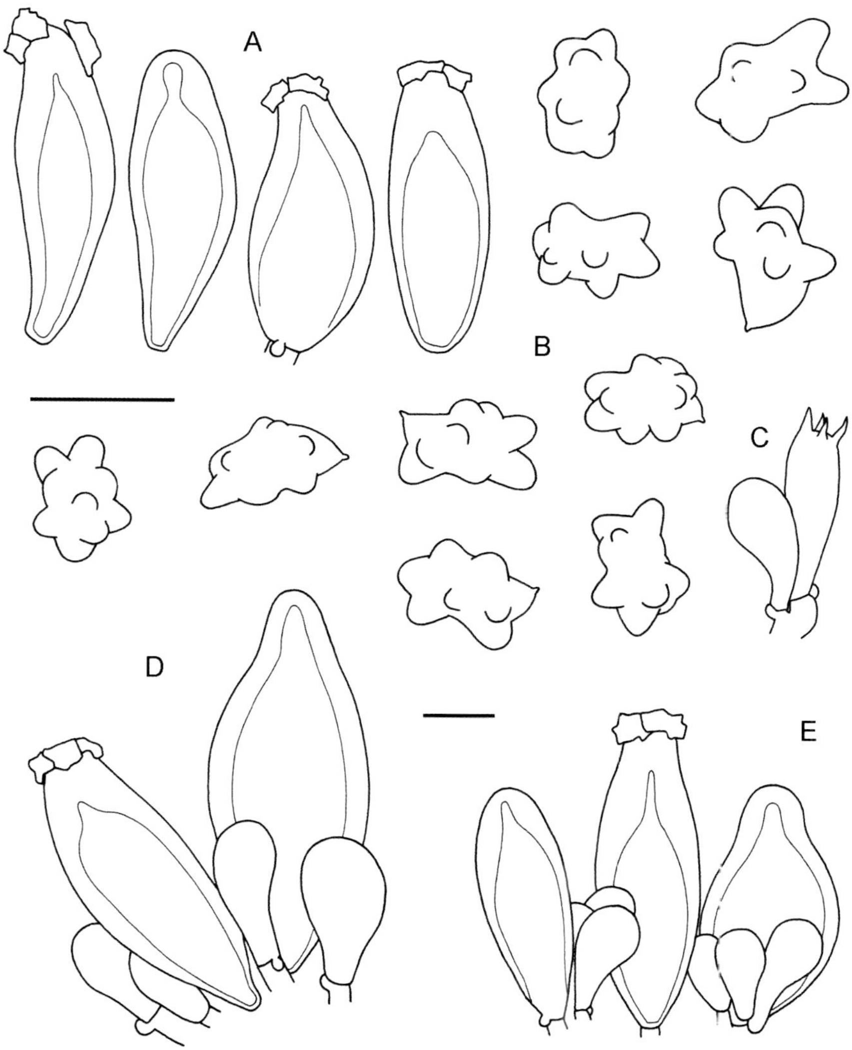

Figure 127. *Inocybe sylvicola.* All holotype (PERTH 08517215). **A,** Pleurocysticia; **B,** Basidiospores; **C,** Basidia; **D,** Caulocystidia and cauloparacystidia. **E,** Cheilocystidia and paracystidia. Scale bars = 10 µm (longer bar for spores only).

INOCYBE VISCATA GROUP

The *I. viscata* group appears to contain three Australian lineages, two of which are documented below. One includes the trans-Tasman species, *I. viscata*, and one of the most common species in northern parts of Australia, *I. torresiae*. Both species exhibit caulocystidia and cauloparacystidia descending at least to the mid-region of the stipe and nodulose spores, but otherwise are quite different in outward appearance and in ecology and habitat preference. From the phylogeny it is clear the cool temperate *I. viscata* is derived from tropical and warm temperate lineages. Three undetermined lineages from Costa Rica and Papua New Guinea are most closely related to the austral species.

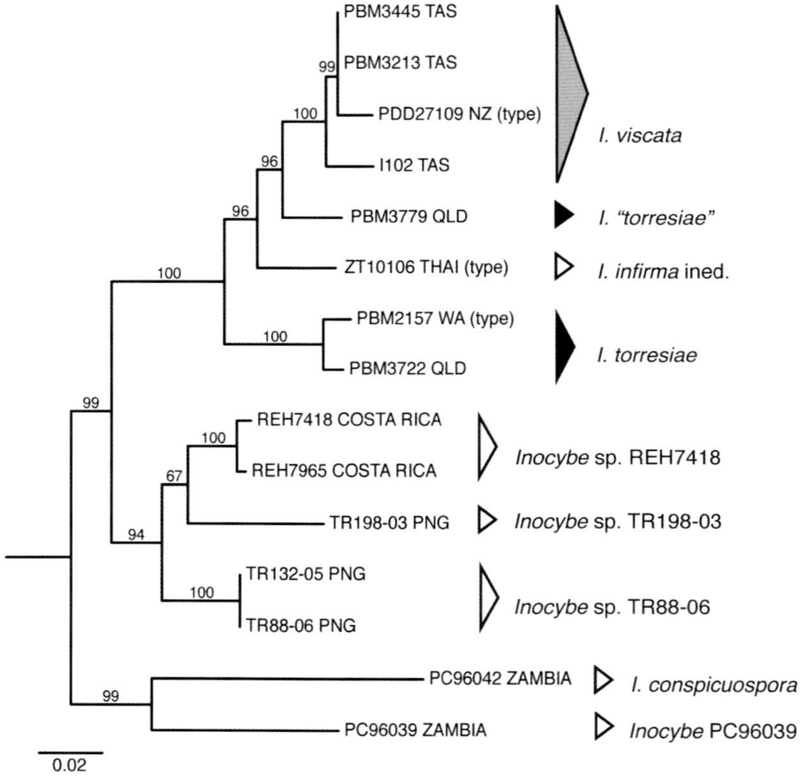

Figure 128. Maximum Likelihood phylogram of the *Inocybe viscata* and *Inocybe torresiae* group. Bootstrap values are shown above branches. The tree was inferred from combined ITS, nLSU, and *rpb2* gene regions and includes a total of 2740 nucleotide sites. One hundred and fifty five sites of the ITS region were omitted before analysis due to alignment ambiguities. Species from Australia are indicated by black-filled triangles. Those indicated with white triangles occur outside Australia. Those with grey-filled triangles occur in both Australia and New Zealand. Fifteen branch tips are included in the tree with 12 represented by ITS, 13 by nLSU, and 12 by *rpb2*. The tree is rooted with two African species of *Inocybe*.

109. Inocybe torresiae Matheny, Bougher & M.D.Barrett, *Nuytsia* 22: 69 (2012)

Type: Beverley Springs Station HS, Kimberley Region, W.A., solitary or rarely caespitose on rich black soil in monsoon forest of *Antidesma ghaesembilla* and *Glochidion disparipes* with *Eucalyptus bigalerita*, *Corymbia bella* and *Albizia procera* more distant, 12 Feb. 1996, *M.D.Barrett MDB F70/96 (E6978, PBM2157)*; holo: PERTH 07678428 (EU600874-nLSU, JQ085937-ITS, EU600872-*rpb1*, EU600873-*rpb2*).

Pileus 28–35 mm wide, conical at first, but becoming campanulate or applanate with an abrupt pronounced umbo; margin incurved to decurved; surface dry, smooth or appressed-fibrillose, excoriate or diffract-scaly with age with an entire margin; scales concolorous with the background; colour pale yellowish brown or isabelline (4B6 to 4B5) to pale tan (5B4), possibly darker or tawny brown by maturity (6D6); context pallid, not changing colour where bruised. *Lamellae* adnate, moderately close, c. 26 L, with several tiers of lamellulae, light yellowish brown (4B4) to fawn-coloured (5B4) maturing to brown (6D7), 4–5 mm deep; edges pallid and indistinctly fimbriate. *Stipe* 40–60 × 4–5 mm, even, cylindrical, central; base not bulbous but occasionally swollen-angular; surface pruinose for entire length with a longitudinally striate appearance; veil absent; colour pallid to pale isabelline (4A3–4B3), lighter than the pileus; context solid. *Odour* variably strong or weak, sharp and sweet, or fruity-soapy (similar to orange extract), occasionally the sweet odour developing slowly after cut. Pl. 210, 211 (see also Pl. 2A).

Basidiospores 6.5–8.5 × 4.5–6.0 µm [mean 7.2 × 5.2 µm; Q (1.17–) 1.27–1.56; Q mean 1.39 (n = 36/3)], nodulose about a pentagonal to subtrapeziform outline, with 7–10 small obtusely and sharply conical nodules (to 2 µm tall), often with a small apical nodule, some spores thick-walled, yellowish brown in KOH; apiculus indistinct. *Basidia* 22–32 × 6–8 µm, narrowly clavate, hyaline, 4-sterigmate. *Pleurocystidia* 34–61 × 11–21 µm, frequent, subcylindrical to subfusiform, sessile at the base, at times rounded-truncate but embedded deeply within the hymenium; necks short or absent; apices obtuse and crystalliferous; walls 2–3 µm thick, hyaline or sometimes with ochraceous contents. *Cheilocystidia* 36–90 × 11–15 µm, subcylindrical, subfusiform, slender-lageniform, sessile or rounded at the base, sometimes ochraceous when collapsed, mixed with basidia. *Caulocystidia* descending the entire length or at least to the centre of the stipe, less frequent on the lower part, mostly lageniform with a tapered or sessile base, or cylindrical, and mixed with clusters of cauloparacystidia, at times the lower part of the stipe with caulocystidioid cells only. *Pileipellis* a cutis of cylindrical hyphae to 11 µm wide, yellowish brown in mass, faintly encrusted, thin-walled. *Clamp connections* present. Fig. 129.

Occurs in northern W.A., northern N.T. and eastern Qld in a broad range of tropical and subtropical vegetation types, including rainforest, eucalypt-sheoak woodland, *Melaleuca* swamp, and disturbed sites cleared of understorey under planted *Eucalyptus*. Fruiting Jan., Feb. and Apr.

Additional specimens examined: N.T.: Fogg Dam, E of Darwin, 12°33'55"S, 131°18'21"E, beside track on raised wall of dam, under *Melaleuca leucadendra* and *Acacia auriculiformis*, 24 Jan. 2014, *M.D.Barrett, T.Lebel, G.M.Bonito & C.N.Barrett MDB F90/14* (MEL, PERTH 08516863). Qld: Bunya Mountains Natl Park, 26°51'S, 151°34'E, in litter in eucalypt forest, 6 May 1989, *A.Young 1373* (BRI AQ808425); Paluma, Mount Spec area, 18.8717°S, 146.1250°E, under *Allocasuarina littoralis* and *Eucalyptus grandis*, 26 Apr. 1990, *N.L.Bougher E4283* (PERTH 07655886); Murray Ra., 5 km from Koombah toward Yarrabah along Yarrabah Rd, Cairns, on side of soil bank in forest dominated by *Allocasuarina littoralis* and *Acacia auriculiformis*, 20 Feb. 1993, *N.L.Bougher E4842* (PERTH 07660529); 6.5 km along Rifle Range Rd, S of Atherton, 17.2625°S, 145.4769°E, in grassy woodland dominated by *Allocasuarina littoralis*, *Syncarpia* and *Eucalyptus grandis*, 23 Feb. 1993, *N.L.Bougher E4867* (PERTH 07554192); 1.5 km from Irvinebank towards Herberton, 17°25'13"S, 145°13'13"E, in woodland of *Corymbia citriodora* and *Eucalyptus crebra* on stony soil, 800 m alt., 20 Feb. 2001, *P.I.Forster PIF26892* (BRI AQ669456); Irvinebank Cemetery, 17°26'11"S, 145°11'28"E, in mown area on stony soil under woodland of *Eucalyptus citriodora*, *E. crebra*, *E. cloeziana* and *Corymbia clarksoniana*, 740 m alt., *P.I.Forster, R.Booth & A.M. Young PIF26874* (BRI AQ 669472); State Forest 185 Danbulla, Lamb Ra., 17°06'57"S, 145°32'17"E, in woodland of *Eucalyptus crebra*, *E. tereticornis*, *E. portuensis* and *Lophostemon confertus* on granite with *Allocasuarina torulosa* dominant in midstorey, 760 m alt., 24 Feb. 2002, *P.I.Forster PIF28333* (BRI AQ 554024); Davies Creek Rd, Dinden Natl Park, 17°00'47.1"S, 145°32'17"E, under *Allocasuarina torulosa* and *Eucalyptus grandis* at woodland-rainforest interface, 20 Feb. 2012, *N.Fechner PBM3722* (PERTH 08362808, TENN 066958; KP171087-nLSU, KP641634-ITS, KM656131-*rpb2*); Davies Creek Rd, Dinden Natl Park, 17°01'58.5"S, 145°36'33.1"E, under *Allocasuarina torulosa* and *Eucalyptus grandis* at woodland-rainforest interface, 20 Feb. 2012, *P.B.Matheny PBM3725* (PERTH 08362823, TENN 066961); *loc. id.*, emerging from side of embankment under *Allocasuarina torulosa* and *Eucalyptus grandis*, 20 Feb. 2012, *N.L.Bougher & M.D.Barrett*

NLB919 (PERTH 08372977, TENN 067765); *loc. id.*, 24 Feb. 2012, *P.B.Matheny PBM3747* (PERTH 08363366, TENN 066948); Clohesy River Rd, at river crossing, 16°55'17.3"S, 145°35'10.6"E, soil on roadside in rainforest mixed with *Eucalyptus* under *Melaleuca leucadendra* and *Corymbia torelliana*, 23 Feb. 2012, *M.D.Barrett & P.B.Matheny PBM3736* (PERTH 08362920, TENN 066971); Mount Whitfield Conservation Park, N of Cairns, 16°53'53.8"S, 145°45'03.1"E, scattered singly on soil in *Melaleuca* swamp in lowland rainforest under *Melaleuca leucadendra*, *Lophostemon suaveolens*, *Corymbia tesselaris* and *Acacia mangium*, 25 Feb. 2012, *P.B.Matheny PBM3757* (TENN 066985); Mt Beerburrum, Glasshouse Mountains Natl Park, 26°57'29.7"S, 152°57'07.3"E, on soil in leaf litter in woodland under *Eucalyptus* spp. (including Black Butt) and *Allocasuarina*, 28 Feb. 2012, *N.Fechner PBM3765* (TENN 066997); Keith and Ivy Boon Memorial Park, corner of Grigor Rd and Maleny–Kenilworth road, E of Conondale Natl Park, 26°39'47"S, 152°27'22"E, scattered singly on soil in grass in park under natural and planted *Eucalyptus* (understorey cleared), 29 Feb. 2012, *P.B.Matheny PBM3771* (PERTH 08363161, TENN 067003); Rainforest Circuit Track, Cunninghams Gap, Main Range Natl Park, 28°02'59"S, 152°23'37.3"E, on soil in dry sclerophyll under *Eucalyptus* and *Allocasuarina*, 29 Feb. 2012, *R.E.Halling & N.Fechner PBM3776* (TENN 067008); Apple Tree Park, Springbrook Natl Park, 28°09'52.1"S, 153°15'37.6"E, 800 m alt., on soil under *Eucalyptus* spp. in wet sclerophyll forest, 2 Mar. 2012, *P.B.Matheny, N.L.Bougher & R.E.Halling PBM3779* (PERTH 08363099, TENN 067011; KP171088-nLSU, KP641635-ITS, KM656132-*rpb2*); Mount Cordeaux Track (near Rainforest Circuit), Cunninghams Gap, Main Range Natl Park, 28°02'59"S, 152°23'37.3"E, on soil among litter in tall sclerophyll forest including eucalypts and *Allocasuarina* sp., 3 Mar. 2012, *N.L.Bougher NLB973* (PERTH 08364796, TENN 067766); Queen Mary Falls, Main Range Natl Park, 28°20'29.5"S, 152°22'16.9"E, scattered singly on soil along bank of track in wet sclerophyll forest under *Eucalyptus*, 4 Mar. 2012, *P.B.Matheny & R.E.Halling PBM3789* (TENN 067021); Boonah–Killarney road, just before Main Range Natl Park (on Boonah side), 28°13'21.4"S, 152°30'51.6"E, on soil under large *Eucalyptus grandis* trees in understorey grazed by cattle, 4 Mar. 2012, *N.L.Bougher NLB982* (PERTH 08364842, TENN 067767).

Diagnostic features: Fruit bodies medium-sized. Pileus pale yellowish brown, campanulate and/or umbonate; stipe pruinose, lacking a distinctive bulb. Spores nodulose; hymenial cystidia sessile or rounded-truncate, deeply embedded in the subhymenium.

Bougher *et al.* (2012) described *I. torresiae* from northern Australia and suggested that it could be characterised by its sweet or citrine odour, noting that this odour was conspicuous in the Western Australian. holotype, but had not been detected in the only other specimens known at the time, *viz.* three collections from Queensland. Subsequently, numerous specimens have been collected in the Northern Territory and, especially, in Queensland where it is probably one of the most common species of *Inocybe*. Field notes of the more recent Queensland material indicate a variable intensity of odour, ranging from absent to weak or slowly becoming sweet after collection.

Inocybe torresiae is also remarkable for the peculiar type of hymenial cystidia that are mostly sessile, rounded-truncate at the base and firmly embedded within the subhymenium. This character, along with the small nodulose spores, unifies all collections cited above. Caulocystidia are present the entire length of the stipe or at least to the mid-region in all collections. Spores of the Queensland collections are slightly more angular with sharper nodules than those of the Western Australian type (compare Fig. 129 A & B).

Several different genotypes of *I. torresiae* have been observed in Australia (Fig. 128). The species occurs in a wide range of tropical and subtropical vegetation types, including rainforest and at the interface between rainforest and eucalypt woodland, in woodland dominated by eucalypts and sheoaks, and in *Melaleuca* swamps surrounded by lowland rainforest. *Inocybe torresiae* also occurs in disturbed sites, such as among grass in areas cleared of understorey under planted *Eucalyptus*.

Inocybe torresiae is most closely related to *I. viscata* from Tasmania and New Zealand, as well as undescribed and unclarified species from Thailand, Papua New Guinea and Costa Rica (Fig. 128).

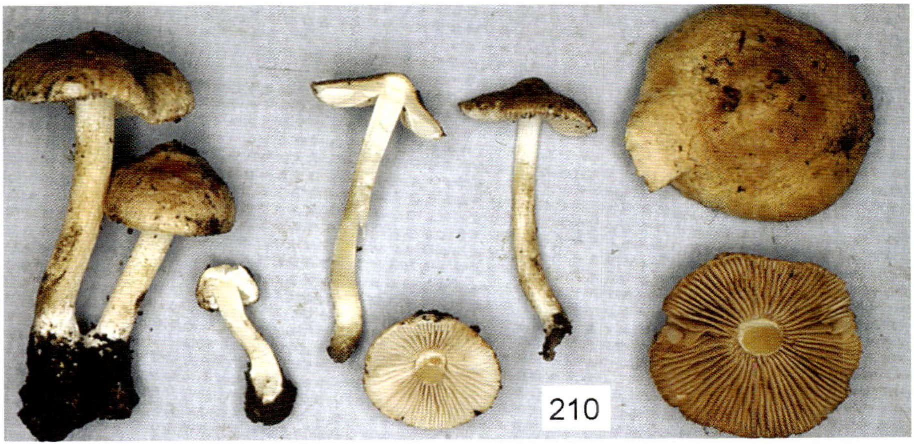

Plate 210. *Inocybe torresiae.* PERTH 07554192. Photograph — N.L.Bougher.

Plate 211. *Inocybe torresiae.* PERTH 08372977. Photograph — N.L.Bougher.

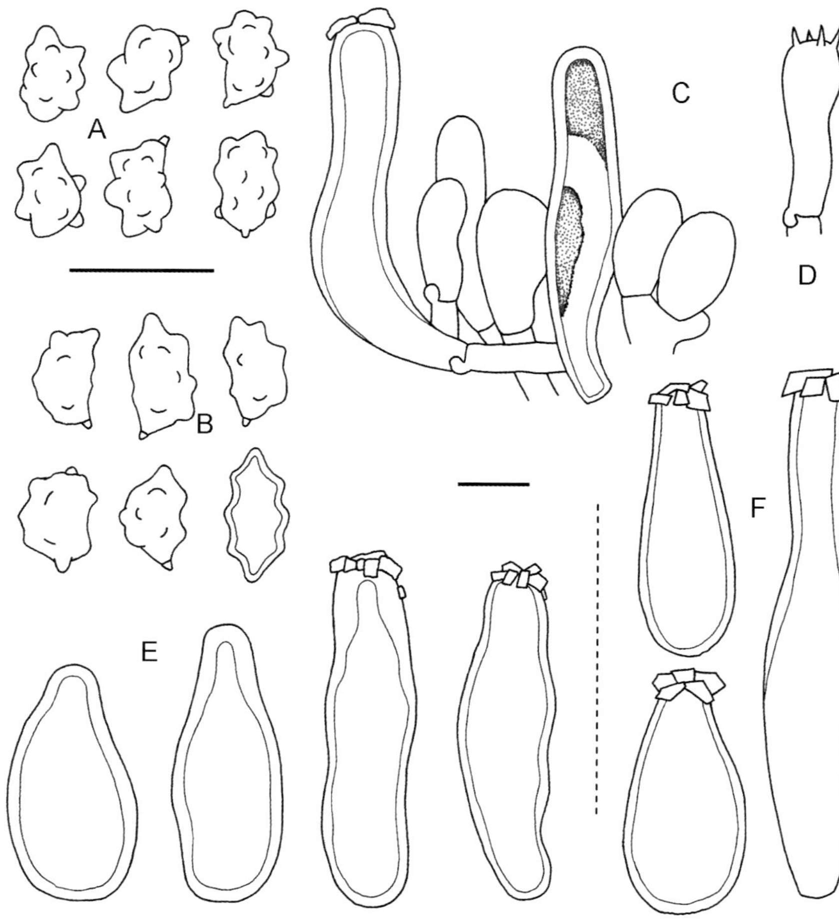

Figure 129. *Inocybe torresiae*. All holotype (PERTH 07678428) except B. **A,** Basidiospores; **B,** Basidiospores (PERTH 07655886); **C,** Caulocystidia and cauloparacystidia; **D,** Basidium; **E,** Pleurocystidia; **F,** Cheilocystidia. Scale bars = 10 µm (longer bar for spores only).

110. Inocybe viscata (E.Horak) Garrido, *Biblioth. Mycol.* 120: 177 (1988)

Astrosporina viscata E.Horak, *New Zealand J. Bot.* 15: 733 ('1978') [1977]. Type: L. Hochstetter, Buller area, West Coast, New Zealand, on ground under *Nothofagus cliffortioides*, 17 Mar. 1968, *E.Horak 68/177;* holo: PDD 27109 (GU233318-ITS).

Pileus 18–60 mm wide, obtusely conical to conico-campanulate, soon expanding to plano-convex to plane, often with a low obtuse umbo that can be pronounced; margin initially incurved, soon decurved and undulate upon expansion, becoming torn or split with age; surface lubricous-viscid (similar to species of *Hebeloma*), often with adhering particles of mud or soil; disc smooth and unbroken, radially fibrillose or rimulose towards the margin, not scaly; colour somewhat complex (almost bicolorous): brown on the disc and around the centre (10YR 4/3) but with a very pale brown margin (10YR 7/3), initially brown (10YR 5/4) at the centre and light yellowish brown to very pale brown (10YR 6/4 to 7/3) towards the margin; centre umber (7.5YR 4/4) with age with a yellowish brown (10YR 5/6) to light yellowish brown (10YR 6/4) margin; context dingy isabelline to pale brown, up to 6 mm thick; colour not changing where cut or bruised, not confluent with the stipe. *Lamellae* adnexed, very close, c. 48–62 L, with several tiers of lamellulae; pallid (2.5Y 8/4; Cartridge Buff) when immature, becoming light yellowish brown (2.5YR6/4) or light olive-brown (2.5Y 4/4) to brown (10YR 5/3) with age; edges pallid and indistinctly fimbriate, ventricose, to 5 mm deep. *Stipe* 35–80 × (3–) 4–10 mm at the apex, even or with a bulbous base, not marginate, more rounded bulbous 6–12 mm wide; terete to compressed; surface densely white pruinose along the upper ½ to ⅔, striatulate, white-fibrillose below to densely so towards the base, rather pallid throughout due to pruinose covering when young but with a light yellowish brown ground color that is more pronounced with age, becoming pale brown to brown (10YR 6/3 to 5/3, 7.5YR 4/4); context firm, solid, pallid to brownish particularly with age, but whitish at the extreme base, no colour changes observed where cut or bruised. *Odour* none. Pl. 212, 213.

Basidiospores 6.5–9.0 (–9.5) × 5.0–6.0 (–6.5) µm [mean 8.0 × 5.4 µm; Q (1.11–) 1.19–1.71 (-1.76); Q mean 1.47 (n = 42/2)], with 6–8 low or moderate-sized obtuse nodules, yellowish brown; apiculus typically not distinctive. *Basidia* 30–35 × 9–10 µm, clavate, hyaline, 4-sterigmate. *Pleurocystidia* 49–53 × 13–17 µm, fusiform, utriform or subcylindrical; upper part usually without a neck, often with a basal pedicel or narrowed lower part but the base mostly rounded (sometimes truncate), very thick-walled (walls to 6 µm), hyaline or pale yellowish. *Cheilocystidia* similar to pleurocystidia, mixed with narrowly clavate paracystidia resembling immature basidia. *Caulocystidia* not frequent, narrowly fusiform to irregularly cylindrical, larger than the pleurocystidia and cheilocystidia (to 80 µm long), at times ochraceous and collapsed, mixed with numerous clavate and variously-shaped cauloparacystidia, descending at least towards the middle region of the stipe; lower part with a vesture of interwoven hyphae or a gelatinized layer of smooth hyaline hyphae 3–5 µm wide similar to hyphae in the epicutis. *Pileipellis* 2-layered with a broad hyaline gelatinised epicutis to 200 µm thick, consisting of loosely interwoven hyaline cylindrical smooth (or faintly wrinkled) thin-walled hyphae 4–7 µm wide. Subcutis not gelatinised, consisting of more regularly arranged hyphae and subcellular elements to 12 µm wide, mostly thin-walled, at times filled with a tawny refractive substance. *Clamp connections* present. Fig. 130 (see also Pl. 12E).

Occurs on soil in wet sclerophyll forest in Tas., and in *Nothofagus* forests in New Zealand. Putative ectomycorrhizal hosts include *Eucalyptus* or *Acacia* and *Pomaderris* (Australia) and *Nothofagus* (New Zealand). Fruiting June–Nov. (Australia); Mar. and Apr. in New Zealand.

Additional specimens examined: Tas.: Truganini Reserve, Hobart, natural bushland, 10 Nov. 2004, *G.M.Gates 197* (TENN 065732,); *loc. id.*, 18 Nov. 2007, *G.M.Gates 1102* (TENN 066121, KP171096-nLSU, KP641647-ITS); Russell Falls Track, Mount Field Natl Park, on disturbed soil along roadside in wet sclerophyll forest under *Eucalyptus regnans*, *Acacia* and *Pomaderris*, 8 June 2009, *P.B.Matheny & D.Ratkowsky PBM3213* (TENN 065733; KP171097-nLSU; KP641648-ITS); *loc. id.*, buried deep in disturbed soil under *Eucalyptus regnans* and *Pomaderris*, 1 June 2010, *P.B.Matheny PBM3445* (PERTH 08516871, TENN 065734; JQ313570-nLSU, KP641649-ITS).

Diagnostic features: Fruit bodies robust. Pileus with a thick viscid pellicle resembling *Hebeloma* spp., (Pl. 12E) not at all scaly, occasionally bicolorous; disc pale brown to brown, margin very pale brown to pale brown; stipe pruinose at least over upper half to two-thirds, striatulate, even or bulbous at the base but not marginate, white, becoming brown with age. Spores nodulose; hymenial cystidia very thick-walled.

We have identified four Tasmanian collections as *I. viscata* based on ITS sequences (Fig. 128) and overall morphological similarity with the New Zealand type collection (PDD 27109). However, we note some variation in hymenial cystidial morphology, the type having more slender cystidia that are tapered towards the base. In contrast, cystidia of *PBM3445* from Tasmania are often more rounded than those of the holotype, but forms with rounded and truncate bases occur in both collections. This latter feature is typical of *I. torresiae*, closely related to *I. viscata* which occurs in tropical and subtropical Australia. We have also observed that Australian collections of *I. viscata* are not pruinose for the entire length of the stipe, unlike those described by Horak (1977). We examined the type and found a stipe that is densely pruinose at the apex but appears to be simply fibrillose on the lower half; indeed, it contains abundant caulocystidia and cauloparacystidia along the upper part of the stipe but only undifferentiated end cells on the lower part. These observations are consistent with what we have seen in Tasmanian material. ITS sequence similarity bears this out as the ITS of Tasmanian collections TENN 065733 and PERTH 08516871 are 99% similar to an ITS sequence produced from the holotype. At least one genetic variant (ITS 94% similar to the holotype) has been detected from Tasmania (*I102*; Fig. 128), and while we tentatively refer this collection to *I. viscata*, further investigations are warranted.

Plate 212. *Inocybe viscata*. PERTH 08516871. Photograph — P.B.Matheny.

Plate 213. *Inocybe viscata*. TENN 065733. Photograph — P.B.Matheny.

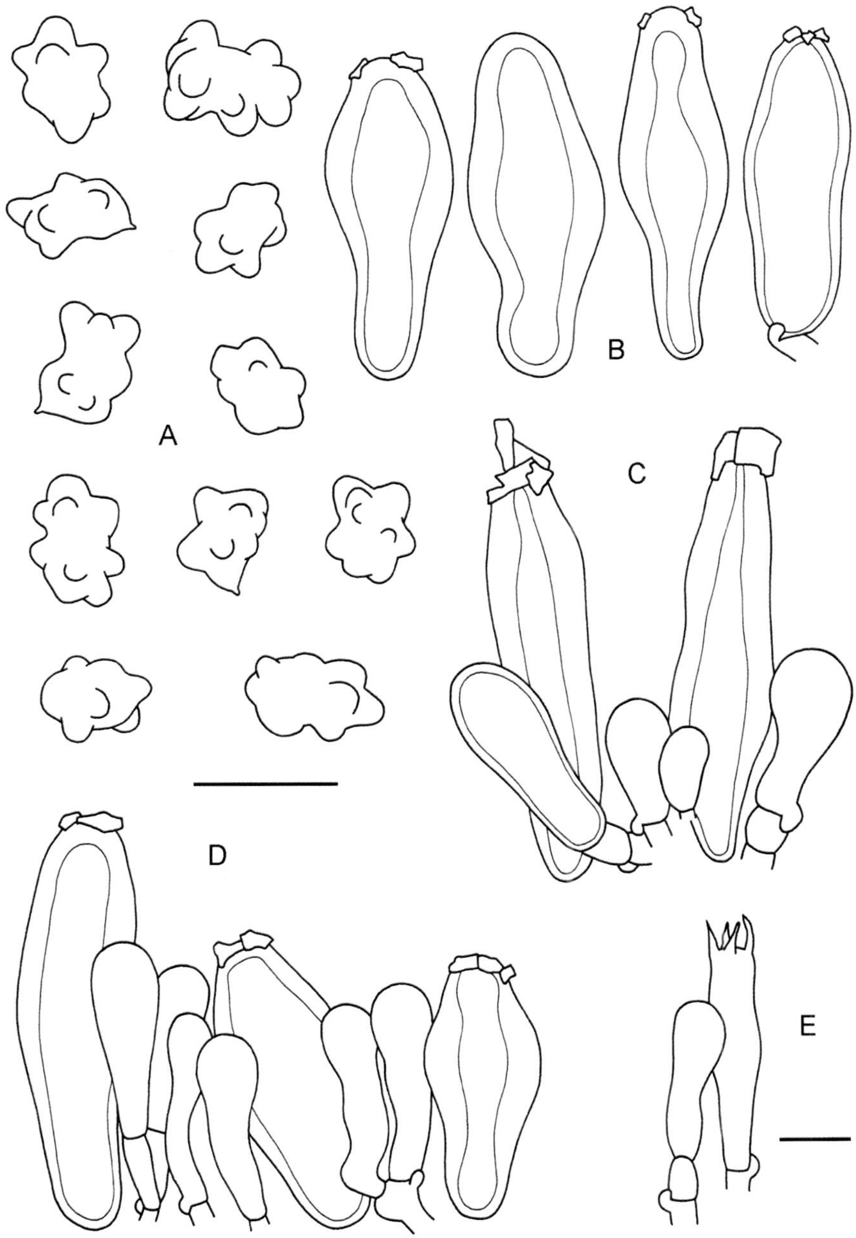

Figure 130. *Inocybe viscata.* All PERTH 08516871. **A,** Basidiospores; **B,** Pleurocystidia; **C,** Caulocystidia and cauloparacystidia; **D,** Cheilocystidia and paracystidia; **E,** Basidia. Scale bars = 10 µm (longer bar for spores only).

This is the first report of *I. viscata* from Australia where, as in New Zealand, it is apparently uncommon or infrequently collected. Presumably PDD 87560, labelled as *I. viscata* from New Zealand, is the collection referred to by Horak (1977) as "one additional collection, the identification of which is tenuous". Our examination of PDD 87560 shows it is probably

I. scissa, based on the very broad rounded bases of the cystidia and larger spores. If so, the holotype of *I. viscata* is the only known collection of this species from New Zealand.

Inocybe viscata appears to be a temperate species nested within a predominantly tropical clade closely related to species from tropical Australia, south-eastern Asia, Papua New Guinea and Costa Rica (Fig. 128).

111. **Inocybe bulbinella** Matheny & Bougher, *sp. nov.*

MycoBank: MB 809270

Type: T. & D.Pedro property, Sunnyglen Rd, Denmark, W.A., 34°57'41"S, 117°26'17"E, on grey sand near *Allocasuarina fraseriana*, *Eucalyptus marginata* and *Eucalyptus staeri*, 16 July 2011, *K.Syme PBM3630*; holo: PERTH 08319596; iso: TENN 066537 (KJ756449-nLSU, KJ756486-ITS, KJ756407-*rpb2*).

Pileus 15–42 mm wide, obtusely conical, expanding to plano-convex or ±plane; umbo absent or low and obtuse to small and subacute; margin decurved to straight, at times undulate and becoming raised; velipellis absent but when young with scattered thin whitish superficial fibrils, these densely accumulated at the margin when near the stipe; surface glossy, dry or at times with adhering grains of sand; disc smooth and unbroken, radially fibrillose, silky-fibrillose or coarsely tomentose towards the margin, rimulose or weakly rimose, not at all scaly; various shades of dark brown, dark yellowish brown (10YR 3/3, 3/4, 4/4 to 7.5YR 3/2), occasionally strong brown (7.5YR 4/4) in places, especially on the disc or dark brown over the disc and shading dark yellowish brown towards the margin, paler between the spreading fibrils; context white to pallid or pale brown, thin except under the disc, not changing colour upon exposure. *Lamellae* adnexed to sinuate, moderately close to close, 30–44 L with several tiers of lamellulae, ventricose, light grey or pale brown to brown, yellowish brown, and eventually dark yellowish brown (10YR 4/3, 4/4, 3/4); edges pallid but indistinctly fimbriate, 3–5 mm deep. *Stipe* 25–50 × 3–8 mm at the apex, enlarged below, typically with a rounded or somewhat angular bulbous base, not marginate; base 6–11 mm wide; cortina fugacious and usually not seen; upper half of immature stipe with a pruinose-like appearance but below with stark pallid vesture of superficial fibrils that wears away with age to reveal the ground colour; with age the apex with a pruinose appearance, at least along the upper quarter to one-third, otherwise finely striatulate, silky-fibrillose with a satiny appearance; ground colour light brown (7.5YR 7/4 to 6/4: Cinnamon-Buff to Cinnamon) to pale brown or brown (10YR 6/3 to 5/3) with age, white above the base due to accumulation of silky fibrils; context solid, light pinkish brown to brown and whitish streaked, white in the bulb. *Odour* absent or at most weakly acidulous. Pl. 214, 215.

Basidiospores 8.0–11.0 (–11.5) × 5.5–7.0 (–7.5) µm [mean 9.6 × 6.4 µm; Q 1.23–1.76 (–1.83); Q mean 1.50 (*n* = 92/4)], angular-nodulose to nodulose, rather variable, ranging from mostly angular with 6–8 small nodules or corners to distinctly nodulose with 8–12 moderate-sized projections, yellowish brown; apiculus small and not very distinct. *Basidia* 29–41 × 9–12 µm, 4-sterigmate, clavate, hyaline. *Pleurocystidia* 67–103 × 18–26 µm, typically fusiform-ventricose or lageniform with a broad ventricose portion near the centre or on the lower ½; necks at times long and cylindrical, occasionally utriform with a long distinct pedicel but without a distinct neck; apices obtuse or occasionally papillate, crystalliferous; mostly thick-walled (2–3 µm), hyaline. *Cheilocystidia* similar to pleurocystidia, occasionally shorter, mixed with paracystidia and basidia. *Caulocystidia* absent or rare and restricted to the apex, when present as metuloids similar to cheilocystidia, but cauloparacystidia not observed; upper part of stipe typically with scattered clusters of largely undifferentiated end cells, these infrequently resembling caulocystidioid hairs. *Pileipellis* bilayered, with a thin interwoven upper layer of cylindrical hyphae that are smooth, hyaline, thin-walled and 4–10 µm wide; lower layer yellowish brown, with cylindrical encrusted hyphae mostly 4–12 µm wide. *Clamp connections* present. Fig. 131.

Occurs in W.A. and Qld, on soil in natural vegetation and disturbed areas with planted trees. In north-eastern Qld it occurs at a rainforest-woodland interface under *Allocasuarina torulosa* and *Eucalyptus grandis*; in south-western W.A. it grows under species of *Corymbia*, *Eucalyptus* and *Acacia*. Fruiting Feb. (Qld) and July (W.A.).

Additional specimens examined: W.A.: Dickson Tower Rd, W of Manjimup, 34°12'49"S, 115°49'40"E, among moss in old growth Jarrah forest, 11 May 2006, *N.L.Bougher NLB186* (PERTH 07698798; JN975002-nLSU, KJ756488-ITS, KJ756409-*rpb2*); Walpole Town, 34°58'33"S, 116°44'01" E, scattered singly on sandy soil in median strip under planted *Corymbia ficifolia*, 14 July 2011, *N.L.Bougher & P.B.Matheny PBM3623* (PERTH 08319634, TENN 066459; KJ756451-nLSU, KJ756489-ITS, KJ756410-*rpb2*); *loc. id.*, *P.B.Matheny PBM3620* (TENN 066603); Jarrahdale Rd, 7.7 km W of Albany Hwy, 32°19'44"S, 116°07'43"E, scattered singly on sandy soil at edge of clearing under *Corymbia calophylla*, *Eucalyptus marginata* and planted *Acacia*, 25 July 2011, *P.B.Matheny PBM3669* (PERTH 08319189, TENN 066521); *loc. id.*, *N.L.Bougher NLB864* (PERTH 08320675). Qld: Dinden Natl Park, 17°00'47.1"S, 145°34'55.4"E, scattered singly on soil on bank of road under *Allocasuarina torulosa* and *Eucalyptus grandis* at rainforest margin, 20 Feb. 2012, *P.B.Matheny PBM3717* (PERTH 08362769, TENN 066953; KJ756450-nLSU, KJ756487-ITS, KJ756408-*rpb2*).

Etymology: From the Latin *bulbinellus* (a small swelling); in reference to the rounded to weakly angular bulb at the base of the stipe.

Diagnostic features: Fruit bodies medium-sized. Pileus dark brown to dark yellowish brown, rimulose; stipe cinnamon-brown with whitish superficial covering, with a pruinose-like appearance along the upper ⅓ to ½; base rounded bulbous to weakly angular; veil fugacious. Spores angular-nodulose; pleurocystidia large.

In the field, the stipe surface of *I. bulbinella* appears pruinose above the mid-region, but upon microscopic examination this appearance must be attributed primarily to clusters of thin-walled caulocystidioid hairs, not metuloids homologous with those on the lamellar surface.

Inocybe bulbinella can be confused with other dark brown or brown species with a bulbous stipe base such as *I. graveolens* (E.Horak) Garrido, *I. calopedes* Matheny & Bougher and allies, but those taxa have an entirely pruinose stipe imparted by the presence of distinct metuloid caulocystidia and a marginate basal bulb. Phylogenetically, *I. bulbinella* is most closely related to eastern North American material referred to as *I. fuligineoatra* Huijsman (*PBM2662*, TENN 062491; results not shown). The latter also has a dark brown pileus that is distinctly rimose but with a reddish brown, densely pruinose stipe. The pruina is caused by the presence of numerous metuloid caulocystidia and cauloparacystidia descending the entire length of the stipe.

Inocybe bulbinella has thus far been found in Australia under *Allocasuarina*, *Corymbia* and *Eucalyptus* in north-eastern Queensland and south-western Western Australia. However, we have noted variation in spore morphology, e.g. the spores of the type specimen are more angular and have fewer nodules than material from Queensland. Moreover, the pleurocystidia of the latter, although fusiform and thick-walled, are not as long and wide as in the type. However, DNA sequences from these collections form a robust monophyletic group with minimal genetic divergence.

Plate 214. *Inocybe bulbinella*, holotype (PERTH 08319596). Photograph — N.L.Bougher.

Plate 215. *Inocybe bulbinella*. PERTH 08319189. Photograph — P.B.Matheny.

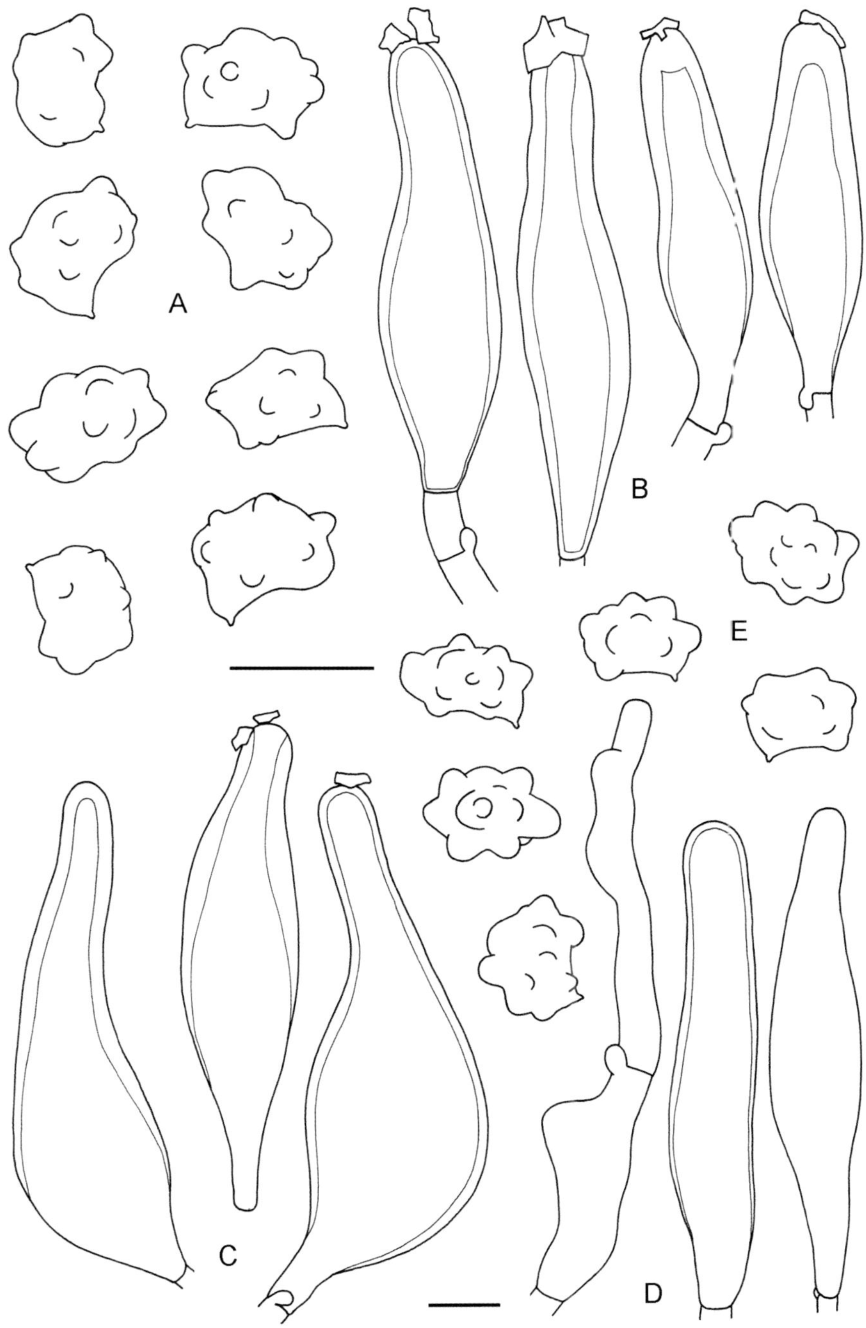

Figure 131. *Inocybe bulbinella.* All holotype (PERTH 08319596), except E (PERTH 08362769). **A,** Basidiospores; **B,** Cheilocystidia; **C,** Pleurocystidia; **D,** Caulocystidioid hair and caulocystidia from stipe apex; **E,** Basidiospores. Scale bars = 10 μm (longer bar for spores only).

112. Inocybe brunneidisca Matheny, Bougher & G.M.Gates, *sp. nov.*

MycoBank: MB 809269

Type: Lady Barron Track, Mount Field Natl Park, Tas., 42°40'S, 146°43'E, scattered singly on soil in wet sclerophyll forest under *Eucalyptus regnans*, *Acacia* and *Pomaderris*, 8 June 2009, *G.M.Gates, D.Ratkowsky & P.B.Matheny PBM3220*; holo: PERTH 08516960; iso: TENN 065798 (JN974959-nLSU; KJ756480-ITS).

Pileus 15–32 mm wide, conical to broadly convex to plano-convex or plane; umbo absent or low and obtuse; margin decurved to straight, upturned with age; surface dry or occasionally moist, dull; velipellis absent; at times with adhering clay particles; centre smooth and unbroken, radially fibrillose towards the margin, not rimose, but occasionally fibrils spreading towards the sometimes torn margin; centre dark brown (7.5YR 4/2, 4/3, 3/2; Sepia or Saccardo's Umber) when young, later very dark greyish brown (10YR 3/2) to dark brown or dark greyish brown (10YR 3/3 to 4/2), fibrils spreading towards the margin giving way to (light) yellowish brown (10YR 6/4 to 5/4) or pale brown to very pale brown (10YR 6/3 to 7/3) ground color, uniformly brown (Tawny Olive to Saccardo's Umber) when young; context watery pale brown to pale greyish brown, several millimeters thick under the disc; colour unchanging where cut. *Lamellae* adnexed, moderately close to very close, c. 40–60 L, with several tiers of lamellulae; light grey (10YR 7/1 to 7/2), very pale brown (10YR 7/3) to brown (10YR 5/3) or clay coloured with age; edges pallid but not distinctly fimbriate, ventricose, 4–5 mm deep. *Stipe* 25–50 × 3–5 mm at apex, terete; base swollen or somewhat bulbous to bulbous, in some cases distinctly marginate and up to 9 mm wide; base typically 6–8 mm wide; cortina absent; white pruinose for its entire length, with a satiny fibrillose-streaked appearance, white or pallid throughout or with a pale yellow (2.5YR 8/4) ground colour; context solid, lacking pink tinges, pallid, nearly white above the bulb. *Odour* none. Pl. 216–218.

Basidiospores (8.0–) 8.5–10.5 (–11.5) × 6.5–7.5 (–8.0) µm [mean 9.8 × 7.2 µm; Q 1.14–1.64; Q mean 1.37 (*n* = 42/3)], mostly with 8–13 conspicuous to moderate-sized nodules, yellowish brown. *Basidia* 31–32 × 10.5–11.5 µm, 4-sterigmate, sphaeropedunculate, especially when young to slender clavate, hyaline. *Pleurocystidia* 62–86 × 14–20 µm, fusiform, cylindrical or ventricose, rarely saccate, necks not delineated, merely tapered at both ends; apices obtuse and crystalliferous, lacking a distinct pedicel; thick-walled (2–3 (–5) µm), hyaline; moderately frequent. *Cheilocystidia* similar to pleurocystidia but shorter and more variable in shape, including cells that are lageniform in outline, at times not as thick-walled, mixed with numerous hyaline paracystidia. *Caulocystidia* similar to cheilocystidia, mixed with cauloparacystidia and descending the entire length of the stipe in dense clusters. *Pileipellis* a cutis of cylindrical to inflated hyphae mostly 7–24 µm wide, encrusted, yellowish brown to cinnamon-brown in mass. *Clamp connections* present. Fig. 132 (see also Pl. 6C).

Occurs in south-western W.A. and Tas.; on soil in wet sclerophyll forest under *Eucalyptus regnans* or *E. obliqua* mixed with *Acacia* and *Pomaderris*. Found in W.A. among litter in open woodland with *Eucalyptus rudis*, *Melaleuca* and *Kunzea*. Fruiting May–July.

Additional specimens examined: W.A.: Modong Nature Reserve, Oakford, among litter in open woodland of *Eucalyptus rudis*, *Melaleuca* and *Kunzea*, 3 July 2007, *N.L.Bougher & J.Keeble E9023* (PERTH 08094217; KJ756448-LSU, KJ756485-ITS); *loc. id.*, *N.L.Bougher & P.Robertson E9024* (PERTH 08094225); *loc. id.*, *N.L.Bougher & P.Davison E9025* (PERTH 08094233). Tas.: type locality, 8 June 2009, *P.B.Matheny, G.M.Gates & D.Ratkowsky, PBM3214* (TENN 065797; KJ756444-LSU, KJ756481-ITS); Halls Falls Track, outside Pyengana, scattered singly on soil under *Acacia*, *Eucalyptus*, *Pomaderris* and *Zieria arborescens*, 21 May 2010, *G.M.Gates, D.Ratkowsky & P.B.Matheny PBM3364* (TENN 065788; KJ756445-nLSU, KJ756482-ITS); *loc. id.*, under *Eucalyptus obliqua* and *Acacia*, *P.B.Matheny & F.Karstedt PBM3365* (TENN065799; KJ756446-LSU, KJ756483-ITS, KJ756405-*rpb2*); *loc. id.*, scattered singly on soil under *Eucalyptus obliqua*, *Acacia* and *Pomaderris*, 23 May 2010, *P.B.Matheny, F.Karstedt & M.Noordeloos PBM3402* (TENN 065800; KJ756447-LSU, KJ756484-ITS, KJ756406-*rpb2*).

Etymology: From the Latin *brunneus* (brown) and *discus* (a circular plate); in reference to the brown disc of the pileus.

216

217

218

Plate 216. *Inocybe brunneidisca*, holotype (PERTH 08516960). Photograph — P.B.Matheny.

Plate 217. *Inocybe brunneidisca*. TENN 065788. Photograph — P.B.Matheny.

Plate 218. *Inocybe brunneidisca*. PERTH 08094225. Photograph — N.L.Bougher.

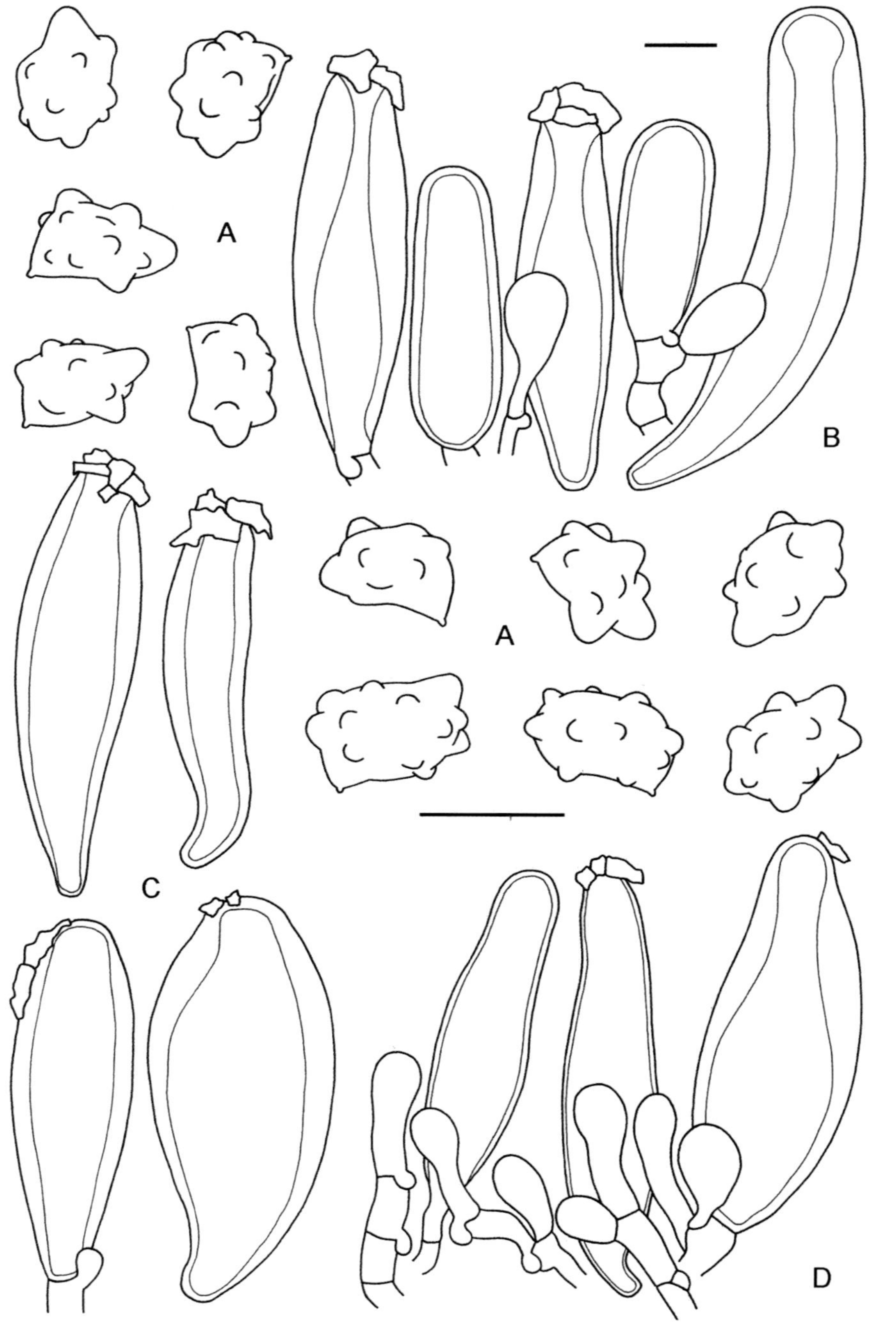

Figure 132. *Inocybe brunneidisca.* All holotype (PERTH 08516960). **A,** Basidiospores; **B,** Cheilocystidia and paracystidia; **C,** Pleurocystidia; **D,** Caulocystidia and cauloparacystidia. Scale bars = 10 µm (longer bar for spores only).

Diagnostic features: Fruit bodies medium-sized. Pileus bicolorous, with a dark brown to brown centre and brown streaks of fibrils against a lighter background; stipe white, entirely pruinose with a somewhat bulbous to marginately bulbous base. Spores with numerous prominent to moderate-sized nodules; pleurocystidia fusiform and thick-walled.

This combination of characteristics is not seen in any known Australian or New Zealand species. Two of the five Tasmanian collections, as well as the Western Australian material, have a distinct marginate bulb, but this trait appears rather variable.

Inocybe brunneidisca may be widely spread in southern Australia as indicated by the current records in Tasmania and Western Australia. However, genetic divergence at the ITS and LSU loci was detected between eastern and western populations of this species and among the Tasmanian populations from Mount Field Natl Park at the centre of the island and Halls Falls in the north-east. Morphologically, the western collections can have broader cauloparacystidia, more abundant paracystidia on the gill edge, and the hymenial cystidia can be more slender than collections from Tasmania. More obviously, the Western Australian collections are not as distinctly bicolorous (see also *in situ* images of PERTH 08094217, 08094225 and 08094233 in Bougher *et al.* (2007d), as *Inocybe* sp.; specimen ID: *3174*, *3179* and *3193*). The Western Australian specimen (*E9023*) sequenced here exhibits greater genetic variation at ITS and LSU loci than is visible among the Tasmanian collections, which may reflect the morphological variation, and sequences of additional collections from additional western populations are needed to determine the extent of their genetic isolation. Presently, we group all specimens together as a broad morphological species. The overall phylogenetic position of *I. brunneidisca* is not resolved in nLSU and *rpb2* gene tree family-level analyses (results not shown). Thus, it can be considered at present as *incertae sedis* within the *Inocybe* clade.

113. **Inocybe humidicola** Matheny & Bougher, *sp. nov.*

MycoBank: MB 809379

Type: Dinden Natl Park, Qld, 17°00'47.1"S, 145°34'55.4"E, on wet disturbed soil at rainforest interface under *Eucalyptus grandis*, 20 Feb. 2012, *P.B.Matheny, M.D.Barrett & N.L.Bougher PBM3720*; holo: PERTH 08362785; iso: TENN 066956 (KJ801180-nLSU, KP171125-ITS, KJ811574-*rpb2*).

Pileus 6–25 mm wide, conical, obtusely conical to plano-convex, often with a small subacute or low obtuse umbo; margin decurved, becoming torn; surface dry; velipellis absent; smooth and unbroken at the centre, finely fibrillose towards the margin, not rimose, not striate, becoming slightly excoriate and breaking up into small ±appressed scales around the disc; centre dark brown (10YR 3/3) shading to dark yellowish brown (10YR 4/4); umbo at times light brown and paler than rest of surface as though hygrophanous; context very thin, c. 1 mm thick under the disc, colour not changing upon exposure. *Lamellae* adnexed, moderately close, 26–30 L, with several tiers of lamellulae, light brown, brown to yellowish brown (10YR 5/4), ventricose, broad, 3–5 mm deep, with edges pallid but indistinctly fimbriate. *Stipe* 15–36 × 1–2 mm at the apex, terete, even, with base slightly swollen to swollen and 2–5 mm wide; cortina fugacious, upper ⅓ to upper ½ pruinose, fibrillose to subfloccose-fibrillose below, when young pallid throughout with a dull yellow tinge or Cartridge Buff, becoming light brown to light yellowish brown above, dark brown at and/or above the base; context pallid in the interior; cortex coloured like the surface, solid, becoming hollow. *Odour* not remarkable. Pl. 219, 220.

Basidiospores 7.5–10.0 × 5.5–7.5 µm [mean 8.8 × 6.7 µm; Q 1.21–1.50; Q mean 1.33 (n = 35/3)], usually with 8–10 obtusely conical nodules to 2 µm tall, yellowish brown. *Basidia* 26–30 × 8–9 µm, clavate, hyaline, 4-sterigmate. *Pleurocystidia* 46–72 × 10–22 µm, fusiform to fusiform-ventricose, typically with obtuse to rounded apices that are bare or crystalliferous, with or without a basal pedicel, thin-walled or at most with a slight apical thickening or plug, hyaline, not frequent. *Cheilocystidia* short-saccate, clavate or utriform, thin-walled, mixed with basidiole-like paracystidia and basidia. *Caulocystidia* similar to pleurocystidia but more narrowly fusiform, irregularly cylindrical or cylindrical, among crowded clavate and variously shaped cauloparacystidia, restricted to the stipe apex, absent towards the centre. *Pileipellis* a cutis of cylindrical repent hyphae mostly 5–12 µm wide, encrusted, dark yellowish brown in mass. *Clamp connections* present. Fig. 133.

Occurs in north-eastern Qld on wet disturbed soil in seasonally inundated *Eucalyptus grandis* forest at interface with rainforest; also in mixed rainforest. Fruiting Feb.

Additional specimens examined: Qld: type locality, on wet disturbed soil at rainforest interface under *Eucalyptus grandis*, 20 Feb. 2012, *P.B.Matheny & M.D.Barrett PBM3719* (PERTH 08362777, TENN 066955; KJ801181-nLSU, KP171126-ITS, KJ811575-*rpb2*); *loc. id.*, scattered singly on wet disturbed soil under *Eucalyptus grandis* at rainforest interface, 20 Feb. 2012, *P.B.Matheny & M.D.Barrett PBM3724* (PERTH 08362815, TENN 066960; KJ801182-nLSU, KP171127-ITS); *loc. id.*, 20 Feb. 2012, *P.B.Matheny & M.D.Barrett PBM3727* (PERTH 08362858, TENN 066963; KJ801183-nLSU, KP171128-ITS); Windy Hollow Rd, Kuranda, 16°50'55.1"S, 145°36'29.3"E, solitary on soil along roadside in mixed rainforest under *Acacia celsa*, *Corymbia torelliana* and rough-barked eucalypts, 23 Feb. 2012, *P.B.Matheny, M.D.Barrett & N.L.Bougher PBM3737A* (PERTH 08362939, TENN 066945; mixed collection); Davies Creek Rd, Dinden Natl Park, 17°01'36.1"S, 145°36'04.3"E, scattered singly on soil along road bank at woodland-rainforest intersection under *Eucalyptus grandis*, *Allocasuarina torulosa* and other eucalypts, 24 Feb. 2012, *M.D.Barrett PBM3744* (PERTH 08363447, TENN 066976; (KJ801184-nLSU, KP171129-ITS, KJ801184-*rpb2*).

Etymology: From the Latin *humidicola* (dweller in moist or damp areas), in reference to the inundated or wet habitat favoured by this species.

Diagnostic features: Fruit bodies small. Pileus dark brown to brown, finely fibrillose, at most excoriate, not scaly; stipe pruinose at the apex only. Spores typically with 8–10 obtuse conical nodules; hymenial cystidia thin-walled. On inundated disturbed soil under eucalypts and/or *Acacia*.

Inocybe humidicola is a small, dull coloured species that lacks outstanding macroscopic features. Microscopically, it can be best recognised by the combination of thin-walled, fusiform to fusiform-ventricose pleurocystidia, caulocystidia confined to the stipe apex, and spores with 8–10 obtuse conical nodules. It appears to have a rather restricted geographic distribution in north-eastern Queensland in mixed rainforest or at the edges of rainforest under eucalypts such as *Eucalyptus grandis* as well as *Corymbia torelliana*, or *Acacia*. Collections should be inspected carefully to detect superficially similar, heterogeneous assemblages of taxa (e.g. PERTH 8362939).

Horak (1979) described two nodulose-spored species similar to *I. humidicola*, viz. *I. incognita* (E.Horak) Garrido, which has a minutely squamulose-squarrose pileus, a strong fruity odour, metuloid (thick-walled) pleurocystidia, and an association with *Nothofagus* in Papua New Guinea. *Inocybe cingulatipes* (E.Horak) Garrido has a squarrose disc and lacks pleurocystidia; it was described from Singapore.

Inocybe humidicola occupies an unresolved position within the *Inocybe* clade based on inspection of nLSU and *rpb2* gene trees (results not shown). At present, it is best regarded as *incertae sedis* within *Inocybe.*

Plate 219. *Inocybe humidicola*, holotype (PERTH 08362785). Photograph — P.B.Matheny.

Plate 220. *Inocybe humidicola*. PERTH 08362777. Photograph — P.B.Matheny.

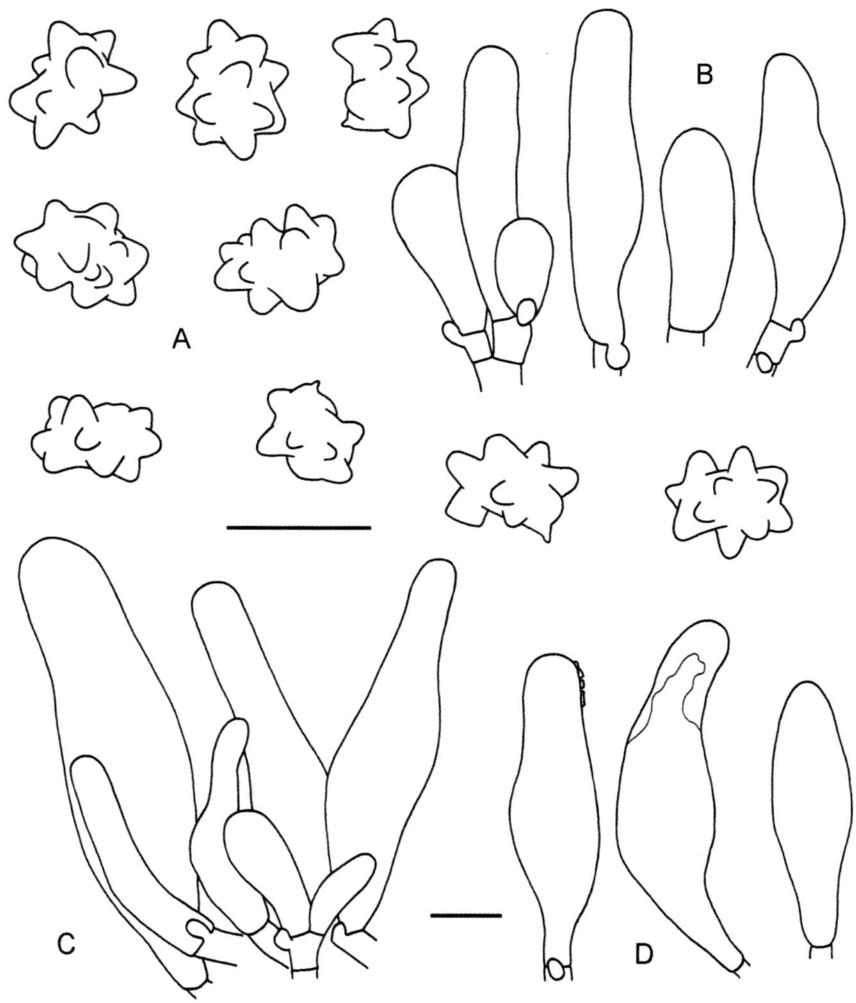

Figure 133. *Inocybe humidicola.* All holotype (PERTH 08362785). **A,** Basidiospores; **B,** Cheilocystidia and paracystidia; **C,** Caulocystidia and cauloparacystidia; **D,** Pleurocystidia. Scale bars = 10 μm (longer bar for spores only).

INOCYBE PARVIFULVA GROUP

This is a small suite of species from Queensland, Papua New Guinea, and New Zealand. *Inocybe parvifulva*, a newly described species, is sister to the New Guinean species *I. papillata* with strong support. The species from New Zealand also appears closely related to this group and is tentatively identified as *I.* aff. *manukanea*. All three of these species share the presence of distinct saccate (thin- or thick-walled) caulocystidia and nodulose spores.

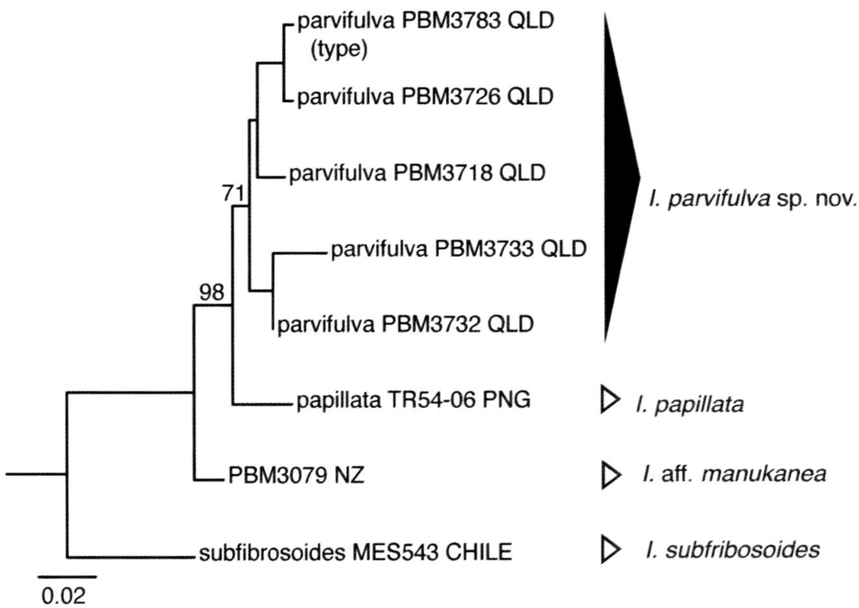

Figure 134. Maximum Likelihood phylogram of the *Inocybe parvifulva* group. Bootstrap values are shown above branches. The tree was inferred from combined ITS, nLSU, and *rpb2* gene regions and includes 2867 total sites. Species from Australia are indicated by black-filled triangles. Non-Australian species are shown as white triangles. The tree is rooted with *I. subfibrosoides*, which appears as sister group to the *I. parvifulva* group in individual family-level nLSU and *rpb2* gene trees (data not shown). Eight branch tips are included, all of which are represented by nLSU, six by ITS, and five by *rpb2*.

114. **Inocybe parvifulva** Matheny & Bougher, *sp. nov.*

MycoBank: MB 809428

Type: Rainforest Circuit, Cunninghams Gap, Main Range Natl Park, Queensland, 28°02’59.0”S, 152°23’37.3”E, singly or in pairs on soil along bank of track in warm temperate rainforest under *Lophostemon confertum*, 3 Mar. 2012, *P.B.Matheny & N.Fechner PBM3783*; holo: PERTH 08363064; iso: TENN 067015 (KP170972-nLSU, KP308805-ITS).

Pileus 6–12 mm wide, conical, plano-convex to plane; centre often with a small nipple-like or obtuse umbo; margin decurved, uplifted with age, translucent-striate when moist and fresh towards the margin; surface dry; velipellis absent; generally smooth and unbroken at the centre but occasionally with small floccose scales or minutely appressed-scaly around the disc, fibrillose towards the margin (or with floccose scales when young but these soon disappearing), not rimose but can be weakly rimulose or torn at the margin, becoming excoriate on the disc with age; edge of margin with sparse pallid fibrils; fulvous at the centre (10YR 5/4 to 5/8; Buckthorn Brown) or brownish yellow to yellowish brown (10YR 6/6 to 5/6), shading to light yellowish brown (10YR 6/4) or almost yellow or light olive-brown (2.5Y 5/6) towards the margin, or fulvous throughout (10YR 5/8) with age, not hygrophanous; floccose fibrils (when present) pallid in contrast to the ground colour; context very thin except under the disc, white or pallid, not changing colour upon exposure. *Lamellae* adnexed, subdistant to moderately close, 16–30 L, with few to several tiers of lamellulae, broad, ventricose and broad for the small size; pale brown, light brown to yellowish brown to dark yellowish brown with age (10YR 6/3 to 4/4), olive tinges absent; edges indistinctly pallid-fimbriate. *Stipe* 10–30 × 1–2 mm, terete, even, at times flexuous; base occasionally slightly swollen, at times slightly flexuous; surface dry; veil (cortina) observed only when very young on buttons, pruinose at the extreme apex if at all, with scattered floccose or indistinct fibrils below, these wearing away with age, appearing finely-fibrillose with age, with a whitish fibrillose vesture at the base or just above it; pallid to light buff, pale ochraceous-buff or pale brown; context solid, becoming hollow, white. *Odour* none. Pl. 221, 222.

Basidiospores 7.5–10.0 × (5.0–) 5.5–7.5 (–8.0) µm [mean 8.5 × 6.5 µm; Q 1.13–1.56 (–1.64); Q mean 1.32 (n = 64/4)], coarsely nodulose with mostly 9–12 obtusely conical nodules that project 1.5–2.5 µm; apiculus small, tapering but distinctive, yellowish brown. *Basidia* 18–33 × 9–12 µm, broadly clavate, hyaline, mostly 4-sterigmate, rarely 2-sterigmate. *Pleurocystidia* 35–55 × 12–25 µm, mostly fusiform or broadly so or ventricose, occasionally subcylindrical; apices obtuse, crystalliferous, thick-walled (mostly (1.5–) 2.0–4.0 µm), hyaline. *Cheilocystidia* similar to pleurocystidia but usually shorter and more variable in shape, broadly clavate, saccate, ventricose or broadly lageniform, some thin-walled, mixed with short clavate hyaline paracystidia. *Caulocystidia* 19–60 × 16–31 µm, in dense clusters at the stipe apex, at times descending to approximately the middle of the stipe, occasionally observed on the lower ½ of the stipe but much more infrequent there, absent above the base and lower ¼, in all cases, variable in shape like cheilocystidia but many sphaeropedunculate, saccate, broadly clavate or cylindrical; often thin-walled and hyaline but metuloids and cauloparacystidia observed at the stipe apex. *Pileipellis* a cutis of cylindrical to inflated hyphae mostly 7–24 µm wide, ochre-tawny in mass, ±thin-walled, weakly encrusted. *Clamp connections* present. Fig. 135.

Occurs in eastern Qld; in warm-temperate rainforest with *Lophostemon*, in mixed rainforest with *Allocasuarina*, *Acacia*, *Eucalyptus*, or edge of rainforest with *Eucalyptus grandis*. Fruiting Jan.–Mar.

Additional specimens examined: Qld: Davies Creek Rd, Dinden Natl Park, 17°00'47.1"S, 145°34'55.4"E, on muddy soil at interface of *Eucalyptus grandis*, *Allocasuarina torulosa* and mixed rainforest, 20 Feb. 2012, *N.L.Bougher NLB916* (PERTH 08372950, TENN 069574); Rainforest Circuit, Cunninghams Gap, Main Range Natl Park, 28°02'59.0"S, 152°23'37.3"E, on soil among litter in warm temperate rainforest including *Lophostemon confertus* (no eucalypts nearby), 3 Mar. 2012, *N.L.Bougher NLB975* (PERTH 08364826, TENN 069568); Dinden Natl Park, 17°00'47.1"S, 145°34'55.4"E, on soil in highly disturbed area (by pigs) under *Eucalyptus grandis* near rainforest edge, 20 Feb. 2012, *P.B.Matheny PBM3718* (TENN 066954; KP170973-nLSU, KP308806-ITS, KM406212-*rpb2*); Black Mountain Rd, near Kuranda, 16°47'07.8"S, 145°37'32.1"E, scattered singly on soil along road bank in mixed rainforest under *Allocasuarina littoralis*, *Acacia mangium* and *Eucalyptus pellita*, 23 Feb. 2012, *P.B.Matheny*, *M.D.Barrett* & *N.L.Bougher PBM3732* (PERTH 08362890, TENN 066967; KP171087-nLSU, KP641634-ITS, KM656131-*rpb2*); *loc. id.*, *PBM3733* (PERTH 08362904, TENN 066968 (PERTH 08362904, TENN 066968; KP170976-nLSU, KP308807-ITS).

Etymology: From the Latin *parvus* (small) and *fulvus* (tawny); in reference to the small fruit bodies and the yellowish brown to brownish yellow pileus.

Diagnostic features: Fruit bodies small. Pileus fulvous, finely floccose-fibrillose or squamulose but sparingly so; stipe even, not bulbous, pruinose above but not below the middle. Spores to 10 µm long, with mostly 9–12 obtusely conical coarse nodules; hymenial cystidia mostly broadly fusiform or ventricose, thick-walled, to 60 µm long; caulocystidia composed of distinctive saccate cells.

Plate 221. *Inocybe parvifulva*, holotype (PERTH 08363064). Photograph — P.B.Matheny.

Plate 222. *Inocybe parvifulva*. PERTH 08364826. Photograph — N.L.Bougher.

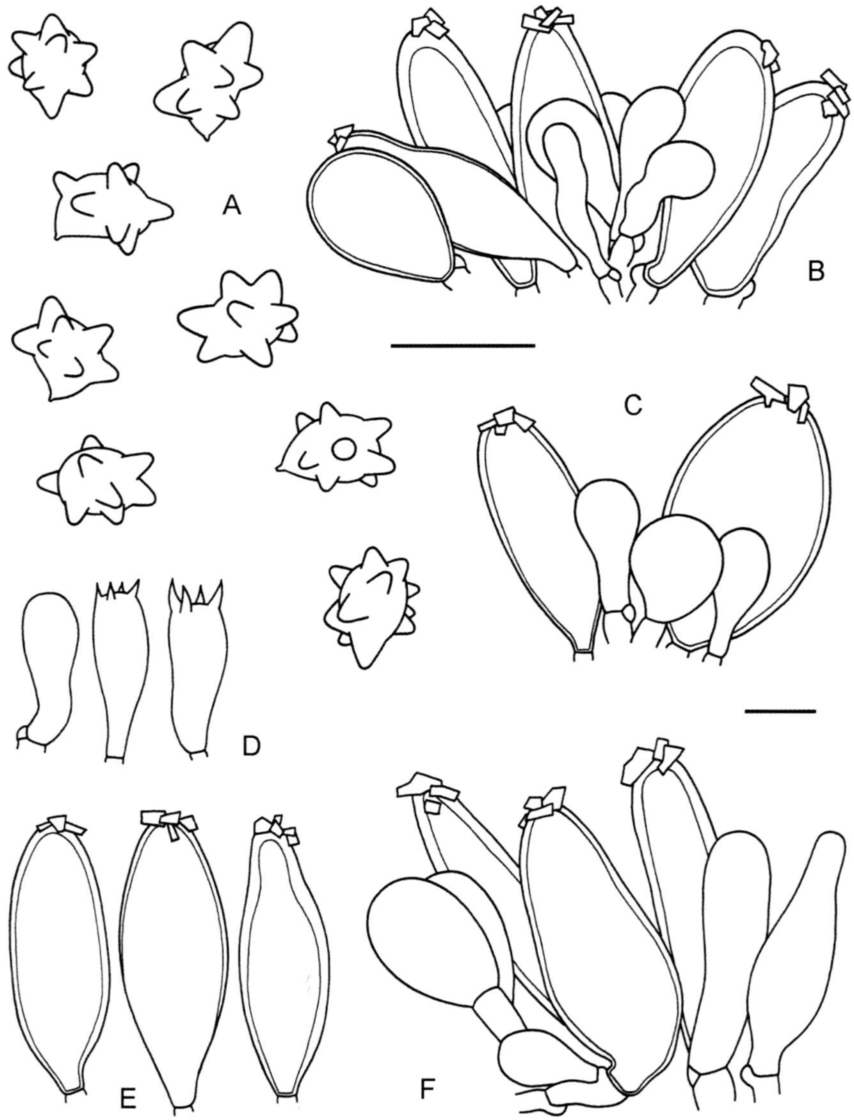

Figure 135. *Inocybe parvifulva.* All holotype (PERTH 08363064) except C (PERTH 08364826). **A,** Basidiospores; **B, C,** Cheilocystidia and paracystidia; **D,** Basidia; **E,** Pleurocystidia; **F,** Caulocystidia and various thin-walled elements from upper stipe. Scale bars = 10 µm (longer bar for spores only).

Inocybe parvifulva is a small species from eastern Queensland with coarsely nodulose spores. The pileus at most is appressed-scaly or finely floccose-fibrillose but never squarrose. Perhaps its most unusual feature is the presence of saccate to sphaeropedunculate, thin-walled caulocystidia at the apex of the stipe; these can be mixed with cylindrical terminal cells. Sequences from multiple collections are somewhat different, but all form a robustly supported single clade (Fig. 134).

Inocybe parvifulva appears to be closely related to two collections from Papua New Guinea and New Zealand (Fig. 134; collection details below). The New Guinea material, which we refer to as *I. papillata* (E.Horak) Garrido, is broadly similar to *I. parvifulva*, but it occurs with *Castanopsis*-*Lithocarpus* and has larger spores: mostly 9.0–12.0 × (5.0–) 5.5–8.5 (–9.0) µm (Horak, 1979a). Like *I. parvifulva* and the unresolved taxon from New Zealand, all three entities share the presence of thin- or thick-walled saccate caulocystidia. The New Zealand collection, a single specimen split between TENN and PDD, has spores like the Australian material (smaller than *I. papillata*) and it lacks a scaly pileus. It may be close phylogenetically to *I. manukanea* (E.Horak) Garrido, but it differs in the coarsely fibrillose (not scaly) pileus and the conspicuous saccate or broadly ventricose caulocystidia, which were depicted as cylindrical to subfusiform in *I. manukanea* (Horak, 1977).

Details of the *I. papillata* and *I. aff. manukanea* collections examined are as follows:

New Zealand: North Island, Waitakere Visitor Centre, just W of Auckland, 36°56'49.2"S, 174°36' 23.04"E, on nearby track, solitary on soil under *Kunzea ericoides*, 6 May 2009, *P.B.Matheny PBM3079* (PDD 97914, TENN 063656; JN974924; *I. aff. manukanea*). Papua New Guinea: Mt Gahavisuka, near carpark, in *Castanopsis* forest, 17 Jan. 2006, *R.Treu TR54-06* (M 0138285; *I. papillata*; KP170971-nLSU, KP308804-ITS).

MISCELLANEOUS GROUPS

(INOCYBE) PSEUDOSPERMA CLADE

Inocybe sect. *Rimosae* (Fr.) Sacc., *Fl. Ital. Crypt.* I, 15: 728 (1916)

Type: *I. rimosa* (Bull. : Fr.) P. Kumm.

Agaricus III. *Rimosae* Fr., *Epicrisis*: 174 (1838)

Inocybe 3. *Rimosae* (Fr.) Quél. in *Mém. Soc. Emul. Montébeliard, sér.* II. 5: 180 (1872)

Pileus fibrillose, often rimose, rarely scaly. *Stipe* scurfy fibrillose, pruinose-furfuraceous, or fibrillose, not smooth. *Basidia* not necropigmented. *Pleurocystidia* absent. *Cheilocystidia* thin-walled, hyaline, continuous along edge of lamellae, originating as modified (sub)hymenial elements. *Caulocystidia* absent or present but poorly differentiated.

Five species of the *Pseudosperma* clade are documented here from Australia, although as many as nine are known. Unlike other groups of Inocybaceae that lack pleurocystidia, species of the *Pseudosperma* clade lack necropigmented basidia.

KEY TO SPECIES OF THE PSEUDOSPERMA CLADE IN AUSTRALIA

1 Pileus areolate-scaly .... **4. I. illudens**

1: Pileus smooth at the disc, rimose towards the margin 2

2 Pileus deeply conical when young; umbo absent; tropical sclerophyll habitats .... **3. I. gracilissima**

2: Pileus conical, convex, or plane, but not deeply conical when young; umbo present; habitats various 3

3 Stipe to 100 mm long; pileus with nipple-like umbo; subtropical rainforest mixed with eucalypts and *Allocasuarina* .... **5. I. pluviorum**

3: Stipe 20–40 mm long; pileus with low broad umbo; lowland tropical sclerophyll or temperate sclerophyll forests 4

4 Lamellae pale yellow to pale olive when young; odour of coconut or cashews; temperate sclerophyll forests .... **2. I. fissurata**

4: Lamellae very pale brown to brown when young; odour absent; lowland tropical sclerophyll swamps .... **1. I. araneosa**

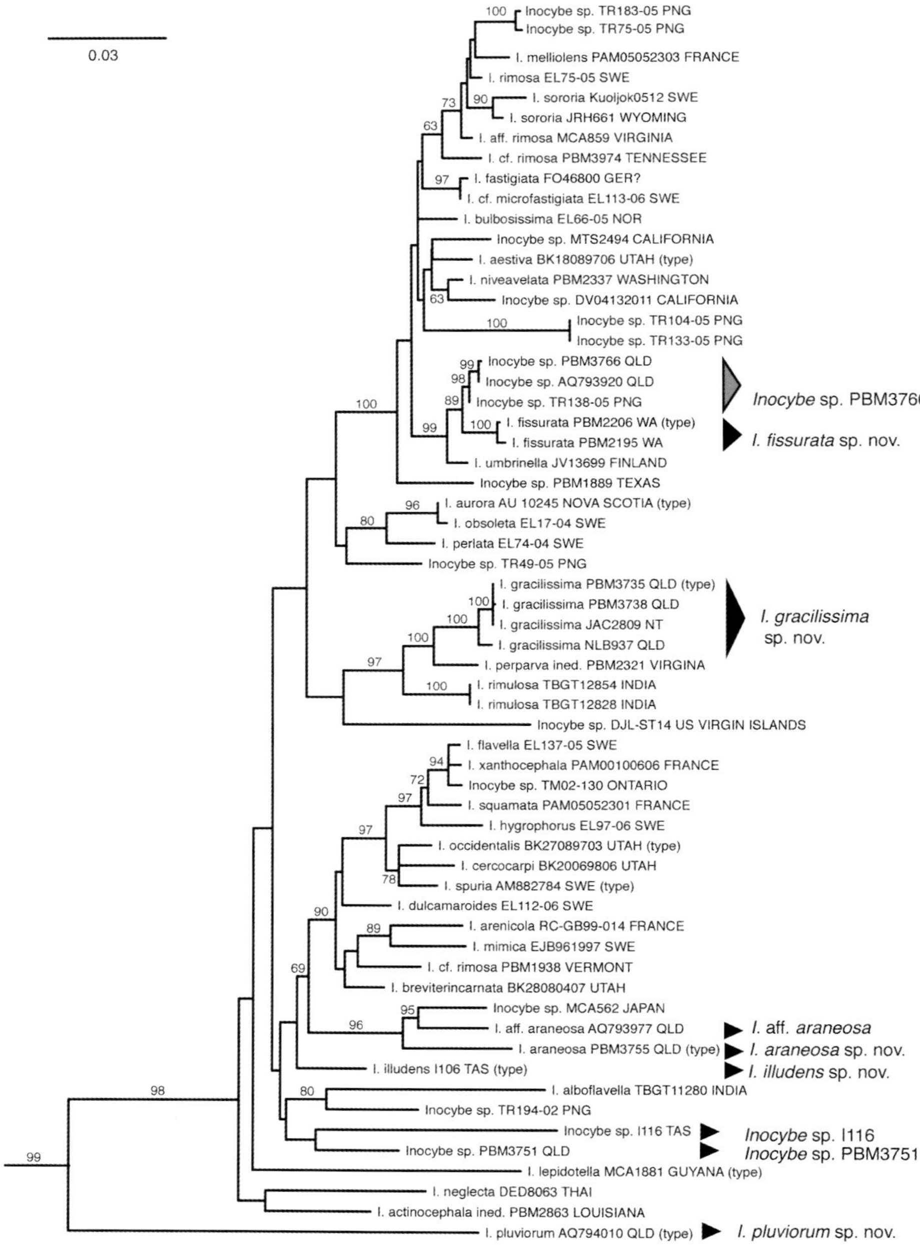

Figure 136. Maximum Likelihood phylogram of the *Pseudosperma* clade. Bootstrap values are indicated above branches. Taxon sampling was guided by results shown in Larsson *et al.* (2009) and Kropp *et al.* (2013). The tree was inferred from combined nLSU and *rpb2* gene coding regions (partitioned separately) and included 2150 total sites. Species known only from Australia are indicated by black-filled triangles. Grey-filled triangles indicate a wider Australasian distribution. The tree is rooted with exemplars of the *Inosperma* and *Mallocybe* clades (*Inocybe viridipes*, *Inocybe isabellina*, *Inocybe sabulosa*, and *Inocybe subtilior*, which have been pruned from the tree figure). All branch tips (66 total) are represented by nLSU sequences and 37 by *rpb2*.

1. Inocybe araneosa Matheny & Bougher, *sp. nov.*

MycoBank: MB 808682

Type: Mount Whitfield Conservation Park, Cairns, Qld, 16°53'53.8"S, 145°45'03.1"E, gregarious to scattered singly on soil in *Melaleuca* swamp in lowland rainforest under *M. leucadendra*, *Lophostemon suaveolens*, *Corymbia tesselaris* and *Acacia mangium*, 25 Feb. 2012, *P.B.Matheny PBM3755*, holo: PERTH 08363285; iso: TENN 066983 (KJ729904-nLSU, KJ729878-ITS, KJ729935-*rpb2*).

Pileus 10–35 mm wide, conical, expanding with age, eventually plane and with a slightly uplifted margin, at times with an obtuse umbo, especially in mature specimens; margin initially incurved, later decurved, with age occasionally with very small scales around the centre, rimulose to rimose towards the margin, long-rimose with age; velipellis absent or present as a greyish cobwebby covering the disc; centre brown (10YR 4/3; Snuff Brown) to strong brown, shading to light yellowish brown (10YR 6/4) towards the margin, at times distinctly bicolorous in youth, with a yellowish brown (10YR 5/4) centre and pale brown (10YR 6/3) margin, paler between spreading fibrils; context pallid, not changing where exposed, up to 2 mm thick under the disc. *Lamellae* adnexed, close to crowded, 50–60 L, with several tiers of lamellulae, medium to broad, up to 4 mm deep, very pale brown (10YR 8/3), at first pale brown to brown (10YR 6/4 to 5/3), with age light yellowish brown with an olive tone (2.5Y 6/4); edges obscurely pallid-fimbriate. *Stipe* 23–40 × 2–5 mm, even, not bulbous; cortina fugacious; surface lightly pruinose-furfuraceous at the apex, lightly fibrillose, glabrescent below; white at first throughout, pallid or almost white above, with pale brown tinges below; context solid, concolorous with the surface. *Odour* not remarkable. Pl. 223, 224 (see also Pl. 2D, 2I, 4H).

Basidiospores (8.0–) 8.5–11.5 × 5.0–7.0 µm [mean 10.2 × 6.1 µm; Q 1.33–1.99 (–2.10); Q mean 1.68 (*n* = 43/2)], smooth, elliptic, with obtuse apices, yellowish brown; apiculus not distinct. *Basidia* 24–36 × 9–10 µm, clavate, hyaline, 4-sterigmate. *Pleurocystidia* absent. *Cheilocystidia* 27–53 × 10–17 µm, cylindrical or nearly so, with rounded to subcapitate apices, at times fusiform, clavate or utriform, thin-walled, hyaline, occasionally multi-septate, in dense clusters arising from long narrow hyphae 3.5–4.0 µm wide. *Caulocystidia* not seen; apex with few scattered bundles of cylindrical hyphae; end cells not differentiated. *Pileipellis* a cutis of coarsely encrusted cylindrical hyphae mostly 5–12 µm wide, thin-walled to slightly thick-walled, russet or dark yellowish brown in mass; velipellis sometimes present, composed of compact, interwoven, hyaline hyphae, narrowly cylindrical, mostly 3–6 µm wide; context hyphae yellow-brown to rust-coloured in KOH. *Clamp connections* present. Fig. 137 (see also Pl. 10G).

Occurs in north-eastern and south-eastern Qld, on soil in *Melaleuca* swamp in lowland tropical forest mixed with *Lophostemon*, *Corymbia* and *Acacia*, or under *Eucalyptus grandis* in warm temperate area. Fruiting Feb.–June.

Additional specimens examined: Qld: type locality, 25 Feb. 2012, *N.L.Bougher NLB950* (PERTH 08364605, TENN067061); end of Cook St, Hallorans Hill Environmental Park, Atherton, on soil among grass under *Eucalyptus grandis* in grassed park, 22 Feb. 2012, *N.L.Bougher NLB922* (PERTH 08373027); Pioneer Park, Tully Falls Rd, gregarious in soil in mixed *Eucalyptus grandis*/riverine vegetation, 18 Apr. 2002, *A.M.Young & R.Booth 3015* (BRI AQ557628); Cliento Park, Nambour, on soil in wet sclerophyll forest near *Lophostemom confertus*, 28 June 2009, *P.Leonard 42609* (BRI AQ793920; KJ729905-nLSU, KJ729879-ITS, KJ729936-*rpb2*); Somerset Track, Mount Mee Natl Park, on soil in wet sclerophyll forest, 24 Apr. 2010, *M.Prance PL58410* (BRI AQ793977; KJ729906-nLSU, KJ729880-ITS, KJ729937-*rpb2*); Tamborine Mountain Rd (Main Western Rd), opposite Witches area near North Tamborine town, 27°56'25.7"S, 153°10'48.4"E, on soil among litter and grass under *Eucalyptus grandis* in grassy area cleared of understory, 1 Mar. 2012, *N.L.Bougher NLB964* (PERTH 08364729 TENN 067962); c. 150 m along unpaved track from carpark at Goomoolahra Picnic Area, Springbrook Natl Park, in tall eucalypt/rainforest with mixed eucalypts, some *Allocasuarina* and possibly *Acacia*, 2 Mar. 2012, *N.L.Bougher NLB965* (PERTH 08364710, TENN 069533); Mount Cordeaux Track (near Rainforest Curcuit), Cunninghams Gap, Main Range Natl Park, 28°02'59.0"S, 152°23'37.3"E, on soil among litter in tall sclerophyll forest (including eucalypts and *Allocasuarina* sp.), 3 Mar. 2012, *N.L.Bougher NLB971* (PERTH 08364761).

Etymology: From the Latin *araneosa* (cobwebby), in reference to the radially rimose pileus resembling a cobweb pattern.

Diagnostic features: Fruit bodies medium-sized. Pileus starkly radially rimose, brown to yellowish brown (umbrinous); stipe whitish, lightly pruinose-furfuraceous at the apex, lightly fibrillose elsewhere, glabrescent towards the base. Spores relatively large, elliptic.

Inocybe araneosa is a distinctive species with a wide warm-temperate and tropical distribution in Qld (Fig. 136). Recent collections made by us are similar to those lodged at BRI labelled as *I. striatula* (Cleland) Grgur. (see listed above). However, *I. striatula* is a species of *Cortinarius* (Matheny & Bougher, 2010).

Inocybe fissurata Matheny & Bougher is more yellowish on the pileus with somewhat longer, oblong-elliptic to oblong-reniform spores and is known only from the south-west of Western Australia. *Inocybe palaeotropica* Turnbull & Watling has smaller spores, a strong odour of 'burnt horn' (we assume this is similar to burnt hair) and an association with dipterocarps or Fagaceae in Indomalaya, Papua New Guinea and the Solomon Islands (Turnbull, 1995).

Inocybe renispora E.Horak, from New Zealand, differs from *I. araneosa* in its phaseoliform spores and squamulose disc (Horak, 1977), while *I. gracilissima* is characterised by its deeply conical pileus when young and a pale olive-yellow to yellowish margin, and by having less cylindrical-oblong spores (in face view). *Inocybe araneosa* and *I. gracilissima* can occur together in Queensland, but the latter appears to be more common.

2. Inocybe fissurata Matheny & Bougher, *sp. nov.*

MycoBank: MB 809325

Type: Floreat Park, Perth, E of Perry Lakes Stadium, along Brookdale Ave., W.A., on soil in lawn under planted *Eucalyptus globulus* and *E. gomphocephala*, 29 Aug. 2001, *P.B.Matheny PBM2206* (*E7054*); holo: PERTH 07676824 (GenBank DNA accession no. AY732213-nLSU, JQ408770-ITS, JQ421069-*rpb2*).

Pileus up to 30 mm wide, plane with a slightly arched margin; umbo low, obtuse; margin decurved; surface dry; centre smooth, at times with a pallid velipellis; margin weakly rimose and fibrillose; centre pallid with an ochraceous undertone (10YR 7/6) or brown when velipellis is absent, elsewhere pale yellow (2.5Y 7/4; Honey Yellow) or in places golden brown, isabelline, or "Chamois" (near 4B5); context white, not changing colour where bruised, up to 5 mm thick under the disc. *Lamellae* adnexed, tightly crowded or convoluted when immature (e.g. PERTH 08690812), close to crowded in age, up to 52 L, with several tiers of lamellulae; pale yellow to pale olive (5Y 7/4 to 6/4), narrow, 2–3 mm deep, subventricose, with edges pallid-fimbriate. *Stipe* up to 30 × 5 mm at the apex, up to 7 mm wide at the gradually enlarged base; cortina not observed; surface at apex scurfy, elsewhere with whitish scurfy fibrils, ground colour pale cream (2.5Y 8/4); context solid, whitish. *Odour* weak, mild or like coconut flakes or cashews, not spermatic. Pl. 225–227.

Basidiospores 9.0–13.0 × 5.5–7.0 µm [mean 11.4 × 6.2 µm; Q 1.50–2.00; Q mean 1.83 (n = 35/3)], smooth, elliptic to oblong-elliptic or oblong-reniform, yellowish brown; apiculus usually indistinct. *Basidia* 32–46 × 11–12 µm, 4-sterigmate, clavate, often with a broadly rounded apex, hyaline. *Pleurocystidia* absent. *Cheilocystidia* 31–55 (–72) × 8–13 µm, a mixture of long cylindrical forms and shorter subfusiform or ventricose forms, all with obtuse swollen or subcapitate apices, thin-walled, hyaline. *Caulocystidia* not well differentiated; terminal cells in clusters, often cylindrical to subclavate, thin-walled, hyaline to ochraceous. *Pileipellis* an ochraceous pigmented layer; hyphae encrusted, cylindrical, mostly 5–12 µm wide; velipellis hyphae mostly 3–6 µm wide, smooth, hyaline, thin-walled. *Clamp connections* present. Fig. 138 (see also Pl. 7A).

Occurs in south-western W.A.; on soil in Karri forest under *Eucalyptus diversicolor* and in Jarrah forest under *E. marginata*, also in lawns under planted *E. gomphocephala* and/or *E. globulus* or *E. botryoides* in parks; also in sand in coastal heath-scrub under *Melaleuca*. Fruiting July and Aug.

Plate 223. *Inocybe araneosa*, holotype (PERTH 08363285). Photograph — P.B.Matheny.

Plate 224. *Inocybe araneosa*. PERTH 08364605. Photograph — N.L.Bougher.

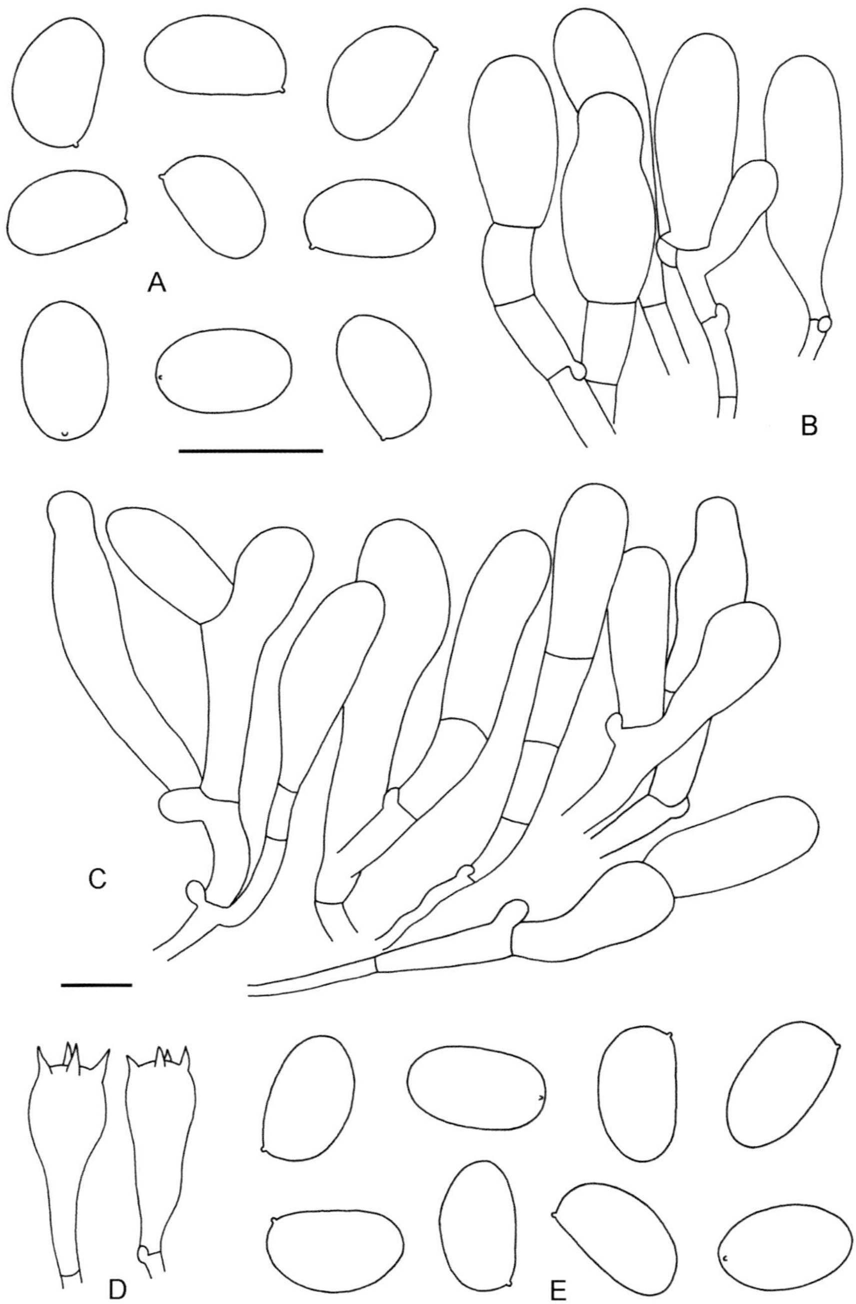

Figure 137. *Inocybe araneosa.* All holotype (PERTH 08363285), except B, E (PERTH 08364729). **A,** Basidiospores; **B,** Cheilocystidia; **C,** Cheilocystidia; **D,** Basidia; **E,** Basidiospores. Scale bars = 10 μm (longer bar for spores only).

Plate 225. *Inocybe fissurata*. PERTH 08690812. Photograph — N.L.Bougher.

Plate 226. *Inocybe fissurata*. PERTH 08517312. Photograph — P.B.Matheny.

Plate 227. *Inocybe fissurata*. PERTH 08690812. Note the crowded distorted pale lamellae in this close-up image of a sectioned young fruit body. Photograph — N.L.Bougher.

Additional specimens examined: W.A.: type locality, 16 Aug. 2001, *P.B.Matheny PBM2175, E7024* (PERTH 08517312, TENN 068363); Two Peoples Bay Nature Reserve, in sand in coastal heath under *Melaleuca* sp., 24 July 1991, *K.Syme & J.Morris KS296/91* (PERTH 05255082); Amphion Block, Dwellingup, in Jarrah forest, 4 July 1985, *R.N.Hilton s.n.* (PERTH 00918881, TENN 066696); Karri Valley Resort, Beedelup Falls Trail, on soil under *Eucalyptus* spp. in Karri forest, 19 Aug. 2001, *P.B.Matheny PBM2195, E7042* (PERTH 07676700, TENN 066680; U555466-nLSU, JQ408771-ITS, E555464-*rpb1*, EU555465-*rpb2*); *loc. id.*, *P.B.Matheny PBM2196, E7043* (PERTH 07676719, TENN 066783); Kings Park, Perth, junction of Saw Ave and May Drive, on lawn near mature *Eucalyptus botryoides* trees planted along May Drive, 31°57'54"S. 115°49'30.1"E, 9 July 2015, *N.L.Bougher NLB1120* (PERTH 08690812).

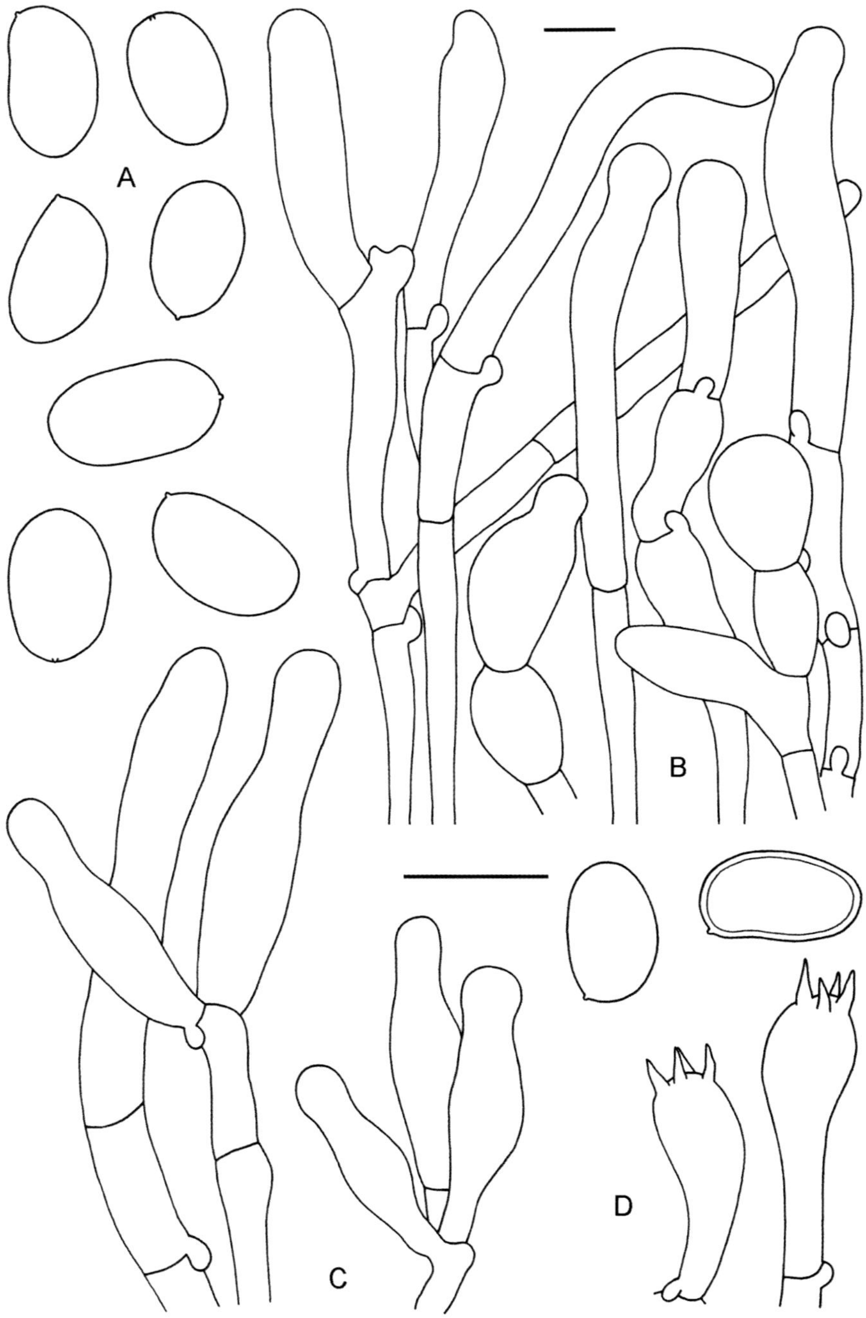

Figure 138. *Inocybe fissurata.* All holotype (PERTH 07676824) except C (PERTH 07676719). **A,** Basidiospores; **B,** Cheilocystidia; **C,** Cheilocystidia; **D,** Basidia. Scale bars = 10 µm (longer bar for spores only).

Etymology: From the Latin *fissuratus* (rimose), in reference to the fibrillose cracks on the pileus surface.

Diagnostic features: Fruit bodies medium-sized. Pileus generally bicolorous with a brownish yellow or yellowish margin contrasting with the brown disc. Odour pleasant to mild, like coconut or cashews. Spores large, mostly 9–13 × 5.5–7 µm, elliptic, oblong-elliptic or oblong-reniform.

Inocybe fissurata is most closely related to an unidentified species from Papua New Guinea and probably Queensland, as well as the northern-temperate *I. umbrinella* Bres. (Fig. 136). In the work by Kropp *et al.* (2013), *I. fissurata* was labelled as *I.* aff. *renispora* E.Horak.

The predominance of long, cylindrical cheilocystidia varies among the collections examined. These are dominant in the holotype, less so in PERTH 07676719, and rare or absent in PERTH 05255082. The shorter, usually subfusoid or ventricose cheilocystidia are present in all collections. In PERTH 05255082 these are especially noticeable. In all collections many of the cheilocystidia are multi-septate and sometimes branched.

Concerning the location of the type collected in 2001: the demolition of Perry Lakes Stadium was completed in 2012, and it has been replaced with a housing estate. The lawn along Brookdale Avenue was then mostly cleared and replaced with woodchips, but some of the planted eucalypt trees that existed there in 2001 have been retained.

3. Inocybe gracilissima Matheny & Bougher, *sp. nov.*

MycoBank: MB 809335

Type: Clohesy River Rd, near turnoff to Cedar Falls Resort, near Kuranda, Qld, 16°55'17.3"S, 145°35'10.6"E, on soil along roadside under *Lophostemon suaveolens* and *Eucalyptus* sp. (rough bark), 23 Feb. 2012, *P.B.Matheny & M.D.Barrett PBM3735*; holo: PERTH 8362912; iso: TENN 066970 (KJ729919-nLSU, KJ729945-*rpb2*).

Pileus 7–30 mm wide, initially deeply conical and 11–17 mm high, expanding with age to conical, obtusely conical or conico-convex; umbo absent; generally rounded at the apex when immature, more acute with age after expansion, margin incurved and initially clasped to the stipe, later decurved, at times somewhat pleated or uneven when joined to the stipe when young, becoming torn with age; surface dry; velipellis absent, smooth and unbroken over the disc; fibrillose to rimose especially with age towards the margin; yellowish brown (10YR 5/6) almost throughout when young; disc at times umber (7.5YR 4/4); generally brownish yellow (10YR 6/6), later with a pale olive-yellow to yellow margin (2.5Y 6/6 to 7/6), pale between the spreading fibrils; context thin, pallid or nearly isabelline. *Lamellae* adnexed to sinuate or subfree, crowded, c. 60–70 L with several tiers of lamellulae, narrow, at most 2 mm deep, at first ivory-yellow, with age light olive-brown (2.5Y 5/4); edges pallid-fimbriate. *Stipe* 25–65 × 1–4 mm, terete, even, not at all bulbous; cortina sparse, whitish, rapidly evanescent; surface somewhat furfuraceous-fibrillose at the extreme apex, fibrillose to almost smooth below; white or whitish throughout but may discolor brown at the base with age; context solid, whitish. *Odour* not remarkable. Pl. 228, 229.

Basidiospores 8.0–12.0 (–14.0) × 5.5–7.0 µm [mean 9.9 × 5.9 µm; Q (1.33–) 1.41–2.00; Q mean 1.68 (*n* = 70/4)], smooth, mostly elliptic to subamygdaliform, less often subphaseoliform, with apices obtuse bluntly pointed, yellowish brown; apiculus not distinct. *Basidia* 24–27 × 8–10 µm, clavate, hyaline, 4-sterigmate, occasionally 2-sterigmate. *Pleurocystidia* absent. *Cheilocystidia* 34–55 × 9–20 µm, variable in shape but usually clavate, subcylindrical or utriform, occasionally fusiform; apices obtuse or subcapitate, thin-walled, hyaline, in dense clusters. *Caulocystidia* absent, but terminal cells from the stipe apex can be clavate or have swollen apices; oleiferous hyphae present in the stipe trama. *Pileipellis* a cutis of yellowish brown cylindrical encrusted hyphae mostly 5–12 (–27) µm wide; trama with conspicuous oleiferous hyphae. *Clamp connections* present. Fig. 139.

Known from northern N.T. and eastern Qld; grows on soil, emerging from leaf litter, in sclerophyll forest, woodland, lowland swamp, or cleared grassy areas, typically under *Acacia*, *Corymbia*, *Eucalyptus*, *Lophostemon*, *Melaleuca* and/or *Allocasuarina*. Fruiting Jan.–Apr.

Additional specimens examined: N.T.: Cutta Cutta Caves, 25 km SE of Katherine, 14°35'S, 132°28'E, alt. 150 m, on limestone outcrop in *Eucalyptus* woodland, 10 Apr. 1989, *J.A.Curnow 2809* (CANB 8916878, TENN 65718; JQ319684-nLSU, JQ408755-ITS); Fogg Dam, E of Darwin, 12°33'55"S, 131°18'21"E, beside track on raised wall of dam, under *Melaleuca ?leucadendra* and *Acacia auriculiformis*, 24 Jan. 2014, *M.D.Barrett, T.Lebel, G.M.Bonito & C.N.Barrett MDB F91/14* (MEL, PERTH 08516561). Qld: Emerald Creek Rd, Dinden S.F., on soil in cleared understory under *Corymbia clarksoniana*, *Eucalyptus tereticornis* and other eucalypts, 24 Feb. 2012, *N.L.Bougher NLB936* (PERTH 08373108, TENN 067739); Davies Creek Rd, Dinden Natl Park, 17°01'36.1"S, 145°36'04.3"E, emerging from litter at a sclerophyll forest-rainforest interface under *Eucalyptus grandis*, *E. tindaliae* and *Allocasuarina torulosa*, 24 Feb. 2012, *N.L.Bougher NLB937* (PERTH 08373116, TENN 067740; KJ801178-nLSU, KP171122-ITS, KJ729946-*rpb2*); Emerald Creek Rd, near Mareeba, Dinden S.F., 17°03'08.5"S, 145°32'24.2"E, on soil under *Corymbia clarksoniana*, *Eucalyptus tereticornis*, and other eucalypts (box trees, smooth barks) in cleared understory, 24 Feb. 2012, *P.B.Matheny PBM3738* (PERTH 08362966, TENN 066946; KJ801179-nLSU, KP171123-ITS, KJ729947-*rpb2*); Mount Whitfield Conservation Park, Cairns, 16°53'53.8"S, 145°45'03.1"E, on soil in leaf litter in lowland tropical *Melaleuca* swamp, under *M. viridiflora*, *Lophostemon suaveolens* and *Corymbia tessularis*, 25 Feb. 2012, *P.B.Matheny PBM3760* (PERTH 8363242, TENN 066988).

Etymology: Superlative form of the Latin *gracilis* (most slender), in reference to the fruit bodies.

Diagnostic features: Fruit bodies small to medium-sized. Pileus strongly conical when young, rimose upon expansion, yellowish brown to brownish yellow with a pale olive-tinged or yellowish margin; velipellis absent; lamellae narrow, ivory-yellow to light olive-brown; odour not distinctive. Spores smooth, 8.0–12.0 (–14.0) × 5.5–7.0 µm, mostly elliptic. Cheilocystidia usually clavate, subcylindrical or utriform; pleurocystidia absent.

Inocybe renispora E.Horak, from New Zealand, differs from *I. gracilissima* in its bean-shaped spores and squamulose pileal disc. *Inocybe palaeotropica* Turnbull & Watling is characterized by its unpleasant odor described as similar to "burnt horn". This latter species has been recorded with dipterocarps in Indomalaya and with fagaceous associates (*Lithocarpus* and *Castanopsis*) in Papua New Guinea and the Solomon Islands (Horak, 1980; Turnbull, 1995). Its spores are smaller (6–9 × 4–6 µm) than those of *I. gracilissima*, but there is some overlap. *Inocybe cutifracta* Petch differs from *I. gracilissima* by its reddish brown pileus and occurrence in Sri Lanka (Latha *et al.* 2016).

Some micromorphological variations among the collections of *I. gracilissima* are worth noting, such as the degree to which spores with an adaxial depression predominate, i.e. less so in the holotype, more so in *PBM3738*. Cheilocystidia are also variable, with PERTH 08373108 (Queensland) having mostly cylindrical or narrowly clavate cystidia similar to the holotype and CANB 8916878 (Northern Territory), whereas PERTH 08373116, collected on the same day from the same area (Dinden National Park, Queensland), has numerous flexuous/contorted cystidia including some with knobby protuberances. A third collection from the same time and locality (PERTH 08362966) has numerous broad cystidia along with some cylindrical forms. Such variability emphasises the necessity of investigating as many collections as possible to circumscribe species in the *Pseudosperma* clade.

Examination of the holotype of *I. renispora* (PDD 27119) confirmed that its spores are predominantly reniform but that some spores are not adaxially depressed. Therefore, its spores are similar to those of some collections of *I. gracilissima*, particularly *PBM3738*. The cheilocystidia of *I. renispora* are predominantly ventricose or clavate rather than predominantly cylindrical and this feature is also in similarity with *PBM3738*.

Phylogenetic analyses confirm that *I. gracilissima* is distinct from other members of the *Pseudosperma* clade; it appears to be most closely related to an undescribed species in the south-eastern United States (Fig. 136; Kropp *et al.*, 2013).

Plate 228. *Inocybe gracilissima*, holotype (PERTH 8362912). Photograph — P.B.Matheny.

Plate 229. *Inocybe gracilissima*. PERTH 08373116. Photograph — N.L.Bougher.

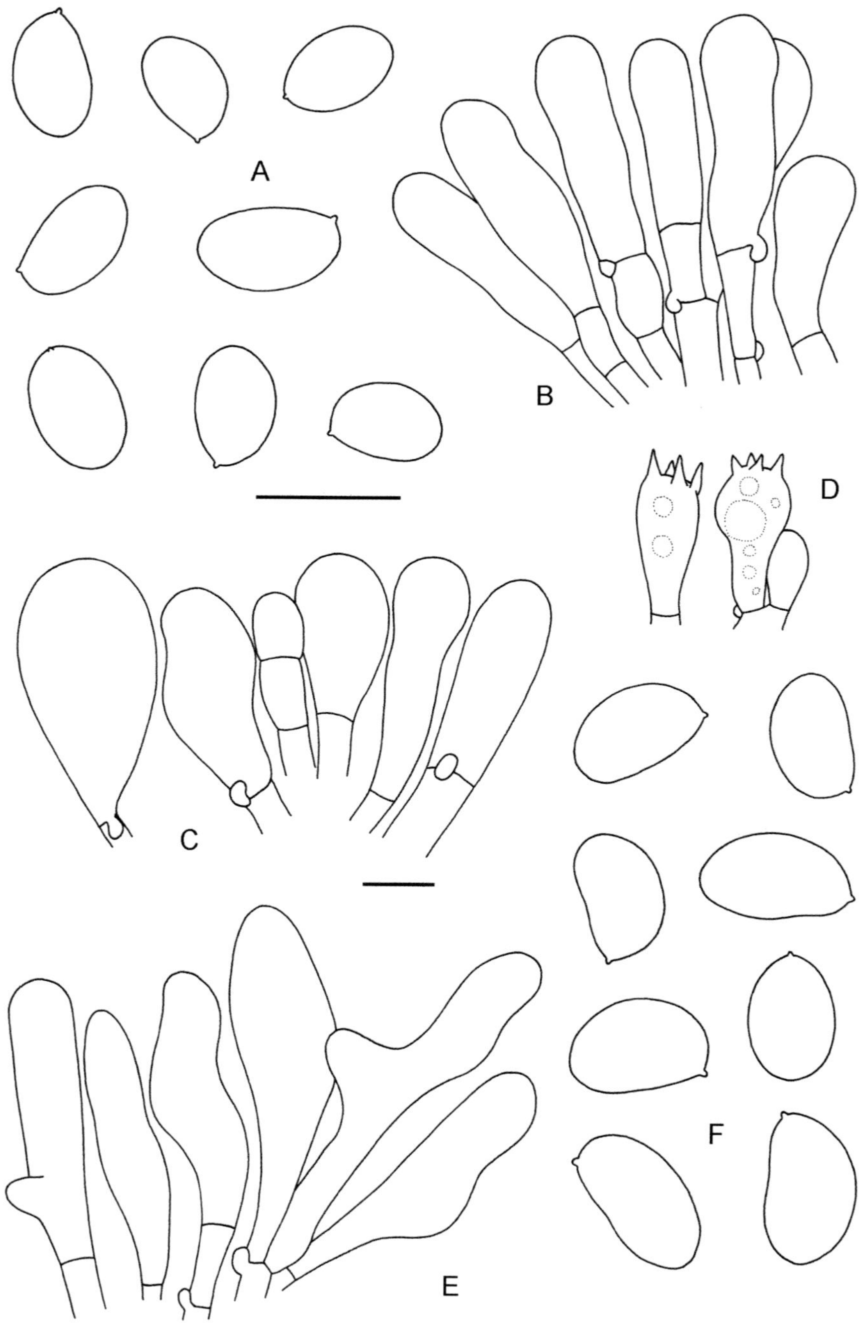

Figure 139. *Inocybe gracilissima.* **A,** Basidiospores (holotype, PERTH 8362912); **B,** Cheilocystidia (holotype, PERTH 8362912); **C,** Cheilocystidia (PERTH 08362966); **D,** Basidia (holotype, PERTH 8362912); **E,** Cheilocystidia (PERTH 08373116); **F,** Basidiospores (PERTH 08362966).

Scale bars = 10 µm (longer bar for spores only).

4. **Inocybe illudens** Matheny, Bougher & G.M.Gates, *sp. nov.*

MycoBank: MB 809381

Type: Pipeline Track, Ferntree, just SW of Hobart, Tas., on soil beside track in wet sclerophyll forest under *Eucalyptus obliqua* and *Pomaderris apetala*, 24 Dec. 2004, *G.M.Gates & D.Ratkowsky 1106*; holo: HO 569026; iso: TENN 065726 (JQ319699-nLSU, JQ408769-ITS, JQ421068-*rpb2*).

Pileus to 60 mm wide, plane, with a slightly raised margin; centre with a broad low obtuse umbo; margin slightly decurved to almost straight; surface dry; velipellis absent; coarsely fibrillose with plaques of blunt squamules on the disc and outwards towards the margin, rimulose in places towards the margin or torn; raw sienna, hazel or rusty brown (5C5) at the centre, ochraceous brown (5C4) towards the margin, much paler between cracks. *Lamellae* sinuate, close to crowded, subventricose, c. 6 mm deep, olive-buff to greyish olive, moderately thin; edges even and concolorous. *Stipe* up to 45 × 16 mm at apex, but tapering to 13 mm at the base, central, terete, dry, long-fibrillose, sandy (4A4) with a satiny or silky sheen, with scattered brownish fibrils at and just above the base. *Odour* not remarkable. Pl. 230.

Basidiospores 9.0–12.0 × 5.5–7.5 µm [mean 10.4 × 6.5 µm; Q 1.40–1.91; Q mean 1.61 (*n* = 20/1), smooth, mostly elliptic to subamygdaliform, less frequently subphaseoliform; apices obtuse, yellowish brown; apiculus small but distinctive. *Basidia* 31–42 × 8–12 µm, clavate, often with a broadly rounded apex, hyaline or often with oily inclusions when mature, developing yellowish granular contents with age, 4-sterigmate. *Pleurocystidia* absent. *Cheilocystidia* 30–45 × 10–12 µm, cylindrical to slender-clavate, hyaline, thin-walled or sometimes slightly thick-walled (especially near the apex), mixed with fertile basidia. *Caulocystidia* absent. *Pileipellis* a trichoderm of dark rusty brown hyphae in mass; hyphae weakly encrusted, but many cells filled with rusty cytoplasmic pigment, thin-walled, mostly 5-12 µm wide. *Clamp connections* present. Fig. 140.

Occurs on soil in wet sclerophyll forest under *Eucalyptus obliqua* and *Pomaderris apetala* in southern Tas.; known only from the type locality. Fruiting Dec.

Plate 230. *Inocybe illudens*, holotype (HO 569026). Photograph — G.M.Gates.

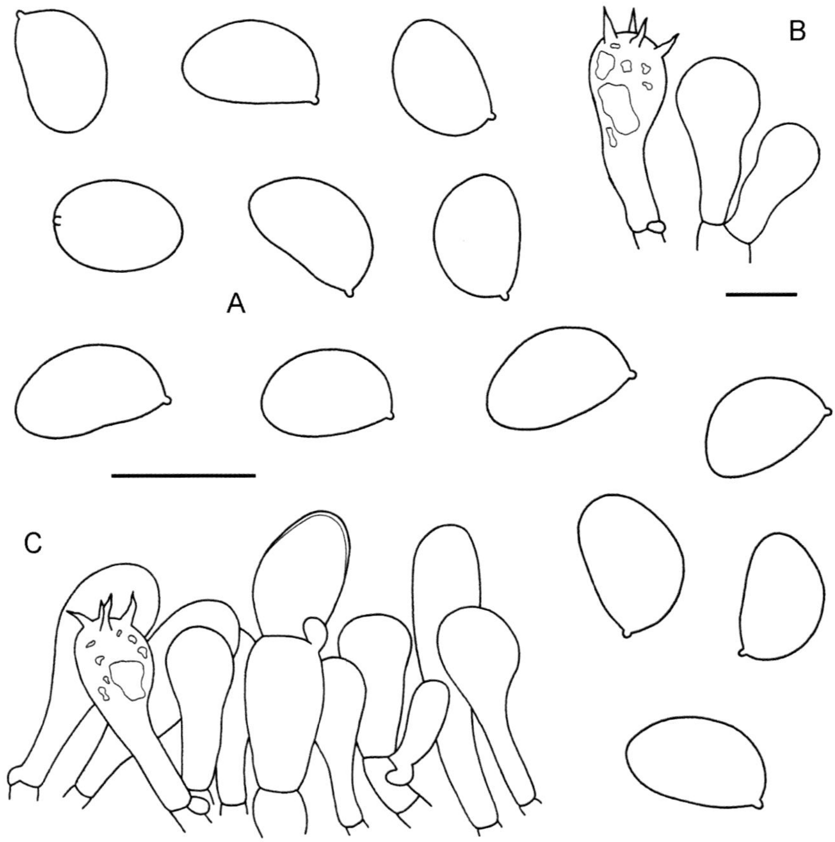

Figure 140. *Inocybe illudens.* All isotype (TENN 065726). **A,** Basidiospores; **B,** Basidia; **C,** Cheilocystidia intermixed with basidia. Scale bars = 10 µm (longer bar for spores only).

Etymology: From the Latin *illudens* (mocking), in reference to the appearance suggestive of *Inocybe squamata* of the Northern Hemisphere.

Diagnostic features: Fruit bodies large and robust. Pileus rusty brown to ochre-brown, with blunt scales; stipe tapered at the base and decorated with brownish fibrils; odour not remarkable. Spores mostly elliptic to subamygdaliform; pleurocystidia absent; cheilocystidia short-cylindrical to slender-clavate.

Inocybe illudens is another new Australian species in the *Pseudosperma* clade (Fig. 136). The scaly pileus and tapered stipe base decorated with brownish fibrils are diagnostic and separate it from other Australian and New Zealand taxa (none are known from southern South America). In contrast, *I. gracilissima* Matheny & Bougher is characterised by its slender habit, conspicuously rimose pileus with an olive-yellow to yellowish margin, and a distribution in eastern Queensland; *I. araneosa* Matheny & Bougher, from tropical Queensland, has a snuff-brown and conspicuously rimose pileus with a non-robust habit. *Inocybe fissurata* Matheny & Bougher, from south-western Western Australia, has a predominantly yellowish bicolorous pileus and slightly longer spores (mean 11.4 × 6.2 µm), while *I. pluviorum* Matheny & Bougher (see below), from subtropical rainforest in Queensland, is characterised by its tall slender fruit bodies with a stipe up to 100 mm long. All four have a smooth and unbroken pileus disc.

Inocybe renispora E.Horak, from New Zealand, shares several traits with *I. illudens*: the comparatively robust habit, somewhat similar pileal colour ("yellowish brownish"), the scaly pileus centre, olive-tinged lamellae, and the absence of a distinctive odour. However, most spores of *I. renispora* are markedly phaseoliform and the stipe is subclavate (rather than tapered), and it lacks the brownish fibrils above the stipe base. Horak (1977) described *I. renispora* as being in association with *Leptospermum* or *Nothofagus*.

5. **Inocybe pluviorum** Matheny & Bougher, *sp. nov.*

MycoBank: MB 809445

Type: Delicia Rd, Linda Garrett Reserve, Mapleton S.F., Blackall Ra., Qld, 26°36'30"S, 152°50'48"E, alt. 425 m, on soil in subtropical rainforest, 5 Apr. 2008, *P.Leonard PL23408*; holo: BRI AQ794010; iso: PERTH 08556456, iso: TENN 069645 (KP170980-nLSU, KM406221-*rpb2*).

Pileus 45 mm wide, nearly plane, greatly expanded; centre with a small but pronounced nipple-like umbo; margin decurved but somewhat raised with age; surface dry; velipellis absent; centre smooth and unbroken over the umbo, soon long-rimose with fibrils spreading strongly towards the margin; centre strong brown or umber (6D8; 7.5YR 4/4; Cinnamon Brown), contrastingly paler or very pale brown between the spreading fibrils; context thin. *Lamellae* adnexed, crowded, c. 45–50 L, medium, golden brown (5D7), edges pallid-fimbriate. *Stipe* 100 × 5 mm, cylindrical, even; base not bulbous; surface dry, finely-fibrillose and almost smooth in appearance over the upper half, more densely fibrillose below, lower one-fifth encased in soil; white with rare ochre tinges; context not observed. *Odour* similar to "flea shampoo for dogs" (notes with type specimen).

Basidiospores 9.0–10.5 (–11.0) × 6.5–7.5 µm [mean 9.6 × 7.0 µm; Q 1.27–1.54; Q mean 1.36 (*n* = 24/1)], smooth, broadly elliptic to elliptic with rounded apices, never phaseoliform, yellowish brown; apiculus very small and indistinct. *Basidia* 34–38 × 10–12 µm, clavate, hyaline, 4-sterigmate. *Pleurocystidia* absent. *Cheilocystidia* 25–36 × 13–14 µm, clavate to subcylindrical, thin-walled, hyaline, mixed with some basidia. *Caulocystidia* absent. *Pileipellis* a cutis of cylindrical hyphae that are light ochraceous in mass, encrusted, mostly 5–10 µm wide, thin-walled. *Clamp connections* present. Fig. 141 (see also Pl. 7B).

Known only from the type locality, a subtropical rainforest in south-eastern Qld that includes *Lophostemon*, *Syncarpia*, *Eucalyptus grandis*, *Eucalyptus* spp., *Corymbia* and *Allocasuarina*. Fruiting Apr.

Etymology: From the Latin *pluviorum* (of wet rainy places) in reference to the habitat of the new species.

Diagnostic features: Fruit body tall, slender. Pileus nearly plane with nipple-like umbo, strong brown, markedly rimose; lamellae crowded; stipe white with rare ochre tinges, long (100 mm). Spores broadly elliptic with very small apiculus; cheilocystidia clavate to subcylindrical.

Inocybe pluviorum is an unusual species broadly similar to those of the *Pseudosperma* clade collected in subtropical forest in south-eastern Queensland. Although known only from a single fruit body, phylogenetic analysis of combined nLSU and *rpb2* data place it with strong support as the sister lineage to all other species of the *Pseudosperma* clade (Fig. 136).

Inocybe araneosa Matheny & Bougher has a similar pileus colour and also shares the pronounced rimose surface. However, the former has a shorter stipe and narrower spores. Other Australian species in the *Pseudosperma* clade can be distinguished from *I. pluviorum* by the characters emphasised in the key and in the discussion of *I. illudens* (see above).

Another collection in BRI (AQ807856), mislabeled "*Inocybe asterospora*", closely resembles *I. pluviorum* in having a similar pileus colour and a long stipe, but it differs in its much larger spores (mostly 10.5–13.0 × 7.0–9.0 µm) and longer (>50 µm) clavate cheilocystidia. It was collected from soil in rainforest at Watagan Forest, near the central coast of New South Wales (32°57'S, 151°13'E), and it is likely to represent an undescribed species.

The odour of *I. pluviorum* is reported to be rather unusual, and notes with the collection in BRI indicate a smell similar to "flea shampoo for dogs"; further collections are required for verification. An unusual smell, described as similar to "burnt horn", has also been ascribed to *I. palaeotropica* by Horak (1980, as *I. umbrina*), and we interpret this smell to be similar to burnt hair or burnt keratin. Turnbull (1995) documented the species from Malaysia but with no description of the odour. Nevertheless, the spores as reported by Horak are smaller than those of *I. pluviorum*, and the cheilocystidia are much longer.

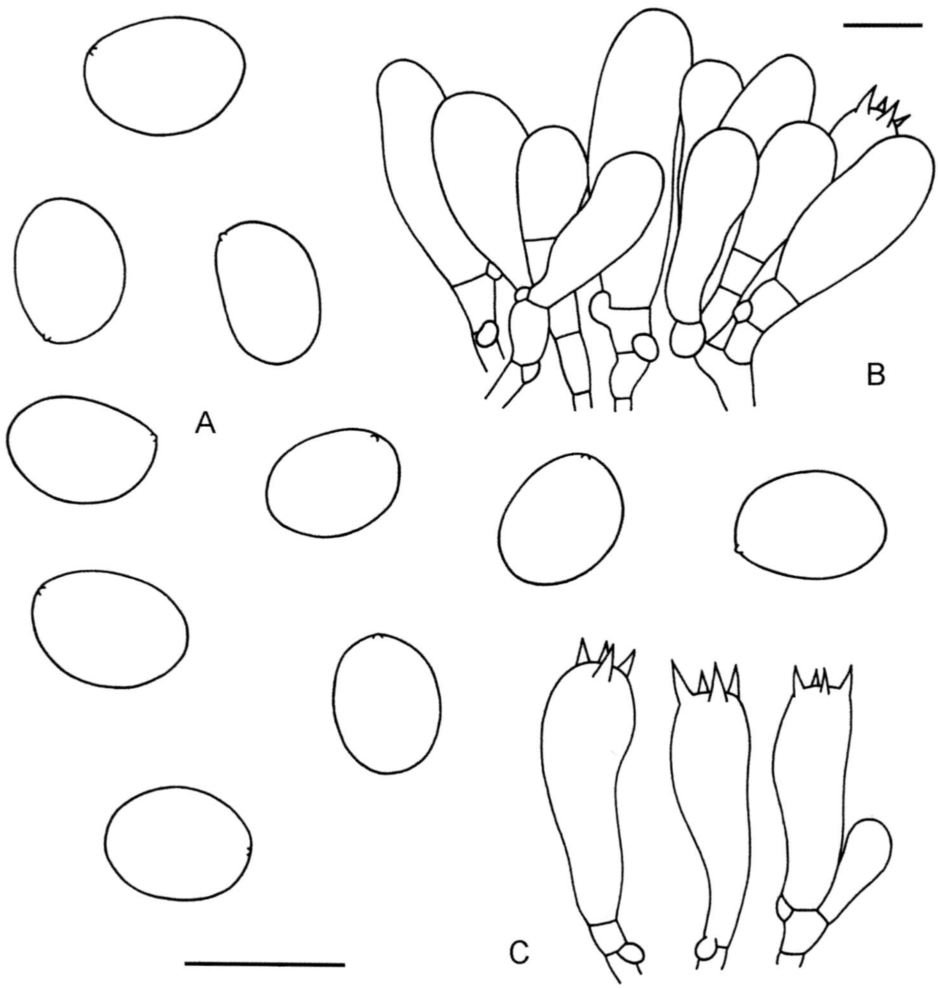

Figure 141. *Inocybe pluviorum.* All holotype (BRI AQ794010). **A,** Basidiospores; **B,** Cheilocystidia; **C,** Basidia. Scale bars = 10 µm (longer bar for spores only).

(INOCYBE) INOSPERMA CLADE

Inocybe subgenus *Inosperma* Kühner in *Bull. mens. Soc. Linn. Lyon* 49 (no.spéc.): 898 (1980)

Type: *I. calamistrata* (Fr. : Fr.) Gillet

Basidia hyaline and non-necropigmented or necropigmented. *Pleurocystidia* absent. *Cheilocystidia* present, continuous along edge of lamellae, originating as modified subhymenial elements. *Caulocystidia* absent. *Spores* with (very) small apiculus and obtuse apex.

The taxonomic circumscription used here is slightly modified from Kuyper (1986). Two sections have been proposed: sect. *Rimosae* and sect. *Cervicolores*. Sect. *Rimosae* is not monophyletic. Species of sect. *Rimosae* in the *Inosperma* clade have been referred to as the *Maculata* clade by Larsson *et al.* (2009). Species of the *Maculata* clade have not yet been recorded as native in Australia; one, *I. erubescens*, is introduced under pines, but is not treated further here. Four species of sect. *Cervicolores* are recognized from Australia. Associating with eucalypts and in habitats with *Acacia* (Fabaceae) and *Pomaderris* (Rhamnaceae) or with *Leptospermum* (Myrtaceae) and *Nothofagus* (Fagaceae). Presently known only from eastern Australia (one species shared with New Zealand).

KEY TO SPECIES OF THE INOSPERMA CLADE IN AUSTRALIA

1 Under introduced north temperate trees, fruit bodies white to yellowish at first .. † **I. erubescens**

1: Under native vegetation, fruit bodies not white 2

2 Pileus recurved squamulose, stipe floccose-squamulose; odour like *Pelargonium* at first, later with a fishy component .... **1. I. calamistratoides**

2: Pileus fibrillose, not scaly, stipe fibrillose, odour absent 3

3 Pileus <15 mm wide; stipe <3 mm thick; green and blue tinges absent........ **3. I. rubricosa**

3: Pileus up to 60 mm wide; stipe up to 8 mm thick; green or blue tinges present 4

4 Pileus often with a central bluish area; spores on average <10 µm long (mean 9.6 × 5.8 µm) .... **2. I. cyanotrichia**

4: Pileus without bluish areas; spores on average >10 µm long (mean 10.7 × 6 µm) .... **4. I. viridipes**

†Not treated further here

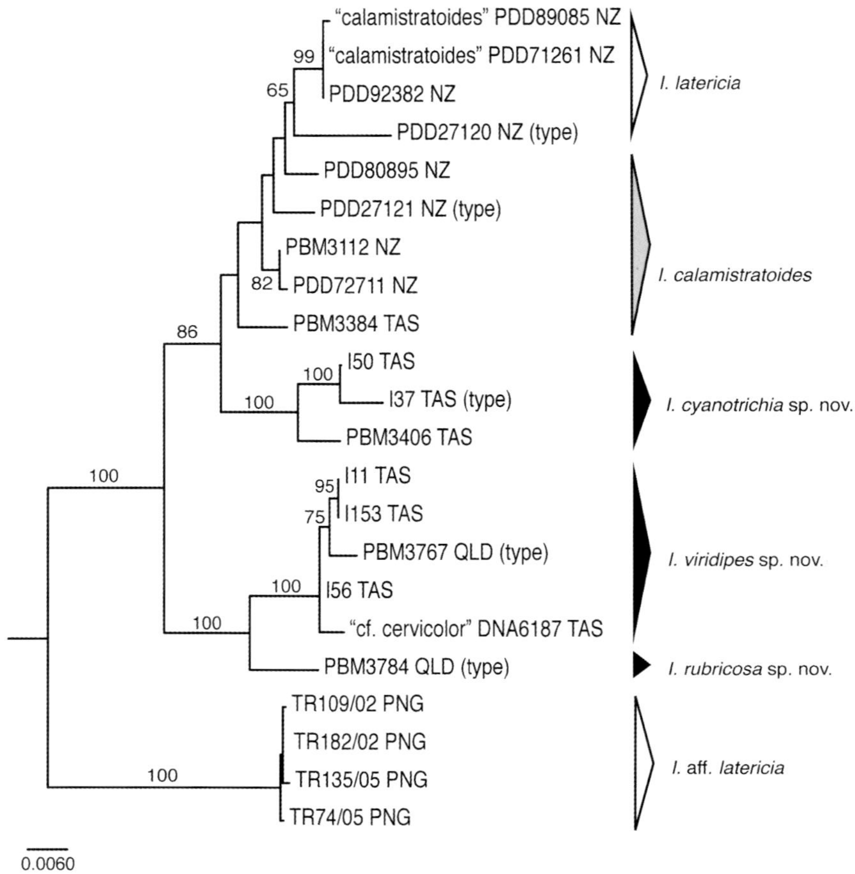

Figure 142. Maximum Likelihood phylogram of Australian species of the *Inosperma* clade. Bootstrap values are indicated above or below branches. Species from the Southern Hemisphere form a monophyletic group based on phylogenetic analyses of nLSU and *rpb2* gene trees (data not shown). The tree was inferred from combined ITS, nLSU, and *rpb2* gene coding regions and included 2846 total sites. Species confined to Australia are indicated by black-filled triangles. Species shared between Australia and New Zealand are indicated by grey-filled triangles. White triangles indicate non-Australian taxa. Of the 22 branch tips in the tree, all are represented by ITS sequences, 18 by nLSU, and 12 by *rpb2*.

1. **Inocybe calamistratoides** E.Horak, *New Zealand J. Bot.* 15: 716 ('1978') [1977]

Type: Granite Ck, Westland District, New Zealand, on the ground under *Nothofagus fusca*, 17 Dec. 1967, *E.Horak*; holo: PDD 27121 (KJ756494-ITS).

Pileus 10–25 mm wide, conical to obtusely conical, expanding with age, lacking an umbo; margin initially incurved, becoming decurved with age; surface dry, recurved-squamulose around the disc and towards the margin, more coarsely fibrillose at edge of margin; velipellis absent, dark reddish brown to chestnut-brown, at times very dark brown over the disc; context not observed, but presumably rubescent. *Lamellae* adnexed to ascending-adnate, c. 38–42 L, close, with few tiers of lamellulae, broad, light reddish brown, becoming dark reddish brown; edges pallid-fimbriate. *Stipe* 25–50 × 4–5 mm at the apex, gradually thickened towards the base (up to 7–8 mm wide); partial veil not directly observed; surface dry; with numerous scattered pinkish to fawn-coloured flocculose-squamulose patches for almost the entire length of the stipe, more densely fibrillose at the base; light reddish brown to reddish brown at least above, whitish mixed with green or blackish green above the base; context not observed. *Odour* distinctly of *Pelargonium* in younger specimens but with a fishy component in older material. Pl. 231.

Basidiospores 7.0–10.0 (–11.0) × 5.0–6.5 µm [mean 8.5 × 5.4 µm; Q 1.40–1.70; Q mean 1.58 (*n* = 25/1)], mostly elliptic, some slightly adaxially concave, apices obtuse, yellowish brown; apiculus small but usually conspicuous, slightly thick-walled. *Basidia* c. 32–40 × 8–10 µm, narrowly clavate, initially hyaline, but becoming necropigmented and collapsing, 4-sterigmate. *Pleurocystidia* absent. *Cheilocystidia* 22–40 × 8–12 µm, clavate to short-clavate or subcylindrical, usually with a slightly or markedly swollen apex, hyaline but becoming brown and collapsing, thin-walled, crowded. *Caulocystidia* absent. *Pileipellis* yellowish brown in mass, composed of cylindrical hyphae mostly 4–10 µm wide but often collapsed, thin-walled, many with intracellular ochre-tawny pigment, encrustations not evident. *Clamp connections* present. Fig. 143.

Known from Tas. and New Zealand; grows on soil under *Nothofagus* mixed with *Leptospermum* (Tas.) or under *Nothofagus*. Fruiting May (Tas.) or Dec.–May (New Zealand).

Additional specimen examined: Tas.: Mount Michael Track, the Blue Tier, alt. 800 m, scattered singly or in pairs on soil under *Nothofagus cunninghami* and *Leptospermum*, 22 May 2010, *M.Noordeloos PBM3384* (TENN 065750; JQ815415-nLSU, JQ801393-ITS, KJ729949-*rpb2*).

Extralimital specimens examined: New Zealand: Gisborne District, Hinerau Track, Urewera Natl Park, L. Waikaremoana, on soil in *Nothofagus* forest, 7 May 2001, *E.Horak & A.Horak ZT9630* (PDD 72711, TENN 065730; AY380369-nLSU (1), JQ815413-nLSU (2), JQ801392-ITS, AY333765-*rpb2*, AY351795-*rpb1*); Lake Taupo District, Taranaki Falls Track, Whakapapa, under *Nothofagus solandri* var *cliffortioides*, 2 Apr. 2005, *C.Shirley JAC9348* (PDD 80783, TENN 065731); Lake Taupo District, Clements Mill Rd, Kaimanawa Forest Park, scattered singly on soil and on road bank under *Nothofagus*, 8 May 2009, *P.B.Matheny PBM3112* (PDD 97855, TENN 063686; JQ815414-LSU, ITS spacers 1 and 2 separately [both unpublished <200 bp], JQ846473-*rpb2*); Nelson District, Bridle Walk, on soil, 8 May 2006, *G.M.Gates s.n.* (PDD 89085; GU222302-ITS).

Other extralimital specimens examined: New Zealand: Nelson District, Murchison, Matakitaki, on the ground under *Nothofagus fusca*, 28 Jan. 1969, *E.Horak ZT69/35* (PDD 27120, holotype of *I. latericia*; KP171150-ITS); Westland District, trail to Stafford Saddle, Jackson Bay, on soil under *Nothofagus menziesii*, 5 Apr. 2000, *E.Horak & A.Horak ZT8481* (PDD 71261, *I.* cf. *latericia*; JQ975024-nLSU, JQ801402-ITS); Nelson District, Canaan Rd, Abel Tasman Natl Park, on soil under *Nothofagus menziesii*, 15 May 2007, *P.Leonard PL61507* (PDD 92382, *I.* cf. *latericia*; GU233413-nLSU, GU233367-ITS).

Diagnostic features: Fruit bodies medium-sized. Pileus dark reddish brown, recurved-scaly around the disc; lamellae light to dark reddish brown; stipe with numerous scattered pinkish to fawn-coloured flocculose-squamulose patches almost its entire length; stipe base greenish or blackish green; odour of *Pelargonium*. Spores smooth, elliptic. Pleurocystidia absent.

The habit description provided here for *I. calamistratoides* is based on Tasmanian material collected in frosty conditions. This might account for the variation in fruit body colour compared to the protologue of Horak (1977), in which the pileus colour was described as pale brown, brown or fuscous, and often tinged greenish or olive. The Tasmanian collection, which is poorly preserved, is dark reddish brown, at times with a dark brown disc, but with a greenish to blackish green-tinged stipe base. Phylogenetic analyses of (*PBM3384*) confirm this material is close related to *I. calamistratoides* (Fig. 142). Based on rRNA sampling (ITS, nLSU), this is a species complex due to the variation in branch lengths among collections from New Zealand and Australia. Furthermore, collections of *I. calamistratoides* form a poorly supported paraphyletic group (including an ITS1 sequence of the holotype) in which *I. latericia* E.Horak (type collection) is derived.

Other collections at PDD also exhibit variation in some microscopic characters. Thus, based on spore morphology, PDD 72519 and PDD 89395 best match *I. calamistratoides*, but both have some encrusted, coloured, thick-walled cheilocystidia, a feature not observed in the Tasmanian material. Three other collections (PDD 71261, PDD 89396, PDD 72910) have smaller, narrower and more cylindrical spores than the type of *I. calamistratoides*, while those of the Tasmanian specimen, although variable in size, are on average somewhat smaller than the type. The spores of the holotype of *I. calamistratoides* [8.5–10.5 × 5.5–6.5 µm; mean 9.4 × 5.8 µm; Q 1.55–1.73; Q mean 1.62 (*n* = 13)] are broader and less cylindrical in face view and less pronounced adaxially concave compared to those of *I. latericia* (type collection). Further morphological comparisons (presence or absence of a greenish stipe base, squarrose *vs* appressed squamulose pileus surface) are needed to better evaluate their taxonomic utility and correlation with any microscopic variation.

Plate 231. *Inocybe calamistratoides*. TENN 065750. Photograph — P.B.Matheny.

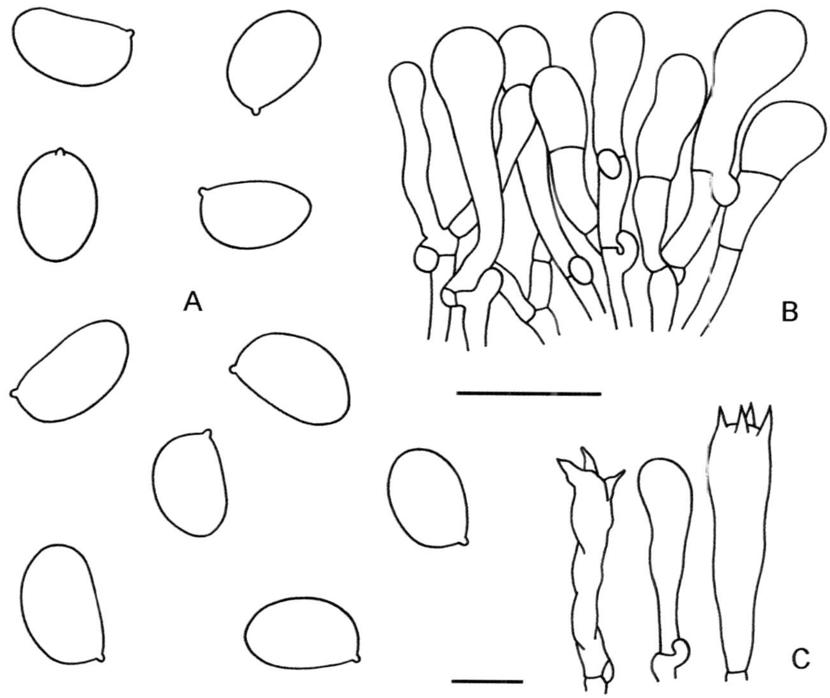

Figure 143. *Inocybe calamistratoides*. TENN 065750. **A,** Basidiospores; **B,** Cheilocystidia; **C,** Basidia. Scale bars = 10 µm (longer bar for spores only).

Inocybe calamistratoides, with its green stipe base, is evidently closely allied to *I. viridipes* Matheny, Bougher & G.M.Gates (see below), but the typically squarrose pileus and flocculose-squamulose stipe surface readily distinguish it from *I. viridipes* and *I. cyanotrichia* Matheny, Bougher & G.M.Gates, while *I. rubricosa* Matheny & Bougher, from Queensland, has a non-scaly pileus and stipe surface, small stature, and it lacks green or blackish green tinges. Based on more inclusive taxon sampling (Kropp *et al.*, 2013), *I. cyanotrichia* is the sister taxon to *I. calamistratoides* with strong support, while *I. viridipes* is sister to this group (Fig. 142). All four temperate Southern Hemisphere species form a robust monophyletic group (Fig. 142).

A GenBank ITS sequence (JX178624) identified as *I. calamistratoides* from New Zealand material is mislabelled. The collection is a representative of the *Inocybe s. str.* clade based on blastn search similarity results (86% similar to *I. fibrillosibrunnea*).

2. **Inocybe cyanotrichia** Matheny, Bougher & G.M.Gates, *sp. nov.*

MycoBank: MB 809319

Type: Clark Cliffs [Tasman Penin.], Tas., on soil in wet sclerophyll forest under *Eucalyptus*, 29 Dec. 2001, *G.M.Gates & D.Ratkowsky 137*; holo: PERTH 08516952; iso: TENN 065728 (JN975033-nLSU, JQ801397-ITS, JQ846476-*rpb2*).

Pileus 17–58 mm wide, hemispherical, convex or plano-convex or conico-convex when immature, becoming convex or plane with age; margin initially slightly incurved, becoming decurved and undulate; velipellis absent; surface dry, finely fibrillose, becoming excoriated-scaly around the disc, dark reddish brown to dark brown (7F8) or clay-brown (5D6) or greyish brown at times, often with a central blue-green irregular patch (24E3), but this not always noticeable; context pallid with fawn-coloured tones beneath the typically reddish brown cuticle, not noticeably rubescent. *Lamellae* adnate to ascending-adnate, seceding with age, close, with several tiers of lamellulae, 3–7 mm deep, moderately thin, initially buff-brown, becoming darker brown (6E6), with conspicuous white fimbriate edges. *Stipe* 14–51 mm × 1.5–8.0 mm at the apex, attenuated downwards, even at the base, attenuated or slightly enlarged; surface dry; partial veil fibrillose, not observed in mature specimens; reddish brown with some loose brown fibrils, blue-green to greyish blue at the base (24E3); context solid, fawn along the cortex but pallid within, rubescent, reddish brown at the base. *Odour* absent when fresh but fishy upon drying (in *137*). Pl. 232–234.

Basidiospores 8.5–10.5 (–11.0) × 5.0–6.5 (–7.0) µm [mean 9.6 × 5.8 µm; Q 1.42–1.91; Q mean 1.67 (*n* = 35/2)], smooth, mostly elliptic to subreniform; apices obtuse, yellowish brown; apiculus small, indistinct. *Basidia* 30–43 × 8–11 µm, 4-sterigmate, cylindroclavate, hyaline to necropigmented. *Pleurocystidia* absent. *Cheilocystidia* (10–) 18–44 × 5–12 µm, cylindrical, subcylindrical, narrowly clavate or shortly clavate, often branched, with narrow subtending hyphae occasionally with minor encrustations, thin-walled, mostly hyaline but some with ochraceous contents. *Caulocystidia* absent. *Pileipellis* a cutis with some emerging cylindrical hyaline or brown hyphae mostly 4–10 µm wide, weakly encrusted, ochraceous in mass. *Clamp connections* present. Fig. 144.

This species is endemic in southern Tas., known only from soil in wet sclerophyll forest under *Eucalyptus*. Fruiting Dec., Jan. and May.

Additional specimens examined: Tas.: Lower Kermandie, on soil in wet sclerophyll forest, 5 Jan. 2002, *G.M.Gates & D.Ratkowsky 150* (TENN 065729; JQ815418-nLSU, JQ801397-ITS, KJ729948-*rpb2*); Cuckoo Falls, near Scottsdale, singly or in pairs on soil under *Acacia*, *Pomaderris* and *Eucalyptus* in wet sclerophyll forest, 24 May 2010, *F.Karstedt PBM3406* (PERTH 08516979; TENN 065719; *I.* cf. *cyanotrichia*; JQ815431-nLSU, JQ486498-*rpb2*).

Etymology: From the Greek *kyaneos* (dark blue) and *trichos* (a hair), in reference to the blue-green patches on the pileus disc and/or the stipe base.

Diagnostic features: Fruit bodies medium-sized or robust. Pileus reddish brown with blue-green tones (not evident in photographs), finely fibrillose to excoriate scaly and lacking distinct recurved scales; stipe fibrillose (not floccose-squamulose) with rubescent context; odour absent when fresh.

Inocybe cyanotrichia is most likely to be confused with *I. viridipes* Matheny, Bougher & G.M.Gates (see below), especially when the blue-green areas of the pileus are not so apparent (as in *PBM3406*). However, the two can be separated by the length of their basidiospores: 8.5–10.5 (–11.0) µm in *I. cyanotrichia*, 9.5–12.0 (–13.5) µm in *I. viridipes*. Phylogenetically, the two Australian species are distinct (Fig. 142; Kropp *et al.*, 2013).

Several species with an alliance to the northern temperate *I. calamistrata* (Fr.:Fr.) Gill. are known from New Zealand, Papua New Guinea and Indonesia (Horak, 1977, 1980), but all can be distinguished morphologically from *I. cyanotricha* by their distinctive squarrose to recurved-scaly pileus and/or stipe surfaces. Thus, *I. latericia* E.Horak has a squarrose pileus surface, a squamulose stipe, and it lacks black-green colouration at the stipe base; it is reported under *Nothofagus* in New Zealand and under *Castanopsis* and *Lithocarpus* in Papua New Guinea. *Inocybe calamistratoides* E.Horak also differs in its recurved-squamulose pileus, squamulose stipe surface, and a blackish green stipe base; it is also associated with *Nothofagus* in New Zealand, with a slight Tasmanian variant recorded under *Nothofagus* and *Leptospermum*. *Inocybe umbrinovirens* E.Horak, from under *Nothofagus* in Papua New Guinea, has a densely recurved-squamulose pileus, larger fruit bodies and broader spores (7.0–8.5 µm). Lastly, *I. fuscospinulosa* Corner & E.Horak, associated with *Castanopsis* and *Lithocarpus* in Java, has a pileus that is densely covered with small spiny scales, as well as smaller spores (6.5–8.0 × 4.0–5.0 µm).

Inocybe cyanotrichia is the sister species to the *I. calamistratoides* complex (including sequences labelled *I. latericia*), but with weak support (Fig. 142), a relationship also indicated by Kropp *et al.* (2013).

Sequences of *I. cyanotrichia* are presented as "*Inocybe cyanotincta* nom. prov." in the phylogenetic tree figures of Kropp *et al.* (2013). However, the epithet "*cyanotincta*" is not appropriate as it merges a Greek prefix with a Latin suffix.

Plates 232, 233. *Inocybe cyanotrichia*, holotype (PERTH 08516952). Photograph — G.M.Gates.

Plate 234. *Inocybe* cf. *cyanotrichia*. PERTH 08516979. Photograph — P.Matheny.

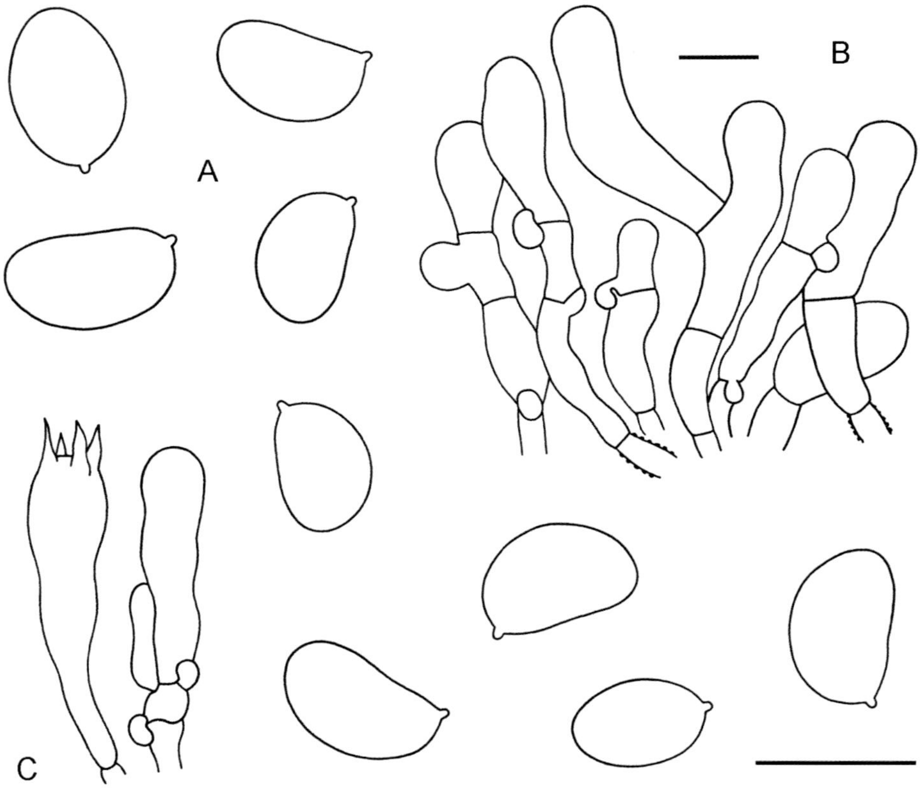

Figure 144. *Inocybe cyanotrichia.* All holotype (PERTH 08516952). **A,** Basidiospores; **B,** Cheilocystidia; **C,** Basidia. Scale bars = 10 µm (longer bar for spores only).

3. **Inocybe rubricosa** Matheny & Bougher, *sp. nov.*

MycoBank: MB 809450

Type: Rainforest Circuit Track, Cunninghams Gap, Main Rain Natl Park, Qld, 28°02'59"S, 152°23'37.3"E, alt. 770 m, on soil in warm-temperate rainforest under *Lophostemon confertum*, 3 Mar. 2012, *P.B.Matheny & N.Fechner PBM3784*: holo: PERTH 08363056; iso: TENN 067016 (KP170990-nLSU, KP308817-ITS, KM406230-*rpb2*).

Pileus 8–9 mm wide, obtusely conical and with an obtuse umbo; margin slightly incurved to decurved; surface fibrillose, glossy, not rimose, not at all scaly, dark brown (7.5YR 3/2) or chestnut-brown, paler in the centre due to greyish fibrils covering the umbo; context white at first, slowly staining red after being cut, thin but up to 2 mm thick under the disc. *Lamellae* adnexed to ascending-adnate, moderately close, c. 24–26 L, with few tiers of lamellulae, dark yellowish brown (Dresden Brown), rubescent in areas; edges conspicuously pallid-fimbriate, broad, ventricose. *Stipe* 35–47 × 2 mm, even, terete; veil not observed; apex with few scattered pallid furfuraceous fibrils, elsewhere finely fibrillose, reddish brown (5YR 5/4), white at the base, lacking green tints; context solid, with a pallid interior; cortex concolorous with the surface. *Odour* none. Pl. 235, 236.

Basidiospores 8.5–10.5 × 5.5–6.5 µm [mean 9.5 × 6.0 µm; Q 1.46–1.75; Q mean 1.59 (n = 25/1)], smooth, elliptic to ovoid or, occasionally, broadly subamygdaliform, mostly not adaxially depressed or subreniform, yellowish brown; apiculus small. *Basidia* 31–38 × 7–10 µm, narrowly clavate, hyaline, becoming necropigmented, 4-sterigmate. *Pleurocystidia* absent. *Cheilocystidia* 15–50 × 5–10 µm, narrowly clavate to subcylindrical, with an obtuse apex, sometimes multiseptate or branched, thin-walled, hyaline, densely arranged in a compact sterile layer on the gill edge, arising from long, narrow, cylindrical subtending hyphae occasionally with encrusted walls. *Pileipellis* a cutis of cylindrical hyphae mostly 6–10 µm wide, light reddish brown to dark rusty brown in mass, thin-walled, with very fine encrusted appearance; many cells with dark tawny contents and collapsed. *Caulocystidia* absent. *Clamp connections* present. Fig. 145.

Plate 235. *Inocybe rubricosa*, holotype (PERTH 08363056). Photograph — P.B.Matheny.

Plate 236. *Inocybe rubricosa*, holotype (PERTH 08363056). Photograph — P.B.Matheny.

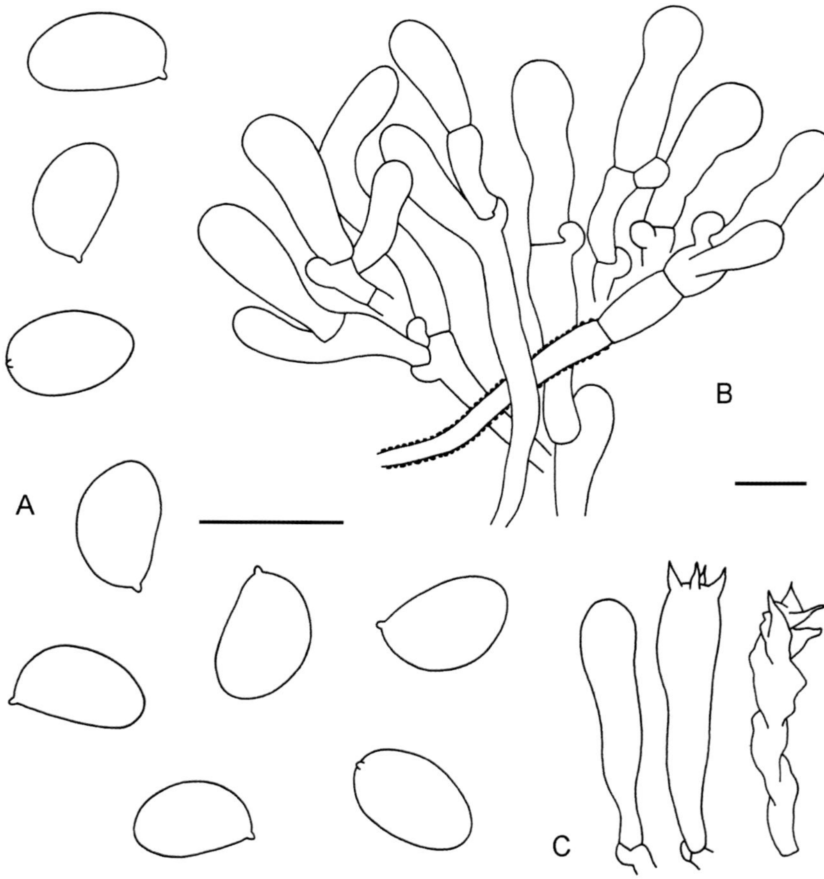

Figure 145. *Inocybe rubricosa.* All holotype (PERTH 08363056). **A,** Basidiospores; **B,** Cheilocystidia; **C,** Basidia. Scale bars = 10 µm (longer bar for spores only).

Known only from the type locality in south-eastern Qld, where it grows on soil in high elevation warm-temperate rainforest. Fruiting Mar.

Etymology: From the Latin *rubricosa* (full of red-earth), in reference to the colour of the pileus (dark brown to chestnut brown) and reddish staining context.

Diagnostic features: Fruit bodies small, but elongate and slender. Pileus dark brown to chestnut-brown with greyish fibrils over the umbo, not scaly; context of pileus slowly staining reddish where cut; edges of lamellae conspicuously fimbriate; stipe finely fibrillose, green tinges absent; odour none. In warm-temperate rainforest under *Lophostemon* (Myrtaceae).

Inocybe rubricosa, a distinctive new species in the *I. calamistratoides* E.Horak alliance (Fig. 142), is readily distinguished from its Australasian relatives by the small slender stature, a lack of an odour and any greenish, olive, or blue tinges, and its occurrence in warm-temperate rainforest in putative association with *Lophostemon. Inocybe latericia* E.Horak also lacks greenish tinges, but it has scaly fruit bodies and and is associated with *Nothofagus* in New Zealand (Horak, 1977). *Inocybe viridipes* Matheny, Bougher & G.M.Gates, also reported from Queensland, as well as Tasmania, is most similar morphologically to *I. rubricosa*. It has larger fruit bodies and greenish or olive tones at the stipe base, while *I. cyanotrichia* Matheny, Bougher & G.M.Gates has larger fruit bodies and frequent bluish tinges to the pileus.

Inocybe rubricosa is known only from the type collection, but its diagnostic features and phylogenetic placement (Fig. 142) strongly support its autonomous status.

4. **Inocybe viridipes** Matheny, Bougher & G.M.Gates, *sp. nov.*

MycoBank: MB 809479

Type: Sunday Creek Rd, just E of Charlie Moreland Camp, SE of Kenilworth, Qld, 26°35'S, 152°44'E, scattered singly in leaf litter on soil beside track in mixed wet sclerophyll forest under *Eucalyptus*, 29 Feb. 2012, *P.B.Matheny PBM3767*; holo: PERTH 08363390; iso: TENN 066999 (KP171094-nLSU, KP641645-ITS, KM656138-*rpb2*).

Pileus 12–35 mm wide, at first conical or broadly convex, conical to obtusely conical or campanulate upon expansion; margin initially incurved, adhering to the stipe in buttons, soon decurved, often torn with age, fibrillose on surface or at first with scattered and appressed white floccose patches of veil remnants, eventually developing appressed squamules around the disc or becoming excoriate; surface dry, initially smooth and unbroken at the centre, fibrillose to coarsely fibrillose towards the margin; colour umber to brown (7.5YR 4/4 to 5/4), becoming dark yellowish brown (10YR 4/3 to 4/4; Snuff Brown) or with yellowish brown (10YR 5/4) margin, when very young dark reddish brown (5YR 3/3); context whitish, slowly rubescent or areas subtly staining pinkish, up to 3 mm thick under the disc. *Lamellae* adnexed to ascending-adnate, subdistant to moderately close, 20–26 L, with several tiers of lamellulae; thickened edges pallid-fimbriate but more distinctive under a hand lens, broad; colour brown, dark yellowish brown, or darker with age (10YR 4/3, 4/4, 3/4; Bay Brown). *Stipe* 36–70 × 3–5 mm, central to excentric, terete, less frequently compressed, even; partial veil not observed; surface not pruinose, fibrillose along its entire length, not at all scaly; pale brown to brown (10YR 6/3 to 5/3), a shade darker below (brown); base with greenish or olive tinges; context solid, becoming hollow, weakly rubescent in places, otherwise concolorous with the stipe surface. *Odour* none. Pl. 237–239.

Basidiospores 9.5–12.0 (–13.5) × 5.0–7.0 µm [mean 10.7 × 6.0 µm; Q 1.54–2.04 (–2.20); Q mean 1.80 (n = 40/2)], smooth, elliptic to subreniform (mostly adaxially flattened rather than concave), yellowish brown, some with walls to 1.5 µm thick; apiculus often indistinct. *Basidia* 38–52 × 8–10 µm, cylindrical to narrowly subclavate, hyaline but often ochraceous and collapsed (necropigmented), usually 4-sterigmate, occasionally 2-sterigmate. *Pleurocystidia* absent. *Cheilocystidia* 16–48 × 6–14 µm, cylindrical to clavate, hyaline except sometimes with golden-granular and amorphous inclusions, thin-walled. *Caulocystidia* absent. *Pileipellis* a cutis of cylindrical hyphae 5–14 µm, encrusted, rusty brown in mass. *Clamp connections* present. Fig. 146.

Known from south-eastern Qld and Tas.; grows on soil in wet or dry sclerophyll forest or in cool- or warm-temperate rainforest. Fruiting Nov.–Mar.

Additional specimens examined: Tas.: Taranna rainforest, SE Tasmania, 43°03'S, 147°54'E, 13 Nov. 2001, *G.M.Gates & D.Ratkowsky 111* (HO 570442, TENN 065720; JQ815430-nLSU, JQ801419-ITS); Myrtle Gully, 17 Jan. 2002, *G.M.Gates & D.Ratkowsky 156* (HO 570444, TENN 065721; JQ815429-nLSU, JQ801418-ITS); Truganini Track, 10 Nov. 2004, *G.M.Gates 196* (HO 570443, TENN 065722); *loc. id.*, 10 Mar. 2011, *G.M.Gates 1153* (HO 562693; KP171095-nLSU, KP641646-ITS, KM656139-*rpb2*).

Etymology: From the Latin *viridis* (green) and *pes* (a foot), in reference to the greenish stipe base.

Diagnostic features: Stipe greenish at the base, fibrillose, not scaly; odour not distinctive; pileus fibrillose to coarsely fibrillose, lacking bluish tinges on the disc. Spores 9.5–12.0 (–13.5) × 5.0–7.0 µm.

Like other species surrounding *I. calamistrata* (Fr. : Fr.) Gillett and its allies (*Inocybe* sect. *Cervicolores* of Larsson *et al.*, 2009), *I. viridipes* has a reddening context, at least some subreniform spores, and elongate-cylindrical to subclavate basidia. The basidia collapse and fill with necropigment as in *I. cyanotrichia* (see above) and *I. misakaensis* Matheny & Watling from Zambia (Matheny & Watling, 2004). The odourless condition differs from most other species in sect. *Cervicolores* which, in the Northern Hemisphere, produce distinctive odours like *Pelargonium*, ripe pears, fish, or have a spermatic smell.

Inocybe cyanotrichia, also recorded from Tasmania, differs rather subtly in the presence of blue-green tinges on the pileus disc. *Inocybe calamistratoides* has a squarrose pileus, squamulose stipe, a fruity odour, and it also occurs in Tasmania under *Nothofagus* mixed with *Leptospermum*, while *I. latericia* E.Horak differs from *I. viridipes* in the squamulose pileus and stipe, an unpleasant (non-spermatic) odour and the absence of a greenish stipe base (Horak, 1977). Comparative examination of their holotypes (PDD 27121 and PDD 27120, respectively) with *I. viridipes* shows that *I. calamistratoides* is more similar to *I. viridipes* than is *I. latericia*. The spores of the last are cylindrical-oblong in face view and more adaxially concave than those of the other two species.

Other species with a probable alliance in the *Cervicolores* clade include *I. umbrinovirens* E.Horak, from under *Nothofagus* in Papua New Guinea, and *I. fuscospinulosa* Corner & E.Horak from mountains in Java Indonesia where it grows in association with *Castanopsis* and *Lithocarpus* (Fagaceae; Horak, 1980). The former has a recurved-squamulose pileus, a strong odour (sweet, fruity, in age like burnt hair), and broadly ovate spores (10–12 × 7–8.5 µm), while the latter bears spiny, erect scales on the pileus and has smaller spores (6.5–8 × 4–5 µm).

A photograph labelled *I. viridipes* and included in a recent Tasmanian field guide by Gates & Ratkowsky (2014) has a greenish blue tinge to the pileus, and most likely represents *I. cyanotrichia*.

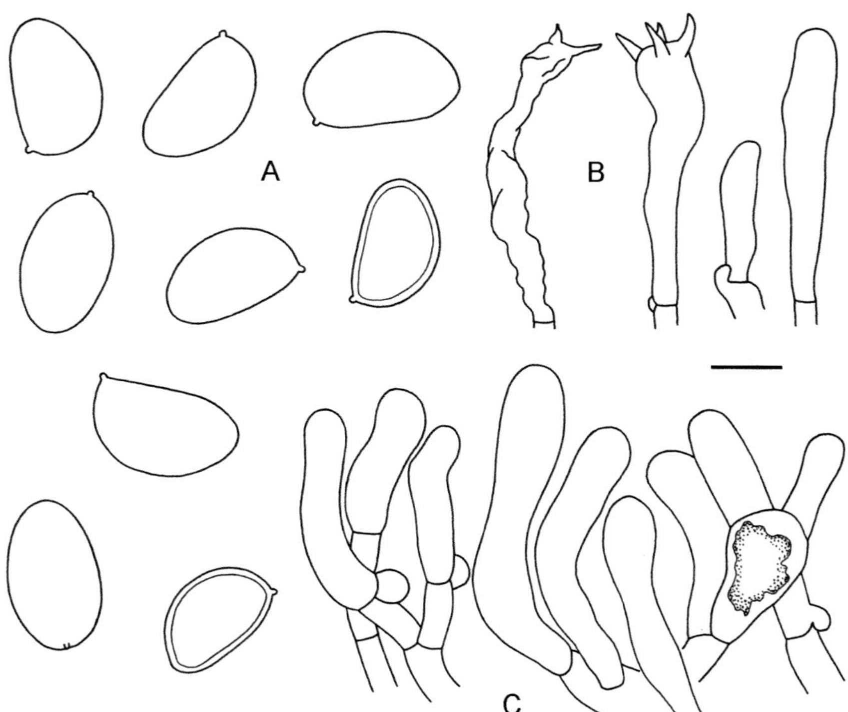

Figure 146. *Inocybe viridipes*. All holotype (PERTH 08363390). **A,** Basidiospores; **B,** Basidia at various stages of development; **C,** Cheilocystidia. Scale bars = 10 µm (longer bar for spores only).

Plate 237. *Inocybe viridipes*, holotype (PERTH 08363390). Note greenish stipe base. Photograph — P.B.Matheny.

Plate 238. *Inocybe viridipes*. HO 562693. Young specimen showing a white velar fringe on the pileus margin. Photograph — G.M.Gates.

Plate 239. *Inocybe viridipes*. HO 562693. Dissected specimens showing the greenish stipe. Photograph — G.M.Gates.

(*INOCYBE*) *MALLOCYBE* CLADE

Inocybe subgenus *Mallocybe* Kuyper, *Persoonia*, Suppl. Vol. 3: 22 (1986)

Type: *I. terrigena* (Fr.) Kuyper

Pileus often woolly-squamulose or tomentose, darkening with KOH. *Lamellae* adnexed, adnate, or subdecurrent.

Basidia necropigmented. *Pleurocystidia* absent. *Cheilocystidia* present, up to 50 µm long in all species except for *I. unicolor*, thin-walled, hyaline, arising as terminal elements of the hymenophoral trama, absent in some collections. Caulocystidia absent.

The taxonomic circumscription is modified from Kuyper (1986). Currently, five species of the *Mallocybe* clade are recognized in Australia, associating with eucalypts, *Allocasuarina* and *Acacia*. Not known from *Nothofagus* rainforests.

KEY TO SPECIES OF THE MALLOCYBE CLADE IN AUSTRALIA

1 Fruit bodies short and squat; cheilocystidia mostly cylindric .... **I. cf. isabellina** (see 1. I. isabellina)

1: Fruit bodies with a long slender stipe; cheilocystidia predominantly short to broadly clavate or pyriform 2

2 Spores long and narrow (mean Q >2.2) .... **5. I. subtilior**

2: Spores wider than above (mean Q <1.90) 3

3 Stipe base with fiery orange aggregated fibrils .... **2. I. pyrrhopoda**

3: Stipe base lacking fiery orange fibrils 4

4 Spores noticeably pale (pale brown in KOH) .... **4. I. subflavospora**

4: Spores darker than above (yellowish brown in KOH) 5

5 Spores mostly elliptic, 9.0–12 × 5.0–6.5 µm; edges of lamellae not entirely sterile; odour indistinct or weakly spermatic .... **1. I. isabellina**

5: Spores mostly phaseoliform and 8.0–11.0 × 4.5–6.0 µm, edges of lamellae entirely sterile; odour distinctly spermatic .... **3. I. sabulosa**

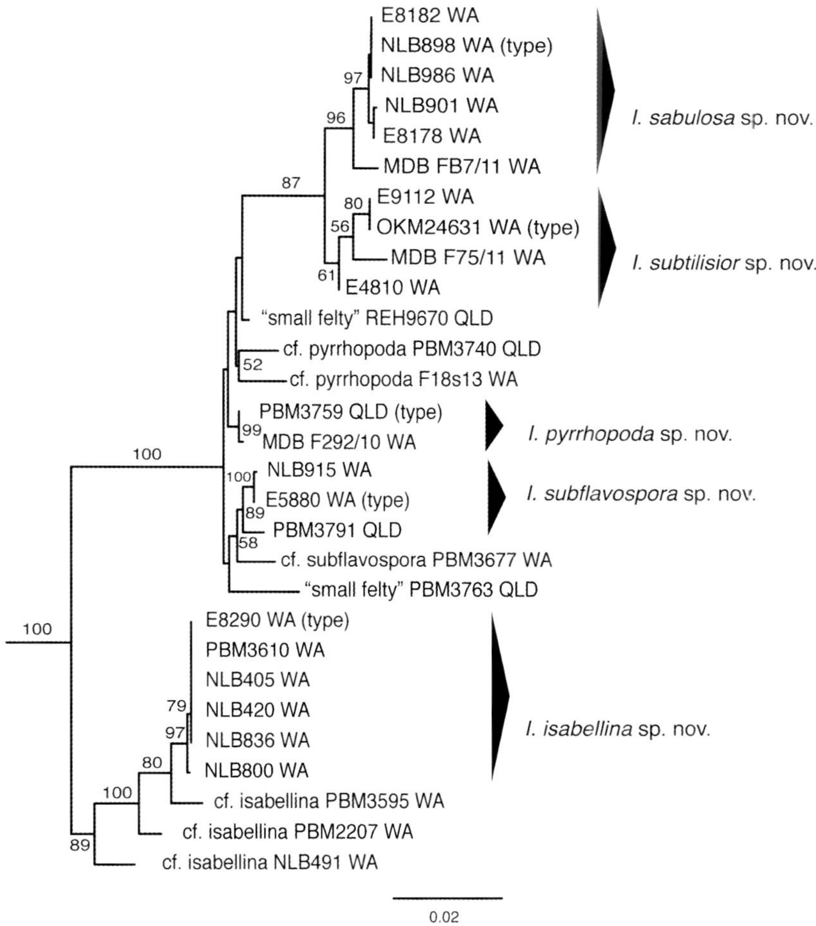

Figure 147. Maximum Likelihood phylogram of Australian species of the *Mallocybe* clade. Bootstrap values are indicated above or below branches or to the right of internodes. Species from the Southern Hemisphere form a monophyletic group based on phylogenetic analysis of nLSU and *rpb2* gene trees (data now shown). The tree was inferred from combined ITS, nLSU, and *rpb2* gene coding regions and included 2881 total sites. Recognised species from Australia are indicated by black-filled triangles. *Inocybe isabellina* was used for rooting purposes based on more inclusive gene tree sampling. Of the 32 branch tips in the tree, 27 are represented by ITS, all by nLSU, and 28 by *rpb2*.

1. **Inocybe isabellina** Matheny & Bougher, *sp. nov.*

MycoBank: MB 809384

Type: Floreat Park, E of Perry Lakes Stadium, Perth, W.A., 31°56'48"S, 115°47'40"E, among grass near *Eucalyptus gomphocephala*, 20 July 2005, *N.L.Bougher E8290*; holo: PERTH 07699255 (KP170915-nLSU, KP171137-ITS, KJ811581-*rpb2*).

Pileus 20–25 mm wide, obtusely conical to convex, becoming flattened-convex; umbo absent; margin inrolled at first and at times with a narrow sterile flap, becoming decurved; surface dry; velipellis absent, at first coarsely and thickly matted-fibrillose to tomentose or felty, becoming radially orientated, at times with minute scurfy appressed scales, not rimose; background colour dull brown to brown with a slight ochre tinge (near 6E8 to 7E8), becoming yellowish brown to dark yellowish brown (7.5YR 4/4, 10YR 4/4 to 5/4; Buckthorn Brown to Dresden Brown); context pallid to pale olive-yellow or yellowish, not changing where bruised, to 4 mm thick under the disc. *Lamellae* adnate to sinuate or weakly subdecurrent, moderately close to close, 28–40 L with several tiers of lamellulae; at first pallid to olive-grey, becoming pale tan with persistent yellowish tinges or olivaceous-yellow (near 4B4; 2.5Y 5/6), eventually dull brownish to olive-brown; not ventricose, 2–3 mm deep; edges pallid-fimbriate but not distinctly so. *Stipe* 22–35 × 2–6 mm, typically longer than the pileus diameter, even, terete, not (or rarely) squat; surface dry, covered with minute longitudinally appressed hyaline fibrils, at times coarsely fibrillose, less often these fibrils aggregate into scattered dark fibrillose squamules (as in NLB800); veil cortinate, evanescent, pallid; dirty cream to pale yellow mixed with olivaceous tones, with brown longitudinal streaks, not pruinose; basal mycelium not conspicuous; context solid, becoming hollow with age. *Odour* usually nondescript or slightly spermatic. Pl. 240, 241.

Basidiospores 9.0–12.0 (–13.0) × 5.0–6.5 (–7.5) µm [mean 10.5 × 6.0 µm; Q 1.50–2.18; Q mean 1.77 (*n* = 55/3)], smooth, variable in shape, elliptic to ovate, oblong-cylindrical to elongate-elliptic, or (sub)amygdaliform, ovate or elongate-ovate in face view; apices rounded to bluntly pointed, with adaxial region often slightly depressed, yellowish brown; apiculus small, conspicuous or not. *Basidia* 32–43 (–50) × 7–9 µm, narrowly clavate, granular, soon becoming necropigmented, 4-sterigmate. *Pleurocystidia* absent. *Cheilocystidia* 17–36 × 6–16 µm, short-clavate or vesiculose, at times cylindrical or pyriform, often more than one terminal cell branching from a common subtending hypha that is encrusted or not, at times forming short concatenated chains, thin-walled, hyaline, lacking crystals, crowded in clusters but the edges of lamellae not entirely sterile. *Caulocystidia* absent. *Pileipellis* a cutis bearing tufts of trichodermally arranged hyphae, these cylindrical to inflated, to 25 µm wide, coarsely encrusted, thin-walled. *Clamp connections* present. Fig. 148 (see also see Pl. 10F, 12C).

Known from south-western W.A., from Perth southward, and east to Two Peoples Bay; in urban bushland, parks, often on calcareous ground but also among burnt debris (charcoal) in Jarrah forest or woodland, or coastal heath-scrub. Usually associated with *Eucalyptus gomphocephala*, *E. marginata*, *Corymbia calophylla* or shrubs of *Spyridium*, *Acacia* and *Melaleuca* mixed with *Agonis flexuosa*; also in association with planted eucalypts. Fruiting June–Sept.

Additional specimens examined: W.A.: Cobb Drabble Reserve, City Beach, Perth, 31°54'36.1"S, 115°45'56.2"E, among grass near *Eucalyptus gomphocephala* trees in urban park, 24 Aug. 2007, *N.L.Bougher NLB405* (PERTH 08096635, TENN 069512; KP170916-nLSU, KP171138-ITS, KJ811582-*rpb2*); Ocean Village park, City Beach, Perth, 31°54'41"S, 115°45'55"E, in grass near *Eucalyptus gomphocephala* trees in an urban park, 8 June 2008, *N.L.Bougher NLB420* (PERTH 08072442; KP170917-nLSU, KP171139-ITS, KJ811583-*rpb2*); Bridgetown Rd, Balingup, 1 km N of Catterick Rd turn-off, 33°50'50.6"S, 116°00'50.3"E, among charcoal on recently burnt ground in Jarrah-Marri forest under *Eucalyptus marginata* and *Corymbia calophylla*, 12 July 2011, *N.L.Bougher NLB800* (PERTH 08318832, TENN 066800; KP170918-nLSU, KP171140-ITS, KJ811584-*rpb2*); Catterick Rd, track BB252, near Greenbushes, 33°51'08.6"S, 116°06'34.4"E, in Blue Gum plantation in Jarrah-Marri forest, 13 July 2011, *P.B.Matheny PBM3610* (PERTH 08319766, TENN 066571; KP170919-nLSU, KJ811585-*rpb2*); Sinker Reef Track, Two Peoples Bay Nature Reserve, 34°56'54.2"S, 118°09'58.8"E, in litter in heath scrub under *Agonis flexuosa*, *Spyridium*, *Acacia* and *Melaleuca*, 19 July 2011, *N.L.Bougher NLB836* (PERTH 08318441, TENN 066691; KP170920-nLSU, KP171141-ITS, KJ811586-*rpb2*).

Additional specimens examined (without molecular confirmation): W.A.: "Markets", Wanneroo, Perth, 31°45'S, 115°48'E, abundant on forest trail, 21 Sept. 1978, *J.Daams s.n.* (PERTH 00777838); L. Joondalup, Yellagonga Regional Park, Edgewater, Perth, 31°45'S, 115°47'E, 1 June 1982, *J.Daams s.n.* (PERTH 00910929); Kalbarri, N of Geraldton, 27°40'59.9"S, 114°13'00.0"E, in sandy soil under *Chamelaucium*, 15 June 1989, *B.Dell OKM23955 (E667)* (PERTH 07547757); Kings Park, Perth, 31°56'S, 115°49'E, on ground under *Eucalyptus marginata*, *Acacia* and *Jacksonia*, 31 May 1991, *O.K.Miller et al. OKM24690* (PERTH 08561222); park at corner of Selby Rd and Grantham Rd, Floreat, Perth, under *Eucalyptus*, 6 July 1991, *L.Bailey OKM25097* (PERTH 08561214); Warwick Open Space, Perth, 31°49'S, 115°48'E, solitary on sandy limestone soil mixed with buried charred remains under *Eucalyptus marginata*, *Corymbia calophylla* and *Banksia attenuata*, 1 July 2001, *P.B.Matheny PBM2141 (E6898)* (PERTH 07667388); off Brookdale Street, E of Perry Lakes Stadium, Floreat, Perth, 31°55'S, 115°46'E, on soil in park under planted eucalypts including *E. gomphocephala*, also introduced species, 6 Sept. 2001, *P.B.Matheny PBM2211 (E7059)* (PERTH 07677154); Ludlow, 33°35'15.6"S, 115°30'02.7"E, on limestone sandy track under *Eucalyptus gomphocephala*, 18 Sept. 2001, *N.L.Bougher & I.C.Tommerup E7063/E4041* (PERTH 07677170); Sinker Reef Track, Two Peoples Bay Nature Reserve (500 m away from *NLB836* [see above]), 34°56'58"S, 118°09'59.6"E, in sand in heath scrub under *Agonis flexuosa*, *Spyridium*, *Acacia* and *Melaleuca*, 19 July 2011, *N.L.Bougher NLB838* (PERTH 0831425); Woodlands Rd,

Porongorup Natl Park, 34°40'06.4"S, 117°52'53.6"E, in tall Karri-Marri forest under *Eucalyptus diversicolor* and *Corymbia calophylla*, 21 July 2011, *N.L.Bougher NLB855* (PERTH 08320772); Forrest Drive, Kings Park, Perth, 31°58'04"S, 115°50'11"E, on limestone track in burnt woodland of *Eucalyptus gomphocephala* and *Acacia*, 5 Aug. 2011, *N.L.Bougher NLB909* (PERTH 08309078).

Other specimens examined (but considered variants based on molecular phylogenetic and/or morphological results): W.A.: In sand dunes along Ritter Rd near L. Yeagerup, W of Vasse Hwy and Pemberton, 23 June 1985, *N.L.Bougher E3730* (PERTH 07712758, TENN 070158); Two Peoples Bay Nature Reserve, near Research Quarters, on side of track near picnic area, 34°59'4"S, 118°10'59"E, gregarious on ground under *Agonis flexuosa*, 27 Mar. 1992, *K.Syme, M.Hart & C.Hart KS370/92* (PERTH 05254728, TENN 066746); Kings Park, Perth, Rokeby Rd entrance, 31°56'59.9"S, 115°49'59.9"E, in grass under planted *Corymbia calophylla*, 3 Sept. 2001, *P.B.Matheny PBM2207 (E7056)* (PERTH 07676840, TENN 066822; AY380381-nLSU, AY337382-*rpb2*, AY351806-*rpb1*); Cobb-Drabble Park, City Beach, Perth, 31°54'30.1"S, 115°45'56.0" E, among grass in parkland near *Eucalyptus gomphocephala* and *Agonis*, 16 Sept. 2008, *N.L.Bougher NLB491* (PERTH 08073287; KP170921-nLSU, KP171142-ITS, KJ811587-*rpb2*); site 3, near the pipeline, Wellington Natl Park, W of Collie, 33°23'34.3"S, 115°58'38.0"E, scattered singly on burnt sandy ground, on charcoal, in Jarrah forest under *Eucalyptus marginata* and *Corymbia calophylla*, 11 July 2011, *N.L.Bougher PBM3595* (TENN 066498; KP170922-nLSU, KP171143-ITS).

Etymology: From the Latin *isabellinus* (soiled tawny yellow), in reference to the yellowish to olivaceous lamellae.

Diagnostic features: Fruit bodies medium-sized. Pileus felty-tomentose, dark reddish brown when fresh; lamellae often olivaceous yellow and adnate or subdecurrent; stipe length longer than pileus width; odour not distinctive or slightly spermatic. Spores smooth, yellowish brown, variable in shape and size but mean Q usually <1.90; cheilocystidia short (<45 µm long), thin-walled; pleurocystidia absent.

Inocybe isabellina is described to accommodate a phylogenetically distinct part of a rather common species complex of the *Mallocybe* clade in the south-west of Western Australia. Due to variation in spore shape and size and ecological observations, we have based our description of *I. isabellina* on a core group of collections for which DNA sequence data are minimally divergent (Fig. 147) and centred around the holotype. Genetic variants are discussed further below.

Based on observation of sequenced specimens, the predominance of inflated cheilocystidia (e.g. in the holotype) and/or narrow ones (e.g. *NLB420*, *PBM3610*) can vary among fruit bodies of the same collection (as in *NLB800*). Therefore, it seems likely that the shape, size, and abundance of cheilocystidia can vary with age (sometimes becoming more inflated with time), and can also be affected by the availability of moisture during development. *NLB491* (as *I.* cf. *isabellina*) has invariably cylindrical cheilocystidia, a feature that can distinguish it taxonomically, and is consistent with its genetic divergence (Fig. 147).

Several lineages (Fig. 147) differ from core *I. isabellina* in various ways, but all are only known from single collections: (1) *PBM3595* (*I.* cf. *isabellina*) lacks the spermatic odour and was found on charcoal on burnt ground in Jarrah forest; the ITS sequence of this specimen is 98% similar to those of core *I. isabellina*; (2) *PBM2207* (*I.* cf. *isabellina*) appears to be a cryptic genetic variant but with a spermatic odour; it too was found among charcoal in urban Jarrah woodland; (3) *NLB491* (*I.* cf. *isabellina*) includes the largest specimens observed to date (pileus up to 50 mm wide), a comparatively short stipe, and mostly cylindrical cheilocystidia. The squat habit is not consistent with *I. isabellina*, and its ITS sequence is 94% similar to core *I. isabellina*. Eventually, *NLB491* could be regarded as a distinct species if it is supported by additional material.

Inocybe sabulosa Matheny & Bougher usually has a distinct spermatic smell, unlike most collections of *I. isabellina*, with more reliable taxonomic differences including the predominantly phaseoliform spores and entirely sterile lamellar edges in *I. sabulosa*. *Inocybe subtilior* Matheny & Bougher has long narrow spores with a mean Q value >2.2, while *I. subflavospora* Matheny & Bougher is characterised by its noticeably pale pigmented spores.

Several broadly similar collections of *Inocybe* from northern Australia (Western Australia, Northern Territory, and Queensland) fail to form discrete groups in our phylogenetic analyses. These are superficially similar to *I. isabellina*, but they generally lack the yellowish or olivaceous lamellae of that species. One of these, the newly described *I. pyrrhopoda* Matheny & Bougher, differs from *I. isabellina* by the presence of rust-coloured fibrils often aggregated above the stipe base. No tropical collections of *I. isabellina* have yet been recorded.

Plate 240. *Inocybe isabellina*, holotype (PERTH 07699255). Photograph — N.L.Bougher.

Plate 241. *Inocybe isabellina*. PERTH 08319766. Photograph — P.B.Matheny.

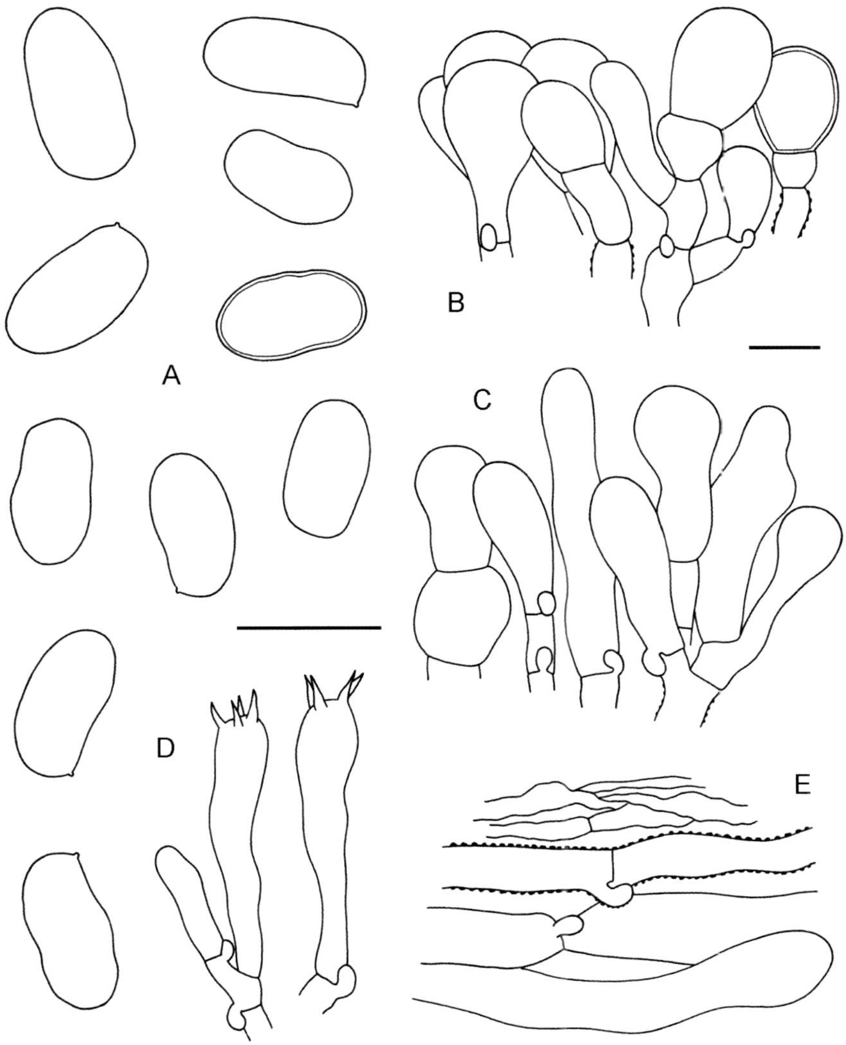

Figure 148. *Inocybe isabellina.* A, B holotype (PERTH 07699255); C, E (PERTH 08072442); D (PERTH 08096635). **A,** Basidiospores; **B, C,** Cheilocystidia; **D,** Basidia.; **E,** Pileipellis. Scale bars = 10 µm (longer bar for spores only).

2. Inocybe pyrrhopoda Matheny & Bougher, *sp. nov.*

MycoBank: MB 809448

Type: Mount Whitfield Conservation Park, Cairns, Qld, 16°53'53.8"S, 145°45'03.1"E, gregarious to scattered or in small clusters on soil among leaf litter in lowland, tropical *Melaleuca* swamp, under *Melaleuca viridiflora*, *Lophostemon suaveolens* and *Corymbia tesselaris*, 25 Feb. 2012, *P.B.Matheny PBM3759*; holo: PERTH 08363250; iso: TENN 066987 (KP170983-nLSU, KP308813-ITS, KM406223-*rpb2*).

Pileus 5–21 mm wide, hemispherical when young, becoming conical to cbtusely so, convex, soon plane; margin at first slightly incurved, soon decurved, uplifted and/or undulate with age; umbo absent or indistinct, less often with a central depression; surface dry; velipellis absent; margin neither striate nor rimose; surface dry, scurfy-furfuraceous when young but noticeably so with age, at times breaking up into small appressed scales around the disc and outwards towards the margin; when fresh strong brown to orange-brown (7.5YR 5/6 to 5/8; 10YR 5/6) or ferruginous (5YR 4/8; Hazel), at times with shades of reddish yellow (7.5YR 6/8) at the centre, with age fading to brownish yellow or yellow (10YR 6/6 to 7/6), generally hazel to fulvous; context dingy pallid, not yellow, not changing colour where exposed, 1–2 mm thick under the disc. *Lamellae* adnexed, moderately close to subdistant, 20–28 L, with several tiers of lamellulae; yellowish tan (10YR 6/6), pale brown (10YR 5/6), to yellowish brown (10YR 5/4) even when young, becoming dark yellowish brown or umber with age (10YR 4/4 to 7.5YR 4/4); edges much paler than faces, indistinctly fimbriate, medium, ventricose, up to 3 mm deep. *Stipe* 10–32 × 1–3 mm at apex, usually longer than pileus width, even or tapered towards a base that is swollen at times, with cortina not directly observed but the surface nowhere pruinose, finely-fibrillose throughout with scattered loose fibrils at times or with age, pallid at the apex and downwards or pale yellow (2.5Y 7/4); base often covered with rusty reddish (ferruginous, 2.5YR 4/8; 7A8) fibrils; context hollow at least with age, ochraceous above, rusty brown below. *Odour* spermatic. Pl. 242–244.

Basidiospores 8.0–10.5 × 5.5–6.5 µm [mean 9.1 × 5.8 µm; Q 1.42–1.82; Q mean 1.57 (n = 35/2), smooth, mostly elliptic or subamygdaliform, with bluntly pointed or rounded apices, not phaseoliform, ovate in face view, yellowish brown; apiculus small, distinct or not. *Basidia* 28–36 × 8–11 µm, narrowly clavate to cylindro-clavate, hyaline, becoming necropigmented, 4-sterigmate. *Pleurocystidia* absent. *Cheilocystidia* 17–32 × 8–11 µm, mostly short-clavate, clavate or subcylindrical, thin-walled, hyaline, arising from long narrow hyphae that are sometimes finely encrusted, densely arranged and forming a sterile lamellar edge. *Caulocystidia* absent; at base of stipe a vesture of narrowly cylindrical superficial hyphae mostly 4–5 µm wide, smooth, mainly thin-walled but sometimes thick-walled, clamped, rusty brown to ochraceous brown in mass. *Pileipellis* a cutis with scattered erect or ascending clusters of hyphae that are cylindrical to inflated, mostly 8–15 µm wide, encrusted, thin-walled, tawny to rusty brown in mass. *Clamp connections* present. Fig. 149 (see also Pl. 10E, 12H).

Occurs in tropical W.A. and Qld; grows on soil in association with eucalypts, *Melaleuca* and/or Phyllanthaceae in lowland habitats. Fruiting Jan., Feb. and Apr.

Additional specimens examined: W.A.: Charnley River Stn HS, West Kimberley, 16°42'48"S, 125°27'40"E, in dense low closed forest on rich black soil, the water table just below surface in wet season, under *Antidesma ghaesembila*, *Glochidion disparipes*, *Melaleuca viridiflora* and *Eucalyptus bigalerita*, 28 Jan. 2010, *M.D.Barrett MDB F292/10* (PERTH 08383278, TENN 069522; KP170984-nLSU, KP308814-ITS, KM406224-*rpb2*). Qld: type locality, 25 Feb. 2012, *N.L.Bougher NLB948* (PERTH 08364583).

Additional specimens examined (genetic variants): W.A.: Lake Argyle Rd, E of Kununurra, East Kimberley, 16°07'22"S, 128°44'31"E, on sandy clay in a flat plain gently sloping toward nearby creek, among scattered *Melaleuca minutifolia* and dense grasses, abundant in area c. 100 m wide, 8 Apr. 2013, *M.D.Barrett, R.L.Barrett & B.M.Anderson MDB F18s/13* (PERTH 08557764; KP170986-nLSU, KP308815-ITS, KM406226-*rpb2*). Qld: end of Emerald Creek Rd, Dinden S.F., near Mareeba, 17°03'08.5"S, 145°32'24.2"E, 24 Feb. 2012, *M.D.Barrett & N.L.Bougher PBM3740* (PERTH 08362963, TENN 066972; KP170985-nLSU, KM406225-*rpb2*).

Etymology From the Greek *pyrrhos* (flame-coloured) and *podos* (a foot), in reference to the rust-coloured fibrils at the stipe base.

Diagnostic features: Fruit bodies small. Pileus furfuraceous to appressed-scaly, generally hazel to fulvous; lamellae pale brown to yellowish brown (not yellow or olivaceous). Stipe pallid, finely fibrillose, often with rust-coloured fibrils at the base. Spores mostly elliptic to subamygdaliform; cheilocystidia short-clavate, clavate or cylindrical, densely arranged.

Inocybe pyrrhopoda is known thus far from tropical lowland regions of north-eastern Queensland and northern Western Australia, where it often produces an abundance of fruit bodies. This species is distinguished genetically (Fig. 147), and also morphologically from other species of the *Mallocybe* clade by the presence of rust-coloured fibrils at the stipe base. We noted however, that this feature is absent in a low proportion of specimens within each of the collections.

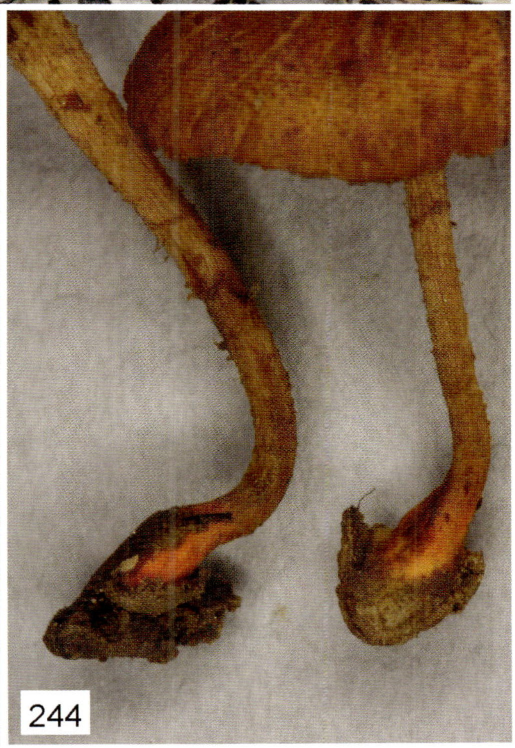

Plate 242. *Inocybe pyrrhopoda*, holotype (PERTH 08363250). Photograph — P.B.Matheny.

Plate 243. *Inocybe pyrrhopoda*. PERTH 08364583. Photograph — N.L.Bougher.

Plate 244. *Inocybe pyrrhopoda*. PERTH 08364583. Note the fiery orange stipe base of these specimens —a striking but apparently inconsistent feature of this species. Photograph — N.L.Bougher.

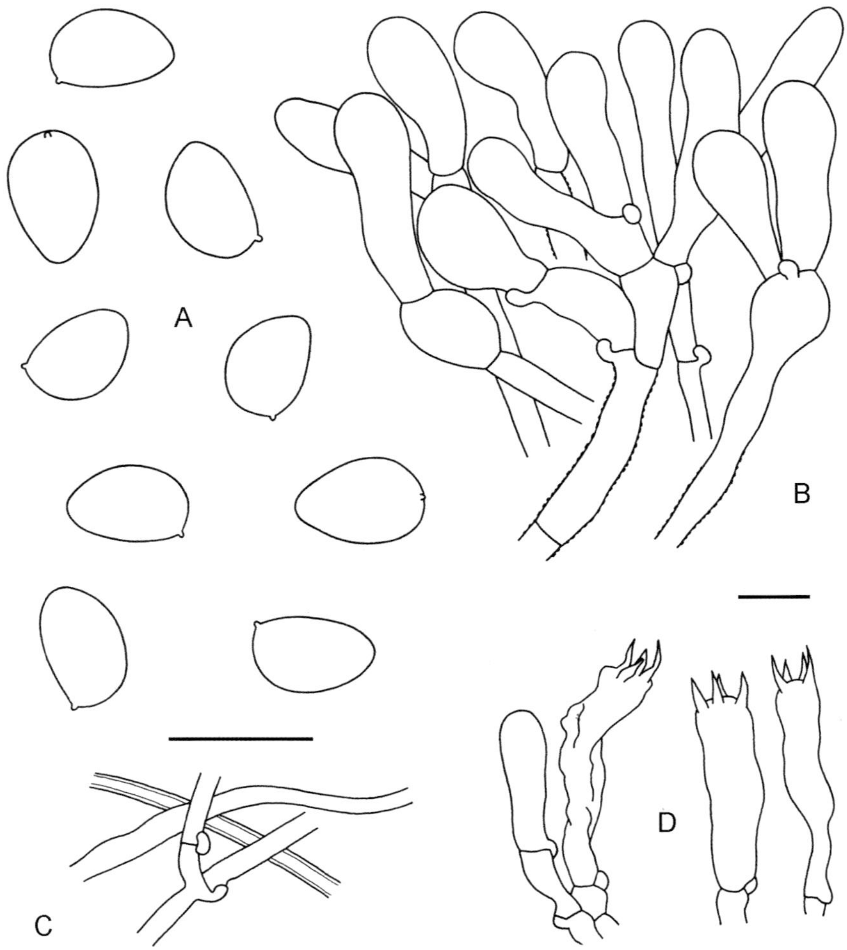

Figure 149. *Inocybe pyrrhopoda.* All holotype (PERTH 08363250). **A,** Basidiospores; **B,** Cheilocystidia; **C,** Superficial hyphae from base of the stipe; **D,** Basidia. Scale bars = 10 μm (longer bar for spores only).

3. **Inocybe sabulosa** Matheny & Bougher, *sp. nov.*

MycoBank: MB 809456

Type: Yarrada Farm, 1644 Great Northern Hwy, Upper Swan, W.A., 31°45'03.6"S, 116°02'41.1"E, singly or in groups of several on sand among scattered grassy weeds and near planted *Eucalyptus camaldulensis* trees, 30 July 2011, *N.L.Bougher NLB898*; holo: PERTH 08320322; iso: TENN 066806 (KP170994-nLSU, KP308821-ITS, KM406234-*rpb2*).

Pileus 10–40 mm wide, initially hemispherical, becoming convex; margin inrolled initially, later incurved, at times undulate; surface dry; velipellis absent; felty or smooth in appearance but with superficial radially arranged to matted minute fibrils or scales that are paler than elsewhere; uniformly dark brown (near 6E6–6F6 or 7E8–7F8), unchanging with age; context dull greyish. *Lamellae* broadly adnexed, close, with several tiers of lamellulae, not ventricose, broad, up to 5 mm deep; (dull) yellowish in youth, becoming yellowish brown (near 5C5) or dark yellowish brown (6D5–6D6); edges smooth, not noticeably fimbriate, slightly paler than the faces. *Stipe* 20–50 × 4–8 mm, tapered towards the base; surface dry; veil whitish cortinate and fugacious; densely longitudinally silky-fibrillose, nowhere pruinose, dull whitish, cream or straw (yellowish); basal mycelium not conspicuous; context dull whitish, solid, becoming hollow. *Odour* spermatic. Pl. 245–248.

Basidiospores 8.0–11.0 × 4.5–6.0 µm [mean 9.5 × 5.2 µm; Q 1.58–2.12 (–2.20); Q mean 1.84 (*n* = 69/4)], smooth, mostly subphaseoliform to phaseoliform (less often elliptic), with rounded (or less often bluntly pointed) apices in profile, elliptic to ovate in face view, yellowish brown; apiculus small and indistinct. *Basidia* 30–37 × 9–11 µm, narrowly clavate, hyaline, becoming necropigmented. *Cheilocystidia* 22–37 × 12–18 µm, clavate, pyriform or sphaeropedunculate (none cylindrical), abundant and densely arranged, imparting a sterile lamellar edge, arising from long, narrow, clamped hyphae. *Pleurocystidia* absent. *Caulocystidia* absent. *Fileipellis* a tangled cutis of irregular clusters of upright or interwoven hyphae, dark tawny to ochre-tawny in mass; hyphae thin-walled, encrusted, cylindrical, 5–12 µm wide. *Clamp connections* present. Fig. 150.

Occurs widely in south-western W.A. and the Kimberley region in northern W.A.; grows in sandy, grassy or weedy areas, at times on calcareous ground, near woodland species of *Acacia*, *Corymbia* and *Eucalyptus*. Fruiting May–Sept. (south-western), Jan. (Kimberley).

Additional specimens examined: W.A.: type locality, 21 Sept. 2013, *N.L.Bougher NLB1676* (PERTH 08577560); Star Swamp (near Hope and Castle Streets), North Beach, Perth, 31°50'59"S, 115°45'29"E, on soil in weedy area on edge of woodland of *Eucalyptus gomphocephala*, *E. marginata* and *Corymbia calophylla*, 19 May 2005, *N.L.Bougher E8178* (PERTH 07680732; TENN 0695155; (JN974916-nLSU, KF308822-ITS, KM406235-*rpb2*); Bold Park, near corner of Elphin Rd and Oceanic Drive, Floreat, Perth, under mature *Eucalyptus gomphocephala*, 26 May 2005, *N.L.Bougher E8182* (PERTH 07680775; TENN 066818; KP170995-nLSU, KP308823-ITS, KM406236-*rpb2*); Port Kennedy Scientific Park, Perth, in sand, coastal dunes in *Acacia rostellifera* shrubland, 17 July 2007, *N.L.Bougher & R.Miles E9057* (PERTH 08094594); bushland on S side of Great Northern Hwy, SW corner of Kununurra, 15°46'22"S, 128°43'52"E, on soil among grassy understorey below *Albizia* and *Corymbia* sp., 10 Jan. 2011, *M.D.Barrett MDB F7/11* (PERTH 08383057; TENN 069516; KP170996-nLSU, KP308824-ITS); Reabold Hill, eastern slope (just N of path), Bold Park, Perth, 31°56'18"S, 115°46'48.7"E, on soil in shrubland with meagre leaf litter and some moss, under *Acacia*, *Templetonia* and scattered *Eucalyptus gomphocephala*, 1 Aug. 2011, *N.L.Bougher NLB901* (PERTH 08320292, TENN 066809; KP170997-nLSU, KP308825-ITS, KM406237-*rpb2*); Woodlake Retreat (Rd), Kingsley, Perth, 31°48'49.8"S, 115°49'11.0"E, in sand near edge of road with *Corymbia calophylla* and *Acacia* sp. (tall shrubs) within 5 m, 5 May 2012, *N.L.Bougher NLB986* (PERTH 08383308; TENN 069514; KP170998-nLSU, KM406238-*rpb1*).

Etymology: From the Latin *sabulosus* (sandy), in reference to these fungi growing in sandy places.

Diagnostic features: Fruit bodies medium-sized. Pileus dark brown, felty; lamellae initially yellowish, becoming dark yellowish brown, with edges indistinctly fimbriate but sterile; stipe tapered, finely fibrillose, nowhere pruinose. Odour spermatic. Spores mostly slightly phaseoliform, yellowish brown; mean Q <2.0. In woodland under eucalypts or shrubs including *Acacia*, often on sandy soil.

Inocybe isabellina Matheny & Bougher is most similar to *I. sabulosa*, but it differs in its indistinct to weakly spermatic smell, predominantly elliptic (non-phaseoliform) spores and partially fertile lamellar edges. Other species in the *Mallocybe* clade differ from *I. sabulosa* by the characters enumerated in the key.

DNA sequences of the Kimberley specimen (MDB F7/11; Fig. 147) confirm a very close relationship with *I. sabulosa*. However, it differs in the comparatively small size of its fruit bodies (pileus c. 4–6 mm wide; stipe 8–11 × 1 mm), the dark orange-brown pileus (5E7), somewhat wider spores (8.5–10.0 × 5.5–6.5 µm), and an isolated northerly geographic location. In spite of these differences, ITS sequence similarity is very high between northern and southern populations, differing only at nine nucleotide positions.

Plate 245. *Inocybe sabulosa*, holotype (PERTH 08320322). Photograph — N.L.Bougher.

Plate 246. *Inocybe sabulosa*. PERTH 08383308. Photograph — N.L.Bougher.

Plate 247. *Inocybe sabulosa*. PERTH 08383057. Photograph — M.D.Barrett.

Plate 248. *Inocybe sabulosa*. PERTH 08320292. Photograph — N.L.Bougher.

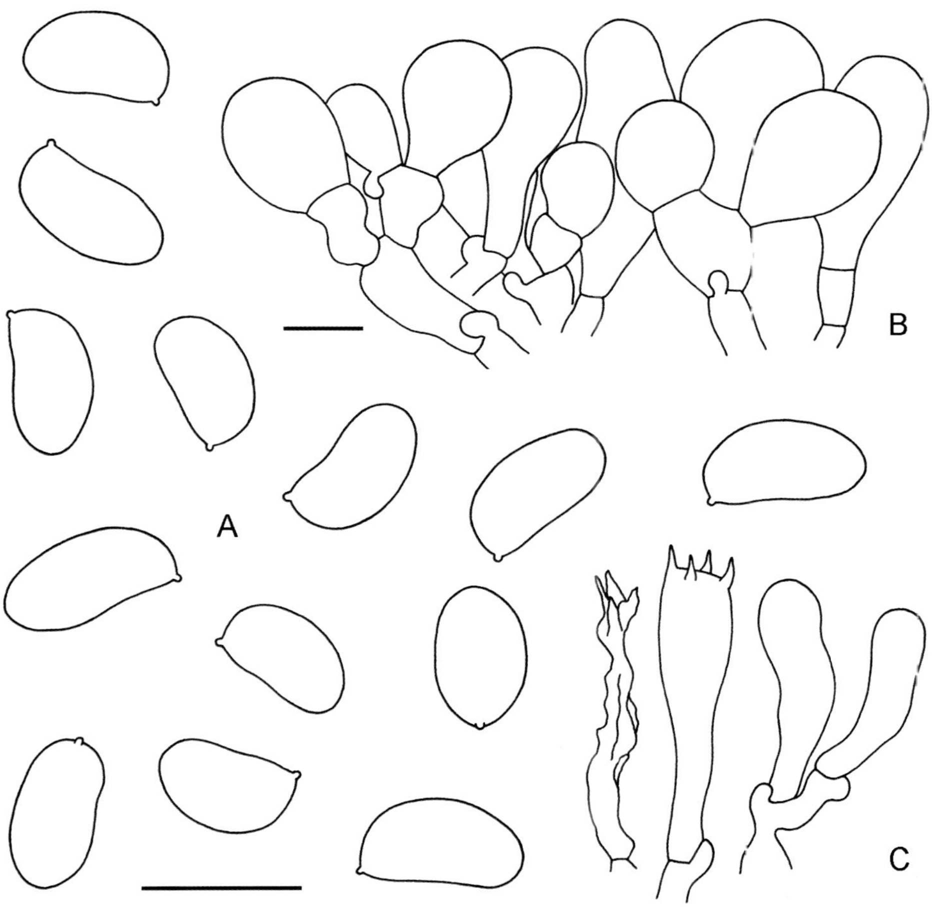

Figure 150. *Inocybe sabulosa.* All holotype (PERTH 08320322). **A,** Basidiospores; **B,** Cheilocystidia; **C,** Basidia. Scale bars = 10 µm (longer bar for spores only).

4. Inocybe subflavospora Matheny & Bougher, *sp. nov.*

MycoBank: MB 809465

Type: S side of Mundaring Weir (on opposite side of road to weir), Mundaring, E of Perth, W.A., Australia, 31°57'S, 116°10'E, singly in troops on mossy grassy ground near *Eucalyptus marginata* in disturbed area, co-occurring with *Laccaria* sp., 24 June 1997, *N.L.Bougher & M.E.Bougher E5880*; holo: PERTH 07609256; iso: TENN 066771 (AY380396l-nLSU, AY337404-*rpb2*, AY351820-21, *rpb1*).

Pileus 8–30 mm wide, initially convex but soon plano-convex, becoming flattened; umbo absent, not depressed; margin at first incurved, later decurved; surface dry, finely fibrillose-squamulose at the centre, appressed when young, or minutely scaly-velvety to matted-fibrillose and breaking into numerous small scales; dark yellowish brown (c. 6E8) to orange-brown (6D8 to 6E8), a shade lighter towards the margin, or reddish brown; context dull watery, not changing colour where bruised. *Lamellae* broadly adnexed to adnate, close, with several tiers of lamellulae; pale brown when young (not yellow), becoming brown (6D6) to dark brown with age, ventricose (or not), broad, up to 7 mm deep; edges concolorous, entire (not appearing fimbriate unless under a hand lens). *Stipe* 10–35 × 1–4 mm, longer than pileus width, even or slightly tapered towards the base; surface dry with veil cortinate and evanescent, densely silky-fibrillose for its entire length, nowhere pruinose; dull cream (c. 4A4); context pale yellowish, solid at first, becoming hollow. *Odour* not distinctive, or like cooked button mushrooms (*Agaricus*), not spermatic. Pl. 249, 250.

Basidiospores 8.0–12.0 × 5.0–6.5 µm [mean 10.0 × 5.7 µm; Q 1.45–2.13 (–2.18); Q mean 1.77 (*n* = 62/5)], smooth, variable in size and shape, *viz.* mostly elliptic to subamygdaliform, less often subphaseoliform or with a slight adaxial depression; apices obtuse to bluntly pointed; pale brown (Chamois, Ochraceous-Buff, Light Ochraceous-Buff); apiculus small but at times distinct. *Basidia* 32–49 × 7–10 µm, cylindroclavate, hyaline at first but becoming golden or necropigmented, 4-sterigmate (occasionally 2-sterigmate). *Pleurocystidia* absent. *Cheilocystidia* 19–42 × 9–13 (–16) µm, clavate to pyriform; lamellar edges sterile, thin-walled, hyaline; at times difficult to locate (e.g. PERTH 08320861). *Caulocystidia* absent, occasionally a few short clavate cells at the extreme apex of the stipe from a caulohymenium. *Pileipellis* a cutis of tangled repent to somewhat interwoven hyphae, at times forming erect fascicles, rusty brown or rusty fulvous in mass; hyphae coarsely encrusted but thin-walled, cylindrical to inflated, mostly 7–20 µm wide. *Clamp connections* present. Fig. 151.

Occurs in south-western W.A. and south-eastern Qld; in disturbed areas on the ground, in grass, often co-occurring with species of *Laccaria* near eucalypts (*Eucalyptus marginata* and *Corymbia calophylla*). Fruiting June–Sept. in W.A.; Feb. and Mar. in Qld.

Additional specimens examined: W.A.: Inglehope, Dwellingup, 32°43'S, 116°04'E, gregarious in open, lightly grassed area mingled with troops of *Laccaria*, 6 June 1980, *R.Hilton s.n.* (PERTH 00903493, TENN 066712); Gleneagle, 32°20'S, 116°04'E, in troops on bare ground on former Bee Farm house site, co-occurring with *Laccaria*, 1 July 1987, *R.Hilton UWA3480* (PERTH 00924202, TENN 066687); rest area, Albany Hwy, 48 km N of Williams, S of Perth, under *Corymbia calophylla*, 5 Sept. 1981, *O.K.Miller & H.Miller OKM19463* (PERTH 08561788); near workshop, CSIRO, Floreat Park, Perth, under *Eucalyptus* sp., 1 July 1991, *I.C.Tommerup E780 / OKM25035* (PERTH 0756449); Strawberry Hill Rd, Gidgegannup, E of Perth, 31°49'53.6"S, 116°08'38.0"E, scattered on ground in litter under *Eucalyptus camaldulensis* and *Hakea* sp. planted along fence-line, 4 Sept. 2011, *N.L.Bougher & M.J.Bougher NLB915* (PERTH 08320861, TENN 066825; KP171074-nLSU, KP641619-ITS, KM656118-*rpb2*). Lightning Swamp, Noranda, Perth, 31°52'15.3"S, 115°54'24.7"E, in sand in open banksia woodland, 8 July 2007, *R.Hart & D.Kidson E9033* (PERTH 08094322); Old Plains Rd, 3.6 km N of Calingiri Rd, 31°7'22.9"S, 116°15'58.7"E, gregarious along weedy, gravelly track in open woodland of *Allocasuarina* sp. and *Eucalyptus wandoo*, 10 Aug. 2014, *N.L.Bougher NLB1077* (PERTH 08578966); *loc. id.*, 10 Aug. 2014, *N.L.Bougher NLB1078* (PERTH 08578931). Qld: Queen Mary Falls, Main Range Natl Park, 28°20'29.5"S, 152°22'16.9"E, scattered singly or in pairs on soil along bank of track in wet sclerophyll forest under *Eucalyptus*, 14 Mar. 2012, *P.B.Matheny PBM3791* (PERTH 08362998, TENN 067023; KP171075-nLSU, KP641620-ITS, KM656119-*rpb2*).

Additional specimens examined (genetic or morphological variants): W.A.: rest area, Albany Hwy, 48 km N of Williams, S of Perth, under *Corymbia calophylla*, 5 Sept. 1981, *O.K.Miller & H.Miller OKM19466* (PERTH 08561761); old dismantled Kunjin railway, 14 km W of Corrigin, 32°21'25.9"S, 117°44'17.1"E, scattered singly on sandy soil under *Allocasuarina* (Black Tamma; no eucalypts nearby), 26 July 2011, *P.B.Matheny PBM3677* (PERTH 08319081, TENN 066457; KP171076-nLSU, KP641621-ITS, KM656120-*rpb2*). Qld: Mt Beerburrum, Glasshouse Mountains Natl Park, 26°57'29.7"S, 152°57'07.3"E, gregarious on soil in leaf litter in woodland under *Eucalyptus* spp., 28 Feb. 2012, *R.E.Halling & P.B.Matheny PBM3763* (PERTH 08363226, TENN 066995); along Kruger Parade, Redbank, 27°36'45.6"S, 152°52'21.4"E, gregarious on soil under *Eucalyptus*, 6 Mar. 2012, *R.E.Halling & N.Fechner REH9670* (BRI, NY 01491110; KP171077-nLSU, KP641621-ITS).

Etymology: From the Latin prefix *sub-* (almost), with *flavus* (yellow) and *spora* (a seed), in reference to the pale brown spores when seen under a light microscope, i.e. paler than the more typical yellowish brown spores of other *Inocybe* species.

Diagnostic features: Fruit bodies small. Pileus often minutely scaly to felty-tomentose; young lamellae pale brown, brown to dark brown (lacking olivaceous or yellow tones); stipe pallid, densely silky-fibrillose; odour not spermatic. Spores pale brown under the light microscope, predominantly elliptic to subamygdaliform (never markedly phaseoliform).

Inocybe subflavospora is similar in outward appearance to *I. isabellina* Matheny & Bougher, both typically with a stipe that is longer than the pileus diameter, a dark yellowish brown to russet-coloured pileus, mostly elliptic to subamygdaliform spores, coarsely encrusted pileipellis hyphae, and occurring in similar disturbed habitats often in association with eucalypts.

Plate 249. *Inocybe subflavospora*, holotype (PERTH 07609256). Photograph — N.L.Bougher.

Plate 250. *Inocybe subflavospora*. PERTH 08320861. Photograph — N.L.Bougher.

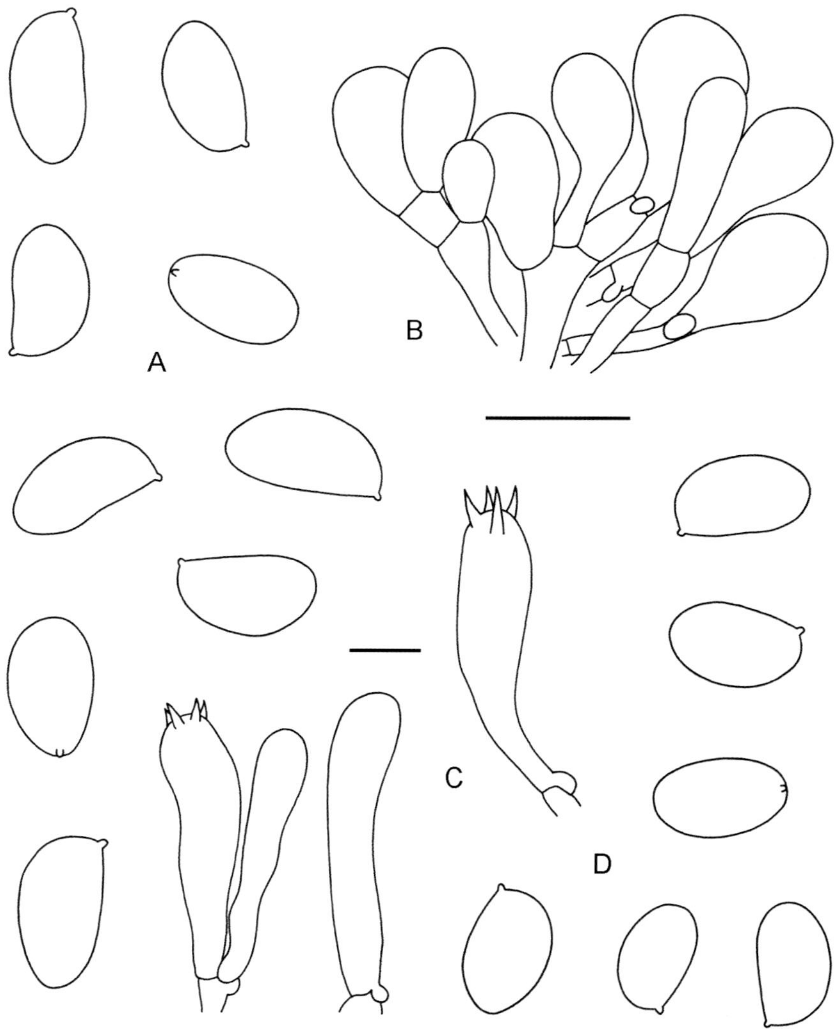

Figure 151. *Inocybe subflavospora.* All holotype (PERTH 07609256) except D, PERTH 08320861. **A,** Basidiospores; **B,** Cheilocystidia; **C,** Basidia; **D,** Basidiospores. Scale bars = 10 µm (longer bar for spores only).

However, *I. subflavospora* can be distinguished from *I. isabellina* Matheny & Bougher by its finely scaly pileus, non-spermatic odour, and the pale brown spores when seen under the microscope. The latter feature is a subtle distinction, but it can be confirmed when compared side-by-side with species of the *Mallocybe* clade having yellowish brown spores. *Inocybe subtilior* differs from *I. subflavospora* by its longer and narrower spores that are yellowish brown, and by the presence of a spermatic odour.

As in the case of *I. isabellina*, a number of collections deviate in various features from typical *I. subflavospora*. One of these (*PBM3791* from southern Queensland) is very close genetically (Fig. 147), forming a strongly supported monophyletic group with *I. subflavospora*, but it has smaller (mostly 7.0–8.5 × 5.0–5.5 µm), yellowish brown, elliptic spores, predominantly cylindrical cheilocystidia, and a felty tomentose-fibrillose pileus surface (lacking small scales).

In several collections (PERTH 08320861, 08578966, 08578931), we could not observe cheilocystidia as the lamellar edges proved to be fertile.

Some variants (cited above separately) fail to form a strongly supported monophyletic group with the type of *I. subflavospora* (Fig. 147). These include: *PBM3763* (from southern Queensland under eucalypts) that shares the similar minutely scaly pileus surface, but differs in its darker (yellowish brown) spores; and *PBM3677* from the wheatbelt region in south-western Western Australia that shares the pale brown spores, but has cylindrical to fusiform cheilocystidia, a short squat habit, and association with *Allocasuarina* (with no nearby eucalypts). Our understanding of speciation and phylogeography in this group remains poor but should improve with further reconciliation of morphological, ecological and genetic differences from additional specimens.

5. **Inocybe subtilior** Matheny & Bougher, *sp. nov.*

MycoBank: MB 809467

Type: Murdoch University Campus Reserve, Perth, W.A., Australia, in sand with *Eucalyptus marginata* and *Pinus pinaster* nearby, 26 May 1991, *O.K.Miller, H.Miller, L.Bailey & M.Bailey OKM24631*; holo: PERTH 08561745; iso: TENN 069564 (AY380398-nLSU, AY337406-*rpb2*, AY351823-*rpb1*).

Pileus 10–30 mm wide, broadly convex, becoming plane, at times with a low obtuse umbo; margin decurved; surface dry, matted felty-fibrillose or with small appressed fibrillose scales; red-brown (7D7 to D6) to brown; context whitish, not changing colour where bruised, firm. *Lamellae* adnate to uncinate, subdistant to moderately close, with c. 26–30 L and few tiers of lamellulae; brown, dark yellowish brown, becoming almost rusty brown with age; edges not conspicuously fimbriate, ±concolorous with the faces (at least in age). *Stipe* 15–37 × 3–5 mm, equal; base at times slightly swollen; surface dry, with tattered loose fibrils or small recurved fibrillose squamules over the upper ¾; context buff to very pale brown. *Odour* spermatic. Pl. 251, 252.

Basidiospores (10.0–) 10.5–13 (–13.5) × (4.0–) 4.5–5.5 (–6.0) µm [mean 11.3 × 4.9 µm; Q 2.00–2.74 (–2.89); Q mean 2.32 (n = 38/2)], smooth, oblong-reniform to cylindrical or oblong-subamygdaliform, yellowish brown; apiculus small and not very distinct. *Basidia* 32–40 × 8–10 µm, clavate to cylindroclavate, hyaline or ochraceous, necropigmented, 4-sterigmate. *Cheilocystidia* 21–45 × 10–18 µm, clavate to short-cylindrical or utriform, predominantly thin-walled (occasional cystidium with a thick wall in PERTH 08095388), hyaline; edges of lamellae sterile. *Pleurocystidia* absent. *Caulocystidia* absent. *Pileipellis* a cutis of repent hyphae, tawny in mass, cylindrical to inflated, 7–23 µm wide, encrusted, thin-walled. *Clamp connections* present. Fig. 152.

Occurs in south-western W.A.; on soil under eucalypts and/or *Chamelaucium*, with genetic and morphological variants in northern W.A. and N.T. Fruiting May and June; northern variants Feb. and Mar.

Additional specimens examined: W.A.: Regans Ford, Brand Hwy, Moore R., 30°57'S, 115°51'E, on open road under *Corymbia calophylla* and *Chamelaucium*, 30 May 1989, *O.K.Miller & H.Miller OKM23761* (PERTH 08561753, TENN 069560); Calingiri Rd, corner with Old Plane Rd, Rica Erickson Nature Reserve, 31°04'59.9"S, 116°25'59.9"E, 2 June 2008, *R.Hart & J.Froudist E9112* (PERTH 08095388, TENN 066769; KP171082-nLSU, KP641628-ITS).

Additional specimens examined (genetic or morphological variants, as cf. *subtilior*): W.A.: 86.2 km E of Broome along the Great Northern Hwy, 17°45'00" S, 122°55'59.9"E, on soil under *Eucalyptus tectifica*, *Brachychiton* and *Lysiphyllum cunninghamii*, 10 Feb. 1993, *N.L.Bougher E4819* (PERTH 07660634, TENN 066823; KP171083-nLSU); Cable Beach Road West, Broome, 17°56'22"S, 122°12'45"E, on red sand near beach dune in low woodland under *Corymbia paractia*, *Acacia* and *Triodia*, 14 Feb. 2011, *M.D.Barrett MDBF75/11* (PERTH 08383081, TENN 069513; KP171084-nLSU, KP641629-ITS, KM656125-*rpb2*). N.T.: Utopia Stn, 22°09'S, 134°37'E, at edge of Coolibah swamp under *Eucalyptus microtheca*, 21 Mar. 1977, *P.K.Latz s.n.* (PERTH 00977403).

Etymology: From the Latin *subtilior* (finer, more slender), in reference to the narrow spores.

Diagnostic features: Fruit bodies medium-sized; stipe as long as or longer than pileus width; odour spermatic. Spores long and narrow, mostly 10.5–13.0 × 4.5–5.5 µm, yellowish brown.

Inocybe subtilior shares a similar habit (long stipe in relation to pileus width) with *I. isabellina* Matheny & Bougher and *I. subflavospora* Matheny & Bougher, but differs in its narrow, oblong spores. Such spores are also characteristic of *Auritella arenicolens*, but the latter is readily distinguished by its larger, squat fruit bodies, pale spores, and narrowly elongate (>50 µm) cheilocystidia. *Inocybe sabulosa* Matheny & Bougher differs from *I. subtilior* in its dark brown felty pileus and shorter and broader spores that are slightly phaseoliform.

Some genetic and morphological variants of *I. subtilior*, collected from the north of Western Australia (see above; Fig. 147), differ in spore size from the type and other southerm populations. Indeed, their shorter and broader spores are more similar to those of *I. isabellina* in size (mostly 8.5–11 × 5.0–6.0 µm; mean Q <2.0), but they are mainly elliptic to oblong-elliptic with obtuse apices. Phylogenetic analyses of DNA sequences suggest the northern collections form a grade (a genetically diverged paraphyletic group) with respect to *I. subtilior* (Fig. 147). Because of the difference from *I. subtilior* in spore size, these northerly and tropical variants will key out either to *I. isabellina* or *I. sabulosa*.

Plate 251. *Inocybe subtilior*. PERTH 08095388. Photograph — N.L.Bougher.

Plate 252. *Inocybe* cf. *subtilior*. PERTH 08383081. Photograph — M.D.Barrett.

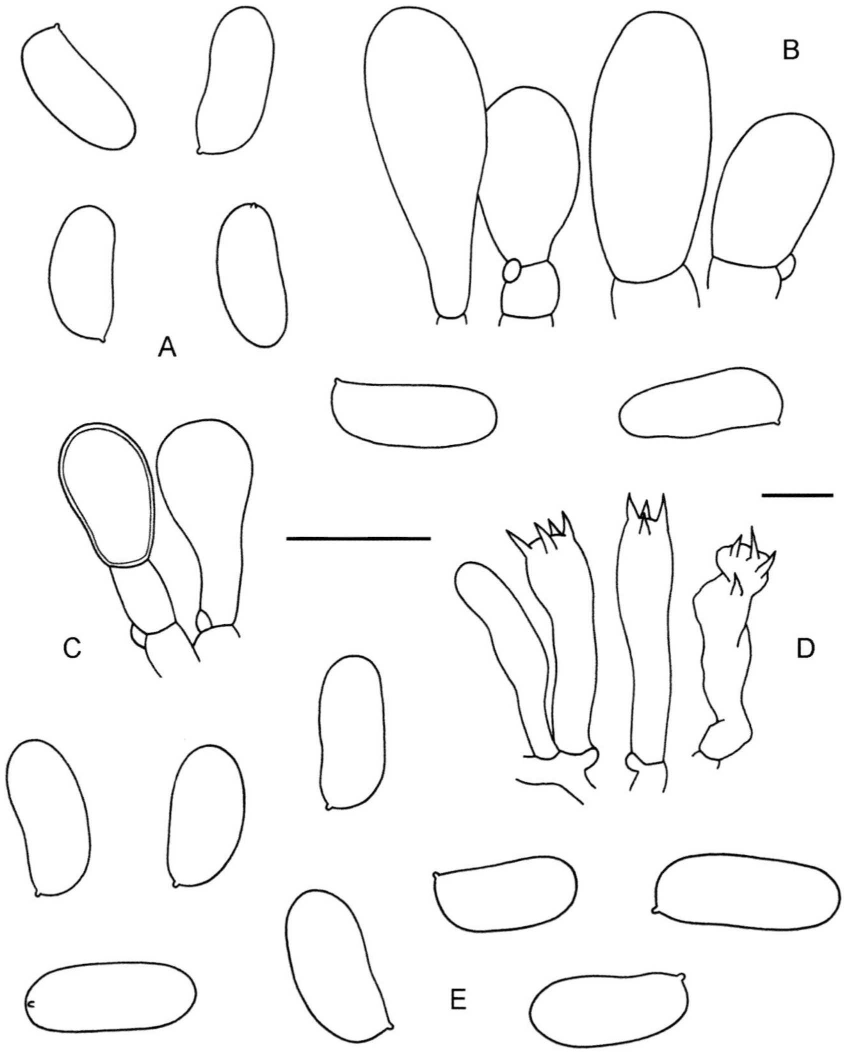

Figure 152. *Inocybe subtilior.* A, B, holotype (PERTH 08561745). C–E, PERTH 08095388. **A,** Basidiospores; **B,** Cheilocystidia; **C,** Cheilocystidia; **D,** Basidia; **E,** Basidiospores. Scale bars = 10 μm (longer bar for spores only).

AURITELLA

Auritella Matheny & Bougher, *Mycotaxon* 97: 232 (2006)

Type: *A. dolichocystis* Matheny, Trappe & Bougher

Auritella Matheny & Bougher, *Mycol. Prog.* 5: 5 (2006), *nom. inval.* (ICN 10.1, 10.3, 40.1)

Fruit bodies agaricoid or secotioid, fleshy, fibrous or tough. *Basidia* necropigmented. *Pleurocystidia* absent. *Cheilocystidia* >50 µm long in agaricoid forms, thin-walled or secondarily thickened (not metuloid); hyaline, pale yellow, or orange-brown. *Caulocystidia* absent or present at the apex, not descending the entire length of the stipe.

Currently, eight species of *Auritella* can be recognized in Australia. Twelve species are known worldwide in tropical forests of Africa, India, and throughout Australia, associating with caesalpinioid legumes (and perhaps other Fabaceae), Dipterocarpaceae, eucalypts and *Allocasuarina*. Not known from *Nothofagus* rainforests.

KEY TO SPECIES OF *AURITELLA* IN AUSTRALIA

1 Fruit bodies secotioid .... **6. A. geoaustralis**

1: Fruit bodies agaricoid 2

2 Pileus 40–75 mm wide; stipe (5–) 10–20 mm wide 3

2: Pileus 10–40 mm wide; stipe 2–10 mm wide 4

3 Spores long and narrow, >10 µm long .... **1. A. arenicolens**

3: Spores elliptic to broadly elliptic, <10 µm long .... **7. A. robusta**

4 Cheilocystidia sinuous 5

4: Cheilocystidia slenderly clavate, sphaeropedunculate, cylindric, or fusiform, not sinuous 6

5 Odour unpleasant; tissues not exuding yellow pigment in KOH .... **2. A. brunnescens**

5: Odour not remarkable; tissues exuding yellow pigment in KOH .... **8. A. serpentinocystis**

6 Stipe 2–4 mm; mean spore Q c. 2.00 .... **5. A. fulvella**

6: Stipe 4–10 mm wide; mean spore Q c. 1.80 or less 7

7 Spores ochraceous-buff in KOH .... **3. A. chamaecephala**

7: Spores yellowish brown in KOH .... **4. A. dolichocystis**

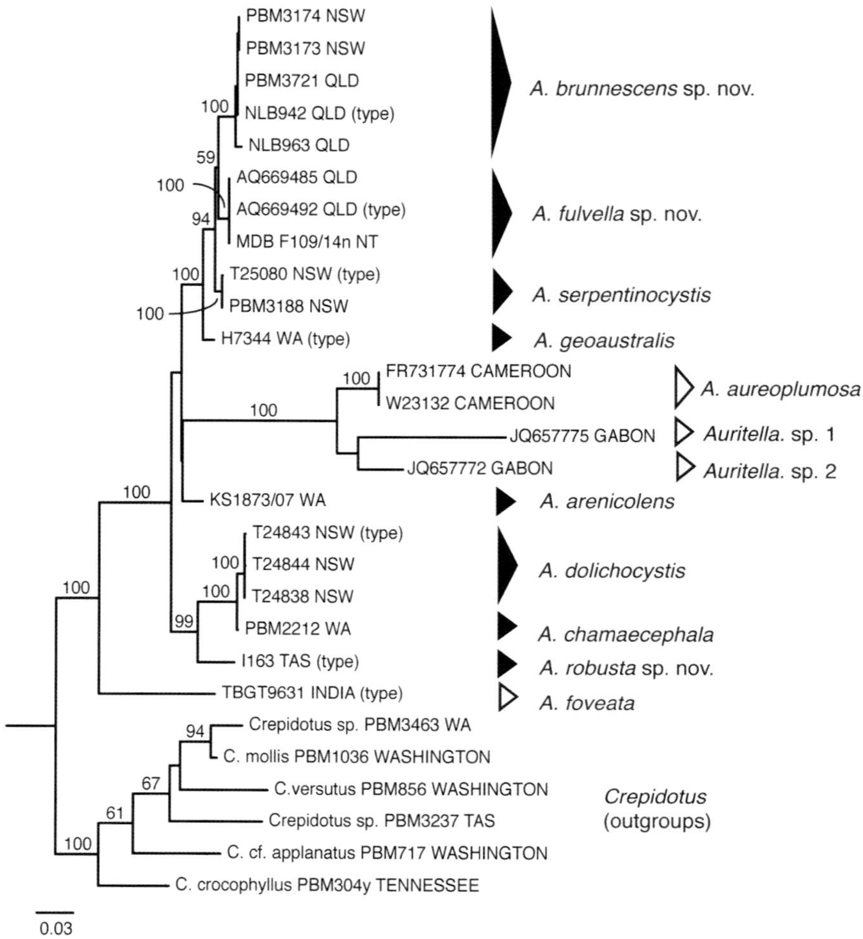

Figure 153. Maximum Likelihood phylogram of the genus *Auritella*. Bootstrap values are indicated above or below branches. The tree was inferred from combined nLSU and *rpb2* gene regions (partitioned separately) and included 2147 sites. Species from Australia are indicated by black-filled triangles. Species outside Australia are indicated by white triangles. The tree is rooted with the genus *Crepidotus* following Matheny *et al.* (2012). All 28 branch tips, are represented by nLSU sequences and 21 by *rpb2*.

1. **Auritella arenicolens** (Cleland) Matheny & Bougher, *Mycotaxon* 97: 232 (2006)

Auritella arenicolens (Cleland) Matheny & Bougher, *Mycol. Prog.* 5: 8 (2006), *nom. inval.*

Naucoria arenicolens Cleland, *Trans. Roy. Soc. S. Australia* 57: 193 (1933); *Inocybe arenicolens* (Cleland) E.Horak, *Persoonia* 11: 6 (1980); *Auritella arenicolens* (Cleland) Matheny & Bougher, *Mycol. Progr.* 5: 8 (2006), *nom. inval.* (ICN Art. 35.1), all as "*arenacolens*". Type: Waitpinga Rd, Encounter Bay, S.A. emerging from sandy soil, 27 May 1932, *J.B.Cleland*; lecto: AD 12222, *fide* P.B.Matheny, N.L.Bougher, *Muelleria* 28: 89 (2010).

Pileus 10–56 mm wide, subovate, convex, becoming plane, incurved to decurved, lightly uplifted and undulate with age; surface dry, velvety-tomentose to appressed-matted-fibrillose, somewhat coarsely so near the entire margin; pale beige (near 6C4), cinnamon to dark brown; context pale rusty ochraceous, hard or tough and breaking cleanly. *Lamellae* adnate or broadly adnexed, close, with several tiers of lamellulae, thin or thickened, dark beige to dark yellowish brown (6E8) or dark brown (6F7) with age, the edges conspicuously pallid-fimbriate, subventricose. *Stipe* 10–40 × 5–17 mm, often short and squat, equal or tapered towards the 6–8 mm diam. base, central, terete; cortina absent or evanescent, finely fibrillose or longitudinally striate with fibrils sometimes breaking up into fibrillose scales; paler than the pileus, pale brown to light yellowish brown; context solid or hollow, concolorous with the pileus, fleshy-fibrous, usually tough or firm. *Odour* none. Pl. 253–255.

Basidiospores (9.5–) 10.0–14.0 × 4.0–5.5 (–6.0) µm [mean 11.7 × 4.7 µm; Q (1.73–) 2.00–3.11 (–3.13); Q mean 2.53 (*n* = 51/4)], smooth, oblong-elliptic or narrowly cylindrical, oblong-cylindrical with a ventral depression appearing elongate-subreniform; apices often obtuse, less often ellipsoidal; pale yellowish brown in KOH, color unchanging in Melzer's solution; walls slightly thickened; apiculus small. *Basidia* 37–63 × 7–11 µm, narrowly clavate to subcylindrical, ochraceous to hyaline, necropigmented basidia often elongate, 4-sterigmate. *Pleurocystidia* absent. *Cheilocystidia* (29–) 47–108 × 6–14 µm, elongate and cylindrical or narrowly subclavate, at times somewhat flexuous, hyaline or ochraceous due to cytoplasmatic pigment, thin-walled, apices obtuse, crowded; edges of lamellae sterile. *Caulocystidia* similar to cheilocystidia, present at apex of stipe. *Pileipellis* with a suprapellis of interwoven cylindrical thin-walled hyphae, up to 10 µm diam.; yellowish brown in mass, smooth in appearance; subpellis interwoven, distinctly ochraceous brown in mass; hyphae 5–10 µm wide; encrusting pigments not observed or very weak; ochraceous refractive hyphae present in trama. *Clamp connections* present in all tissues. Fig. 154 (see also Pl. 7G, 8A, 11B).

Occurs in W.A. and S.A.; on sandy soil under *Corymbia calophylla*, *Eucalyptus marginata*, *Melaleuca thymoides* (Myrtaceae), *Spyridium globulosum* (Rhamnaceae), *Acacia* (Fabaceae) and *Hakea* spp. (Proteaceae). Fruiting May to July.

Additional specimens examined: S.A.: Encounter Bay, in sand, 27 May 1933, *J.B.Cleland s.n.* (AD 12261); *loc. id.*, in sand, 22 May 1930, *J.B.Cleland s.n.* (AD 12262). W.A.: Lot 536, Scotsdale Rd, Denmark, in litter near boundary firebreak under *Corymbia calophylla*, *Eucalyptus marginata*, *Hakea* spp. and associated understorey, 25 May 1993, *K.Syme KS21/91* (PERTH 04988728, TENN 066785); Rocky Point Track, Two Peoples Bay Nature Reserve, gregarious in sandy soil under *Spyridium globulosum* and *Melaleuca thymoides*, 28 May 1992, *K.Syme KS540/92* (PERTH 05429692, TENN 066775); 20 km N of Kellerberrin along Higginson Rd, partially submerged or pileus even with surface, under *Allocasuarina* and *Acacia*, 15 June 1995, *I.Tommerup, W.Dunstan & N.L.Bougher E5465* (PERTH 07710070, TENN 070157); Worsley Alumina Pty Ltd. Bauxite lease, Boddington, in litter in *Eucalyptus marginata* forest, 2 July 2002, *N.L.Bougher E7236* (PERTH 07675852); Rockhole Rd, Dallinup Ck, Munglinup, in sand, 10 June 2006, *K.Syme KS1470/06* (PERTH 07575939); Management Track 1 off Springvale Rd, Lake Shaster Nature Reserve, near mallee *Melaleuca* sp., *Leucopogon* sp., 16 June 2006, *K.Syme KS1666/06* (PERTH 08118981); Kau Rock Nature Reserve, Esperance area, in sand, 18 June 2006, *K.Syme KS1527/06* (PERTH 07899181); Norwood Rd, Esperance-Grass Patch region, gregarious or scattered near mallee eucalypts, *Hakea* sp., *Callitris* sp., *Melaleuca* sp., 13 July 2006, *K.Syme KS1635/06* (PERTH 08118957); Burdett Rd, Mount Burdett Nature Reserve, near *Allocasuarina* sp., *Melaleuca* sp., 13 July 2006, *K.Syme KS1644/06* (PERTH 08118965); Goodga River Track, Two Peoples Bay Nature Reserve, near *Allocasuarina fraseriana*, *Eucalyptus marginata* and *Beaufortia squarrosa*, 23 May 2007, *K.Syme KS1873/07* (PERTH 08047081, TENN 066678; KJ702338-nLSU, KJ729857-ITS, KJ729920-*rpb2*); White Hill Rd, near corner of Quail Rd, Yalgorup Natl Park, Yalgorup, 32°21'41"S, 115°38'19"E, in open area under shrubs including *Hibbertia hypercoides*, *Melaleuca* sp. and *Phyllanthus* sp. with *Eucalyptus gomphocephala* trees 30 m away, 23 May 2008, *N.L.Bougher NLB412* (PERTH 08072639; JX258833-ITS as "*Auritella* sp."); corner of Calingiri Rd and Old Plane Rd, Rica Erickson Nature Reserve, Calingiri, 31°05'S, 116° 26'E, 2 June 2008, *R.Hart & J.Froudist E9112* (PERTH 08095388, TENN 066769).

Diagnostic features: Fruit bodies medium to large. Stipe generally shorter than pileus width; odour absent. Spores pale brown, oblong-cylindrical or oblong with an adaxial depression (oblong-subreniform); basidia very long and narrow, narrowly clavate; cheilocystidia very long, mostly cylindrical and very narrow.

The habit of *A. arenicolens* is similar to other Australian species of the genus which typically have tough basidiomata, sombre brown colours, the lack of a rimose or squamulose pileus, distinctly fimbriate lamellar edges, and the absence of fruity or spermatic odours (Matheny & Bougher, 2006a, 2010). Microscopically, however, *A. arenicolens* is quite distinct due to the combination of long, narrowly cylindrical spores and long, slender basidia and cheilocystidia.

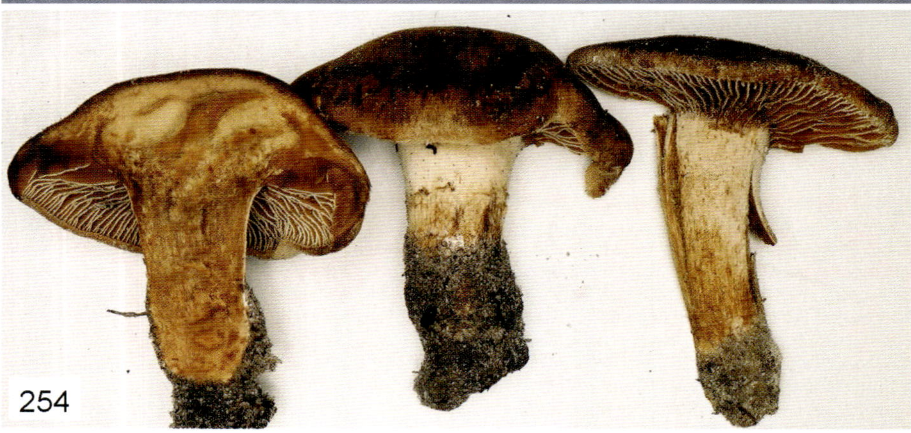

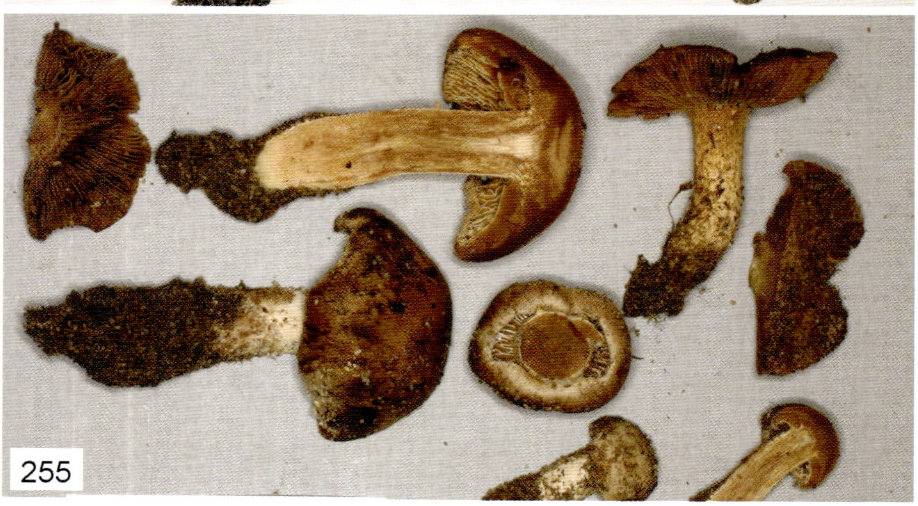

Plate 253. *Auritella arenicolens*. PERTH 07710070. Photograph — N.L.Bougher.

Plate 254. *Auritella arenicolens*. PERTH 08047081. Photograph — K.Syme.

Plate 255. *Auritella arenicolens*. PERTH 08072639. Photograph — N.L.Bougher.

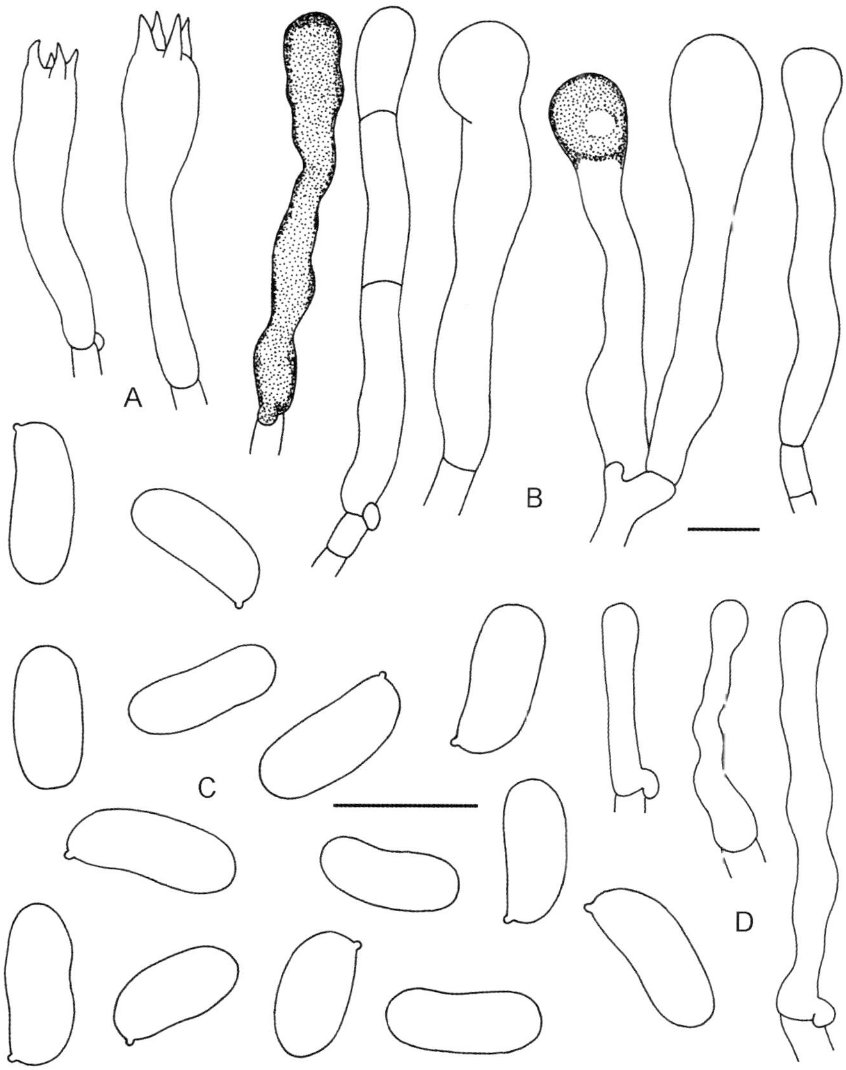

Figure 154. *Auritella arenicolens.* All PERTH 05429692. **A,** Basidia; **B,** Cheilocystidia; **C,** Basidiospores; **D,** Caulocystidia. Scale bars = 10 µm (longer bar for spores only).

The spores of *A. arenicolens* (including the lectotype) do not appear to be predominantly attenuated at the apex or "subfusoid" as described and illustrated by Horak (1980). Rather, the majority have an obtuse apex, and many are broader at the apex than at the base, similar to the spores illustrated by Grgurinovic (1997) who commented that Horak (1980) mistakenly recorded AD 12262 as the holotype, whereas Cleland's protologue for *A. arenicolens* was based on AD 12222. In the collections seen by us, spore shape is quite variable, with at least some ellipsoidal spores present. Horak (1980) and Grgurinovic (1997) reported the absence of caulocystidia, but we have seen them in clusters at the apex of the stipe in PERTH 05429692 and other comparatively recent collections.

Auritella arenicolens was originally described (as *Naucoria arenacolens*) from South Australia (Cleland, 1933), transferred to *Inocybe* by Horak (1980), and it was retained in that genus by

Grgurinovic (1997). Later, it was assigned to *Auritella* by Matheny & Bougher (2006a) under the specific epithet "*arenacolens*", an orthographic error subsequently corrected to "*arenicolens*" (Matheny & Bougher, 2006b).

Auritella arenicolens is in need of additional study. While sequence data for one collection from Western Australia (*KS1873/07*) is included here (Fig. 153), its identity is in doubt given that preliminary ITS sequence data from other specimens (*E5465*, PERTH 07710070; *NLB412*, PERTH 08072639; *KS540/92*, PERTH 05429692) are highly divergent from *KS1873/07* (data not shown).

2. **Auritella brunnescens** Matheny & Bougher, *sp. nov.*

MycoBank: MB 808677

Type: Davies Creek Rd, Dinden Natl Park, Qld, 17°00.1'36.1"S, 145°36'04.3"E, in litter under *Eucalyptus grandis*, *Allocasuarina torulosa*, *E. tindaliae* and other eucalypts at woodland-rainforest interface, 25 Feb. 2012, *M.D.Barrett & N.L.Bougher NLB942*; holo: PERTH 08364567; iso: TENN 067055 (KJ702339-nLSU, KJ702342-ITS, KJ702347-*rpb2*).

Pileus 11–40 mm wide, hemispherical, obtusely conical, convex to plano-convex; umbo absent, margin decurved; surface dry; velipellis absent; surface finely matted-fibrillose to fibrillose or minutely felty, not at all scaly; pale brown, light yellowish brown to brown (10YR 6/3, 6/4, 5/3) or drab orange-brown (7.5YR 5/6 but duller), in age yellowish brown (10YR 5/6) or ochraceous and with a slightly depressed centre; context pallid or with isabelline tones just above the lamellae, firm, colour unchanging where exposed. *Lamellae* ascending-adnate to adnexed, close to crowded or moderately so, 28–32 L, with several tiers of lamellulae, ventricose, up to 5 mm deep, warm brown (7.5YR 4/4) to dark yellowish brown with distinct pallid-fimbriate edges. *Stipe* 12–70 × 3–7 mm, even, often flexuous, cylindrical, at times subclavate towards the base, but base not at all bulbous, surface nowhere pruinose, fibrillose or with scattered fibrils imparting a tattered fibrillose appearance throughout, very pale brown mixed at times with pallid streaks; partial veil not observed; context solid, firm, fleshy-fibrous, often staining or bruising brown after cut. *Odour* usually unpleasant but at times not remarkable, unpleasant after drying. Pl. 256, 257 (see also Pl. 3A, I).

Basidiospores 7.0–9.0 × (4.0–) 4.5–5.5 µm [mean 7.9 × 5.0 µm; Q 1.36–1.88; Q mean 1.58 (*n* = 32/2)], smooth, mostly elliptic with blunt apices, at times subamygdaliform, pale yellowish brown; apiculus small but usually distinct under oil. *Basidia* 28–48 × 8–10 µm, 4-sterigmate, narrowly clavate, hyaline but necropigmented with age. *Pleurocystidia* absent. *Cheilocystidia* 31–73 × 5–8 µm, irregularly cylindrical, often flexuous or serpentine in outline, less often short-clavate, hyaline or more often amber to ochraceous brown, thin-walled or the walls slightly thickened; apices smooth and usually swollen or obtuse, sometimes coated with a brown mucilaginous material. *Caulocystidia* absent. *Pileipellis* a loose turf of interwoven to almost trichodermally arranged hyphae that are fulvous or brownish yellow in mass, darker or rust-colored with age; terminal hyphae often cylindrical or at times short-clavate, some frequently septate, smooth or weakly encrusted in appearance, mostly 7–20 µm wide. *Clamp connections* present. Fig. 155.

Known from eastern Qld and south-eastern N.S.W.; grows on soil under *Eucalyptus* in temperate dry sclerophyll forest (N.S.W.), or under *E. grandis*, *E. tindaliae* and *Allocasuarina torulosa* in tropical wet sclerophyll forest at a sclerophyll-rainforest interface (Qld). Fruiting Feb.–May.

Additional specimens examined: Qld: type locality, *M.D.Barrett & N.L.Bougher NLB943* (PERTH 08364575); Davies Creek Rd, Dinden Natl Park, 17°00'47.1"S, 145°34'55.4"E, in soil on roadbank under *Allocasuarina torulosa* and *Eucalyptus grandis* at rainforest margin, 20 Feb. 2012, *M.D.Barrett, N.L.Bougher & P.B.Matheny PBM3721* (TENN 066957; KJ702340-nLSU, KJ702345-ITS, KJ702350-*rpb2*); near North Tamborine, Tamborine Mountain Rd, opposite 'At the Polish Place', 27°57'10.2"S, 153°10'51.6"E, on soil among litter and grass under *Eucalyptus* sp. (marri-like bark) and *E. grandis* in an area cleared of understorey, 1 Mar. 2012, *N.L.Bougher NLB963* (PERTH 08364745; KJ702341-nLSU, KJ702346-ITS, KJ702351-*rpb2*). N.S.W.: 1 km E of Fitzroy Falls, above Kangaroo Valley, Southern Tablelands, on soil on track under *Eucalyptus* in dry sclerophyll forest, 18 May 2009, *P.B.Matheny PBM3173* (TENN 065742; JQ313558-nLSU, KJ702343-ITS, KJ702348-*rpb2*); *loc. id.*, *P.B.Matheny PBM3174* (TENN 065354; JQ313571-nLSU, KJ702344-ITS, KJ702349-*rpb2*).

Plate 256. *Auritella brunnescens*, holotype (PERTH 08364567). Photograph — N.L.Bougher.

Plate 257. *Auritella brunnescens*, holotype (PERTH 08364567). Photograph — N.L.Bougher.

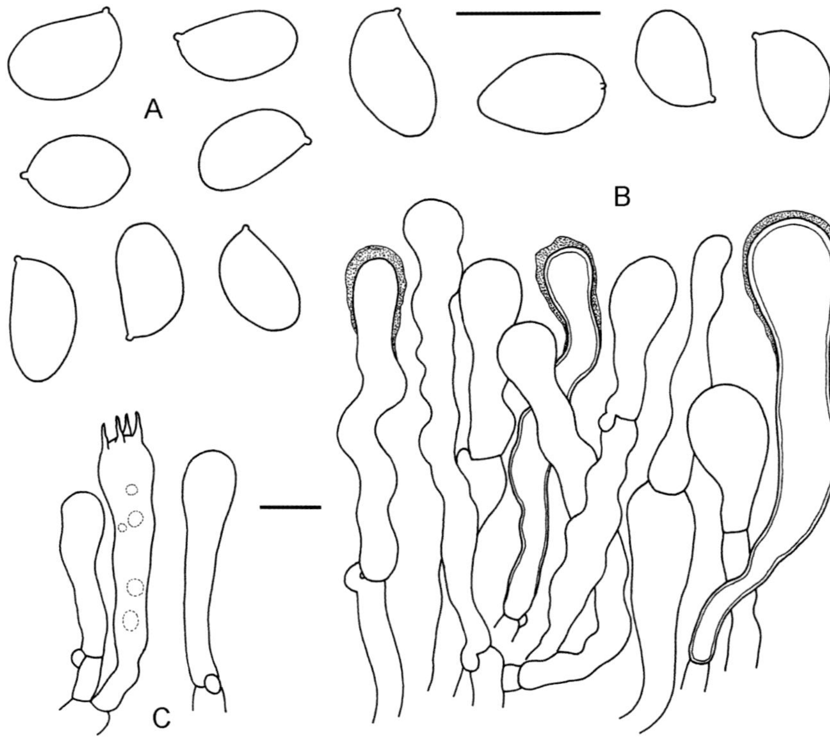

Figure 155. *Auritella brunnescens.* All holotype (PERTH 08364567). **A,** Basidiospores; **B,** Cheilocystidia; **C,** Basidia. Scale bars = 10 µm (longer bar for spores only).

Etymology: From the Latin *brunnescens* (becoming brown), in reference to the brownish-staining context.

Diagnostic features: Odour usually unpleasant; lamellae with distinct pallid-fimbriate edges; stipe context solid, firm, fleshy-fibrous, often staining or bruising brown after cut. Cheilocystidia often flexuous-cylindrical to long-clavate; spores mostly elliptic with blunt apices.

Auritella brunnescens is one of few species of Australian Inocybaceae to be found in both tropical and temperate habitats. In Queensland, it was observed in wet sclerophyll forest under *Allocasuarina* and *Eucalyptus* at the edge of rainforest, while in the Southern Tablelands of New South Wales, it was collected in a sclerophyll forest under *Eucalyptus*. Initially, we identified the specimens as *A. serpentinocystis* Matheny, Trappe & Bougher, given the similar colouration, texture, spore size and narrow, flexuous cheilocystidia. That species was described by Matheny & Bougher (2006a, 2006b) based on herbarium collections and field notes provided by the collector (J.Trappe). Among these notes, the odour is recorded as musty or indistinctive. However, the odour of *A. brunnescens*, observed directly by us in the field is decidedly unpleasant, remaining so even after drying. The context of *A. brunnescens* also stains brown after cutting (Pl. 3A), and while we assume *A. serpentinocystis* does not stain brown, this trait needs to be observed more carefully. Dried sections of *A. serpentinocystis* exude yellow pigment into solutions of KOH (similar to species of the genus *Pleuroflammula* Singer); but we have not observed this in *A. brunnescens*. Both species typically feature a stipe length longer than the pileus diameter, which distinguishes them from other species of *Auritella*. Phylogenetic data clearly separate *A. brunnescens* and *A. serpentinocystis* (Fig. 153).

Auritella dolichocystis Matheny, Trappe & Bougher, described from New South Wales, can be distinguished from *A. brunnescens* by its non-sinuous cheilocystidia and squat habit. *Auritella fulvella* Matheny & Bougher differs in its smaller size (stipe 2–4 mm wide), narrowly clavate cheilocystidia, and comparatively narrow spores with a mean Q-value of 2.00.

3. **Auritella chamaecephala** Matheny, O.K.Mill. & Bougher, *Mycotaxon* 97: 232 (2006)

Auritella chamaecephala Matheny, O.K.Mill. & Bougher, *Mycol. Progr.* 5: 11 (2006), *nom. inval.* (ICN Art. 36.1). Type: Gully Rd, near Nornalup, W.A., under *Eucalyptus jacksonii*, 10 June 1989, *O.K. & H.Miller OKM23901*; holo: PERTH 08561206 (formerly VPI); iso: WTU.

Pileus 20–35 mm wide, ovate or hemispherical-convex, becoming broadly convex, incurved to decurved and undulate with age at the margin; surface dull, dry, indistinctly tomentose, with a smooth appearance or with dense tufts of fibrils, light brown (5C4) with a darker centre (5C6) to brown (6D6 to 6D7), brown with 3% KOH, dark green with $FeSO_4$; context firm, beige (near 5B5 to 5A4), becoming mottled after cut, 3.5–6 mm thick under the centre. *Lamellae* adnate to narrowly adnate, close to subdistant, with several tiers of lamellulae, pale cream than light brown (near 5C5) maturing to brown (5E7 to 6D5) or dark brown (6F5 to 6F6), 2–5 mm deep; edges pallid and fimbriate. *Stipe* 15–30 × 4–8 mm, terete, equal or only slightly enlarged at the base, cortina not observed, fibrillose, in places minutely punctate, floccose near the base, or with tufts of fibrils, light brown (slightly paler than 5B3) to brown (6D5); basal mycelium white but sparse; partial veil a rapidly evanescent whitish cortina; context fleshy fibrous, solid, coloured like the pileal context. *Odour* indistinct or unpleasant [rotten mushroom-like when fresh, strongly peppery when dried (PERTH08578915)]. Pl. 258, 259.

Basidiospores (7.0–) 7.5–9.5 × 4.0–5.5 µm [mean 8.3 × 4.7 µm; Q 1.60–2.00 (–2.11); Q mean 1.77 (*n* = 42/3)], smooth, mostly elliptic, occasionally subamygdaliform to subphaseoliform, pale brownish yellow (Ochraceous Buff to Light Ochraceous Buff; 10YR 6/6), only slightly thick-walled; apiculus small and indistinct. *Basidia* 30–45 × 7–9 µm, narrowly clavate, clavate or cylindroclavate, 4-sterigmate, hyaline to ochraceous. *Pleurocystidia* absent. *Cheilocystidia* 39–88 × 7–14 µm, often elongated into a narrowly clavate outline, or skittle-shaped to rarely irregularly cylindrical, at times slightly flexuous but not serpentine, often with subcapitate or swollen apices, thin-walled, hyaline to ochraceous; lamellar edges sterile. *Caulocystidia* 20–55 × 7–10 µm, numerous at the apex of stipe, clavate to cylindrical or subutriform, rather similar to cheilocystidia but the apices often obtuse. *Lamellar trama* regular, hyphae often inflated, 6–20 µm wide, pale yellowish brown in mass, smooth, frequently septate. *Pileipellis* an irregular cutis of cylindrical hyphae, 5–11 µm wide, thin-walled, ochre-tawny in mass, encrusted, many cells also with rusty-ochraceous refractive content; terminal cells undifferentiated; tramal hyphae more pale in mass, smooth, cylindrical; refractive hyphae present. *Clamp connections* present. Fig. 156.

Occurs in south-western W.A.; gregarious on ground under *Eucalyptus diversicolor*, *E. jacksonii*, *Corymbia calophylla*, *Allocasuarina decussata* (Casuarinaceae) and *Agonis flexuosa*; also *E. wandoo* and *E. marginata*. Fruiting May, June, Aug. and Sept.

Additional specimens examined: W.A.: Denmark, lot 406, SE corner, in old pasture and moss above granite outcrop under *Allocasuarina decussata*, *Corymbia calophylla*, *Agonis flexuosa* and *Eucalyptus diversicolor*, 20 May 1993, *K.Syme KS645/93* (PERTH 05439353, WTU); John Forrest Natl Park, E of Perth, near trail to the lake, solitary on soil under *Eucalyptus wandoo*, *E. marginata* and *Corymbia calophylla*, 8 Sept. 2001, *P.B.Matheny PBM2212/E7060* (PERTH 07676867, WTU; AY635781–rpb2, AY635765–nLSU); Lights Rd, Hallowell Railway Reserve Track, 34°59'17.4"S, 117°16'30.6"E, in forest with *Eucalyptus marginata* and *Allocasuarina* sp., 16 Sept. 2006, *K.Syme 1805/06* (PERTH 08106401); Old Plains Rd, 3.4 km N of Calingiri Rd, 31°07'30.6"S, 116°16'09.1"E, under sparse litter on side of gravelly track in open woodland of *Allocasuarina* sp. and *Eucalyptus wandoo*, 11 Aug. 2014, *N.L.Bougher NLB1079* (PERTH 08578915).

Diagnostic features: Fruit bodies medium-sized, squat. Stipe generally shorter than pileus width. Odour absent or unpleasant. Spores short and mostly elliptic; cheilocystidia narrowly clavate or skittle-shaped.

Matheny & Bougher (2006a) discussed the close similarity between *A. chamaecephala* (from Western Australia) and *A. dolichocystis* Matheny, Trappe & Bougher (from eastern Australia), e.g. gross morphology (light brown to yellowish brown pileus, squat habit), spore size and cheilocystidia. Differences noted between the species included basidiospore colour (pale for *A. chamaecephala*, ochraceous tawny for *A. dolichocystis*), and nucleotide differences at *rpb2* and nLSU loci. The nucleotide differences between the two species cited by Matheny & Bougher (2006a) are based on data for only one collection of *A. chamaecephala* (PERTH 07676867), and it is uncertain whether its nucleotide sequence matches that of the holotype. Furthermore, the cheilocystidia of PERTH 07676867 are predominantly multi-septate and have variable apices, whereas those of the holotype are mostly sparingly or basally septate and have consistently subcapitate or swollen apices. The forms of cheilocystidia overlap between these collections, and therefore we follow Matheny & Bougher (2006a) and accept PERTH 07676867 under the published concept of *A. chamaecephala.* The form of cheilocystidia which predominates in the holotype (and in PERTH 05439353) is more similar to that of *A. dolichocystis* than to PERTH 05439353. We await further collections and nucleotide sequences to clarify the distinction between *A. chamaecephala* and *A. dolichocystis*, and to confirm the taxonomic status of PERTH 05439353.

Plate 258. *Auritella chamaecephala* (PERTH 08578915). Photograph — N.L.Bougher.

Plate 259. *Auritella chamaecephala* (PERTH 08106401). Photograph — K.Syme.

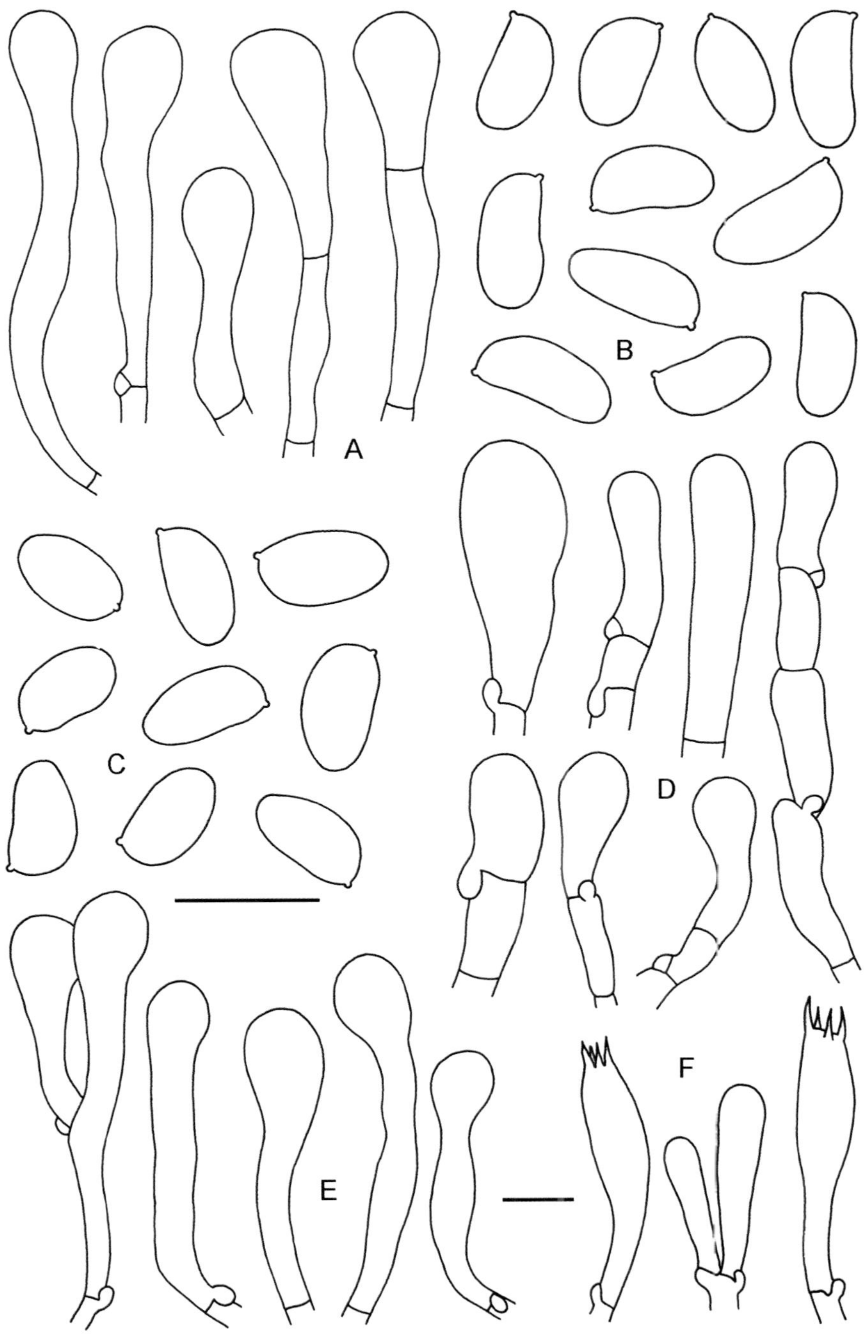

Figure 156. *Auritella chamaecephala.* **A,** Cheilocystidia, holotype (PERTH 08561206); **B,** Basidiospores (PERTH 05439353); **C,** Basidiospores (PERTH 07676867); **D,** Cheilocystidia (PERTH 07676867); **E,** Cheilocystidia (PERTH 05439353); **F,** Basidia (PERTH 05439353). Scale bars = 10 µm (longer bar for spores only).

Auritella arenicolens (Cleland) Matheny & Bougher and *A. chamaecephala* Matheny, O.K.Mill. & Bougher are the only agaricoid species of *Auritella* currently known to occur in Western Australia; *A. geoaustralis* Matheny & Bougher also occurs in Western Australia, but is an independent secotioid taxon (Fig. 153; Matheny & Bougher 2006a). *Auritella arenicolens* and *A. chamaecephala* can be readily distinguished by their spores, i.e. 10–14 µm long and predominantly cylindrical in the former, (7.0–) 7.5–9.5 µm long and mostly elliptic in the latter. In addition, the cheilocystidia of *A. arenicolens* are often elongate and cylindrical, whereas those of *A. chamaecephala* are usually narrowly clavate or skittle-shaped.

4. **Auritella dolichocystis** Matheny, Trappe & Bougher, *Mycotaxon* 97: 232 (2006)

Auritella dolichocystis, Matheny, Trappe & Bougher, *Mycol. Progr.* 5: 9 (2006), *nom. inval.* (ICN Art. 36.1). Type: Holy Camp Rd, Glenrock, Weddin Shire, N.S.W., among sheep-grazed regrowth of *Eucalyptus albens* with bare soil and some grass, 25 Oct. 1999, *J.Trappe 24838*; holo: PERTH 07536569; iso: WTU (AY635763-nLSU, AY635767-*rpb2*).

Pileus 12–40 mm wide, convex to plane; surface dry; disc glabrous to obscurely fibrillose or finely felty, finely-fibrillose to fibrillose towards the margin, at times slightly sulcate but not striate; edge of pileus appressed against the stipe when young; light yellowish brown to light brown; context brownish white or pale brown, up to 5 mm thick. *Lamellae* adnexed, becoming adnate to adnate-subdecurrent; lamellulae infrequent, brownish grey to light greyish brown when young, dull sombre brown with age; edges pallid and coarsely fimbriate. *Stipe* 7–27 × 4–10 mm, cylindrical or tapering towards the unswollen base, longitudinally fibrillose, subsquamulose when young, with upper half or apical 1–2 mm sparsely pruinose, white at apex, brownish white to pale brown below, or concolorous with the pileus, sometimes with darker brown fibrils and other areas after being handled. Cortina absent even on young basidiomata. *Odour* not distinctive, faintly sweetly pungent.

Basidiospores (7.0–) 7.5–10.5 × 4.0–5.5 µm [mean 8.6 × 4.8 µm; Q (1.40–) 1.50–2.25 (–2.63); Q mean 1.81 (n = 57/5)], smooth, elliptic, less frequently subamygdaliform or phaseoliform; Ochraceous-Tawny to Ochraceous-Buff, with slightly thickened walls; apiculus distinct or not. *Basidia* 30–45 × 7–10 µm, narrowly clavate, 4-sterigmate, ochraceous to hyaline. *Pleurocystidia* absent. *Cheilocystidia* (42–) 45–75 (–82) × 9–14 (–16) µm, subfusiform, sublageniform to narrowly clavate, with a median to basal subventricose swelling; apices swollen to subcapitate; thin-walled, hyaline to ochraceous. *Caulocystidia* frequent at the stipe apex, usually similar to cheilocystidia, at times lacking subcapitate apices or shorter, in clusters, irregularly cylindrical, fusiform to clavate; stipe hyphae brownish yellow to hyaline, 3–10 µm wide; refractive hyphae frequent. *Lamellar trama* parallel. *Pileipellis* a cutis of encrusted cylindrical hyphae, 4–13 µm wide, ochraceous tawny in mass; tramal hyphae brownish yellow in mass, cylindrical. *Clamp connections* present in all tissues. Fig. 157.

Occurs in the central-west of N.S.W.; grows on the ground under *Eucalyptus* (Myrtaceae) and with *Callitris* (Cupressaceae) and *Acacia* and *Dillwynia* (Fabaceae). Fruiting Apr., Oct.–Nov.

Additional specimens examined: N.S.W.: Parkes Shire, Bumberry L., alt. 400 m, on flat with ungrazed *Eucalyptus sideroxylon* (Large Mugga Ironbark) and a few *E. microcarpa* (Grey Box) with a grass groundcover and a few *Callitris endlicheri*, *Acacia parvifolia*, *Dillwynia juniperina* and several other shrubs, 11 Nov. 1999, *J.Trappe 24893* (MEL); inside of Genaren Hill Sanctuary near south-central gate, Genaren, Parkes Shire, flat area in ungrazed woodland of *E. sideroxylon*, *E. microcarpa* and *E. dealbata*, 21 Apr. 2000, *J.Trappe 25067* (MEL); *loc. id.*, 24 Oct. 1999, *J. Trappe 24843* (MEL, WTU; AY635764-nLSU, AY635780-*rpb2*); *loc. id.*, 24 Oct. 1999, *J.Trappe 224844* (MEL, WTU; AY380371-nLSU, AY337371-*rpb2*).

Diagnostic features: Fruit bodies medium-sized to somewhat robust. Pileus yellowish brown to light brown, fibrillose to felty, flesh tough; lamellae with coarsely fimbriate edges; cortina absent; dried tissues not exuding yellow pigment in alkali solution. Spores elliptic, subamygdaliform or phaseoliform, not elongate; cheilocystidia with a median to basal subventricose swelling and swollen to subcapitate apices.

Dried tissues of *A. dolichocystis* do not exude yellow pigment in KOH, unlike *A. serpentinocystis* Matheny, Trappe & Bougher, and they do not have a strong, unpleasant odour, unlike *A. brunnescens*. The narrow, elongate cheilocystidia resemble those of *A. arenicolens* (Cleland) Matheny & Bougher, although the latter lacks subcapitate cystidial apices and has longer and much paler spores. *Auritella serpentinocystis* occurs in the same region of New South Wales, but it differs in its narrow, undulate or serpentine cheilocystidia.

Observations of young material suggest *A. dolichocystis* lacks a partial veil. We have seen an evanescent structure in very immature fruit bodies of *A. brunnescens*, an additional feature that could be used to distinguish the two taxa, both of which have been recorded in New South Wales.

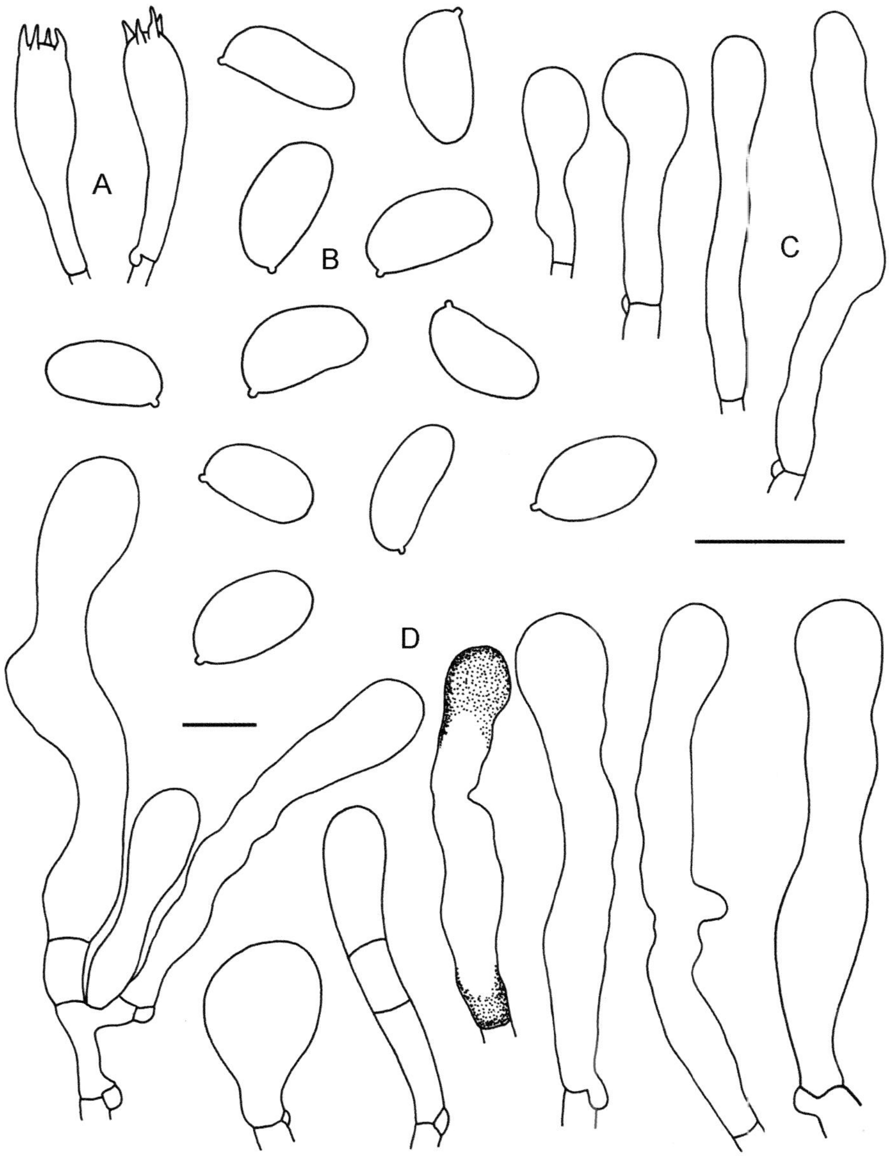

Figure 157. *Auritella dolichocystis.* All holotype (PERTH 07536569). **A,** Basidia; **B,** Basidiospores; **C,** Caulocystidia; **D,** Cheilocystidia. Scale bars = 10 µm (longer bar for spores only).

Auritella chamaecephala Matheny, O.K.Mill. & Bougher is most similar to *A. dolichocystis* as they have a similar squat habit, a light brown to yellowish brown pileus, similar spore size and cheilocystidia. However, the former has paler spores, is known only from Western Australia and has different *rpb2* and nLSU nucleotide sequences (Matheny & Bougher, 2006a). The two species form a well-supported monophyletic group with *A. robusta* Matheny, Bougher & G.M.Gates (Fig. 153), which is characterised by the larger size of its fruit bodies and somewhat shorter spores.

5. **Auritella fulvella** Matheny & Bougher, *sp. nov.*

MycoBank: MB 808678

Type: Herberton Cemetery, Herberton, Qld, 17°23'47"S, 145°22'39"E, on granodiorite in woodland under *Corymbia citriodora* and *Eucalyptus crebra*, 19 Feb. 2001, *P.I.Forster, R.Booth & A.M.Young PIF26850*; holo: BRI AQ669492; iso: PERTH 08517266, TENN 068485 (KJ702352-nLSU, KJ702354-ITS, KJ702356-*rpb2*).

Pileus 8–17 mm wide, pulvinate, becoming broadly convex and flat across the top, lacking an umbo and velipellis; margin incurved (at first strongly incurved), becoming decurved; surface dry, densely fibrillose, occasionally becoming excoriate-scaly around the disc (scales ±appressed); margin entire, not rimose; tan-brown (5D7) to brown (6E7), lighter in N.T. material (5C4), fulvous (6C7 to D7) in the dried state. *Lamellae* usually free, moderately close with several tiers of lamellulae, thick, sometimes fused, broad, 4–5 mm deep, brown to very dark brown; edges conspicuously fimbriate. *Stipe* 12–22 × 2–4 (–5) mm wide, even, smooth, dry, appressed-silky-fibrillose, veil (cortina) not observed even in immature stages, light brown or concolorous with the pileus, light brown to pale brown (5B5 to B4) in dried state and lighter in colour compared to the dried fulvous pileus; context solid, becoming hollow, brown above, pale brown below, appearing brunnescent after cutting. *Odour* not recorded. Pl. 260–262.

Basidiospores 8.0–11.0 × 4.0–5.5 µm [mean 9.5 × 4.8 µm; Q 1.60–2.40 (–2.50); Q mean 2.00 (n = 40/2)], cylindrical or subreniform with a ventral depression, with obtuse apices, pale brown to yellowish brown; apiculus small, inconspicuous; shape more variable and more often elliptic from lamellar mounts of BRI AQ669485 (genetically the same as the holotype). *Basidia* 31–43 × 7–10 µm, narrowly clavate to subcylindrical, hyaline, becoming necropigmented, 4-sterigmate. *Pleurocystidia* absent. *Cheilocystidia* 29–58 × 7–15 µm, narrowly clavate, sphaeropedunculate, cylindrical or, less often, fusiform, occasionally with a flexuous outline; apices often swollen, thin-walled or occasionally with a thickened wall especially near the apex, hyaline, crowded, arising from long cylindrical hyphae sometimes with encrusted walls. *Caulocystidia* not observed at least below the pileal insertion; at extreme apex some disorganised patches of cystidioid end-cells with swollen apices (but not sphaeropedunculate) resembling some forms of the cheilocystidia. *Pileipellis* a compact interwoven cutis, ochre-tawny in mass, composed of cylindrical hyphae mostly 5–12 µm wide, thin-walled, smooth or encrusted, many filled with ochraceous pigment, some collapsed. *Clamp connections* present. Fig. 158.

Occurs in northern N.T. and north-eastern Qld; on the ground in tropical woodland dominated by eucalypts. Fruiting Jan. and Feb.

Additional specimens examined: N.T.: 800 m W of turnoff to Florence Falls and Buley Rockhole from main road, Lichfield Natl Park, 13°07'27"S, 130°47'55"E, in rather dense woodland of *Eucalyptus miniata*, *E. tetrodonta* (Myrtaceae), *Erythrophleum chlorostachys* (Fabaceae) and *Grevillea pteridifolia* (Proteaceae) (not burnt for several years), 25 Jan. 2014, *M.D.Barrett, T.Lebel, G.M.Bonito & C.N.Barrett MDB F109/14* (MEL 2382701). Qld: type locality, 20 Feb. 2001, *P.I.Forster, R.Booth & A.M.Young PIF26860* (BRI AQ669485, TENN 068468; KJ702353-nLSU, KJ702355-ITS, KJ702357-*rpb2*).

Etymology: From the Latin *fulvellus* (little tawny or reddish yellow), in reference to the small size of the fulvous coloured fruit bodies.

Diagnostic features: Fruit bodies small. Pileus fulvous, glabrous. Stipe 2–4 (–5) mm wide. Spores cylindrical, subreniform or elliptic, generally narrow (mean Q: 2.00) and 4.0–5.5 µm wide; cheilocystidia mostly narrowly clavate. Dried tissues not exuding pigment in 5% KOH.

Morphologically, *A. fulvella* appears most similar to *A. dolichocystis* Matheny, Trappe & Bougher and *A. chamaecephala* Matheny, O.K.Mill. & Bougher, but it differs in its smaller fruit bodies and narrower spores. *Auritella serpentinocystis* Matheny, Trappe & Bougher from New South Wales has serpentine and narrowly sinuous cheilocystidia and yellow or ochraceous exudation of pigment of dried tissues in KOH. Phylogenetic data (Fig. 153) suggest *A. fulvella* is most closely related to the cluster of *A. serpentinocystis* and *A. geoaustralis* Matheny & Bougher, the latter a secotioid species. *Auritella brunnescens* Matheny & Bougher, which also occurs in north-eastern Queensland under eucalypts and *Allocasuarina* and under eucalypts in New South Wales, is larger, has an unpleasant odour, elliptic to subamgydaliform spores with a mean Q value of c. 1.60, and sinuous cheilocystidia.

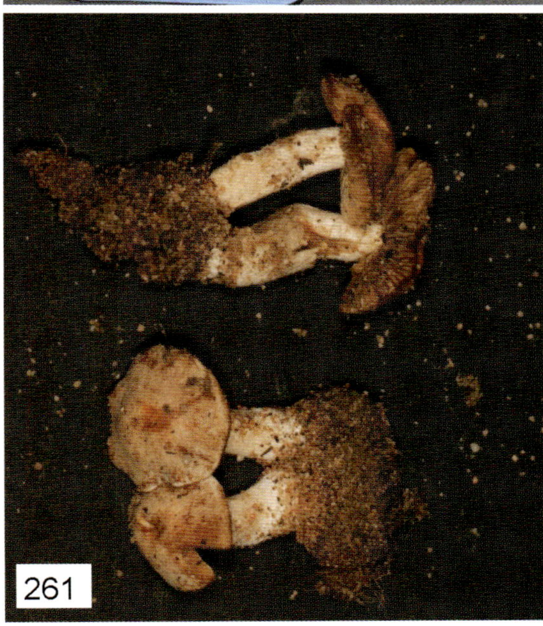

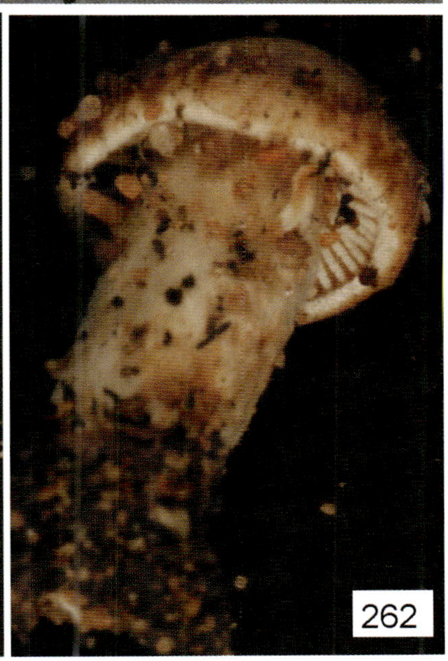

Plate 260. *Auritella fulvella.* BRI AQ 669485. Photograph — unknown photographer, image from BRI.

Plates 261, 262. *Auritella fulvella.* MEL 2382701. Photographs — M.D.Barrett.

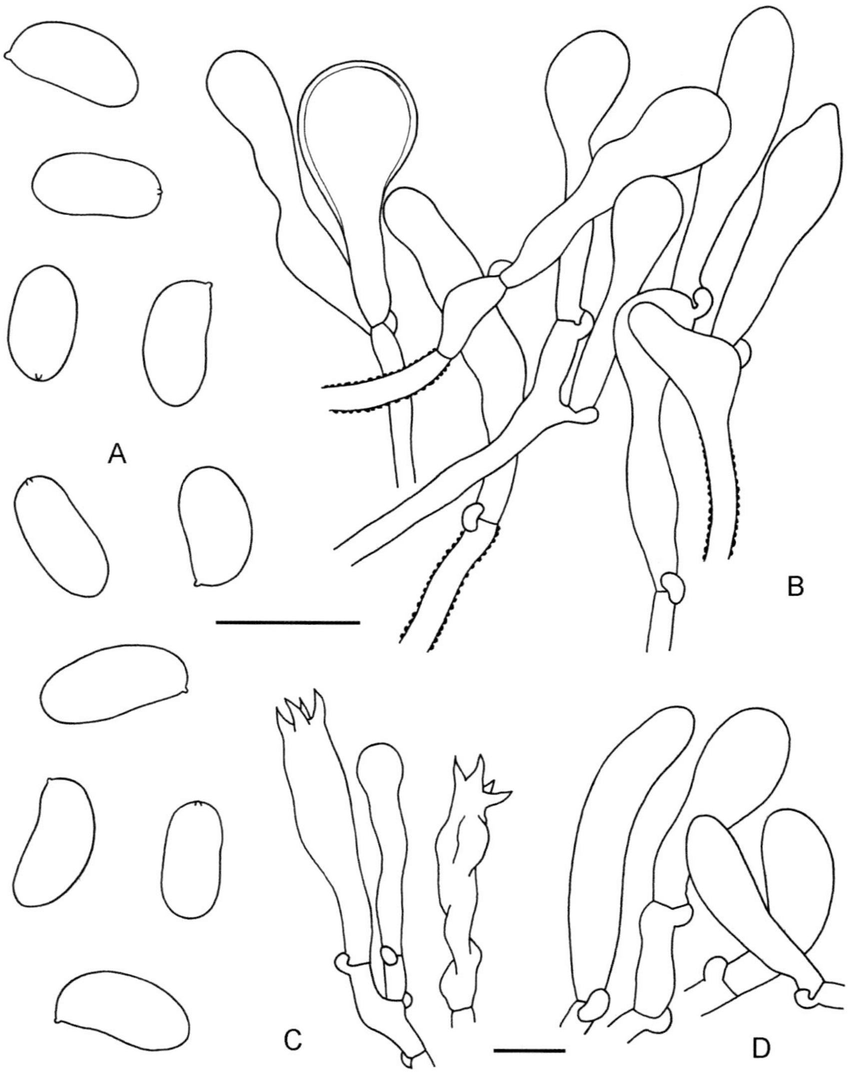

Figure 158. *Auritella fulvella.* All holotype (BRI AQ669492). **A,** Basidiospores; **B,** Cheilocystidia; **C,** Basidia; **D,** Cystidioid elements from extreme upper stipe. Scale bars = 10 µm (longer bar for spores only).

6. Auritella geoaustralis Matheny & Bougher, *Mycotaxon* 97: 232 (2006)

Auritella geoaustralis Matheny & Bougher, *Mycol. Progr.* 5: 7 (2006), *nom. inval.* (ICN Art. 36.1).
Type: 30 km N of Kellerberrin, W.A., emerging from soil under *Eucalyptus wandoo* and *Gastrolobium parviflorum*, 31 July 1996, *S.Bolsenbroek H7344*; holo: PERTH 07664605; iso: WTU (AY333774–rpb2, AY380395–nLSU).

Fruit bodies to 20 × 12 mm, secotioid, pulvinate or roughly elliptic, cream, not bruising. Peridium thin, to 1 mm wide, cream in section, dry, minutely silky-fibrillose. *Gleba* composed of irregularly empty locules, some radial pattern evident, dark brown (near 7E7). *Columella* truncate, cream in section, emerging from fruit bodies only as a small bump. Other minor sterile intrusions present in other parts of the fruit body. Basal mycelium not visible. *Odour* unpleasant. Pl. 263.

Basidiospores (8.5–) 9.0–10.5 (–11.0) × 5.5–6.0 (–6.5) µm [mean 9.7 × 5.9 µm; Q (1.42–) 1.50–1.75 (–1.83); Q mean 1.63 (n = 20/1)], smooth, ovate-elliptic; wall slightly thickened, pale in KOH (Ochraceous-Buff to Ochraceous-Tawny), not reacting in Melzer's reagent; apiculus small and often indistinct, but occasionally conspicuous. *Basidia* 40–60 × 8–10 µm, necropigmented, statismosporic, irregularly cylindrical to cylindro-clavate, at times widest just below the apex, 4-sterigmate; sterigmata often up to 6 µm long with an acute apex; forming hymenia. *Cystidia* not seen. *Peridial hyphae* cylindrical to somewhat inflated, at times femur-shaped, thin-walled, wrinkled or with bands of possibly encrusting pigment, at times some hyphae decorated with sparse exterior encrusting granules. Hyphae and encrustations not reacting in Melzer's reagent. *Clamp connections* present in all tissues. Fig. 159.

Known only from the type locality in the wheatbelt region of W.A.; scattered, emerging from soil under *Eucalyptus wandoo* (Myrtaceae) and *Gastrolobium parviflorum* (Fabaceae). Fruiting July.

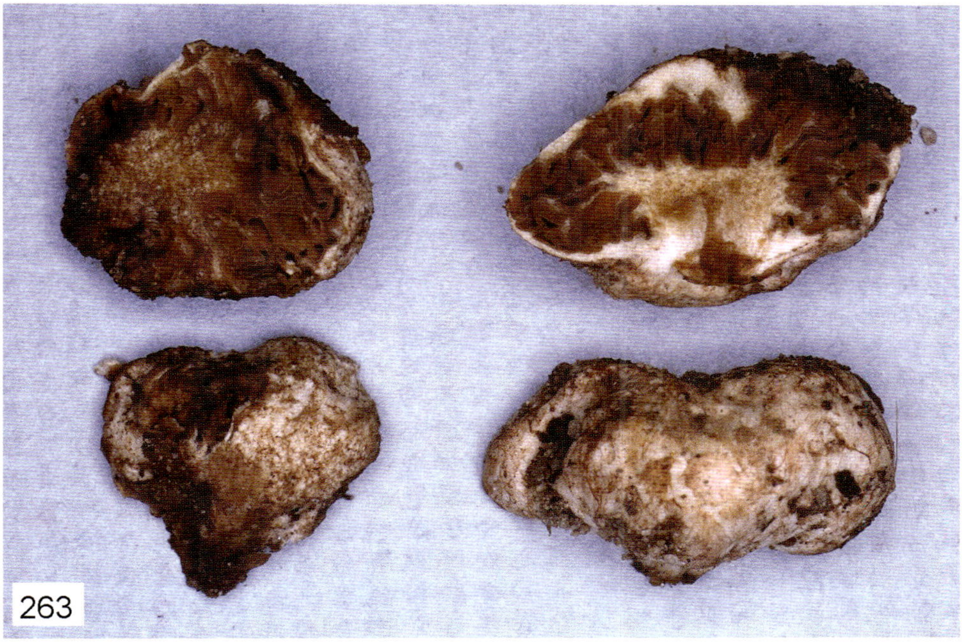

Plate 263. *Auritella geoaustralis*, holotype. (PERTH 07664605). Photograph — N.L.Bougher.

Etymology: From the Latin *geo-* (earth) and *australis* (southern), in reference to its subterranean habit and southern distribution.

Diagnostic features: Fruit bodies secotioid. Peridium dry, cream-coloured, silky-fibrillose; gleba dark brown, composed of irregularly empty locules, not gelatinised; columella present. Spores smooth, ovate-elliptic; apiculus typically very small; ochre-buff to ochre-tawny, nondextrinoid; basidia necropigmented; cystidia not observed.

Aside from its sequestrate form, *A. geoaustralis* exhibits characters shared with other species of *Auritella* such as smooth spores, necropigmented basidia, the absence of pleurocystidia, and an association with Myrtaceae and Fabaceae. In the past *A. geoaustralis* had been considered to represent the informal "*Genus B*" by Bougher & Lebel (2001) and as "*Geoinocybe*" by Francis & Bougher (2004). However, the inclusion of *A. geoaustralis* in *Auritella* has been firmly supported by nucleotide sequences for three genes, *rpb1*, *rpb2*, and nLSU (Matheny, 2005;

Matheny & Watling, 2004). In a phylogeny including this and four other species of *Auritella* presented by Matheny & Bougher (2006a), *A. geoaustralis* is the sister species to the agaricoid *A. serpentinocystis* Matheny, Trappe & Bougher. In the most recent phylogenetic work presented here (Fig. 153), it forms a well-supported sister lineage to a clade of three other Australian species, *viz. A. brunnescens* Matheny & Bougher, *A. fulvella* Matheny & Bougher and *A. serpentinocystis.*

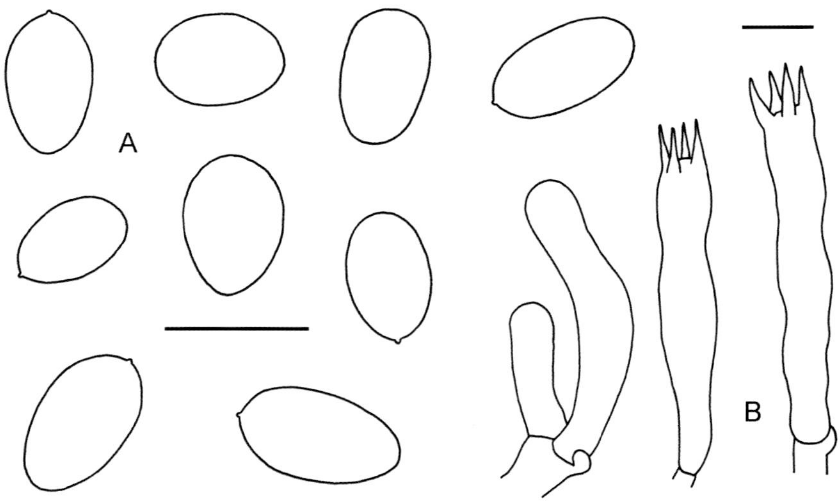

Figure 159. *Auritella geoaustralis*. All PERTH 07664605. **A,** Basidiospores; **B,** Basidia. Scale bars = 10 µm (longer bar for spores only).

7. Auritella robusta Matheny, Bougher & G.M.Gates, *sp. nov.*

MycoBank: MB 809260

Type: Pipeline Track, Mt Wellington, Hobart, Tas., 42°53'S, 147°13'E, on soil in wet sclerophyll forest under *Eucalyptus obliqua*, *Pomaderris* and *Olearia argophylla*, 13 Jan. 2014, *G.M.Gates 1163*; holo: HO 572018; iso: PERTH 08557756, TENN 066593 (KJ702358-nLSU, KJ702359-ITS, KJ702360-*rpb2*).

Pileus 39–73 mm wide, initially plano-convex to gibbous, becoming plane with a depressed centre; margin incurved to decurved; surface dry; velipellis absent, not rimose, finely fibrillose, at times forming fine appressed scales towards the margin; burnt yellow-ochre (5C6), almost rusty, discoloring to blackish brown in areas with age. *Lamellae* sinuate, close to crowded, with several tiers of lamellulae; dark greyish brown to yellowish brown, becoming dark brown with age; edges white-fimbriate, up to 6 mm deep. *Stipe* 43–62 × 11–15 mm, somewhat enlarged below and 15–20 mm wide towards the base, stout, terete; surface dry, fibrous, whitish with overlain longitudinal isabelline to brown fibrils imparting a streaked appearance; base with white mycelium; veil not observed; context solid, becoming hollow, bruising pinkish brown. *Odour* like perfumed rose. Pl. 264.

Basidiospores (6.0–) 6.5–8.5 × 5.0–6.0 µm [mean 7.6 × 5.3 µm; Q 1.20–1.60; Q mean 1.43 (*n* = 28/1)], smooth, elliptical to broadly elliptical or ovate, with rounded or obtuse apices, yellowish brown; apiculus small but conspicuous. *Basidia* 38–55 × 7–10 µm, slender-clavate to cylindroclavate, hyaline but rapidly becoming necropigmented, 4-sterigmate (rarely with 1 or 2 sterigmata). *Pleurocystidia* absent. *Cheilocystidia* 31–115 (–154) × 7.0–11.0 µm apically, up to 17 µm wide elsewhere, variable in shape (clavate, fusiform or cylindrical), with or without a swollen apex, long narrow forms sometimes sinuous, thin-walled, hyaline or becoming necropigmented and granular, with subtending hyphae sometimes swollen at their apex or with a finely encrusted wall. *Caulocystidia* absent, but a caulohymenium can be present at the extreme apex of the stipe. *Pileipellis* a cutis of tangled and interwoven hyphae, these yellowish brown or fulvous in mass; individual hyphae hyaline or with dark ochraceous contents, mostly 5–9 μm wide, smooth, thin-walled. *Clamp connections* present. Fig. 160.

Occurs in Tas.; on soil in wet-sclerophyll forest under *Eucalyptus obliqua*, *Pomaderris apetala* and *Olearia argophylla*, and also in *Eucalyptus delegatensis* forest. Fruiting Jan. and May.

Additional specimen examined: Tas.: Mt Maurice, plot 5, 41°22'2"S, 147°36'6"E, under *Eucalyptus delegatensis*, 14 May 2007, *B.M.Horton T526* (MEL, *n.v.*).

Etymology: From the Latin *robustus* (large), in reference to the large fruit bodies.

Diagnostic features: Fruit bodies large and robust; context bruising pinkish brown; odour of perfumed rose. Spores elliptical to broadly elliptical or ovate; cheilocystidia very long and highly variable in shape.

Auritella robusta is the only Tasmanian species of the genus, and although known from just two collections, its diagnostic characters justify its novel status. The very long cheilocystidia are somewhat similar to those of *A. arenicolens* (Cleland) Matheny & Bougher, but the spores of the latter are considerably longer and narrower; *A. robusta* also becomes pinkish brown where bruised and has an aromatic odour, unlike *A. arenicolens*. The large size of *A. robusta*, with a pileus sometimes exceeding 70 mm in width, readily separates it from the other Australian taxa. Phylogenetic analysis also supports its recognition at species level (Fig. 153). ITS blastn analysis of the type collection reveals 99% sequence similarity with an ITS sequence produced from a fruit body (*T526*) previously sampled from *Eucalyptus delegatensis* forest in Tasmania (Horton, 2011).

Plate 264. *Auritella robusta*, holotype (HO 572018). Photograph — G.M.Gates.

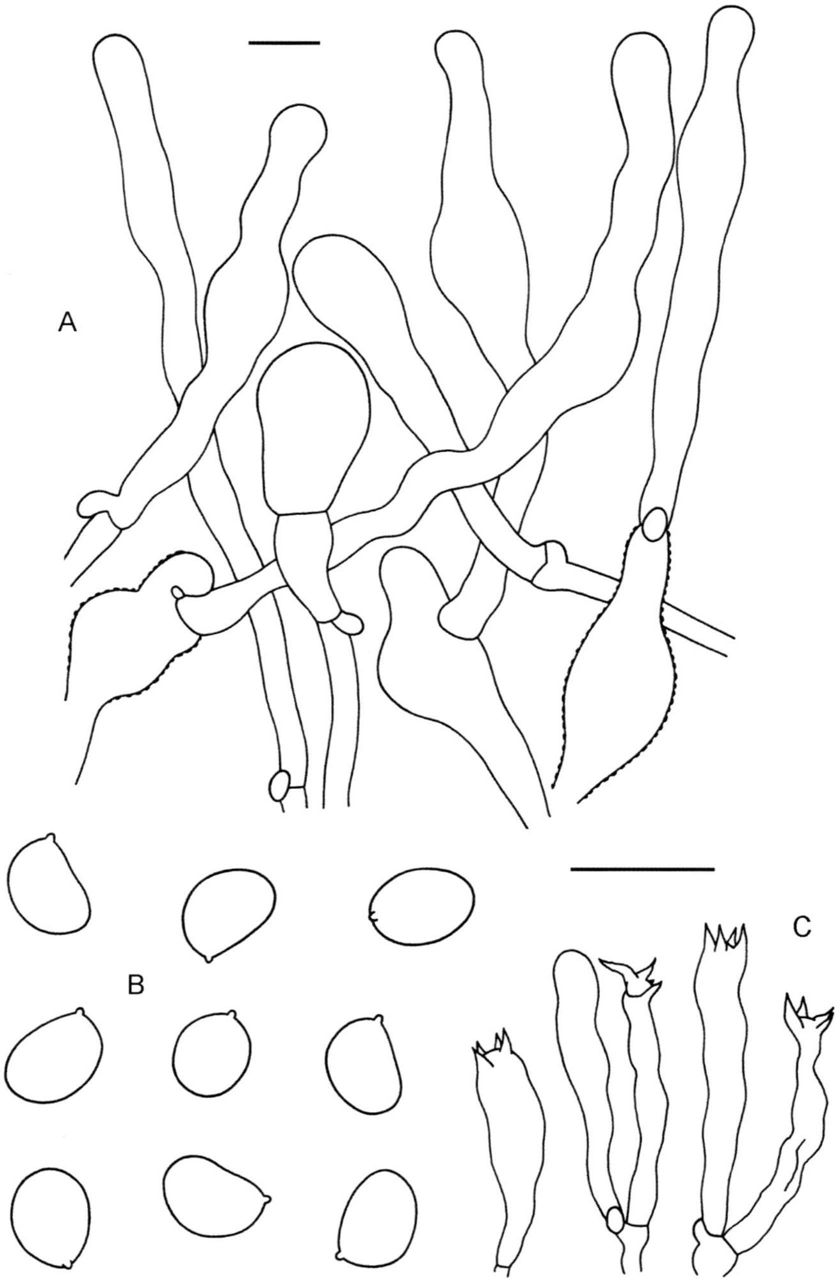

Figure 161. *Auritella robusta.* All holotype (HO 572018). **A,** Cheilocystidia. **B,** Basidiospores. **C,** Basidia. Scale bars – 10μm (longer bar for spores only).

8. Auritella serpentinocystis Matheny & Bougher, *Mycotaxon* 97: 232 (2006)

Auritella serpentinocystis Matheny, Trappe & Bougher, *Mycol. Progr.* 5: 9 (2006), *nom. inval.* (ICN Art. 36.1). Type: near south-central gate, Genaren Hill Sanctuary, Parkes Shire, N.S.W., just E of ridgetop with skeletal rocky soil, under *Eucalyptus sideroxylon*, *E. dealbata* and *E. microcarpa*, 22 Apr. 2000, *J.Trappe 25080*; holo: PERTH 07536755; iso: WTU (AY038325-nLSU, AY333773-*rpb2*, AF389547-*rpb1*).

Pileus 13–40 mm wide, convex to obtusely conical, lacking an umbo; margin decurved, entire; dark yellowish brown to brown (10YR 3/4 to 4/3) when young, generally darker at the centre, browner (10YR 4/3) to pale brown (10YR 6/3) with age, mostly pale brown over the centre, when dry Amber Brown, Sudan Brown to Antique Brown, minutely velvety, occasionally cracking circumscissily, dry; context brownish white or pale brown. *Lamellae* adnexed to ascending-adnate (not subdecurrent), seceding, closely spaced, with numerous lamellulae; edges distinctly pallid and fimbriate, pale brown, becoming dark brown to dark yellowish brown or bay-brown (10YR 3/3 to 3/4). *Stipe* 13–40 × 2–9 mm, often longer than the pileus diameter, cylindrical to tapered, not squat, pale brown to brownish white, longitudinally fibrillose, not pruinose or only at the extreme apex; cortina rapidly evanescent; context fleshy-fibrous, not particularly tough, staining pinkish to reddish brown after exposure to air; tissues exuding yellow or ochraceous pigment in 3–5% KOH solutions. *Odour* musty or unpleasant.

Basidiospores 8.0–9.5 (–10.0) × 5.0–5.5 (–6.0) µm [mean 8.6 × 5.4 µm; Q 1.45–1.73; Q mean 1.59 (n = 10/1; holotype)], smooth, usually elliptic to subphaseoliform or subamygdaliform to elliptic, brownish yellow (ochraceous-buff); apices obtuse; apiculus prominent but small. *Basidia* 28–39 × 8–9 µm, clavate to narrowly so, 4-sterigmate, occasionally 1- or 2-sterigmate, necropigmented, hyaline becoming ochraceous brown. *Pleurocystidia* absent. *Cheilocystidia* (25–) 35–72 (–82) × 5–10 µm, narrowly cylindrical to narrowly subclavate or clavate, serpentine or flexuous, thin-walled or less often secondarily thickened in the apical region, hyaline or less often pale yellow or ochraceous; apices obtuse, not capitate, occasionally mixed with a few clavate cells; entire edge sterile. *Caulocystidia* 25–43 × 9–15 µm, in dense clusters, restricted to the extreme apex, unlike cheilocystidia, (sub)clavate to cylindrical, but not long and serpentine, thin-walled, hyaline; caulobasidia at times scattered at the apex. *Pileipellis* a cutis of subparallel to somewhat upturned cylindrical hyphae 5–11 µm wide, encrusted, ochraceous to tawny in mass; tramal hyphae pale yellow to subhyaline, to 13 µm wide, frequently septate. Mounts of context tissue dissolving pale yellow to ochraceous pigment in KOH. *Clamp connections* present in all tissues. Fig. 162 (see also Pl. 11A, 12A).

Occurs in eastern and central-western N.S.W.; scattered in woodland under *Eucalyptus*, *Acacia* and *Micromyrtus*, or in wet sclerophyll forest under *Eucalyptus* and *Leptospermum*. Fruiting Apr. and May.

Additional specimens examined: N.S.W.: Watagan Forest, c. 100 km N of Sydney, 32°57'S, 151°13'E, gregarious on soil in litter in eucalypt forest, 11 Apr. 1982, *A.Young 732* (BRI AQ808415); type locality, 21 Apr. 2000, *J.Trappe 25066* (MEL); Yengo Natl Park, 32°48'29"S, 150°53'53"E, alt. 283 m, scattered in sandy soil, in wet sclerophyll forest under *Eucalyptus* and *Leptospermum* with no understorey, 21 May 2009, *N.Fechner & R.Halling PBM3188* (MEL 2335650A, TENN 063641; JQ313559-nLSU, KJ729858-ITS, KJ756402-*rpb2*).

Diagnostic features: Fruit bodies medium-sized, not squat; context staining pinkish to reddish brown upon exposure to air. Cheilocystidia mostly narrowly cylindrical to narrowly clavate and highly flexuous in outline. Dried tissues exuding yellow to ochraceous pigment in KOH.

Auritella serpentinocystis is characterised by the abundant, long and narrow, crowded, serpentine cheilocystidia, a velvety to felty-tomentose or fibrillose pileus and occurrence under *Eucalyptus* in eastern New South Wales. Unlike some other *Auritella* species, it often has a stipe that is longer than pileus diameter, and therefore it does not have a squat habit. Rehydrated context tissue exudes a yellow or ochraceous pigment in alkali solution, similar to the genus *Pleuroflammula* Singer (Singer, 1986). This peculiar reaction to KOH separates it from *A. brunnescens* Matheny & Bougher. Dried material of *A. serpentinocystis* also lacks a strong odour, unlike *A. brunnescens*.

Auritella arenicolens (Cleland) Matheny & Bougher from Western Australia and South Australia has similar but less sinuous cheilocystidia and also differs in its much longer spores and squat habit. *Auritella dolichocystis* Matheny, Trappe & Bougher, also from New South

Wales, is light yellowish brown and has elongated clavate cheilocystidia with subcapitate apices. Nucleotide differences at *rpb2* and nLSU loci also separate *A. dolichocystis* from *A. serpentinocystis* (Fig. 153; Matheny & Bougher, 2006a).

Matheny & Bougher (2006a) referred to a collection from Western Australia (*OKM 23824/CSIRO E587*) with a well-developed velipellis and longer basidiospores. This specimen cannot be located and may have been misplaced in the transfer of collections from CSIRO to PERTH in 2004.

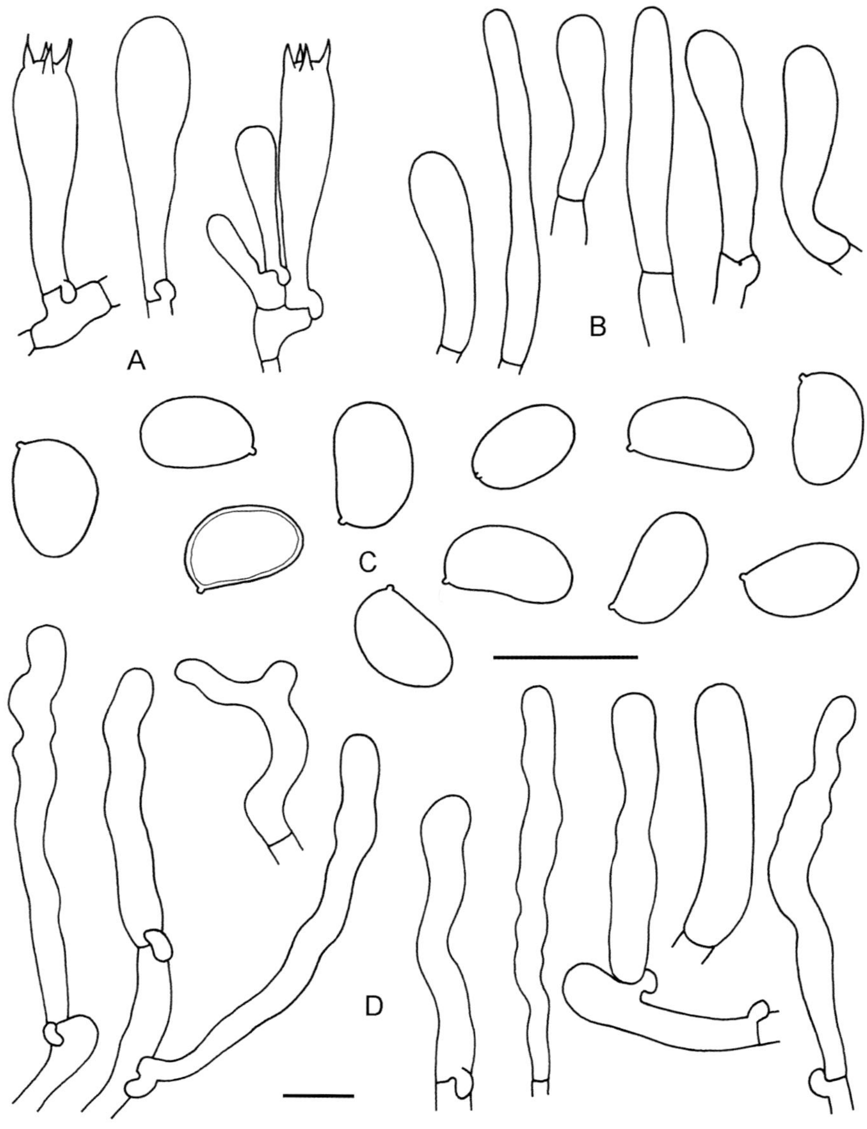

Figure 162. *Auritella serpentinocystis*. All holotype (PERTH 07536755). **A,** Basidia; **B,** Caulocystidia; **C,** Basidiospores; **D,**. Cheilocystidia. Scale bars: 10 μm (longer bar for spores only).

TUBARIOMYCES

Tubariomyces Esteve-Rav. & Matheny, *Mycologia* 102(6): 1390 (2010)

Type: *T. inexpectatus* (M. Villarreal, Esteve-Rav., Keykoop & E. Horak) Esteve-Rav. & Matheny

Fruit bodies very small to small, omphalinoid or tubarioid in habit. *Pileus* <20 mm wide, squamulose, felty, velvety, or granular. *Lamellae* adnate, arcuate, or shortly decurrent, distant or subdistant. *Stipe* pruinose to fibrillose for entire length or with minute scattered scales. *Basidia* necropigmented. *Pleurocystidia* absent. *Cheilocystidia* present, thin-walled or with slightly thickened walls. *Caulocystidia* present, descending entire length of stipe, often pigmented. *Pileipellis* a palisade of clavate to subcyclindrical terminal hyphae, forming a trichoderm or subhymeniform-like covering.

Six species of *Tubariomyces* are now known worldwide. Four are formally described (Alvarado *et al.*, 2010, Vizzini *et al.*, 2013). Two species are recorded from tropical Australia. Others are known from Mediterranean Europe and Zambia. Associating with Cistaceae, Fagaceae, Fabaceae, Phyllanthaceae, and Myrtaceae. Not known from *Nothofagus* rainforest. The genus was earlier referred to as the "*Mallocybella* clade" in works by Matheny *et al.*, (2009) and Matheny (2009).

KEY TO SPECIES OF *TUBARIOMYCES* IN AUSTRALIA

1 Pileus conical to obtusely conical, lamellae short decurrent .... **1. T. australiensis**

1: Pileus plano-convex and umbilicate, lamellae adnate .... **2. T. umbilicatus**

1. **Tubariomyces australiensis** Matheny, Bougher & M.D.Barrett, *sp. nov.*

MycoBank: MB 808679

Type: Bachsten Creek Campsite, on S arm of Bachsten Ck, Edkins Ra., NW Kimberley region, W.A., Australia, 15°59'28"S, 125°19'41"E, in sandy soil in relatively dense eucalypt woodland under *Eucalyptus miniata*, *E. tetrodonta*, *Corymbia polycarpa*, *Melaleuca leucadendra* and *Petalostigma pubescens*, beside river with tall grassy understory, 15 Jan. 2010, *M.D.Barrett MDB F62/10*; holo: PERTH 08383189 (KP641654-ITS).

Pileus 3–5 mm wide, conical to obtusely conical; margin slightly decurved to almost straight; umbo absent; surface pubescent to smooth; margin sometimes obscurely translucent-striate, not obviously hygrophanous; strong brown to light yellowish brown or yellowish brown (7.5YR 5/6, 10YR 5/4 or 10YR5/6), light yellowish brown between striae; context very thin. *Lamellae* short-decurrent, distant, 12–16 L, with few tiers of lamellulae, pale brown to yellowish brown (10YR 6/3 to 10YR 5/4), edges distinctly white-fimbriate. *Stipe* 15–20 × 0.5–1.0 mm, terete, even, indistinctly pruinose or otherwise nearly smooth, no partial veil observed, light yellowish brown to yellowish brown (like the pileus); context not observed. *Odour* not known. Pl. 265, 266.

Basidiospores 7.5–9.0 × 4.0–5.5 µm [mean 8.4 × 4.7 µm; Q 1.60–2.16 (–2.25); Q mean 1.82 (*n* = 23/1)], smooth, elliptic with obtuse apices, less often subamygdaliform, occasionally with a slight ventral depression, yellowish brown; apiculus small, indistinct. *Basidia* 18–32 × 6–7 µm, clavate, ochraceous and collapsing with age, 4-sterigmate. *Pleurocystidia* absent. *Cheilocystidia* 29–51 × 8–12 µm, cylindrical to irregularly cylindrical or less often ventricose with a slight swelling at the centre or above the base, with apices obtuse or slightly swollen, thin- to slightly thick-walled (to 1 µm), hyaline, but many becoming rusty-ochraceous, densely arranged and crowded on the lamellar edge. *Caulocystidia* 29–46 × 6–8 µm, similar to cheilocystidia but some contorted and with swollen apices, visible throughout the stipe, with walls 1–2 µm thick, mostly rusty-ochraceous. *Pileipellis* a crowded palisade of cylindrical to clavate rusty-ochraceous, smooth, thick-walled terminal cells 20–47 × 5–12 µm, subtended by polygonal and irregular swollen cells up to 20 µm wide. *Clamp connections* present. Fig. 163 (see also Pl. 11C, 12B).

Known only from the type locality in the north-western Kimberley, W.A.; grows in monsoonal eucalypt woodland. Fruiting Jan.

Etymology: The epithet *australiensis* is used in reference to the occurrence of the new species in Australia.

Plate 265. *Tubariomyces australiensis*, holotype (PERTH 08383189). Photograph — M.D.Barrett.

Plate 266. *Tubariomyces australiensis*, holotype (PERTH 08383189). Photograph — M.D.Barrett.

Diagnostic features: Fruit bodies very small, with an omphalinoid or tubarioid habit. Pileus conical to obtusely so; lamellae short-decurrent and distant. Spores smooth, mostly elliptic and <10.0 µm long; caulocystidia descending entire the length of the stipe, with crowded cylindrical cheilocystidia and a pileipellis of terminal cells.

Tubariomyces australiensis exhibits morphological traits typical of *Tubariomyces* Esteve-Rav. & Matheny (Alvarado *et al.*, 2010), e.g. an omphalinoid or tubarioid habit, and cheilocystidia and pileipellis structure. Three species have been described from Mediterranean Europe: *T. inexpectatus* (M.Villarreal, Esteve-Rav., Heykoop & E.Horak) Esteve-Rav. & Matheny, *T. hygrophoroides* Esteve-Rav., P.A.Moreau & E.E.Hermos. and *T. similis* Della Maggiora, Tolaini & Vizzini (Vizzini *et al.*, 2013). An additional, as yet undescribed species has also been acknowledged from Zambia (Matheny *et al.*, 2009; Alvarado *et al.*, 2010). The description of *T. australiensis* raises the number of confirmed named species to four and is the first report of the genus from Australia.

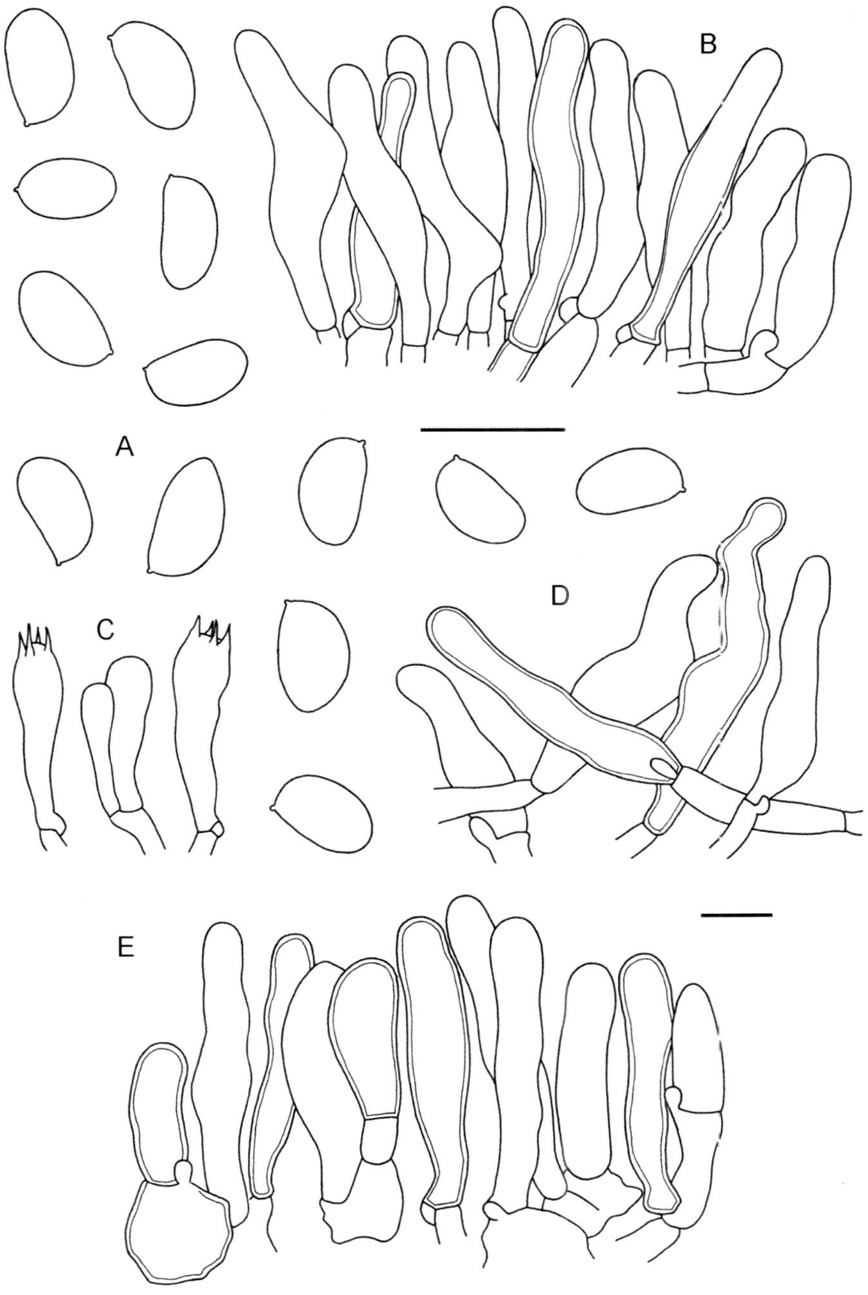

Figure 163. *Tubariomyces australiensis,* holotype (PERTH 08383189). **A,** Basidiospores; **B,** Cheilocystidia; **C,** Basidia; **D,** Caulocystidia (from upper stipe); **E,** Pileipellis. Scale bars = 10 µm (longer bar for spores only).

Plate 267. *Tubariomyces "umbilicatus"*. TENN 066973. Photograph — P.B.Matheny.

Plate 268. *Tubariomyces "umbilicatus"*. TENN 066973. Photograph — P.B.Matheny.

An additional collection from north-eastern Queensland, almost certainly represents a different species of *Tubariomyces*. This entity (known only from very scant and inadequate material and thus only provisionally described below as *T. "umbilicatus"*) is distinguished by its granulose pileus surface with a central depression, adnate lamellae and entirely pruinose stipe surface. Microscopically, it is very similar to *T. australiensis*.

Tubariomyces australiensis differs morphologically, ecologically, and geographically from other species of *Tubariomyces* outside Australia, *viz*, having much shorter and elliptic spores (<10 µm long, and growing in association with myrtaceous plants and *Petalostigma* (Picrodendraceae). The caulocystidia of *T. australiensis* are found in scattered dense clusters and are often slightly thick-walled (walls 1–2 µm thick). The end cells of the pileipellis are clavate to subcylindrical and form a palisade or hymeniform arrangement. In the European species of *Tubariomyces*, the pileipellis is described as a trichoderm with caulocystidioid (cylindrical to clavate) end cells.

Tubariomyces australiensis is a small, rather inconspicuous fungus that can readily be overlooked by collectors, or perhaps lodged in herbaria as misidentified or unidentified material.

2. Tubariomyces "umbilicatus" Matheny & Bougher, *nom. prov.*

Pileus up to 8 mm wide, plano-convex, with a distinct central depression; margin decurved; surface dry, almost granular under a hand lens, drying pleated somewhat at the margin; velipellis absent; hazel to strong brown (5YR 4/8 to 7.5YR 5/6); context very thin. *Lamellae* adnate, seceding, subdistant, 20–22 L, with several tiers of lamellulae, dark yellowish brown (Dresden Brown), with paler and fimbriate edges, medium. *Stipe* 10–11 × 1 mm, tapered at the base, uniformly pruinose, reddish yellow to yellowish red or ochraceous (7.5YR 6/6 to 5Y 5/8), curved or flexuous; veil not observed. *Odour* none. Pl. 267, 268.

Basidiospores 7.5–9.0 × 4.0–5.0 µm [mean 8.4 × 4.7 µm; Q 1.50–2.00; Q mean 1.79 (*n* = 20/1)], smooth, mostly amygdaliform with bluntly pointed apices, occasionally elliptic or cylindrical, at times a with slight adaxial depression but not truly phaseoliform, yellowish brown; apiculus small, indistinct. *Basidia* 28–35 × 8–9 µm, clavate, hyaline, becoming necropigmented, 4-sterigmate. *Pleurocystidia* absent. *Cheilocystidia* 30–50 × 7–11 µm, mostly cylindrical to subclavate, apices obtuse or occasionally indistinctly subcapitate, with slightly thickened walls, hyaline, densely arranged, imparting sterile lamellar edges. *Caulocystidia* 40–50 × 7–9 µm, in scattered dense clusters descending the entire length of the stipe, mostly cylindrical to subclavate, at times somewhat flexuous; walls 1–2 µm thick; ochraceous to rusty brown in mass, ochraceous when viewed individually. *Pileipellis* a palisade of clavate to subcylindrical terminal cells arising from occasionally branched or often inflated hyphae, these encrusted or not; end cells 26–65 × 6–15 µm, slightly thick-walled, smooth, ochraceous in mass. *Clamp connections* present.

Specimen examined: Qld: end of Emerald Creek Rd, Dinden S.F., near Mareeba, 17°03'08.5"S, 145°32'24.2"E, scattered singly on bank of shallow ditch in cleared understorey under *Corymbia clarksoniana*, *Eucalyptus tereticornis* and *Eucalyptus* sp. (box trees), 24 Feb. 2012, *M.D.Barrett PBM3741* (TENN 066973).

In the field, this provisionally recognized species resembles a member of the *Mallocybe* clade, but the pruinose stipe clearly indicates *Tubariomyces*. Microscopically, it is very similar to *T. australiensis* but that species is distinguished by short, decurrent lamellae and a conical to obtusely conical pileus.

The single collection cited here is inadequate to designate as a type and we did not attempt to sample it for DNA sequencing. However, we provide a description and images here of this apparently undescribed species to bring it to the attention of future collectors.

APPENDIX 1: NEW TAXA

New Species

APPENDIX 1

APPENDIX 1

Provisional Names

New names

APPENDIX 1

APPENDIX 2: EXCLUDED AND DOUBTFUL TAXA

Excluded Taxa

Inocybe austrofibrillosa Grgur., *Austral. Mycol.* 21: 35 (2002)

Cortinarius fibrillosus Cleland, *Trans. Roy. Soc. S. Aust.* 52: 222 (1928); *Inocybe fibrillosa* (Cleland) Grgur., *Larger Fungi of S. Aust.*: 190 (1997), *non I. fibrillosa* Peck, *Ann. Rep. N.Y. St. Mus.* 41: 65 (1888).

Type: Mt. Lofty, S.A., 23 June 1928, *J.B.Cleland*; lectotype: AD 4169.

Inocybe cystidiocatenata Grgur., *Larger Fungi of S. Aust.* 188 (1997); *Cortinarius cystidiocatenatus* (Grgur.) Gasparini, *Austral. Mycol.* 25: 25 (2006).

Type: Morialta Falls, S.A.,caespitose on burnt soil under *Eucalyptus* sp., 28 June 1984, *C.A.Grgurinovic & R.E.Halling 35684*; holotype: AD 12264.

Pileus 1 to 1½ inches [25–37 mm diam.], deeply convex, then convex and gibbous, hoary from whitish mealy fibrils sometimes forming a white edge to the pileus, when moist reddish brown [near Russet (xv.) or darker], when dry between Cinnamon and Cinnamon Buff (xxix). *Lamellae* adnexed, moderately close, yellower than Ochraceous Tawny (xv.). *Stipe* up to 1¾ inches [4.3 cm] high, slender, fibrillose, pallid whitish, hollow; context reddish, thin. Spores pale brown, oblique 9.0 × 4.5 µm. Caespitose on the ground near stumps.

Basidiospores (7.5–) 8.0–9.0 × 4.5–5.0 (–5.0) µm (n = 20/1), smooth in appearance including when in Melzer's reagent, elliptic to subamygdaliform with obtuse apices, noticeably pale yellowish (2.5Y 7/6; Warm Buff) in KOH; dextrinoid, reddish brown (near 2.5Y 4/6; Tawny) in Melzer's reagent; thick-walled; germ pore and plage absent; apiculus small but distinct. *Basidia* 22–28 × 6–8 µm, 4-sterigmate, clavate, mostly hyaline but at times ochraceous. *Pleurocystidia* absent. *Cheilocystidia* 17–21 × 14–15 µm, saccate to pyriform or broadly clavate, thin-walled, hyaline, difficult to revive but edges of lamellae sterile under the stereomicroscope. *Caulocystidia* absent; vesture hyphae interwoven, often thick-walled at least from the lower part of the stipe, hyaline, cylindric, 4–12 µm diam. *Lamellar trama* regular, reddish brown (light Tawny) pigmented in mass; hyphae cylindric to inflated, up to 25 µm wide. *Pileipellis* an interwoven suprapellis or ascending layer of cylindric to inflated hyphae, which are hyaline, thin- to thick-walled, walls up to 1.5 (–2.0) µm thick; subpellis ochraceous pigmented in mass, with hyphae primarily inflated, up to 20 µm wide, lightly incrusted. *Clamp connections* present.

Other specimens examined (see also below in *Inocybe cystidiocatenata*): Tas. Mount Wellington picnic area, near *Leptospermum* sp., 17 May 1991, *N.L.Bougher E4439* (PERTH 07630964) as "*Cortinarius fibrillosus* - frosty cortinarius". W.A. Amphion Forest Block, off Murray River Road, Dwellingup, in *Eucalyptus marginata* & *Corymbia calophylla* forest, 5 Aug. 1996, *N.L.Bougher E5774* (PERTH 07598483) as "*Cortinarius fibrillosus* - frosty cortinarius"; Plavins Forest Block, Murray River Road, Dwellingup, in *Eucalyptus marginata* & *Corymbia calophylla* forest, 26 June 1997, *M.Glen & S.Q.Bolsenbroek E5889* (PERTH 07608527) as "*Cortinarius fibrillosus* - frosty cortinarius"; Alcoa Mine, Nettleton Road, Dwellingup, in *Eucalyptus marginata* & *Corymbia calophylla* forest, 4 July 2000, *M.Glen & I.C.Tommerup E6584* (PERTH 07670613), as "*Inocybe austrofibrillosa*"; Alcoa Mine, Nettleton Road, Dwellingup, in litter in rehabilitated forest, 8 July 2002, *G.Mullins & N.Kerp E7299* (PERTH 07676115), as "*Cortinarius fibrillosus* - frosty cortinarius"; Ryedene Property, near Two Peoples Bay, in litter in woodland of *Eucalyptus marginata*, *Corymbia calophylla* and *Allocasuarina*, 20 June 2007, *N.L.Bougher & L.Bell NLB285* (PERTH 08074070), as "*Inocybe cystidiocatenata*".

Given the presence of cheilocystidia and smooth pale brownish spores, Grgurinovic (1997) transferred this species to *Inocybe*. Later, realizing *Inocybe fibrillosa* (Cleland) Grgur. was a homonym of *Inocybe fibrillosa* Peck, the new name, *Inocybe austrofibrillosa* Grgur., was proposed (Grgurinovic, 2002). Several lines of evidence, however, demonstrate that *I. austrofibrillosa* is not an *Inocybe*: the *Telamonia*-like habit, the absence of pleurocystidia but presence of a distinct apiculus on the spores, and the reddish brown colour or dextrinoid reaction of the spores in Melzer's reagent. We consider Cleland's original generic disposition, *Cortinarius*, as the best placement of this unusual species despite the smooth spore walls, pale colored spore deposit and spore wall reaction in Melzer's reagent. The presence of cheilocystidia does not in itself exclude *Cortinarius* from consideration as a small number of species in that genus have them. Furthermore, pale pigmented spores that appear smooth under light microscopy, although highly unusual, can exceptionally be found in *Cortinarius*, as, for example, in *C. aureifolius* Peck.

Recent collections of this species, identified as *Inocybe cystidiocatenata* Grgur., are not uncommon in Western Australia. One of these (PERTH 07670613) was sequenced (AY333317, AY380361, AY333306: accessions at GenBank) and confirmed as closely related to other species of *Cortinarius* (Matheny & Ammirati, 2003). There is little doubt that *I. cystidiocatenata* is conspecific with *C. fibrillosus*, and that the latter has nomenclatural priority. Grgurinovic maintained their separation chiefly on cystidial characters, which we were unable to discern after comparison of the two types. Both types also share the thick-walled hyphae of the pileipellis and vesture hyphae of the stipe. The spores of the type of *I. cystidiocatenata*, although largely unreactive in Melzer's, feature occasional spores that are dextrinoid. The following features stand out for this species: the bright brown lamellae and flesh (near 6C8), pale spores, and cheilocystidia of inflated elements in chains in a broad layer on the lamellar edge.

Inocybe cystidiocatenata Grgur., *Larger Fungi of S. Aust*: 188 (1997)

Cortinarius cystidiocatenatus (Grgur.) Gasparini, *Austral. Mycol.* 25: 25 (2006).

Type: Morialta Falls, S.A.,caespitose on burnt soil under *Eucalyptus* (Myrtaceae), 28 June 1984, *C.A.Grgurinovic & R.E.Halling 35684*; holotype: AD 12264.

Cortinarius fibrillosus Cleland, *Trans. Roy. Soc. S. Aust.* 52: 222 (1928); *Inocybe fibrillosa* (Cleland) Grgur., *Larger Fungi of S. Aust.* 190 (1997), *non I. fibrillosa* Peck, *Ann. Rep. N.Y. St. Mus.* 41: 65 (1888).

Inocybe austrofibrillosa Grgur., *Austral. Mycol.* 21: 35 (2002).

For a macroscopic description and illustrations see Grgurinovic (1997, Plates 8a and 29c).

Basidiospores 8.0–9.0 (–10.5) × 5.0–5.5 µm (n = 20/1), elliptic to subamygdaliform with mostly obtuse apices, smooth in KOH and Melzer's, noticeably quite pale in KOH, very pale brown (Warm Buff), mostly unchanging in Melzer's but occasional spores dextrinoid, with thick wall and a small but distinct apiculus, germ pore and plage absent. *Basidia* 25–31 × 6–8 µm, 4-sterigmate, clavate, hyaline or occasionally ochraceous, not necropigmented. *Pleurocystidia* absent. *Cheilocystidia* elliptic or saccate, quite small (as depicted by the type line drawings (Grgurinovic, 1997)) in relation to the size of the basidia. *Caulocystidia* absent; stipe sheathed by vesture hyphae, which are interwoven, filiform, smooth, usually thick-walled, similar to those of the suprapellis on the pileus; terminal cells undifferentiated; stipe hyphae pale ochraceous in mass. *Lamellar trama* regular, bright cinnamon pigmented in mass; hyphae cylindrical to inflated, up to 25 µm wide, encrusted. *Clamp connections* present.

Specimens examined (see also above in *Inocybe austrofibrillosa*)**:** Only *C.A.Grgurinovic & R.E.Halling 35684*, see above.

This species is synonymous with *Cortinarius fibrillosus* Cleland. For a discussion of its status, see the commentary under *I. austrofibrillosa* above.

Inocybe marangania Grgur., *Larger Fungi of S. Austral.*: 179 (1997)

Pholiota marangania (Grgur.) Matheny & Bougher, *Muelleria* 28: 101 (2010).

Type: Mt. Lofty Botanic Gardens below "Carmino", S.A., gregarious in troops among *Funaria*, 9 May 1984, *C.A.Grgurinovic 2A584* & *R.J.Chinnock*; holotype AD 12263.

For a macroscopic description, see Grgurinovic (1997).

Basidiospores 7.0–8.0 × 4.0–5.0 µm (n = 15/1), smooth, (sub)amygdaliform, not phaseoliform, with conical apices in profile; apices with an indistinct germ pore; Tawny or Hazel in KOH, thick-walled (walls up to 0.4 µm thick), with an indistinct apiculus, not reacting in Melzer's. *Basidia* 20–27 × 6–7 µm, 4-sterigmate, cylindrico-clavate to cylindric, hyaline or ochraceous. *Pleurocystidia* 43–80 × 10–14 µm, mostly lageniform with elongated cylindrical necks; apices obtuse or at times swollen or subacute, not capitate; also intermediate shapes tapered distally, thin-walled; apices not crystalliferous but necks may be granular encrusted, hyaline or ochraceous or with vacuolated portions with yellow contents. *Cheilocystidia* 28–55 × 10–14 µm, a few similar to pleurocystidia, but mostly fusiform or broadly so, merely tapered distally to an obtuse apex, thin-walled, hyaline or ochraceous; paracystidia not observed. Extreme apex of stipe with clusters of caulocystidia, usually cylindrical and bright ochraceous to hyaline, or parts of the cells with ochraceous contents; bright tawny refractive hyphae frequent in stipe trama. *Lamellar trama* regular, pale yellowish in mass, hyphae cylindrical, up to 16 µm wide. *Pileipellis* an exocutis of loosely interwoven narrowly cylindrical hyphae 2.5-4.0 µm wide, often branching, ochraceous or yellowish to subhyaline, thin-walled; subpellis a bright ochraceous-tawny repent layer in mass; hyphae cylindrical to inflated, 5–22 µm wide, encrusted with pigment. *Clamp connections* present.

Specimens examined: Only *C.A.Grgurinovic 2A584* & *R.J.Chinnock*, see above.

The presence of a gelatinised pileipellis, an indistinct germ pore, and what appears to be growth on a charred substrate indicate *Inocybe marangania* is a species of *Pholiota*, likely related to the *P. highlandensis* group.

Inocybe striatula (Cleland) Grgur., *Larger Fungi S. Austral.*: 190 (1997)

Cortinarius striatulus Cleland, *Trans. Roy. Soc. S. Austral.* 52: 221 (1928).

Type: Mt. Lofty, S.A.,28 July 1928, *J.B.Cleland*; holotype (AD 4660);

Pileus up to 2.5 cm diam., somewhat convex to nearly plane, more or less gibbous, wavy on the margin, sometimes slightly repand, very finely fibrillose, submembranaceous; when moist striate at the periphery and between Russet and Cinnamon Brown (xv.), with a pale ring around the edge; when dry paler than Cinnamon Buff (xxix.). *Lamellae* sinuate, moderately close, slightly ventricose, near Sayal Brown (xxix.), with narrow finely serrate whitish edges. *Stipe* 5 cm high, moderately slender, a little flexuous, sometimes flattened, attenuated at the base and sometimes apically as well, fibrillose, markedly hollow, pallid, with a brownish tint above; context pallid with a slight reddish-brown tint. Spores yellow-brown, oblique, 7.5 × 4 µm. Subcaespitose amongst shrubs under eucalypts.

Basidiospores 7.0–8.5 × (4.0) 4.5–5.5 µm (n = 20/1), faintly verrucose in Melzer's, elliptic to subamygdaliform with obtuse or bluntly pointed apices, germ pore absent; "Light Ochraceous Buff" or "Ochraceous Buff" in KOH, thick-walled, with a small but distinct apiculus, not reacting in Melzer's. *Basidia* 22–26 × 6–7 µm, 4-sterigmate, cylindrico-clavate to clavate, hyaline to ochraceous, many collapsed. *Pleurocystidia* none. *Cheilocystidia* not demonstrable with confidence. *Caulocystidia* absent; stipe surface with sparse layer of vesture hyphae at apex down to the base, similar to suprapellis hyphae described below. *Lamellar trama* regular, Ochraceous-Buff to hyaline in mass, not brightly pigmented; hyphae cylindric, up to 13 µm wide. *Pileipellis* with a suprapellis composed of an interwoven layer of narrowly cylindrical, smooth, thin-walled, hyaline hyphae up to 7.0 µm wide, mostly collapsed; subpellis, "Ochraceous-Tawny" pigmented and composed of mostly inflated hyphae up to 27 µm diam, faintly encrusted. *Clamp connections* present.

Specimens examined: S.A. Waterfall Gully, 1 July 1945, *J.B.Cleland* (AD 4656); Belair National Park, 11 August 1945, *J.B.Cleland* (AD4655). Tas. Cradle Mountain, Tas., on soil near *Nothofagus cunninghamii*, 12 May 1991, *N.L.Bougher E4379* (PERTH 07630530). W.A. Plantagenet, 12 km west of Denmark, along old railway line road, in regrowth farmland, some overgrown tracks, medium to tall forest of *Eucalyptus diversicolor*, *E. megacarpa*, *E. patens*, *Corymbia calophylla* and *Agonis* sp., 24 June 2001, *P.B.Matheny E6783/PBM2137* (PERTH 07651775).

This is a species of *Cortinarius* and should be recognized as originally proposed by Cleland: *C. striatulus*. The spores are faintly verrucose, possess a distinct apiculus, and the holotype and other authentic materials display a *Telamonia*-like habit, including the downwards attenuated stipe and striate pileus carefully emphasized in the protologue. Relatively recently collected material of what we interpret to represent this species (e.g. PERTH 07651775) demonstrates the presence of globose, pyriform to subcylindric, small cheilocystidia 16–40 × 15–24 µm. The other characters enumerated above do not warrant the placement of this species in Inocybaceae.

APPENDIX 2

Doubtful Taxa

Inocybe albidipes Cleland & Cheel, *Trans. Roy. Soc. S. Austral.* 42: 107 (1918).

The type specimens of *I. albidipes* have yet to be located although a watercolour is extant as per Horak (1980). The protologue describes a species with nodulose spores, stipe with a swollen to marginate bulb, and light to yellowish brown fibrillose and rimose pileus with an acute umbo when young. The stipe would be white given Cleland's choice of specific epithet. This description would suggest an affinity with north-temperate taxa such as *Inocybe mixtilis* (Britz.) Sacc., or *Inocybe praetervisa* Quél.

Inocybe crassipes (Cooke & Massee) Pegler, *Austr. J. Bot.* 13: 342 (1965).

Agaricus ozes var. *crassipes* Cooke & Mass., *Grevillea* 15: 93 (1887); *Collybia ozes* (Fr.) Quél. var. *crassipes* (Cooke & Mass.) Sacc., *Syll. Fung.* 9: 34 (1891).

Type material of *I. crassipes* is not located at AD. Grgurinovic (1997) indicates a study of the type but presents only a statement made by Pegler (1965), who studied type material at Kew (K) and considered the species in section *Rimosae*. We concur with Horak (1980) that *I. crassipes* is best regarded as doubtful given its fragmentary state and lack of data on cheilocystidia. We were able to view a copy of a painting of the holotype of *Agaricus ozes* var. *crassipes* at MEL. The icon features a pileus that is dark reddish brown, somewhat rimose with a few tears along the margin, more or less avellaneous to fawn coloured lamellae that are medium in breadth with darker pigmented edges, and a dark reddish brown stipe that is fibrillose-striate and enlarged towards the base. No other critical diagnostic features were discernible. In our opinion, it could be consistent with a member of section *Rimosae*, but the darker lamellar edges would be unusual for this group. The rather somber colours and darker lamellar edges could also point to *Auritella*, but without complete microscopic data, it is not possible to apply the name.

Inocybe subasterospora Cleland & Cheel, *Trans. Roy. Soc. S. Aust.* 42: 106 (1918).

Type material was not located at AD. However, two watercolour icons are referenced in the protologue (below), which we have not seen. Cleland & Cheel (1918) state the species is common in the Sydney district, so it would seem likely material corresponding to the following description may be encountered there.

[Description from protologue] *Pileus* occasionally up to 1¼ inch [c. 31 mm] in diameter, slightly convex, sometimes depressed, subgibbous to occasionally papillately umbonate, dark brown, occasionally lighter or rusty brown and often more chestnut at the periphery, usually fibrously streaked and splitting, occasionally more scaly. *Lamellae* adnate to adnexed, separating from the stem, moderately crowded, pale milk-coffee coloured, then cinnamon, sometimes with a white edge. *Stipe* up to 1½ inches [c. 37 mm] high, moderately stout, solid, slightly fibrillose, pale brownish, slightly mealy above, base sometimes a little swollen.

Basidiospores 7–9 × 5.2–7 um, irregularly knobby. *Cystidia* numerous, ventricose, apices usually knobby 42–70 × 14–17 um.

Specimens cited by Cleland: N.S.W. Sydney district, Mar. to July, Oct., Dec.; Pittwater, April; Parramatta, July; Milson Island, Hawkesbury River, July and Nov.; Hill Top (E.C.), May; localities not noted (several); locality not stated (stipe white but pileus dark); Miss Clarke, Watercolour No. 33; D.I.C., Watercolor No. 63, Herb., J.B.C., Nov. 1914).

APPENDIX 3. TABLE OF DNA SEQUENCES USED IN THIS STUDY

Species	Field no.	Herbarium accession no.	State or Country	Plant associates	GenBank Accession No.		
					ITS[a]	nLSU-rRNA	*rpb2*
Auritella arenicolens	KS1873/07	PERTH 08047081	W.A.	*Allocasuarina, Eucalyptus, Beaufortia*	KJ729857	KJ702338	KJ729920
A. brunnescens	NLB942 (holotype)	PERTH 08364567	Qld.	*Eucalyptus, Allocasuarina*	KJ702342	KJ702339	KJ702347
	PBM3173	TENN 065742	N.S.W.	*Eucalyptus*	KJ702343	JQ313558	KJ702348
	PBM3174	TENN 065354	N.S.W.	*Eucalyptus*	KJ702344	JQ313571	KJ702349
	PBM3721	TENN 066957	Qld.	*Allocasuarina, Eucalyptus*	KJ702345	KJ702340	KJ702350
	NLB963	PERTH 08364745	Qld.	*Eucalyptus*	KJ702346	KJ702341	KJ702351
A. chamaecephala	PBM2212	PERTH E7060	W.A.	*Eucalyptus, Corymbia*	—	AY635765	AY635781
A. dolichocystis	T24843	MEL	N.S.W.	*Eucalyptus, Callitris, Acacia, Dillwynia*	—	AY635764	AY635780
	T24844	MEL	N.S.W.	*Eucalyptus, Callitris, Acacia, Dillwynia*	—	AY380371	AY33731
	T24838 (holotype)	PERTH 07536569	N.S.W.	*Eucalyptus*	—	AY635763	AY635767
A. fulvella	PIF26850 (holotype)	BRI 669492	Qld.	*Corymbia, Eucalyptus*	KJ702354	KJ702352	KJ702356
	PIF26860	BRI 669485	Qld.	*Corymbia, Eucalyptus*	KJ702355	KJ702353	KJ702357
A. geoaustralis	H7344 (holotype)	PERTH	W.A.		—	AY380395	AY333774
A. robusta	I163 (holotype)	HO 572018	Tas.	*Eucalyptus, Pomaderris, Olearia*	KJ702359	KJ702358	KJ702360
A. serpentinocystis	T25080 (holotype)	PERTH 07536755	N.S.W.	*Eucalyptus*	—	AY333773	AY038325
	PBM3188	MEL 2335650A	N.S.W.	*Eucalyptus, Leptospermum*	KJ729858	JQ313559	KJ756402
Inocybe acaciae	NLB356 (holotype)	PERTH 08073813	W.A.	*Acacia*	KJ729859	JN974921	KJ729921
	T35064	OSC	N.S.W.	*Eucalyptus*	KJ729860	KJ729888	KJ729922
I. acanthosperma	PBM3773 (holotype)	PERTH 08363145	Qld.	*Eucalyptus, Allocasuarina*	KJ729861	KJ729889	KJ729923
	MDB F5/10	PERTH 08363102	N.T.	*Corymbia, Lophostemon, Pandanus*, rainforest trees	KJ729862	KJ729890	—
I. albofloccosa	PBM3645 (holotype)	PERTH 08319413	W.A.	*Melaleuca* shrubs, *Agonis* woodland	KJ729863	KJ729891	KJ729924
	PBM3423	TENN	Tas.	*Eucalyptus,*	KJ729864	KJ729892	KJ729925

APPENDIX 3

Species	Field no.	Herbarium accession no.	State or Country	Plant associates	GenBank Accession No.		
					ITS[a]	nLSU-rRNA	*rpb2*
		067737		*Nothofagus*			
	PBM3487	PERTH 08241775	W.A.	*Eucalyptus, Allocasuarina*	KJ729865	KJ729893	—
	PBM3488	PERTH 08241783	W.A.	*Eucalyptus, Allocasuarina*	KJ729866	KJ729894	KJ729926
	PBM3500	PERTH 08242259	W.A.	*Eucalyptus, Corymbia, Allocasuarina*	KJ729867	KJ729895	KJ729927
	PBM3604	PERTH 08319820	W.A.	*Eucalyptus, Corymbia*	KJ729868	KJ729896	KJ729928
	PBM3605	PERTH 08319812	W.A.	*Eucalyptus, Corymbia*	KJ729869	KJ729897	KJ729929
	PBM3609	PERTH 08319774	W.A.	*Eucalyptus, Corymbia*	KJ729870	KJ729898	KJ729930
	NLB813	PERTH 08318697	W.A.	*Eucalyptus, Corymbia*	KJ729871	KJ729899	KJ729931
I. "actinocephala".	PBM2863	WTU	Louisiana, USA	*Quercus, Carya, Fagus*	JQ408789	JN975012	JQ846495
I. "actinospora"	D25	WTU	Argentina	*Alnus*	—	AY380363	AY337361
	GDc	TENN 063762	Argentina	*Alnus*	HQ586875	HQ641117	—
I. acuta.	JV10770	TURA 5066	Finland	*Picea, Pinus, Betula, Populus, Salix*	KP171102	KM197208	—
	ELL5505	GB	Norway		AM882923 (*I. "acutella"*)	AM882923 (*I. "acutella"*)	—
I. aestiva	BK18089706 (holotype)	UTC	Utah, USA	*Populus, Abies, Pinus*	EU600847	EU600846	—
I. albovestita		PDD 71309	N.Z.	*Nothofagus*	KP171103	EU555443	—
I. alienospora	PBM3743	PERTH 08362947	Qld.	*Allocasuarina, Eucalyptus*	KP171104	KM197209	KM245970
	REH9667	NY	Qld.	*Acacia*	KP171105	KM197210	KM245971
I. aff. *alienospora*	PBM3728	PERTH 08362866	Qld.	*Corymbia, Eucalyptus, Acacia*	KP171106	KM197211	KM245972
	PBM3758	PERTH 08363269	Qld.	*Melaleuca, Corymbia, Eucalyptus*	KP171107	KM197212	KM245973
I. alloumbrina	PBM3775 (holotype)	PERTH 08363129	Qld.	*Eucalyptus, Allocasuarina*	KJ778848	KJ801158	KM245974
	PBM3372	TENN 068358	Tas.	*Eucalyptus, Pomaderris, Acacia*	KJ778849	—	—
	REH9669	BRI	Qld.	*Acacia*	KJ778850	KJ801159	—
I. alnea	PAM8082901	TENN 063832	France	*Quercus, Carpinus*	HQ586856	HQ641097	—
	PAM7090608	TENN 063831	France	*Alnus*	HQ586854	HQ641095	—
	PAM8090501	LIP	France		—	KM197213	—
I. amblyspora	JFA12538	WTU	Sweden	*Betula*	KJ729872	KJ729900	—
	BJ900831	GB	Sweden		AM882908	AM882908	—
		21670 (VMNH)	Italy?		JF908261	—	—

APPENDIX 3

Species	Field no.	Herbarium accession no.	State or Country	Plant associates	GenBank Accession No.		
					ITS[a]	nLSU-rRNA	*rpb2*
I. amygdalina	I107	HO 568205	Tas.	Wet sclerophyll forest (*Eucalyptus*, *Acacia*, *Pomaderris*)	KJ729876	KJ729902	KJ729933
	PBM3359	TENN 066003	Tas.	*Eucalyptus*, *Acacia*, *Pomaderris*	KJ729877	KJ729903	KJ729934
I. aff. *amygdalina*	PBM3700	PERTH 08318840	W.A.	*Allocasuarina*	KJ756497	KJ756467	KJ756403
I. araneosa	PBM3755 (holotype)	PERTH 08363285	Qld.	*Melaleuca*, *Lophostemon*, *Corymbia*, *Acacia*	KJ729878	KJ729904	KJ729935
I. aff. *araneosa*	PL58410	BRI AQ793977	Qld.	Wet sclerophyll forest	KJ729880	KJ729906	KJ729937
I. arenicola	RC_GB99_01 4	GB	France	*Pinus*, *Salix*, sand dune	FJ904134	FJ904134	—
I. assimilata	EL14805	GB	Sweden		AM882929	AM882929	—
I. cf. *assimilata*	EL16206	GB	Sweden		FN550881	FN550881	—
	EL2304	GB	Sweden		AM882930	AM882930	—
I. aff. *assimilata*	PBM3328	TENN 063958	Tennessee, USA	*Quercus*, *Carya*, *Fagus*	KP171108	KM197214	KM245975
I. aff. *asterospora*	PBM3309	TENN	Tennessee, USA	*Tsuga*, *Carya*	—	KM197215	KM245976
	DJL-TN06-58	TENN	Tennessee, USA		KP171109	KM197216	—
I. cf. *asterospora*	PBM2453	TENN 065796	New York, USA		DQ404390	AY702015	—
	MR00015	GB	Sweden		AM882897	AM882897	—
I. atrisquamosa	TJB10483 (holotype)	BRI AQ79833	Qld.	*Eucalyptus*, *Syncarpia*	KJ729881	KJ729907	KJ729938
	T35059	OSC	N.S.W.	*Eucalyptus*	KJ729882	KJ729908	KJ729939
	TJB10553	BRI AQ798334	Qld.	*Eucalyptus*	KJ729883	KJ729909	KJ729940
	I29	TENN 065712	Tas.	*Eucalyptus*, *Acacia*	KJ729884	KJ729910	—
	RFLP19		N.S.W.	*Eucalyptus*	DQ388826	—	—
	RFLP30		N.S.W.	*Eucalyptus*	DQ388837	—	—
	RFLP36		N.S.W.	*Eucalyptus*	DQ388843	—	—
I. atrorubripes	PBM3690 (holotype)	PERTH 08318956	W.A.	*Allocasuarina*	KF830037	KF808349	KF830053
	MR00281	TENN 066456	W.A.	*Allocasuarina*	KF830038	KF808350	KF830054
I. aureorimosa	PBM3598 (holotype)	PERTH 08319871	W.A.	*Eucalyptus*, *Corymbia*	KF977198	KF915274	KF991371
	MR00224	PERTH 08320136	W.A.	*Eucalyptus*, *Corymbia*	—	KF915273	KF991370
	PBM3646	PERTH	W.A.	*Melaleuca*,	KF977197	KF915272	KF991369

APPENDIX 3

Species	Field no.	Herbarium accession no.	State or Country	Plant associates	GenBank Accession No.		
					ITS[a]	nLSU-rRNA	*rpb2*
		08319405		*Agonis, Spyridium, Acacia*			
I. auritelloides	PBM3411 (holotype)	PERTH 08516928	Tas.	*Nothofagus*	KJ729885	KJ729911	KJ729941
	I108	TENN 068292	Tas.		KJ729886	KJ729912	—
	PBM3410	TENN 068293	Tas.	*Nothofagus*	KJ729887	KJ729913	KJ729942
I. aurora	AU 10245 (isotype)	WTU-F-043332	Nova Scotia, Canada	*Picea, Tsuga*	HQ201337	HQ201338	—
I. australiensis	PBM3246	TENN 065714	Tas.	*Eucalyptus, Acacia, Pomaderris*	KJ756468	KJ756433	KJ729953
	PBM3351	TENN 065715	Tas.	*Eucalyptus, Acacia, Pomaderris*	KJ756469	KJ756434	KJ729954
	PBM3493	PERTH 08241724	W.A.	*Eucalyptus, Corymbia, Allocasuarina*	KJ756470	KJ756435	KJ729955
	PBM3634	PERTH 08319529	W.A.	*Melaleuca, Agonis*	KJ756471	KJ756436	KJ729956
	NLB1007	PERTH 08383359	W.A.	*Corymbia, Agonis*	KJ756472	KJ801160	—
	PBM3659	PERTH 08318905	W.A.	*Eucalyptus, Acacia*	KJ756473	KJ756437	
I. cf. *australiensis*	PBM3629	TENN 066599	W.A.	*Lepidosperma, Astartea, Eucalyptus, Corymbia*	KP171110	KM197217	KM245977
I. austrosquarrosa	H. Lepp 546 (holotype)	CANB 574568	A.C.T.	*Eucalyptus*	—	KJ756438	—
	I35	HO 567551	Tas.	Wet sclerophyll forest	KJ756474	KJ756439	—
I. avellaneifolia	PBM3648 (holotype)	PERTH 08319391	W.A.	Shrubs of *Melaleuca, Spyridium, Acacia, Agonis*	KJ756475	KJ756440	KJ729951
I. bella	I98 (holotype)	HO 571334	Tas.	*Eucalyptus*, *Pomaderris*, rainforest understorey	KJ756476	KJ756441	—
I. bicornis f. *bicornis*	E5563 (holotype)	PERTH 07605684	W.A.	*Eucalyptus*	KC305486	EU555447	KC305448
	PBM3676	PERTH 08319103	W.A.	*Allocasuarina*	KC305487	KC305409	KC305448
	PBM3691	PERTH 08318948	W.A.	*Allocasuarina*	KJ756477	KJ756442	KJ729952
I. bicornis f. *secotioides*	NLB890 (holotype)	PERTH 08320403	W.A.	*Allocasuarina*	KC305488	KC305410	KC305447

APPENDIX 3

Species	Field no.	Herbarium accession no.	State or Country	Plant associates	GenBank Accession No.		
					ITS[a]	nLSU-rRNA	*rpb2*
I. boreoaustralis	PBM3780 (holotype)	PERTH 08363080	Qld.	*Lophostemon*	KF830044	KF808359	KF830060
	PBM3777	PERTH 08363110	Qld.	*Eucalyptus, Allocasuarina*	KF830045	KF808358	KF830061
I. brevicula	NLB256 (holotype)	PERTH 08092338	W.A.	*Eucalyptus, Corymbia*	KJ756478	JN974925	KJ729950
	NLB1073	PERTH 08516634	W.A.	*Eucalyptus, Corymbia*	KJ756479	KJ756443	KJ756404
I. "brevi-squamulosa"	ZT10102	SFSU	Thailand	*Tectona, Dipterocarpus*	GQ893019	GQ892974	—
I. breviterincarnata	BK2808047	UTC	Utah, USA	*Populus, Abies, Pinus*	JQ408749	EU555451	EU555450
I. cf. *bridgesiana*		CONC F0252	Chile	*Nothofagus*	KP171111	KP170898	KM245978
I. brunneidisca	PBM3220 (holotype)	PERTH 08516960	Tas.	*Eucalyptus, Acacia, Pomaderris*	KJ756480	JN974959	—
	PBM3214	TENN 065797	Tas.	*Eucalyptus, Acacia, Pomaderris*	KJ756481	KJ756444	—
	PBM3364	TENN 065788	Tas.	*Eucalyptus, Acacia, Pomaderris*	KJ756482	KJ756445	—
	PBM3365	TENN 065799	Tas.	*Eucalyptus, Acacia, Pomaderris*	KJ756483	KJ756446	KJ756405
	PBM3402	TENN 065800	Tas.	*Eucalyptus, Acacia, Pomaderris*	KJ756484	KJ756447	KJ756406
	E9023	PERTH 08094217	W.A.	*Eucalyptus, Melaleuca, Kunzea*	KJ756485	KJ756448	
I. bulbinella	PBM3630 (holotype)	PERTH 08319596	W.A.	*Allocasuarina, Eucalyptus*	KJ756486	KJ756449	KJ756407
	PBM3717	PERTH 08362769	Qld.	*Allocasuarina, Eucalyptus*	KJ756487	KJ756450	KJ756408
	NLB186	PERTH 07698798	W.A.	*Eucalyptus*	KJ756488	JN975002	KJ756409
	PBM3623	PERTH 08319634	W.A.	*Corymbia* (planted)	KJ756489	KJ756451	KJ756410
I. bulbobissima	EL66-05	GB	Norway	*Salix*	AM882765	AM882765	—
I. cacaocolor	PBM3790 (holotype)	PERTH 08363005	Qld.	*Eucalyptus*	KJ778845	KJ756464	KJ756421
	H. Lepp 3475	CANB 638191	N.S.W.	*Eucalyptus, Callitris*	KJ778846	KJ756465	—
	PBM3670	PERTH 08319170	W.A.	*Corymbia, Eucalyptus*	KJ778847	KJ756466	KJ756423
I. caerulata	PBM3127 (holotype)	PDD 97860	N.Z.	*Kunzea*	KJ756490	JQ313560	KJ756411
	I57a ("I16")	HO 567548	Tas.	Wet sclerophyll forest	KJ756491	KJ756452	KJ756412
	I57	HO 567549	Tas.	Wet sclerophyll	KJ756492	KJ756453	—

APPENDIX 3

Species	Field no.	Herbarium accession no.	State or Country	Plant associates	GenBank Accession No.		
					ITS[a]	nLSU-rRNA	*rpb2*
				forest			
	I141	HO 567550	Tas.	Wet sclerophyll forest	KJ756493	KJ756454	—
I. calamistratoides	ZT 67/262 (holotype)	PDD 27121	N.Z.	*Nothofagus*	KJ756494	—	—
	PBM3384	TENN 065750	Tas.	*Nothofagus, Leptospermum*	JQ801393	JQ815415	KJ729949
	ZT 9630	PDD 72711	N.Z.	*Nothofagus*	JQ801392	AY380369 JQ815413	AY333765
	PBM3112	PDD 97855	N.Z.	*Nothofagus*	ITS1 and ITS2 (both <200 bp)	JQ815414	JQ846473
		PDD 89085	N.Z.		GU222302	—	—
I. calida	TAA185175	GB	Estonia		AM882760	AM882760	—
I. calocephala	PBM3600 (holotype)	PERTH 08319855	W.A.	*Eucalyptus, Corymbia*	—	KJ756455	KJ756413
	REH9133	NY, BRI	Qld.	*Eucalyptus, Syncarpia, Leptospermum*	KJ756495	KJ756456	—
	PBM3478	PERTH 08241708	W.A.	*Eucalyptus, Allocasuarina*	—	KJ801161	—
	PBM2125	PERTH 07652097	W.A.	*Corymbia, Agonis*	—	EU555456	—
	PBM2118	PERTH 07671393	W.A.	*Corymbia, Agonis*	KJ756496	—	—
I. calopedes	TJB10555	CORT, BRI	W.A.	Wet sclerophyll forest	KJ778838	KJ756457	KJ756414
	PBM3385	TENN 068947	Tas.	*Nothofagus, Leptospermum*	KJ778839	JQ313572	KJ756415
I. calospora group I	JFA12539	WTU	Sweden	*Betula*	—	AY038313	AY337365
	PAM 08092808	LIP	France		—	KP170899	—
	PAM 99082905	TENN 063577	France	*Betula*	HQ586871	HQ641114	—
I. calospora f. *pectinata*	PAM 10082903	LIP	France	*Alnus*	KP171112	KP170900	—
I. calospora group II	EL9505	GB	Finland		AM882759	AM882759	—
	PAM 10092501	LIP	France	*Alnus, Quercus, Corylus*	KP171113	KP170901	—
	PAM 00073102	TENN 063837	France	*Quercus, Castanea, Betula*	HQ586853	HQ641093	—
	PAM 03082401	TENN 063836	France	*Alnus*	HQ586852	HQ641094	
I. campestris	PBM3682 (holotype)	PERTH 08319030	W.A.	*Allocasuarina, Eucalyptus*	—	KJ756458	KJ756416
	PBM3668	PERTH 08319197	W.A.	*Eucalyptus*	KJ778840	—	—
	NLB891	PERTH 08320381	W.A.	*Acacia, Allocasuarina*	KJ778841	KJ756459	KJ756417
	PBM3689	PERTH 08318964	W.A.	*Acacia, Allocasuarina*	KJ778842	KJ756460	KJ756418

APPENDIX 3

Species	Field no.	Herbarium accession no.	State or Country	Plant associates	GenBank Accession No.		
					ITS[a]	nLSU-rRNA	*rpb2*
	PBM3698	PERTH 08318867	W.A.	*Allocasuarina, Acacia*	—	KJ756461	KJ756419
	PBM3699	PERTH 08318859	W.A.	*Allocasuarina, Acacia*	KJ778843	KJ756462	KJ756420
	MR00287	TENN 066547	W.A.	*Eucalyptus, Acacia*	KJ778844	KJ756463	KJ756421
I. cerasphora	BSI 01/184	WTU	Chile	*Nothofagus*	—	AY380370	AY337367
I. cercocarpi	BK20069806	UTC	Utah, USA	*Cercocarpus, Pinus*	—	EU600890	EU600889
I. cf. "*cervicolor*"	DNA6187		Tas.		FN550942	—	—
I. chalcoceps	I47 (holotype)	PERTH 08556474	Tas.	Wet sclerophyll forest	—	KJ729914	—
I. clypeata	NLB905 (holotype)	PERTH 08320268	W.A.	Shrubs of *Acacia, Templetonia, Melaleuca, Spyridium*, also *Eucalyptus*	KF871781	KF853405	KF891958
	NLB1061	PERTH 08363706	W.A.	*Eucalyptus* and shrubs of *Acacia, Templetonia, Melaleuca, Spyridium*	KF871782	KF853406	KF891959
I. conspicuospora		PC 96042	Zambia	Miombo woodland	—	EU555471	EU555470
I. curvipes	E. Cooper 4	CANB 404918	A.C.T.	*Pinus radiata*	JN035289	JN035293	—
	TR37-03	M 0138268	P.N.G.	*Pinus*	JN035290	JN035294	—
	PBM2401	WTU	Washington, USA	*Betula*	—	QY239022	AY337414
I. cyanotrichia	I37 (holotype)	PERTH 08516952	Tas.	*Eucalyptus*	JQ801397	JN975033	JQ846476
	I50	TENN 065729	Tas.	*Eucalyptus, Acacia, Pomaderris*	JQ801397	JQ815418	KJ729948
I. cf. *cyanotrichia*	PBM3406	PERTH 08516979	Tas.	*Eucalyptus, Acacia, Pomaderris*	—	JQ815431	JQ486498
I. dewrangia.	PBM3660	PERTH 08319286	W.A.	*Eucalyptus, Corymbia*	KP171114	KP170902	—
	NLB1066	PERTH 08561184	W.A.	*Eucalyptus, Gastrolobium*	KP171115	KP170903	KM245979
	T26597		N.S.W.	*Eucalyptus*	KP171116	KP170904	KM245980
I. diabolica	JV5172F (isotype)	WTU	Norway	*Betula*	HQ201350	HQ201350	—
	EL9006	GB	Sweden		FN550896	FN550896	—
I. cf. *diminuta*	PBM3287	TENN 064301	North Carolina, USA	*Pinus, Fagus, Quercus, Tsuga*	KP171117	JN974973	KM245981

APPENDIX 3

Species	Field no.	Herbarium accession no.	State or Country	Plant associates	GenBank Accession No.		
					ITS[a]	nLSU-rRNA	*rpb2*
I. dulcamaroides	EL112_06	GB	Sweden	*Dryas*, alpine	FJ904126	FJ904126	—
I. eburnea	PBM3367 (holotype)	PERTH 08516987	Tas.	*Eucalyputs, Acacia, Pomaderris*	KJ778851	KJ801162	KJ756424
	PBM3370	PERTH 08516995	Tas.	*Eucalyputs, Acacia, Pomaderris*	KJ778852	KJ801163	KJ756425
I. emergens	E6203	PERTH 07659902	W.A.	*Eucalyptus, Acacia*	JQ085935	EU569838	—
	NLB893	PERTH 08320365	W.A.	*Eucalyptus, Acacia*	—	KP170905	KM245982
	NLB896	PERTH 08320349	W.A.	*Acacia, Allocasuarina*	JX258837	KJ729915	KJ729943
	MR00291	TENN 066564	W.A.	*Allocasuarina*	—	KJ729916	KJ729944
I. eriocaulis	PBM2132	PERTH 07651619	W.A.	*Corymbia, Eucalyptus, Allocasuarina*	KJ778853	EU569843	EU569842
I. cf. *eriocaulis*	PBM3401	TENN 065711	Tas.	*Eucalyptus, Acacia, Pomaderris*	KP171118	KP170906	KM245983
	PBM3631	PERTH 08319588	W.A.	*Eucalyptus, Corymbia, Allocasuarina*	KP171119	KP170907	KM245984
	RT105-1	ECM root tip	NZ	*Nothofagus*	EF634112	—	—
I. exigens	NLB1018 (type)	PERTH 08383367	W.A.	*Agonis, Allocasuarina,* myrtaceous shrubs	KJ778854	KJ801164	KJ756426
	NLB1019b	PERTH 08383375	W.A.	*Agonis, Spyridium, Leucopogon*	KJ778855	KJ801165	KJ756427
I. exigua	T24874 (epitype)	PERTH 08517290	N.S.W.	*Eucalyptus, Acacia, Callitris, Micromyrtus, Helichrysum*	KJ778856	AY380389	AY351809
	T25090	TENN 068379	N.S.W.	*Eucalyptus, Acacia, Callitris, Micromyrtus, Helichrysum*	—	KP170908	KM245985
I. farinosipes	MR00226 (holotype)	PERTH 08320101	W.A.	*Eucalyptus, Corymbia*	KC305483	KC305405	KC305445
	PBM3471	PERTH 08242054	W.A.	*Eucalyptus, Corymbia, Acacia, Agonis*	—	KC305407	—
	MR00228	TENN 066595	W.A.	*Eucalyptus, Corymbia*	KC305485	KC305408	KC305446
	PBM3590	PERTH 0831944	W.A.	*Eucalyptus, Corymbia*	KC305481	KC305403	—
	PBM3611	PERTH 08319758	W.A.	*Eucalyptus*	KC305484	KC305406	KC305444
	NLB1070	PERTH	W.A.	*Eucalyptus,*	KJ778857	KJ801166	KJ756428

APPENDIX 3

Species	Field no.	Herbarium accession no.	State or Country	Plant associates	GenBank Accession No.		
					ITS[a]	nLSU-rRNA	*rpb2*
		08516642		*Corymbia*			
I. fastigiata	FO46800	TUB	Germany?		—	DQ071697	—
I. fibrillosibrunnea	E5971	PERTH 07607032	W.A.	*Allocasuarina*	KC305472	EU307848	EU307850
	NLB802	PERTH 07607032	W.A.	*Eucalyptus, Corymbia*	KC305474	KC305397	KC307849
I. fibrillosipes	PBM3224 (holotype)	PERTH 08517002	Tas.	*Eucalyptus, Acacia, Pomaderris*	KJ778858	KJ801167	—
	MR00074	GB	Tas.	*Eucalyptus, Nothofagus, Pomaderris*	FN550917	FN550917	—
	Gates 200060803	GB	Tas.	*Eucalyptus, Nothofagus, Pomaderris*	FN550932	FN550932	—
I. fissurata	PBM2206 (holotype)	PERTH 07676824	W.A.	*Eucalyptus*	JQ408770	AY732213	JQ421069
	PBM2195	PERTH 07676700	W.A.	*Eucalyptus*	JQ408771	EU555466	EU555464
I. flavella	EL137-05	GB	Sweden	*Corylus, Alnus, Quercus*	AM882776	AM882776	—
	TM02-130	TRTC	Ontario, Canada	*Tsuga*-dominated forest	—	EU522733	—
I. flavoalbida	PBM3768 (holotype)	PERTH 08363196	Qld.	*Eucalyptus*	KJ729873	KJ729901	KJ729932
	TR104-02	M 0138262	P.N.G.	*Castanopsis*	KJ729874	—	—
	TR44-06	M 0138283	P.N.G.	*Castanopsis*	KJ729875	JN974960	—
I. floccosibrunnea	NLB273 (holotype)	PERTH 8073937	W.A.	*Eucalyptus*	KC305477	JN974918	KC305440
	PBM3495	PERTH 08242305	W.A.	*Eucalyptus, Allocasuarina*	KC305476	KC305499	—
	PBM3617	PERTH 08319685	W.A.	*Eucalyptus, Corymbia*	KC305479	KC305401	KC305441
	PBM3484	PERTH 08241813	W.A.	*Eucalyptus, Allocasuarina*	KC305478	KC305400	KC305442
I. fragilis	PBM3662 (holotype)	PERTH 08319251	W.A.	*Corymbia, Eucalyptus*	KF830029	KF808341	KF830047
	PBM3665	PERTH 08319227	W.A.	*Corymbia, Eucalyptus*	KF830030	KF808342	KF830048
I. fragrans	I149 (holotype)	HO 562696	Tas.	*Nothofagus, Eucalyptus, Eucryphia*	KF977202	KF915278	KF991375
	PBM3200	TENN 063642	Tas.	*Nothofagus*	KF977203	KF915279	—
I. froudistii	NLB1060 (holotype)	PERTH 08363714	W.A.	*Melaleuca, Acacia, Templetonia*	—	KJ801168	KJ756429
	I14	TENN 067788	Tas.	*Acacia, Eucalyptus*	—	KC305386	—
	PBM3239	TENN 067789	Tas.	*Acacia, Eucalyptus*	—	KC305385	KJ756430
	PBM3243	TENN	Tas.	*Acacia*	—	KC305384	KJ756431

APPENDIX 3

Species	Field no.	Herbarium accession no.	State or Country	Plant associates	GenBank Accession No.		
					ITS[a]	nLSU-rRNA	*rpb2*
		067790					
	PBM3245	TENN 067771 (formerly 065741)	Tas.	*Acacia*	KC305462	KC305383	KC305431
	E6103	PERTH 07589069	W.A.	*Acacia*	KC305461	—	—
	PBM2151 (E6944)	PERTH 07678118	W.A.	*Eucalyptus, Corymbia*	—	KJ801169	KJ811567
	Green55 (E8369)	PERTH 07700261	W.A.	*Acacia*	—	GB305374	KC305432
	PBM3462	PERTH 08242143	W.A.	*Eucalyptus, Corymbia, Acacia, Agonis*	KC305460	KC305382	KC305433
	PBM3465	PERTH 08242119	W.A.	*Eucalyptus, Corymbia, Acacia, Agonis*	—	KJ729917	—
	PBM3467	PERTH 08242097	W.A.	*Eucalyptus, Corymbia, Acacia, Agonis*	KC305457	KC305379	KC305424
	PBM3468	PERTH 08242089	W.A.	*Eucalyptus, Corymbia, Acacia, Agonis*	—	KJ729918	—
	PBM3603	TENN 066607	W.A.	*Eucalyptus, Corymbia*	—	KC305376	KC305428
	PBM3612	PERTH 08319731	W.A.	*Eucalyptus, Corymbia*	KC305456	KC305378	KC305426
	PBM3614	PERTH 08319715	W.A.	*Eucalyptus, Corymbia*	KC305455	KC305377	KC305425
	NLB799	PERTH 08318123	W.A.	*Eucalyptus, Corymbia*	KC305454	KC305375	KC305423
	NLB902	PERTH 0830284	W.A.	*Acacia, Melaleuca, Eucalyptus*	KC305458	KC305380	KC305427
	NLB1043	PERTH 08383405	W.A.	*Eucalyptus, Corymbia*	KJ778859	KJ801170	KJ756432
I. fulvilubrica	PBM3352 (holotype)	TENN 065756	Tas.	*Eucalyptus, Acacia, Pomaderris*	—	JQ085922	—
	PBM3218	TENN 065760	Tas.	*Eucalyptus, Acacia, Pomaderris*	—	JQ085923	—
	PBM3407	TENN 065755	Tas.	*Eucalyptus, Acacia, Pomaderris*	JQ085924	JQ085925	—
	PBM3435	TENN 065758	Tas.	*Eucalyptus, Acacia*	JQ085926	JQ085927	—
	PBM3440	TENN 065759	Tas.	*Eucalyptus, Acacia, Pomaderris, Nothofagus*	JQ085928	JQ085929	JQ085930
	I3a	HO 563719	Tas.		—	JQ085931	—
	PBM2134 (E6776)	PERTH 07651708	W.A.	*Eucalyptus, Corymbia*	JQ085932	EU569850	EU569849
I. fuscata	MES544	FLAS	Chile	*Nothofagus*	KP171120	KP170909	KM245986
I. fuscosquarrosa	PBM3447 (holotype)	PERTH 08556458	Tas.	*Eucalyptus, Acacia,*	KC305465	KC305389	KC305437

APPENDIX 3

Species	Field no.	Herbarium accession no.	State or Country	Plant associates	GenBank Accession No.		
					ITS[a]	nLSU-rRNA	*rpb2*
				Pomaderris			
	I4	TENN 067787	Tas.	*Eucalyptus, Acacia*	KC305475	KC305398	—
	I49	TENN 069528	Tas.	*Eucalyptus, Acacia, Pomaderris*	KC305372	KC305396	—
	I52	TENN 069529	Tas.		KC305466	KC305390	—
	I65	TENN 069530	Tas.		KC305467	KC305391	—
	I75	TENN 069531	Tas.	Wet sclerophyll forest	KC305547	KC305395	KC305435
	I95	TENN 069532	Tas.		KC305470	KC305394	—
	PBM3426	TENN 069526	Tas.	*Eucalyptus, Acacia, Pomaderris*	KC305468	KC305392	KC305436
	PBM3442	TENN 065787	Tas.	*Eucalyptus, Acacia*	KC305469	KC305393	KC305438
I. geniculata	PBM3622 (holotype)	PERTH 08319642	W.A.	*Eucalyptus, Agonis*	KJ778860	KJ801171	—
	PBM3241	TENN 067071	Tas.	*Eucalyptus, Pomaderris*	KJ778861	KJ801172	KJ811568
	PBM3353	TENN 067072	Tas.	*Eucalyptus, Pomaderris*	KJ778862	KJ801173	KJ811569
	I154	HO 562694	Tas.	*Pomaderris*	KJ778863	KJ801174	KJ811570
	PBM2184 (E7033)	PERTH 07676646	W.A.	*Eucalyptus*	KJ778864	EU555468	EU555467
	NLB262	PERTH 08092257	W.A.	*Eucalyptus, Corymbia*	KJ778865	KJ801175	KJ811571
	PBM3664	PERTH 08319235	W.A.	*Eucalyptus, Corymbia*	KJ778866	KJ801176	KJ811572
I. gilviceps	REH9327 (holotype)	BRI 794277	Qld.	*Eucalyptus*	KP171121	KJ801177	KJ811573
I. gomphocephala	PBM3652 (holotype)	TENN 066646	W.A.	*Agonis*, shrubs of Myrtaceae and Fabaceae	KC305463	KC305387	KC305434
	PBM3651	PERTH 08319375	W.A.	*Agonis, Spyridium*, shrubs of Myrtaceae and Fabaceae	KC305464	KC305388	—
I. gracilissima	PBM3735 (holotype)	PERTH 08362912	Qld.	*Lophostemon, Eucalyptus*	—	KJ729919	KJ729945
	JAC2809	CBG 8916878	N.T.	*Eucalyptus*	JQ408755	JQ319684	J.C. s.n.
	NLB937	PERTH 08373116	Qld.	*Eucalyptus, Allocasuarina*	KP171122	KJ801178	KJ729946
	PBM3738	PERTH 08362966	Qld.	*Corymbia, Eucalyptus*	KP171123	KJ801179	KJ729947
I. horakomyces	ZT8702	PDD 71140	N.Z.	*Nothofagus*	—	EU600854	—
	J.C. s.n.	PDD 80935	N.Z.	*Nothofagus*	—	KP170910	—
I. "horakomyces"	PL54104	PDD 77535	N.Z.	*Nothofagus*	KP171124	—	—
I. humidicola	PBM3720 (holotype)	PERTH 08362785	Qld.	*Eucalyptus*	KP171125	KJ801180	KJ811574

APPENDIX 3

Species	Field no.	Herbarium accession no.	State or Country	Plant associates	GenBank Accession No.		
					ITS[a]	nLSU-rRNA	*rpb2*
	PBM3719	PERTH 08362777	Qld.	*Eucalyptus*	KP171126	KJ801181	KJ811575
	PBM3724	PERTH 08362815	Qld.	*Eucalyptus*	KP171127	KJ801182	—
	PBM3727	PERTH 08362858	Qld.	*Eucalyptus*	KP171128	KJ801183	—
	PBM3744	PERTH 08363447	Qld.	*Eucalyptus, Allocasuarina*	KP171129	KJ801184	KJ811576
I. hydrocybiformis	ZT10077	ZT	Thailand	*Pinus, Dipterocarpus*	—	GQ892971	—
	ZT9879	ZT	Thailand	*Tectona, Dipterocarpus*	—	GQ892972	—
	TBGT12318	TBGT	India	*Hoppea*	KP171130	KP170911	KM245987
I. aff. *hydro-cybiformis*	DED8165	SFSU	Thailand	*Castanopsis, Pinus, Dipterocarpus*	—	GQ892973	—
	TBGT10480	TBGT	India	*Hoppea, Vatteria*	KP171131	KP170912	KM245988
I. hygrophorus	EL97-06	GB	Sweden	*Betula, Salix*	FJ904137	FJ904137	—
I. hymenocarpa	PBM2181 (E7030) (holotype)	PERTH 07676603	W.A.	*Eucalyptus, Trymalium, Chorilaena*	KP171132	EU307851	EU307853
I. illudens	I106 (holotype)	HO 569026	Tas.	*Eucalyptus, Pomaderris*	JQ408769	JQ319699	JQ421068
I. "infirma".	ZT10106 (holotype)	SFSU	Thailand	*Tectona, Dipterocarpus*	GQ892987	GQ89294	—
I. insignis	MK070101101	TENN	Illinois, USA	*Carya, Quercus*	—	KP170913	KM245989
I. intricata	PBM2600	TENN 062438	Tennessee, USA	*Betula, Picea*	EU523561	EU307835	—
I. cf. *intricata*	LAS200609	GB	Sweden		FN550895	FN550895	—
	PAM 08071405	LIP	France		—	KP170914	—
I. invadens	PBM3494 (holotype)	PERTH 08241716	W.A.	*Eucalyptus, Allocasuarina*	—	KJ801185	—
	PBM3217	TENN 069535	Tas.	*Eucalyptus, Acacia, Pomaderris*	KP171133	KJ801186	KJ811577
	PBM3746	PERTH 08363374	Qld.	*Eucalyptus, Allocasuarina*	JX258832	KJ801187	KJ811578
I. ionocaulis	PBM3486 (holotype)	PERTH 08241791	W.A.	*Eucalyptus, Allocasuarina*	KP171134	KJ801188	KJ811579
	PBM3496	PERTH 08242291	W.A.	*Eucalyptus, Allocasuarina*	KP171135	—	—
	PBM3626	PERTH 08319561	W.A.	*Eucalyptus, Agonis*	KP171136	KJ801189	KJ811580
I. isabellina	E8290 (holotype)	PERTH 07699255	W.A.	*Eucalyptus*	KP171137	KP170915	KJ811581
	NLB405	PERTH 08096635	W.A.	*Eucalyptus*	KP171138	KP170916	KJ811582
	NLB420	PERTH 08072442	W.A.	*Eucalyptus*	KP171139	KP170917	KJ811583

APPENDIX 3

Species	Field no.	Herbarium accession no.	State or Country	Plant associates	GenBank Accession No.		
					ITS[a]	nLSU-rRNA	*rpb2*
	NLB800	PERTH 08318832	W.A.	*Eucalyptus, Corymbia*	KP171140	KP170918	KJ811584
	PBM3610	PERTH 08319766	W.A.	*Eucalyptus, Corymbia*	—	KP170919	KJ811585
	NLB836	PERTH 08318441	W.A.	*Agonis, Spyridium, Acacia, Melaleuca*	KP171141	KP170920	KJ811586
I. cf. *isabellina*	NLB491	PERTH 08073287	W.A.	*Eucalyptus, Acacia*	KP171142	KP170921	KJ811587
	PBM2207	PERTH07676840	W.A.	*Eucalyptus*	—	AY380381	AY337382
	PBM3595	TENN 066498	W.A.	*Eucalyptus, Corymbia*	KP171143	KP170922	—
I. kriegIsteineri	JV10514F	WTU	Finland	*Betula, Pinus*	KP308782	KP170953	KM406198
	SJ88034	GB	Sweden		AM882715 (*I.* "*rennyi*")	AM882715 (*I.* "*rennyi*")	—
	EC300	WTU	Washington USA	*Pseudotsuga*	—	JN975029	—
I. lacera	PBM1462	WTU	Washington USA		—	AY038318	AY337383
	PBM2541	TENN 062386	Massachusetts, USA	Mixed deciduous forest (no conifers)	KP171144	JN974993	KM245991
I. lanuginosa	F50698 (epitype)	WTU	Finland	*Pinus*	HQ201356	HQ201356	—
	REM5119	GB	Sweden		AM882843	AM882843	—
	PBM3023	TENN 062780	Pennsylvania, USA	*Tsuga, Pinus*	HQ232480	KP170923	KM245992
	PBM956	WTU	Washington USA	*Pseudotsuga, Tsuga*		AY038319	AY337384
I. lanuginosa var. *alpina*	EL14006	GB	Sweden		FN550879	FN550879	—
I. lasseri	MCA1971	BRG	Guyana	*Dicymbe*	—	EU569857	KJ811580
I. lasseroides	PBM3749	PERTH 08363323	Qld.	*Allocasuarina, Eucalyptus*	KP171145	KP170924	KM245993
	PBM3750	PERTH 08363307	Qld.	*Allocasuarina, Eucalyptus*	KP171146	KP170925	—
I. aff. *lasseroides*	PBM3786	PERTH 08363048	Qld.	*Allocasuarina*	KP171147	KP170926	KM245994
	PBM3787	PERTH 08363021	Qld.	*Allocasuarina*	KP171148	KP170927	KM245995
	TJB10466	CORT	Qld.		KP171149	KP170928	KM245996
I. latericia	(holotype)	PDD 27120	N.Z.	*Nothofagus*	KP171150	—	—
		PDD 89085	N.Z.		GU222302	—	—
		PDD 71261	N.Z.	*Nothofagus*	JQ801402	JN975024	—
		PDD 92382	N.Z.	*Nothofagus*	GU233367	GU233413	—
I. aff. *latericia*	TR109/02	M 0138263	P.N.G.	*Castanopsis*	JQ801405	JN975023	JQ846487
	TR182/02	M 0138266	P.N.G.	*Castanopsis*	JQ801406	JQ815423	—
	TR135/05	M 0138278	P.N.G.	*Castanopsis*	JQ801404	JQ815422	—
	TR74/05	M 0138273	P.N.G.	*Castanopsis*	JQ801403	JQ815421	JQ846486

APPENDIX 3

Species	Field no.	Herbarium accession no.	State or Country	Plant associates	GenBank Accession No.		
					ITS[a]	nLSU-rRNA	*rpb2*
I. laureata	NLB833 (holotype)	PERTH 08318476	W.A.	*Agonis, Spyridium, Acacia, Melaleuca, Olax*	KP308751	KP170929	KM245997
	PBM3649	TENN 66477	W.A.	*Agonis, Spyridium, Acacia, Melaleuca*	KP308752	KP170930	KM245998
I. lepidotella	MCA1881	TENN 066442	Guyana	*Dicymbe*	JN642233	JN642234	—
I. leptophylla	BK07099719	UTC	Utah, USA	Pinaceae	—	AY038320	AY337387
I. leptospermi	H. Lepp 540	CANB 574560	A.C.T.	*Eucalyptus*	KP308753	KP170931	KJ811588
	I2	TENN 067033	Tas.		KP308754	KP170932	—
	PBM3398	TENN 065746	Tas.	*Nothofagus*	KP308755	JQ313561	KJ811589
	PBM3400	TENN 067032	Tas.	*Leptospermum*	KP308756	KP170933	KJ811590
	PBM3608	PERTH 08319782	W.A.	*Corymbia, Eucalyptus*	KP308757	KP170934	KJ811591
	PBM3628	TENN 06641	W.A.	*Eucalyptus, Melaleuca*	KP308757	KP170935	KJ811592
	PBM3642	PERTH 08319456	W.A.	*Melaleuca, Agonis*	KP308759	KP170936	KJ811593
	PBM3080	TENN 063625	N.Z.	*Kunzea*	KP308760	JN974983	KJ811594
	JAC s.n.	PDD 80003	N.Z.	*Kunzea, Leptospermum*	KP308761	KP170937	—
	JAC s.n.	PDD 80267	N.Z.	*Kunzea*	KP308762	KP170938	—
I. leucotaenia	NLB1064 (holotype)	PERTH 0836384	W.A.	*Eucalyptus, Corymbia*	KF871784	KF853408	KF891961
	NLB423	PERTH 08072477	W.A.	*Eucalyptus, Corymbia*	KF871785	KF853409	KF891962
	NLB482	PERTH 08073171	W.A.	*Eucalyptus*	KF871783	KF853407	KF891960
I. "lineata"	DED8019	SFSU	Thailand	*Dipterocarpus*	EU569861	EU569861	—
	DED8048	SFSU	Thailand		—	GQ892958	KM245999
I. lucifera	PBM3419 (holotype)	PERTH 08517029	Tas.	*Nothofagus*	KP308763	KP170939	KJ811595
	I110	HO 571944	Tas.	*Nothofagus*	KP308764	KP170940	—
I. luteobulbosa	PBM3425	TENN 067034	Tas.	*Eucalyptus, Acacia, Pomaderris*	KP308765	KP170941	KJ811596
	I155	HO 562695	Tas.	*Pomaderris, Eucalyptus*	KP308766	KP170942	KJ811597
	PL s.n.	PDD 75276	N.Z.	*Nothofagus*	KP308767	JQ313573	KJ811598
	PBM3088	PDD 97897	N.Z.	*Nothofagus*	KP308768	KP170943	KJ811599
	PBM3130	TENN 063872	N.Z.	*Nothofagus*	KP308769	KP170944	KJ811600
	PBM3138	PDD 97902	N.Z.	*Nothofagus*	KP308770	KP170945	KJ811601
		PDD 72618	N.Z.	*Nothofagus*	HM060318	—	—

APPENDIX 3

Species	Field no.	Herbarium accession no.	State or Country	Plant associates	GenBank Accession No.		
					ITS[a]	nLSU-rRNA	*rpb2*
I. luteo-olivacea	REH8914 (holotype)	BRI 794189	Qld.	*Acacia*	KP308771	JN974991	KJ811602
	REH9627	NY 01491109	Qld.	*Acacia*	KP308772	KP170946	KJ811603
I. magnifolia	MCA2441 (holotype)	BRG	Guyana	*Dicymbe*	JN642228	JN642244	EU600899
I. aff. *manukanea*	PBM3079	PDD 97914	N.Z.	*Kunzea*	—	JN974924	—
I. marginata	MCA1881	BRG	Guyana	*Dicymbe*	—	AY509115	AY509114
I. mallocyboides	PBM3501 (holotype)	PERTH 08242240	W.A.	*Eucalyptus*, *Kunzea*, other myrtaceous shrubs	KP308773	KP170947	—
I. mallopoda	NLB258 (holotype)	PERTH 08092214	W.A.	*Eucalyptus*, *Corymbia*, *Leptospermum*	KP308774	JN974995	KM246000
I. aff. *margaritispora*	PBM2976	TENNN 062734	Tennessee, USA	*Quercus*	—	KP170948	KM246001
	MR00198	TENN 063960	Tennessee, USA	*Quercus*	KP308775	JN974998	KM246002
	Clone 7a environmental sequence	DUKE	Georgia, USA	*Carya*	HQ541862	—	—
I. melliolens	PAM0505230 3	GB	France	*Tilia*	FJ904148	FJ904148	
I. mellita	MR00288 (holotype)	PERTH 08306621	W.A.	*Eucalyptus*, *Acacia*	KP308776	KP170949	KM406190
	NLB892	PERTH 08320373	W.A.	*Eucalyptus*	KP308777	KP170950	KM406191
I. memorialis	NLB494 (holotype)	PERTH 08073317	W.A.	*Eucalyptus*	KP308778	—	—
	NLB352B	PERTH 08073775	W.A.	*Eucalyptus*	KP308779	JN974919	KM406192
I. mendica	PBM3101	PDD 97864	N.Z.	*Nothofagus*	KP308780	KP170951	KM406193
I. meridionalis	PBM3413 (holotype)	PERTH 08517037	Tas.	*Nothofagus*	KF977204	KF915281	KF991377
	PBM3382	TENN 065709	Tas.	*Nothofagus*, *Leptospermum*	KF977207	KF915284	KF991380
	PBM3390	TENN 065705	Tas.	*Nothofagus*	KF977205	KF915282	KF991378
	PBM3391	TENN 065705	Tas.	*Nothofagus*	KF977206	KF915283	KF991379
	PBM3450	TENN 065707	Tas.	*Nothofagus*, *Eucalyptus*	KF977208	KF915285	KF991381
I. cf. *meridionalis* 1	PBM3100	PDD 97872	N.Z.	*Nothofagus*	—	KF915280	KF991376
	PBM3115	PDD 97871	N.Z.	*Nothofagus*	—	JN974926	KM406194
I. cf. *meridionalis* 2	PBM3189	TENN 063642	N.S.W.	*Eucalyptus*	KF977209	KF915286	KF991382
I. cf. *micro-fastigiata*	EL113-06	GB	Sweden	*Dryas*	FJ904156	FJ904156	—

APPENDIX 3

Species	Field no.	Herbarium accession no.	State or Country	Plant associates	GenBank Accession No.		
					ITS[a]	nLSU-rRNA	*rpb2*
I. mimica	EBJ961997	GB	Sweden	*Pinus, Picea*, calcareous	FJ904124	FJ904124	—
I. mixtilis	H. Lepp 2002	CANB 574539	A.C.T.	*Pinus, Ulmus*	KP308781	KP170952	KM406195
	PAM 07110104	TENN 063575	France	*Pinus, Betula, Populus*	HQ586872	HQ641115	KM406196
	PAM 05103003	TENN 063752	France	*Pseudotsuga*	HQ586870	HQ641113	KM406197
	PBM1315	WTU	Washington USA	*Pseudotsuga, Alnus*	—	AY380387	AY337395
I. napipes	PBM2555	TENN 062398	New Hampshire, USA	*Betula, Abies*	KP308783	KP170954	—
	PBM2376	WTU	Norway	*Picea*	—	AY239024	AY337390
	PBM3296	TENN 064312	North Carolina, USA	*Picea, Abies, Tsuga, Betula*	KP308784	KP170955	—
	EL7005	GB	Norway		AM882927	AM882927	—
I. "neglecta"	DED8063	SFSU	Thailand	*Lithocarpus, Castanopsis*	EU600829	EU600829	—
I. cf. *neoumbrina*	I103	TENN 069570	Tas.		KP308785	—	—
	I13	TENN 069571	Tas.		KP308786	—	—
I. cf. *neuquenensis*		CONC F0831	Chile	*Nothofagus*	KF977201	KF915277	KF991374
I. nigricans	PBM3697 (holotype)	PERTH 08318875	W.A.	*Allocasuarina*	KP308787	KP170956	KM406199
	MR00278	TENN 066463	W.A.	*Allocasuarina*	KP308788	KP170957	KM406200
	MR00282	PERTH 08306850	W.A.	*Allocasuarina*	—	KP170958	KM406201
I. niveivelata	PBM2337	WTU	Washington , USA	*Abies, Pinus*	JQ313566	AY333776	AY333776
I. nobilissima	PBM3772 (holotype)	PERTH 08363153	Qld.	*Eucalyptus, Allocasuarina*	KP308789	KP170959	KM406202
	NLB972	PERTH 08364788	Qld.	*Allocasuarina*	KP308790	KP170960	KM406203
	NLB974	PERTH 08364818	Qld.	*Allocasuarina*	KP308791	KP170961	KM406204
I. nothopus	T25060 (holotype)	PERTH 08517088	N.S.W.	*Eucalyptus, Acacia, Micromyrtus*	KP308792	AY380388	AY337391
I. nudiuscula f. *nudiuscula*	PBM3602 (holotype)	PERTH 08319839	W.A.	*Eucalyptus, Corymbia*	KP308793	KP170962	KC305451
	PBM3774	PERTH 08363137	Qld.	*Eucalyptus, Allocasuarina*	KP308794	KP170963	KC305449
	PBM3782	PERTH 08363072	Qld.	*Lophostemon*	KP308795	KP170964	KC305450
I. nudiuscula secotioid form	T18260	OSC	Vic.	*Acacia, Eucalyptus*	KP308796	EU555457	—

APPENDIX 3

Species	Field no.	Herbarium accession no.	State or Country	Plant associates	GenBank Accession No.		
					ITS[a]	nLSU-rRNA	*rpb2*
I. obsoleta	EL17-03	GB	Sweden	*Pinus, Corylus*	AM882769	AM882769	—
I. "obtusiuscula"	PAM 02081710	TENN 063753	France	*Salix*	HQ586869	HQ641112	—
I. occidentalis	BK27089703 (holotype)	UTC	Utah, USA	*Picea, Abies*	EU600893	EU600893	EU600892
I. ochracea	L3X29 environmental isolate		Estonia	*Alnus*	FM993253	FM993253	—
I. olivaceohinnulea	PBM3624 (holotype)	PERTH 08319626	W.A.	*Eucalyptus, Agonis*	KP308797	KP170965	KM406205
I. ornatipes	NLB456 (holotype)	PERTH 08072906	W.A.	*Eucalyptus*	KP308798	KP170966	KM406206
I. pallidiflavipes	PBM3434 (holotype)	TENN 067778	Tas.	*Eucalyptus, Acacia*	KP308799	JQ313574	KM406207
	PBM3357	TENN 067779	Tas.	*Eucalyptus, Pomaderris*	KP308800	KP170967	KM406208
	PBM3453	TENN 067777	Tas.	*Eucalyptus, Acacia, Pomaderris, Doryphora*	KP308801	KP170968	KM406209
I. papillata	TR54-06	M 0138285	P.N.G.	*Castanopsis*	KP308804	KP170971	—
I. "parvibulbosa"	ZT10099	SFSU	Thailand	*Tectona, Dipterocarpus*	GQ893000	GQ892955	—
	ZT10078	SFSU	Thailand	*Pinus, Dipterocarpus*	GQ893001	GQ892956	—
I. parvifulva	PBM3783 (holotype)	PERTH 08363064	Qld.	*Lophostemon*	KP308805	KP170972	—
	PBM3718	TENN 066954	Qld.	*Eucalyptus*	KP308806	KP170973	KM406212
	PBM3726	TENN 066962	Qld.	*Eucalyptus*	—	KP170974	KM406213
	PBM3732	TENN 066967	Qld.	*Allocasuarina, Acacia, Eucalyptus*	—	KP170975	KM406214
	PBM3733	TENN 066968	Qld.	*Allocasuarina, Acacia, Eucalyptus*	KP308807	KP170976	—
I. perlata	EL74-04	GB	Sweden	*Corylus, Betula*	AM882771	AM882771	—
I. "perparva"	PBM2321	WTU	Louisiana, USA	*Quercus, Carya, Juglans, Robinia*	JQ408788	JQ319708	JQ846494
I. persicinipes	PBM2197 (E7044) (holotype)	PERTH 07676727	W.A.	*Eucalyptus*	KF977215	EU600837	EU600836
I. aff. *persicinipes*	E6640	PERTH 07674422	W.A.	*Eucalyptus, Acacia, Templetonia*	KP308808	KP170977	KM406215
	MR00290	TENN	W.A.	*Allocasuarina*	KP308809	KP170978	KM406216

APPENDIX 3

Species	Field no.	Herbarium accession no.	State or Country	Plant associates	GenBank Accession No.		
					ITS[a]	nLSU-rRNA	*rpb2*
		066565					
I. "pileosulcata"	DED8058	SFSU	Thailand	*Dipterocarpus*	EU600838	EU600838	—
	DED8164	SFSU	Thailand	*Pinus, Castanopsis, Dipterocarpus*	GQ892996	GQ892951	KM406217
		TBG T10742	India	*Hoppea, Viteria*	KP308810	KP170979	KM406218
I. aff. *"pileo-sulcata"*	DED8163	SFSU	Thailand	*Pinus, Castanopsis, Dipterocarpus*	GQ892997	GQ892952	KM406219
I. phaeosquarrosa	PBM3087	PDD 97856	N.Z.	*Nothofagus*	KP308811	JQ313575	KM406220
I. pluvialis	PBM3227 (holotype)	PERTH 08517150	Tas.	*Eucalyptus, Acacia, Pomaderris*	KF871774	KF853398	KF891951
	PBM3228	TENN 067042	Tas.	*Eucalyptus*	KF853401	KF853401	KF891954
	PBM3455	TENN 067045	Tas.	*Eucalyptus, Acacia, Pomaderris*	KF871776	KF853400	KF891953
I. pluviorum	PL23408 (holotype)	PERTH 08556466	Qld.	Subtropical rainforest	—	KP170980	KM406221
I. pluviosilvestris	PBM3389 (holotype)	PERTH 08517134	Tas.	*Nothofagus*	KP308802	KP170969	KM406210
	PBM3418	TENN 068284	Tas.	*Nothofagus*	KP308803	KP170970	KM406211
I. praecox	20003325-019	UBC F19023	B.C., Canada		HQ604502 (*I.*"*glab-rescens*")	HQ604502 (*I.*"*glab-rescens*")	—
	2050404-006	UBC F19288	B.C., Canada		HQ604509 (*I.*"*flocc-ulosa*")	HQ604509 (*I.*"*flocc-ulosa*")	—
	2090420-002		B.C., Canada		HQ604500 (*I.* cf. "*glab-rescens*")	HQ604500 (*I.* cf. "*glab-rescens*")	—
	2000311-019	UBC F19019	B.C., Canada		HQ604511 (*I.* cf. "*glab-rescens*")	HQ604511 (*I.* cf. "*glab-rescens*")	—
	2010429-002	UBC F19095	B.C., Canada		HQ604503 (*I.* "*splendens* v. *phaeo-leuca*")	HQ604503 (*I.* "*splendens* v. *phaeo-leuca*")	—
	2020502-008	UBC F19121	B.C., Canada		HQ604501 (*I.*"*ambly-spora*")	HQ604501 (*I.*"*ambly-spora*")	—
	PBM1402	WTU	Washington, USA	*Pseudotsuga, Tsuga, Alnus*	—	AY038311	AY337360
I. praetervisa (group I)	EL12904	GB	Sweden		AM882720	AM882720	—
	PAM 01043008	TENN 063895	France	*Pinus*	HQ586860	HQ641103	—
I. praetervisa	EC329	WTU	Washington	*Pseudotsuga*	—	KP170981	—

APPENDIX 3

Species	Field no.	Herbarium accession no.	State or Country	Plant associates	GenBank Accession No.		
					ITS[a]	nLSU-rRNA	*rpb2*
(group II)							
	EC332	WTU	Washington	*Pseudotsuga*	—	KP170982	—
	Clone A6 environmental isolate		California, USA	*Notholithocarpus*	—	DQ273514	—
	PBM3050	TENN 062800	California, USA	*Pseudotsuga*	KP308812	JN974999	KM406222
I. praetervisa (group III)	SJ95027	GB	Sweden		AM882721	AM882721	—
I. praetervisa (group IV)	EL8506	GB	Sweden		FN550892	FN550892	—
I. pygmaeformis	PBM3233 (holotype)	PERTH 08556431	Tas.	*Eucalyptus, Pomaderris, Acacia*	KF830036	KF808348	—
I. aff. *pusio*	PBM3408	TENN 068282	Tas.	*Pinus*	—	KC305373	—
I. pyrrhopoda	PBM3759 (holotype)	PERTH 08363250	Qld.	*Melaleuca, Lophostemon, Corymbia*	KP308813	KP170983	KM406223
	MDB F292/10	PERTH 08383278	W.A.	*Antidesma, Glochidion, Melaleuca, Eucalyptus*	KP308814	KP170984	KM406224
I. cf. *pyrrhopoda*	PBM3740	PERTH 08362963	Qld.	*Corymbia, Eucalyptus*	ITS1 and ITS2 (both <200 bp)	KP170985	KM406225
	MDB F18s/13	PERTH 08557764	W.A.	*Melaleuca*	KP308815	KP170986	KM406226
I. redolens	PBM2185 (E7034) (holotype)	PERTH 07676654	W.A.	*Eucalyptus*	JQ085933	EU600844	—
	I18	HO 563720	Tas.	*Eucalyptus, Melaleuca, Nothofagus, Phyllocladus*	—	JQ085934	—
	I160	TENN 067763	Tas.	Wet sclerophyll forest	ITS1 and ITS2 (both <200 bp)	KP170987	KM406227
I. aff. *redolens*	PBM3762	TENN 066994	Qld.	*Eucalyptus*	KP308816	KP170988	KM406228
I. relicina	JV10258	WTU	Finland	*Picea, Betula*	—	AY038324	AY333778
I. rhodella	MCA3917 (holotype)	BRG	Guyana	*Dicymbe*	JN642230	JN642234	—
I. rimosa	EL75-05	GB	Sweden	*Fagus, Quercus*	AM882762	AM882762	—
I. aff. *rimosa*	MCA859	WTU	Virginia, USA	*Quercus*	JQ408772	JQ319700	—
I. cf. *rimosa*	PBM3974	TENN 069510	Tennessee, USA	*Betula, Quercus*	—	KP170989	KM406229
	PMB1938	WTU	Vermont, USA	*Fagus, Quercus,*	JQ408775	JQ319701	JQ846491

APPENDIX 3

Species	Field no.	Herbarium accession no.	State or Country	Plant associates	GenBank Accession No.		
					ITS[a]	nLSU-rRNA	*rpb2*
				Pinus, Betula			
I. roseola	I23 (holotype)	HO 569186	Tas.	*Eucalyptus, Acacia*	—	KF808357	—
I. rubricosa	PBM3784 (holotype)	PERTH 08363056	Qld.	*Lophostemon*	KP308817	KP170990	KM406230
I. rufescens	NLB834 (holotype)	PERTH 08318468	W.A.	*Agonis, Spyridium, Acacia, Melaleuca*	KP308818	KP170991	KM406231
I. cf. *rufescens*	NLB1003	PERTH 08383324	W.A.	*Eucalyptus*	KP308819	KP170992	KM406232
I. rufobadia	NLB885 (holotype)	PERTH 08320454	W.A.	*Allocasuarina,* heath plants	KF977213	KF915290	KF991385
	PBM3694	PERTH 08319006	W.A.	*Allocasuarina, Acacia*	KF977214	KF915291	KF991386
I. rufoflavipes	NLB1002 (holotype)	PERTH 08383316	W.A.	*Eucalyptus*	KP308820	KP170993	KM406233
I. rufuloides	E8353	PERTH 07700598	W.A.	*Pinus, Melaleuca, Adenanthos*	JN035292	JN035295	—
	NLB618	PERTH 08305978	W.A.	*Pinus*	JN035291	—	—
I. sabulosa	NLB898 (holotype)	PERTH 08320322	W.A.	*Eucalyptus*	KP308821	KP170994	KM406234
	E8178	PERTH 07680732	W.A.	*Eucalyptus, Corymbia*	KP308822	JN974916	KM406235
	E8182	PERTH 07680775	W.A.	*Eucalyptus*	KP308823	KP170995	KM406236
	MDB F7/11	PERTH 08383057	W.A.	*Albizia, Corymbia*	KP308824	KP170996	—
	NLB901	PERTH 08320292	W.A.	*Acacia, Templetonia, Eucalyptus*	KP308825	KP170997	KM406237
	NLB986	PERTH 08383308	W.A.	*Corymbia, Acacia*	—	KP170998	KM406238
I. saliceticola	BJ900827O	GB	Sweden		AM882717	AM882717	—
I. salicis	EL7105	GB	Norway		AM882724	AM882724	—
I. salicis-herbaceae	ADW0060	TENN 063513	Idaho	Pinaceae	HQ586876	KP170999	KM406239
	EL1406	GB	Sweden		FN550893	FN550893	—
	EL5807	GB	Sweden		FN550891	FN550891	
	PBM1021	WTU	Washington , USA	*Abies, Picea, Pseudotsuga, Pinus*	—	AY038322	—
	PAM 95082312	LIP	France	*Picea*	—	HQ641100	—
I. scabriuscula	E. Horak s.n.	PDD 88624	N.Z.	*Nothofagus*	GU233365	GU233425	—
	G.S. Ridley s.n.	PDD 74240	N.Z.	*Nothofagus* forest, with *Leptospermum*	KP308826	—	—
I. scissa	ZT8660	PDD 71173	N.Z.	*Nothofagus*	JX258835	KP171000	

APPENDIX 3

Species	Field no.	Herbarium accession no.	State or Country	Plant associates	GenBank Accession No.		
					ITS[a]	nLSU-rRNA	*rpb2*
	PL12102	PDD 75859	N.Z.	*Nothofagus*	KP308827	JN974985	
I. sejuncta	MDB F97/11 (holotype)	PERTH 08383103	W.A.	*Eucalyptus, Corymbia, Melaleuca*	—	KP171001	KM555102
	PBM3752	PERTH 08363293	Qld.	*Eucalyptus*	—	KP171002	KM555103
	REH9668	BRI	Qld.	*Acacia*	KP308828	KP171003	KM555104
I. serrata complex (clade I)	PBM3222 "granulosipes"	TENN 069652	Tas.	*Eucalyptus, Acacia, Pomaderris*	KP636802	KP171004	—
	PBM3221 "granulosipes"	TENN 069651	Tas.	*Eucalyptus, Acacia, Pomaderris*	KP636803	KP171005	KM555105
	PBM3223 "granulosipes"	TENN 069653	Tas.	*Eucalyptus, Acacia, Pomaderris*	KP636804	KP171006	KM555106
	PBM3178 "fulvo-olivacea"	TENN 069656	Tas.	*Eucalyptus*	—	KP171007	KM555107
	HL967 undet.	CANB 574596	A.C.T.	*Eucalyptus*	KP636805	KP171008	—
	PBM3240 "granulosipes"	TENN 065744	Tas.	*Eucalyptus, Acacia, Pomaderris*	KP636806	JQ313563	KM555108
	PBM3234 "granulosipes"	TENN 069654	Tas.	*Eucalyptus, Acacia, Pomaderris*	KP636807	KP171009	—
	PBM3361 "granulosipes"	TENN 069655	Tas.	*Eucalyptus, Pomaderris*	KP636808	KP171010	KM555109
	PBM3231 "granulosipes"	TENN 069660	Tas.	*Eucalyptus, Acacia, Pomaderris*	KP636809	KP171011	KM555110
	PBM3235 "granulosipes"	TENN 069659	Tas.	*Eucalyptus, Acacia, Pomaderris*	KP636810	KP171012	KM555111
	Gates 20060805 "Inocybe sp."	Gates pers. herb.	Tas.		FN550880	FN550880	—
	PBM3477 "flavopus ined."	PERTH 08241988	W.A.	*Eucalyptus, Corymbia, Bossiaea*	KP636811	KP171013	KM555112
	NLB364 "fulvo-olivacea"	PERTH 08096198	W.A.	*Melaleuca, Andersonia*	KP636812	KP171014	—
	PBM3491 "fulvo-olivacea"	PERTH 08241740	W.A.	*Eucalyptus, Corymbia*	KP636813	KP171015	KM555113
	PBM3643 "serrata"	TENN 066579	W.A.	*Melaleuca, Taxandria, Agonis*	KP636814	KP171016	KM555114
	PBM3176 "fulvo-olivacea"	TENN 069657	N.S.W.	*Eucalyptus*	—	KP171017	KM555115
	PBM3479 "granulosipes"	PERTH 08241961	W.A.	*Eucalyptus, Allocasuarina*	KP636815	KP171018	KM555116
	NLB338 "fulvo-olivacea"	PERTH 08073627	W.A.	*Eucalyptus, Corymbia, Allocasuarina*	KP636816	KP171019	KM555117
	PBM3490 "fulvo-olivacea"	PERTH 08241759	W.A.	*Eucalyptus, Allocasuarina*	KP636817	KP171020	KM555118
	E8260	PERTH	W.A.	*Kunzea,*	KP636818	JN974917	KM555119

APPENDIX 3

Species	Field no.	Herbarium accession no.	State or Country	Plant associates	GenBank Accession No.		
					ITS[a]	nLSU-rRNA	*rpb2*
	"serrata"	07681216		*Astartea*			
	PBM3177 "fulvo-olivacea"	TENN 065355	N.S.W.	*Eucalyptus*	—	KP171021	—
	HL577 undet.	CANB 574547	N.S.W.	*Eucalyptus*	—	JN974966	—
I. serrata complex (clade II)	PBM3191 "fulvo-olivacea"	TENN 063044	N.S.W.	*Eucalyptus*	KP636819	KP171022	KM555120
	PBM3244 "fulvo-olivacea"	TENN 069665	Tas.	*Eucalyptus*	KP636820	KP171023	KM555121
	T25079 "crassocystis ined."	TENN 069663	N.S.W.	*Eucalyptus, Acacia, Micromyrtus, Callitris*	—	AY380392	AY337399
	PBM3671 "serrata"	TENN 066455	W.A.	*Corymbia, Eucalyptus*, planted *Acacia*	KP636821	KP171024	KM555122
	NLB814 "serrata"	PERTH 08318689	W.A.	*Eucalyptus, Corymbia*	KP636822	KP171025	KM555123
	E6320 "serrata"	PERTH 07612052	W.A.	*Gastrolobium, Allocasuarina*	—	EU600856	EU600855
	NLB331 "serrata"	PERTH 08073554	W.A.	*Eucalyptus, Corymbia*	KP636823	KP171026	KM555124
	PBM3615 "serrata"	TENN 066590	W.A.	*Eucalyptus, Corymbia*	ITS1 only (<200 bp)	KP171027	—
I. serrata complex (clade III)	PBM3432 "fulvo-olivacea"	TENN 069629	Tas.	*Eucalyptus, Acacia*	KP636824	KP171028	KM555125
	PBM3354 "fulvo-olivacea"	TENN 069628	Tas.	*Eucalyptus, Pomaderris, Acacia*	KP636825	KP171029	KM555126
	PBM3362 "serrata"	TENN 069626	Tas.	*Eucalyptus, Pomaderris, Acacia*	KP636826	KP171030	KM555127
	PBM3457 "serrata"	PERTH 08242208	W.A.	*Eucalyptus, Corymbia*	KP636827	KP171031	—
	PBM3430 "serrata"	TENN 069627	Tas.	*Eucalyptus, Acacia*	KP636828	KP171032	KM555128
I. serratoides	PBM3422 (holotype)	PERTH 08557748	Tas.	*Eucalyptus*	KP636829	KP171033	KM555129
	PBM3072b	PDD 97910	N.Z.	*Kunzea*	KP636830	KP171034	KM555130
	PBM3175	TENN 063978	N.S.W.	*Eucalyptus*	KP636831	KP171035	—
	PBM3356	TENN 068701	Tas.	*Eucalyptus, Pomaderris*	KP636832	KP171036	KM555131
	Uncultured fungus	Clone RFLP37	Qld.	*Eucalyptus*	DQ388844	—	—
	Uncultured fungus	Clone RFPL25	Qld.	*Eucalyptus*	DQ388832	—	—
	Uncultured fungus	Clone RFLP1	Qld.	*Eucalyptus*	DQ388808	—	—
I. sindonia	J. Cooper s.n.	PDD 80239	N.Z.	*Pinus*	KP636833	KC305363	—
I. sinuospora	E5980	PERTH 07606532	W.A.	*Allocasuarina*	JQ085936	EU600861	—
	PBM3686A	PERTH 08318999	W.A.	*Allocasuarina*	JX258838	KP171037	KM555132

APPENDIX 3

Species	Field no.	Herbarium accession no.	State or Country	Plant associates	GenBank Accession No.		
					ITS[a]	nLSU-rRNA	*rpb2*
I. solitaria	PBM3366 (holotype)	PERTH 08516774	Tas.	*Eucalyptus, Acacia, Pomaderris*	ITS1 and ITS2 (both <200 bp)	KP171038	KM555133
	PBM3404	TENN 066001	Tas.	*Nothofagus*	KP636834	KP171039	—
I. sororia	Kuoljok0512	GB	Sweden	*Salix*	FJ904150	FJ904150	—
	JRH661	WTU	Wyoming, USA	Pinaceae	—	EU600863	—
Inocybe sp. 1 (ECM)			Argentina	*Nothofagus*	UDB 007079		
Inocybe sp. 3 (ECM)			Argentina	*Nothofagus*	UDB 007037		
Inocybe sp. 6, 8, 14 (ECM)			Argentina	*Nothofagus*	UDB 007187		
			Argentina	*Nothofagus*	UDB 007138		
			Argentina	*Nothofagus*	UDB 007183		
Inocybe sp. 7 (ECM)			Argentina	*Nothofagus*	UDB 007067	—	—
			Argentina	*Nothofagus*	UDB 007168	—	—
Inocybe sp. 9 (ECM)			Argentina	*Nothofagus*	UDB 007068		
			Argentina	*Nothofagus*	UDB 007079		
Inocybe sp. 12 (ECM)			Argentina	*Nothofagus*	UDB 007177		
Inocybe sp. 13 (ECM)			Argentina	*Nothofagus*	UDB 007182		
Inocybe sp. DED8021	DED8021	SFSU	Thailand	*Tectona, Dipterocarpus*	GQ892999	GQ892954	KM555134
Inocybe sp. DJL-SJ14	DJL-SJ14	CFMR	U.S. Virgin Islands	*Coccoloba, Neea*	JQ408784	EU600851	—
Inocybe sp. DV04132011	DV04132011	TENN 066033	California, USA	*Quercus*	KP636835	KP171040	—
Inocybe sp. ECV3648	ECV3648	SFSU	Thailand	*Dipterocarpus*	GQ893012	GQ892967	KM555135
Inocybe sp. EL14305	EL14305	GB	Sweden		AM882722	AM882722	—
Inocybe sp. environmental sample	RT55-1		N.Z.	*Nothofagus*	EF634110	—	—
	RT60-3		N.Z.	*Nothofagus*	EF634111	—	—
Inocybe sp. environmental sample	M0303		Japan	*Fagus*	AB218084	AB218084	—
Inocybe sp. H. Lepp	H. Lepp 647	CANB	N.S.W.	*Eucalyptus*	KP636836	KP171041	KM555136

APPENDIX 3

Species	Field no.	Herbarium accession no.	State or Country	Plant associates	GenBank Accession No.		
					ITS[a]	nLSU-rRNA	*rpb2*
647		574551					
Inocybe sp. I116	I116	Gates pers. herb.	Tas.	Sclerophyll forest	FJ904142	FJ904142	—
Inocybe sp. I134	I134	TENN 069625	Tas.		KP636837	KP171042	KM555137
Inocybe sp. JAC2809b	JAC2809b	CBG 8916878 (CANB)	N.T.	*Eucalyptus*	KP636838	—	—
Inocybe sp. MCA562	MCA562	WTU	Japan	*Castanopsis* and mixed conifers	JQ408785	JN975016	JQ421077
Inocybe sp. MDB F128/10	MDB F128/10	PERTH 08383235	W.A.	*Eucalyptus, Corymbia, Heteropogon*	KP636839	KP171043	KM555138
Inocybe sp. MR00254	MR00254	TENN 066468	W.A.	Myrtaceous shrubs and trees	KP636840	KP171044	KM555139
Inocybe sp. MTS2494	MTS2494	UC	California, USA	*Quercus*	JQ408786	JN975008	—
Inocybe sp. NLB1074	NLB1074	PERTH 08561176	W.A.		—	KP171045	KM555140
Inocybe sp. PC96039		PC 96039	Zambia	Miombo woodland	—	EU555474	EU555473
Inocybe sp. PBM1889	PBM1889	WTU	Texas, USA	*Fagus, Quercus, Pinus*	JQ408787	JQ319707	JQ846493
Inocybe sp. PBM2166	PBM2166	PERTH 07671873	W.A.	*Eucalyptus*	KP636841	KP171046	—
	PBM2178	PERTH 07672004	W.A.	*Eucalyptus*	KP636842	KP171047	—
	PBM2180	PERTH 07672020	W.A.	*Eucalyptus*	KP636843	EU555461	EU555460
Inocybe sp. PBM3086	PBM3806	PDD 97895	N.Z.	*Nothofagus*	KP636844	JN974923	KM555141
Inocybe sp. PBM3103	PBM3103	PDD 97866	N.Z.	*Nothofagus*	KP636845	KP171048	KM555142
	PBM3107	PDD 97865	N.Z.	*Nothofagus*	KP636846	KP171049	—
Inocybe sp. PBM3136	PBM3136	PDD 97905	N.Z.	*Nothofagus, Kunzea*	KP636847	KP171050	KM555143
	PBM3139	PDD 97908	N.Z.	*Nothofagus, Kunzea*	KP636848	KP171051	KM555144
Inocybe sp. PBM3242	PBM3242	TENN 068381	Tas.	*Eucalyptus, Acacia*	KF830035	KF808347	—
Inocybe sp. PBM3335	PBM3335	TENN 063942	Tennessee	*Quercus*	KP636849	JN974990	—
	CO4327	CSU	Oklahoma	*Quercus, Carya*	KP636850	KP171052	—

APPENDIX 3

Species	Field no.	Herbarium accession no.	State or Country	Plant associates	GenBank Accession No.		
					ITS[a]	nLSU-rRNA	*rpb2*
Inocybe sp. PBM3448	PBM3448	TENN 068382	Tas.	*Eucalyptus*	KF830033	KF808345	KF830051
Inocybe sp. PBM3601	PBM3601	TENN 066647	W.A.	*Eucalyptus, Corymbia*	KF830032	KF808344	KF830050
	MR00219	TENN 066605	W.A.	*Eucalyptus, Corymbia*	KF830032	KF808343	KF830049
Inocybe sp. PBM3679	PBM3679	TENN 066638	W.A.	*Allocasuarina*	KF977210	KF915287	KB991383
Inocybe sp. PBM3686B	PBM3686B	TENN 066548	W.A.	*Allocasuarina*	KF977200	KF915276	KF991373
Inocybe sp. PBM3745	PBM3745	TENN 066977	Qld.	*Eucalyptus, Allocasuarina*	KF977211	KF915288	—
Inocybe sp. PBM3751	PBM3751	TENN 066980	Qld.	*Allocasuarina, Eucalyptus*	KP636851	KP171053	KM555145
Inocybe sp. PBM3766	PBM3766	TENN 066998	Qld.	*Eucalyptus, Allocasuarina*	KP636852	KP171054	KM555146
	TR138-05	M 0137870	P.N.G.	Mixed *Castanopsis* forest	JN408792	JN975009	—
	PL42609	BRI AQ793920	Qld.	*Lophostemon*	KJ729879	KJ729905	KJ729936
Inocybe sp. PBM3792	PBM3792	TENN 067024	Qld.	*Eucalyptus*	KF830034	KF808346	KF830052
Inocybe sp. PBM3906	PBM3906	TENN 067756	NC	*Picea, Abies*	KP636853	KP171055	KM555147
Inocybe sp. PL4203	PL4203	PDD 77716	N.Z.	*Nothofagus*	—	JN974956	KM555148
Inocybe sp. PL42609	PL42609	BRI AQ793920	Qld.	*Lophostemon*, wet sclerophyll forest	KJ729879	KJ729905	KJ729936
	PBM3766	TENN 066998	Qld.	*Eucalyptus, Allocasuarina*	KJ729880	KJ729906	KJ729937
	TR138-05	M 0137870	P.N.G.	*Castanopsis*	JQ408792	JN975009	—
Inocybe sp. REH7418	REH7418	NY	Costa Rica	*Quercus*	KP636854	JN974931	KM555149
	REH7965	NY	Costa Rica	*Quercus*	KP636855	JN973932	KM555150
Inocybe sp. REH9538	REH9538	NY	Qld.	*Syncarpia, Lophostemon*	KP636856	KP171056	KM555151
Inocybe sp. SAT07-328-06	SAT07-328-06	TENN	Washington USA	*Pseudotsuga, Tsuga, Alnus*	KP636857	KP171057	—
Inocybe sp. SA100602A	SA100602A	SA	Slovakia		KP636858	EU307837	—
	Clone A2YN18RM1		France	Alpine environmental sample	FJ567238	—	—
Inocybe sp. T1300	T1300	MEL	Tas.	*Eucalyptus*	JF960792	—	—
Inocybe sp. T35054	T35054	OSC	N.S.W.	*Eucalyptus*	KF977212	KF915289	KF991384
Inocybe sp.		TBGT	India	*Vitex*	KP636859	KP171058	KM656097

APPENDIX 3

Species	Field no.	Herbarium accession no.	State or Country	Plant associates	GenBank Accession No.		
					ITS[a]	nLSU-rRNA	*rpb2*
TBGT11280		11280					
Inocybe sp. TBGT12854		TBGT 12854	India	*Vateria, Ficus, Calphyllum*	KP636860	KP171059	KM656098
		TBGT 12828	India	*Vateria, Ficus, Calophyllum*	KP636861	KP171060	KM656099
Inocybe sp. TR49-05	TR49-05	M 0138271	P.N.G.	*Castanopsis*	JQ408790	JN975014	JQ421079
Inocybe sp. TR88-06	TR88-06	M 0138287	P.N.G.	*Castansopsis*	KP636862	JN974992	KM656100
	TR132-05	M 0138276	P.N.G.	*Castanopsis*	KP636863	KP171061	KM656101
Inocybe sp. TR104-05	TR104-05	M 0138275	P.N.G.	*Castanopsis*	KP636864	JN975011	—
	TR133-05	M 0138277	P.N.G.	*Castanopsis*	JQ408791	JQ319709	—
Inocybe sp. TR183-05	TR183-05	M 0137871	P.N.G.	*Castanopsis*	JQ408773	JN975005	JQ421070
	TR75-05	M 0138274	P.N.G.	*Castanopsis*	JQ408774	JQ815425	—
Inocybe sp. TR194-02	TR194-02	M 0138267	P.N.G.	*Castanopsis*	JQ408793	JN975032	JQ421080
Inocybe sp. TR198-03	TR198-03	M 0138269	P.N.G.	*Castanopsis*	KP636865	KP171062	—
Inocybe sp. ZT10031	ZT10031	ZT	Thailand		GQ893020	GQ892976	—
Inocybe sp. ZT10105	ZT10105	ZT	Thailand	*Tectona, Dipterocarpus*	GQ893003	GQ892957	—
I. spadicea	PBM2203 (E7051) (paratype)	PERTH 07676794	W.A.	*Eucalyptus, Agonis*	KP636866	EU600865	—
I. spiniformis	PBM3756 (holotype)	PERTH 08363277	Qld.	*Melaleuca, Lophostemon, Corymbia, Acacia*	KP636867	KP171063	KM656102
	PBM3748	PERTH 08363331	Qld.	*Allocasuarina, Eucalyptus*	KP636868	KP171064	KM656103
I. squamata	PAM 05052301	GB	France	*Populus, Picea*	FJ904132	FJ904132	—
I. "stellata"	ZT10097 (holotype)	SFSU	Thailand	*Castanopsis, Quercus, Lithocarpus, Dipterocarpus*	GQ893008	GQ892963	—
	DED8162	SFSU	Thailand	*Castanopsis, Pinus, Dipterocarpus*	GQ893006	GQ892961	KM656104
	ECV3651	SFSU	Thailand	*Dipterocarpus*	GQ893007	GQ892962	KM656105
I. aff. "*stellata*"	DED8015	SFSU	Thailand	*Dipterocarpus*	GQ893011	GQ892966	KM656106
	DED8060	SFSU	Thailand	*Dipterocarpus*	GQ893010	GQ892965	KM656107
	ZT10123	ZT	Thailand	*Quercus, Lithocarpus, Castanopsis*	GQ893009	GQ892965	—
I. straminipes	SA100602B	SAV	Slovakia		—	EU307838	EU307839

APPENDIX 3

Species	Field no.	Herbarium accession no.	State or Country	Plant associates	GenBank Accession No.		
					ITS[a]	nLSU-rRNA	*rpb2*
	PAM 01042904	TENN 063560	France	*Alnus, Salix*	HQ586868	HQ641111	KM656108
	PAM 95081306	TENN 063864	France	*Polygonum*	HQ586858	HQ641101	KM656109
	PAM 08090402	LIP	France		—	KP171065	—
	EL904	GB	Sweden		AM882718	AM882718	—
	PBM2782	TENN 062580	Massa-chusetts	*Fagus, Quercus, Pinus*	KP636869	KP171066	KM656110
	AWW269	TENN 063767	Rhode Island	Deciduous woodlands	HQ586874	HQ641111 6	KM656111
	EL7206	GB	Sweden		FN550890	FN550890	—
I. subangustifolia	REH9319 (holotype)	BRI 794269	Qld.	*Eucalyptus*	KP636870	KP171067	KM656112
	NLB983	PERTH 08364850	Qld.	*Eucalyptus*, rainforest plants	KP636871	KP171068	KM656113
I. subclavata	PBM3123	PDD 97858	N.Z.	*Kunzea*	KP636872	JN974974	—
	PBM3147	TENN 063878	N.Z.	*Kunzea*	KP636873	KP171069	KM656114
		PDD 77794	N.Z.	*Nothofagus*	GU233360	—	—
I. subferruginea	NLB319 (holotype)	PERTH 08074437	W.A.	*Eucalyptus, Corymbia, Allocasuarina*	KP636874	KP171070	KM656115
	E7066	PERTH 07676905	W.A.	*Eucalyptus*	KP636875	EU600894	—
	NLB226	PERTH 07699700	W.A.	*Corymbia*	KP636876	—	—
	PBM3621	PERTH 08319650	W.A.	*Corymbia*	KP636877	KP171071	KM656116
	PBM3674	TENN 066497	W.A.	*Corymbia, Eucalyptus, Acacia*	KP636878	KP171072	—
I. subfibrosoides	MES543	FLAS	Chile	*Nothofagus*	KP636879	KP171073	KM656117
I. subflavospora	E5880 (holotype)	PERTH 07609256	W.A.	*Eucalyptus*	—	AY380396	AY337404
	NLB915	PERTH 08320861	W.A.	*Eucalyptus, Hakea*	KP641619	KP171074	KM656118
	PBM3791	PERTH 0862998	Qld.	*Eucalyptus*	KP641620	KP171075	KM656119
I. cf. *subflavospora*	PBM3677	PERTH 08319081	W.A.	*Allocasuarina*	KP641621	KP171076	KM656120
	REH9670	BRI	Qld.	*Eucalyptus*	KP641621	KP171077	—
I. subfulva	PBM1482	WTU	Missouri, USA	*Quercus, Carya, Juniperus*	KP641623	JN974989	—
I. "sublineata"	DED8044	SFSU	Thailand	*Castanopsis*	GQ892995	GQ892950	KM656121
I. subpulverulenta	H. Lepp 1966 (holotype)	CANB 574535	A.C.T.	*Eucalyptus*	KP641624	KP171078	KM656122
	H. Lepp 929	CANB 574588	A.C.T.	*Acacia*	KP641625	KP171079	—
	H. Lepp 953	CANB 574592	A.C.T.		KP641626	KP171080	KM656123

APPENDIX 3

Species	Field no.	Herbarium accession no.	State or Country	Plant associates	GenBank Accession No.		
					ITS[a]	nLSU-rRNA	*rpb2*
I. cf *subpulverulenta*	PBM3416	TENN 067775	Tas.	*Nothofagus*	KP641627	KP171081	KM656124
I. subtilior	OKM24631 (holotype)	PERTH 08561745	W.A.	*Eucalyptus, Pinus*	—	AY380398	AY337406
	E9112	PERTH 08095388	W.A.		KP641628	KP171082	—
I. cf. *subtilior*	E4810	PERTH 07660634	W.A.	*Eucalyptus, Brachychiton, Lysiphyllum*	—	KP171083	—
	MDB F75/11	PERTH 08383081	W.A.	*Corymbia, Acacia, Triodia*	KP641629	KP171084	KM656125
I. sylvicola	PMB3387 (holotype)	PERTH 08517215	Tas.	*Nothofagus*	KP641630	JQ313576 (*I.* "*scissa*")	KM656126
	I64	TENN 065737	Tas.	Rainforest	KP641631	KP171085	—
	PBM3394	TENN 065745	Tas.	*Nothofagus*	KP641632	JQ313568 (*I.* "*scissa*")	KM656127
I. tahquamenonensis	PBM1142	WTU	New Hampshire, USA	*Fagus, Quercus, Betula, Tsuga*	—	AY380399	AY337407
I. tasmanica	PBM3395 (holotype)	PERTH 08517231	Tas.	*Nothofagus*	KF830028	KF808340	KF830046
I. tenax	PBM3377	TENN 067034	Tas.	*Nothofagus, Leptospermum*	KF871768	KF853392	KF871768
	PBM3417	TENN 067037	Tas.	*Nothofagus*	KF871771	KF853395	KF891948
	PBM3443	TENN 067040	Tas.	*Nothofagus, Pomaderris*	KF871770	KF853394	KF891947
	PBM3201	MEL 23356645A (as *I. umbrosa*)	Vic.	*Nothofagus*	KF871769	KF843393	KF891946
	PBM3132	PDD 97903	N.Z.	*Nothofagus, Kunzea*	KF871773	KF853397	KF891950
	PBM3140	PDD 97907	N.Z.	*Nothofagus*	KF871772	KF853396	KF891949
	Horak s.n.	PDD 71260	N.Z.	*Nothofagus*	KF977216	—	—
I. tenuis	PBM3360 (holotype)	PERTH 08517258	Tas.	*Eucalyptus, Pomaderris*	KF871780	KF853404	KF891957
I. teraturgus	SJ79017	GB	Sweden		AM882789	AM882789	—
I. tertia	PBM3730 (holotype)	PERTH 08362882	Qld.	*Allocasuarina, Acacia, Eucalyptus*	KP641633	KP171086	KM656128
I. "*thailandica*"	DED8049	SFSU	Thailand	*Castanopsis*	GQ893013	GQ892968	KM656129
	DED8008	SFSU	Thailand	*Castanopsis*	EU600871	EU600871	—
	DED8020	SFSU	Thailand	*Castanopsis*	GQ893014	GQ892969	KM656130
I. tomentipes	E8486 (holotype)	PERTH 08093334	W.A.	*Corymbia*	KF871787	KF853411	—

APPENDIX 3

Species	Field no.	Herbarium accession no.	State or Country	Plant associates	GenBank Accession No.		
					ITS[a]	nLSU-rRNA	*rpb2*
	PBM3618	PERTH 08319677	W.A.	*Allocasuarina, Agonis, Astartea*	KF871786	KF853410	—
I. torresiae	MDB F70/96 (E6978) (PBM2157) (holotype)	PERTH 07678428	W.A.	*Antidesma, Glochidion, Eucalyptus, Corymbia, Albizia*	JQ085937	EU600874	EU600873
	PBM3722	PERTH 08362808	Qld.	*Allocasuarina, Eucalyptus*	KP641634	KP171087	KM656131
	PBM3779	PERTH 08363099	Qld.	*Eucalyptus*	KP641635	KP171088	KM656132
I. trachysperma	PBM2127 (E6760)	PERTH 07677669	W.A.	*Eucalyptus, Corymbia*	KP641636	—	—
I. tubarioides	PBM2550	TENN 062394	North Carolina, USA	*Tsuga, Quercus, Pinus, Carya*	—	AY732211	EU307855
I. tumidula	PBM3764 (holotype)	PERTH 08557772	Qld.	*Eucalyptus*	KP641637	KP171089	KM656133
	PBM3770	PERTH 08363188	Qld.	*Eucalyptus*	KP641638	KP171090	KM656134
I. umbrinella	JV13699	GB	Finland	*Pinus, Populus, Salix*	FJ904165	FJ904165	FJ904165
I. vagata	PBM3187 (paratype)	CANB 782347	N.S.W.	*Eucalyptus*	KP641639	KP171091	KM656135
		PDD 72861	N.Z.	*Kunzea, Leptospermum*	HM060326	HM060325	—
I. vestipedes	PBM3470 (holotype)	PERTH 08242062	W.A.	*Eucalyptus, Corymbia, Agonis*	KP641640	KF171092	KM656136
	PBM2199 (E7046)	PERTH 07676743	W.A.	*Eucalyptus*	KP641641	EU600859	EU600858
	PBM3482	PERTH 08241848	W.A.	*Allocasuarina, Eucalyptus*	KP641642	KP171093	KM656137
I. violaceocaulis	PBM2164 (E7013) (holotype)	PERTH 07671857	W.A.	*Eucalyptus*	—	AY732208	—
	I86	TENN 066846	Tas.		—	KC305368	—
	REH9202	NY	Qld.	*Allocasuarina, Eucalyptus*	KP641643	KC305367	KC305417
	PBM3215	TENN 066847	TAS	*Eucalyptus, Acacia, Pomaderris*	KP641644	KC305366	KC305418
	PBM2198	PERTH 07676735	W.A.	*Eucalyptus*	—	AY380404	—
I. viridipes	PBM3767 (holotype)	PERTH 0836390	Tas.	*Eucalyptus*	KP641645	KP171094	KM656138
	I11	HO 570442	Tas.	Rainforest	JQ801419	JQ815430	—
	I56	HO 570444	Tas.		JQ801418	JQ815429	—
	I153	HO 562693	Tas.		KP641646	KP171095	KM656139
I. viscata	E. Horak 68/177 (holotype)	PDD 27109	Tas.	*Nothofagus*	GU233318	—	—

APPENDIX 3

Species	Field no.	Herbarium accession no.	State or Country	Plant associates	GenBank Accession No.		
					ITS[a]	nLSU-rRNA	*rpb2*
	I102	TENN 066121	Tas.		KP641647	KP171096	—
	PBM3213	TENN 065733	Tas.	*Eucalyptus, Acacia, Pomaderris*	KP641648	KP171097	—
	PBM3445	TENN 065734	Tas.	*Eucalyptus, Acacia*	KP641649	JQ313570	—
I. vittata	PBM3683 (holotype)	PERTH 08319022	W.A.	*Allocasuarina* (eucalypts in vicinity)	KF871779	KF853403	KF891956
	PBM3693	PERTH 08318913	W.A.	*Allocasuarina, Acacia*	KP641650	KP171098	KM656140
	PBM3696	PERTH 08318883	W.A.	*Allocasuarina, Acacia*	KF871778	KF853402	KF891955
I. xanthocephala	PAM 00100606	GB	France	*Salix*	FJ904130	FJ904130	—
I. xanthocystis	NLB797 (holotype)	PERTH 08318158	W.A.	*Eucalyptus, Corymbia*	KF830040	KF808352	KF830056
	PBM3466	PERTH 08242100	W.A.	*Eucalyptus, Corymbia, Acacia, Agonis*	KF830042	KF808355	KF830058
	PBM3469	PERTH 08242070	W.A.	*Eucalyptus, Corymbia, Acacia, Agonis*	KF830041	KF808353	KF830057
	PBM3589	PERTH 08319952	W.A.	*Eucalyptus, Corymbia*	KF830039	KF808351	KF830055
	NLB455	PERTH 08072892	W.A.	*Eucalyptus*	—	KF808354	—
	NLB798	PERTH 08318131	W.A.	*Eucalyptus, Corymbia*	KF830043	KF808356	KF830059
I. aff. *xanthomelas*	PAM 07062202	TENN 063835	France	*Quercus, Carpinus*	HQ586861	HQ641104	KM656141
I. "cf. *xanthomelas*"	EL3505	GB	Norway		AM882989	AM882989	—
	PAM 06060405	TENN 063834	France	*Salix*	HQ586867	HQ641109	KM245990
	PAM 10082807	LIP	France	*Alnus*	KP641651	KP171099	—
	PAM 10082808	LIP	France	*Alnus*	KP641652	KP171100	—
	PAM 10082305	LIP	France	*Alnus*	KP641653	KP171101	—
Tubariomyces australiensis	MDB F62/10 (holotype)	PERTH 0838319	W.A.	*Eucalyptus, Corymbia, Melaleuca, Petalostigma*	KP641654	—	—

[a] NCBI GenBank will not accession DNA sequences <200 bp in length as these could be unreliable if used in phylogenetic analysis.
Names in "quotes" are unpublished manuscript names.
For authorities of published names, see the Index.

GLOSSARY

abaxial: *of basidiospores*, the opposite side to that with the apiculus when seen in side (profile) view. cf. adaxial.

abrupt: appearing as if cut off transversely; truncate.

acidulous: odour similar to that of species in *Suillus* or *Lycoperdon*; odour similar to that of a tin can owing to presence of acetic acid. A common odour in many species of Inocybaceae, particularly from temperate regions of the southern hemisphere.

aculeate: prickly; with narrow spines.

acuminate: tapering gradually to a protracted point.

acute: terminating in a distinct but not protracted point, the converging edges separated by an angle less than 90 degrees.

adaxial: *of basidiospores*, the side that the apiculus is on when seen in side (profile) view. cf. abaxial.

adnate: *of lamellae*, attached to the stipe by at least part of the lamellar width; also described as broadly or narrowly adnate depending upon the degree of attachment.

adnexed: *of lamellae*, partially attached to the stipe. cf. emarginate.

agaric: common name for a fungus which produces gilled fruit bodies, usually a member of the Agaricales. cf. agaricoid.

agaricoid: like an agaric; refers to fungi with a distinct pileus and stipe but not belonging to the Agaricales (e.g. some polypores). **cf. agaric**.

agglutinate: fixed together as if with glue.

allantoid: *of basidiospores*, slightly curved with rounded ends; sausage-shaped. cf. **phaseoliform** which is a broader shape.

alveola: a small cavity or pit on a surface.

alveolate: *of a surface*, having deep honeycomb-like pits.

amorphous: shapeless or structureless.

ampulliform: flask-shaped.

amygdaliform: *of basidiospores*, almond-shaped.

amyloid: staining grey-blue to blue-black in iodine-containing solutions such as Melzer's reagent. cf. dextrinoid, inamyloid.

anastomosing: *of lamellae*, with interconnections (anastomoses); also called interveining.

angiocarpy: *of fruit bodies*, a type of development whereby the hymenium is protected within a cavity during part, most, or all of its development. adj. **angiocarpic**, **angiocarpous**. cf. **bivelangiocarpy**, **gymnocarpy**, **monovelangiocarpy**.

angular: *of basidiospores*, having various angular shapes; *of fruit bodies*, angular swollen base of stipe.

annulus: ring-like band or bands of tissue on the stipe usually derived from the partial veil after the expansion of the pileus; may occur near the stipe apex (superior position), centrally, or near the base (inferior position).

apiculus (hilar appendage): a short projection at the base of basidiospores where the spore is attached to the sterigma. adj. **apiculate**.

appendiculate: *of fruit bodies*, with a rim or flap of overlapping or overhanging tissue on the pileal margin.

applanate: flat or flattened; also referred to as plane.

appressed: pressed closely against but not united with.

arachnoid: cobweb-like, e.g. the partial veil of some fruit bodies. cf. **cortina, cortinoid**.

arcuate: *of lamellae*, curved like a bow.

arenicolous: living or growing in sand.

areolate: *of surfaces*, split or cracked into polygonal or irregular shaped blocks.

aseptate: *of hyphae*, having no cross walls or septa. cf. **septum**.

attachment: *of the stipe to the pileus*, may be central, eccentric or lateral.

attenuate: gradually narrowed towards the end.

avellaneous: dull grayish brown or gray tinged with pink; "Avellaneous" of Ridgway (1912), a greyish brown with a vinaceous or pinkish tint.

ballistospore: *of a basidiospore*, a spore that is forcibly discharged. adj. **ballistosporic**. cf. **statismospore**.

basal mycelium: mycelium attached and immediately below fruit bodies.

basidiole: a basidium-like hymenial element, lacking sterigmata because it is either immature or sterile.

basidioma: a basidium-producing structure; a fruit body of the Basidiomycota containing basidia. pl. **basidiomata**. cf. **fruit body**.

basidiomycete: common name for a member of the Basidiomycota, the basidium being the diagnostic character.

basidiospore: sexual spore (post-meiotic product) borne on basidia; basidiospores of the Inocybaceae are typically yellowish brown (rarely lighter), smooth, angular, or nodulose.

basidium: cell or organ diagnostic for the Basidiomycota in which karyogamy and meiosis occur, giving rise to basidiospores. The Inocybaceae typically produce four (less often two) sterigmata, each producing one basidiospore. pl. **basidia**. cf. **necropigmented basidium**.

bivelangiocarpy: *of fruit bodies*, a type of angiocarpic development whereby the hymenium is protected within a cavity during part, most, or all of its development by both a partial veil (of lipsanenchyma tissue origin) and universal veil. adj. **bivelangiocarpic**, **bivelangiocarpous**. cf. **angiocarpy**, **monovelangiocarpy**.

blastn: Basic Local Alignment Search Tool, nucleotides; a tool to find regions of similarity between nucleotide sequences

bruising: *of the surface and flesh of some fruit bodies*, changing colour when damaged or exposed to air. cf. **brunnescent**, **rubescent**.

brunnescent: *of the surface and flesh of fruit bodies*, turning a brown colour where cut or bruised.

bulbous: *of stipes*, swollen at the base.

button: a young basidioma before expansion of the pileus and stipe. cf. **primordium**.

caespitose: growing in tufts; the habit whereby numerous fruit bodies are fused together at their base just above the substrate. cf. **gregarious**.

campanulate: *of a pileus*, bell-shaped.

canescent: *of surfaces*, softly downy.

capitate: *of cystidia*, with a knob-like to globose head.

catenate: in chains, *e.g.* of a row or chain of cells or spores.

catenulate: in short chains of two or three.

caulocystidium: a differentiated sterile end-cell on the stipe resembling and developmentally analogous with cheilocystidia. Metuloid or not. Often occurring in clusters with cauloparacystidia. pl. **caulocystidia**.

caulocystidioid hair: differentiated sterile end-cell (cystidium) on the stipe not developmentally homologous with the cheilocystidia. Metuloid or not. Often occurring scattered and not occurring with cauloparacystidia.

caulohymenium: area on the upper stipe of some Inocybaceae, usually comprised of basidia and cystidia arranged in a similar way to their arrangement on the lamellar edge.

cauloparacystidium: a differentiated sterile end-cell (cystidium) on the stipe resembling and developmentally homologous to the paracystidia on lamellae edges. Often occurring with caulocystidia, and usually smaller than them and typically thin-walled, hyaline, often clavate. Thick-walled and variously shaped forms not resembling the paracystidia may not be developmentally analogous to the paracystidia. Such forms are referred to as paracystidia-like cells. pl. **cauloparacystidia**.

chamois: dull yellow; "Chamois" of Ridgway (1912).

cheilocystidium: a differentiated sterile end-cell (cystidium) on the margin (edge) of the lamellae. In the Inocybaceae often similar to pleurocystidia, and often occurring with paracystidia. pl. **cheilocystidia**.

chronogram: a phylogenetic tree calibrated to geological time.

citriniform (limoniform): *of basidiospores*, with a beak-like projection at each end (lemon-shaped).

clade: a group of taxa, the descendants of which all share the same common ancestor; monophyletic.

clamp connection: a swollen or loop-like hyphal outgrowth which, at cell division, makes a connection between the resulting two cells by fusing with the lower and acting as a passage for nuclear transfer. Present in all basidioma tissues of almost all the Inocybaceae.

clavate: club-shaped, greatly widened at one end.

close: *of lamellae*, neither crowded nor well-spaced but arbitrarily somewhat between. cf. **crowded**, **distant**, **subdistant**.

co-evolution: the interdependent evolution of two or more taxa having an obvious ecological relationship.

collybioid: resembling *Collybia*, i.e. with a convex pileus, non-decurrent lamella attachment, strongly incurved or inrolled young pileal margin and no annulus or volva.

columella: *of secotioid fruit bodies*, sterile tissue representing a rudimentary stipe. May extend to the top of the gleba (percurrent columella), partial way (truncate columella), be branched into many thin intrusions throughout the gleba (dendritic columella), or emergent outside below the fruit body (stipitate columella). It may be reduced to a small basal pad at the base of the gleba.

concolorous: having the same colour. cf. **versicolorous**.

context: *of fruit bodies*, macroscopic term for the sterile mass of mycelium composing fruit bodies. cf. **trama**.

cortina: cobweb-like (arachnoid) partial veil of some fruit bodies. When present in some of the Inocybaceae, the cortina usually soon disappears (fugacious or evanescent). adj. **cortinoid**. cf. **partial veil**.

crenate: *of a lamella or pileus*, having the margin scalloped.

crenulate: minutely crenate.

crowded: *of lamellae*, crammed together. cf. **close**, **distant**, **subdistant**.

cutis: *of a pileus or stipe*, a pellis with hyphae arranged predominantly parallel to the surface. May be divided into epicutis and subcutis if the outer most layer of the cutis is distinguished from that below.

cyanescent: *of the surface and flesh of fruit bodies*, turning a blue colour.

cylindrical: circular in cross-section, with parallel sides.

cystidiate: *of edges or surfaces*, obviously glistening or differentiated due to the presence of cystidia, e.g. the lamellae margins of many Inocybaceae or the stipes of some Inocybaceae.

cystidium: a sterile end-cell of distinct shape, colour and/or size in the hymenium or other locations. cf. types categorised on location: **caulocystidium**, **cheilocystidium**, **hymenial cystidium**, **pileocystidium**, **pleurocystidium**. cf. types categorised on appearance: **leptocystidium**, **metuloid cystidium**. pl. **cystidia**.

dacryform (lacryform): tear-drop shaped.

decurrent: *of lamellae*, extending along the stipe beyond the point of insertion.

decurved: *of pilei margins,* (incurved) bent downwards.

dentate: *of edges or surfaces*, toothed. cf. **denticulate**.

denticulate: *of edges or surfaces*, finely toothed. cf. **dentate**.

depauperate: lacking, in reference to clades of Inocybaceae that lack pleurocystidia.

depressed: *of pilei or spores*, a sunken or concave area lower than the remainder of a structure.

dextrinoid: staining reddish brown in iodine-containing solutions such as Melzer's reagent. cf. **amyloid**, **inamyloid**.

digitate: shaped like the fingers of a hand.

dimorphic: *of basidia or basidiospores,* having two different forms, especially of disjunct size ranges.

disc: the central area of a pileus.

distant: *of lamellae*, spaced well apart. cf. **close**, **crowded**.

divergent (bilateral) trama: *of hyphae of lamellae*, having lateral strata angled away obliquely from a central strand (mediostratum) of vertically oriented hyphae. cf. **convergent (inverse) trama**, **regular trama**.

diverticulate: *of cystidia*, with many apical branched or contorted protrusions.

dry sclerophyll: hard-leaved vegetation with short internodes used to describe plant communities in Australia dominated by a eucalypt overstory between 10–30 metres in height.

echinate: *of basidiospores*, with long sharp pointed spines. cf. **echinulate**.

echinulate: *of basidiospores*, bearing small, pointed spines. cf. **echinate**.

ectomycorrhiza: see **mycorrhiza**.

ellipsoidal: oval in outline, bilaterally symmetrical, three-dimensional.

emarginate: *of lamellae*, sharply adnexed to the stipe. cf. **sinuate**.

emergent: *of fruit bodies*, partially or completely rising above the soil surface. cf. **epigeous**, **hypogeous**, **sequestrate**.

endophyte: an organism which completes its life cycle in a plant that shows no external sign of the infection.

entire: *of edges or surfaces*, having a smooth margin, e.g. margin of pileus that is not rimose or striate.

epigeous: *of fruit bodies,* growing on or developing on the surface of the soil. cf. **emergent**, **hypogeous**.

equal: *of stipes,* having equal diameter from the apex to the base.

erect: perpendicular to a surface.

eroded: *of pilei or lamellae margins,* irregularly degraded.

erumpent: *of fruit bodies*, bursting through the surface.

evanescent: present briefly, soon vanishing. cf. **fugacious**.

excentric (=eccentric): off-centre, to one side, any attachment intermediate between central and lateral.

exudation: droplet of clear, white or coloured liquid. May change colour when dry, or stain the underlying tissue.

fabaceous: having the odour or taste of raw beans.

facultative: *of organisms*, having the capacity to derive nutrition as a saprotroph or as a parasite. cf. **obligative**.

farinaceous: *of odour*, with the odour or taste of flour or fresh ground meal from whole grain.

farinaceous: *of surfaces*, covered with mealy flour-like particles. cf. **mealy**.

fasciculate: having many fruit bodies crowded close together, often fused below ground but not above ground. cf. **caespitose**, **gregarious**.

fibrillose: *of surfaces*, having visible silk-like fibrils.

fibrous: *of stipes*, usually rather thick and leaving a ragged edge when broken in two.

filiform: filamentous; thread-like in shape.

fimbriate: *of edges or surfaces*, finely fringed, e.g. the lamellar margin of many Inocybaceae.

flesh (context): macroscopic term for the sterile tissue comprising the fruit body of larger fungi. Trama is usually used as the equivalent microscopic term. cf. **context**, **stroma**, **trama**.

floccose: *of surfaces*, with easily removed cotton-like fibrils: having the appearance of cotton flannel; with a soft cottony texture.

flocculose: *of surfaces*, finely woolly or cottony.

foveate: pitted.

foveolate: delicately pitted.

free: *of lamellae*, not attached to the stipe.

fruit body: multi-cellular reproductive structure of larger fungi (and micro fungi), producing and disseminating usually the sexual spores (less often asexual spores). cf. **basidioma**.

fugacious: present briefly, soon vanishing. cf. **evanescent**.

fulvous: brownish yellow to yellowish brown.

furfuraceous (scurfy): with bran-like scales similar to dandruff.

fuscescent: turning the colour of smoky drab.

fuscous: smoky drab; combinations of grey, brown, purple, or black; "Fuscous" of Ridgway (1912).

fusiform: spindle-shaped; tapered at both ends and widest at the centre.

fusoid: somewhat fusiform.

geniculate: bent abruptly like a knee joint.

gibbous: refers to nodulose-spored condition of basidiospores. cf. **nodulose**.

glabrescent: *of surfaces*, becoming bald or smooth.

glabrous: *of surfaces*, bald, wax-like, smooth, without hairs or raised fibres or scales or raised patches.

gleba: the interior of sequestrate fungi, containing the hymenium and various sterile components.

globose: spherical; nearly so (subglobose).

glutinous: *of surfaces*, covered with a jelly-like gluten, slightly or not sticky. cf. **viscid**.

granular: *of surfaces*, coarsely powdery.

granulose: *of surfaces*, covered with a powder composed of particles as big as grains of salt.

gregarious: growing in groups; the habit whereby numerous fruit bodies are grouped close to each other but not joined together. cf. **caespitose**.

guttulate: *of a spore*, containing one or more oil-like globules.

gymnocarpy: *of fruit bodies*, a type of development whereby the hymenium is exposed from initiation to maturity. adj. **gymnocarpic**, **gymnocarpous**. cf. **angiocarpy**.

habit: the manner of growth.

hirsute: *of surfaces*, covered with coarse, rough hairs. cf. **villous**.

hispid: *of surfaces*, covered with long rough hairs or bristles.

hispid-squarrose: *of surfaces*, covered with stiff erect scales.

hoary: *of surfaces*, covered with a very short, closely interwoven hairs.

host: other living organism on or in which fungi live. A host plant for the Inocybaceae is its mycorrhizal partner.

humicolous (terricolous): living or growing on or in soil.

hyaline: *of structures observed under the microscope*, transparent to translucent and colourless or almost so.

hygrophanous: *of fruit bodies*, changing colour on drying.

hymenial cystidium: a sterile end-cell (cystidium) occurring with the hymenium on the lamellae. cf. **caulocystidium**, **pleurocystidium**.

hymenium: fertile layer bearing basidia in fruit bodies. In the Inocybaceae occurring on the lamellae; infrequently on the stipe. cf. **caulohymenium**. pl. **hymenia**. adj. **hymenial**.

hymenophoral trama: tissue comprised of sterile hyphae which support the hymenium; located between the hymenial layers of the lamellae. cf. **trama**.

hymenophore: the spore-bearing part of a fruit body (hymenium), e.g. the lamellae of a mushroom.

hypha: an individual fungal filament of a mycelium. pl. **hyphae**. cf. **mycelium**.

hypogeous: *of fruit bodies*, growing on or developing below the surface of the soil. cf. **emergent**, **epigeous**.

imbricate: partly overlapping, such as squamules on a pileus.

inamyloid: not reacting in iodine-containing solutions such as Melzer's Reagent. cf. **amyloid**, **dextrinoid**.

infundibuliform: funnel-shaped.

innate: *of surfaces*, being an integral part of the tissue; not readily removable from the adjacent tissue. cf. **superficial**.

involute: *of pilei margins,* having the edge rolled inwards or downwards. cf. **revolute**.

isabelline: dingy yellowish brown with an olive tone ; "Isabella Color" of Ridgway (1912).

ITS: internal transcribed spacer; the ribosomal RNA region that includes internal transcribed spacer 1 (ITS1), the 5.8S ribosomal RNA gene, and internal transcribed spacer 2 (ITS2); the ITS1-5.8S-ITS2 region; both ITS1 and ITS2 are highly variable (ITS1 more so) - gene regions used to distinguish many species of fungi at the species level.

ivory: creamy-white.

KOH: potassium hydroxide, usually prepared as a weak 3–5% solution.

lacerate: *of edges or surfaces*, appearing as if roughly cut or torn.

lacrymoid: tear-drop shaped.

lageniform: *of cystidia*, gourd-shaped; flask-shaped; widest at the base and tapered towards the apex; narrower with a more distinct neck than utriform.

lamella: a usually radially-arranged, vertically-orientated, plate-like structure on the under-side of fruit bodies. Supports the hymenium. pl. **lamellae**.

lamellula: a short lamella of varying lengths that extends from the margin of the pileus towards the stipe but do not reach the stipe. pl. **lamellulae**.

lanceolate: *of cystidia*, lance-shaped; narrow, slightly swollen at or below the middle and tapered towards the apex, or sometimes, each end.

leptocystidium: a smooth, thin-walled cystidium without any conspicuous contents.

limoniform (citriniform): *of basidiospores,* lemon-shaped.

locule: *of sequestrate fungi*, a cavity or chamber in the gleba. adj. **loculate**.

lubricous: greasy or slippery or oily but not viscid (sticky) or slimy.

marginate: *of bulb at the base of stipes*, having a circular ridge that forms the top of the bulb.

matted fibrillose: *of surfaces*, with matted or interwoven fibrils.

mealy: *of surfaces*, covered with mealy flour-like particles. cf. **farinaceous**.

membranous: parchment-like or membrane-like.

metuloid cystidium: modified cystidium apically bearing calcium-oxalate crystals; typically thick-walled.

monophyletic: a taxon that includes the common ancestor and all of its descendants.

monovelangiocarpy: *of fruit bodies*, a type of angiocarpic development whereby the hymenium is protected only by the universal veil. adj. **monovelangiocarpic**, **monovelangiocarpous** cf. **angiocarpy**, **bivelangiocarpy**.

mucronate (rostrate): *of basidiospores, of cystidia,* with a short, beak-like extension (mucro or rostrum).

muscarine: small quaternary ammonium compound that stimulates the parasympathetic nervous system, produced in varying quantities by numerous species of Inocybaceae and other fungi; and which produces the SLUDGE syndrome, symptoms of which include salivation, lacrymation, urge to urinate and defecate, gastrointestinal distress, and emesis.

mycelium: the vegetative body of a fungus, comprised of hyphae. pl. **mycelia**. cf. **hypha**.

mycorrhiza: a mutualistic or weakly pathogenic association of a fungus and the roots of a plant. Some types are: ectomycorrhiza, where the fungus is on the surface of the roots and forms a Hartig net; endomycorrhiza, where the fungus invades the cells of roots where it is frequently confined to well-defined layers; ectendomycorrhiza, where there is a Hartig net but the fungus also penetrates the root cells; and arbuscular, where the hyphae in the cortical cells of the root may be coiled or branched. pl. **mycorrhizae**.

necropigmented basidium: a basidium that becomes ochraceous and collapses after spore discharge.

nLSU: nuclear large subunit ribosomal RNA (nLSU-rRNA), comprises part of the ribosome, a subcellular unit involved in protein translation; refers to a relatively conserved region of the nuclear ribosomal RNA tandem repeat encoded by the 25S ribosomal RNA gene.

nodulose: *of basidiospores*, ornamented with small or large, blunt knobs.

obligative: *of organisms*, wholly dependent for its nutrition on another living organism. cf. facultative.

oblong: *of basidiospores*, having rounded ends and parallel sides.

obovate: reverse egg-shaped, i.e with the widest diameter remote from the point of attachment.

obtuse: with blunt or rounded ends.

oleiferous hypha: an oily or resinous hypha. cf. **refractive hypha**.

omphalinoid: resembling *Omphalina*, i.e. small, fragile, with decurrent lamella attachment, umbilicate pileus, and no annulus or volva.

ovoid: egg-shaped in three dimensions.

palisade: with hyphae that are erect and parallel with each other.

paracystidium: a differentiated sterile end-cell (cystidium) on the lamellar margin occurring among cheilocystidia; usually short, clavate. pl. **paracystidia**.

partial veil: *of fruit bodies,* a protective layer of tissue which extends from the stipe to the pileal margin during the development of the hymenium; after rupturing may or may not leave discernible remnants on the pileal margin and stipe. cf. **universal veil**.

PDAB: macrochemical known as p-dimethylamino-benzaldehyde composed of PDAB dissolved in a solution of conc. HCl acid and 95% ethyl alcohol.

pedicel: *of cystidia*, a slender stalk. adj. **pedicellate**.

pellis: *of fruit bodies*, anatomical structure of hyphae that compose the outer layer of the pileus (pileipellis), or stipe (stipitipellis), not including any veil.

peridium: the outer layer (pellis) of sequestrate fungi. Consisting of one to several layers: an innermost endoperidium, middle mesoperidium, and external exoperidium.

phaseoliform: *of basidiospores*, bean-shaped. cf. **reniform**; cf. **allantoid** which is a narrower shape.

phylogeny: the evolutionary history of organisms often depicted like branches on a tree.

pigment: a coloured material. Pigments of the larger fungi may occur within a hypha (intracellular, plasmatic, vacuolar), in the hyphal walls (intraparietal), on the hyphae (incrusting), or between the hyphae (intercellular).

pileipellis: an anatomical structure of hyphae that compose the outer layer (pellis) of the pileus, not including any veil.

pileocystidium: a differentiated sterile end-cell (cystidium) in the pileipellis. pl. **pileocystidia**.

pileus: the pileus-shaped, hymenium-bearing part of fruit bodies. adj. **pileate**.

pilose: *of surfaces*, hairy.

pleurocystidia: sterile terminal cells found on the sides of lamellae that are differentiated in size or shape from basidia or immature basidia.

pleurocystidium: a differentiated sterile end-cell (cystidium) on the side (face) of the lamellae. In the Inocybaceae often similar to cheilocystidia. pl. **pleurocystidia**.

polymorphic (versiform): variably shaped.

primordium: a young basidioma before expansion of the pileus and stipe. cf. **button**.

pruinose: *of surfaces*, finely frost-like, flour-like or powdered appearance. In the Inocybaceae, often referred to the stipe surface and reflecting the presence of caulocystidia.

pubescent: *of surfaces*, covered with very short, soft hairs.

pulverulent: *of surfaces*, appearing as though covered with a fine powder.

pulvinate: cushion-like.

pyriform: *of cystidia*, pear-shaped.

Q-value: *of basidiospores*, quotient of length and width, the former divided by the latter.

receding: *of lamellae*, attached to the stipe at first but becoming free later.

recurved - curved back: curved back upward.

recurved: *of pileus margin or squamules on pileus or stipe*, having the base appressed on a surface but the tips turned upwards.

reflexed: *of pileus margin or squamules on pileus or stipe*, turned abruptly upwards or backwards.

refractive hypha: a conspicuous hypha found in various tissues of Inocybaceae fruit bodies that are yellow, ochraceous, or brown and opaque, typically long and bearing irregular knobs. cf. **oleiferous hypha**, cf. **thromboplera** (Clémençon, 2004).

reniform: *of basidiospores*, kidney-shaped. cf. **phaseoliform**.

revolute: *of pilei margins,* having the edge rolled upwards or backwards. cf. **involute**.

rimose: *of pileus surfaces*, condition where fibrils have split and cracked usually in a radial manner and often revealing paler flesh beneath.

rimulose: *of pileus surfaces*, having tiny cracks between fibrils or weakly rimose.

rostrate (mucronate): *of basidiospores, of cystidia*, with a short, beak-like extension (mucro or rostrum).

rpb1: gene that encodes the largest subunit of RNA polymerase II, enzyme used to transcribe DNA into messenger RNA; in the Inocybaceae the most variable region of *rpb1* (between conserved domains A and C) is used for molecular phylogenetic purposes and to verify rRNA sequence differences between species.

rpb2: gene that encodes the second largest subunit of RNA polymerase II, enzyme used to transcribe DNA into messenger RNA; in the Inocybaceae the most variable region of *rpb2* (between conserved domains 6 and 7) is used for molecular phylogenetic purposes and to verify rRNA sequence differences between species.

rRNA: nuclear or mitochondrial ribosomal RNA; composed of gene regions most often used in fungal systematics and detection of fungi from the environment.

rubescent: *of the surface and flesh of fruit bodies*, turning a red colour, usually where cut or bruised.

rugose: *of surfaces*, having coarse wrinkles.

rugulose: *of surfaces*, having undulating low wrinkles.

saccate: *of cystidia*, sac- or bag-like.

scabrous: *of surfaces*, rough due to scales or minute projections.

scurfy (furfuraceous): with bran-like scales similar to dandruff.

secotioid: *of fruit bodies*, having hypogeous or emergent fruit bodies in which the basidiospores are not forcibly discharged (statismosporic) and remain enclosed by the basidioma during most or all stages of development, including at maturity. May be referred to as truffle-like or sequestrate.

septum: a dividing wall delimiting adjacent cells of a hypha or spore. pl. **septa**. adj. **septate**.

sericeous: *of surfaces,* silky.

serrate: *of edges or surfaces*, notched or toothed like a saw.

serrulate: *of edges or surfaces*, finely notched or toothed.

sinuate: *of lamellae*, having a deep notch near the point of attachment to the stipe. cf. emarginate.

solid: *of stipes*, without any hollows or gaps.

spathulate: spatula-shaped.

spermatic: odour similar to piperidine or the male flowers (catkins) of *Castanea* (chestnut).

spinose: *of basidiospores*, with spines.

spinulose: *of basidiospores*, with minute spines.

spore print: deposition of basidiospores discharged from a fruit body accumulated over sufficient time so as to become visible.

squamose: *of surfaces*, covered with scales.

squamule: a small scale. adj. **squamulose**.

squamulose: *of surfaces*, covered with small scales.

squarrose: *of surfaces*, covered with erect or recurved scales.

STAC: Smooth-spored Temperate Austral Clade. A hyper-diverse clade of *Inocybe* species with smooth spores and metuloids, found in Australia and other southern hemisphere regions.

statismospore: *of a basidiospore*, a spore that is not forcibly discharged. adj. **statismosporic**. cf. **ballistospore**.

stellate: star-shaped.

sterigma: a projection on a basidium bearing a terminal basidiospore. pl. **sterigmata**.

sterigmate: refers to number of sterigmata on basidia.

stipe: stem of fruit bodies, adj. **stipitate**.

stipe trama: tissue of sterile hyphae comprising a stipe, cf. **trama**.

stipitipellis: an anatomical structure of hyphae that compose the outer layer (pellis) of the stipe, not including any veil.

stramineous: straw-coloured, yellowish.

striate: *of pilei margin, of stipe surface*, having a finely lined appearance.

striatulate: *of stipe surface*, having a longitudinal lined appearance.

strigose: *of surfaces*, covered with coarse, bristle-like hairs.

stuffed: *of stipes*, composed of loosely arranged tissue/fibres.

subhymenium: a layer of hyphal cells subtending the hymenium.

subulate: *of cystidia*, slender, more or less cylindrical, and tapering to a point; awl-shaped.

sulcate: *of pilei margin*, having radial grooves.

superficial: *of surfaces*, material readily removable from the adjacent tissue. cf. **innate**.

sympatric: taxa with overlapping geographic distributions.

tawny: brownish orange; "Tawny" of Ridgway (1912).

terete: circular in cross-section.

terricolous (humicolous): living or growing on or in soil.

tomentose: *of surfaces*, covered with soft, densely matted, woolly hairs.

trama: any tissue of sterile hyphae within a basidioma. adj. **tramal**. cf. **context**, **hymenophoral trama**, **stipe trama**.

translucent-striate: *of a pileus,* having the lines of the lamellae visible through the pileus.

trapeziform: *of basidiospores*, four sides, two of which are usually parallel, or no sides parallel.

trichoderm: *of a pileus or stipe*, a pellis with hyphae arranged predominantly at right angles to the surface.

tricholomatoid: resembling *Tricholoma*, i.e. with a fleshy stipe, sinuate lamella attachment, and no annulus or volva.

tubarioid: resembling *Tubaria*, i.e. small, brownish pileus, with subdecurrent to adnate lamella attachment, and no annulus or volva.

turbinate: top-shaped; flattened at the apex and abruptly tapered at the base.

umber: a deep dull dark brown, smoky brown; earth brown sometimes with a very slight reddish tinge.

umbilicate: *of a pileus*, having a central navel-like depression, e.g. as in *Tubariomyces umbilicatus*.

umbonate: *of a pileus*, having a central protrusion (umbo).

umbrinous: raw umber or brown or "Snuff-Brown" to "Saccardo's Umber" of Ridgway (1912); burnt umber refers to a reddish brown colour.

undulate: wavy, not flat.

universal veil: *of fruit bodies,* a protective layer of tissue which initially envelopes a basidioma; after rupturing may or may not leave discernible remnants on the pileus surface and lower stipe. cf. **partial veil**, **volva**.

utriform: *of cystidia*, bladder-shaped; broader with a less distinct neck than lageniform.

veil: a protective layer of tissue enveloping fruit bodies during their development. cf. **partial veil**, **universal veil**.

velipellis: a distinct outer-most layer on the pileus of some species of Inocybaceae. Typically forming an evanescent or persistent, superficial, white or greyish, matted-fibrillose or filmy patch on the pileus (less often viscid). Comprised of hyphae morphologically distinct from those of the pileipellis, e.g. hyaline without any wall incrustations. Although assumed to be derived from the universal veil, the ontogeny is uncertain as velipellis hyphae may be entirely separate or merge and inter-mix with the pileipellis.

velutinous: *of surfaces*, velvety; with dense short, fine flexible hairs.

ventricose: *of cystidia*, swollen in the middle; *of lamellae*, broader midway between the stipe and the margin of the pileus.

vermiform: worm-like.

verrucose: *of basidiospores,* warty-roughened, with the surface ornamented by small rounded warts; *of hyphae*, distinctly and regularly rough-walled.

verruculose: *of basidiospores,* finely warted.

versicolorous: having different colours. cf. **concolorous**.

versiform (polymorphic): variably shaped.

vesiculose: *of cystidia*, swollen, bladder-like but more sac-shaped.

vesture: a superficial layer of usually whitish, silky, fibrils conspicuous on the stipe of some species of Inocybaceae. Usually comprised of interwoven cylindric hyphae with undifferentiated or cystidioid terminal cells.

villose: *of surfaces*, covered with long fine hairs.

viscid: *of surfaces*, sticky, or slimy and sticky. cf. **glutinous**.

volva: cup-like remnants of the universal veil at or near the base of the stipe. cf. **universal veil**. adj. **volvate**.

wet sclerophyll: hard-leaved vegetation with short internodes; used to describe plant communities in Australia dominated by a tall eucalypt overstory (>30 meters) and annual rainfall of at least 1000 mm.

yellowish brown: "Buckthorn Brown" of Ridgway (1912).

zonate: *of pilei*, having concentric bands of different colour or texture.

BIBLIOGRAPHY

Alessio, C.L. (1980), *Inocybe. Iconographia Mycologica* 29. Trento.

Alvarado, P., Manjón, J.L., Matheny, P.B. & Esteve-Raventós, F. (2010), *Tubariomyces*, a new genus of Inocybaceae from the Mediterranean region, *Mycologia* 102: 1389–1397.

Baker, R. (1907), Contributions to a knowledge of the flora of Australia. Part V, *Proc. Linn. Soc. New South Wales* 31: 711–721.

Baroni, T.J. & Matheny, P.B. (2011), A re-evaluation of gasteroid and cyphelloid species of Entolomataceae from eastern North America, *Harvard Pap. Bot.* 16: 293–310.

Bastias, B.A., Xu, Z. & Cairney, J.W. (2006), Influence of long-term repeated prescribed burning on mycelial communities of ectomycorrhizal fungi, *New Phytol.* 172: 149–158.

Bon, M. (1984), Macromycetes de la zone maritime Picarde (8ème suppl): les Inocybes sabulicoles, *Doc. Mycol.* 14(53): 9–40.

Bon, M. (1997), Clé monographique du genre *Inocybe* (Fr.) Fr. (2ème partie: sous-genre *Inocybe = Inocybium* (Earle) Sing.), *Doc. Mycol.* 27(108): 1–69.

Bougher, N.L. (2009), *Fungi of the Perth Region and Beyond: a self-managed field book*. Western Australian Naturalists' Club Inc., Perth.

Bougher, N.L. (2010), History of the study of fungi at Kings Park, Perth, Western Australia, *Western Australian Naturalist* 27: 61–90.

Bougher, N.L., Hart, R. & Weaver, J. (2005), *Acourt Road Bush, Jandakot Regional Park Fungi Report 2005*. Perth Urban Bushland Fungi Project Report for the Department of Environment and Conservation, Perth.

Bougher, N.L., Hart, R. & De Bueger, S. (2006a), *Modong Nature Reserve Fungi Report 2006*. Perth Urban Bushland Fungi Project Report for the Jandakot Regional Park, Department of Environment and Conservation, Perth.

Bougher, N.L., Hart, R. & De Bueger, S. (2006b), *Brownman Swamp Bushland (Beeliar Regional Park) Fungi Report 2006*. Perth Urban Bushland Fungi Project Report for the Department of Environment and Conservation, Perth.

Bougher, N.L., Hart, R., De Bueger, S., Froudist, J. & Glossop, B. (2007a), *Fungi of West Bay Bushland, Augusta, Western Australia*. Perth Urban Bushland Fungi Project Report for the Environmental Research Group Augusta, Department of Environment and Conservation, Perth.

Bougher, N.L., Hart, R., De Bueger, S. & Glossop, B. (2007b), *Port Kennedy Scientific Park Fungi Report 2007, Rockingham Lakes Regional Park*. Perth Urban Bushland Fungi Project Report for the Department of Environment and Conservation, Perth.

Bougher, N.L., Hart, R., De Bueger, S., Sarti, K. & Glossop, B. (2007c), *Fungi of Bungendore Park, Perth, Western Australia*. Perth Urban Bushland Fungi Project Report for Bungendore Management, Department of Environment and Conservation, Perth.

Bougher, N.L., Hart, R., De Bueger, S. & Glossop, B. (2007d), *Modong Nature Reserve Fungi Report 2007, Jandakot Regional Park*. Perth Urban Bushland Fungi Project Report for the Department of Environment and Conservation, Perth.

Bougher, N.L. & Lebel, T. (2001), Sequestrate (truffle-like) fungi of Australia and New Zealand, *Austral. Syst. Bot.* 14: 439–484.

Bougher, N.L. & Matheny, P.B. (2011), Two species of *Inocybe* (fungi) introduced into Western Australia, *Nuytsia* 21: 139–148.

Bougher, N.L., Matheny, P.B. & Gates, G.M. (2012), Five new species and records of *Inocybe* (Agaricales) from temperate and tropical Australia, *Nuytsia* 22: 57–74.

BIBLIOGRAPHY

Bougher, N.L. & Syme, K. (1998), *Fungi of Southern Australia*. University of Western Australia Press, Nedlands.

Braaten, C.C., Matheny, P.B., Viess, D.L., Wood, M.G., Williams, J.H. & Bougher, N.L. (2014), Two new species of *Inocybe* from Australia and North America that include novel secotioid forms, *Botany* 92: 9–22.

Byrne, M., Stean, D.A., Joseph, L., Yeates, D.K., Jordan, G.J., Crayn, D., Aplin, K., Cantrill, D.J., Cook, L.G., Crisp, M.D., Keogh, J.S., Melville, J., Moritz, C., Porch, N., Sniderman, J.M.K., Stunnucks, P. & Weston, P.H. (2011), Decline of a biome: evolution, contraction, fragmentation, extinction and invasion of the Australian mesic zone biota, *J. Biogeogr.* 38: 1635–1656.

Catcheside, P. & May, T. (2010), *Inocybe violaceocaulis* in South Australia, *Fungimap Newslett.* 40: 5–6.

Cleland, J.B. (1933), Australian fungi: notes and descriptions – No. 9, *Trans. Roy. Soc. S. Australia* 57: 187–194.

Cleland, J.B. (1934–1935), *Toadstools and Mushrooms and other Larger Fungi of South Australia*, parts 1 and 2. Government Printer, Adelaide.

Cleland, J.B. & Cheel, E. (1915), Notes on Australian fungi, No. 1, *J. & Proc. Roy. Soc. New South Wales* 48: 433–443.

Cleland, J.B. & Cheel, E. (1918), Australian fungi: notes and descriptions. No. 1, *Trans. & Proc. Roy. Soc. S. Australia* 42: 88–138.

Clémençon, H. (2004), Cytology and plectology of the Hymenomycetes. *Biblioth. Mycol.* 199: 1–488.

Cooke, M.C. (1881), Australian fungi, *Grevillea* 9: 142–149.

Cooke, M.C. (1892), *Handbook of Australian Fungi*. Williams & Norgate, London.

Cortez, V.G. & Coelho, G. (2005), Additions to the mycobiota (Agaricales, Basidiomycetes) of Rio Grande do Sul, Brazil, *Iheringia*, Sér. Bot. 60: 69–75.

Cribb, A.B. (1987), Mushrooms and toadstools, *In*: J.Covacevich, P.Davie & J.Pearn (eds), *Toxic Plants and Animals. a guide for Australia*: 22–31. Queensland Museum, South Brisbane.

Crisp, M.D. & Cook, L.G. (2007), A congruent molecular signature of vicariance across multiple plant lineages, *Molec. Phylogen. Evol.* 43: 1106–1117.

Crisp, M.D., Cook, L. & Steane, D. (2004), Radiation of the Australian flora: what can comparisons of molecular phylogenies across multiple taxa tell us about the evolution of diversity in present-day communities? *Philos. Trans. Roy. Soc. London*, Ser. B, 359: 1551–1571.

Donk, M.A. (1962), The generic names proposed for Agaricaceae, *Beih. Nova Hedwigia* 5: 1–320.

Dunstan, W.A., Dell, B. & Malajczuk, N. (1998), The diversity of ectomycorrhizal fungi associated with introduced *Pinus* spp. in the southern hemisphere, with particular reference to Western Australia, *Mycorrhiza* 8: 71–79.

Earle, F.S. (1909), Genera of North American gill fungi, *Bull. New York Bot. Gard.* 5: 373–462.

Esteve-Raventós, F. & Villarreal, M. (2001), *Inocybe lavandulochlora*, una nuova specie dell sezione *Lilacinae* R. Heim., *Revista Micol.* 3: 215–223.

Fan, Y.G. & Bau, T. (2014), *Inocybe hainanensis*, a new lilac-stiped species from tropical China, *Mycosystema* 33: 954–960.

Fries, E. (1821), *Systema Mycologicum*. Vols I & II. Officina Berlingiana, Lund.

Fries, E. (1863), *Monographia Hymenomycetum Sueciae*. Uppsala. Reprint. A.Asher & Co., Amsterdam [1963].

Ford, S. & May, T.W. (1998), A cryptogamic extravaganza: report of the FNCV expedition 22–25 May 1998 to commemorate the centenary of Wilson's Promontory National Park, Field Naturalists Club of Victoria, Blackburn.

Francis, A.A. & Bougher, N.L. (2004), Cortinarioid sequestrate (truffle-like) fungi of Western Australia, *Australas. Mycol.* 23: 1–26.

Fuhrer, B. (2005), *A Field Guide to Australian Fungi*. Bloomings Books, Melbourne (reprinted with revisions 2009).

Gates, G.M. & Ratkowsky, D. (2014), *A Field Guide to Tasmanian Fungi.* Tasmanian Field Naturalists' Club Inc., Hobart.

Grgurinovic, C.A. (1997), *Larger Fungi of South Australia*. The Botanic Gardens of Adelaide and State Herbarium and The Flora and Fauna of South Australia Handbooks Committee, Adelaide.

Grgurinovic, C.A. (2002), Fungal names needing correction. *Australas. Mycol.* 21: 35–37.

Grgurinovic, C.A. & Simpson, J.A. (2001), Conservation status of the known Agaricales, Boletales, Cantharellales, Lycoperdales, Phallales and Russulales of South Australia, *Fungal Diversity* 8: 97–127.

Heim, R. (1931), *Le Genre* Inocybe, *Encyclopedie Mycologique 1*. Paul Lechevalier & Fils, Paris.

Hilton, R.N. (1983), The Drummond collection of Western Australian fungi at the Royal Botanic Gardens, Kew, *Nuytsia* 4: 333–357.

Hopper, S.D. & Gioia, P. (2004), The south-west Australian floristic region: evolution and conservation of a global hotspot of biodiversity, *Ann. Rev. Ecol. Evol. Syst.* 35: 623–650.

Horak, E. (1968), Synopsis generum Agaricalium (Die Gattungstypen der Agaricales). *Beitr. Kryptogamen Fl. Schweiz* 13: 1–741.

Horak, E. (1977), Fungi Agaricini Novaezelandiae. VI. *Inocybe* (Fr.) Fr. and *Astrosporina* Schroeter, *New Zealand J. Bot.* 15: 713–747.

Horak, E. (1979a), *Astrosporina* (Agaricales) in Indomalaya and Australia, *Persoonia* 10: 157–205.

Horak, E. (1979b), Fungi, Basidiomycetes. Agaricales y Gasteromycetes secotioides, *Fl. Criptog. Tierra del Fuego* 11(6): 1–524.

Horak, E. (1980), *Inocybe* (Agaricales) in Indomalaya and Australasia, *Persoonia* 11: 1–37.

Horak, E., Matheny, P.B., Desjardin, D.E. & Soytong, K. (2015) The genus *Inocybe* (Inocybaceae, Agaricales, Basidiomycota) in Thailand and Malaysia. *Phytotaxa* 230 (3): 201-238.

Horton, B.M. (2011), *Eucalyptus Decline and Ectomycorrhizal Fungal Community Ecology of* Eucalyptus delegatensis *Forest, Tasmania, Australia*. Ph.D. thesis, University of Tasmania, Hobart.

Iotti, M., Marchetti, M., Bonuso, E. & Zambonelli, A. (2004), Morphological and molecular characterization of the mycorrhizas of *Inocybe rufuloides* and *I. splendens*, *Mycotaxon* 94: 75–84.

Judge, B.S., Ammirati, J.F., Lincoff, G.H., Trestrail, J.H. & Matheny, P.B. (2010), Ingestion of a newly described North American mushroom species from Michigan resulting in chronic renal failure: *Cortinarius orellanosus*, *Clin. Toxicol.* 48: 545–549.

Jülich, W. ('1982') [1981], Higher taxa of Basidiomycetes, *Biblioth. Mycol.* 85: 1–485.

Kobayashi, T. (2002), The taxonomic studies of the genus *Inocybe*, *Nova Hedwigia* 124: 1–246.

Kobayasi, Y. (1952), On the genus *Inocybe* from Japan, *Nagaoa* 2: 76–115.

Kornerup, A. & Wanscher, J.H. (1967), *Methuen Handbook of Colour*, 2nd edn. Methuen & Co. Ltd, London.

Kosentka, P., Sprague, S.L., Ryberg, M., Gartz, J., May, A.L., Campagna, S.R. & Matheny, P.B. (2013), Evolution of the toxins muscarine and psilocybin in a family of mushroom-forming fungi, *PLoS ONE* 8(5): e64646.

Kropp, B.R., Matheny, P.B. & Hutchison, L.J. (2013), *Inocybe* section *Rimosae* in Utah: phylogenetic affinities and new species, *Mycologia* 105: 728–747.

Kuyper, T.W. (1986), A revision of the genus *Inocybe* in Europe: I. Subgenus *Inosperma* and the smooth-spored species of subgenus *Inocybe*, *Persoonia* (Suppl. Vol.) 3: 1–247.

Larkin, M.A., Blackshields, G., Brown, N.P., Chenna, R., McGettigan, P.A., McWilliams, H., Valentin, F., Wallace, I.M., Wilm, A., Lopez, R., Thompson, J.D., Gibson T.J. & Higgins, D.G. (2007), Clustal W and Clustal X version 2.0, *Bioinformatics* 23: 2947–2948.

Latha, K.P., Manimohan, P. & Matheny, P.B. (2016) A new species of *Inocybe* representing the *Nothocybe* lineage. *Phytotaxa* 267: 040-050.

Maddison, D.R. & Maddison, W.P. (2005), *MacClade 4: version 4.08 for OS X*. Sinauer Associates, Sunderland, Massachusetts.

Matheny, P.B. (2005), Improving phylogenetic inference of mushrooms using RPB1 and RPB2 sequences (*Inocybe*, Agaricales), *Molec. Phylogen. Evol.* 35: 1–20.

Matheny, P.B. (2009), A phylogenetic classification of the Inocybaceae, *McIlvainea* 18: 11–21.

Matheny, P.B., Aime, M.C., Bougher, N.L., Buyck, B., Desjardin, D.E., Horak, E., Kropp, B.R., Lodge, D.J., Trappe, J.M. & Hibbett, D.S. (2009), Out of the paleotropics? Historical biogeography and diversification of the cosmopolitan mushroom family Inocybaceae, *J. Biogeogr.* 36: 577–592.

Matheny, P.B., Aime, M.C., Smith, M.E. & Henkel, T.W. (2012a), New species and reports of *Inocybe* (Agaricales) from Guyana, *Kurtziana* 37: 23–39.

Matheny, P.B. & Bougher, N.L. (2005), A new violet species of *Inocybe* (Agaricales) from urban and rural landscapes in Western Australia, *Australas. Mycol.* 24: 7–12.

Matheny, P.B. & Bougher, N.L. (2006a), The new genus *Auritella* from Africa and Australia (Inocybaceae, Agaricales): molecular systematics, taxonomy and historical biogeography, *Mycol. Progr.* 5: 2–17.

Matheny, P.B. & Bougher, N.L. (2006b), Validation of *Auritella*, *Mycotaxon* 98: 231–233.

Matheny, P.B. & Bougher, N.L. (2010), Type studies of *Inocybe* from Australia, *Muelleria* 28: 87–104.

Matheny, P.B. & Kropp, B.R. (2001), A revision of the *Inocybe lanuginosa* group and allied species in North America, *Sydowia* 53: 93–139.

Matheny, P.B., Liu, Y.J., Ammirati, J.F. & Hall, B.D. (2002), Using RPB1 sequences to improve phylogenetic inference among mushrooms (*Inocybe*, Agaricales), *Amer. J. Bot.* 89: 688–698.

Matheny, P.B., Norvell, L.L. & Giles, E.C. (2013), A common new species of *Inocybe* in the Pacific Northwest with a diagnostic PDAB reaction, *Mycologia* 105: 436-446.

Matheny, P.B., Pradeep, C.K., Vrinda, K.B. & Varghese, S.P. (2012b), *Auritella foveata*, a new species of Inocybaceae (Agaricales) from tropical India, *Kew Bull.* 67: 119–125.

Matheny, P.B. & Watling, R. (2004), A new and unusual species of *Inocybe* (Inosperma clade) from tropical Africa, *Mycotaxon* 89: 497–503.

Mathieson, J. (1946), Antibiotics from Victorian Basidiomycetes, *Austral. J. Exper. Biol. Med. Sci.* 24: 57–62.

May, T.W. (1989), Report of the F.N.C.V. fungal excursions: 1986–1988, *Victorian Naturalist* 106: 48–58.

May, T.W. & Wood, A.E. (1997), *Fungi of Australia. Vol. 2A Catalogue and Bibliography of Australian Macrofungi 1. Basidiomycota p.p.* Australian Biological Resources Study, Canberra.

McAlpine, D. (1895), *Systematic Arrangement of Australian Fungi.* R.S.Brain, Government Printer, Melbourne.

McMullan-Fisher, S., Leonard, P. & Guard, F. (2014), *Australian Subtropical Fungi.* Suncoast Fungi, Buderim.

Miller, O.K., Jr. & Hilton, R.N. (1987), New and interesting agarics from Western Australia, *Sydowia* 39: 126–137.

Moser, M. (1978), *Fungorum Rariorum Icones Coloratae.* 7: 1–48. Cramer, Vaduz.

Munsell (1954), *Munsell Soil Color Charts.* Munsell Color Company Inc., Baltimore.

Nouhra, E., Urcelay, C., Longo, S. & Tedersoo, L. (2013), Ectomycorrhizal fungal communities associated to *Nothofagus* species in Northern Patagonia, *Mycorrhiza* 23: 487–496.

Obase, K., Lee, J.K., Lee, S.Y. & Chun, K.W. (2011), Diversity and community structure of ectomycorrhizal fungi in *Pinus thunbergii* coastal forests in the eastern region of Korea, *Mycoscience* 52: 383–391.

Pegler, D.N. (1965), Studies on Australasian Agaricales, *Austral. J. Bot.* 13: 323–356.

Pegler, D.N. (1986), Agaric flora of Sri Lanka, *Kew Bull., Addit. Ser.* 12: 1–519.

Pradeep, C.K., Vrinda, K.B., Varghese, S. & Matheny, P.B. (2016) New and noteworthy species of *Inocybe* (Agaricales) from tropical India. *Mycol. Prog.* 15: 24.

Ridgway, R. (1912), *Color Standards and Color Nomenclature.* Privately Published, Washington.

Roy, M.H., Rochet, J., Manzi, S., Jargeat, P., Gryta, H., Moreau, P.A. & Gardes, M. (2013), What determines *Alnus*-associated ectomycorrhizal community diversity and specificity? A comparison of host and habitat effects at a regional scale, *New Phytol.* 198: 1228–1238.

Ryberg, M., Larsson, E. & Jacobsson, S. (2010), An evolutionary perspective on morphological and ecological characters in the mushroom family Inocybaceae (Agaricomycotina, Fungi), *Molec. Phylogen. Evol.* 55: 431–442.

Ryberg, M. & Matheny, P.B. (2012), Asynchronous origins of ectomycorrhizal clades of Agaricales, *Proc. Roy. Soc.*, Ser. B, 279: 2003–2011.

Ryberg, M., Nilsson, R.H., Kristiansson, E., Topel, M., Jacobsson, S. & Larsson, E. (2008), Mining metadata from unidentified ITS sequences in GenBank: a case study in *Inocybe* (Basidiomycota), *BMC Evol. Biol.* 8: 50.

Shepherd, C.J. & Totterdell, C.J. (1988), *Mushrooms and Toadstools of Australia.* Inkata Press, Melbourne.

Singer, R. (1953), The agarics of the Argentine sector of Tierra del Fuego and limitrophous regions of the Magallanes area. II. The brown-spored genera (except *Cortinarius*), *Sydowia* 7: 206–265.

Singer, R. (1986), *The Agaricales in Modern Taxonomy*, 4th edn. Koeltz Scientific Books, Koenigstein.

Singer, R. & Digilio, A.P.L. (1952), Pródromo de la flora Agaricina Argentina, *Lilloa* 25: 5–462.

Sinnott, N.H. (1976), *Victorian Agaricales and Cantharellaceae.* Privately Published, Essendon.

Stamatakis, A. (2006), RAxML-VI-HPC: maximum likelihood-based phylogenetic analyses with thousands of taxa and mixed models, *Bioinformatics* 22: 2688–2690.

Stamatakis, A. (2014), RAxML version 8: a tool for phylogenetic analysis and post-analysis of large phylogenies, *Bioinformatics* 30: 1312–1313.

Stangl, J. (1984), *Inocybe mixtilis* (Britz.) Sacc., *Int. J. Mycol. Lichenol.* 1: 327–334.

Stangl, J. (1989), Die Gattung *Inocybe* in Bayern, *Hoppea* 46: 5–388.

Stangl, J. & Veselsky, J. (1982), Risspilze der section *Lilacinae* Heim., *Ceská Mykol.* 36: 85–99.

Syme, K. (1992), *Survey of the Larger Fungi of the Two Peoples Bay Nature Reserve*. Denmark Environment Centre, Denmark.

Tait, R. (2007), *Fungi Out West: some fungi of southern inland Queensland.* Chinchilla Field Naturalists' Club Inc. & Cranbrook Press, Toowoomba.

Thiers, B. [continuously updated], Index herbariorum: a global directory of public herbaria and associated staff. New York Botanical Garden's Virtual Herbarium. http://sweetgum.nybg.org/ih/

Turnbull, E. (1995), *Inocybe* in Peninsular Malaysia, *Edinburgh. J. Bot.* 52: 351–359.

Turnbull, E. & Watling, R. (1999), Taxonomic and floristic notes on Malaysian larger fungi III, *Malayan Nat. J.* 53: 189–200.

Vesper, A. & Richter, T. (2010), Studien zu zwei wenig bekannten Risspilzen, *Inocybe rufuloides* and *I. subporospora*, *Boletus* 32: 100–105.

Villarreal, M., Esteve-Raventós, F., Heykoop, M. & Horak, E. (1998), *Inocybe inexpectata*, a new and unusual species of subgenus *Mallocybe*, *Mycol. Res.* 102: 479–482.

Vizzini, A., Della Maggiora, M., Tolaini, F. & Ercole, E. (2013), A new cryptic species in the genus *Tubariomyces* (Inocybaceae, Agaricales), *Mycol. Progr.* 12: 375–381.

Walbert, K., Ramsfield, T.D., Ridgway, H.J. & Jones, E.E. (2010), Ectomycorrhizal species associated with *Pinus radiata* in New Zealand including novel associations determined by molecular analysis, *Mycorrhiza* 20: 209–215.

Young, A.M. (1982), *Common Australian Fungi*. University of New South Wales Press, Kensington. (and revised edition 1994.)

Young, A.M. (2005), *A Field Guide to the Fungi of Australia*. University of New South Wales Press, Sydney.

ABBREVIATIONS AND CONTRACTIONS

Literature

Author abbreviations follow R.K.Brummitt & C.E.Powell, *Authors of Plant Names* (Royal Botanic Gardens, Kew, 1992).

Journal titles are abbreviated in accordance with G.D.R.Bridson, *Botanico-Periodicum-Huntianum*, 2nd edn (Hunt Institute for Botanical Documentation, Pittsburgh, 2004). Other literature is abbreviated in accordance with F.A.Stafleu & R.S.Cowan, *Taxonomic Literature*, 2nd edn (Bohn, Scheltema & Holkema, Utrecht, 1976–1987), except that upper case initial letters are used for proper names and significant words.

Herbaria

Abbreviations of herbaria are in accordance with P.K.Holmgren, N.H.Holmgren & L.C.Barnett, *Index Herbariorum* Part I, 8th edn (New York Botanical Garden, 1990). Those most commonly cited are:

AD	State Herbarium of South Australia, Adelaide
ADW	Herbarium, Waite Agricultural Research Institute, Adelaide (transferred to DAR)
BM	The Natural History Museum, London
BRI	Queensland Herbarium, Brisbane
BRIP	Plant Pathology Branch, Department of Primary Industries, Indooroopilly, Brisbane
CANB	Australian National Herbarium, Canberra
DAR	Plant Pathology Herbarium, Orange Agricultural Institute, Orange
HO	Tasmanian Herbarium, Hobart
HUV	Herbarium Ustilaginales Vánky, Tübingen
IMI	Herbarium, CAB International Mycological Institute, Kew
K	Royal Botanic Gardens, Kew
MEL	National Herbarium of Victoria, Melbourne
NSW	National Herbarium of New South Wales, Sydney
NTPPMH	Northern Territory Plant Pathology and Mycological Herbarium
PDD	Landcare Research Herbarium, Auckland
PERTH	Western Australian Herbarium, Perth
TENN	University of Tennessee Herbarium, Knoxville
VPRI	Herbarium, Institute of Horticultural Development, Melbourne

States, Territories

Abbreviations of Australian States and Territories as used in statements of distribution and citation of collections are:

A.C.T.	Australian Capital Territory
N.S.W.	New South Wales
N.T.	Northern Territory
Qld	Queensland
S.A.	South Australia
Tas.	Tasmania
Vic.	Victoria
W.A.	Western Australia

General abbreviations

add.	addendum
alt.	altitude
app.	appendix

ABBREVIATIONS AND CONTRACTIONS

Art.	Article (of the International Code of Botanical Nomenclature)
auct.	*auctoris/auctorum* (of an author or authors)
auct. mult.	*auctorum multorum* (of many authors)
auct. non	*auctorum non* (of authors [but] not ...), used for misapplied names
c.	*circa* (about)
cf.	*confer* (compare)
Ck	Creek
cm	centimetre
coll.	collected by
comb.	*combinatio* (combination)
cons.	*conservandus* (conserved)
det.	determined by
diam.	diameter
E	east
edn	edition
ed./eds	editor/editors
e.g.	*exempli gratia* (for example)
et al.	*et alii/et aliorum* (and others/and of others)
f.	*forma* (form)
fig./figs	figure/figures (in other works)
Fig.	Figure (referring to a Figure in this volume)
f.sp.	form species
holo	holotype
Hwy	Highway
i.e.	*id est* (that is)
ined.	*ineditus* (unpublished)
Is.	Island/s
iso	isotype
isolecto	isolectotype
km	kilometre
L.	Lake
lecto	lectotype
loc. cit.	*loco citato* (in bibliographic citations: in the same work and page as just cited)
loc. id.	*loco idem* (in specimen citations: in the same place as just cited)
m	metre
mm	millimetre
Mt/Mts	Mount/Mounts
Mtn/Mtns	Mountain/Mountains
MYP	malt yeast extract peptone
N	north
Natl	National
n.d.	no date
NE	north-east (ern)
nom. cons.	*nomen conservandum* (conserved name)
nom. cons. prop.	*nomen conservandum propositus* (proposed conserved name)
nom. dub.	*nomen dubium* (doubtful name)
nom. illeg.	*nomen illegitimum* (illegitimate name)
nom. inval.	*nomen invalidum* (name not validly published)
nom. nov.	*nomen novum* (new name)
nom. nud.	*nomen nudum* (name published without a description or reference to a published description)
nom. rej.	*nomen rejiciendum* (rejected name)
non	not
nov.	*novus* (new)
n. ser.	new series

n.v.	*non vidi* (not seen)
NW	north-west (ern)
op. cit.	*opere citato* (in the work cited above)
p./pp.	page/pages
Penin.	Peninsula
pers. comm.	by personal communication
pl./pls	plate/plates
p.p.	*pro parte* (in part)
pro. syn.	as a synonym
Q	Ratio of length to width.
q.v.	*quod vide* (which see)
R.	River
Ra.	Range
Rd	Road
S	south
SE	south-east (ern)
sect.	*sectio*/section
sensu	in the sense of
ser./sér.	series/série
s. lat.	*sensu lato* (in a wide sense)
s.n.	*sine numero* (without number)
sp./spp.	species (singular/plural)
sp. nov.	*species nova* (new species)
s. str.	*sensu stricto* (in a narrow sense)
St	Saint/Street
subg.	subgenus
subsp./ subspp.	subspecies (singular/plural)
subtrib.	*subtribus* (subtribe)
suppl.	supplement
SW	south-west (ern)
syn	syntype
synon.	synonym
T	Type (collection)
trib.	*tribus* (tribe)
var.	*varietas* (variety)
viz.	*videlicet* (namely)
vol./vols	volume/s
W	west

Symbols

±	*in species descriptions*, more-or-less
<	less than
>	more than
µm	micrometre

PUBLICATION DATES OF PREVIOUS VOLUMES OF *FUNGI OF AUSTRALIA*

Volume 1A	9 August 1996
Volume 1B	2 October 1996
Volume 2A	23 April 1997
Volume 2B	22 August 2003
Hygrophoraceae	21 October 2005
Septoria	26 September 2006
The Smut Fungi	19 March 2008

The publication date of *Fungi of Australia: Inocybaceae* will be listed in the next volume of the series, or see the ABRS website.

INDEX

Accepted names are in roman, synonyms and doubtful names in *italic*.

Principal page references are in **bold**, figures and plates in *italic*.